Geometric Properties of

D0215164

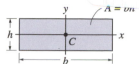

Rectangular area

$I_x = \frac{1}{12} bh^3$

$I_y = \frac{1}{12} hb^3$

Material Property Relations

Poisson's ratio

$$\nu = -\frac{\epsilon_{lat}}{\epsilon_{long}}$$

Generalized Hooke's Law

$$\epsilon_x = \frac{1}{E}\left[\sigma_x - \nu(\sigma_y + \sigma_z)\right]$$

$$\epsilon_y = \frac{1}{E}\left[\sigma_y - \nu(\sigma_x + \sigma_z)\right]$$

$$\epsilon_z = \frac{1}{E}\left[\sigma_z - \nu(\sigma_x + \sigma_y)\right]$$

$$\gamma_{xy} = \frac{1}{G}\tau_{xy}, \ \gamma_{yz} = \frac{1}{G}\tau_{yz}, \ \gamma_{zx} = \frac{1}{G}\tau_{zx}$$

where

$$G = \frac{E}{2(1 + \nu)}$$

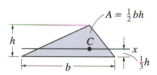

Triangular area

$I_x = \frac{1}{36} bh^3$

Relations Between *w, V, M*

$$\frac{dV}{dx} = w(x), \qquad \frac{dM}{dx} = V$$

Elastic Curve

$$\frac{1}{\rho} = \frac{M}{EI}$$

$$EI\frac{d^4v}{dx^4} = w(x)$$

$$EI\frac{d^3v}{dx^3} = V(x)$$

$$EI\frac{d^2v}{dx^2} = M(x)$$

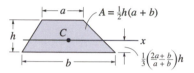

Trapezoidal area

$A = \frac{1}{2}h(a + b)$

$\frac{1}{3}\left(\frac{2a + b}{a + b}\right)h$

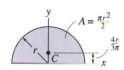

Semicircular area

$A = \frac{\pi r^2}{2}$

$I_x = \frac{1}{8}\pi r^4$

$I_y = \frac{1}{8}\pi r^4$

Buckling

Critical axial load

$$P_{cr} = \frac{\pi^2 EI}{(KL)^2}$$

Critical stress

$$\sigma_{cr} = \frac{\pi^2 E}{(KL/r)^2}, \ r = \sqrt{I/A}$$

Secant formula

$$\sigma_{max} = \frac{P}{A}\left[1 + \frac{ec}{r^2}\sec\left(\frac{L}{2r}\sqrt{\frac{P}{EA}}\right)\right]$$

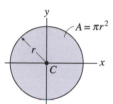

Circular area

$A = \pi r^2$

$I_x = \frac{1}{4}\pi r^4$

$I_y = \frac{1}{4}\pi r^4$

Energy Methods

Conservation of energy

$$U_e = U_i$$

Strain energy

$$U_i = \frac{N^2 L}{2AE} \quad \text{constant axial load}$$

$$U_i = \int_0^L \frac{M^2 dx}{2EI} \quad \text{bending moment}$$

$$U_i = \int_0^L \frac{f_s V^2 dx}{2GA} \quad \text{transverse shear}$$

$$U_i = \int_0^L \frac{T^2 dx}{2GJ} \quad \text{torsional moment}$$

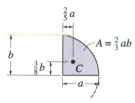

Semiparabolic area

$A = \frac{2}{3}ab$

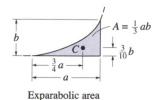

Exparabolic area

$A = \frac{1}{3}ab$

MECHANICS OF MATERIALS

MECHANICS OF MATERIALS

TENTH EDITION

R. C. HIBBELER

PEARSON

Boston   Columbus   Indianapolis   New York   San Francisco   Hoboken   Amsterdam
Cape Town   Dubai   London   Madrid   Milan   Munich   Paris   Montréal   Toronto   Delhi
Mexico City   São Paulo   Sydney   Hong Kong   Seoul   Singapore   Taipei   Tokyo

Vice President and Editorial Director, ECS: *Marcia J. Horton*
Senior Editor: *Norrin Dias*
Editorial Assistant: *Michelle Bayman*
Program/Project Management Team Lead: *Scott Disanno*
Program Manager: *Sandra L. Rodriguez*
Project Manager: *Rose Kernan*
Director of Operations: *Nick Sklitsis*
Operations Specialist: *Maura Zaldivar-Garcia*
Product Marketing Manager: *Bram Van Kempen*
Field Marketing Manager: *Demetrius Hall*
Marketing Assistant: *Jon Bryant*
Cover Designer: *Black Horse Designs*
Cover Image: *Fotosearch/Getty Images*

Credits and acknowledgments borrowed from other sources and reproduced, with permission, in this textbook appear on appropriate page within text. Unless otherwise specified, all photos provided by R.C. Hibbeler.

Pearson Education Ltd., *London*
Pearson Education Singapore, Pte. Ltd
Pearson Education Canada, Inc.
Pearson Education—Japan
Pearson Education Australia PTY, Limited
Pearson Education North Asia, Ltd., *Hong Kong*
Pearson Educación de Mexico, S.A. de C.V.
Pearson Education Malaysia, Pte. Ltd.
Pearson Education, Inc., *Hoboken, New Jersey*

Library of Congress Cataloging-in-Publication Data
Hibbeler, R. C.,
Mechanics of materials / R.C. Hibbeler.
Tenth edition. | Hoboken, NJ: Pearson, 2015. | Includes index.
LCCN 2015044964 | ISBN 9780134319650
Materials. | Mechanics, Applied. | Strength of materials. | Structural analysis
  (Engineering) | Materials—Problems, exercises, etc. | Mechanics, Applied—Problems,
  exercises, etc. | Strength of materials—Problems, exercises, etc. | Structural analysis
  (Engineering)—Problems, exercises, etc.
LCC TA405.H47 2015 | DDC 620.1/123—dc23
LC record available at http://lccn.loc.gov/2015044964

6 2021

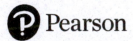

ISBN 10: 0-13-431965-6
ISBN 13: 978-0-13-431965-0

To the Student

With the hope that this work will stimulate
an interest in Mechanics of Materials
and provide an acceptable guide to its understanding.

PREFACE

It is intended that this book provide the student with a clear and thorough presentation of the theory and application of the principles of mechanics of materials. To achieve this objective, over the years this work has been shaped by the comments and suggestions of hundreds of reviewers in the teaching profession, as well as many of the author's students. The tenth edition has been significantly enhanced from the previous edition, and it is hoped that both the instructor and student will benefit greatly from these improvements.

NEW TO THIS EDITION

• **Updated Material.** Many topics in the book have been re-written in order to further enhance clarity and to be more succinct. Also, some of the artwork has been enlarged and improved throughout the book to support these changes.

• **New Layout Design.** Additional design features have been added to this edition to provide a better display of the material. Almost all the topics are presented on a one or two page spread so that page turning is minimized.

• **Improved Preliminary and Fundamental Problems.** These problems sets are located just after each group of example problems. They offer students basic applications of the concepts covered in each section, and they help provide the chance to develop their problem-solving skills before attempting to solve any of the standard problems that follow. The problems sets may be considered as extended examples, since in this edition their complete solutions are given in the back of the book. Additionally, when assigned, these problems offer students an excellent means of preparing for exams, and they can be used at a later time as a review when studying for the Fundamentals of Engineering Exam.

• **New Photos.** The relevance of knowing the subject matter is reflected by the real-world application of the 14 new or updated photos placed throughout the book. These photos generally are used to explain how the principles apply to real-world situations and how materials behave under load.

• **New Problems.** There are approximately 30%, or about 430 problems added to this edition, which involve applications to many different fields of engineering.

• **New Review Problems.** Updated review problems have been placed at the end of each chapter so that instructors can assign them as additional preparation for exams.

CONTENTS

The subject matter is organized into 14 chapters. Chapter 1 begins with a review of the important concepts of statics, followed by a formal definition of both normal and shear stress, and a discussion of normal stress in axially loaded members and average shear stress caused by direct shear.

In Chapter 2 normal and shear strain are defined, and in Chapter 3 a discussion of some of the important mechanical properties of materials is given. Separate treatments of axial load, torsion, and bending are presented in Chapters 4, 5, and 6, respectively. In each of these chapters, both linear-elastic and plastic behavior of the material covered in the previous chapters, where the state of stress results from combined loadings. In Chapter 9 the concepts for transforming multiaxial states of stress are presented. In a similar manner, Chapter 10 discusses the methods for strain transformation, including the application of various theories of failure. Chapter 11 provides a means for a further summary and review of previous material by covering design applications of beams and shafts. In Chapter 12 various methods for computing deflections of beams and shafts are covered. Also included is a discussion for finding the reactions on these members if they are statically indeterminate. Chapter 13 provides a discussion of column buckling, and lastly, in Chapter 14 the problem of impact and the application of various energy methods for computing deflections are considered.

Sections of the book that contain more advanced material are indicated by a star (*). Time permitting, some of these topics may be included in the course. Furthermore, this material provides a suitable reference for basic principles when it is covered in other courses, and it can be used as a basis for assigning special projects.

Alternative Method of Coverage. Some instructors prefer to cover stress and strain transformations *first,* before discussing specific applications of axial load, torsion, bending, and shear. One possible method for doing this would be first to cover stress and its transformation, Chapter 1 and Chapter 9, followed by strain and its transformation, Chapter 2 and the first part of Chapter 10. The discussion and example problems in these later chapters have been styled so that this is possible. Also, the problem sets have been subdivided so that this material can be covered without prior knowledge of the intervening chapters. Chapters 3 through 8 can then be covered with no loss in continuity.

HALLMARK ELEMENTS

Organization and Approach. The contents of each chapter are organized into well-defined sections that contain an explanation of specific topics, illustrative example problems, and a set of homework problems. The topics within each section are placed into subgroups defined by titles. The purpose of this is to present a structured method for introducing each new definition or concept and to make the book convenient for later reference and review.

Chapter Contents. Each chapter begins with a full-page illustration that indicates a broad-range application of the material within the chapter. The "Chapter Objectives" are then provided to give a general overview of the material that will be covered.

Procedures for Analysis. Found after many of the sections of the book, this unique feature provides the student with a logical and orderly method to follow when applying the theory. The example problems are solved using this outlined method in order to clarify its numerical application. It is to be understood, however, that once the relevant principles have been mastered and enough confidence and judgment have been obtained, the student can then develop his or her own procedures for solving problems.

Photographs. Many photographs are used throughout the book to enhance conceptual understanding and explain how the principles of mechanics of materials apply to real-world situations.

Important Points. This feature provides a review or summary of the most important concepts in a section and highlights the most significant points that should be realized when applying the theory to solve problems.

Example Problems. All the example problems are presented in a concise manner and in a style that is easy to understand.

Homework Problems. Apart from of the preliminary, fundamental, and conceptual problems, there are numerous standard problems in the book that depict realistic situations encountered in engineering practice. It is hoped that this realism will both stimulate the student's interest in the subject and provide a means for developing the skill to reduce any such problem from its physical description to a model or a symbolic representation to which principles may be applied. Throughout the book there is an approximate balance of problems using either SI or FPS units. Furthermore, in any set, an attempt has been made to arrange the problems in order of increasing difficulty. The answers to all but every fourth problem are listed in the back of the book. To alert the user to a problem without a reported answer, an asterisk (*) is placed before the problem number. Answers are reported to three significant figures, even though the data for material properties may be known with less accuracy. Although this might appear to be a poor practice, it is done simply to be consistent, and to allow the student a better chance to validate his or her solution.

Appendices. The appendices of the book provide a source for review and a listing of tabular data. Appendix A provides information on the centroid and the moment of inertia of an area. Appendices B and C list tabular data for structural shapes, and the deflection and slopes of various types of beams and shafts.

Accuracy Checking. The Tenth Edition has undergone a rigorous Triple Accuracy Checking review. In addition to the author's review of all art pieces and pages, the text was checked by the following individuals:

- Scott Hendricks, Virginia Polytechnic University
- Karim Nohra, University of South Florida
- Kurt Norlin, Bittner Development Group
- Kai Beng Yap, Engineering Consultant

ACKNOWLEDGMENTS

Over the years, this text has been shaped by the suggestions and comments of many of my colleagues in the teaching profession. Their encouragement and willingness to provide constructive criticism are very much appreciated and it is hoped that they will accept this anonymous recognition. A note of thanks is given to the reviewers.

S. Apple, *Arkansas Tech University*
A. Bazar, *University of California, Fullerton*
M. Hughes, *Auburn University*
R. Jackson, *Auburn University*
E. Tezak, *Alfred State College*
H. Zhao, *Clemson University*

There are a few people that I feel deserve particular recognition. A long-time friend and associate, Kai Beng Yap, was of great help to me in preparing the problem solutions. A special note of thanks also goes to Kurt Norlin in this regard. During the production process I am thankful for the assistance of Rose Kernan, my production editor for many years, and to my wife, Conny, for her help in proofreading and typing, that was needed to prepare the manuscript for publication.

I would also like to thank all my students who have used the previous edition and have made comments to improve its contents; including all those in the teaching profession who have taken the time to e-mail me their comments, but in particular G. H. Nazari.

I would greatly appreciate hearing from you if at any time you have any comments or suggestions regarding the contents of this edition.

Russell Charles Hibbeler
hibbeler@bellsouth.net

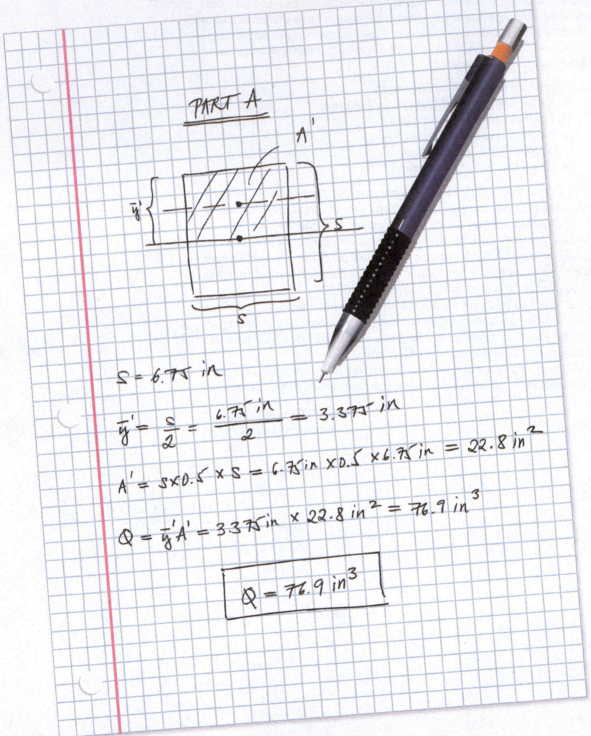

PART A

$$S = 6.75 \text{ in}$$

$$\bar{y}' = \frac{S}{2} = \frac{6.75 \text{ in}}{2} = 3.375 \text{ in}$$

$$A' = S \times 0.5 \times S = 6.75 \text{ in} \times 0.5 \times 6.75 \text{ in} = 22.8 \text{ in}^2$$

$$Q = \bar{y}'A' = 3.375 \text{ in} \times 22.8 \text{ in}^2 = 76.9 \text{ in}^3$$

$$\boxed{Q = 76.9 \text{ in}^3}$$

your answer specific feedback

Express your answer to three significant figures and include appropriate units.

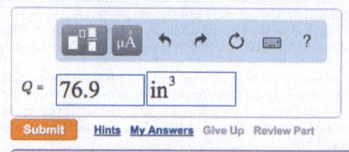

$Q =$ | 76.9 | in^3

Submit   **Hints** **My Answers** Give Up Review Part

Incorrect; Try Again; 5 attempts remaining

The distance between the horizontal centroidal axis of area A' and the neutral axis of the beam's cross section is half the distance between the top of the shaft and the neutral axis.

www.MasteringEngineering®.com

RESOURCES FOR INSTRUCTORS

• **MasteringEngineering.** This online Tutorial Homework program allows you to integrate dynamic homework with automatic grading and adaptive tutoring. MasteringEngineering allows you to easily track the performance of your entire class on an assignment-by-assignment basis, or the detailed work of an individual student.

• **Instructor's Solutions Manual.** An instructor's solutions manual was prepared by the author. The manual includes homework assignment lists and was also checked as part of the accuracy checking program. The Instructor Solutions Manual is available at www.pearsonhighered.com.

• **Presentation Resources.** All art from the text is available in PowerPoint slide and JPEG format. These files are available for download from the Instructor Resource Center at www.pearsonhighered.com. If you are in need of a login and password for this site, please contact your local Pearson representative.

• **Video Solutions.** Developed by Professor Edward Berger, Purdue University, video solutions located on the Pearson Engineering Portal offer step-by-step solution walkthroughs of representative homework problems from each section of the text. Make efficient use of class time and office hours by showing students the complete and concise problem solving approaches that they can access anytime and view at their own pace. The videos are designed to be a flexible resource to be used however each instructor and student prefers. A valuable tutorial resource, the videos are also helpful for student self-evaluation as students can pause the videos to check their understanding and work alongside the video. Access the videos at pearsonhighered.com/engineering-resources/ and follow the links for the *Mechanics of Materials* text.

RESOURCES FOR STUDENTS

• **Mastering Engineering.** Tutorial homework problems emulate the instrutor's office-hour environment.

• **Engineering Portal**—The Pearson Engineering Portal, located at pearsonhighered.com/engineering-resources/includes opportunities for practice and review including:

• **Video Solutions**—Complete, step-by-step solution walkthroughs of representative homework problems from each section of the text. Videos offer fully worked solutions that show every step of the representative homework problems—this helps students make vital connections between concepts.

CONTENTS

MECHANICS OF MATERIALS

CHAPTER 1

(© alexskopje/Fotolia)

The bolts used for the connections of this steel framework are subjected to stress. In this chapter we will discuss how engineers design these connections and their fasteners.

STRESS

CHAPTER OBJECTIVES

- In this chapter we will review some of the important principles of statics and show how they are used to determine the internal resultant loadings in a body. Afterwards the concepts of normal and shear stress will be introduced, and specific applications of the analysis and design of members subjected to an axial load or direct shear will be discussed.

1.1 INTRODUCTION

Mechanics of materials is a branch of mechanics that studies the internal effects of stress and strain in a solid body. Stress is associated with the strength of the material from which the body is made, while strain is a measure of the deformation of the body. A thorough understanding of the fundamentals of this subject is of vital importance for the design of any machine or structure, because many of the formulas and rules of design cited in engineering codes are based upon the principles of this subject.

Historical Development.   The origin of mechanics of materials dates back to the beginning of the seventeenth century, when Galileo Galilei performed experiments to study the effects of loads on rods and beams made of various materials. However, it was not until the beginning of the nineteenth century when experimental methods for testing materials were vastly improved. At that time many experimental and theoretical studies in this subject were undertaken, primarily in France, by such notables as Saint-Venant, Poisson, Lamé, and Navier.

Through the years, after many fundamental problems had been solved, it became necessary to use advanced mathematical and computer techniques to solve more complex problems. As a result, mechanics of materials has expanded into other areas of mechanics, such as the *theory of elasticity* and the *theory of plasticity*.

1.2   EQUILIBRIUM OF A DEFORMABLE BODY

Since statics plays an important role in both the development and application of mechanics of materials, it is very important to have a good grasp of its fundamentals. For this reason we will now review some of the main principles of statics that will be used throughout the text.

Loads.   A body can be subjected to both surface loads and body forces. *Surface loads* that act on a small area of contact are reported by *concentrated forces*, while *distributed loadings* act over a larger surface area of the body. When the loading is coplanar, as in Fig. 1–1a, then a resultant force $\mathbf{F}_R$ of a distributed loading is equal to the area under the distributed loading diagram, and this resultant acts through the geometric center or centroid of this area.

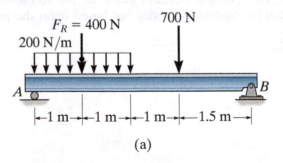

(a)

Fig. 1–1

A *body force* is developed when one body exerts a force on another body without direct physical contact between the bodies. Examples include the effects caused by the earth's gravitation or its electromagnetic field. Although these forces affect all the particles composing the body, they are normally represented by a single concentrated force acting on the body. In the case of gravitation, this force is called the *weight W* of the body and acts through the body's center of gravity.

Support Reactions. For bodies subjected to coplanar force systems, the supports most commonly encountered are shown in Table 1–1. As a general rule, *if the support prevents translation in a given direction, then a force must be developed on the member in that direction. Likewise, if rotation is prevented, a couple moment must be exerted on the member.* For example, the roller support only prevents translation perpendicular or normal to the surface. Hence, the roller exerts a normal force **F** on the member at its point of contact. Since the member can freely rotate about the roller, a couple moment cannot be developed on the member.

Many machine elements are pin connected in order to enable free rotation at their connections. These supports exert a force on a member, but no moment.

TABLE 1–1			
Type of connection	Reaction	Type of connection	Reaction
Cable	One unknown: F	External pin	Two unknowns: F_x, F_y
Roller	One unknown: F	Internal pin	Two unknowns: F_x, F_y
Smooth support	One unknown: F	Fixed support	Three unknowns: F_x, F_y, M
Journal bearing	One unknown: F	Thrust bearing	Two unknowns: F_x, F_y

In order to design the members of this building frame, it is first necessary to find the internal loadings at various points along their length.

Equations of Equilibrium.
Equilibrium of a body requires both a **balance of forces**, to prevent the body from translating or having accelerated motion along a straight or curved path, and a **balance of moments**, to prevent the body from rotating. These conditions are expressed mathematically as the equations of equilibrium:

$$\Sigma \mathbf{F} = \mathbf{0}$$
$$\Sigma \mathbf{M}_O = \mathbf{0}$$
(1–1)

Here, $\Sigma \mathbf{F}$ represents the sum of all the forces acting on the body, and $\Sigma \mathbf{M}_O$ is the sum of the moments of all the forces about any point O either on or off the body.

If an x, y, z coordinate system is established with the origin at point O, the force and moment vectors can be resolved into components along each coordinate axis, and the above two equations can be written in scalar form as six equations, namely,

$$\Sigma F_x = 0 \quad \Sigma F_y = 0 \quad \Sigma F_z = 0$$
$$\Sigma M_x = 0 \quad \Sigma M_y = 0 \quad \Sigma M_z = 0$$
(1–2)

Often in engineering practice the loading on a body can be represented as a system of *coplanar forces* in the x–y plane. In this case equilibrium of the body can be specified with only three scalar equilibrium equations, that is,

$$\Sigma F_x = 0$$
$$\Sigma F_y = 0$$
$$\Sigma M_O = 0$$
(1–3)

Successful application of the equations of equilibrium must include all the known and unknown forces that act on the body, and the best way to account for these loadings is to draw the body's free-body diagram before applying the equations of equilibrium. For example, the free-body diagram of the beam in Fig. 1–1a is shown in Fig. 1–1b. Here each force is identified by its magnitude and direction, and the body's dimensions are included in order to sum the moments of the forces.

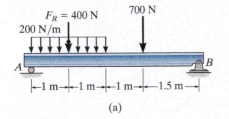

(a)

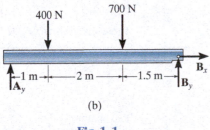

(b)

Fig. 1–1

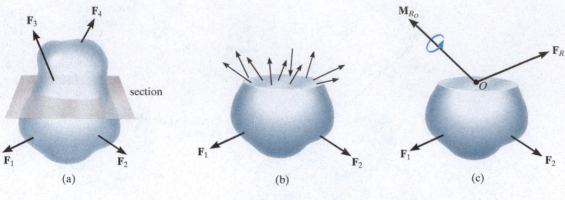

Fig. 1–2

Internal Resultant Loadings.

In mechanics of materials, statics is primarily used to determine the resultant loadings that act within a body. This is done using the ***method of sections.*** For example, consider the body shown in Fig. 1–2*a*, which is held in equilibrium by the four external forces.* In order to obtain the internal loadings acting on a specific region within the body, it is necessary to pass an imaginary section or "cut" through the region where the internal loadings are to be determined. The two parts of the body are then separated, and a free-body diagram of one of the parts is drawn. When this is done, there will be a distribution of internal force acting on the "exposed" area of the section, Fig. 1–2*b*. These forces actually represent the effects of the material of the top section of the body acting on the bottom section.

Although the exact distribution of this internal loading may be *unknown*, its resultants $\mathbf{F}_R$ and $\mathbf{M}_{R_O}$, Fig. 1–2*c*, are determined by applying the equations of equilibrium to the segment shown in Fig. 1–2*c*. Here these loadings act at point *O*; however, this point is often chosen at the centroid of the sectioned area.

---

*The body's weight is not shown, since it is assumed to be quite small, and therefore negligible compared with the other loads.

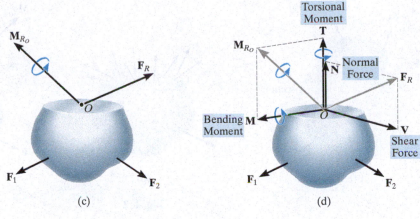

(c)   (d)

Fig. 1–2 (cont.)

The weight of this sign and the wind loadings acting on it will cause normal and shear forces and bending and torsional moments in the supporting column.

Three Dimensions.

For later application of the formulas for mechanics of materials, we will consider the components of $\mathbf{F}_R$ and $\mathbf{M}_{R_O}$ acting both normal and tangent to the sectioned area, Fig. 1–2d. Four different types of resultant loadings can then be defined as follows:

Normal force, N. This force acts perpendicular to the area. It is developed whenever the external loads tend to push or pull on the two segments of the body.

Shear force, V. The shear force lies in the plane of the area, and it is developed when the external loads tend to cause the two segments of the body to slide over one another.

Torsional moment or torque, T. This effect is developed when the external loads tend to twist one segment of the body with respect to the other about an axis perpendicular to the area.

Bending moment, M. The bending moment is caused by the external loads that tend to bend the body about an axis lying within the plane of the area.

Notice that graphical representation of a moment or torque is shown in three dimensions as a vector (arrow) with an associated curl around it. By the *right-hand rule*, the thumb gives the arrowhead sense of this vector and the fingers or curl indicate the tendency for rotation (twisting or bending).

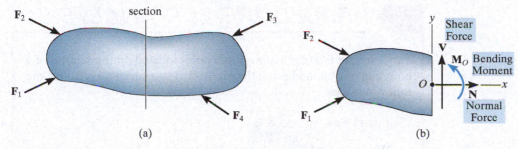

Fig. 1–3

Coplanar Loadings.   If the body is subjected to a *coplanar system of forces*, Fig. 1–3a, then only normal-force, shear-force, and bending-moment components will exist at the section, Fig. 1–3b. If we use the x, y, z coordinate axes, as shown on the left segment, then **N** can be obtained by applying $\Sigma F_x = 0$, and **V** can be obtained from $\Sigma F_y = 0$. Finally, the bending moment $\mathbf{M}_O$ can be determined by summing moments about point O (the z axis), $\Sigma M_O = 0$, in order to eliminate the moments caused by the unknowns **N** and **V**.

IMPORTANT POINTS

- *Mechanics of materials* is a study of the relationship between the external loads applied to a body and the stress and strain caused by the internal loads within the body.

- External forces can be applied to a body as *distributed* or *concentrated surface loadings*, or as *body forces* that act throughout the volume of the body.

- Linear distributed loadings produce a *resultant force* having a *magnitude* equal to the *area* under the load diagram, and having a *location* that passes through the *centroid* of this area.

- A support produces a *force* in a particular direction on its attached member if it *prevents translation* of the member in that direction, and it produces a *couple moment* on the member if it *prevents rotation*.

- The equations of equilibrium $\Sigma \mathbf{F} = \mathbf{0}$ and $\Sigma \mathbf{M} = \mathbf{0}$ must be satisfied in order to prevent a body from translating with accelerated motion and from rotating.

- The method of sections is used to determine the internal resultant loadings acting on the surface of a sectioned body. In general, these resultants consist of a normal force, shear force, torsional moment, and bending moment.

PROCEDURE FOR ANALYSIS

The resultant *internal* loadings at a point located on the section of a body can be obtained using the method of sections. This requires the following steps.

Support Reactions.

- When the body is sectioned, decide which segment of the body is to be considered. If the segment has a support or connection to another body, then *before* the body is sectioned, it will be necessary to determine the reactions acting on the chosen segment. To do this, draw the free-body diagram of the *entire body* and then apply the necessary equations of equilibrium to obtain these reactions.

Free-Body Diagram.

- Keep all external distributed loadings, couple moments, torques, and forces in their *exact locations*, before passing the section through the body at the point where the resultant internal loadings are to be determined.

- Draw a free-body diagram of one of the "cut" segments and indicate the unknown resultants N, V, M, and T at the section. These resultants are normally placed at the point representing the geometric center or *centroid* of the sectioned area.

- If the member is subjected to a *coplanar* system of forces, only N, V, and M act at the centroid.

- Establish the x, y, z coordinate axes with origin at the centroid and show the resultant internal loadings acting along the axes.

Equations of Equilibrium.

- Moments should be summed at the section, about each of the coordinate axes where the resultants act. Doing this eliminates the unknown forces N and V and allows a direct solution for M and T.

- If the solution of the equilibrium equations yields a negative value for a resultant, the *directional sense* of the resultant is *opposite* to that shown on the free-body diagram.

The following examples illustrate this procedure numerically and also provide a review of some of the important principles of statics.

EXAMPLE 1.1

Determine the resultant internal loadings acting on the cross section at C of the cantilevered beam shown in Fig. 1–4a.

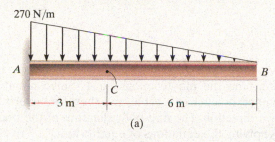

270 N/m

A

B

C

3 m

6 m

(a)

Fig. 1–4

SOLUTION

Support Reactions. The support reactions at A do not have to be determined if segment CB is considered.

Free-Body Diagram. The free-body diagram of segment CB is shown in Fig. 1–4b. It is important to keep the distributed loading on the segment until *after* the section is made. Only then should this loading be replaced by a single resultant force. Notice that the intensity of the distributed loading at C is found by proportion, i.e., from Fig. 1–4a, $w/6\text{ m} = (270\text{ N/m})/9\text{ m}, w = 180\text{ N/m}$. The magnitude of the resultant of the distributed load is equal to the area under the loading curve (triangle) and acts through the centroid of this area. Thus, $F = \frac{1}{2}(180\text{ N/m})(6\text{ m}) = 540\text{ N}$, which acts $\frac{1}{3}(6\text{ m}) = 2\text{ m}$ from C as shown in Fig. 1–4b.

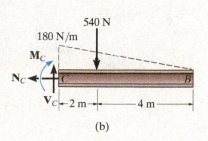

540 N

180 N/m

$\mathbf{M}_C$

N_C

C

B

$\mathbf{V}_C$ 2 m   4 m

(b)

Equations of Equilibrium. Applying the equations of equilibrium we have

$\xrightarrow{+}\ \Sigma F_x = 0;\qquad\qquad -N_C = 0$

$\qquad\qquad\qquad\qquad N_C = 0$ *Ans.*

$+\uparrow\Sigma F_y = 0;\qquad\quad V_C - 540\text{ N} = 0$

$\qquad\qquad\qquad\qquad V_C = 540\text{ N}$ *Ans.*

$\zeta+\Sigma M_C = 0;\qquad -M_C - 540\text{ N}(2\text{ m}) = 0$

$\qquad\qquad\qquad\quad M_C = -1080\text{ N}\cdot\text{m}$ *Ans.*

The negative sign indicates that $\mathbf{M}_C$ acts in the opposite direction to that shown on the free-body diagram. Try solving this problem using segment AC, by first checking the support reactions at A, which are given in Fig. 1–4c.

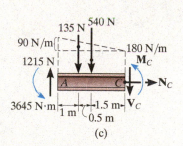

135 N   540 N

90 N/m   180 N/m

1215 N $\mathbf{M}_C$

A C N_C

3645 N·m   1.5 m $\mathbf{V}_C$

1 m   0.5 m

(c)

1

EXAMPLE 1.2

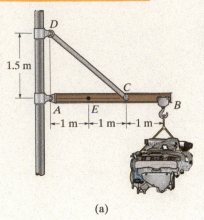

(a)

The 500-kg engine is suspended from the crane boom in Fig. 1–5a. Determine the resultant internal loadings acting on the cross section of the boom at point E.

SOLUTION

Support Reactions.   We will consider segment AE of the boom, so we must first determine the pin reactions at A. Since member CD is a two-force member, it acts like a cable, and therefore exerts a force F_{CD} having a known direction. The free-body diagram of the boom is shown in Fig. 1–5b. Applying the equations of equilibrium,

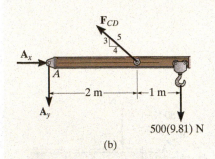

(b)

$$\zeta + \Sigma M_A = 0; \qquad F_{CD}\left(\tfrac{3}{5}\right)(2\text{ m}) - [500(9.81)\text{ N}](3\text{ m}) = 0$$

$$F_{CD} = 12\,262.5\text{ N}$$

$$\xrightarrow{+}\Sigma F_x = 0; \qquad A_x - (12\,262.5\text{ N})\left(\tfrac{4}{5}\right) = 0$$

$$A_x = 9810\text{ N}$$

$$+\uparrow\Sigma F_y = 0; \qquad -A_y + (12\,262.5\text{ N})\left(\tfrac{3}{5}\right) - 500(9.81)\text{ N} = 0$$

$$A_y = 2452.5\text{ N}$$

Free-Body Diagram.   The free-body diagram of segment AE is shown in Fig. 1–5c.

Equations of Equilibrium.

$$\xrightarrow{+}\Sigma F_x = 0; \qquad N_E + 9810\text{ N} = 0$$

$$N_E = -9810\text{ N} = -9.81\text{ kN} \qquad\qquad Ans.$$

$$+\uparrow\Sigma F_y = 0; \qquad -V_E - 2452.5\text{ N} = 0$$

$$V_E = -2452.5\text{ N} = -2.45\text{ kN} \qquad\qquad Ans.$$

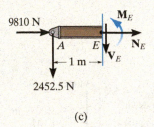

(c)

Fig. 1–5

$$\zeta + \Sigma M_E = 0; \qquad M_E + (2452.5\text{ N})(1\text{ m}) = 0$$

$$M_E = -2452.5\text{ N}\cdot\text{m} = -2.45\text{ kN}\cdot\text{m} \qquad\qquad Ans.$$

EXAMPLE 1.3

Determine the resultant internal loadings acting on the cross section at C of the beam shown in Fig. 1–6a.

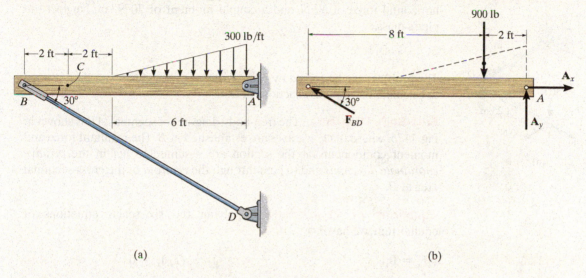

(a)

(b)

Fig. 1–6

SOLUTION

Support Reactions.   Here we will consider segment BC, but first we must find the force components at pin A. The free-body diagram of the *entire* beam is shown in Fig. 1–6b. Since member BD is a two-force member, like member CD in Example 1.2, the force at B has a known direction, Fig. 1–6b. We have

$$\zeta + \Sigma M_A = 0; \quad (900 \text{ lb})(2 \text{ ft}) - (F_{BD} \sin 30°) \, 10 \text{ ft} = 0 \quad F_{BD} = 360 \text{ lb}$$

Free-Body Diagram.   Using this result, the free-body diagram of segment BC is shown in Fig. 1–6c.

Equations of Equilibrium.

$$\xrightarrow{+} \Sigma F_x = 0; \qquad N_C - (360 \text{ lb}) \cos 30° = 0$$

$$N_C = 312 \text{ lb} \qquad\qquad Ans.$$

$$+\uparrow \Sigma F_y = 0; \qquad (360 \text{ lb}) \sin 30° - V_C = 0$$

$$V_C = 180 \text{ lb} \qquad\qquad Ans.$$

$$\zeta + \Sigma M_C = 0; \qquad M_C - (360 \text{ lb}) \sin 30°(2 \text{ ft}) = 0$$

$$M_C = 360 \text{ lb} \cdot \text{ft} \qquad\qquad Ans.$$

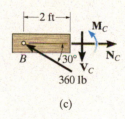

(c)

EXAMPLE 1.4

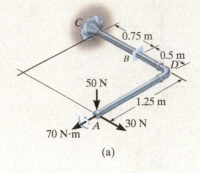

(a)

(b)

Fig. 1-7

Determine the resultant internal loadings acting on the cross section at B of the pipe shown in Fig. 1–7a. End A is subjected to a vertical force of 50 N, a horizontal force of 30 N, and a couple moment of 70 N · m. Neglect the pipe's mass.

SOLUTION

The problem can be solved by considering segment AB, so we do not need to calculate the support reactions at C.

Free-Body Diagram.   The free-body diagram of segment AB is shown in Fig. 1–7b, where the x, y, z axes are established at B. The resultant force and moment components at the section are assumed to act in the *positive coordinate directions* and to pass through the *centroid* of the cross-sectional area at B.

Equations of Equilibrium.   Applying the six scalar equations of equilibrium, we have*

$\Sigma F_x = 0;$ $(F_B)_x = 0$ *Ans.*

$\Sigma F_y = 0;$ $(F_B)_y + 30\text{ N} = 0$ $(F_B)_y = -30\text{ N}$ *Ans.*

$\Sigma F_z = 0;$ $(F_B)_z - 50\text{ N} = 0$ $(F_B)_z = 50\text{ N}$ *Ans.*

$\Sigma (M_B)_x = 0;$ $(M_B)_x + 70\text{ N} \cdot \text{m} - 50\text{ N}\,(0.5\text{ m}) = 0$

$$(M_B)_x = -45\text{ N} \cdot \text{m} \qquad \textit{Ans.}$$

$\Sigma (M_B)_y = 0;$ $(M_B)_y + 50\text{ N}\,(1.25\text{ m}) = 0$

$$(M_B)_y = -62.5\text{ N} \cdot \text{m} \qquad \textit{Ans.}$$

$\Sigma (M_B)_z = 0;$ $(M_B)_z + (30\text{ N})(1.25) = 0$ *Ans.*

$$(M_B)_z = -37.5\text{ N} \cdot \text{m}$$

NOTE: What do the negative signs for $(F_B)_y$, $(M_B)_x$, $(M_B)_y$, and $(M_B)_z$ indicate? The normal force $N_B = |(F_B)_y| = 30\text{ N}$, whereas the shear force is $V_B = \sqrt{(0)^2 + (50)^2} = 50\text{ N}$. Also, the torsional moment is $T_B = |(M_B)_y| = 62.5\text{ N} \cdot \text{m}$, and the bending moment is $M_B = \sqrt{(45)^2 + (37.5)^2} = 58.6\text{ N} \cdot \text{m}$.

---

*The *magnitude* of each moment about the x, y, or z axis is equal to the magnitude of each force times the perpendicular distance from the axis to the line of action of the force. The *direction* of each moment is determined using the right-hand rule, with positive moments (thumb) directed along the positive coordinate axes.

It is suggested that you test yourself on the solutions to these examples, by covering them over and then trying to think about which equilibrium equations must be used and how they are applied in order to determine the unknowns. Then before solving any of the problems, build your skills by first trying to solve the Preliminary Problems, which actually require little or no calculations, and then do some of the Fundamental Problems given on the following pages. The solutions and answers to all these problems are given in the back of the book. **Doing this throughout the book will help immensely in understanding how to apply the theory, and thereby develop your problem-solving skills.**

PRELIMINARY PROBLEMS

P1–1.   In each case, explain how to find the resultant internal loading acting on the cross section at point A. Draw all necessary free-body diagrams, and indicate the relevant equations of equilibrium. Do not calculate values. The lettered dimensions, angles, and loads are assumed to be known.

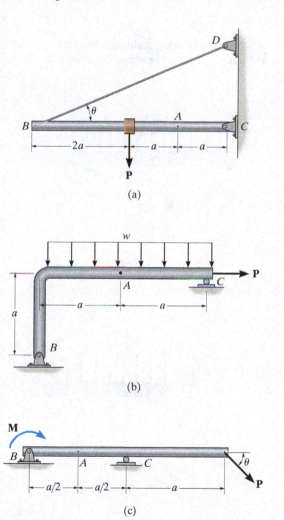

(a)

(b)

(c)

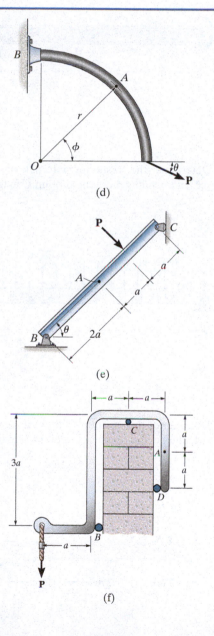

(d)

(e)

(f)

FUNDAMENTAL PROBLEMS

F1–1. Determine the resultant internal normal force, shear force, and bending moment at point C in the beam.

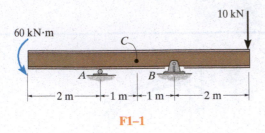

F1–1

F1–2. Determine the resultant internal normal force, shear force, and bending moment at point C in the beam.

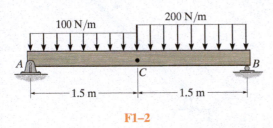

F1–2

F1–3. Determine the resultant internal normal force, shear force, and bending moment at point C in the beam.

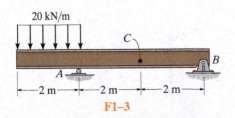

F1–3

F1–4. Determine the resultant internal normal force, shear force, and bending moment at point C in the beam.

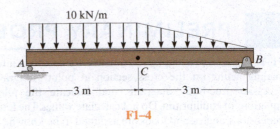

F1–4

F1–5. Determine the resultant internal normal force, shear force, and bending moment at point C in the beam.

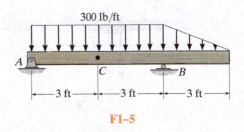

F1–5

F1–6. Determine the resultant internal normal force, shear force, and bending moment at point C in the beam.

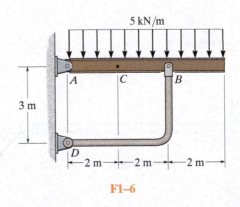

F1–6

PROBLEMS

1–1. The shaft is supported by a smooth thrust bearing at B and a journal bearing at C. Determine the resultant internal loadings acting on the cross section at E.

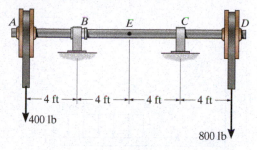

Prob. 1–1

1–2. Determine the resultant internal normal and shear force in the member at (a) section a–a and (b) section b–b, each of which passes through the centroid A. The 500-lb load is applied along the centroidal axis of the member.

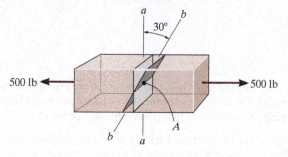

Prob. 1–2

1–3. Determine the resultant internal loadings acting on section b–b through the centroid C on the beam.

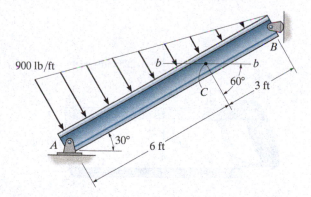

Prob. 1–3

*1–4. The shaft is supported by a smooth thrust bearing at A and a smooth journal bearing at B. Determine the resultant internal loadings acting on the cross section at C.

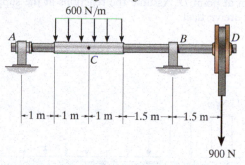

Prob. 1–4

1–5. Determine the resultant internal loadings acting on the cross section at point B.

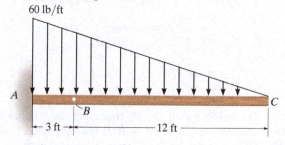

Prob. 1–5

1–6. Determine the resultant internal loadings on the cross section at point D.

1–7. Determine the resultant internal loadings at cross sections at points E and F on the assembly.

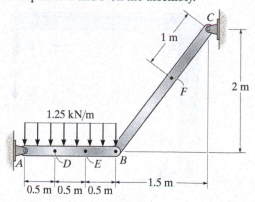

Probs. 1–6/7

*1–8. The beam supports the distributed load shown. Determine the resultant internal loadings acting on the cross section at point *C*. Assume the reactions at the supports *A* and *B* are vertical.

1–9. The beam supports the distributed load shown. Determine the resultant internal loadings acting on the cross section at point *D*. Assume the reactions at the supports *A* and *B* are vertical.

1–11. Determine the resultant internal loadings acting on the cross sections at points *D* and *E* of the frame.

*1–12. Determine the resultant internal loadings acting on the cross sections at points *F* and *G* of the frame.

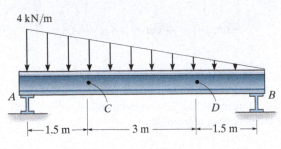

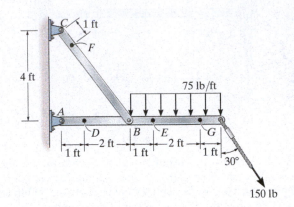

Probs. 1–8/9

Probs. 1–11/12

1–10. The boom *DF* of the jib crane and the column *DE* have a uniform weight of 50 lb/ft. If the supported load is 300 lb, determine the resultant internal loadings in the crane on cross sections at points *A*, *B*, and *C*.

1–13. The blade of the hacksaw is subjected to a pretension force of *F* = 100 N. Determine the resultant internal loadings acting on section *a–a* that passes through point *D*.

1–14. The blade of the hacksaw is subjected to a pretension force of *F* = 100 N. Determine the resultant internal loadings acting on section *b–b* that passes through point *D*.

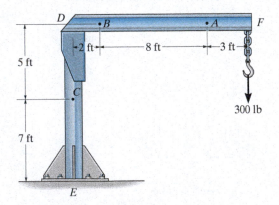

Prob. 1–10

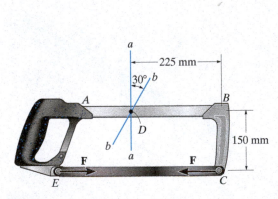

Probs. 1–13/14

1–15. The beam supports the triangular distributed load shown. Determine the resultant internal loadings on the cross section at point C. Assume the reactions at the supports A and B are vertical.

*1–16. The beam supports the distributed load shown. Determine the resultant internal loadings on the cross section at points D and E. Assume the reactions at the supports A and B are vertical.

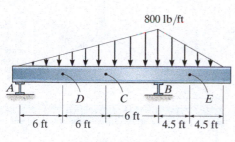

Probs. 1–15/16

1–17. The shaft is supported at its ends by two bearings A and B and is subjected to the forces applied to the pulleys fixed to the shaft. Determine the resultant internal loadings acting on the cross section at point D. The 400-N forces act in the −z direction and the 200-N and 80-N forces act in the +y direction. The journal bearings at A and B exert only y and z components of force on the shaft.

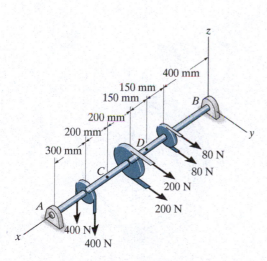

Prob. 1–17

1–18. The shaft is supported at its ends by two bearings A and B and is subjected to the forces applied to the pulleys fixed to the shaft. Determine the resultant internal loadings acting on the cross section at point C. The 400-N forces act in the −z direction and the 200-N and 80-N forces act in the +y direction. The journal bearings at A and B exert only y and z components of force on the shaft.

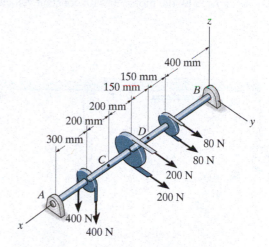

Prob. 1–18

1–19. The hand crank that is used in a press has the dimensions shown. Determine the resultant internal loadings acting on the cross section at point A if a vertical force of 50 lb is applied to the handle as shown. Assume the crank is fixed to the shaft at B.

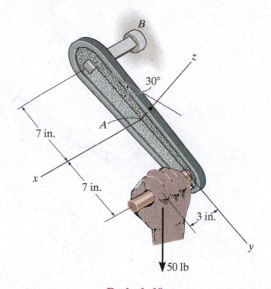

Prob. 1–19

*1–20.   Determine the resultant internal loadings acting on the cross section at point C in the beam. The load D has a mass of 300 kg and is being hoisted by the motor M with constant velocity.

1–21.   Determine the resultant internal loadings acting on the cross section at point E. The load D has a mass of 300 kg and is being hoisted by the motor M with constant velocity.

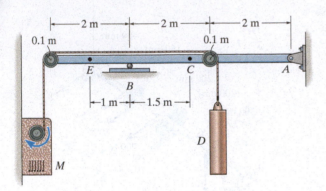

Probs. 1–20/21

1–22.   The metal stud punch is subjected to a force of 120 N on the handle. Determine the magnitude of the reactive force at the pin A and in the short link BC. Also, determine the resultant internal loadings acting on the cross section at point D.

1–23.   Determine the resultant internal loadings acting on the cross section at point E of the handle arm, and on the cross section of the short link BC.

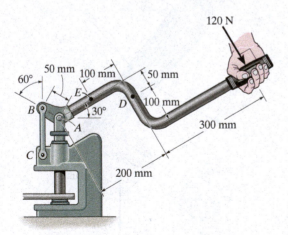

Probs. 1–22/23

*1–24.   Determine the resultant internal loadings acting on the cross section at point C. The cooling unit has a total weight of 52 kip and a center of gravity at G.

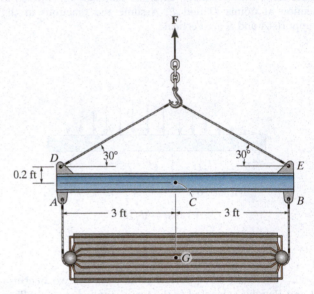

Prob. 1–24

1–25.   Determine the resultant internal loadings acting on the cross section at points B and C of the curved member.

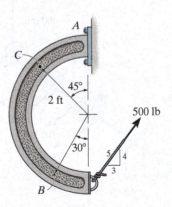

Prob. 1–25

1–26. Determine the resultant internal loadings acting on the cross section of the frame at points F and G. The contact at E is smooth.

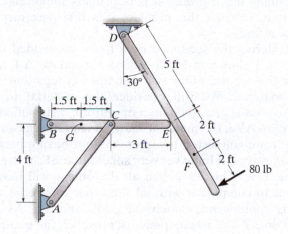

Prob. 1–26

1–27. The pipe has a mass of 12 kg/m. If it is fixed to the wall at A, determine the resultant internal loadings acting on the cross section at B.

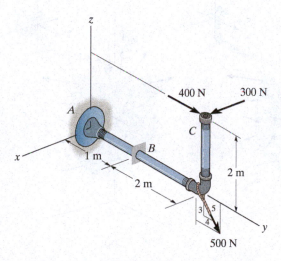

Prob. 1–27

*1–28 The brace and drill bit is used to drill a hole at O. If the drill bit jams when the brace is subjected to the forces shown, determine the resultant internal loadings acting on the cross section of the drill bit at A.

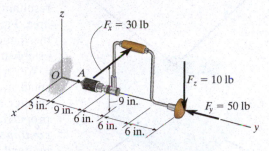

Prob. 1–28

1–29. The curved rod AD of radius r has a weight per length of w. If it lies in the horizontal plane, determine the resultant internal loadings acting on the cross section at point B. *Hint:* The distance from the centroid C of segment AB to point O is $CO = 0.9745r$.

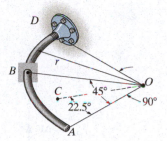

Prob. 1–29

1–30. A differential element taken from a curved bar is shown in the figure. Show that $dN/d\theta = V$, $dV/d\theta = -N$, $dM/d\theta = -T$, and $dT/d\theta = M$.

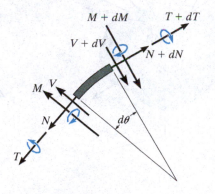

Prob. 1–30

1

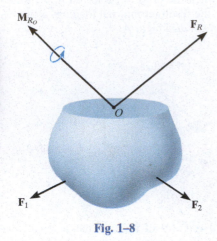

Fig. 1–8

1.3 STRESS

It was stated in Section 1.2 that the force and moment acting at a specified point O on the sectioned area of the body, Fig. 1–8, represents the resultant effects of the *distribution of loading* that acts over the sectioned area, Fig. 1–9a. Obtaining this *distribution* is of primary importance in mechanics of materials. To solve this problem it is first necessary to establish the concept of stress.

We begin by considering the sectioned area to be subdivided into small areas, such as ΔA shown in Fig. 1–9a. As we reduce ΔA to a smaller and smaller size, we will make two assumptions regarding the properties of the material. We will consider the material to be **continuous**, that is, to consist of a *continuum* or uniform distribution of matter having no voids. Also, the material must be **cohesive**, meaning that all portions of it are connected together, without having breaks, cracks, or separations. A typical finite yet very small force $\Delta \mathbf{F}$, acting on ΔA, is shown in Fig. 1–9a. This force, like all the others, will have a unique direction, but to compare it with all the other forces, we will replace it by its *three components*, namely, $\Delta \mathbf{F}_x$, $\Delta \mathbf{F}_y$, and $\Delta \mathbf{F}_z$. As ΔA approaches zero, so do $\Delta \mathbf{F}$ and its components; however, the quotient of the force and area will approach a finite limit. This quotient is called **stress**, and it describes the *intensity of the internal force* acting on a *specific plane* (area) passing through a point.

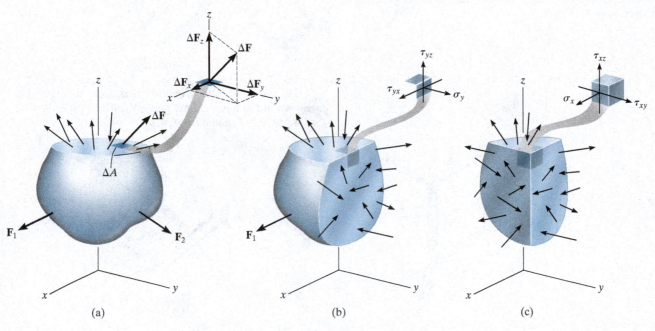

(a)   (b)   (c)

Fig. 1–9

Normal Stress.

The *intensity* of the force acting normal to ΔA is referred to as the **normal stress**, σ (sigma). Since $\Delta \mathbf{F}_z$ is normal to the area then

$$\sigma_z = \lim_{\Delta A \to 0} \frac{\Delta F_z}{\Delta A} \tag{1-4}$$

If the normal force or stress "pulls" on ΔA as shown in Fig. 1–9a, it is *tensile stress*, whereas if it "pushes" on ΔA it is *compressive stress*.

Shear Stress.

The intensity of force acting tangent to ΔA is called the **shear stress**, τ (tau). Here we have two shear stress components,

$$\tau_{zx} = \lim_{\Delta A \to 0} \frac{\Delta F_x}{\Delta A}$$
$$\tag{1-5}$$
$$\tau_{zy} = \lim_{\Delta A \to 0} \frac{\Delta F_y}{\Delta A}$$

The subscript notation z specifies the orientation of the area ΔA, Fig. 1–10, and x and y indicate the axes along which each shear stress acts.

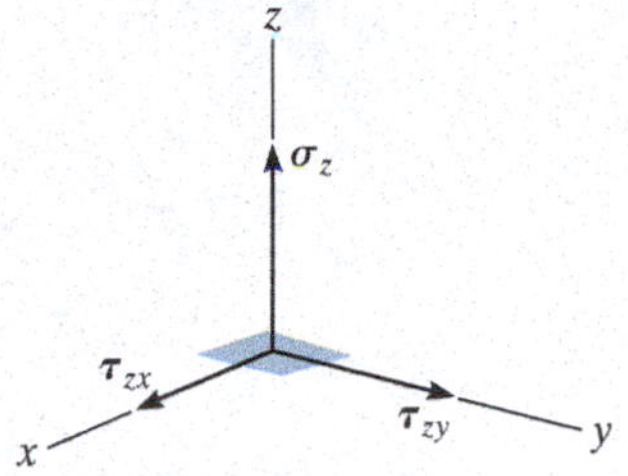

Fig. 1–10

General State of Stress.

If the body is further sectioned by planes parallel to the x–z plane, Fig. 1–9b, and the y–z plane, Fig. 1–9c, we can then "cut out" a cubic volume element of material that represents the **state of stress** acting around a chosen point in the body. This state of stress is then characterized by three components acting on each face of the element, Fig. 1–11.

Units.

Since stress represents a force per unit area, in the International Standard or SI system, the magnitudes of both normal and shear stress are specified in the basic units of newtons per square meter (N/m^2). This combination of units is called a pascal ($1\ Pa = 1\ N/m^2$), and because it is rather small, prefixes such as kilo- (10^3), symbolized by k, mega- (10^6), symbolized by M, or giga- (10^9), symbolized by G, are used in engineering to represent larger, more realistic values of stress.* In the Foot-Pound-Second system of units, engineers usually express stress in pounds per square inch (psi) or kilopounds per square inch (ksi), where 1 kilopound (kip) = 1000 lb.

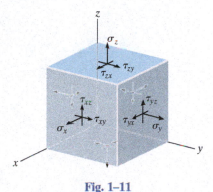

Fig. 1–11

*Sometimes stress is expressed in units of N/mm^2, where $1\ mm = 10^{-3}\ m$. However, in the SI system, prefixes are not allowed in the denominator of a fraction, and therefore it is better to use the equivalent $1\ N/mm^2 = 1\ MN/m^2 = 1\ MPa$.

1.4   AVERAGE NORMAL STRESS IN AN AXIALLY LOADED BAR

We will now determine the average stress distribution acting over the cross-sectional area of an axially loaded bar such as the one shown in Fig. 1–12*a*. Specifically, the **cross section** is the section taken *perpendicular* to the longitudinal axis of the bar, and since the bar is prismatic all cross sections are the same throughout its length. Provided the material of the bar is both **homogeneous** and **isotropic**, that is, it has the same physical and mechanical properties throughout its volume, and it has the same properties in all directions, then when the load *P* is applied to the bar through the centroid of its cross-sectional area, the bar will deform uniformly throughout the central region of its length, Fig. 1–12*b*.

Realize that many engineering materials may be approximated as being both homogeneous and isotropic. Steel, for example, contains thousands of randomly oriented crystals in each cubic millimeter of its volume, and since most objects made of this material have a physical size that is very much larger than a single crystal, the above assumption regarding the material's composition is quite realistic.

Note that **anisotropic materials**, such as wood, have different properties in different directions; and although this is the case, if the grains of wood are oriented along the bar's axis (as for instance in a typical wood board), then the bar will also deform uniformly when subjected to the axial load *P*.

Average Normal Stress Distribution.
If we pass a section through the bar, and separate it into two parts, then equilibrium requires the resultant normal force *N* at the section to be equal to *P*, Fig. 1–12*c*. And because the material undergoes a *uniform* deformation, it is necessary that the cross section be subjected to a *constant normal stress distribution*.

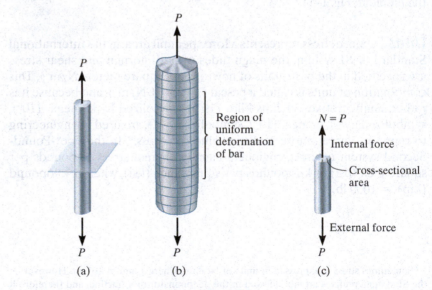

Region of uniform deformation of bar

$N = P$

Internal force

Cross-sectional area

External force

(a)   (b)   (c)

Fig. 1–12

As a result, each small area ΔA on the cross section is subjected to a force $\Delta N = \sigma \, \Delta A$, Fig. 1–12d, and the *sum* of these forces acting over the entire cross-sectional area must be equivalent to the internal resultant force **P** at the section. If we let $\Delta A \rightarrow dA$ and therefore $\Delta N \rightarrow dN$, then, recognizing σ is *constant*, we have

$$+\uparrow F_{Rz} = \Sigma F_z; \qquad \int dN = \int_A \sigma \, dA$$

$$N = \sigma A$$

$$\boxed{\sigma = \frac{N}{A}} \qquad (1\text{–}6)$$

Here

$\sigma =$ average normal stress at any point on the cross-sectional area

$N =$ *internal resultant normal force*, which acts through the *centroid* of the cross-sectional area. N is determined using the method of sections and the equations of equilibrium, where for this case $N = P$.

$A =$ cross-sectional area of the bar where σ is determined

Equilibrium.   The stress distribution in Fig. 1–12 indicates that only a normal stress exists on any small volume element of material located at each point on the cross section. Thus, if we consider vertical equilibrium of an element of material and then apply the equation of force equilibrium to its free-body diagram, Fig. 1–13,

$$\Sigma F_z = 0; \qquad \sigma(\Delta A) - \sigma'(\Delta A) = 0$$

$$\sigma = \sigma'$$

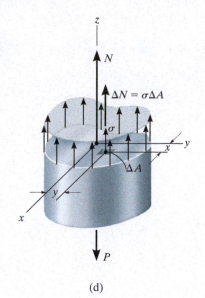

(d)

Fig. 1–12 (cont.)

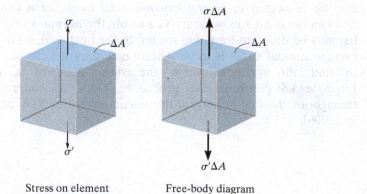

Stress on element   Free-body diagram

Fig. 1–13

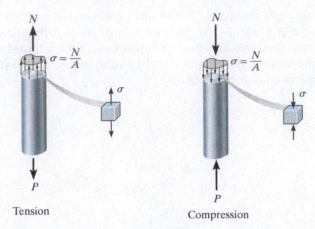

Tension   Compression

Fig. 1–14

This steel tie rod is used as a hanger to suspend a portion of a staircase. As a result it is subjected to tensile stress.

In other words, the normal stress components on the element must be equal in magnitude but opposite in direction. Under this condition the material is subjected to *uniaxial stress*, and this analysis applies to members subjected to either tension or compression, as shown in Fig. 1–14.

Although we have developed this analysis for *prismatic* bars, this assumption can be relaxed somewhat to include bars that have a *slight taper*. For example, it can be shown, using the more exact analysis of the theory of elasticity, that for a tapered bar of rectangular cross section, where the angle between two adjacent sides is 15°, the average normal stress, as calculated by $\sigma = N/A$, is only 2.2% *less* than its value found from the theory of elasticity.

Maximum Average Normal Stress.   For our analysis, both the internal force N and the cross-sectional area A were *constant* along the longitudinal axis of the bar, and as a result the normal stress $\sigma = N/A$ is also *constant* throughout the bar's length. Occasionally, however, the bar may be subjected to *several external axial loads*, or a change in its cross-sectional area may occur. As a result, the normal stress within the bar may be different from one section to the next, and, if the *maximum* average normal stress is to be determined, then it becomes important to find the location where the ratio N/A is a *maximum*. Example 1.5 illustrates the procedure. Once the internal loading throughout the bar is known, the maximum ratio N/A can then be identified.

IMPORTANT POINTS

- When a body subjected to external loads is sectioned, there is a distribution of force acting over the sectioned area which holds each segment of the body in equilibrium. The intensity of this internal force at a point in the body is referred to as *stress*.

- Stress is the limiting value of force per unit area, as the area approaches zero. For this definition, the material is considered to be continuous and cohesive.

- The magnitude of the stress components at a point depends upon the type of loading acting on the body, and the orientation of the element at the point.

- When a prismatic bar is made of homogeneous and isotropic material, and is subjected to an axial force acting through the centroid of the cross-sectional area, then the center region of the bar will deform uniformly. As a result, the material will be subjected *only to normal stress*. This stress is uniform or *averaged* over the cross-sectional area.

PROCEDURE FOR ANALYSIS

The equation $\sigma = N/A$ gives the *average* normal stress on the cross-sectional area of a member when the section is subjected to an internal resultant normal force N. Application of this equation requires the following steps.

Internal Loading.

- Section the member *perpendicular* to its longitudinal axis at the point where the normal stress is to be determined, and draw the free-body diagram of one of the segments. Apply the force equation of equilibrium to obtain the *internal axial force* N at the section.

Average Normal Stress.

- Determine the member's cross-sectional area at the section and calculate the average normal stress $\sigma = N/A$.

- It is suggested that σ be shown acting on a small volume element of the material located at a point on the section where stress is calculated. To do this, first draw σ on the face of the element coincident with the sectioned area A. Here σ acts in the *same direction* as the internal force N since all the normal stresses on the cross section develop this resultant. The normal stress σ on the opposite face of the element acts in the opposite direction.

EXAMPLE 1.5

The bar in Fig. 1–15a has a constant width of 35 mm and a thickness of 10 mm. Determine the maximum average normal stress in the bar when it is subjected to the loading shown.

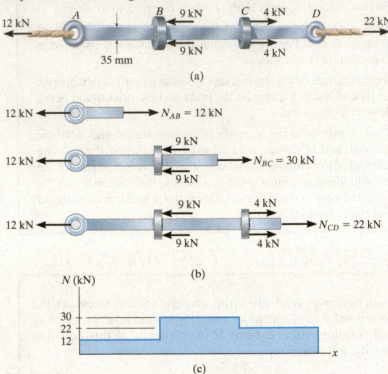

(a)

(b)

(c)

SOLUTION

Internal Loading.   By inspection, the internal axial forces in regions *AB, BC,* and *CD* are all constant yet have different magnitudes. Using the method of sections, these loadings are shown on the free-body diagrams of the left segments shown in Fig. 1–15b.* The *normal force diagram*, which represents these results graphically, is shown in Fig. 1–15c. The largest loading is in region *BC*, where $N_{BC} = 30$ kN. Since the cross-sectional area of the bar is *constant*, the largest average normal stress also occurs within this region of the bar.

Average Normal Stress.   Applying Eq. 1–6, we have

$$\sigma_{BC} = \frac{N_{BC}}{A} = \frac{30(10^3)\ \text{N}}{(0.035\ \text{m})(0.010\ \text{m})} = 85.7\ \text{MPa} \qquad \textit{Ans.}$$

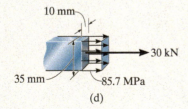

(d)

Fig. 1–15

The stress distribution acting on an arbitrary cross section of the bar within region *BC* is shown in Fig. 1–15d.

*Show that you get these *same results* using the right segments.

EXAMPLE 1.6

The 80-kg lamp is supported by two rods AB and BC as shown in Fig. 1–16a. If AB has a diameter of 10 mm and BC has a diameter of 8 mm, determine the average normal stress in each rod.

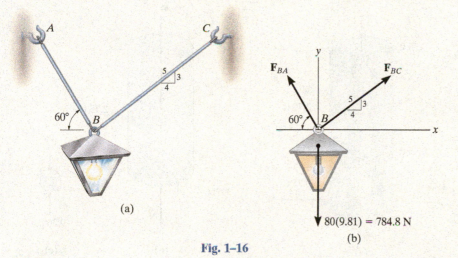

(a)

(b)

Fig. 1–16

SOLUTION

Internal Loading.   We must first determine the axial force in each rod. A free-body diagram of the lamp is shown in Fig. 1–16b. Applying the equations of force equilibrium,

$$\xrightarrow{+}\ \Sigma F_x = 0; \qquad F_{BC}\left(\tfrac{4}{5}\right) - F_{BA}\cos 60° = 0$$

$$+\uparrow \Sigma F_y = 0; \qquad F_{BC}\left(\tfrac{3}{5}\right) + F_{BA}\sin 60° - 784.8\ \text{N} = 0$$

$$F_{BC} = 395.2\ \text{N}, \qquad F_{BA} = 632.4\ \text{N}$$

By Newton's third law of action, equal but opposite reaction, these forces subject the rods to tension throughout their length.

Average Normal Stress.   Applying Eq. 1–6,

$$\sigma_{BC} = \frac{F_{BC}}{A_{BC}} = \frac{395.2\ \text{N}}{\pi(0.004\ \text{m})^2} = 7.86\ \text{MPa} \qquad \textit{Ans.}$$

$$\sigma_{BA} = \frac{F_{BA}}{A_{BA}} = \frac{632.4\ \text{N}}{\pi(0.005\ \text{m})^2} = 8.05\ \text{MPa} \qquad \textit{Ans.}$$

The average normal stress distribution acting over a cross section of rod AB is shown in Fig. 1–16c, and at a point on this cross section, an element of material is stressed as shown in Fig. 1–16d.

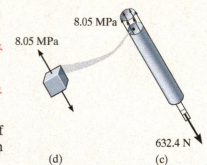

(d)

(c)

EXAMPLE   1.7

The cylinder shown in Fig. 1–17a is made of steel having a specific weight of $\gamma_{st} = 490 \text{ lb/ft}^3$. Determine the average compressive stress acting at points A and B.

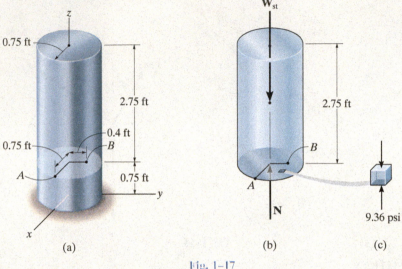

Fig. 1–17

SOLUTION

Internal Loading. A free-body diagram of the top segment of the cylinder is shown in Fig. 1–17b. The weight of this segment is determined from $W_{st} = \gamma_{st} V_{st}$. Thus the internal axial force N at the section is

$$+\uparrow \Sigma F_z = 0; \qquad N - W_{st} = 0$$

$$N - (490 \text{ lb/ft}^3)(2.75 \text{ ft})\left[\pi(0.75 \text{ ft})^2\right] = 0$$

$$N = 2381 \text{ lb}$$

Average Compressive Stress. The cross-sectional area at the section is $A = \pi(0.75 \text{ ft})^2$, and so the average compressive stress becomes

$$\sigma = \frac{N}{A} = \frac{2381 \text{ lb}}{\pi(0.75 \text{ ft})^2} = 1347.5 \text{ lb/ft}^2$$

$$\sigma = 1347.5 \text{ lb/ft}^2 \,(1 \text{ ft}^2/144 \text{ in}^2) = 9.36 \text{ psi} \qquad \textit{Ans.}$$

The stress shown on the volume element of material in Fig. 1–17c is representative of the conditions at either point A or B. Notice that this stress acts *upward* on the bottom or shaded face of the element since this face forms part of the bottom surface area of the section, and on this surface, the resultant force **N** is pushing upward.

EXAMPLE 1.8

Member *AC* shown in Fig. 1–18*a* is subjected to a vertical force of 3 kN. Determine the position *x* of this force so that the average compressive stress at the smooth support *C* is equal to the average tensile stress in the tie rod *AB*. The rod has a cross-sectional area of 400 mm^2 and the contact area at *C* is 650 mm^2.

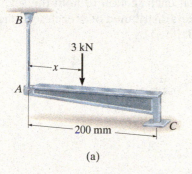

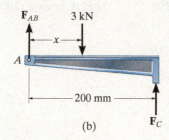

(a)   (b)

Fig. 1–18

SOLUTION

Internal Loading. The forces at *A* and *C* can be related by considering the free-body diagram of member *AC*, Fig. 1–18*b*. There are three unknowns, namely, F_{AB}, F_C, and *x*. To solve we will work in units of newtons and millimeters.

$$+\uparrow \Sigma F_y = 0; \qquad\qquad F_{AB} + F_C - 3000 \text{ N} = 0 \qquad (1)$$

$$\zeta +\Sigma M_A = 0; \qquad -3000 \text{ N}(x) + F_C(200 \text{ mm}) = 0 \qquad (2)$$

Average Normal Stress. A necessary third equation can be written that requires the tensile stress in the bar *AB* and the compressive stress at *C* to be equivalent, i.e.,

$$\sigma = \frac{F_{AB}}{400 \text{ mm}^2} = \frac{F_C}{650 \text{ mm}^2}$$

$$F_C = 1.625 F_{AB}$$

Substituting this into Eq. 1, solving for F_{AB}, then solving for F_C, we obtain

$$F_{AB} = 1143 \text{ N}$$

$$F_C = 1857 \text{ N}$$

The position of the applied load is determined from Eq. 2,

$$x = 124 \text{ mm} \qquad\qquad Ans.$$

As required, $0 < x < 200$ mm.

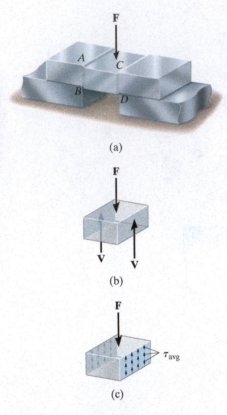

(a)

(b)

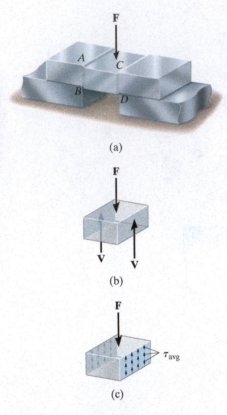

τ_{avg}

(c)

Fig. 1–19

The pin *A* used to connect the linkage of this tractor is subjected to *double shear* because shearing stresses occur on the surface of the pin at *B* and *C*. See Fig. 1–21*c*.

1.5   AVERAGE SHEAR STRESS

Shear stress has been defined in Section 1.3 as the stress component that acts *in the plane* of the sectioned area. To show how this stress can develop, consider the effect of applying a force **F** to the bar in Fig. 1–19*a*. If **F** is large enough, it can cause the material of the bar to deform and fail along the planes identified by *AB* and *CD*. A free-body diagram of the unsupported center segment of the bar, Fig. 1–19*b*, indicates that the shear force $V = F/2$ must be applied at each section to hold the segment in equilibrium. The ***average shear stress*** distributed over each sectioned area that develops this shear force is defined by

$$\tau_{avg} = \frac{V}{A} \tag{1–7}$$

Here

τ_{avg} = average shear stress at the section, which is assumed to be the *same* at each point on the section

V = internal resultant shear force on the section determined from the equations of equilibrium

A = area of the section

The distribution of average shear stress acting over the sections is shown in Fig. 1–19*c*. Notice that τ_{avg} is in the *same direction* as **V**, since the shear stress must create associated forces, all of which contribute to the internal resultant force **V**.

The loading case discussed here is an example of ***simple or direct shear***, since the shear is caused by the *direct action* of the applied load **F**. This type of shear often occurs in various types of simple connections that use bolts, pins, welding material, etc. In all these cases, however, application of Eq. 1–7 is *only approximate*. A more precise investigation of the shear-stress distribution over the section often reveals that much larger shear stresses occur in the material than those predicted by this equation. Although this may be the case, application of Eq. 1–7 is generally acceptable for many problems involving the design or analysis of small elements. For example, engineering codes allow its use for determining the size or cross section of fasteners such as bolts, and for obtaining the bonding strength of glued joints subjected to shear loadings.

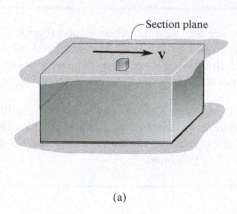

(a)

Fig. 1–20

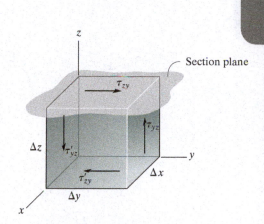

(b)

Shear Stress Equilibrium.

Let us consider the block in Fig. 1–20a, which has been sectioned and is subjected to the internal shear force V. A volume element taken at a point located on its surface will be subjected to a direct shear stress τ_{zy}, as shown in Fig. 1–20b. However, force and moment equilibrium of this element will also require shear stress to be developed on three other sides of the element. To show this, it is first necessary to draw the free-body diagram of the element, Fig. 1–20c. Then force equilibrium in the y direction requires

$$\Sigma F_y = 0; \qquad \tau_{zy}\underbrace{(\underbrace{\Delta x\, \Delta y}_{\text{area}})}_{} - \tau'_{zy}\, \Delta x\, \Delta y = 0$$

$$\tau_{zy} = \tau'_{zy}$$

In a similar manner, force equilibrium in the z direction yields $\tau_{yz} = \tau'_{yz}$. Finally, taking moments about the x axis,

$$\Sigma M_x = 0; \qquad -\tau_{zy}\underbrace{(\Delta x\, \Delta y)}_{}\,\Delta z + \tau_{yz}(\Delta x\, \Delta z)\,\Delta y = 0$$

$$\tau_{zy} = \tau_{yz}$$

In other words,

$$\tau_{zy} = \tau'_{zy} = \tau_{yz} = \tau'_{yz} = \tau$$

and so, ***all four shear stresses must have equal magnitude and be directed either toward or away from each other at opposite edges of the element***, Fig. 1–20d. This is referred to as the **complementary property of shear**, and the element in this case is subjected to *pure shear*.

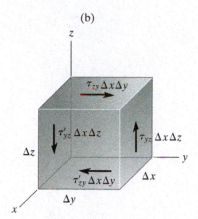

Free-body diagram

(c)

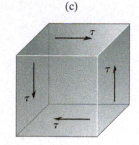

Pure shear

(d)

IMPORTANT POINTS

- If two parts are *thin or small* when joined together, the applied loads may cause shearing of the material with negligible bending. If this is the case, it is generally assumed that an *average shear stress* acts over the cross-sectional area.

- When shear stress τ acts on a plane, then equilibrium of a volume element of material at a point on the plane requires associated shear stress of the same magnitude act on the three other sides of the element.

PROCEDURE FOR ANALYSIS

The equation $\tau_{avg} = V/A$ is used to determine the *average shear stress* in the material. Application requires the following steps.

Internal Shear.

- Section the member at the point where the average shear stress is to be determined.

- Draw the necessary free-body diagram, and calculate the internal shear force $\mathbf{V}$ acting at the section that is necessary to hold the part in equilibrium.

Average Shear Stress.

- Determine the sectioned area A, and then calculate the average shear stress $\tau_{avg} = V/A$.

- It is suggested that τ_{avg} be shown on a small volume element of material located at a point on the section where it is determined. To do this, first draw τ_{avg} on the face of the element, coincident with the sectioned area A. This stress acts in the *same direction* as $\mathbf{V}$. The shear stresses acting on the three adjacent planes can then be drawn in their appropriate directions following the scheme shown in Fig. 1–20d.

EXAMPLE 1.9

Determine the average shear stress in the 20-mm-diameter pin at A and the 30-mm-diameter pin at B that support the beam in Fig. 1–21a.

SOLUTION

Internal Loadings.   The forces on the pins can be obtained by considering the equilibrium of the beam, Fig. 1–21b.

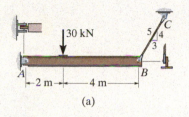

(a)

$$\zeta + \Sigma M_A = 0;$$

$$F_B\left(\frac{4}{5}\right)(6 \text{ m}) - 30 \text{ kN}(2 \text{ m}) = 0 \qquad F_B = 12.5 \text{ kN}$$

$$\xrightarrow{+} \Sigma F_x = 0; \quad (12.5 \text{ kN})\left(\frac{3}{5}\right) - A_x = 0 \qquad A_x = 7.50 \text{ kN}$$

$$+\uparrow \Sigma F_y = 0; \quad A_y + (12.5 \text{ kN})\left(\frac{4}{5}\right) - 30 \text{ kN} = 0 \quad A_y = 20 \text{ kN}$$

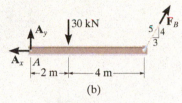

(b)

Thus, the resultant force acting on pin A is

$$F_A = \sqrt{A_x^2 + A_y^2} = \sqrt{(7.50 \text{ kN})^2 + (20 \text{ kN})^2} = 21.36 \text{ kN}$$

The pin at A is supported by two fixed "leaves" and so the free-body diagram of the center segment of the pin shown in Fig. 1–21c has *two* shearing surfaces between the beam and each leaf. Since the force of the beam (21.36 kN) acting on the pin is supported by shear force on each of two surfaces, it is called **double shear**. Thus,

$$V_A = \frac{F_A}{2} = \frac{21.36 \text{ kN}}{2} = 10.68 \text{ kN}$$

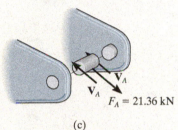

(c)

In Fig. 1–21a, note that pin B is subjected to **single shear**, which occurs on the section between the cable and beam, Fig. 1–21d. For this pin segment,

$$V_B = F_B = 12.5 \text{ kN}$$

Average Shear Stress.

$$(\tau_A)_{\text{avg}} = \frac{V_A}{A_A} = \frac{10.68(10^3) \text{ N}}{\frac{\pi}{4}(0.02 \text{ m})^2} = 34.0 \text{ MPa} \qquad \textit{Ans.}$$

$$(\tau_B)_{\text{avg}} = \frac{V_B}{A_B} = \frac{12.5(10^3) \text{ N}}{\frac{\pi}{4}(0.03 \text{ m})^2} = 17.7 \text{ MPa} \qquad \textit{Ans.}$$

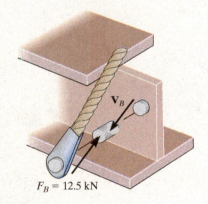

(d)

Fig. 1–21

EXAMPLE   1.10

If the wood joint in Fig. 1–22a has a thickness of 150 mm, determine the average shear stress along shear planes a–a and b–b of the connected member. For each plane, represent the state of stress on an element of the material.

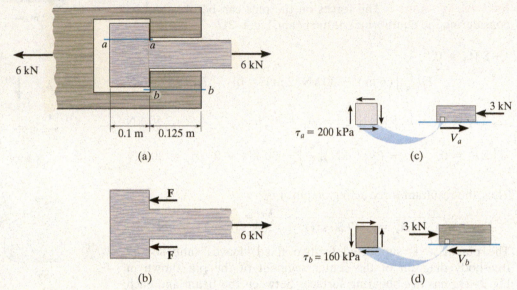

Fig. 1–22

SOLUTION

Internal Loadings.   Referring to the free-body diagram of the member, Fig. 1–22b,

$$\xrightarrow{+} \Sigma F_x = 0; \qquad 6 \text{ kN} - F - F = 0 \qquad F = 3 \text{ kN}$$

Now consider the equilibrium of segments cut across shear planes a–a and b–b, shown in Figs. 1–22c and 1–22d.

$$\xrightarrow{+} \Sigma F_x = 0; \qquad V_a - 3 \text{ kN} = 0 \qquad V_a = 3 \text{ kN}$$

$$\xrightarrow{+} \Sigma F_x = 0; \qquad 3 \text{ kN} - V_b = 0 \qquad V_b = 3 \text{ kN}$$

Average Shear Stress.

$$(\tau_a)_{\text{avg}} = \frac{V_a}{A_a} = \frac{3(10^3) \text{ N}}{(0.1 \text{ m})(0.15 \text{ m})} = 200 \text{ kPa} \qquad \textit{Ans.}$$

$$(\tau_b)_{\text{avg}} = \frac{V_b}{A_b} = \frac{3(10^3) \text{ N}}{(0.125 \text{ m})(0.15 \text{ m})} = 160 \text{ kPa} \qquad \textit{Ans.}$$

The state of stress on elements located on sections a–a and b–b is shown in Figs. 1–22c and 1–22d, respectively.

EXAMPLE 1.11

The inclined member in Fig. 1–23a is subjected to a compressive force of 600 lb. Determine the average compressive stress along the smooth areas of contact at *AB* and *BC*, and the average shear stress along the horizontal plane *DB*.

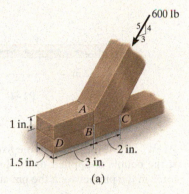

(a)

Fig. 1–23

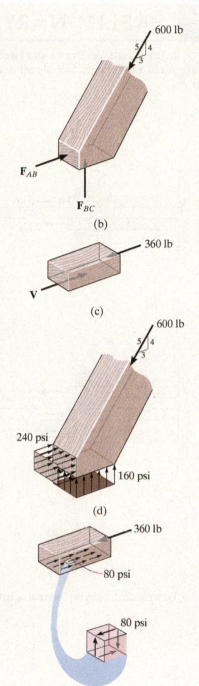

(b)

(c)

(d)

(e)

SOLUTION

Internal Loadings. The free-body diagram of the inclined member is shown in Fig. 1–23b. The compressive forces acting on the areas of contact are

$$\xrightarrow{+} \Sigma F_x = 0; \qquad F_{AB} - 600 \text{ lb}\left(\tfrac{3}{5}\right) = 0 \qquad F_{AB} = 360 \text{ lb}$$

$$+\uparrow \Sigma F_y = 0; \qquad F_{BC} - 600 \text{ lb}\left(\tfrac{4}{5}\right) = 0 \qquad F_{BC} = 480 \text{ lb}$$

Also, from the free-body diagram of the segment *ABD*, Fig. 1–23c, the shear force acting on the sectioned horizontal plane *DB* is

$$\xrightarrow{+} \Sigma F_x = 0; \qquad V - 360 \text{ lb} = 0 \qquad V = 360 \text{ lb}$$

Average Stress. The average compressive stresses along the vertical and horizontal planes of the inclined member are

$$\sigma_{AB} = \frac{F_{AB}}{A_{AB}} = \frac{360 \text{ lb}}{(1 \text{ in.})(1.5 \text{ in.})} = 240 \text{ psi} \qquad \textit{Ans.}$$

$$\sigma_{BC} = \frac{F_{BC}}{A_{BC}} = \frac{480 \text{ lb}}{(2 \text{ in.})(1.5 \text{ in.})} = 160 \text{ psi} \qquad \textit{Ans.}$$

These stress distributions are shown in Fig. 1–23d.

The average shear stress acting on the horizontal plane defined by *DB* is

$$\tau_{\text{avg}} = \frac{360 \text{ lb}}{(3 \text{ in.})(1.5 \text{ in.})} = 80 \text{ psi} \qquad \textit{Ans.}$$

This stress is shown uniformly distributed over the sectioned area and on an element of material in Fig. 1–23e.

PRELIMINARY PROBLEMS

P1–2. In each case, determine the largest internal shear force resisted by the bolt. Include all necessary free-body diagrams.

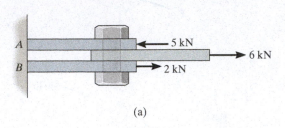

(a)

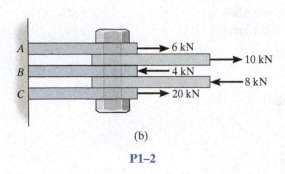

(b)

P1–2

P1–3. Determine the largest internal normal force in the bar.

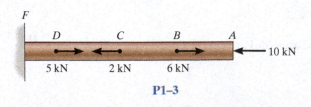

P1–3

P1–4. Determine the internal normal force at section A if the rod is subjected to the external uniformly distributed loading along its length of 8 kN/m.

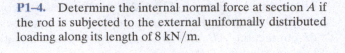

P1–4

P1–5. The lever is held to the fixed shaft using the pin AB. If the couple is applied to the lever, determine the shear force in the pin between the pin and the lever.

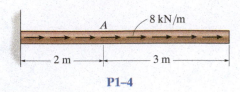

P1–5

P1–6. The single-V butt joint transmits the force of 5 kN from one bar to the other. Determine the resultant normal and shear force components on the face of the weld, section AB.

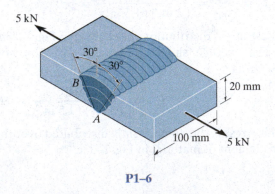

P1–6

FUNDAMENTAL PROBLEMS

F1–7. The uniform beam is supported by two rods AB and CD that have cross-sectional areas of 10 mm^2 and 15 mm^2, respectively. Determine the intensity w of the distributed load so that the average normal stress in each rod does not exceed 300 kPa.

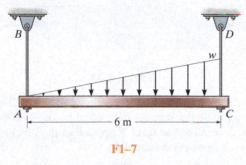

F1–7

F1–8. Determine the average normal stress on the cross section. Sketch the normal stress distribution over the cross section.

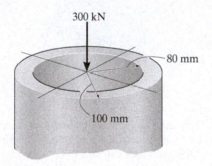

F1–8

F1–9. Determine the average normal stress on the cross section. Sketch the normal stress distribution over the cross section.

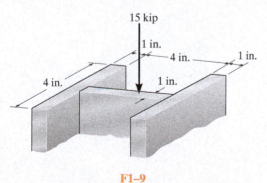

F1–9

F1–10. If the 600-kN force acts through the centroid of the cross section, determine the location $\bar{y}$ of the centroid and the average normal stress on the cross section. Also, sketch the normal stress distribution over the cross section.

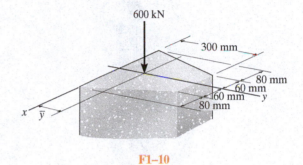

F1–10

F1–11. Determine the average normal stress at points A, B, and C. The diameter of each segment is indicated in the figure.

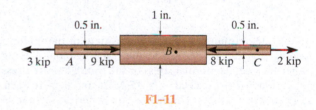

F1–11

F1–12. Determine the average normal stress in rod AB if the load has a mass of 50 kg. The diameter of rod AB is 8 mm.

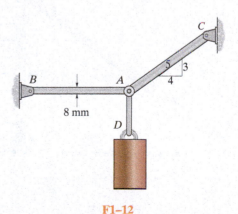

F1–12

PROBLEMS

1–31. The supporting wheel on a scaffold is held in place on the leg using a 4-mm-diameter pin. If the wheel is subjected to a normal force of 3 kN, determine the average shear stress in the pin. Assume the pin only supports the vertical 3-kN load.

3 kN

Prob. 1–31

*1–32. Determine the largest intensity w of the uniform loading that can be applied to the frame without causing either the average normal stress or the average shear stress at section b–b to exceed $\sigma = 15$ MPa and $\tau = 16$ MPa, respectively. Member CB has a square cross section of 30 mm on each side.

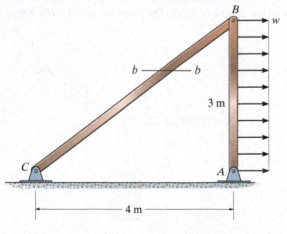

Prob. 1–32

1–33. The bar has a cross-sectional area A and is subjected to the axial load P. Determine the average normal and average shear stresses acting over the shaded section, which is oriented at θ from the horizontal. Plot the variation of these stresses as a function of θ $(0 \leq \theta \leq 90°)$.

Prob. 1–33

1–34. The small block has a thickness of 0.5 in. If the stress distribution at the support developed by the load varies as shown, determine the force **F** applied to the block, and the distance d to where it is applied.

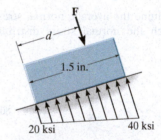

20 ksi   40 ksi

Prob. 1–34

1–35. If the material fails when the average normal stress reaches 120 psi, determine the largest centrally applied vertical load **P** the block can support.

*1–36. If the block is subjected to a centrally applied force of $P = 6$ kip, determine the average normal stress in the material. Show the stress acting on a differential volume element of the material.

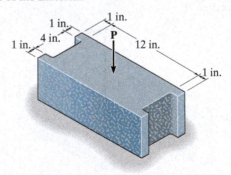

Probs. 1–35/36

1–37. The plate has a width of 0.5 m. If the stress distribution at the support varies as shown, determine the force **P** applied to the plate and the distance d to where it is applied.

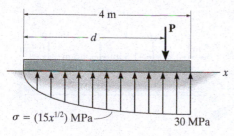

$\sigma = (15x^{1/2})$ MPa — 30 MPa

Prob. 1–37

1–38. The board is subjected to a tensile force of 200 lb. Determine the average normal and average shear stress in the wood fibers, which are oriented along plane a–a at 20° with the axis of the board.

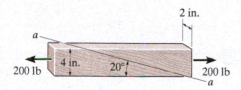

Prob. 1–38

1–39. The boom has a uniform weight of 600 lb and is hoisted into position using the cable BC. If the cable has a diameter of 0.5 in., plot the average normal stress in the cable as a function of the boom position θ for $0° \le \theta \le 90°$.

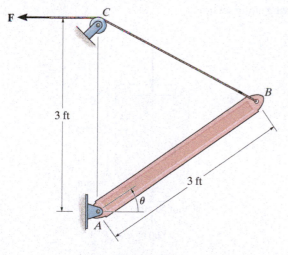

Prob. 1–39

*1–40. Determine the average normal stress in each of the 20-mm-diameter bars of the truss. Set $P = 40$ kN.

1–41. If the average normal stress in each of the 20-mm-diameter bars is not allowed to exceed 150 MPa, determine the maximum force **P** that can be applied to joint C.

1–42. Determine the maximum average shear stress in pin A of the truss. A horizontal force of $P = 40$ kN is applied to joint C. Each pin has a diameter of 25 mm and is subjected to double shear.

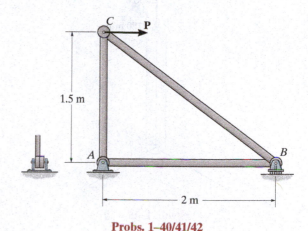

Probs. 1–40/41/42

1–43. If $P = 5$ kN, determine the average shear stress in the pins at A, B, and C. All pins are in double shear, and each has a diameter of 18 mm.

*1–44. Determine the maximum magnitude P of the loads the beam can support if the average shear stress in each pin is not to exceed 80 MPa. All pins are in double shear, and each has a diameter of 18 mm.

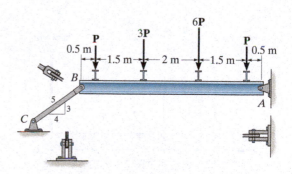

Probs. 1–43/44

1–45. The column is made of concrete having a density of 2.30 Mg/m³. At its top B it is subjected to an axial compressive force of 15 kN. Determine the average normal stress in the column as a function of the distance z measured from its base.

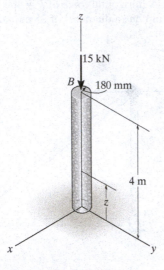

Prob. 1–45

1–46. The beam is supported by two rods AB and CD that have cross-sectional areas of 12 mm² and 8 mm², respectively. If $d = 1$ m, determine the average normal stress in each rod.

1–47. The beam is supported by two rods AB and CD that have cross-sectional areas of 12 mm² and 8 mm², respectively. Determine the position d of the 6-kN load so that the average normal stress in each rod is the same.

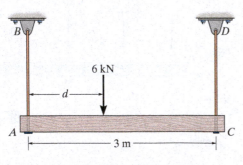

Probs. 1–46/47

*1–48. If $P = 15$ kN, determine the average shear stress in the pins at A, B, and C. All pins are in double shear, and each has a diameter of 18 mm.

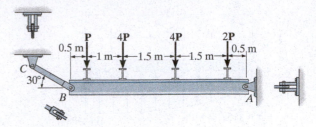

Prob. 1–48

1–49. The railcar docklight is supported by the $\frac{1}{8}$-in.-diameter pin at A. If the lamp weighs 4 lb, and the extension arm AB has a weight of 0.5 lb/ft, determine the average shear stress in the pin needed to support the lamp. *Hint:* The shear force in the pin is caused by the couple moment required for equilibrium at A.

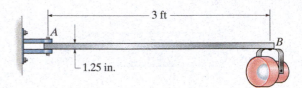

Prob. 1–49

1–50. The plastic block is subjected to an axial compressive force of 600 N. Assuming that the caps at the top and bottom distribute the load uniformly throughout the block, determine the average normal and average shear stress acting along section a–a.

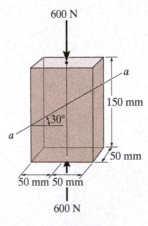

Prob. 1–50

1–51. The two steel members are joined together using a 30° scarf weld. Determine the average normal and average shear stress resisted in the plane of the weld.

1–54. The two members used in the construction of an aircraft fuselage are joined together using a 30° fish-mouth weld. Determine the average normal and average shear stress on the plane of each weld. Assume each inclined plane supports a horizontal force of 400 lb.

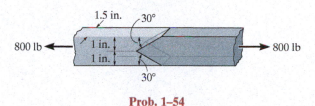

Prob. 1–54

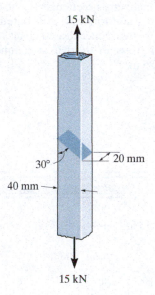

Prob. 1–51

1–55. The 2-Mg concrete pipe has a center of mass at point *G*. If it is suspended from cables *AB* and *AC*, determine the average normal stress in the cables. The diameters of *AB* and *AC* are 12 mm and 10 mm, respectively.

***1–56.** The 2-Mg concrete pipe has a center of mass at point *G*. If it is suspended from cables *AB* and *AC*, determine the diameter of cable *AB* so that the average normal stress in this cable is the same as in the 10-mm-diameter cable *AC*.

***1–52.** The bar has a cross-sectional area of $400(10^{-6})$ m^2. If it is subjected to a triangular axial distributed loading along its length which is 0 at $x = 0$ and 9 kN/m at $x = 1.5$ m, and to two concentrated loads as shown, determine the average normal stress in the bar as a function of x for $0 \leq x < 0.6$ m.

1–53. The bar has a cross-sectional area of $400(10^{-6})$ m^2. If it is subjected to a uniform axial distributed loading along its length of 9 kN/m, and to two concentrated loads as shown, determine the average normal stress in the bar as a function of x for 0.6 m $< x \leq 1.5$ m.

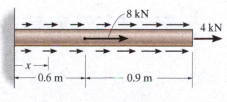

Probs. 1–52/53

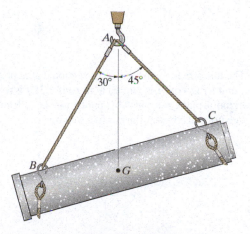

Probs. 1–55/56

1–57. The pier is made of material having a specific weight γ. If it has a square cross section, determine its width w as a function of z so that the average normal stress in the pier remains constant. The pier supports a constant load $\mathbf{P}$ at its top where its width is w_1.

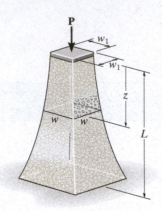

Prob. 1–57

1–58. Rods AB and BC have diameters of 4 mm and 6 mm, respectively. If the 3 kN force is applied to the ring at B, determine the angle θ so that the average normal stress in each rod is equivalent. What is this stress?

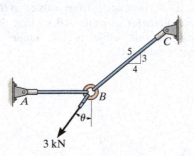

Prob. 1–58

1–59. The uniform bar, having a cross-sectional area of A and mass per unit length of m, is pinned at its center. If it is rotating in the horizontal plane at a constant angular rate of ω, determine the average normal stress in the bar as a function of x.

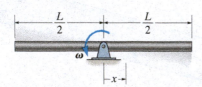

Prob. 1–59

*1–60. The bar has a cross-sectional area of $400(10^{-6})$ m². If it is subjected to a uniform axial distributed loading along its length and to two concentrated loads, determine the average normal stress in the bar as a function of x for $0 < x \leq 0.5$ m.

1–61. The bar has a cross-sectional area of $400(10^{-6})$ m². If it is subjected to a uniform axial distributed loading along its length and to two concentrated loads, determine the average normal stress in the bar as a function of x for 0.5 m $< x \leq 1.25$ m.

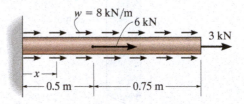

Probs. 1–60/61

1–62. The prismatic bar has a cross-sectional area A. If it is subjected to a distributed axial loading that increases linearly from $w = 0$ at $x = 0$ to $w = w_0$ at $x = a$, and then decreases linearly to $w = 0$ at $x = 2a$, determine the average normal stress in the bar as a function of x for $0 \leq x < a$.

1–63. The prismatic bar has a cross-sectional area A. If it is subjected to a distributed axial loading that increases linearly from $w = 0$ at $x = 0$ to $w = w_0$ at $x = a$, and then decreases linearly to $w = 0$ at $x = 2a$, determine the average normal stress in the bar as a function of x for $a < x \leq 2a$.

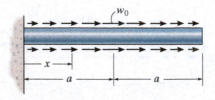

Probs. 1–62/63

*1–64. The bars of the truss each have a cross-sectional area of 1.25 in². Determine the average normal stress in members AB, BD, and CE due to the loading $P = 6$ kip. State whether the stress is tensile or compressive.

1–65. The bars of the truss each have a cross-sectional area of 1.25 in². If the maximum average normal stress in any bar is not to exceed 20 ksi, determine the maximum magnitude P of the loads that can be applied to the truss.

1–67. Determine the greatest constant angular velocity ω of the flywheel so that the average normal stress in its rim does not exceed $\sigma = 15$ MPa. Assume the rim is a thin ring having a thickness of 3 mm, width of 20 mm, and a mass of 30 kg/m. Rotation occurs in the horizontal plane. Neglect the effect of the spokes in the analysis. *Hint:* Consider a free-body diagram of a semicircular segment of the ring. The center of mass for this segment is located at $\hat{r} = 2r/\pi$ from the center.

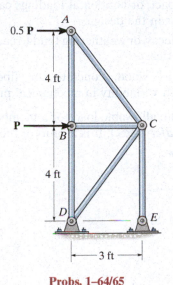

Probs. 1–64/65

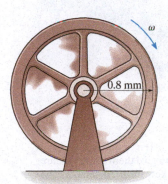

Prob. 1–67

1–66. Determine the largest load P that can be applied to the frame without causing either the average normal stress or the average shear stress at section a–a to exceed $\sigma = 150$ MPa and $\tau = 60$ MPa, respectively. Member CB has a square cross section of 25 mm on each side.

*1–68. The radius of the pedestal is defined by $r = (0.5e^{-0.08y^2})$ m, where y is in meters. If the material has a density of 2.5 Mg/m³, determine the average normal stress at the support.

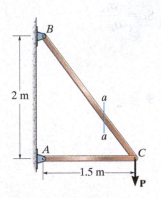

Prob. 1–66

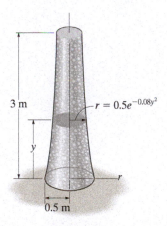

Prob. 1–68

1.6 ALLOWABLE STRESS DESIGN

To ensure the safety of a structural or mechanical member, it is necessary to restrict the applied load to one that is *less than* the load the member can fully support. There are many reasons for doing this.

- The intended measurements of a structure or machine may not be exact, due to errors in fabrication or in the assembly of its component parts.
- Unknown vibrations, impact, or accidental loadings can occur that may not be accounted for in the design.
- Atmospheric corrosion, decay, or weathering tend to cause materials to deteriorate during service.
- Some materials, such as wood, concrete, or fiber-reinforced composites, can show high variability in mechanical properties.

One method of specifying the allowable load for a member is to use a number called the *factor of safety* (F.S.). It is a ratio of the failure load F_{fail} to the allowable load F_{allow},

Cranes are often supported using bearing pads to give them stability. Care must be taken not to crush the supporting surface, due to the large bearing stress developed between the pad and the surface.

$$\text{F.S.} = \frac{F_{fail}}{F_{allow}} \qquad (1-8)$$

Here F_{fail} is found from experimental testing of the material.

If the load applied to the member is *linearly related* to the stress developed within the member, as in the case of $\sigma = N/A$ and $\tau_{avg} = V/A$, then we can also express the factor of safety as a ratio of the failure stress σ_{fail} (or τ_{fail}) to the *allowable stress* σ_{allow} (or τ_{allow}). Here the area A will cancel, and so,

$$\text{F.S.} = \frac{\sigma_{fail}}{\sigma_{allow}} \qquad (1-9)$$

or

$$\text{F.S.} = \frac{\tau_{fail}}{\tau_{allow}} \qquad (1-10)$$

Specific values of F.S. depend on the types of materials to be used and the intended purpose of the structure or machine, while accounting for the previously mentioned uncertainties. For example, the F.S. used in the design of aircraft or space vehicle components may be close to 1 in order to reduce the weight of the vehicle. Or, in the case of a nuclear power plant, the factor of safety for some of its components may be as high as 3 due to uncertainties in loading or material behavior. Whatever the case, the factor of safety or the allowable stress for a specific case can be found in design codes and engineering handbooks. Design that is based on an allowable stress limit is called ***allowable stress design*** (ASD). Using this method will ensure a balance between both public and environmental safety on the one hand and economic considerations on the other.

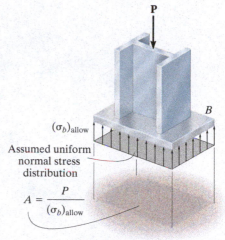

$$A = \frac{P}{(\sigma_b)_{\text{allow}}}$$

The area of the column base plate B is determined from the allowable bearing stress for the concrete.

Simple Connections.

By making simplifying assumptions regarding the behavior of the material, the equations $\sigma = N/A$ and $\tau_{\text{avg}} = V/A$ can often be used to analyze or design a simple connection or mechanical element. For example, if a member is subjected to normal force at a section, its required area at the section is determined from

$$A = \frac{N}{\sigma_{\text{allow}}} \tag{1–11}$$

or if the section is subjected to an average shear force, then the required area at the section is

$$A = \frac{V}{\tau_{\text{allow}}} \tag{1–12}$$

Three examples of where the above equations apply are shown in Fig. 1–24. The first figure shows the normal stress acting on the bottom of a base plate. This compressive stress caused by one surface that bears against another is often called ***bearing stress***.

Assumed uniform shear stress

$$l = \frac{P}{\tau_{\text{allow}} \pi d}$$

The embedded length l of this rod in concrete can be determined using the allowable shear stress of the bonding glue.

The area of the bolt for this lap joint is determined from the shear stress, which is largest between the plates.

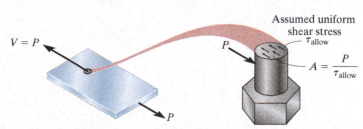

$$A = \frac{P}{\tau_{\text{allow}}}$$

Fig. 1–24

1.7   LIMIT STATE DESIGN

We have stated in the previous section that a properly designed member must account for the uncertainties resulting from the variability of *both* the material's properties and the applied loading. Each of these uncertainties can be investigated using statistics and probability theory, and so in structural engineering there has been an increasing trend to *separate* load uncertainty from material uncertainty.* This method of design is called *limit state design* (LSD), or more specifically, in the United States it is called *load and resistance factor design* (LRFD). We will now discuss how this method is applied.

Load Factors.   Various types of loads R can act on a structure or structural member, and each can be multiplied by a **load factor** γ (gamma) that accounts for its variability. The loads include *dead load*, which is the fixed weight of the structure, and *live loads*, which involve people or vehicles that move about. Other types of live loads include *wind*, *earthquake*, and *snow loads*. The dead load D is multiplied by a relatively small factor such as $\gamma_D = 1.2$, since it can be determined with greater certainty than, for example, the live load L caused by people, which can have a load factor of $\gamma_L = 1.6$.

Building codes often require a structure to be designed to support various *combinations* of the loads, and when applied in combination, each type of load will have a unique load factor. For example, the load factor of one load combination of dead (D), live (L), and snow (S) loads gives a total load R of

$$R = 1.2D + 1.6L + 0.5S$$

The load factors for this combined loading reflect the *probability* that R will occur for all the events stated. In this equation, notice that the load factor $\gamma_S = 0.5$ is small, because of the low probability that a maximum snow load will occur *simultaneously* with the maximum dead and live loads.

Resistance Factors.   **Resistance factors** ϕ (phi) are determined from the probability of material failure as it relates to the material's quality and the consistency of its strength. These factors will differ for different types of materials. For example, concrete has smaller factors than steel, because engineers have more confidence about the behavior of steel under load than they do about concrete. A typical resistance factor $\phi = 0.9$ is used for a steel member in tension.

---

* ASD combines these uncertainties by using the factor of safety or defining the allowable stress.

Design Criteria. Once the load and resistance factors γ and ϕ have been specified using a code, then proper design of a structural member requires that its predicted strength, ϕP_n, be greater than the predicted load it is intended to support. Thus, the LRFD criterion can be stated as

$$\phi P_n \geq \Sigma \gamma_i R_i \qquad (1-13)$$

Here P_n is the **nominal strength** of the member, meaning the load, when applied to the member, causes it either to fail (ultimate load), or deform to a state where it is no longer serviceable. In summary then, the resistance factor ϕ reduces the nominal strength of the member and requires it to be equal to or greater than the applied load or combination of loads calculated using the load factors γ.

IMPORTANT POINT

- Design of a member for strength is based on selecting either an allowable stress or a factor of safety that will enable it to safely support its intended load (ASD), or using load and resistance factors to modify the strength of the material and the load, respectively (LRFD).

PROCEDURE FOR ANALYSIS

When solving problems using the average normal and average shear stress equations, careful consideration should first be given to finding the section over which the critical stress is acting. Once this section is determined, the member must then be designed to have a sufficient cross-sectional area at the section to resist the stress that acts on it. This area is determined using the following steps.

Internal Loading.

- Section the member through the area and draw a free-body diagram of a segment of the member. The internal resultant force at the section is then determined using the equations of equilibrium.

Required Area.

- Provided either the allowable stress or the load and resistance factors are known or can be determined, then the required area needed to sustain the calculated load or factored load at the section is determined from $A = N/\sigma$ or $A = V/\tau$.

Appropriate factors of safety must be considered when designing cranes and cables used to transfer heavy loads.

EXAMPLE 1.12

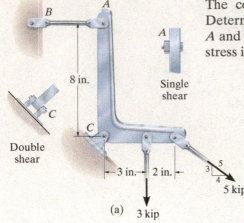

8 in.

Single shear

Double shear

3 in. 2 in.

5 kip

(a)   3 kip

The control arm is subjected to the loading shown in Fig. 1–25a. Determine to the nearest $\frac{1}{4}$ in. the required diameters of the steel pins at A and C if the factor of safety for shear is F.S. = 1.5 and the failure shear stress is $\tau_{fail} = 12$ ksi.

SOLUTION

Pin Forces.   A free-body diagram of the arm is shown in Fig. 1–25b. For equilibrium we have

$$\zeta + \Sigma M_C = 0;$$
$$F_{AB}(8 \text{ in.}) - 3 \text{ kip}(3 \text{ in.}) - 5 \text{ kip}\left(\tfrac{3}{5}\right)(5 \text{ in.}) = 0$$
$$F_{AB} = 3 \text{ kip}$$

$$\xrightarrow{+} \Sigma F_x = 0; \quad -3 \text{ kip} - C_x + 5 \text{ kip}\left(\tfrac{4}{5}\right) = 0 \quad C_x = 1 \text{ kip}$$
$$+\uparrow \Sigma F_y = 0; \quad C_y - 3 \text{ kip} - 5 \text{ kip}\left(\tfrac{3}{5}\right) = 0 \quad C_y = 6 \text{ kip}$$

The pin at C resists the resultant force at C, which is

$$F_C = \sqrt{(1 \text{ kip})^2 + (6 \text{ kip})^2} = 6.083 \text{ kip}$$

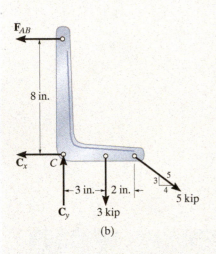

F_{AB}

8 in.

C_x C

3 in. 2 in.

5 kip

C_y 3 kip

(b)

Allowable Shear Stress.   We have

$$\text{F.S.} = \frac{\tau_{fail}}{\tau_{allow}}; \quad 1.5 = \frac{12 \text{ ksi}}{\tau_{allow}}; \quad \tau_{allow} = 8 \text{ ksi}$$

Pin A.   This pin is subjected to *single shear*, Fig. 1–25c, so that

$$A = \frac{V}{\tau_{allow}}; \quad \pi\left(\frac{d_A}{2}\right)^2 = \frac{3 \text{ kip}}{8 \text{ kip/in}^2}; \quad d_A = 0.691 \text{ in.}$$

3 kip

3 kip

Pin at A

(c)

Use $d_A = \frac{3}{4}$ in.   *Ans.*

Pin C.   Since this pin is subjected to *double shear*, a shear force of 3.041 kip acts over its cross-sectional area *between* the arm and each supporting leaf for the pin, Fig. 1–25d. We have

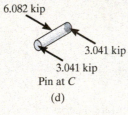

6.082 kip

3.041 kip

3.041 kip

Pin at C

(d)

$$A = \frac{V}{\tau_{allow}}; \quad \pi\left(\frac{d_C}{2}\right)^2 = \frac{3.041 \text{ kip}}{8 \text{ kip/in}^2}; \quad d_C = 0.696 \text{ in.}$$

Use $d_C = \frac{3}{4}$ in.   *Ans.*

Fig. 1–25

EXAMPLE   1.13

The suspender rod is supported at its end by a fixed-connected circular disk as shown in Fig. 1–26a. If the rod passes through a 40-mm-diameter hole, determine the minimum required diameter of the rod and the minimum thickness of the disk needed to support the 20-kN load. The allowable normal stress for the rod is $\sigma_{allow} = 60$ MPa, and the allowable shear stress for the disk is $\tau_{allow} = 35$ MPa.

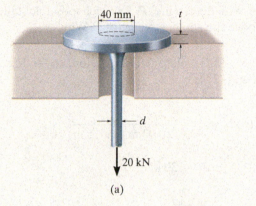

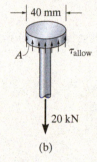

(a)   (b)

Fig. 1–26

SOLUTION

Diameter of Rod.   By inspection, the axial force in the rod is 20 kN. Thus the required cross-sectional area of the rod is

$$A = \frac{N}{\sigma_{allow}}; \qquad \frac{\pi}{4}d^2 = \frac{20(10^3)\ \text{N}}{60(10^6)\ \text{N/m}^2}$$

so that

$$d = 0.0206\ \text{m} = 20.6\ \text{mm} \qquad \qquad Ans.$$

Thickness of Disk.   As shown on the free-body diagram in Fig. 1–26b, the material at the sectioned area of the disk must resist *shear stress* to prevent movement of the disk through the hole. If this shear stress is *assumed* to be uniformly distributed over the sectioned area, then, since $V = 20$ kN, we have

$$A = \frac{V}{\tau_{allow}}; \qquad 2\pi(0.02\ \text{m})(t) = \frac{20(10^3)\ \text{N}}{35(10^6)\ \text{N/m}^2}$$

$$t = 4.55(10^{-3})\ \text{m} = 4.55\ \text{mm} \qquad \qquad Ans.$$

EXAMPLE 1.14

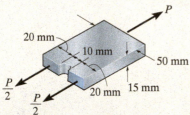

(a)

Determine the largest load P that can be applied to the bars of the lap joint shown in Fig. 1–27a. The bolt has a diameter of 10 mm and an allowable shear stress of 80 MPa. Each plate has an allowable tensile stress of 50 MPa, an allowable bearing stress of 80 MPa, and an allowable shear stress of 30 MPa.

SOLUTION

To solve the problem we will determine P for each possible failure condition; then we will choose the smallest value of P. Why?

Failure of Plate in Tension.   If the plate fails in tension, it will do so at its smallest cross section, Fig. 1–27b.

$$(\sigma_{\text{allow}})_t = \frac{N}{A}; \qquad 50(10^6)\ \text{N/m}^2 = \frac{P}{2(0.02\ \text{m})(0.015\ \text{m})}$$

$$P = 30\ \text{kN}$$

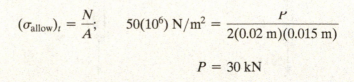

Failure of plate in tension

(b)

Failure of Plate by Bearing.   A free-body diagram of the top plate, Fig. 1–27c, shows that the bolt will exert a complicated distribution of stress on the plate along the curved central area of contact with the bolt.* To simplify the analysis for small connections having pins or bolts such as this, design codes allow the *projected area* of the bolt to be used when calculating the bearing stress. Therefore,

$$(\sigma_{\text{allow}})_b = \frac{N}{A}; \qquad 80(10^6)\text{N/m}^2 = \frac{P}{(0.01\ \text{m})(0.015\ \text{m})}$$

$$P = 12\ \text{kN}$$

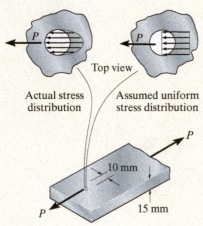

Actual stress   Assumed uniform
distribution   stress distribution

Failure of plate in bearing caused by bolt

(c)

Fig. 1–27

*The material strength of a bolt or pin is generally greater than that of the plate material, so bearing failure of the member is of greater concern.

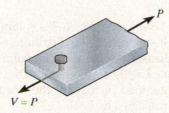

$V = P$

Failure of bolt by shear

(e)

Failure of Plate by Shear.   There is the possibility for the bolt to tear through the plate along the section shown on the free-body diagram in Fig. 1–27d. Here the shear is $V = P/2$, and so

$$(\tau_{\text{allow}})_p = \frac{V}{A}; \qquad 30(10^6) \text{ N/m}^2 = \frac{P/2}{(0.02 \text{ m})(0.015 \text{ m})}$$

$$P = 18 \text{ kN}$$

Failure of Bolt by Shear.   The bolt can fail in shear along the plane between the plates. The free-body diagram in Fig. 1–27e indicates that $V = P$, so that

$$(\tau_{\text{allow}})_b = \frac{V}{A}; \qquad 80(10^6) \text{ N/m}^2 = \frac{P}{\pi(0.005 \text{ m})^2}$$

$$P = 6.28 \text{ kN}$$

Comparing the above results, the largest allowable load for the connections depends upon the bolt shear. Therefore,

$$P = 6.28 \text{ kN} \qquad\qquad Ans.$$

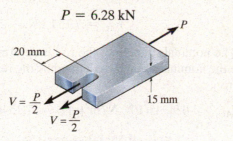

20 mm

$V = \dfrac{P}{2}$

$V = \dfrac{P}{2}$

15 mm

Failure of plate by shear

(d)

Fig. 1–27 (cont.)

EXAMPLE   1.15

The 400-kg uniform bar AB shown in Fig. 1–28a is supported by a steel rod AC and a roller at B. If it supports a live distributed loading of 3 kN/m, determine the required diameter of the rod. The failure stress for the steel is $\sigma_{fail} = 345$ MPa. Use the LRFD method, where the resistance factor for tension is $\phi = 0.9$ and the load factors for the dead and live loads are $\gamma_D = 1.2$ and $\gamma_L = 1.6$, respectively.

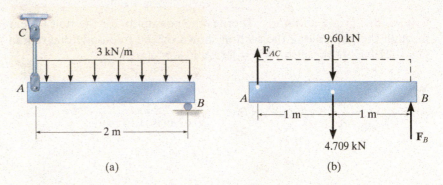

(a)   (b)

Fig. 1–28

SOLUTION

Factored Loads. Here the dead load is the bar's weight $D = 400(9.81)$ N $= 3.924$ kN. Therefore, the factored dead load is $1.2D = 4.709$ kN. The live load resultant is $L = (3\,\text{kN/m})(2\,\text{m}) = 6\,\text{kN}$, so that the factored live load is $1.6L = 9.60$ kN.

From the free-body diagram of the bar, Fig. 1–28b, the factored load in the rod can now be determined.

$$\zeta + \Sigma M_B = 0; \quad 9.60 \text{ kN}(1\,\text{m}) + 4.709 \text{ kN}(1\,\text{m}) - F_{AC}(2\,\text{m}) = 0$$

$$F_{AC} = 7.154 \text{ kN}$$

Area. The nominal strength of the rod is determined from $P_n = \sigma_{fail}\,A$, and since the nominal strength is defined by the resistance factor $\phi = 0.9$, we require

$$\phi P_n \geq F_{AC}; \quad 0.9[345(10^6)\text{ N/m}^2]\,A_{AC} = 7.154(10^3)\text{ N}$$

$$A_{AC} = 23.04(10^{-6})\text{ m}^2 = 23.04\text{ mm}^2 = \frac{\pi}{4}d_{AC}^2$$

$$d_{AC} = 5.42\text{ mm} \qquad\qquad Ans.$$

FUNDAMENTAL PROBLEMS

F1–13. Rods *AC* and *BC* are used to suspend the 200-kg mass. If each rod is made of a material for which the average normal stress can not exceed 150 MPa, determine the minimum required diameter of each rod to the nearest mm.

F1–15. Determine the maximum average shear stress developed in each 3/4-in.-diameter bolt.

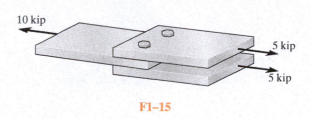

F1–15

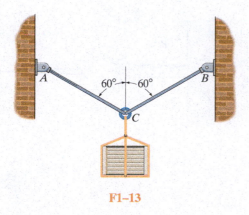

F1–13

F1–14. The pin at *A* has a diameter of 0.25 in. If it is subjected to double shear, determine the average shear stress in the pin.

F1–16. If each of the three nails has a diameter of 4 mm and can withstand an average shear stress of 60 MPa, determine the maximum allowable force **P** that can be applied to the board.

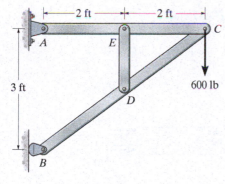

F1–14

F1–16

F1–17. The strut is glued to the horizontal member at surface AB. If the strut has a thickness of 25 mm and the glue can withstand an average shear stress of 600 kPa, determine the maximum force **P** that can be applied to the strut.

F1–19. If the eyebolt is made of a material having a yield stress of $\sigma_Y = 250$ MPa, determine the minimum required diameter d of its shank. Apply a factor of safety F.S. = 1.5 against yielding.

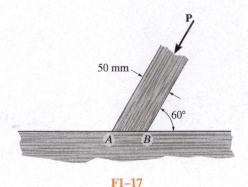

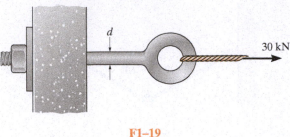

F1–17

F1–19

F1–18. Determine the maximum average shear stress developed in the 30-mm-diameter pin.

F1–20. If the bar assembly is made of a material having a yield stress of $\sigma_Y = 50$ ksi, determine the minimum required dimensions h_1 and h_2 to the nearest 1/8 in. Apply a factor of safety F.S. = 1.5 against yielding. Each bar has a thickness of 0.5 in.

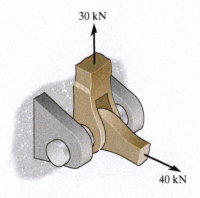

F1–18

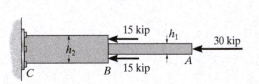

F1–20

F1–21. Determine the maximum force **P** that can be applied to the rod if it is made of material having a yield stress of $\sigma_Y = 250$ MPa. Consider the possibility that failure occurs in the rod and at section *a–a*. Apply a factor of safety of F.S. = 2 against yielding.

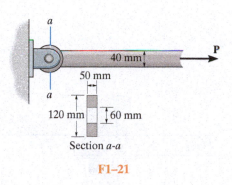

Section *a-a*

F1–21

F1–22. The pin is made of a material having a failure shear stress of $\tau_{fail} = 100$ MPa. Determine the minimum required diameter of the pin to the nearest mm. Apply a factor of safety of F.S. = 2.5 against shear failure.

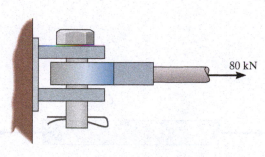

F1–22

F1–23. If the bolt head and the supporting bracket are made of the same material having a failure shear stress of $\tau_{fail} = 120$ MPa, determine the maximum allowable force **P** that can be applied to the bolt so that it does not pull through the plate. Apply a factor of safety of F.S. = 2.5 against shear failure.

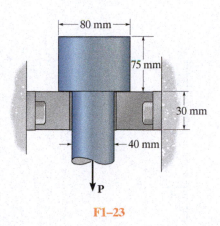

F1–23

F1–24. Six nails are used to hold the hanger at *A* against the column. Determine the minimum required diameter of each nail to the nearest 1/16 in. if it is made of material having $\tau_{fail} = 16$ ksi. Apply a factor of safety of F.S. = 2 against shear failure.

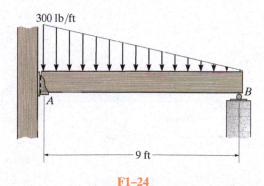

F1–24

PROBLEMS

1–69. If A and B are both made of wood and are $\frac{3}{8}$ in. thick, determine to the nearest $\frac{1}{4}$ in. the smallest dimension h of the vertical segment so that it does not fail in shear. The allowable shear stress for the segment is $\tau_{\text{allow}} = 300$ psi.

1–71. The connection is made using a bolt and nut and two washers. If the allowable bearing stress of the washers on the boards is $(\sigma_b)_{\text{allow}} = 2$ ksi, and the allowable tensile stress within the bolt shank S is $(\sigma_t)_{\text{allow}} = 18$ ksi, determine the maximum allowable tension in the bolt shank. The bolt shank has a diameter of 0.31 in., and the washers have an outer diameter of 0.75 in. and inner diameter (hole) of 0.50 in.

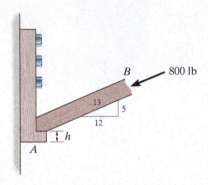

Prob. 1–69

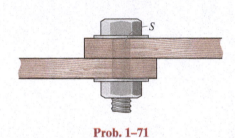

Prob. 1–71

1–70. The lever is attached to the shaft A using a key that has a width d and length of 25 mm. If the shaft is fixed and a vertical force of 200 N is applied perpendicular to the handle, determine the dimension d if the allowable shear stress for the key is $\tau_{\text{allow}} = 35$ MPa.

*1–72. The tension member is fastened together using *two* bolts, one on each side of the member as shown. Each bolt has a diameter of 0.3 in. Determine the maximum load P that can be applied to the member if the allowable shear stress for the bolts is $\tau_{\text{allow}} = 12$ ksi and the allowable average normal stress is $\sigma_{\text{allow}} = 20$ ksi.

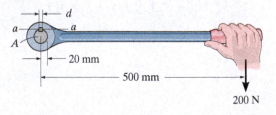

Prob. 1–70

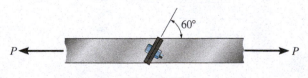

Prob. 1–72

1–73. The steel swivel bushing in the elevator control of an airplane is held in place using a nut and washer as shown in Fig. (a). Failure of the washer A can cause the push rod to separate as shown in Fig. (b). If the maximum average shear stress is $\tau_{max} = 21$ ksi, determine the force **F** that must be applied to the bushing. The washer is $\frac{1}{16}$ in. thick.

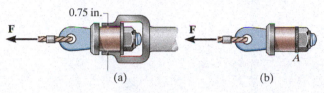

0.75 in.

F   F   A

(a)   (b)

Prob. 1–73

1–74. The spring mechanism is used as a shock absorber for a load applied to the drawbar AB. Determine the force in each spring when the 50-kN force is applied. Each spring is originally unstretched and the drawbar slides along the smooth guide posts CG and EF. The ends of all springs are attached to their respective members. Also, what is the required diameter of the shank of bolts CG and EF if the allowable stress for the bolts is $\sigma_{allow} = 150$ MPa?

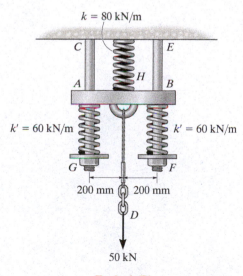

$k = 80$ kN/m

C   E
A   H   B

$k' = 60$ kN/m $k' = 60$ kN/m

G   F

200 mm   200 mm

D

50 kN

Prob. 1–74

1–75. Determine the size of *square* bearing plates A′ and B′ required to support the loading. Take $P = 1.5$ kip. Dimension the plates to the nearest $\frac{1}{2}$ in. The reactions at the supports are vertical and the allowable bearing stress for the plates is $(\sigma_b)_{allow} = 400$ psi.

*1–76. Determine the maximum load **P** that can be applied to the beam if the bearing plates A′ and B′ have square cross sections of 2 in. × 2 in. and 4 in. × 4 in., respectively, and the allowable bearing stress for the material is $(\sigma_b)_{allow} = 400$ psi.

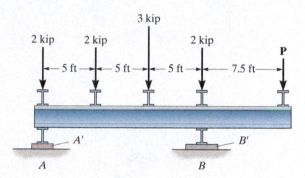

3 kip

2 kip   2 kip   2 kip   P

5 ft   5 ft   5 ft   7.5 ft

A′   B′
A   B

Probs. 1–75/76

1–77. Determine the required diameter of the pins at A and B to the nearest $\frac{1}{16}$ in. if the allowable shear stress for the material is $\tau_{allow} = 6$ ksi. Pin A is subjected to double shear, whereas pin B is subjected to single shear.

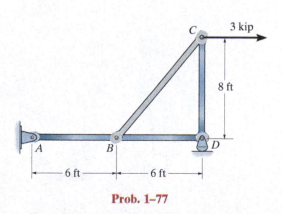

C   3 kip
8 ft
A   B   D

6 ft   6 ft

Prob. 1–77

1–78. If the allowable tensile stress for wires AB and AC is $\sigma_{\text{allow}} = 200$ MPa, determine the required diameter of each wire if the applied load is $P = 6$ kN.

1–79. If the allowable tensile stress for wires AB and AC is $\sigma_{\text{allow}} = 180$ MPa, and wire AB has a diameter of 5 mm and AC has a diameter of 6 mm, determine the greatest force P that can be applied to the chain.

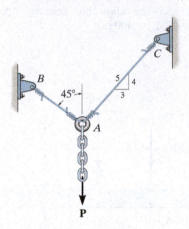

Probs. 1–78/79

*1–80. The cotter is used to hold the two rods together. Determine the smallest thickness t of the cotter and the smallest diameter d of the rods. All parts are made of steel for which the failure normal stress is $\sigma_{\text{fail}} = 500$ MPa and the failure shear stress is $\tau_{\text{fail}} = 375$ MPa. Use a factor of safety of $(\text{F.S.})_t = 2.50$ in tension and $(\text{F.S.})_s = 1.75$ in shear.

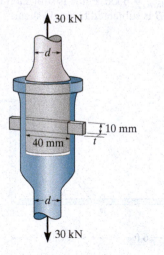

Prob. 1–80

1–81. Determine the required diameter of the pins at A and B if the allowable shear stress for the material is $\tau_{\text{allow}} = 100$ MPa. Both pins are subjected to double shear.

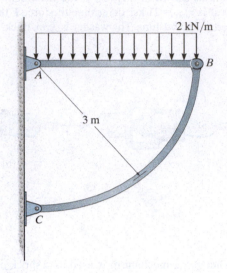

Prob. 1–81

1–82. The steel pipe is supported on the circular base plate and concrete pedestal. If the thickness of the pipe is $t = 5$ mm and the base plate has a radius of 150 mm, determine the factors of safety against failure of the steel and concrete. The applied force is 500 kN, and the normal failure stresses for steel and concrete are $(\sigma_{\text{fail}})_{\text{st}} = 350$ MPa and $(\sigma_{\text{fail}})_{\text{con}} = 25$ MPa, respectively.

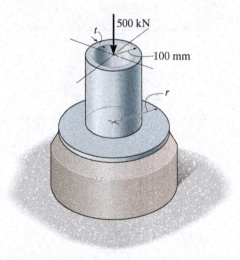

Prob. 1–82

1–83. The boom is supported by the winch cable that has a diameter of 0.25 in. and an allowable normal stress of $\sigma_{allow} = 24$ ksi. Determine the greatest weight of the crate that can be supported without causing the cable to fail if $\phi = 30°$. Neglect the size of the winch.

*1–84. The boom is supported by the winch cable that has an allowable normal stress of $\sigma_{allow} = 24$ ksi. If it supports the 5000 lb crate when $\phi = 20°$, determine the smallest diameter of the cable to the nearest $\frac{1}{16}$ in.

1–86. The two aluminum rods support the vertical force of $P = 20$ kN. Determine their required diameters if the allowable tensile stress for the aluminum is $\sigma_{allow} = 150$ MPa.

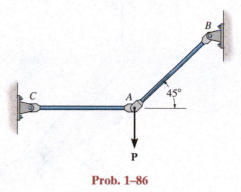

Prob. 1–86

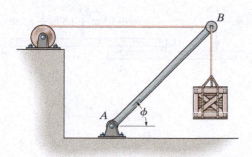

Probs. 1–83/84

1–85. The assembly consists of three disks A, B, and C that are used to support the load of 140 kN. Determine the smallest diameter d_1 of the top disk, the largest diameter d_2 of the opening, and the largest diameter d_3 of the hole in the bottom disk. The allowable bearing stress for the material is $(\sigma_b)_{allow} = 350$ MPa and allowable shear stress is $\tau_{allow} = 125$ MPa.

1–87. The two aluminum rods AB and AC have diameters of 10 mm and 8 mm, respectively. Determine the largest vertical force $\mathbf{P}$ that can be supported. The allowable tensile stress for the aluminum is $\sigma_{allow} = 150$ MPa.

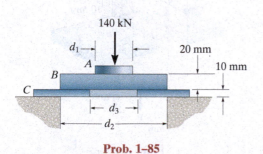

Prob. 1–85

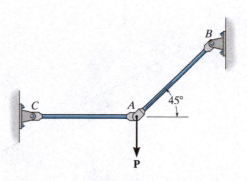

Prob. 1–87

*1–88.   Determine the required minimum thickness t of member AB and edge distance b of the frame if $P = 9$ kip and the factor of safety against failure is 2. The wood has a normal failure stress of $\sigma_{fail} = 6$ ksi, and a shear failure stress of $\tau_{fail} = 1.5$ ksi.

1–90.   The compound wooden beam is connected together by a bolt at B. Assuming that the connections at A, B, C, and D exert only vertical forces on the beam, determine the required diameter of the bolt at B and the required outer diameter of its washers if the allowable tensile stress for the bolt is $(\sigma_t)_{allow} = 150$ MPa and the allowable bearing stress for the wood is $(\sigma_b)_{allow} = 28$ MPa. Assume that the hole in the washers has the same diameter as the bolt.

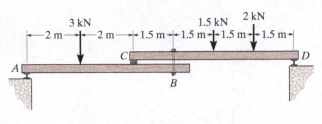

Prob. 1–90

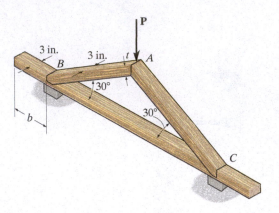

Prob. 1–88

1–89.   Determine the maximum allowable load P that can be safely supported by the frame if $t = 1.25$ in. and $b = 3.5$ in. The wood has a normal failure stress of $\sigma_{fail} = 6$ ksi, and a shear failure stress of $\tau_{fail} = 1.5$ ksi. Use a factor of safety against failure of 2.

1–91.   The hanger is supported using the rectangular pin. Determine the magnitude of the allowable suspended load P if the allowable bearing stress is $(\sigma_b)_{allow} = 220$ MPa, the allowable tensile stress is $(\sigma_t)_{allow} = 150$ MPa, and the allowable shear stress is $\tau_{allow} = 130$ MPa. Take $t = 6$ mm, $a = 5$ mm and $b = 25$ mm.

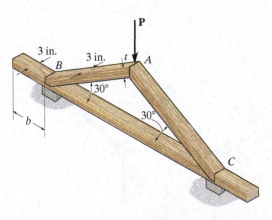

Prob. 1–89

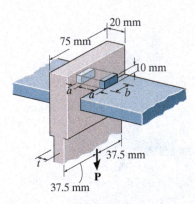

Prob. 1–91

*1–92. The hanger is supported using the rectangular pin. Determine the required thickness t of the hanger, and dimensions a and b if the suspended load is $P = 60$ kN. The allowable tensile stress is $(\sigma_t)_{\text{allow}} = 150$ MPa, the allowable bearing stress is $(\sigma_b)_{\text{allow}} = 290$ MPa, and the allowable shear stress is $\tau_{\text{allow}} = 125$ MPa.

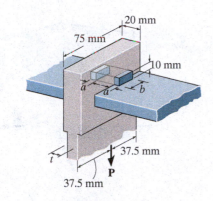

Prob. 1–92

1–93. The rods AB and CD are made of steel. Determine their smallest diameter so that they can support the dead loads shown. The beam is assumed to be pin connected at A and C. Use the LRFD method, where the resistance factor for steel in tension is $\phi = 0.9$, and the dead load factor is $\gamma_D = 1.4$. The failure stress is $\sigma_{\text{fail}} = 345$ MPa.

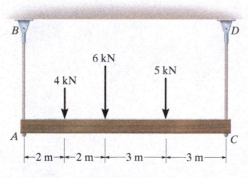

Prob. 1–93

1–94. The aluminum bracket A is used to support the centrally applied load of 8 kip. If it has a thickness of 0.5 in., determine the smallest height h in order to prevent a shear failure. The failure shear stress is $\tau_{\text{fail}} = 23$ ksi. Use a factor of safety for shear of F.S. = 2.5.

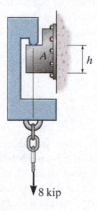

Prob. 1–94

1–95. If the allowable tensile stress for the bar is $(\sigma_t)_{\text{allow}} = 21$ ksi, and the allowable shear stress for the pin is $\tau_{\text{allow}} = 12$ ksi, determine the diameter of the pin so that the load P will be a maximum. What is this load? Assume the hole in the bar has the same diameter d as the pin. Take $t = \frac{1}{4}$ in. and $w = 2$ in.

*1–96. The bar is connected to the support using a pin having a diameter of $d = 1$ in. If the allowable tensile stress for the bar is $(\sigma_t)_{\text{allow}} = 20$ ksi, and the allowable bearing stress between the pin and the bar is $(\sigma_b)_{\text{allow}} = 30$ ksi, determine the dimensions w and t so that the gross area of the cross section is $wt = 2$ in^2 and the load P is a maximum. What is this maximum load? Assume the hole in the bar has the same diameter as the pin.

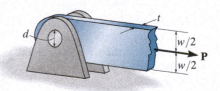

Probs. 1–95/96

CHAPTER REVIEW

The internal loadings in a body consist of a normal force, shear force, bending moment, and torsional moment. They represent the resultants of both a normal and shear stress distribution that act over the cross section. To obtain these resultants, use the method of sections and the equations of equilibrium.	$\Sigma F_x = 0$ $\Sigma F_y = 0$ $\Sigma F_z = 0$ $\Sigma M_x = 0$ $\Sigma M_y = 0$ $\Sigma M_z = 0$	
If a bar is made from homogeneous isotropic material and it is subjected to a series of external axial loads that pass through the centroid of the cross section, then a uniform normal stress distribution will act over the cross section. This average normal stress can be determined from $\sigma = N/A$, where N is the internal axial load at the section.	$\sigma = \dfrac{N}{A}$	
The average shear stress can be determined using $\tau_{avg} = V/A$, where V is the shear force acting on the cross section. This formula is often used to find the average shear stress in fasteners or in parts used for connections.	$\tau_{avg} = \dfrac{V}{A}$	
The ASD method of design of any simple connection requires that the average stress along any cross section not exceed an allowable stress of σ_{allow} or τ_{allow}. These values are reported in codes and are considered safe on the basis of experiments or through experience. Sometimes a factor of safety is reported provided the failure stress is known.	$\text{F.S.} = \dfrac{\sigma_{fail}}{\sigma_{allow}} = \dfrac{\tau_{fail}}{\tau_{allow}}$	
The LRFD method of design is used for the design of structural members. It modifies the load and the strength of the material separately, using load and resistance factors.	$\phi P_n \geq \Sigma \gamma_i R_i$	

CONCEPTUAL PROBLEMS

C1–1. Hurricane winds have caused the failure of this highway sign. Assuming the wind creates a uniform pressure on the sign of 2 kPa, use reasonable dimensions for the sign and determine the resultant shear and moment at each of the two connections where the failure occurred.

C1–1

C1–2. High-heel shoes can often do damage to soft wood or linoleum floors. Using a reasonable weight and dimensions for the heel of a regular shoe and a high-heel shoe, determine the bearing stress under each heel if the weight is transferred down only to the heel of one shoe.

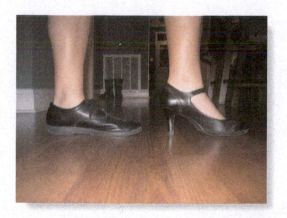

C1–2

C1–3. Here is an example of the single shear failure of a bolt. Using appropriate free-body diagrams, explain why the bolt failed along the section between the plates, and not along some intermediate section such as *a–a*.

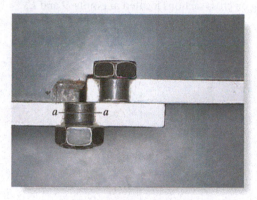

C1–3

C1–4. The vertical load on the hook is 1000 lb. Draw the appropriate free-body diagrams and determine the maximum average shear force on the pins at *A*, *B*, and *C*. Note that due to symmetry four wheels are used to support the loading on the railing.

C1–4

REVIEW PROBLEMS

R1–1. The beam *AB* is pin supported at *A* and supported by a cable *BC*. A *separate* cable *CG* is used to hold up the frame. If *AB* weighs 120 lb/ft and the column *FC* has a weight of 180 lb/ft, determine the resultant internal loadings acting on cross sections located at points *D* and *E*.

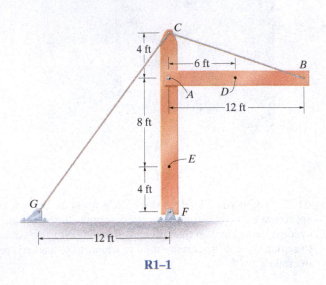

R1–1

R1–2. The long bolt passes through the 30-mm-thick plate. If the force in the bolt shank is 8 kN, determine the average normal stress in the shank, the average shear stress along the cylindrical area of the plate defined by the section lines *a–a*, and the average shear stress in the bolt head along the cylindrical area defined by the section lines *b–b*.

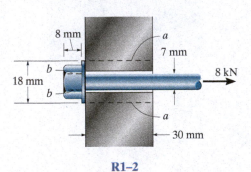

R1–2

R1–3. Determine the required thickness of member *BC* to the nearest $\frac{1}{16}$ in., and the diameter of the pins at *A* and *B* if the allowable normal stress for member *BC* is $\sigma_{allow} = 29\,\text{ksi}$ and the allowable shear stress for the pins is $\tau_{allow} = 10\,\text{ksi}$.

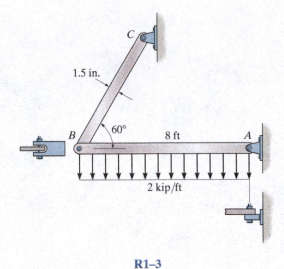

R1–3

*R1–4. The circular punch *B* exerts a force of 2 kN on the top of the plate *A*. Determine the average shear stress in the plate due to this loading.

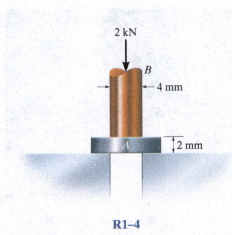

R1–4

R1–5. Determine the average punching shear stress the circular shaft creates in the metal plate through section AC and BD. Also, what is the average bearing stress developed on the surface of the plate under the shaft?

R1–7. The yoke-and-rod connection is subjected to a tensile force of 5 kN. Determine the average normal stress in each rod and the average shear stress in the pin A between the members.

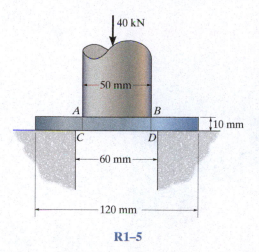

40 kN

50 mm

A B

10 mm

C D

60 mm

120 mm

R1–5

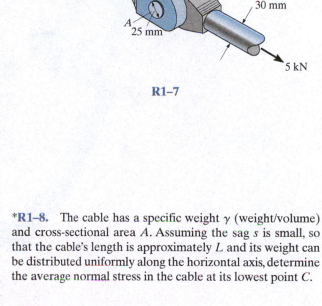

5 kN

40 mm

30 mm

A

25 mm

5 kN

R1–7

R1–6. The 150 mm by 150 mm block of aluminum supports a compressive load of 6 kN. Determine the average normal and shear stress acting on the plane through section a–a. Show the results on a differential volume element located on the plane.

*R1–8. The cable has a specific weight γ (weight/volume) and cross-sectional area A. Assuming the sag s is small, so that the cable's length is approximately L and its weight can be distributed uniformly along the horizontal axis, determine the average normal stress in the cable at its lowest point C.

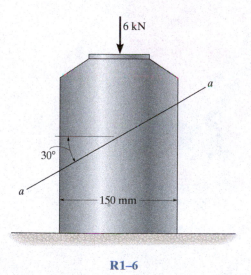

6 kN

a

30°

150 mm

a

R1–6

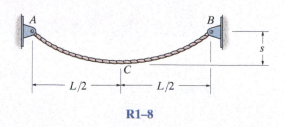

A B

C

$L/2$ $L/2$

s

R1–8

CHAPTER 2

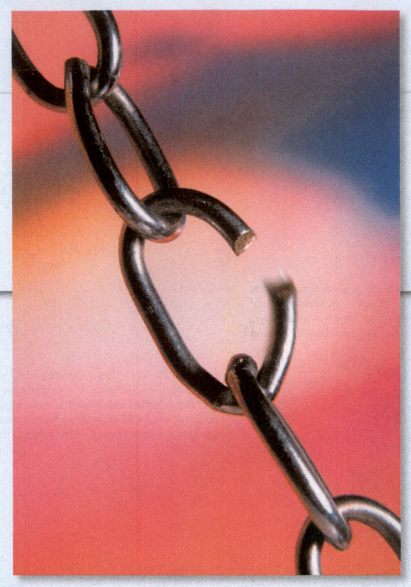

(© Eyebyte/Alamy)

Noticeable deformation occurred in this chain link just before excessive stress caused it to fracture.

STRAIN

■ In engineering the deformation of a body is specified using the concepts of normal and shear strain. In this chapter we will define these quantities and show how they can be determined for various types of problems.

2.1 DEFORMATION

Whenever a force is applied to a body, it will tend to change the body's shape and size. These changes are referred to as *deformation*, and they may be highly visible or practically unnoticeable. For example, a rubber band will undergo a very large deformation when stretched, whereas only slight deformations of structural members occur when a building is occupied. Deformation of a body can also occur when the temperature of the body is changed. A typical example is the thermal expansion or contraction of a roof caused by the weather.

In a general sense, the deformation will not be uniform throughout the body, and so the change in geometry of any line segment within the body may vary substantially along its length. Hence, to study deformation, we will consider line segments that are very short and located in the neighborhood of a point. Realize, however, that the deformation will also depend on the orientation of the line segment at the point. For example, as shown in the adjacent photos, a line segment may elongate if it is oriented in one direction, whereas it may contract if it is oriented in another direction.

Note the before and after positions of three different line segments on this rubber membrane which is subjected to tension. The vertical line is lengthened, the horizontal line is shortened, and the inclined line changes its length and rotates.

2.2   STRAIN

In order to describe the deformation of a body by changes in the lengths of line segments and changes in the angles between them, we will develop the concept of strain. Strain is actually measured by experiment, and once the strain is obtained, it will be shown in the next chapter how it can be related to the stress acting within the body.

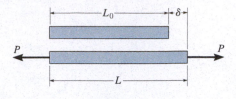

Fig. 2–1

Normal Strain.   If an axial load P is applied to the bar in Fig. 2–1, it will change the bar's length L_0 to a length L. We will define the *average normal strain* ϵ (epsilon) of the bar as the change in its length δ (delta) $= L - L_0$ divided by its original length, that is

$$\epsilon_{avg} = \frac{L - L_0}{L_0} \qquad (2\text{--}1)$$

The normal strain at a point in a body of arbitrary shape is defined in a similar manner. For example, consider the very small line segment Δs located at the point, Fig. 2–2. After deformation it becomes $\Delta s'$, and the change in its length is therefore $\Delta s' - \Delta s$. As $\Delta s \to 0$, in the limit the normal strain at the point is therefore

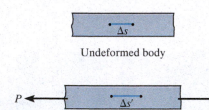

Undeformed body

Deformed body

Fig. 2–2

$$\epsilon = \lim_{\Delta s \to 0} \frac{\Delta s' - \Delta s}{\Delta s} \qquad (2\text{--}2)$$

In both cases ϵ (or ϵ_{avg}) is a change in length per unit length, and it is positive when the initial line elongates, and negative when the line contracts.

Units.   As shown, normal strain is a *dimensionless quantity*, since it is a ratio of two lengths. However, it is sometimes stated in terms of a ratio of length units. If the SI system is used, where the basic unit for length is the meter (m), then since ϵ is generally very small, for most engineering applications, measurements of strain will be in micrometers per meter (μm/m), where $1\ \mu$m $= 10^{-6}$ m. In the Foot-Pound-Second system, strain is often stated in units of inches per inch (in./in.), and for experimental work, strain is sometimes expressed as a percent. For example, a normal strain of $480(10^{-6})$ can be reported as $480(10^{-6})$ in./in., $480\ \mu$m/m, or 0.0480%. Or one can state the strain as simply $480\ \mu$ (480 "micros").

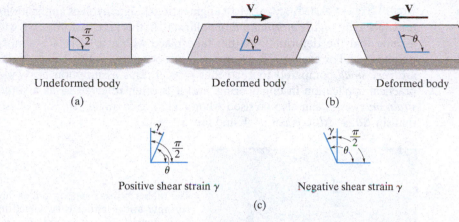

Positive shear strain γ Negative shear strain γ

(c)

Fig. 2–3

Shear Strain.

Deformations not only cause line segments to elongate or contract, but they also cause them to change direction. If we select two line segments that are originally perpendicular to one another, then the *change in angle* that occurs between them is referred to as **shear strain**. This angle is denoted by γ (gamma) and is always measured in radians (rad), which are dimensionless. For example, consider the two perpendicular line segments at a point in the block shown in Fig. 2–3*a*. If an applied loading causes the block to deform as shown in Fig. 2–3*b*, so that the angle between the line segments becomes θ, then the shear strain at the point becomes

$$\gamma = \frac{\pi}{2} - \theta \qquad (2\text{–}3)$$

Notice that if θ is smaller than $\pi/2$, Fig. 2–3*c*, then the shear strain is *positive*, whereas if θ is larger than $\pi/2$, then the shear strain is *negative*.

Cartesian Strain Components.

We can generalize our definitions of normal and shear strain and consider the undeformed element at a point in a body, Fig. 2–4*a*. Since the element's dimensions are very small, its deformed shape will become a parallelepiped, Fig. 2–4*b*. Here the *normal strains* change the sides of the element to

$$(1 + \epsilon_x)\Delta x \qquad (1 + \epsilon_y)\Delta y \qquad (1 + \epsilon_z)\Delta z$$

which produces a *change in the volume of the element*. And the *shear strain* changes the angles between the sides of the element to

$$\frac{\pi}{2} - \gamma_{xy} \qquad \frac{\pi}{2} - \gamma_{yz} \qquad \frac{\pi}{2} - \gamma_{xz}$$

which produces a *change in the shape of the element*.

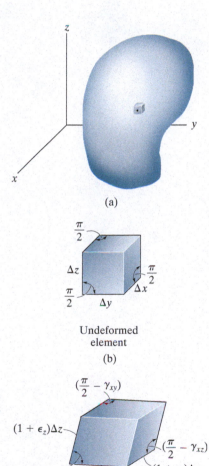

(a)

Undeformed element

(b)

Deformed element

(c)

Fig. 2–4

Small Strain Analysis.   Most engineering design involves applications for which only *small deformations* are allowed. In this text, therefore, we will assume that the deformations that take place within a body are almost infinitesimal. For example, the *normal strains* occurring within the material are *very small* compared to 1, so that $\epsilon \ll 1$. This assumption has wide practical application in engineering, and it is often referred to as a *small strain analysis*. It can also be used when a change in angle, $\Delta\theta$, is small, so that $\sin \Delta\theta \approx \Delta\theta$, $\cos \Delta\theta \approx 1$, and $\tan \Delta\theta \approx \Delta\theta$.

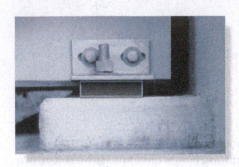

The rubber bearing support under this concrete bridge girder is subjected to both normal and shear strain. The normal strain is caused by the weight and bridge loads on the girder, and the shear strain is caused by the horizontal movement of the girder due to temperature changes.

▶ *IMPORTANT POINTS*

- Loads will cause all material bodies to deform and, as a result, points in a body will undergo *displacements or changes in position*.

- *Normal strain* is a measure per unit length of the elongation or contraction of a small line segment in the body, whereas *shear strain* is a measure of the change in angle that occurs between two small line segments that are originally perpendicular to one another.

- The state of strain at a point is characterized by six strain components: three normal strains ϵ_x, ϵ_y, ϵ_z and three shear strains γ_{xy}, γ_{yz}, γ_{xz}. These components all depend upon the original orientation of the line segments and their location in the body.

- Strain is the geometrical quantity that is measured using experimental techniques. Once obtained, the stress in the body can then be determined from material property relations, as discussed in the next chapter.

- Most engineering materials undergo very small deformations, and so the normal strain $\epsilon \ll 1$. This assumption of "small strain analysis" allows the calculations for normal strain to be simplified, since first-order approximations can be made about its size.

EXAMPLE 2.1

Determine the average normal strains in the two wires in Fig. 2–5 if the ring at A moves to A'.

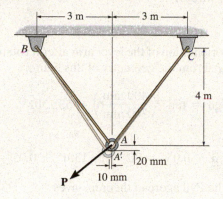

Fig. 2–5

SOLUTION

Geometry. The original length of each wire is

$$L_{AB} = L_{AC} = \sqrt{(3\ m)^2 + (4\ m)^2} = 5\ m$$

The final lengths are

$$L_{A'B} = \sqrt{(3\ m - 0.01\ m)^2 + (4\ m + 0.02\ m)^2} = 5.01004\ m$$

$$L_{A'C} = \sqrt{(3\ m + 0.01\ m)^2 + (4\ m + 0.02\ m)^2} = 5.02200\ m$$

Average Normal Strain.

$$\epsilon_{AB} = \frac{L_{A'B} - L_{AB}}{L_{AB}} = \frac{5.01004\ m - 5\ m}{5\ m} = 2.01(10^{-3})\ m/m \quad \textit{Ans.}$$

$$\epsilon_{AC} = \frac{L_{A'C} - L_{AC}}{L_{AC}} = \frac{5.02200\ m - 5\ m}{5\ m} = 4.40(10^{-3})\ m/m \quad \textit{Ans.}$$

EXAMPLE 2.2

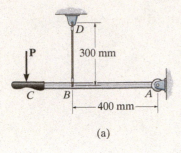

(a)

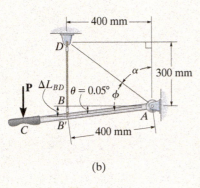

(b)

Fig. 2–6

When force **P** is applied to the rigid lever arm ABC in Fig. 2–6a, the arm rotates counterclockwise about pin A through an angle of 0.05°. Determine the normal strain in wire BD.

SOLUTION I

Geometry. The orientation of the lever arm after it rotates about point A is shown in Fig. 2–6b. From the geometry of this figure,

$$\alpha = \tan^{-1}\left(\frac{400 \text{ mm}}{300 \text{ mm}}\right) = 53.1301°$$

Then

$$\phi = 90° - \alpha + 0.05° = 90° - 53.1301° + 0.05° = 36.92°$$

For triangle ABD the Pythagorean theorem gives

$$L_{AD} = \sqrt{(300 \text{ mm})^2 + (400 \text{ mm})^2} = 500 \text{ mm}$$

Using this result and applying the law of cosines to triangle $AB'D$,

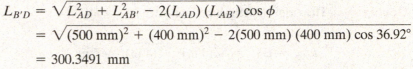

$$L_{B'D} = \sqrt{L_{AD}^2 + L_{AB'}^2 - 2(L_{AD})(L_{AB'})\cos\phi}$$
$$= \sqrt{(500 \text{ mm})^2 + (400 \text{ mm})^2 - 2(500 \text{ mm})(400 \text{ mm})\cos 36.92°}$$
$$= 300.3491 \text{ mm}$$

Normal Strain.

$$\epsilon_{BD} = \frac{L_{B'D} - L_{BD}}{L_{BD}}$$

$$= \frac{300.3491 \text{ mm} - 300 \text{ mm}}{300 \text{ mm}} = 0.00116 \text{ mm/mm} \qquad Ans.$$

SOLUTION II

Since the strain is small, this same result can be obtained by approximating the elongation of wire BD as ΔL_{BD}, shown in Fig. 2–6b. Here,

$$\Delta L_{BD} = \theta L_{AB} = \left[\left(\frac{0.05°}{180°}\right)(\pi \text{ rad})\right](400 \text{ mm}) = 0.3491 \text{ mm}$$

Therefore,

$$\epsilon_{BD} = \frac{\Delta L_{BD}}{L_{BD}} = \frac{0.3491 \text{ mm}}{300 \text{ mm}} = 0.00116 \text{ mm/mm} \qquad Ans.$$

EXAMPLE   2.3

The plate shown in Fig. 2–7a is fixed connected along AB and held in the horizontal guides at its top and bottom, AD and BC. If its right side CD is given a uniform horizontal displacement of 2 mm, determine (a) the average normal strain along the diagonal AC, and (b) the shear strain at E relative to the x, y axes.

SOLUTION

Part (a).   When the plate is deformed, the diagonal AC becomes AC', Fig. 2–7b. The lengths of diagonals AC and AC' can be found from the Pythagorean theorem. We have

$$AC = \sqrt{(0.150 \text{ m})^2 + (0.150 \text{ m})^2} = 0.21213 \text{ m}$$

$$AC' = \sqrt{(0.150 \text{ m})^2 + (0.152 \text{ m})^2} = 0.21355 \text{ m}$$

Therefore the average normal strain along AC is

$$(\epsilon_{AC})_{\text{avg}} = \frac{AC' - AC}{AC} = \frac{0.21355 \text{ m} - 0.21213 \text{ m}}{0.21213 \text{ m}}$$

$$= 0.00669 \text{ mm/mm} \qquad \qquad Ans.$$

Part (b).   To find the shear strain at E relative to the x and y axes, which are 90° apart, it is necessary to find the change in the angle at E. After deformation, Fig. 2–7b,

$$\tan\left(\frac{\theta}{2}\right) = \frac{76 \text{ mm}}{75 \text{ mm}}$$

$$\theta = 90.759° = \left(\frac{\pi}{180°}\right)(90.759°) = 1.58404 \text{ rad}$$

Applying Eq. 2–3, the shear strain at E is therefore the change in the angle AED,

$$\gamma_{xy} = \frac{\pi}{2} - 1.58404 \text{ rad} = -0.0132 \text{ rad} \qquad Ans.$$

The *negative sign* indicates that the once 90° angle becomes larger.

NOTE: If the x and y axes were horizontal and vertical at point E, then the 90° angle between these axes would not change due to the deformation, and so $\gamma_{xy} = 0$ at point E.

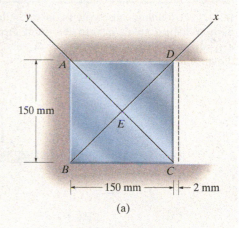

(a)

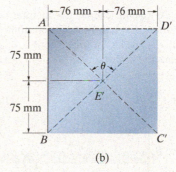

(b)

Fig. 2–7

2

PRELIMINARY PROBLEMS

P2–1. A loading causes the member to deform into the dashed shape. Explain how to determine the normal strains ϵ_{CD} and ϵ_{AB}. The displacement Δ and the lettered dimensions are known.

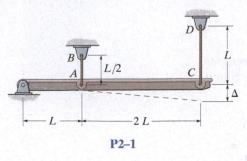

P2–1

P2–2. A loading causes the member to deform into the dashed shape. Explain how to determine the normal strains ϵ_{CD} and ϵ_{AB}. The displacement Δ and the lettered dimensions are known.

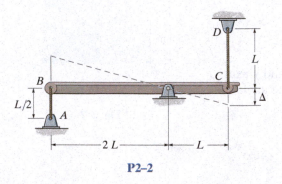

P2–2

P2–3. A loading causes the wires to elongate into the dashed shape. Explain how to determine the normal strain ϵ_{AB} in wire AB. The displacement Δ and the distances between all lettered points are known.

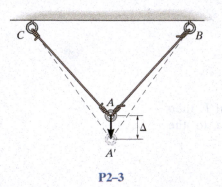

P2–3

P2–4. A loading causes the block to deform into the dashed shape. Explain how to determine the strains ϵ_{AB}, ϵ_{AC}, ϵ_{BC}, $(\gamma_A)_{xy}$. The angles and distances between all lettered points are known.

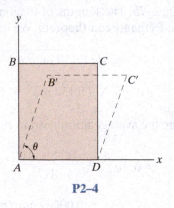

P2–4

P2–5. A loading causes the block to deform into the dashed shape. Explain how to determine the strains $(\gamma_A)_{xy}$, $(\gamma_B)_{xy}$. The angles and distances between all lettered points are known.

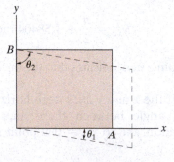

P2–5

FUNDAMENTAL PROBLEMS

F2–1. When force **P** is applied to the rigid arm *ABC*, point *B* displaces vertically downward through a distance of 0.2 mm. Determine the normal strain in wire *CD*.

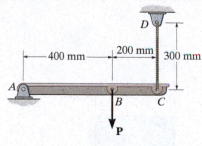

F2–1

F2–2. If the force **P** causes the rigid arm *ABC* to rotate clockwise about pin *A* through an angle of 0.02°, determine the normal strain in wires *BD* and *CE*.

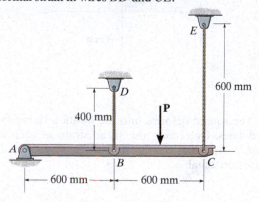

F2–2

F2–3. The rectangular plate is deformed into the shape of a parallelogram shown by the dashed line. Determine the average shear strain at corner *A* with respect to the *x* and *y* axes.

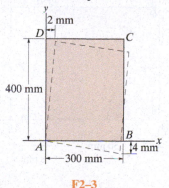

F2–3

F2–4. The triangular plate is deformed into the shape shown by the dashed line. Determine the normal strain along edge *BC* and the average shear strain at corner *A* with respect to the *x* and *y* axes.

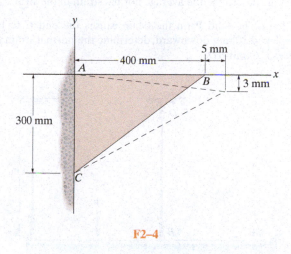

F2–4

F2–5. The square plate is deformed into the shape shown by the dashed line. Determine the average normal strain along diagonal *AC* and the shear strain at point *E* with respect to the *x* and *y* axes.

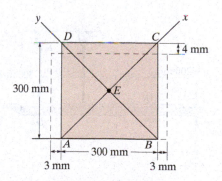

F2–5

PROBLEMS

2–1. An air-filled rubber ball has a diameter of 6 in. If the air pressure within the ball is increased until the diameter becomes 7 in., determine the average normal strain in the rubber.

2–2. A thin strip of rubber has an unstretched length of 15 in. If it is stretched around a pipe having an outer diameter of 5 in., determine the average normal strain in the strip.

2–3. If the load **P** on the beam causes the end *C* to be displaced 10 mm downward, determine the normal strain in wires *CE* and *BD*.

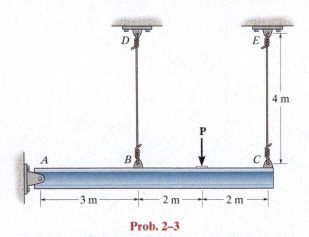

Prob. 2–3

***2–4.** The force applied at the handle of the rigid lever causes the lever to rotate clockwise about the pin *B* through an angle of 2°. Determine the average normal strain in each wire. The wires are unstretched when the lever is in the horizontal position.

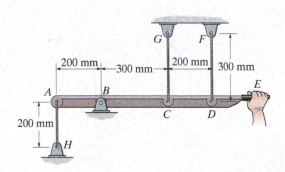

Prob. 2–4

2–5. The rectangular plate is subjected to the deformation shown by the dashed line. Determine the average shear strain γ_{xy} in the plate.

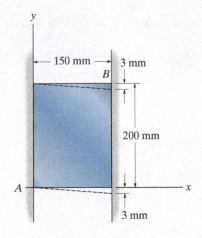

Prob. 2–5

2–6. The square deforms into the position shown by the dashed lines. Determine the shear strain at each of its corners, *A*, *B*, *C*, and *D*, relative to the *x*, *y* axes. Side *D'B'* remains horizontal.

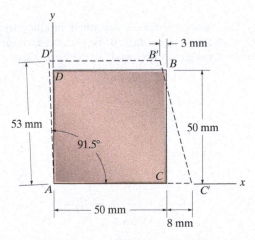

Prob. 2–6

2–7. The pin-connected rigid rods AB and BC are inclined at $\theta = 30°$ when they are unloaded. When the force P is applied θ becomes 30.2°. Determine the average normal strain in wire AC.

2–10. Determine the shear strain γ_{xy} at corners A and B if the plastic distorts as shown by the dashed lines.

2–11. Determine the shear strain γ_{xy} at corners D and C if the plastic distorts as shown by the dashed lines.

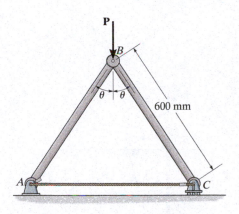

Prob. 2–7

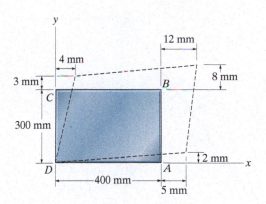

Probs. 2–10/11

*2–8. The wire AB is unstretched when $\theta = 45°$. If a load is applied to the bar AC, which causes θ to become 47°, determine the normal strain in the wire.

2–9. If a horizontal load applied to the bar AC causes point A to be displaced to the right by an amount ΔL, determine the normal strain in the wire AB. Originally, $\theta = 45°$.

*2–12. The material distorts into the dashed position shown. Determine the average normal strains ϵ_x, ϵ_y and the shear strain γ_{xy} at A, and the average normal strain along line BE.

2–13. The material distorts into the dashed position shown. Determine the average normal strains along the diagonals AD and CF.

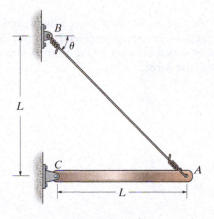

Probs. 2–8/9

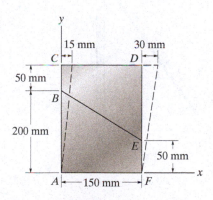

Probs. 2–12/13

2–14. Part of a control linkage for an airplane consists of a rigid member CB and a flexible cable AB. If a force is applied to the end B of the member and causes it to rotate by $\theta = 0.5°$, determine the normal strain in the cable. Originally the cable is unstretched.

*2–16.** The nylon cord has an original length L and is tied to a bolt at A and a roller at B. If a force **P** is applied to the roller, determine the normal strain in the cord when the roller is at C, and at D. If the cord is originally unstrained when it is at C, determine the normal strain ϵ'_D when the roller moves to D. Show that if the displacements Δ_C and Δ_D are small, then $\epsilon'_D = \epsilon_D - \epsilon_C$.

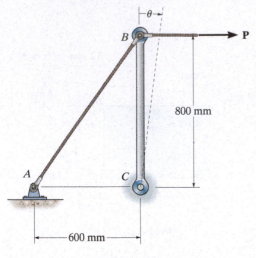

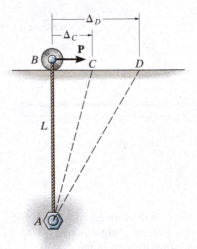

Prob. 2–14

Prob. 2–16

2–15. Part of a control linkage for an airplane consists of a rigid member CB and a flexible cable AB. If a force is applied to the end B of the member and causes a normal strain in the cable of 0.004 mm/mm, determine the displacement of point B. Originally the cable is unstretched.

2–17. A thin wire, lying along the x axis, is strained such that each point on the wire is displaced $\Delta x = kx^2$ along the x axis. If k is constant, what is the normal strain at any point P along the wire?

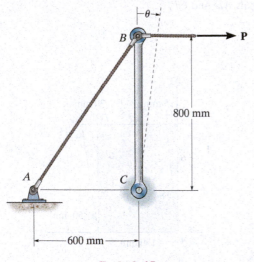

Prob. 2–15

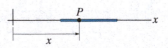

Prob. 2–17

2–18. Determine the shear strain γ_{xy} at corners A and B if the plate distorts as shown by the dashed lines.

2–19. Determine the shear strain γ_{xy} at corners D and C if the plate distorts as shown by the dashed lines.

*2–20. Determine the average normal strain that occurs along the diagonals AC and DB.

2–22. The triangular plate is fixed at its base, and its apex A is given a horizontal displacement of 5 mm. Determine the shear strain, γ_{xy}, at A.

2–23. The triangular plate is fixed at its base, and its apex A is given a horizontal displacement of 5 mm. Determine the average normal strain ϵ_x along the x axis.

*2–24. The triangular plate is fixed at its base, and its apex A is given a horizontal displacement of 5 mm. Determine the average normal strain $\epsilon_{x'}$ along the x' axis.

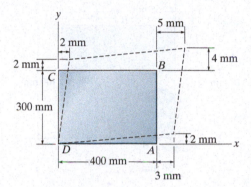

Probs. 2–18/19/20

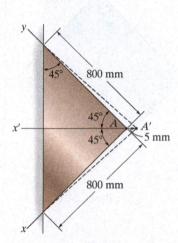

Probs. 2–22/23/24

2–21. The corners of the square plate are given the displacements indicated. Determine the average normal strains ϵ_x and ϵ_y along the x and y axes.

2–25. The polysulfone block is glued at its top and bottom to the rigid plates. If a tangential force, applied to the top plate, causes the material to deform so that its sides are described by the equation $y = 3.56\,x^{1/4}$, determine the shear strain at the corners A and B.

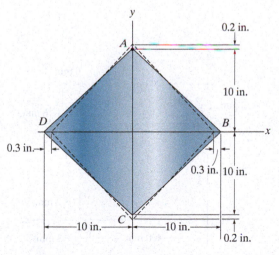

Prob. 2–21

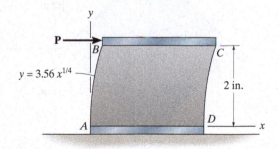

Prob. 2–25

2–26. The corners of the square plate are given the displacements indicated. Determine the shear strain at A relative to axes that are directed along AB and AD, and the shear strain at B relative to axes that are directed along BC and BA.

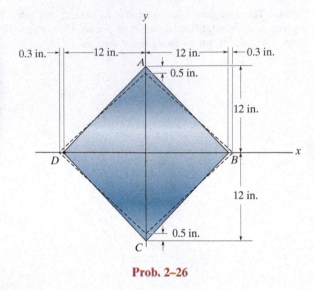

Prob. 2–26

2–27. The corners of the square plate are given the displacements indicated. Determine the average normal strains along side AB and diagonals AC and BD.

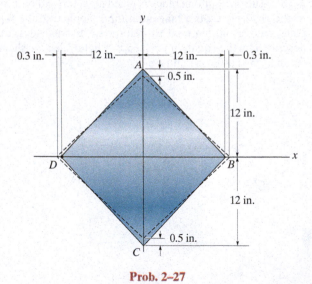

Prob. 2–27

***2–28.** The block is deformed into the position shown by the dashed lines. Determine the average normal strain along line AB.

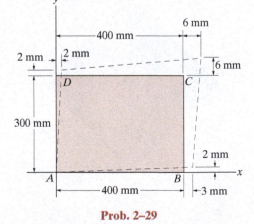

Prob. 2–28

2–29. The rectangular plate is deformed into the shape shown by the dashed lines. Determine the average normal strain along diagonal AC, and the average shear strain at corner A relative to the x, y axes.

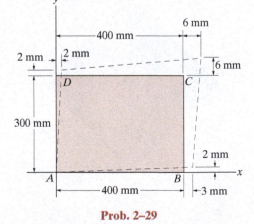

Prob. 2–29

2–30. The rectangular plate is deformed into the shape shown by the dashed lines. Determine the average normal strain along diagonal BD, and the average shear strain at corner B relative to the x, y axes.

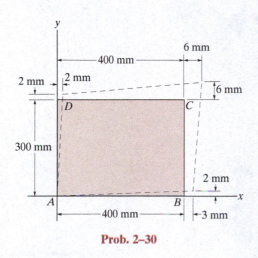

Prob. 2–30

2–31. The nonuniform loading causes a normal strain in the shaft that can be expressed as $\epsilon_x = k \sin\left(\dfrac{\pi}{L}x\right)$, where k is a constant. Determine the displacement of the center C and the average normal strain in the entire rod.

Prob. 2–31

*2–32. The rectangular plate undergoes a deformation shown by the dashed lines. Determine the shear strain γ_{xy} and $\gamma_{x'y'}$ at point A.

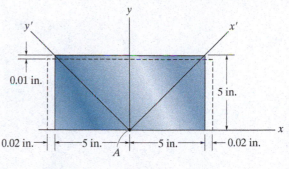

Prob. 2–32

2–33. The fiber AB has a length L and orientation θ. If its ends A and B undergo very small displacements u_A and v_B respectively, determine the normal strain in the fiber when it is in position $A' B'$.

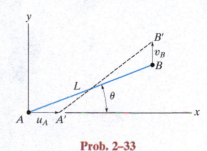

Prob. 2–33

2–34. If the normal strain is defined in reference to the final length $\Delta s'$, that is,

$$\epsilon' = \lim_{\Delta s' \to 0} \left(\frac{\Delta s' - \Delta s}{\Delta s'}\right)$$

instead of in reference to the original length, Eq. 2–2, show that the difference in these strains is represented as a second-order term, namely, $\epsilon - \epsilon' = \epsilon \epsilon'$.

CHAPTER 3

(© Tom Wang/Alamy)

Horizontal ground displacements caused by an earthquake produced fracture of this concrete column. The material properties of the steel and concrete must be determined so that engineers can properly design the column to resist the loadings that caused this failure.

MECHANICAL PROPERTIES OF MATERIALS

CHAPTER OBJECTIVES

- Having discussed the basic concepts of stress and strain, in this chapter we will show how stress can be related to strain by using experimental methods to determine the stress–strain diagram for a specific material. Other mechanical properties and tests that are relevant to our study of mechanics of materials also will be discussed.

3.1 THE TENSION AND COMPRESSION TEST

The strength of a material depends on its ability to sustain a load without undue deformation or failure. This strength is inherent in the material itself and must be determined by *experiment*. One of the most important tests to perform in this regard is the **tension or compression test**. Once this test is performed, we can then determine the relationship between the average normal stress and average normal strain in many engineering materials such as metals, ceramics, polymers, and composites.

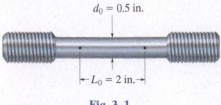

$d_0 = 0.5$ in.

$L_0 = 2$ in.

Fig. 3–1

To perform a tension or compression test, a specimen of the material is made into a "standard" shape and size, Fig. 3–1. As shown it has a constant circular cross section with enlarged ends, so that when tested, failure will occur somewhere within the central region of the specimen. Before testing, two small punch marks are sometimes placed along the specimen's uniform length. Measurements are taken of both the specimen's initial cross-sectional area, A_0, and the *gage-length* distance L_0 between the punch marks. For example, when a metal specimen is used in a tension test, it generally has an initial diameter of $d_0 = 0.5$ in. (13 mm) and a gage length of $L_0 = 2$ in. (51 mm), Fig. 3–1. A testing machine like the one shown in Fig. 3–2 is then used to stretch the specimen at a very slow, constant rate until it fails. The machine is designed to read the load required to maintain this uniform stretching.

At frequent intervals, data is recorded of the applied load P. Also, the elongation $\delta = L - L_0$ between the punch marks on the specimen may be measured, using either a caliper or a mechanical or optical device called an *extensometer*. Rather than taking this measurement and then calculating the strain, it is also possible to read the normal strain *directly* on the specimen by using an *electrical-resistance strain gage*, which looks like the one shown in Fig. 3–3. As shown in the adjacent photo, the gage is cemented to the specimen along its length, so that it becomes an integral part of the specimen. When the specimen is strained in the direction of the gage, both the wire and specimen will experience the same deformation or strain. By measuring the change in the electrical resistance of the wire, the gage may then be calibrated to directly read the normal strain in the specimen.

Typical steel specimen with attached strain gage

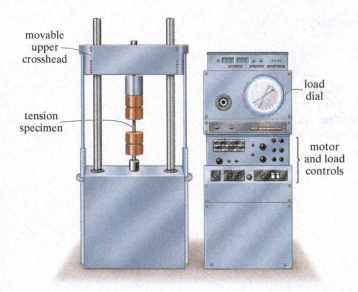

movable upper crosshead

tension specimen

load dial

motor and load controls

Fig. 3–2

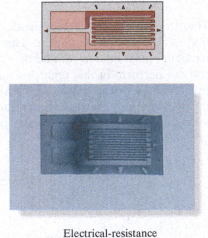

Electrical-resistance strain gage

Fig. 3–3

3.2 THE STRESS–STRAIN DIAGRAM

Once the stress and strain data from the test are known, then the results can be plotted to produce a curve called the **stress–strain diagram**. This diagram is very useful since it applies to a specimen of the material made of *any* size. There are two ways in which the stress–strain diagram is normally described.

Conventional Stress–Strain Diagram. The **nominal** or **engineering stress** is determined by dividing the applied load P by the specimen's *original* cross-sectional area A_0. This calculation assumes that the stress is *constant* over the cross section and throughout the gage length. We have

$$\sigma = \frac{P}{A_0} \tag{3–1}$$

Likewise, the **nominal** or **engineering strain** is found directly from the strain gage reading, or by dividing the change in the specimen's gage length, δ, by the specimen's *original gage length* L_0. Thus,

$$\epsilon = \frac{\delta}{L_0} \tag{3–2}$$

When these values of σ and ϵ are plotted, where the vertical axis is the stress and the horizontal axis is the strain, the resulting curve is called a **conventional stress–strain diagram**. A typical example of this curve is shown in Fig. 3–4. Realize, however, that two stress–strain diagrams for a particular material will be quite similar, but will never be exactly the same. This is because the results actually depend upon such variables as the material's composition, microscopic imperfections, the way the specimen is manufactured, the rate of loading, and the temperature during the time of the test.

From the curve in Fig. 3–4, we can identify four different regions in which the material behaves in a unique way, depending on the amount of strain induced in the material.

Elastic Behavior.   The initial region of the curve, indicated in light orange, is referred to as the elastic region. Here the curve is a *straight line* up to the point where the stress reaches the **proportional limit**, σ_{pl}. When the stress slightly exceeds this value, the curve bends until the stress reaches an elastic limit. For most materials, these points are very close, and therefore it becomes rather difficult to distinguish their exact values. What makes the elastic region unique, however, is that after reaching σ_Y, if the load is removed, the specimen will recover its original shape. In other words, no damage will be done to the material.

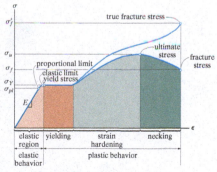

Conventional and true stress–strain diagram for ductile material (steel) (not to scale)

Fig. 3–4

Because the curve is a straight line up to σ_{pl}, any increase in stress will cause a proportional increase in strain. This fact was discovered in 1676 by Robert Hooke, using springs, and is known as **Hooke's law.** It is expressed mathematically as

$$\sigma = E\epsilon \qquad (3-3)$$

Here E represents the constant of proportionality, which is called the **modulus of elasticity** or **Young's modulus**, named after Thomas Young, who published an account of it in 1807.

As noted in Fig. 3–4, the modulus of elasticity represents the *slope* of the straight line portion of the curve. Since strain is dimensionless, from Eq. 3–3, E will have the same units as stress, such as psi, ksi, or pascals.

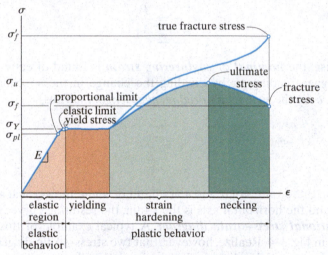

Conventional and true stress–strain diagram
for ductile material (steel) (not to scale)

Fig. 3–4 (Repeated)

Yielding.   A slight increase in stress above the elastic limit will result in a breakdown of the material and cause it to *deform permanently*. This behavior is called **yielding**, and it is indicated by the rectangular dark orange region in Fig. 3–4. The stress that causes yielding is called the **yield stress** or **yield point**, σ_Y, and the deformation that occurs is called **plastic deformation**. Although not shown in Fig. 3–4, for low-carbon steels or those that are hot rolled, the yield point is often distinguished by two values. The **upper yield point** occurs first, followed by a sudden decrease in load-carrying capacity to a **lower yield point**. Once the yield point is reached, then as shown in Fig. 3–4, *the specimen will continue to elongate (strain) **without** any increase in load*. When the material behaves in this manner, it is often referred to as being **perfectly plastic.**

Strain Hardening.   When yielding has ended, any load causing an increase in stress will be supported by the specimen, resulting in a curve that rises continuously but becomes flatter until it reaches a maximum stress referred to as the **ultimate stress**, σ_u. The rise in the curve in this manner is called **strain hardening**, and it is identified in Fig. 3–4 as the region in light green.

Necking.   Up to the ultimate stress, as the specimen elongates, its cross-sectional area will decrease in a fairly *uniform* manner over the specimen's entire gage length. However, just after reaching the ultimate stress, the cross-sectional area will then begin to decrease in a *localized* region of the specimen, and so it is here where the stress begins to increase. As a result, a constriction or "neck" tends to form with further elongation, Fig. 3–5a. This region of the curve due to necking is indicated in dark green in Fig. 3–4. Here the stress–strain diagram tends to curve downward until the specimen breaks at the **fracture stress**, σ_f, Fig. 3–5b.

Typical necking pattern which has occurred on this steel specimen just before fracture.

True Stress–Strain Diagram.

Instead of always using the *original* cross-sectional area A_0 and specimen length L_0 to calculate the (engineering) stress and strain, we could have used the *actual* cross-sectional area A and specimen length L at the *instant* the load is measured. The values of stress and strain found from these measurements are called **true stress** and **true strain**, and a plot of their values is called the **true stress–strain diagram**. When this diagram is plotted, it has a form shown by the upper blue curve in Fig. 3–4. Note that the conventional and true σ–ϵ diagrams are practically coincident when the strain is small. The differences begin to appear in the strain-hardening range, where the magnitude of strain becomes more significant. From the conventional σ–ϵ diagram, the specimen appears to support a *decreasing* stress (or load), since A_0 is constant, $\sigma = N/A_0$. In fact, the true σ–ϵ diagram shows the area A within the necking region is always *decreasing* until fracture, σ_f', and so the material *actually* sustains *increasing stress*, since $\sigma = N/A$.

Although there is this divergence between these two diagrams, we can neglect this effect since most engineering design is done only within the elastic range. This will generally restrict the deformation of the material to very small values, and when the load is removed the material will restore itself to its original shape. The conventional stress–strain diagram can be used in the elastic region because the true strain up to the elastic limit is small enough, so that the error in using the engineering values of σ and ϵ is very small (about 0.1%) compared with their true values.

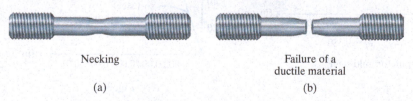

Necking	Failure of a ductile material
(a)	(b)

Fig. 3–5

This steel specimen clearly shows the necking that occurred just before the specimen failed. This resulted in the formation of a "cup-cone" shape at the fracture location, which is characteristic of ductile materials.

Steel.   A typical conventional stress–strain diagram for a mild steel specimen is shown in Fig. 3–6. In order to enhance the details, the elastic region of the curve has been shown in green using an exaggerated strain scale, also shown in green. Following this curve, as the load (stress) is increased, the proportional limit is reached at $\sigma_{pl} = 35$ ksi (241 MPa), where $\epsilon_{pl} = 0.0012$ in./in. When the load is further increased, the stress reaches an upper yield point of $(\sigma_Y)_u = 38$ ksi (262 MPa), followed by a drop in stress to a lower yield point of $(\sigma_Y)_l = 36$ ksi (248 MPa). The end of yielding occurs at a strain of $\epsilon_Y = 0.030$ in./in., which is 25 times greater than the strain at the proportional limit! Continuing, the specimen undergoes strain hardening until it reaches the ultimate stress of $\sigma_u = 63$ ksi (434 MPa); then it begins to neck down until fracture occurs, at $\sigma_f = 47$ ksi (324 MPa). By comparison, the strain at failure, $\epsilon_f = 0.380$ in./in., is 317 times greater than ϵ_{pl}!

Since $\sigma_{pl} = 35$ ksi and $\epsilon_{pl} = 0.0012$ in./in., we can determine the modulus of elasticity. From Hooke's law, it is

$$E = \frac{\sigma_{pl}}{\epsilon_{pl}} = \frac{35 \text{ ksi}}{0.0012 \text{ in./in.}} = 29(10^3) \text{ ksi}$$

Although steel alloys have different carbon contents, most grades of steel, from the softest rolled steel to the hardest tool steel, have about this same modulus of elasticity, as shown in Fig. 3–7.

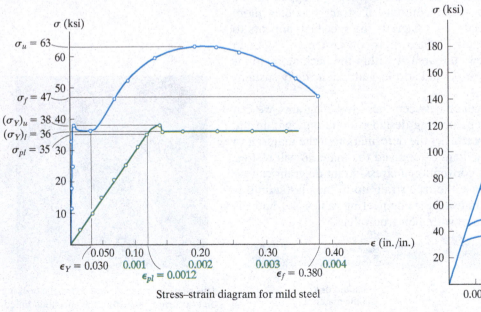

Stress–strain diagram for mild steel

Fig. 3–6

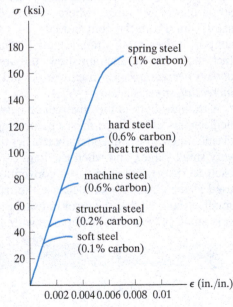

Fig. 3–7

3.3   STRESS–STRAIN BEHAVIOR OF DUCTILE AND BRITTLE MATERIALS

Materials can be classified as either being ductile or brittle, depending on their stress–strain characteristics.

Ductile Materials.   Any material that can be subjected to large strains before it fractures is called a ***ductile material***. Mild steel, as discussed previously, is a typical example. Engineers often choose ductile materials for design because these materials are capable of absorbing shock or energy, and if they become overloaded, they will usually exhibit large deformation before failing.

One way to specify the ductility of a material is to report its percent elongation or percent reduction in area at the time of fracture. The ***percent elongation*** is the specimen's fracture strain expressed as a percent. Thus, if the specimen's original gage length is L_0 and its length at fracture is L_f, then

$$\text{Percent elongation} = \frac{L_f - L_0}{L_0}(100\%) \qquad (3\text{–}4)$$

For example, as in Fig. 3–6, since $\epsilon_f = 0.380$, this value would be 38% for a mild steel specimen.

The ***percent reduction in area*** is another way to specify ductility. It is defined within the region of necking as follows:

$$\text{Percent reduction of area} = \frac{A_0 - A_f}{A_0}(100\%) \qquad (3\text{–}5)$$

Here A_0 is the specimen's original cross-sectional area and A_f is the area of the neck at fracture. Mild steel has a typical value of 60%.

Besides steel, other metals such as brass, molybdenum, and zinc may also exhibit ductile stress–strain characteristics similar to steel, whereby they undergo elastic stress–strain behavior, yielding at constant stress, strain hardening, and finally necking until fracture. In most metals and some plastics, however, constant yielding will *not occur* beyond the elastic range. One metal where this is the case is aluminum, Fig. 3–8. Actually, this metal often does not have a well-defined *yield point*, and consequently it is standard practice to define a ***yield strength*** using a graphical procedure called the ***offset method***. Normally for structural design a 0.2% strain (0.002 in./in.) is chosen, and from this point on the ϵ axis a line parallel to the initial straight line portion of the stress–strain diagram is drawn. The point where this line intersects the curve defines the yield strength. From the graph, the yield strength is $\sigma_{YS} = 51$ ksi (352 MPa).

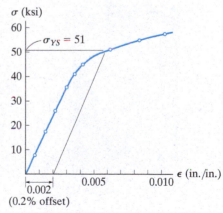

Yield strength for an aluminum alloy

Fig. 3–8

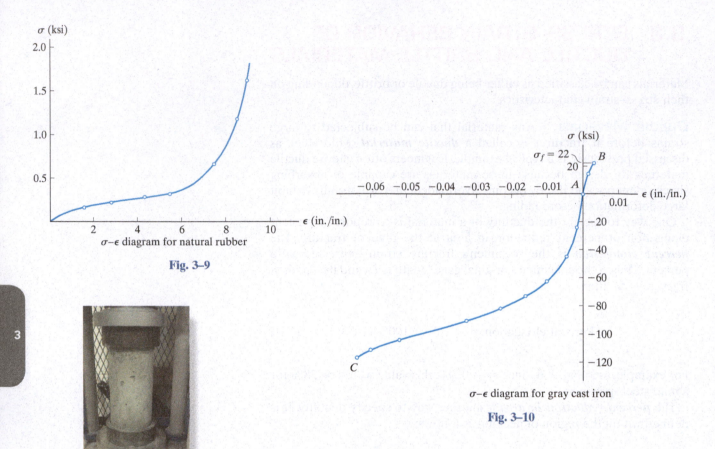

$\sigma-\epsilon$ diagram for natural rubber

Fig. 3–9

$\sigma-\epsilon$ diagram for gray cast iron

Fig. 3–10

Concrete used for structural purposes must be tested in compression to be sure it reaches its ultimate design stress after curing for 30 days.

Realize that the yield strength is not a physical property of the material, since it is a stress that causes a *specified* permanent strain in the material. In this text, however, we will assume that the yield strength, yield point, elastic limit, and proportional limit all *coincide* unless otherwise stated. An exception would be natural rubber, which in fact does not even have a proportional limit, since stress and strain are *not* linearly related. Instead, as shown in Fig. 3–9, this material, which is known as a polymer, exhibits *nonlinear elastic behavior*.

Wood is a material that is often moderately ductile, and as a result it is usually designed to respond only to elastic loadings. The strength characteristics of wood vary greatly from one species to another, and for each species they depend on the moisture content, age, and the size and arrangement of knots in the wood. Since wood is a fibrous material, its tensile or compressive characteristics parallel to its grain will differ greatly from these characteristics perpendicular to its grain. Specifically, wood splits easily when it is loaded in tension perpendicular to its grain, and consequently tensile loads are almost always intended to be applied parallel to the grain of wood members.

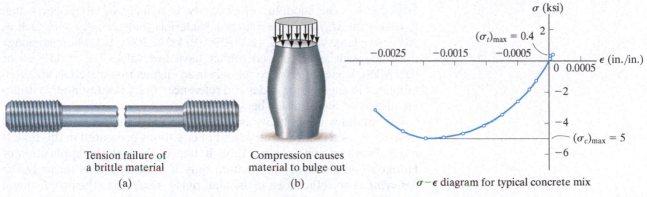

Tension failure of
a brittle material

(a)

Compression causes
material to bulge out

(b)

Fig. 3–11

$\sigma-\epsilon$ diagram for typical concrete mix

Fig. 3–12

Brittle Materials.

Materials that exhibit little or no yielding before failure are referred to as ***brittle materials***. Gray cast iron is an example, having a stress–strain diagram in tension as shown by the curve AB in Fig. 3–10. Here fracture at $\sigma_f = 22$ ksi (152 MPa) occurred due to a microscopic crack, which then spread rapidly across the specimen, causing complete fracture. Since the appearance of initial cracks in a specimen is quite random, brittle materials do not have a well-defined tensile fracture stress. Instead the *average* fracture stress from a set of observed tests is generally reported. A typical failed specimen is shown in Fig. 3–11*a*.

Compared with their behavior in tension, brittle materials exhibit a much higher resistance to axial compression, as evidenced by segment AC of the gray cast iron curve in Fig. 3–10. For this case any cracks or imperfections in the specimen tend to close up, and as the load increases the material will generally bulge or become barrel shaped as the strains become larger, Fig. 3–11*b*.

Like gray cast iron, concrete is classified as a brittle material, and it also has a low strength capacity in tension. The characteristics of its stress–strain diagram depend primarily on the mix of concrete (water, sand, gravel, and cement) and the time and temperature of curing. A typical example of a "complete" stress–strain diagram for concrete is given in Fig. 3–12. By inspection, its maximum compressive strength is about 12.5 times greater than its tensile strength, $(\sigma_c)_{\max} = 5$ ksi (34.5 MPa) versus $(\sigma_t)_{\max} = 0.40$ ksi (2.76 MPa). For this reason, concrete is almost always reinforced with steel bars or rods whenever it is designed to support tensile loads.

It can generally be stated that most materials exhibit both ductile and brittle behavior. For example, steel has brittle behavior when it contains a high carbon content, and it is ductile when the carbon content is reduced. Also, at low temperatures materials become harder and more brittle, whereas when the temperature rises they become softer and more ductile. This effect is shown in Fig. 3–13 for a methacrylate plastic.

Steel rapidly loses its strength when heated. For this reason engineers often require main structural members to be insulated in case of fire.

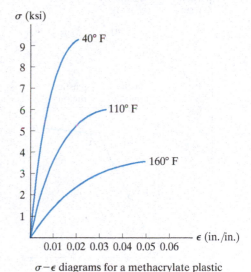

$\sigma-\epsilon$ diagrams for a methacrylate plastic

Fig. 3–13

Stiffness.

The modulus of elasticity is a mechanical property that indicates the **stiffness** of a material. Materials that are very stiff, such as steel, have large values of E [$E_{st} = 29(10^3)$ ksi or 200 GPa], whereas spongy materials such as vulcanized rubber have low values [$E_r = 0.10$ ksi or 0.69 MPa]. Values of E for commonly used engineering materials are often tabulated in engineering codes and reference books. Representative values are also listed on the inside back cover.

The modulus of elasticity is one of the most important mechanical properties used in the development of equations presented in this text. It must always be remembered, though, that E, through the application of Hooke's law, Eq. 3–3, can be used only if a material has *linear elastic behavior*. Also, if the stress in the material is *greater* than the proportional limit, the stress–strain diagram ceases to be a straight line, and so Hooke's law is no longer valid.

Strain Hardening.

If a specimen of ductile material, such as steel, is loaded into the *plastic region* and then unloaded, *elastic strain is recovered* as the material returns to its equilibrium state. The *plastic strain remains*, however, and as a result the material will be subjected to a **permanent set**. For example, a wire when bent (plastically) will spring back a little (elastically) when the load is removed; however, it will not fully return to its original position. This behavior is illustrated on the stress–strain diagram shown in Fig. 3–14a. Here the specimen is loaded beyond its yield point A to point A'. Since interatomic forces have to be overcome to elongate the specimen *elastically*, then these same forces pull the atoms back together when the load is removed, Fig. 3–14a. Consequently, the modulus of elasticity, E, is the same, and therefore the slope of line $O'A'$ is the same as line OA. With the load removed, the permanent set is OO'.

If the load is reapplied, the atoms in the material will again be displaced until yielding occurs at or near the stress A', and the stress–strain diagram continues along the same path as before, Fig. 3–14b. Although this new stress–strain diagram, defined by $O'A'B$, now has a *higher* yield point (A'), a consequence of strain hardening, it also has *less ductility*, or a smaller plastic region, than when it was in its original state.

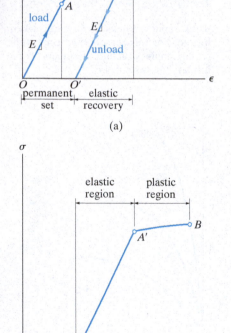

(a)

(b)

Fig. 3–14

This pin was made of a hardened steel alloy, that is, one having a high carbon content. It failed due to brittle fracture.

3.4   STRAIN ENERGY

As a material is deformed by an external load, the load will do external work, which in turn will be stored in the material as internal energy. This energy is related to the strains in the material, and so it is referred to as **strain energy**. To show how to calculate strain energy, consider a small volume element of material taken from a tension test specimen, Fig. 3–15. It is subjected to the uniaxial stress σ. This stress develops a force $\Delta F = \sigma \, \Delta A = \sigma (\Delta x \, \Delta y)$ on the top and bottom faces of the element, which causes the element to undergo a vertical displacement $\epsilon \, \Delta z$, Fig. 3–15b. By definition, **work** is determined by the product of a force and displacement in the direction of the force. Here the force is increased uniformly from *zero* to its final magnitude ΔF when the displacement $\epsilon \, \Delta z$ occurs, and so during the displacement the work done on the element by the force is equal to the *average* force magnitude $(\Delta F/2)$ times the displacement $\epsilon \, \Delta z$. The conservation of energy requires this "external work" on the element to be equivalent to the "internal work" or strain energy stored in the element, assuming that no energy is lost in the form of heat. Consequently, the strain energy is $\Delta U = (\frac{1}{2}\Delta F) \, \epsilon \, \Delta z = (\frac{1}{2} \sigma \, \Delta x \, \Delta y) \, \epsilon \, \Delta z$. Since the volume of the element is $\Delta V = \Delta x \, \Delta y \, \Delta z$, then $\Delta U = \frac{1}{2} \sigma \epsilon \, \Delta V$.

For engineering applications, it is often convenient to specify the strain energy per unit volume of material. This is called the **strain energy density**, and it can be expressed as

$$u = \frac{\Delta U}{\Delta V} = \frac{1}{2} \sigma \epsilon \qquad (3\text{--}6)$$

Finally, if the material behavior is *linear elastic*, then Hooke's law applies, $\sigma = E\epsilon$, and therefore we can express the **elastic strain energy density** in terms of the uniaxial stress σ as

$$u = \frac{1}{2} \frac{\sigma^2}{E} \qquad (3\text{--}7)$$

Modulus of Resilience.

When the stress in a material reaches the proportional limit, the strain energy density, as calculated by Eq. 3–6 or 3–7, is referred to as the **modulus of resilience**. It is

$$u_r = \frac{1}{2} \sigma_{pl} \, \epsilon_{pl} = \frac{1}{2} \frac{\sigma_{pl}^2}{E} \qquad (3\text{--}8)$$

Here u_r is equivalent to the shaded *triangular area* under the elastic region of the stress–strain diagram, Fig. 3–16a. Physically the modulus of resilience represents the largest amount of strain energy per unit volume the material can absorb without causing any permanent damage to the material. Certainly this property becomes important when designing bumpers or shock absorbers.

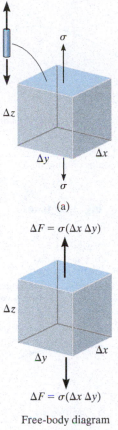

(a)

$\Delta F = \sigma(\Delta x \, \Delta y)$

$\Delta F = \sigma(\Delta x \, \Delta y)$

Free-body diagram

(b)

Fig. 3–15

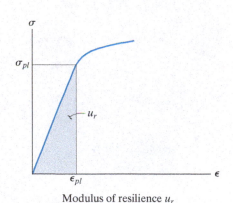

Modulus of resilience u_r

(a)

Fig. 3–16

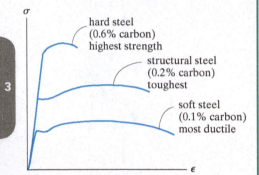

Modulus of toughness u_t

(b)

Fig. 3–16 (cont.)

Fig. 3–17

This nylon specimen exhibits a high degree of toughness as noted by the large amount of necking that has occurred just before fracture.

Modulus of Toughness.

Another important property of a material is its ***modulus of toughness***, u_t. This quantity represents the *entire area* under the stress–strain diagram, Fig. 3–16b, and therefore it indicates the maximum amount of strain energy per unit volume the material can absorb just before it fractures. Certainly this becomes important when designing members that may be accidentally overloaded. By alloying metals, engineers can change their resilience and toughness. For example, by changing the percentage of carbon in steel, the resulting stress–strain diagrams in Fig. 3–17 show how its resilience and toughness can be changed.

IMPORTANT POINTS

- A *conventional stress–strain diagram* is important in engineering since it provides a means for obtaining data about a material's tensile or compressive strength without regard for the material's physical size or shape.

- *Engineering stress and strain* are calculated using the *original* cross-sectional area and gage length of the specimen.

- A *ductile material*, such as mild steel, has four distinct behaviors as it is loaded. They are *elastic behavior, yielding, strain hardening,* and *necking*.

- A material is *linear elastic* if the stress is proportional to the strain within the elastic region. This behavior is described by *Hooke's law*, $\sigma = E\epsilon$, where the *modulus of elasticity E* is the slope of the line.

- Important points on the stress–strain diagram are the *proportional limit, elastic limit, yield stress, ultimate stress*, and *fracture stress*.

- The *ductility* of a material can be specified by the specimen's *percent elongation* or the *percent reduction in area*.

- If a material does not have a distinct yield point, a *yield strength* can be specified using a graphical procedure such as the *offset method*.

- *Brittle materials*, such as gray cast iron, have very little or no yielding and so they can fracture suddenly.

- *Strain hardening* is used to establish a higher yield point for a material. This is done by straining the material beyond the elastic limit, then releasing the load. The modulus of elasticity remains the same; however, the material's ductility *decreases*.

- *Strain energy* is energy stored in a material due to its deformation. This energy per unit volume is called *strain energy density*. If it is measured up to the proportional limit, it is referred to as the *modulus of resilience*, and if it is measured up to the point of fracture, it is called the *modulus of toughness*. It can be determined from the area under the $\sigma - \epsilon$ diagram.

EXAMPLE 3.1

A tension test for a steel alloy results in the stress–strain diagram shown in Fig. 3–18. Calculate the modulus of elasticity and the yield strength based on a 0.2% offset. Identify on the graph the ultimate stress and the fracture stress.

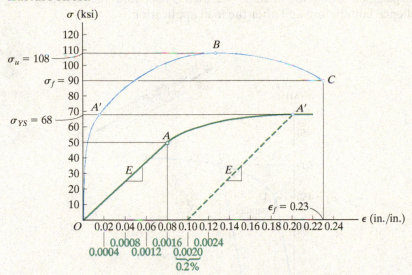

Fig. 3–18

SOLUTION

Modulus of Elasticity.   We must calculate the *slope* of the initial straight-line portion of the graph. Using the magnified curve and scale shown in green, this line extends from point O to point A, which has coordinates of approximately (0.0016 in./in., 50 ksi). Therefore,

$$E = \frac{50 \text{ ksi}}{0.0016 \text{ in./in.}} = 31.2(10^3) \text{ ksi} \qquad \textit{Ans.}$$

Note that the equation of line OA is thus $\sigma = 31.2(10^3)\epsilon$.

Yield Strength.   For a 0.2% offset, we begin at a strain of 0.2% or 0.0020 in./in. and graphically extend a (dashed) line parallel to OA until it intersects the σ–ϵ curve at A'. The yield strength is approximately

$$\sigma_{YS} = 68 \text{ ksi} \qquad \textit{Ans.}$$

Ultimate Stress.   This is defined by the peak of the σ–ϵ graph, point B in Fig. 3–18.

$$\sigma_u = 108 \text{ ksi} \qquad \textit{Ans.}$$

Fracture Stress.   When the specimen is strained to its maximum of $\epsilon_f = 0.23$ in./in., it fractures at point C. Thus,

$$\sigma_f = 90 \text{ ksi} \qquad \textit{Ans.}$$

EXAMPLE 3.2

The stress–strain diagram for an aluminum alloy that is used for making aircraft parts is shown in Fig. 3–19. If a specimen of this material is stressed to $\sigma = 600$ MPa, determine the permanent set that remains in the specimen when the load is released. Also, find the modulus of resilience both before and after the load application.

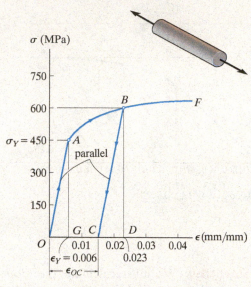

Fig. 3–19

SOLUTION

Permanent Strain.   When the specimen is subjected to the load, it strain hardens until point B is reached on the σ–ϵ diagram. The strain at this point is approximately 0.023 mm/mm. When the load is released, the material behaves by following the straight line BC, which is parallel to line OA. Since both of these lines have the same slope, the strain at point C can be determined analytically. The slope of line OA is the modulus of elasticity, i.e.,

$$E = \frac{450 \text{ MPa}}{0.006 \text{ mm/mm}} = 75.0 \text{ GPa}$$

From triangle *CBD*, we require

$$E = \frac{BD}{CD}; \qquad 75.0(10^9) \text{ Pa} = \frac{600(10^6) \text{ Pa}}{CD}$$

$$CD = 0.008 \text{ mm/mm}$$

This strain represents the amount of *recovered elastic strain*. The permanent set or strain, ϵ_{OC}, is thus

$$\epsilon_{OC} = 0.023 \text{ mm/mm} - 0.008 \text{ mm/mm}$$

$$= 0.0150 \text{ mm/mm} \qquad\qquad \textit{Ans.}$$

NOTE: If gage marks on the specimen were originally 50 mm apart, then after the load is *released* these marks will be 50 mm + $(0.0150)(50 \text{ mm}) = 50.75 \text{ mm}$ apart.

Modulus of Resilience.   Applying Eq. 3–8, the areas under *OAG* and *CBD* in Fig. 3–19 are*

$$(u_r)_{\text{initial}} = \frac{1}{2}\, \sigma_{pl}\, \epsilon_{pl} = \frac{1}{2}(450 \text{ MPa})(0.006 \text{ mm/mm})$$

$$= 1.35 \text{ MJ/m}^3 \qquad\qquad \textit{Ans.}$$

$$(u_r)_{\text{final}} = \frac{1}{2}\, \sigma_{pl}\, \epsilon_{pl} = \frac{1}{2}(600 \text{ MPa})(0.008 \text{ mm/mm})$$

$$= 2.40 \text{ MJ/m}^3 \qquad\qquad \textit{Ans.}$$

NOTE: By comparison, the effect of strain hardening the material has caused an increase in the modulus of resilience; however, note that the modulus of toughness for the material has decreased, since the area under the original curve, *OABF*, is larger than the area under curve *CBF*.

*Work in the SI system of units is measured in joules, where $1 \text{ J} = 1 \text{ N} \cdot \text{m}$.

EXAMPLE 3.3

The aluminum rod, shown in Fig. 3–20a, has a circular cross section and is subjected to an axial load of 10 kN. If a portion of the stress–strain diagram is shown in Fig. 3–20b, determine the approximate elongation of the rod when the load is applied. Take $E_{al} = 70$ GPa.

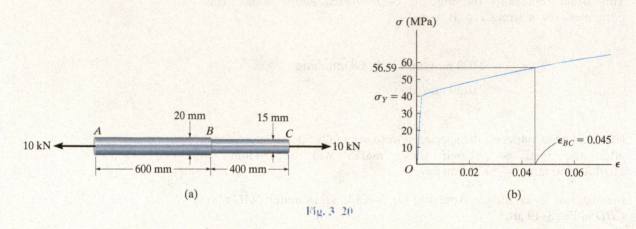

Fig. 3–20

SOLUTION

In order to find the elongation of the rod, we must first obtain the strain. This is done by calculating the stress, then using the stress–strain diagram. The normal stress within each segment is

$$\sigma_{AB} = \frac{N}{A} = \frac{10(10^3)\ \text{N}}{\pi(0.01\ \text{m})^2} = 31.83\ \text{MPa}$$

$$\sigma_{BC} = \frac{N}{A} = \frac{10(10^3)\ \text{N}}{\pi(0.0075\ \text{m})^2} = 56.59\ \text{MPa}$$

From the stress–strain diagram, the material in segment AB is strained *elastically* since $\sigma_{AB} < \sigma_Y = 40$ MPa. Using Hooke's law,

$$\epsilon_{AB} = \frac{\sigma_{AB}}{E_{al}} = \frac{31.83(10^6)\ \text{Pa}}{70(10^9)\ \text{Pa}} = 0.0004547\ \text{mm/mm}$$

The material within segment BC is strained plastically, since $\sigma_{BC} > \sigma_Y = 40$ MPa. From the graph, for $\sigma_{BC} = 56.59$ MPa, $\epsilon_{BC} \approx 0.045$ mm/mm. The approximate elongation of the rod is therefore

$$\delta = \Sigma \epsilon L = 0.0004547(600\ \text{mm}) + 0.0450(400\ \text{mm})$$

$$= 18.3\ \text{mm} \qquad \qquad \textit{Ans.}$$

FUNDAMENTAL PROBLEMS

F3–1. Define a homogeneous material.

F3–2. Indicate the points on the stress–strain diagram which represent the proportional limit and the ultimate stress.

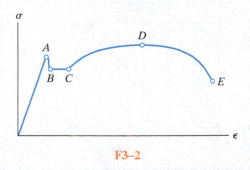

F3–2

F3–3. Define the modulus of elasticity E.

F3–4. At room temperature, mild steel is a ductile material. True or false?

F3–5. Engineering stress and strain are calculated using the *actual* cross-sectional area and length of the specimen. True or false?

F3–6. As the temperature increases the modulus of elasticity will increase. True or false?

F3–7. A 100-mm-long rod has a diameter of 15 mm. If an axial tensile load of 100 kN is applied, determine its change in length. Assume linear elastic behavior with $E = 200$ GPa.

F3–8. A bar has a length of 8 in. and cross-sectional area of 12 in². Determine the modulus of elasticity of the material if it is subjected to an axial tensile load of 10 kip and stretches 0.003 in. The material has linear elastic behavior.

F3–9. A 10-mm-diameter rod has a modulus of elasticity of $E = 100$ GPa. If it is 4 m long and subjected to an axial tensile load of 6 kN, determine its elongation. Assume linear elastic behavior.

F3–10. The material for the 50-mm-long specimen has the stress–strain diagram shown. If $P = 100$ kN, determine the elongation of the specimen.

F3–11. The material for the 50-mm-long specimen has the stress–strain diagram shown. If $P = 150$ kN is applied and then released, determine the permanent elongation of the specimen.

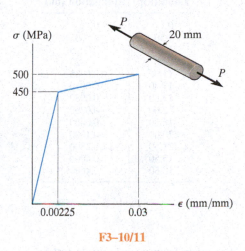

F3–10/11

F3–12. If the elongation of wire BC is 0.2 mm after the force **P** is applied, determine the magnitude of **P**. The wire is A-36 steel and has a diameter of 3 mm.

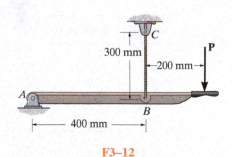

F3–12

PROBLEMS

3–1. A tension test was performed on a steel specimen having an original diameter of 0.503 in. and gage length of 2.00 in. The data is listed in the table. Plot the stress–strain diagram and determine approximately the modulus of elasticity, the yield stress, the ultimate stress, and the fracture stress. Use a scale of 1 in. = 20 ksi and 1 in. = 0.05 in./in. Redraw the elastic region, using the same stress scale but a strain scale of 1 in. = 0.001 in./in.

Load (kip)	Elongation (in.)
0	0
1.50	0.0005
4.60	0.0015
8.00	0.0025
11.00	0.0035
11.80	0.0050
11.80	0.0080
12.00	0.0200
16.60	0.0400
20.00	0.1000
21.50	0.2800
19.50	0.4000
18.50	0.4600

Prob. 3–1

3–2. Data taken from a stress–strain test for a ceramic are given in the table. The curve is linear between the origin and the first point. Plot the diagram, and determine the modulus of elasticity and the modulus of resilience.

3–3. Data taken from a stress–strain test for a ceramic are given in the table. The curve is linear between the origin and the first point. Plot the diagram, and determine approximately the modulus of toughness. The fracture stress is $\sigma_f = 53.4$ ksi.

σ (ksi)	ϵ (in./in.)
0	0
33.2	0.0006
45.5	0.0010
49.4	0.0014
51.5	0.0018
53.4	0.0022

Probs. 3–2/3

*3–4. The stress–strain diagram for a steel alloy having an original diameter of 0.5 in. and a gage length of 2 in. is given in the figure. Determine approximately the modulus of elasticity for the material, the load on the specimen that causes yielding, and the ultimate load the specimen will support.

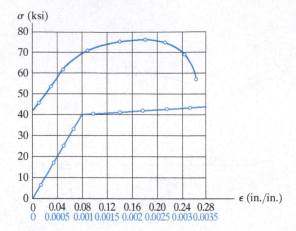

Prob. 3–4

3–5. The stress–strain diagram for a steel alloy having an original diameter of 0.5 in. and a gage length of 2 in. is given in the figure. If the specimen is loaded until it is stressed to 70 ksi, determine the approximate amount of elastic recovery and the increase in the gage length after it is unloaded.

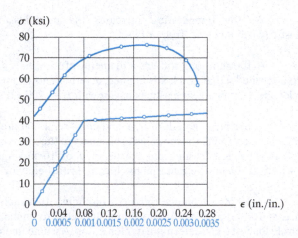

Prob. 3–5

3–6. The stress–strain diagram for a steel alloy having an original diameter of 0.5 in. and a gage length of 2 in. is given in the figure. Determine approximately the modulus of resilience and the modulus of toughness for the material.

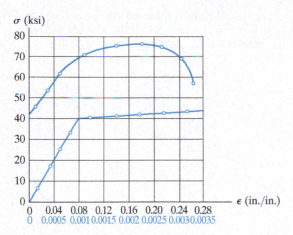

Prob. 3–6

3–7. The rigid beam is supported by a pin at C and an A-36 steel guy wire AB. If the wire has a diameter of 0.2 in., determine how much it stretches when a distributed load of $w = 100$ lb/ft acts on the beam. The material remains elastic.

*3–8. The rigid beam is supported by a pin at C and an A-36 steel guy wire AB. If the wire has a diameter of 0.2 in., determine the distributed load w if the end B is displaced 0.75 in. downward.

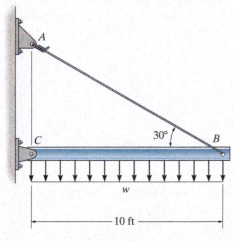

Probs. 3–7/8

3–9. Acetal plastic has a stress-strain diagram as shown. If a bar of this material has a length of 3 ft and cross-sectional area of 0.875 in^2, and is subjected to an axial load of 2.5 kip, determine its elongation.

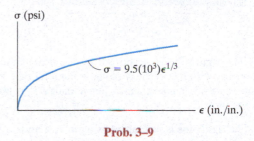

$$\sigma = 9.5(10^3)\epsilon^{1/3}$$

Prob. 3–9

3–10. The stress–strain diagram for an aluminum alloy specimen having an original diameter of 0.5 in. and a gage length of 2 in. is given in the figure. Determine approximately the modulus of elasticity for the material, the load on the specimen that causes yielding, and the ultimate load the specimen will support.

3–11. The stress–strain diagram for an aluminum alloy specimen having an original diameter of 0.5 in. and a gage length of 2 in. is given in the figure. If the specimen is loaded until it is stressed to 60 ksi, determine the approximate amount of elastic recovery and the increase in the gage length after it is unloaded.

*3–12. The stress–strain diagram for an aluminum alloy specimen having an original diameter of 0.5 in. and a gage length of 2 in. is given in the figure. Determine approximately the modulus of resilience and the modulus of toughness for the material.

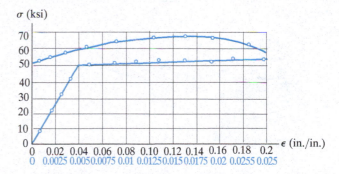

Probs. 3–10/11/12

3–13.   A bar having a length of 5 in. and cross-sectional area of 0.7 in.² is subjected to an axial force of 8000 lb. If the bar stretches 0.002 in., determine the modulus of elasticity of the material. The material has linear elastic behavior.

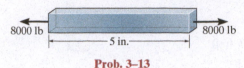

8000 lb   **8000 lb**

5 in.

Prob. 3–13

3–14.   The rigid pipe is supported by a pin at *A* and an A-36 steel guy wire *BD*. If the wire has a diameter of 0.25 in., determine how much it stretches when a load of *P* = 600 lb acts on the pipe.

3–15.   The rigid pipe is supported by a pin at *A* and an A-36 guy wire *BD*. If the wire has a diameter of 0.25 in., determine the load *P* if the end *C* is displaced 0.075 in. downward.

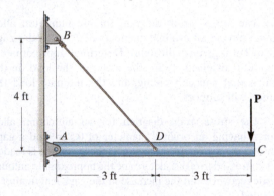

B

4 ft

A   *D*   *C*

P

3 ft   3 ft

Probs. 3–14/15

3–16.   Direct tension indicators are sometimes used instead of torque wrenches to ensure that a bolt has a prescribed tension when used for connections. If a nut on the bolt is tightened so that the six 3-mm high heads of the indicator are strained 0.1 mm/mm, and leave a contact area on each head of 1.5 mm², determine the tension in the bolt shank. The material has the stress–strain diagram shown.

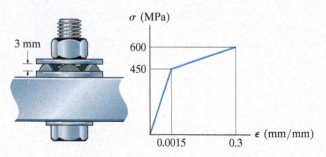

3 mm

σ (MPa)

600

450

0.0015   0.3

ϵ (mm/mm)

Prob. 3–16

3–17.   The rigid beam is supported by a pin at *C* and an A992 steel guy wire *AB* of length 6 ft. If the wire has a diameter of 0.2 in., determine how much it stretches when a distributed load of *w* = 200 lb/ft acts on the beam. The wire remains elastic.

3–18.   The rigid beam is supported by a pin at *C* and an A992 steel guy wire *AB* of length 6 ft. If the wire has a diameter of 0.2 in., determine the distributed load *w* if the end *B* is displaced 0.12 in. downward. The wire remains elastic.

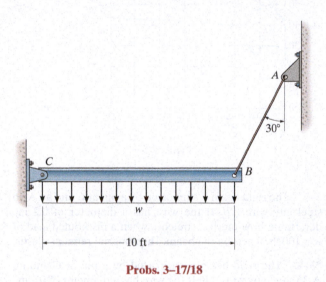

A

30°

C   *B*

w

10 ft

Probs. 3–17/18

3–19.   The stress–strain diagram for a bone is shown, and can be described by the equation $\epsilon = 0.45(10^{-6})\,\sigma + 0.36(10^{-12})\,\sigma^3$, where σ is in kPa. Determine the yield strength assuming a 0.3% offset.

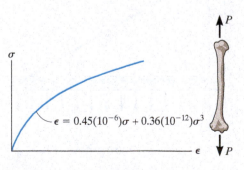

P

σ

$\epsilon = 0.45(10^{-6})\sigma + 0.36(10^{-12})\sigma^3$

ϵ

P

Prob. 3–19

*3–20. The stress–strain diagram for a bone is shown and can be described by the equation $\epsilon = 0.45(10^{-6})\,\sigma + 0.36(10^{-12})\,\sigma^3$, where σ is in kPa. Determine the modulus of toughness and the amount of elongation of a 200-mm-long region just before it fractures if failure occurs at $\epsilon = 0.12$ mm/mm.

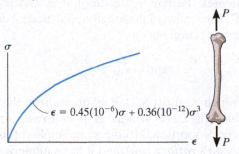

$$\epsilon = 0.45(10^{-6})\sigma + 0.36(10^{-12})\sigma^3$$

Prob. 3–20

3–21. The two bars are made of a material that has the stress–strain diagram shown. If the cross-sectional area of bar AB is 1.5 in² and BC is 4 in², determine the largest force P that can be supported before any member fractures. Assume that buckling does not occur.

3–22. The two bars are made of a material that has the stress–strain diagram shown. Determine the cross-sectional area of each bar so that the bars fracture simultaneously when the load $P = 3$ kip. Assume that buckling does not occur.

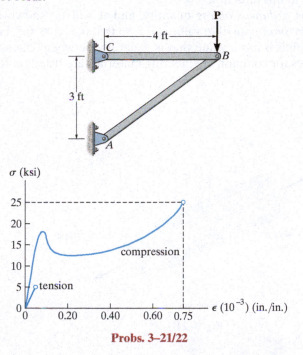

Probs. 3–21/22

3–23. The pole is supported by a pin at C and an A-36 steel guy wire AB. If the wire has a diameter of 0.2 in., determine how much it stretches when a horizontal force of 2.5 kip acts on the pole.

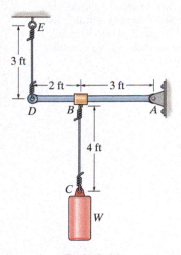

Prob. 3–23

*3–24. The bar DA is rigid and is originally held in the horizontal position when the weight W is supported from C. If the weight causes B to displace downward 0.025 in., determine the strain in wires DE and BC. Also, if the wires are made of A-36 steel and have a cross-sectional area of 0.002 in², determine the weight W.

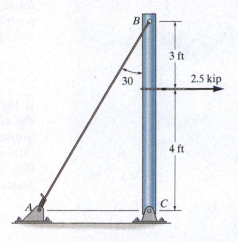

Prob. 3–24

3.5   POISSON'S RATIO

When a deformable body is subjected to a force, not only does it elongate but it also contracts laterally. For example, consider the bar in Fig. 3–21 that has an original radius r and length L, and is subjected to the tensile force P. This force elongates the bar by an amount δ, and its radius contracts by an amount δ'. The strains in the longitudinal or axial direction and in the lateral or radial direction become

$$\epsilon_{long} = \frac{\delta}{L} \quad \text{and} \quad \epsilon_{lat} = \frac{\delta'}{r}$$

In the early 1800s, the French scientist S. D. Poisson realized that *within the elastic range* the *ratio* of these strains is a *constant*, since the displacements δ and δ' are proportional to the same applied force. This ratio is referred to as **Poisson's ratio**, ν (nu), and it has a numerical value that is unique for any material that is both *homogeneous and isotropic*. Stated mathematically it is

$$\nu = -\frac{\epsilon_{lat}}{\epsilon_{long}} \tag{3–9}$$

The negative sign is included here since *longitudinal elongation* (positive strain) causes *lateral contraction* (negative strain), and vice versa. Keep in mind that these strains are caused only by the single axial or longitudinal force P; i.e., no force acts in a lateral direction in order to strain the material in this direction.

Poisson's ratio is a *dimensionless* quantity, and it will be shown in Sec. 10.6 that its *maximum* possible value is 0.5, so that $0 \le \nu \le 0.5$. For most nonporous solids it has a value that is generally between 0.25 and 0.355. Typical values for common engineering materials are listed on the inside back cover.

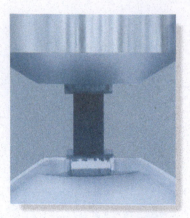

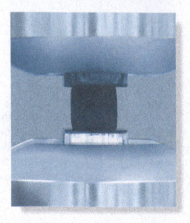

When the rubber block is compressed (negative strain), its sides will expand (positive strain). The ratio of these strains remains constant.

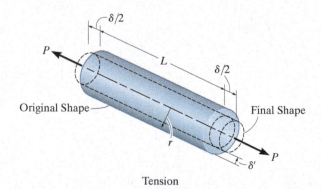

Tension

Fig. 3–21

EXAMPLE 3.4

A bar made of A-36 steel has the dimensions shown in Fig. 3–22. If an axial force of $P = 80$ kN is applied to the bar, determine the change in its length and the change in the dimensions of its cross section. The material behaves elastically.

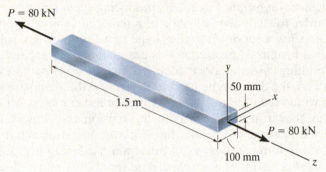

$P = 80$ kN

1.5 m

50 mm

y

x

$P = 80$ kN

100 mm

z

Fig. 3–22

SOLUTION

The normal stress in the bar is

$$\sigma_z = \frac{N}{A} = \frac{80(10^3)\ \text{N}}{(0.1\ \text{m})(0.05\ \text{m})} = 16.0(10^6)\ \text{Pa}$$

From the table on the inside back cover for A-36 steel $E_{\text{st}} = 200$ GPa, and so the strain in the z direction is

$$\epsilon_z = \frac{\sigma_z}{E_{\text{st}}} = \frac{16.0(10^6)\ \text{Pa}}{200(10^9)\ \text{Pa}} = 80(10^{-6})\ \text{mm/mm}$$

The axial elongation of the bar is therefore

$$\delta_z = \epsilon_z\, L_z = [80(10^{-6})](1.5\ \text{m}) = 120\ \mu\text{m} \qquad \textit{Ans.}$$

Using Eq. 3–9, where $\nu_{\text{st}} = 0.32$ as found from the inside back cover, the lateral contraction strains in *both* the x and y directions are

$$\epsilon_x = \epsilon_y = -\nu_{\text{st}}\, \epsilon_z = -0.32[80(10^{-6})] = -25.6\ \mu\text{m/m}$$

Thus the changes in the dimensions of the cross section are

$$\delta_x = \epsilon_x\, L_x = -[25.6(10^{-6})](0.1\ \text{m}) = -2.56\ \mu\text{m} \qquad \textit{Ans.}$$

$$\delta_y = \epsilon_y\, L_y = -[25.6(10^{-6})](0.05\ \text{m}) = -1.28\ \mu\text{m} \qquad \textit{Ans.}$$

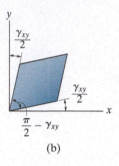

(a)

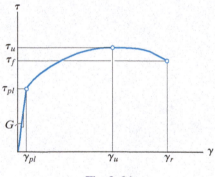

(b)

Fig. 3–23

3.6   THE SHEAR STRESS–STRAIN DIAGRAM

In Sec. 1.5 it was shown that when a small element of material is subjected to *pure shear*, equilibrium requires that equal shear stresses must be developed on four faces of the element, Fig. 3–23a. Furthermore, if the material is *homogeneous* and *isotropic*, then this shear stress will distort the element *uniformly*, Fig. 3–23b, producing shear strain.

In order to study the behavior of a material subjected to pure shear, engineers use a specimen in the shape of a thin tube and subject it to a torsional loading. If measurements are made of the applied torque and the resulting angle of twist, then by the methods to be explained in Chapter 5, the data can be used to determine the shear stress and shear strain within the tube and thereby produce a shear stress–strain diagram such as shown in Fig. 3–24. Like the tension test, this material when subjected to shear will exhibit linear elastic behavior and it will have a defined *proportional limit* τ_{pl}. Also, strain hardening will occur until an *ultimate shear stress* τ_u is reached. And finally, the material will begin to lose its shear strength until it reaches a point where it fractures, τ_f.

For most engineering materials, like the one just described, the elastic behavior is *linear*, and so Hooke's law for shear can be written as

$$\tau = G\gamma \qquad\qquad (3\text{–}10)$$

Here G is called the **shear modulus of elasticity** or the **modulus of rigidity**. Its value represents the slope of the line on the τ–γ diagram, that is, $G = \tau_{pl}/\gamma_{pl}$. Units of measurement for G will be the *same* as those for τ (Pa or psi), since γ is measured in radians, a dimensionless quantity. Typical values for common engineering materials are listed on the inside back cover.

Later it will be shown in Sec. 10.6 that the three material constants, E, ν, and G can all be *related* by the equation

$$G = \frac{E}{2(1 + \nu)} \qquad\qquad (3\text{–}11)$$

Therefore, if E and G are known, the value of ν can then be determined from this equation rather than through experimental measurement.

Fig. 3–24

EXAMPLE 3.5

A specimen of titanium alloy is tested in torsion and the shear stress–strain diagram is shown in Fig. 3–25a. Determine the shear modulus G, the proportional limit, and the ultimate shear stress. Also, determine the maximum distance d that the top of a block of this material, shown in Fig. 3–25b, could be displaced horizontally if the material behaves elastically when acted upon by a shear force **V**. What is the magnitude of **V** necessary to cause this displacement?

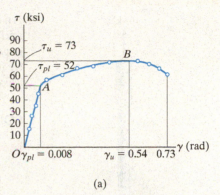

(a)

SOLUTION

Shear Modulus.   This value represents the slope of the straight-line portion OA of the τ–γ diagram. The coordinates of point A are (0.008 rad, 52 ksi). Thus,

$$G = \frac{52 \text{ ksi}}{0.008 \text{ rad}} = 6500 \text{ ksi} \qquad Ans.$$

The equation of line OA is therefore $\tau = G\gamma = 6500\gamma$, which is Hooke's law for shear.

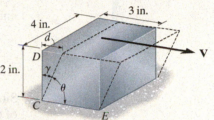

(b)

Fig. 3–25

Proportional Limit.   By inspection, the graph ceases to be linear at point A. Thus,

$$\tau_{pl} = 52 \text{ ksi} \qquad Ans.$$

Ultimate Stress.   This value represents the maximum shear stress, point B. From the graph,

$$\tau_u = 73 \text{ ksi} \qquad Ans.$$

Maximum Elastic Displacement and Shear Force.   The shear strain at the corner C of the block in Fig. 3–25b is determined by finding the difference in the 90° angle DCE and the angle θ. This angle is $\gamma = 90° - \theta$ as shown. From the τ–γ diagram the maximum elastic shear strain is 0.008 rad, a very small angle. The top of the block in Fig. 3–25b will therefore be displaced horizontally a distance d given by

$$\tan(0.008 \text{ rad}) \approx 0.008 \text{ rad} = \frac{d}{2 \text{ in.}}$$

$$d = 0.016 \text{ in.}$$

The corresponding *average* shear stress in the block is $\tau_{pl} = 52$ ksi. Thus, the shear force V needed to cause the displacement is

$$\tau_{avg} = \frac{V}{A}; \qquad 52 \text{ ksi} = \frac{V}{(3 \text{ in.})(4 \text{ in.})}$$

$$V = 624 \text{ kip} \qquad Ans.$$

EXAMPLE 3.6

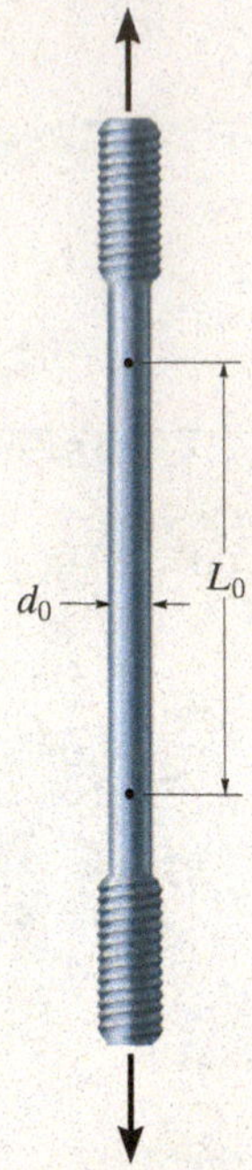

165 kN

d_0 → ← L_0

165 kN

Fig. 3–26

An aluminum specimen shown in Fig. 3–26 has a diameter of $d_0 = 25$ mm and a gage length of $L_0 = 250$ mm. If a force of 165 kN elongates the gage length 1.20 mm, determine the modulus of elasticity. Also, determine by how much the force causes the diameter of the specimen to contract. Take $G_{al} = 26$ GPa and $\sigma_Y = 440$ MPa.

SOLUTION

Modulus of Elasticity.   The average normal stress in the specimen is

$$\sigma = \frac{N}{A} = \frac{165\,(10^3)\,\text{N}}{(\pi/4)\,(0.025\,\text{m})^2} = 336.1\,\text{MPa}$$

and the average normal strain is

$$\epsilon = \frac{\delta}{L} = \frac{1.20\,\text{mm}}{250\,\text{mm}} = 0.00480\,\text{mm/mm}$$

Since $\sigma < \sigma_Y = 440$ MPa, the material behaves elastically. The modulus of elasticity is therefore

$$E_{al} = \frac{\sigma}{\epsilon} = \frac{336.1\,(10^6)\,\text{Pa}}{0.00480} = 70.0\,\text{GPa} \qquad \textit{Ans.}$$

Contraction of Diameter.   First we will determine Poisson's ratio for the material using Eq. 3–11.

$$G = \frac{E}{2(1+\nu)}$$

$$26\,\text{GPa} = \frac{70.0\,\text{GPa}}{2(1+\nu)}$$

$$\nu = 0.347$$

Since $\epsilon_{long} = 0.00480$ mm/mm, then by Eq. 3–9,

$$\nu = -\frac{\epsilon_{lat}}{\epsilon_{long}}$$

$$0.347 = -\frac{\epsilon_{lat}}{0.00480\,\text{mm/mm}}$$

$$\epsilon_{lat} = -0.00166\,\text{mm/mm}$$

The contraction of the diameter is therefore

$$\delta' = (0.00166)\,(25\,\text{mm})$$

$$= 0.0416\,\text{mm} \qquad \textit{Ans.}$$

*3.7 FAILURE OF MATERIALS DUE TO CREEP AND FATIGUE

The mechanical properties of a material have up to this point been discussed only for a static or slowly applied load at constant temperature. In some cases, however, a member may have to be used in an environment for which loadings must be sustained over long periods of time at elevated temperatures, or in other cases, the loading may be repeated or cycled. We will not cover these effects in this book, although we will briefly mention how one determines a material's strength for these conditions, since in some cases they must be considered for design.

Creep. When a material has to support a load for a very long period of time, it may continue to deform until a sudden fracture occurs or its usefulness is impaired. This time-dependent permanent deformation is known as *creep*. Normally creep is considered when metals and ceramics are used for structural members or mechanical parts that are subjected to high temperatures. For some materials, however, such as polymers and composite materials–including wood or concrete–temperature is *not* an important factor, and yet creep can occur strictly from long-term load application. As a typical example, consider the fact that a rubber band will not return to its original shape after being released from a stretched position in which it was held for a very long period of time.

For practical purposes, when creep becomes important, a member is usually designed to resist a specified creep strain for a given period of time. An important mechanical property that is used in this regard is called the *creep strength*. This value represents the highest stress the material can withstand during a specified time without exceeding an allowable creep strain. The creep strength will vary with temperature, and for design, a temperature, duration of loading, and allowable creep strain must all be specified. For example, a creep strain of 0.1% per year has been suggested for steel used for bolts and piping.

Several methods exist for determining the allowable creep strength for a particular material. One of the simplest involves testing several specimens simultaneously at a constant temperature, but with each subjected to a different axial stress. By measuring the length of time needed to produce the allowable creep strain for each specimen, a curve of stress versus time can be established. Normally these tests are run to a maximum of 1000 hours. An example of the results for stainless steel at a temperature of 1200°F and prescribed creep strain of 1% is shown in Fig. 3–27. As noted, this material has a yield strength of 40 ksi (276 MPa) at room temperature (0.2% offset) and the creep strength at 1000 h is found to be approximately $\sigma_c = 20$ ksi (138 MPa).

The long-term application of the cable loading on this pole has caused the pole to deform due to creep.

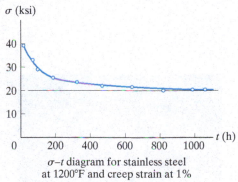

σ–t diagram for stainless steel
at 1200°F and creep strain at 1%

Fig. 3–27

For longer periods of time, extrapolations from the curves must be made. To do this usually requires a certain amount of experience with creep behavior, and some supplementary knowledge about the creep properties of the material. Once the material's creep strength has been determined, however, a factor of safety is applied to obtain an appropriate allowable stress for design.

Fatigue.   When a metal is subjected to repeated cycles of stress or strain, it causes its internal structure to break down, ultimately leading to fracture. This behavior is called *fatigue*, and it is usually responsible for a large percentage of failures in connecting rods and crankshafts of engines; steam or gas turbine blades; connections or supports for bridges, railroad wheels, and axles; and other parts subjected to cyclic loading. In all these cases, fracture will occur at a stress that is *less* than the material's yield stress.

The nature of this failure apparently results from the fact that there are microscopic imperfections, usually on the surface of the member, where the localized stress becomes *much greater* than the average stress acting over the cross section. As this higher stress is cycled, it leads to the formation of minute cracks. Occurrence of these cracks causes a further increase of stress at their tips, which in turn causes a further extension of the cracks into the material as the stress continues to be cycled. Eventually the cross-sectional area of the member is reduced to the point where the load can no longer be sustained, and as a result sudden fracture occurs. The material, even though known to be ductile, behaves as if it were brittle.

In order to specify a safe strength for a metallic material under repeated loading, it is necessary to determine a limit below which no evidence of failure can be detected after applying a load for a specified number of cycles. This limiting stress is called the *endurance* or *fatigue limit*. Using a testing machine for this purpose, a series of specimens are each subjected to a specified stress and cycled to failure. The results are plotted as a graph representing the stress S (or σ) on the vertical axis and the number of cycles-to-failure N on the horizontal axis. This graph is called an *S–N diagram* or *stress–cycle diagram*, and most often the values of N are plotted on a logarithmic scale since they are generally quite large.

Examples of S–N diagrams for two common engineering metals are shown in Fig. 3–28. The endurance limit is usually identified as the stress for which the S–N graph becomes horizontal or asymptotic. As noted, it has a well-defined value of $(S_{el})_{st} = 27$ ksi (186 MPa) for steel. For aluminum, however, the endurance limit is not well defined, and so here it may be specified as the stress having a limit of, say, 500 million cycles, $(S_{el})_{al} = 19$ ksi (131 MPa). Once a particular value is obtained, it is often assumed that for any stress below this value the fatigue life will be infinite, and therefore the number of cycles to failure is no longer given consideration.

The design of members used for amusement park rides requires careful consideration of cyclic loadings that can cause fatigue.

Engineers must account for possible fatigue failure of the moving parts of this oil-pumping rig.

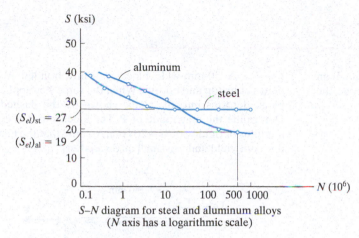

S–N diagram for steel and aluminum alloys
(N axis has a logarithmic scale)

Fig. 3–28

IMPORTANT POINTS

- *Poisson's ratio*, ν, is a ratio of the lateral strain of a homogeneous and isotropic material to its longitudinal strain. Generally these strains are of opposite signs, that is, if one is an elongation, the other will be a contraction.

- The *shear stress–strain diagram* is a plot of the shear stress versus the shear strain. If the material is homogeneous and isotropic, and is also linear elastic, the slope of the straight line within the elastic region is called the modulus of rigidity or the shear modulus, G.

- There is a mathematical relationship between G, E, and ν.

- *Creep* is the time-dependent deformation of a material for which stress and/or temperature play an important role. Members are designed to resist the effects of creep based on their material creep strength, which is the largest initial stress a material can withstand during a specified time without exceeding a specified creep strain.

- *Fatigue* occurs in metals when the stress or strain is cycled. This phenomenon causes brittle fracture of the material. Members are designed to resist fatigue by ensuring that the stress in the member does not exceed its *endurance* or *fatigue limit*. This value is determined from an *S–N* diagram as the maximum stress the material can resist when subjected to a specified number of cycles of loading.

FUNDAMENTAL PROBLEMS

F3–13. A 100-mm-long rod has a diameter of 15 mm. If an axial tensile load of 10 kN is applied to it, determine the change in its diameter. $E = 70$ GPa, $\nu = 0.35$.

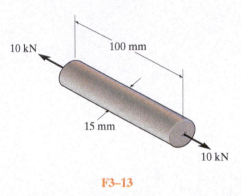

F3–13

F3–14. A solid circular rod that is 600 mm long and 20 mm in diameter is subjected to an axial force of $P = 50$ kN. The elongation of the rod is $\delta = 1.40$ mm, and its diameter becomes $d' = 19.9837$ mm. Determine the modulus of elasticity and the modulus of rigidity of the material. Assume that the material does not yield.

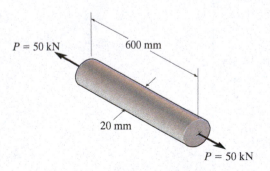

F3–14

F3–15. A 20-mm-wide block is firmly bonded to rigid plates at its top and bottom. When the force **P** is applied the block deforms into the shape shown by the dashed line. Determine the magnitude of **P**. The block's material has a modulus of rigidity of $G = 26$ GPa. Assume that the material does not yield and use small angle analysis.

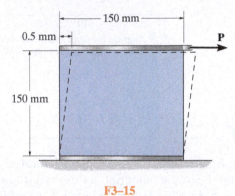

F3–15

F3–16. A 20-mm-wide block is bonded to rigid plates at its top and bottom. When the force **P** is applied the block deforms into the shape shown by the dashed line. If $a = 3$ mm and **P** is released, determine the permanent shear strain in the block.

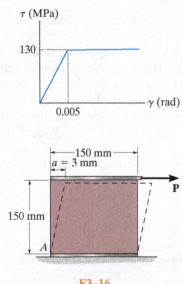

F3–16

PROBLEMS

3–25. The acrylic plastic rod is 200 mm long and 15 mm in diameter. If an axial load of 300 N is applied to it, determine the change in its length and the change in its diameter. $E_p = 2.70$ GPa, $\nu_p = 0.4$.

300 N   300 N

|← 200 mm →|

Prob. 3–25

3–26. The plug has a diameter of 30 mm and fits within a rigid sleeve having an inner diameter of 32 mm. Both the plug and the sleeve are 50 mm long. Determine the axial pressure p that must be applied to the top of the plug to cause it to contact the sides of the sleeve. Also, how far must the plug be compressed downward in order to do this? The plug is made from a material for which $E = 5$ MPa, $\nu = 0.45$.

p

Prob. 3–26

3–27. The elastic portion of the stress–strain diagram for an aluminum alloy is shown in the figure. The specimen from which it was obtained has an original diameter of 12.7 mm and a gage length of 50.8 mm. When the applied load on the specimen is 50 kN, the diameter is 12.67494 mm. Determine Poisson's ratio for the material.

*3–28. The elastic portion of the stress–strain diagram for an aluminum alloy is shown in the figure. The specimen from which it was obtained has an original diameter of 12.7 mm and a gage length of 50.8 mm. If a load of $P = 60$ kN is applied to the specimen, determine its new diameter and length. Take $\nu = 0.35$.

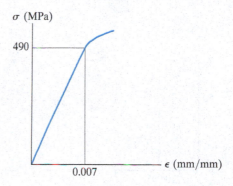

σ (MPa)

490

0.007

ϵ (mm/mm)

Probs. 3–27/28

3–29. The brake pads for a bicycle tire are made of rubber. If a frictional force of 50 N is applied to each side of the tires, determine the average shear strain in the rubber. Each pad has cross-sectional dimensions of 20 mm and 50 mm. $G_r = 0.20$ MPa.

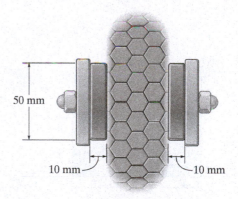

50 mm

10 mm   10 mm

Prob. 3–29

3–30. The lap joint is connected together using a 1.25 in. diameter bolt. If the bolt is made from a material having a shear stress–strain diagram that is approximated as shown, determine the shear strain developed in the shear plane of the bolt when $P = 75$ kip.

3–31. The lap joint is connected together using a 1.25 in. diameter bolt. If the bolt is made from a material having a shear stress–strain diagram that is approximated as shown, determine the permanent shear strain in the shear plane of the bolt when the applied force $P = 150$ kip is removed.

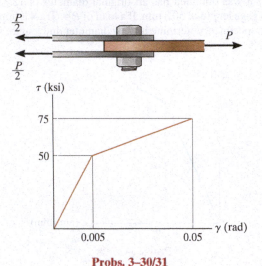

Probs. 3–30/31

*3–32. The rubber block is subjected to an elongation of 0.03 in. along the x axis, and its vertical faces are given a tilt so that $\theta = 89.3°$. Determine the strains ϵ_x, ϵ_y and γ_{xy}. Take $\nu_r = 0.5$.

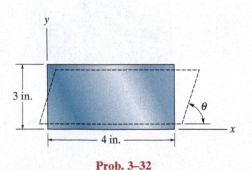

Prob. 3–32

3–33. The shear stress–strain diagram for an alloy is shown in the figure. If a bolt having a diameter of 0.25 in. is made of this material and used in the lap joint, determine the modulus of elasticity E and the force P required to cause the material to yield. Take $\nu = 0.3$.

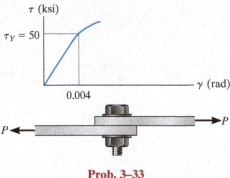

Prob. 3–33

3–34. A shear spring is made from two blocks of rubber, each having a height h, width b, and thickness a. The blocks are bonded to three plates as shown. If the plates are rigid and the shear modulus of the rubber is G, determine the displacement of plate A when the vertical load $\mathbf{P}$ is applied. Assume that the displacement is small so that $\delta = a \tan \gamma \approx a\gamma$.

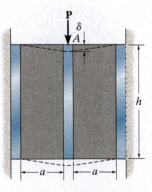

Prob. 3–34

CHAPTER REVIEW

One of the most important tests for material strength is the tension test. The results, found from stretching a specimen of known size, are plotted as normal stress on the vertical axis and normal strain on the horizontal axis.

Many engineering materials exhibit initial linear elastic behavior, whereby stress is proportional to strain, defined by Hooke's law, $\sigma = E\epsilon$. Here E, called the modulus of elasticity, is the slope of this straight line on the stress–strain diagram.

$$\sigma = E\epsilon$$

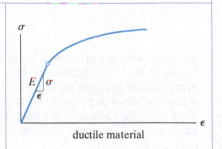

ductile material

When the material is stressed beyond the yield point, permanent deformation will occur. In particular, steel has a region of yielding, whereby the material will exhibit an increase in strain with no increase in stress. The region of strain hardening causes further yielding of the material with a corresponding increase in stress. Finally, at the ultimate stress, a localized region on the specimen will begin to constrict, forming a neck. It is after this that the fracture occurs.

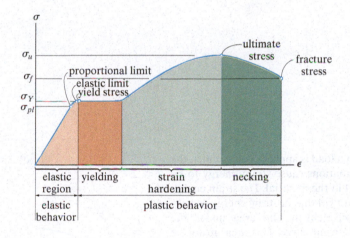

Ductile materials, such as most metals, exhibit both elastic and plastic behavior. Wood is moderately ductile. Ductility is usually specified by the percent elongation to failure or by the percent reduction in the cross-sectional area.

$$\text{Percent elongation} = \frac{L_f - L_0}{L_0} \, (100\%)$$

$$\text{Percent reduction of area} = \frac{A_0 - A_f}{A_0} \, (100\%)$$

Brittle materials exhibit little or no yielding before failure. Cast iron, concrete, and glass are typical examples.

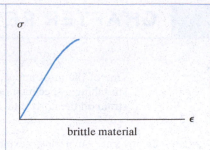

brittle material

The yield point of a material at A can be increased by strain hardening. This is accomplished by applying a load that causes the stress to be greater than the yield stress, then releasing the load. The larger stress A' becomes the new yield point for the material.

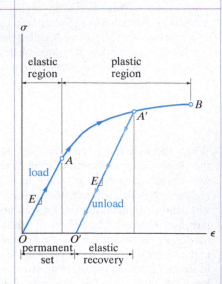

When a load is applied to a member, the deformations cause strain energy to be stored in the material. The strain energy per unit volume, or strain energy density, is equivalent to the area under the stress–strain curve. This area up to the yield point is called the modulus of resilience. The entire area under the stress–strain diagram is called the modulus of toughness.

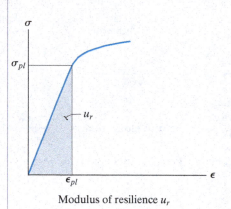

Modulus of resilience u_r

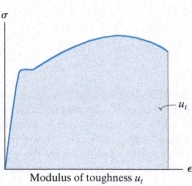

Modulus of toughness u_t

Poisson's ratio ν is a dimensionless material property that relates the lateral strain to the longitudinal strain. Its range of values is $0 \leq \nu \leq 0.5$.

$$\nu = -\frac{\epsilon_{lat}}{\epsilon_{long}}$$

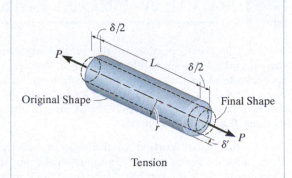

Original Shape   Final Shape

Tension

Shear stress–strain diagrams can also be established for a material. Within the elastic region, $\tau = G\gamma$, where G is the shear modulus, found from the slope of the line. The value of ν can be obtained from the relationship that exists between G, E, and ν.

$$G = \frac{E}{2(1 + \nu)}$$

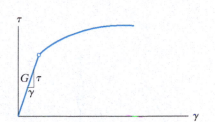

When materials are in service for long periods of time, considerations of creep become important. Creep is the time rate of deformation, which occurs at high stress and/or high temperature. Design requires that the stress in the material not exceed an allowable stress which is based on the material's creep strength.

Fatigue can occur when the material undergoes a large number of cycles of loading. This effect will cause microscopic cracks to form, leading to a brittle failure. To prevent fatigue, the stress in the material must not exceed a specified endurance or fatigue limit.

REVIEW PROBLEMS

R3–1. The elastic portion of the tension stress–strain diagram for an aluminum alloy is shown in the figure. The specimen used for the test has a gage length of 2 in. and a diameter of 0.5 in. When the applied load is 9 kip, the new diameter of the specimen is 0.49935 in. Calculate the shear modulus G_{al} for the aluminum.

R3–2. The elastic portion of the tension stress–strain diagram for an aluminum alloy is shown in the figure. The specimen used for the test has a gage length of 2 in. and a diameter of 0.5 in. If the applied load is 10 kip, determine the new diameter of the specimen. The shear modulus is $G_{al} = 3.8(10^3)$ ksi.

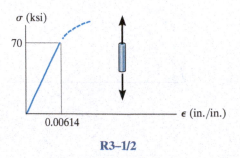

R3–1/2

R3–3. The rigid beam rests in the horizontal position on two 2014-T6 aluminum cylinders having the *unloaded* lengths shown. If each cylinder has a diameter of 30 mm, determine the placement x of the applied 80-kN load so that the beam remains horizontal. What is the new diameter of cylinder A after the load is applied? $\nu_{al} = 0.35$.

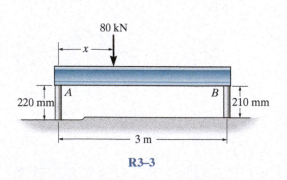

R3–3

*R3–4. The wires each have a diameter of $\frac{1}{2}$ in., length of 2 ft, and are made from 304 stainless steel. If $P = 6$ kip, determine the angle of tilt of the rigid beam AB.

R3–5. The wires each have a diameter of $\frac{1}{2}$ in., length of 2 ft, and are made from 304 stainless steel. Determine the magnitude of force **P** so that the rigid beam tilts 0.015°.

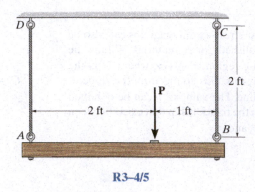

R3–4/5

R3–6. The head H is connected to the cylinder of a compressor using six $\frac{3}{16}$-in. diameter steel bolts. If the clamping force in each bolt is 800 lb, determine the normal strain in the bolts. If $\sigma_Y = 40$ ksi and $E_{st} = 29(10^3)$ ksi, what is the strain in each bolt when the nut is unscrewed so that the clamping force is released?

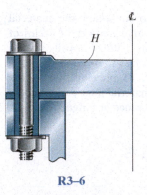

R3–6

R3–7. The stress–strain diagram for polyethylene, which is used to sheath coaxial cables, is determined from testing a specimen that has a gage length of 10 in. If a load P on the specimen develops a strain of $\epsilon = 0.024$ in./in., determine the approximate length of the specimen, measured between the gage points, when the load is removed. Assume the specimen recovers elastically.

R3–9. The 8-mm-diameter bolt is made of an aluminum alloy. It fits through a magnesium sleeve that has an inner diameter of 12 mm and an outer diameter of 20 mm. If the original lengths of the bolt and sleeve are 80 mm and 50 mm, respectively, determine the strains in the sleeve and the bolt if the nut on the bolt is tightened so that the tension in the bolt is 8 kN. Assume the material at A is rigid. $E_{al} = 70$ GPa, $E_{mg} = 45$ GPa.

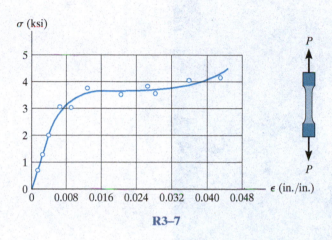

R3–7

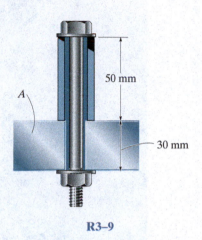

R3–9

*R3–8. The solid rod, or radius r, with two rigid caps attached to its ends is subjected to an axial force P. If the rod is made from a material having a modulus of elasticity E and Poisson's ratio ν, determine the change in volume of the material.

R3–10. An acetal polymer block is fixed to the rigid plates at its top and bottom surfaces. If the top plate displaces 2 mm horizontally when it is subjected to a horizontal force $P = 2$ kN, determine the shear modulus of the polymer. The width of the block is 100 mm. Assume that the polymer is linearly elastic and use small angle analysis.

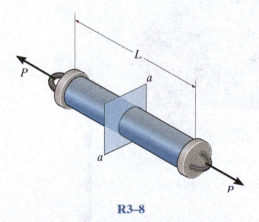

R3–8

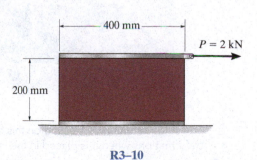

R3–10

CHAPTER 4

(© Hazlan Abdul Hakim/Getty Images)

The string of drill pipe stacked on this oil rig will be subjected to large axial deformations when it is placed in the hole.

AXIAL LOAD

CHAPTER OBJECTIVES

■ In this chapter we will discuss how to determine the deformation of an axially loaded member, and we will also develop a method for finding the support reactions when these reactions cannot be determined strictly from the equations of equilibrium. An analysis of the effects of thermal stress, stress concentrations, inelastic deformations, and residual stress will also be discussed.

4.1 SAINT-VENANT'S PRINCIPLE

In the previous chapters, we have developed the concept of stress as a means of measuring the force distribution within a body and strain as a means of measuring a body's deformation. We have also shown that the mathematical relationship between stress and strain depends on the type of material from which the body is made. In particular, if the material behaves in a linear elastic manner, then Hooke's law applies, and there is a proportional relationship between stress and strain.

Using this idea, consider the manner in which a rectangular bar will deform elastically when the bar is subjected to the force **P** applied along its centroidal axis, Fig. 4–1a. The once horizontal and vertical grid lines drawn on the bar become distorted, and *localized deformation* occurs at each end. Throughout the midsection of the bar, the lines remain horizontal and vertical.

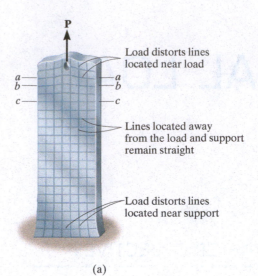

(a)

Fig. 4–1

If the material remains elastic, then the *strains* caused by this deformation are directly related to the *stress* in the bar through Hooke's law, $\sigma = E\epsilon$. As a result, a profile of the variation of the stress distribution acting at sections *a–a*, *b–b*, and *c–c*, will look like that shown in Fig. 4–1*b*. By comparison, the stress tends to reach a uniform value at section *c–c*, which is sufficiently removed from the end since the localized deformation caused by **P** *vanishes*. The minimum distance from the bar's end where this occurs can be determined using a mathematical analysis based on the theory of elasticity. It has been found that this distance should at least be equal to the *largest dimension* of the loaded cross section. Hence, section *c–c* should be located at a distance at least equal to the width (not the thickness) of the bar.*

In the same way, the stress distribution at the support in Fig. 4–1*a* will also even out and become uniform over the cross section located the same distance away from the support.

The fact that the localized stress and deformation behave in this manner is referred to as **Saint-Venant's principle**, since it was first noticed by the French scientist Barré de Saint-Venant in 1855. Essentially it states that the stress and strain produced at points in a body *sufficiently removed* from the region of external load application will be *the same* as the stress and strain produced by *any other applied external loading* that has the same statically equivalent resultant and is applied to the body within the same region. For example, if two symmetrically applied forces $P/2$ act on the bar, Fig. 4–1*c*, the stress distribution at section *c–c* will be uniform and therefore equivalent to $\sigma_{avg} = P/A$ as in Fig. 4–1*c*.

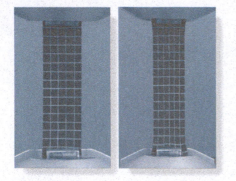

Notice how the lines on this rubber membrane distort after it is stretched. The localized distortions at the grips smooth out as stated by Saint-Venant's principle.

*When section *c–c* is so located, the theory of elasticity predicts the maximum stress to be $\sigma_{max} = 1.02\,\sigma_{avg}$.

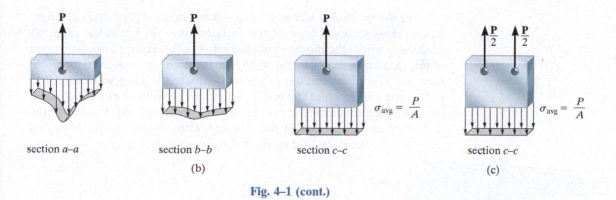

$$\sigma_{avg} = \frac{P}{A}$$

$$\sigma_{avg} = \frac{P}{A}$$

section *a–a*   section *b–b*   section *c–c*   section *c–c*

(b)   (c)

Fig. 4–1 (cont.)

4.2   ELASTIC DEFORMATION OF AN AXIALLY LOADED MEMBER

Using Hooke's law and the definitions of stress and strain, we will now develop an equation that can be used to determine the *elastic* displacement of a member subjected to axial loads. To generalize the development, consider the bar shown in Fig. 4–2a, which has a cross-sectional area that gradually varies along its length *L*, and is made of a material that has a variable stiffness or modulus of elasticity. The bar is subjected to concentrated loads at its ends and a variable external load distributed along its length. This distributed load could, for example, represent the weight of the bar if it is in the vertical position, or friction forces acting on the bar's surface.

Here we wish to find the ***relative displacement*** δ (delta) of one end of the bar with respect to the other end as caused by the loading. We will *neglect* the localized deformations that occur at points of concentrated loading and where the cross section suddenly changes. From Saint-Venant's principle, these effects occur within small regions of the bar's length and will therefore have only a slight effect on the final result. For the most part, the bar will deform uniformly, so the normal stress will be uniformly distributed over the cross section.

The vertical displacement of the rod at the top floor *B* only depends upon the force in the rod along length *AB*. However, the displacement at the bottom floor *C* depends upon the force in the rod along its entire length, *ABC*.

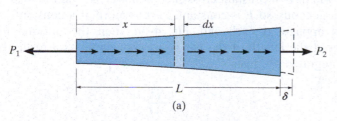

(a)

Fig. 4–2

Using the method of sections, a differential element (or wafer) of length dx and cross-sectional area $A(x)$ is isolated from the bar at the arbitrary position x, where the modulus of elasticity is $E(x)$. The free-body diagram of this element is shown in Fig. 4–2b. The resultant internal axial force will be a function of x since the external distributed loading will cause it to vary along the length of the bar. This load, $N(x)$, will deform the element into the shape indicated by the dashed outline, and therefore the displacement of one end of the element with respect to the other end becomes $d\delta$. The stress and strain in the element are therefore

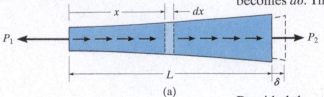

(a)

$$\sigma = \frac{N(x)}{A(x)} \quad \text{and} \quad \epsilon = \frac{d\delta}{dx}$$

Provided the stress does not exceed the proportional limit, we can apply Hooke's law; i.e., $\sigma = E(x)\epsilon$, and so

$$\frac{N(x)}{A(x)} = E(x)\left(\frac{d\delta}{dx}\right)$$

$$d\delta = \frac{N(x)dx}{A(x)E(x)}$$

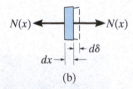

(b)

Fig. 4–2 (Repeated)

For the entire length L of the bar, we must integrate this expression to find δ. This yields

$$\boxed{\delta = \int_0^L \frac{N(x)dx}{A(x)E(x)}} \qquad (4\text{–}1)$$

Here

δ = displacement of one point on the bar relative to the other point

L = original length of bar

$N(x)$ = internal axial force at the section, located a distance x from one end

$A(x)$ = cross-sectional area of the bar expressed as a function of x

$E(x)$ = modulus of elasticity for the material expressed as a function of x

Constant Load and Cross-Sectional Area.   In many cases the bar will have a constant cross-sectional area A; and the material will be homogeneous, so E is constant. Furthermore, if a constant external force is applied at each end, Fig. 4–3a, then the internal force N throughout the length of the bar is also constant. As a result, Eq. 4–1 when integrated becomes

$$\boxed{\delta = \frac{NL}{AE}} \qquad (4\text{–}2)$$

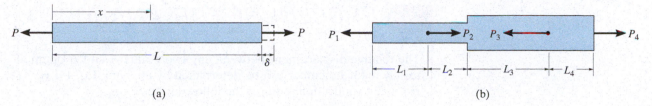

(a)   (b)

Fig. 4–3

If the bar is subjected to several different axial forces along its length, or the cross-sectional area or modulus of elasticity changes abruptly from one region of the bar to the next, as in Fig. 4–3b, then the above equation can be applied to each *segment of the bar* where these quantities remain *constant*. The displacement of one end of the bar with respect to the other is then found from the *algebraic addition* of the relative displacements of the ends of each segment. For this general case,

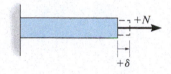

$$\delta = \sum \frac{NL}{AE} \qquad (4\text{--}3)$$

Sign Convention.   When applying Eqs. 4–1 through 4–3, it is best to use a consistent sign convention for the internal axial force and the displacement of the bar. To do so, we will consider both the force and displacement to be *positive* if they cause *tension and elongation*, Fig. 4–4; whereas a *negative* force and displacement will cause *compression* and *contraction*.

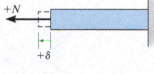

Fig. 4–4

IMPORTANT POINTS

- *Saint-Venant's principle* states that both the localized deformation and stress which occur within the regions of load application or at the supports tend to "even out" at a distance sufficiently removed from these regions.

- The displacement of one end of an axially loaded member relative to the other end is determined by relating the applied *internal* load to the stress using $\sigma = N/A$ and relating the displacement to the strain using $\epsilon = d\delta/dx$. Finally these two equations are combined using Hooke's law, $\sigma = E\epsilon$, which yields Eq. 4–1.

- Since Hooke's law has been used in the development of the displacement equation, it is important that no internal load causes yielding of the material, and that the material behaves in a linear elastic manner.

PROCEDURE FOR ANALYSIS

The relative displacement between any two points A and B on an axially loaded member can be determined by applying Eq. 4–1 (or Eq. 4–2). Application requires the following steps.

Internal Force.

- Use the method of sections to determine the internal axial force N within the member.

- If this force varies along the member's length due to an *external distributed loading*, a section should be made at the arbitrary location x from one end of the member, and the internal force represented as a function of x, i.e., $N(x)$.

- If several *constant external forces* act on the member, the internal force in each *segment* of the member between any two external forces must be determined.

- For any segment, an internal *tensile force* is *positive* and an internal *compressive force* is *negative*. For convenience, the results of the internal loading throughout the member can be shown graphically by constructing the normal-force diagram.

Displacement.

- When the member's cross-sectional area *varies* along its length, the area must be expressed as a function of its position x, i.e., $A(x)$.

- If the cross-sectional area, the modulus of elasticity, or the internal loading *suddenly changes*, then Eq. 4–2 should be applied to each segment for which these quantities are constant.

- When substituting the data into Eqs. 4–1 through 4–3, be sure to account for the proper sign of the internal force N. Tensile forces are positive and compressive forces are negative. Also, use a consistent set of units. For any segment, if the result is a *positive* numerical quantity, it indicates *elongation*; if it is *negative*, it indicates a *contraction*.

EXAMPLE 4.1

The uniform A-36 steel bar in Fig. 4–5a has a diameter of 50 mm and is subjected to the loading shown. Determine the displacement at D, and the displacement of point B relative to C.

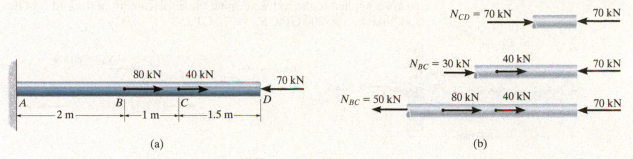

(a)   (b)

Fig. 4-5

SOLUTION

Internal Forces.   The internal forces within the bar are determined using the method of sections and horizontal equilibrium. The results are shown on the free-body diagrams in Fig. 4–5b. The normal-force diagram in Fig. 4–5c shows the variation of these forces along the bar.

Displacement.   From the table on the inside back cover, for A-36 steel, $E = 200$ GPa. Using the established sign convention, the displacement of the end of the bar is therefore

$$\delta_D = \Sigma\frac{NL}{AE} = \frac{[-70(10^3)\text{ N}](1.5\text{ m})}{\pi(0.025\text{ m})^2[200(10^9)\text{ N/m}^2]}$$

$$+ \frac{[-30(10^3)\text{ N}](1\text{ m})}{\pi(0.025\text{ m})^2[200(10^9)\text{ N/m}^2]} + \frac{[50(10^3)\text{ N}](2\text{ m})}{\pi(0.025\text{ m})^2[200(10^9)\text{ N/m}^2]}$$

$$\delta_D = -89.1(10^{-3})\text{ mm} \qquad\qquad\qquad\qquad\quad \textit{Ans.}$$

This negative result indicates that point D moves to the left.

The displacement of B relative to C, $\delta_{B/C}$, is caused only by the internal load within region BC. Thus,

$$\delta_{B/C} = \frac{NL}{AE} = \frac{[-30(10^3)\text{ N}](1\text{ m})}{\pi(0.025\text{ m})^2[200(10^9)\text{ N/m}^2]} = -76.4(10^{-3})\text{ mm} \quad \textit{Ans.}$$

Here the negative result indicates that B will move towards C.

EXAMPLE 4.2

The assembly shown in Fig. 4–6a consists of an aluminum tube AB having a cross-sectional area of 400 mm². A steel rod having a diameter of 10 mm is attached to a rigid collar and passes through the tube. If a tensile load of 80 kN is applied to the rod, determine the displacement of the end C of the rod. Take $E_{st} = 200$ GPa, $E_{al} = 70$ GPa.

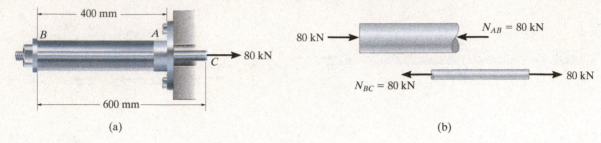

(a)   (b)

Fig. 4–6

SOLUTION

Internal Force. The free-body diagrams of the tube and rod segments in Fig. 4–6b show that the rod is subjected to a tension of 80 kN, and the tube is subjected to a compression of 80 kN.

Displacement. We will first determine the displacement of C with respect to B. Working in units of newtons and meters, we have

$$\delta_{C/B} = \frac{NL}{AE} = \frac{[+80(10^3)\text{ N}](0.6\text{ m})}{\pi(0.005\text{ m})^2[200(10^9)\text{ N/m}^2]} = +0.003056\text{ m} \rightarrow$$

The positive sign indicates that C moves *to the right* relative to B, since the bar elongates.

The displacement of B with respect to the *fixed* end A is

$$\delta_B = \frac{NL}{AE} = \frac{[-80(10^3)\text{ N}](0.4\text{ m})}{[400\text{ mm}^2(10^{-6})\text{ m}^2/\text{mm}^2][70(10^9)\text{ N/m}^2]}$$

$$= -0.001143\text{ m} = 0.001143\text{ m} \rightarrow$$

Here the negative sign indicates that the tube shortens, and so B moves to the *right* relative to A.

Since both displacements are to the right, the displacement of C relative to the fixed end A is therefore

$$(\xrightarrow{+}) \qquad \delta_C = \delta_B + \delta_{C/B} = 0.001143\text{ m} + 0.003056\text{ m}$$

$$= 0.00420\text{ m} = 4.20\text{ mm} \rightarrow \qquad \textit{Ans.}$$

EXAMPLE 4.3

Rigid beam AB rests on the two short posts shown in Fig. 4–7a. AC is made of steel and has a diameter of 20 mm, and BD is made of aluminum and has a diameter of 40 mm. Determine the displacement of point F on AB if a vertical load of 90 kN is applied over this point. Take $E_{st} = 200$ GPa, $E_{al} = 70$ GPa.

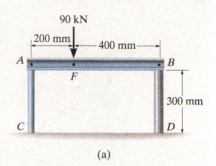

(a)

SOLUTION

Internal Force.   The compressive forces acting at the top of each post are determined from the equilibrium of member AB, Fig. 4–7b. These forces are equal to the internal forces in each post, Fig. 4–7c.

Displacement.   The displacement of the top of each post is

Post AC:

$$\delta_A = \frac{N_{AC}L_{AC}}{A_{AC}E_{st}} = \frac{[-60(10^3)\text{ N}](0.300\text{ m})}{\pi(0.010\text{ m})^2[200(10^9)\text{ N/m}^2]} = -286(10^{-6})\text{ m}$$

$$= 0.286\text{ mm} \downarrow$$

Post BD:

$$\delta_B = \frac{N_{BD}L_{BD}}{A_{BD}E_{al}} = \frac{[-30(10^3)\text{ N}](0.300\text{ m})}{\pi(0.020\text{ m})^2[70(10^9)\text{ N/m}^2]} = -102(10^{-6})\text{ m}$$

$$= 0.102\text{ mm} \downarrow$$

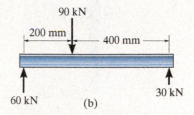

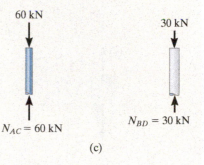

(c)

A diagram showing the centerline displacements at A, B, and F on the beam is shown in Fig. 4–7d. By proportion of the blue shaded triangle, the displacement of point F is therefore

$$\delta_F = 0.102\text{ mm} + (0.184\text{ mm})\left(\frac{400\text{ mm}}{600\text{ mm}}\right) = 0.225\text{ mm} \downarrow \qquad \textit{Ans.}$$

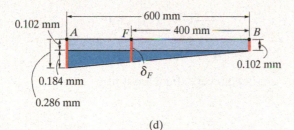

(d)

Fig. 4–7

EXAMPLE 4.4

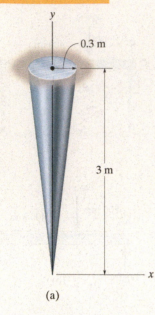

(a)

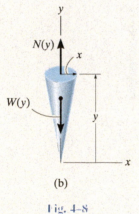

(b)

Fig. 4–8

A member is made of a material that has a specific weight of $\gamma = 6 \text{ kN/m}^3$ and modulus of elasticity of 9 GPa. If it is in the form of a *cone* having the dimensions shown in Fig. 4–8a, determine how far its end is displaced due to gravity when it is suspended in the vertical position.

SOLUTION

Internal Force.   The internal axial force varies along the member, since it is dependent on the weight $W(y)$ of a segment of the member below any section, Fig. 4–8b. Hence, to calculate the displacement, we must use Eq. 4–1. At the section located a distance y from the cone's free end, the radius x of the cone as a function of y is determined by proportion; i.e.,

$$\frac{x}{y} = \frac{0.3 \text{ m}}{3 \text{ m}}; \qquad x = 0.1y$$

The volume of a cone having a base of radius x and height y is

$$V = \frac{1}{3} \pi y x^2 = \frac{\pi(0.01)}{3} y^3 = 0.01047 y^3$$

Since $W = \gamma V$, the internal force at the section becomes

$$+\uparrow \Sigma F_y = 0; \qquad N(y) = 6(10^3)(0.01047y^3) = 62.83y^3$$

Displacement.   The area of the cross section is also a function of position y, Fig. 4–8b. We have

$$A(y) = \pi x^2 = 0.03142\, y^2$$

Applying Eq. 4–1 between the limits of $y = 0$ and $y = 3$ m yields

$$\delta = \int_0^L \frac{N(y)\, dy}{A(y)\, E} = \int_0^3 \frac{(62.83y^3)\, dy}{(0.03142y^2)\, 9(10^9)}$$

$$= 222.2(10^{-9}) \int_0^3 y\, dy$$

$$= 1(10^{-6}) \text{ m} = 1\ \mu\text{m} \qquad \qquad \textit{Ans.}$$

NOTE: This is indeed a very small amount.

PRELIMINARY PROBLEMS

P4–1. In each case, determine the internal normal force between lettered points on the bar. Draw all necessary free-body diagrams.

P4–4. The rod is subjected to an external axial force of 800 N and a uniform distributed load of 100 N/m along its length. Determine the internal normal force in the rod as a function of x.

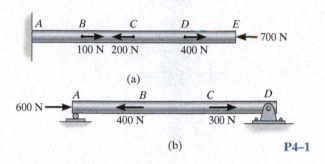

(a)

(b)   **P4–1**

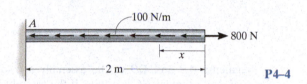

P4–4

P4–2. Determine the internal normal force between lettered points on the cable and rod. Draw all necessary free-body diagrams.

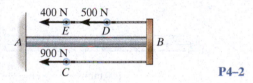

P4–2

P4–5. The rigid beam supports the load of 60 kN. Determine the displacement at B. Take $E = 60$ GPa, and $A_{BC} = 2\,(10^{-3})\ \text{m}^2$.

P4–3. The post weighs 8 kN/m. Determine the internal normal force in the post as a function of x.

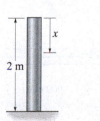

P4–3

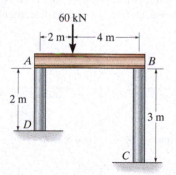

P4–5

FUNDAMENTAL PROBLEMS

F4–1. The 20-mm-diameter A-36 steel rod is subjected to the axial forces shown. Determine the displacement of end C with respect to the fixed support at A.

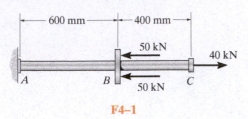

F4–1

F4–2. Segments AB and CD of the assembly are solid circular rods, and segment BC is a tube. If the assembly is made of 6061-T6 aluminum, determine the displacement of end D with respect to end A.

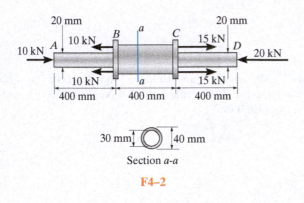

Section *a-a*

F4–2

F4–3. The 30-mm-diameter A992 steel rod is subjected to the loading shown. Determine the displacement of end C.

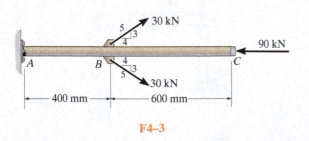

F4–3

F4–4. If the 20-mm-diameter rod is made of A-36 steel and the stiffness of the spring is $k = 50$ MN/m, determine the displacement of end A when the 60-kN force is applied.

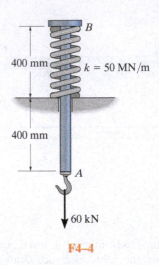

F4–4

F4–5. The 20-mm-diameter 2014-T6 aluminum rod is subjected to the uniform distributed axial load. Determine the displacement of end A.

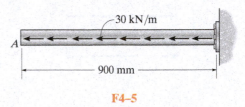

F4–5

F4–6. The 20-mm-diameter 2014-T6 aluminum rod is subjected to the triangular distributed axial load. Determine the displacement of end A.

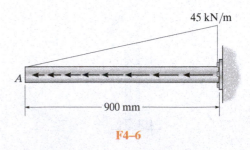

F4–6

PROBLEMS

4–1. The A992 steel rod is subjected to the loading shown. If the cross-sectional area of the rod is 80 mm^2, determine the displacement of B and A. Neglect the size of the couplings at B and C.

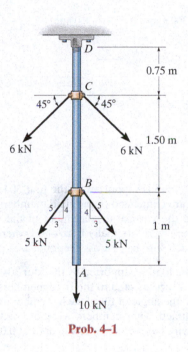

Prob. 4–1

4–2. The copper shaft is subjected to the axial loads shown. Determine the displacement of end A with respect to end D if the diameters of each segment are $d_{AB} = 0.75$ in., $d_{BC} = 1$ in., and $d_{CD} = 0.5$ in. Take $E_{cu} = 18(10^3)$ ksi.

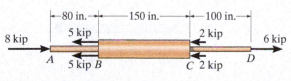

Prob. 4–2

4–3. The composite shaft, consisting of aluminum, copper, and steel sections, is subjected to the loading shown. Determine the displacement of end A with respect to end D and the normal stress in each section. The cross-sectional area and modulus of elasticity for each section are shown in the figure. Neglect the size of the collars at B and C.

*4–4. The composite shaft, consisting of aluminum, copper, and steel sections, is subjected to the loading shown. Determine the displacement of B with respect to C. The cross-sectional area and modulus of elasticity for each section are shown in the figure. Neglect the size of the collars at B and C.

Aluminum	Copper	Steel
$E_{al} = 10(10^3)$ ksi	$E_{cu} = 18(10^3)$ ksi	$E_{st} = 29(10^3)$ ksi
$A_{AB} = 0.09$ in^2	$A_{BC} = 0.12$ in^2	$A_{CD} = 0.06$ in^2

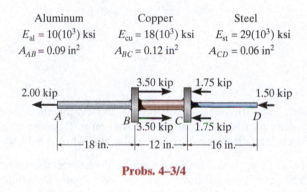

Probs. 4–3/4

4–5. The 2014-T6 aluminium rod has a diameter of 30 mm and supports the load shown. Determine the displacement of end A with respect to end E. Neglect the size of the couplings.

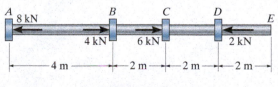

Prob. 4–5

4–6. The A-36 steel drill shaft of an oil well extends 12 000 ft into the ground. Assuming that the pipe used to drill the well is suspended freely from the derrick at A, determine the maximum average normal stress in each pipe string and the elongation of its end D with respect to the fixed end at A. The shaft consists of three different sizes of pipe, AB, BC, and CD, each having the length, weight per unit length, and cross-sectional area indicated.

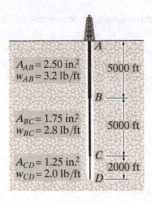

$A_{AB} = 2.50$ in.²
$w_{AB} = 3.2$ lb/ft
5000 ft

$A_{BC} = 1.75$ in.²
$w_{BC} = 2.8$ lb/ft
5000 ft

$A_{CD} = 1.25$ in.²
$w_{CD} = 2.0$ lb/ft
2000 ft

Prob. 4–6

4–7. The truss is made of three A-36 steel members, each having a cross-sectional area of 400 mm². Determine the horizontal displacement of the roller at C when $P = 8$ kN.

*4–8. The truss is made of three A-36 steel members, each having a cross-sectional area of 400 mm². Determine the magnitude P required to displace the roller to the right 0.2 mm.

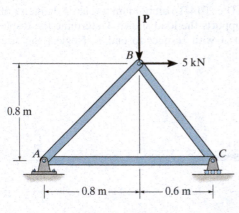

Probs. 4–7/8

4–9. The assembly consists of two 10-mm diameter red brass C83400 copper rods AB and CD, a 15-mm diameter 304 stainless steel rod EF, and a rigid bar G. If $P = 5$ kN, determine the horizontal displacement of end F of rod EF.

4–10. The assembly consists of two 10-mm diameter red brass C83400 copper rods AB and CD, a 15-mm diameter 304 stainless steel rod EF, and a rigid bar G. If the horizontal displacement of end F of rod EF is 0.45 mm, determine the magnitude of P.

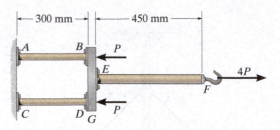

Probs. 4–9/10

4–11. The load is supported by the four 304 stainless steel wires that are connected to the rigid members AB and DC. Determine the vertical displacement of the 500-lb load if the members were originally horizontal when the load was applied. Each wire has a cross-sectional area of 0.025 in².

*4–12. The load is supported by the four 304 stainless steel wires that are connected to the rigid members AB and DC. Determine the angle of tilt of each member after the 500-lb load is applied. The members were originally horizontal, and each wire has a cross-sectional area of 0.025 in².

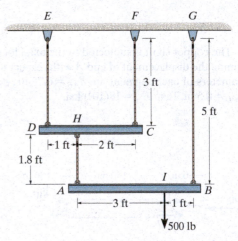

Probs. 4–11/12

4–13. The rigid bar is supported by the pin-connected rod *CB* that has a cross-sectional area of 14 mm² and is made from 6061-T6 aluminum. Determine the vertical deflection of the bar at *D* when the distributed load is applied.

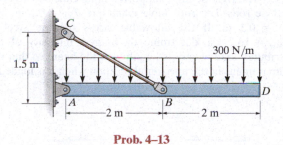

1.5 m

300 N/m

A   *B*   *D*

2 m   2 m

Prob. 4–13

4–14. The post is made of Douglas fir and has a diameter of 100 mm. If it is subjected to the load of 20 kN and the soil provides a frictional resistance distributed around the post that is triangular along its sides; that is, it varies from $w = 0$ at $y = 0$ to $w = 12$ kN/m at $y = 2$ m, determine the force **F** at its bottom needed for equilibrium. Also, what is the displacement of the top of the post *A* with respect to its bottom *B*? Neglect the weight of the post.

4–15. The post is made of Douglas fir and has a diameter of 100 mm. If it is subjected to the load of 20 kN and the soil provides a frictional resistance that is distributed along its length and varies linearly from $w = 4$ kN/m at $y = 0$ to $w = 12$ kN/m at $y = 2$ m, determine the force **F** at its bottom needed for equilibrium. Also, what is the displacement of the top of the post *A* with respect to its bottom *B*? Neglect the weight of the post.

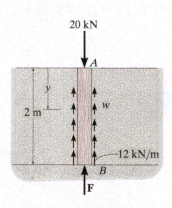

20 kN

A

y

2 m

w

12 kN/m

B

F

Probs. 4–14/15

*****4–16.** The coupling rod is subjected to a force of 5 kip. Determine the distance *d* between *C* and *E* accounting for the compression of the spring and the deformation of the bolts. When no load is applied the spring is unstretched and $d = 10$ in. The material is A-36 steel and each bolt has a diameter of 0.25 in. The plates at *A*, *B*, and *C* are rigid and the spring has a stiffness of $k = 12$ kip/in.

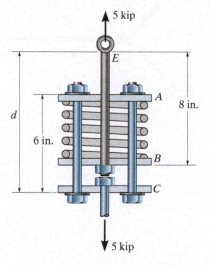

5 kip

E

A   8 in.

d

6 in.

B

C

5 kip

Prob. 4–16

4–17. The pipe is stuck in the ground so that when it is pulled upward the frictional force along its length varies linearly from zero at *B* to f_{max} (force/length) at *C*. Determine the initial force *P* required to pull the pipe out and the pipe's elongation just before it starts to slip. The pipe has a length *L*, cross-sectional area *A*, and the material from which it is made has a modulus of elasticity *E*.

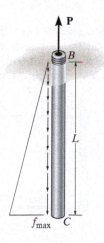

P

B

L

f_{max} *C*

Prob. 4–17

4–18. The linkage is made of three pin-connected A992 steel members, each having a diameter of $1\frac{1}{4}$ in. If a horizontal force of $P = 60$ kip is applied to the end B of member AB, determine the displacement of point B.

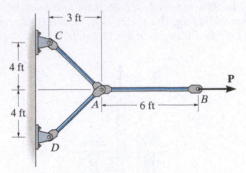

Prob. 4–18

4–19. The linkage is made of three pin-connected A992 steel members, each having a diameter of $1\frac{1}{4}$ in. Determine the magnitude of the force $\mathbf{P}$ needed to displace point B 0.25 in. to the right.

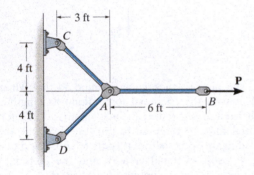

Prob. 4–19

*4–20. The assembly consists of three titanium (Ti-6A1-4V) rods and a rigid bar AC. The cross-sectional area of each rod is given in the figure. If a force of 60 kip is applied to the ring F, determine the horizontal displacement of point F.

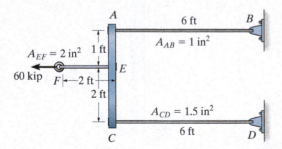

Prob. 4–20

4–21. The rigid beam is supported at its ends by two A-36 steel tie rods. If the allowable stress for the steel is $\sigma_{\text{allow}} = 16.2$ ksi, the load $w = 3$ kip/ft, and $x = 4$ ft, determine the smallest diameter of each rod so that the beam remains in the horizontal position when it is loaded.

4–22. The rigid beam is supported at its ends by two A-36 steel tie rods. The rods have diameters $d_{AB} = 0.5$ in. and $d_{CD} = 0.3$ in. If the allowable stress for the steel is $\sigma_{\text{allow}} = 16.2$ ksi, determine the largest intensity of the distributed load w and its length x on the beam so that the beam remains in the horizontal position when it is loaded.

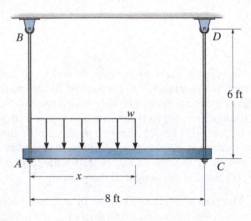

Probs. 4–21/22

4–23. The steel bar has the original dimensions shown in the figure. If it is subjected to an axial load of 50 kN, determine the change in its length and its new cross-sectional dimensions at section a–a. $E_{\text{st}} = 200$ GPa, $\nu_{\text{st}} = 0.29$.

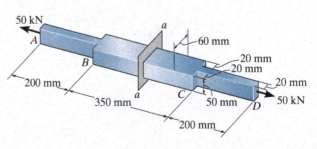

Prob. 4–23

*4–24. Determine the relative displacement of one end of the tapered plate with respect to the other end when it is subjected to an axial load P.

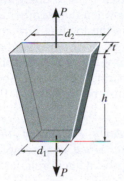

Prob. 4–24

4–25. The assembly consists of two rigid bars that are originally horizontal. They are supported by pins and 0.25-in.-diameter A-36 steel rods. If the vertical load of 5 kip is applied to the bottom bar AB, determine the displacement at C, B, and E.

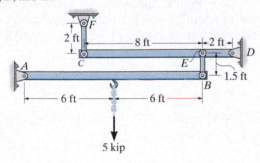

Prob. 4–25

4–26. The truss consists of three members, each made from A-36 steel and having a cross-sectional area of 0.75 in². Determine the greatest load P that can be applied so that the roller support at B is not displaced more than 0.03 in.

4–27. Solve Prob. 4–26 when the load **P** acts vertically downward at C.

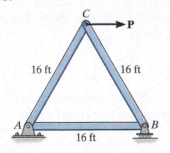

Probs. 4–26/27

*4–28. The observation cage C has a weight of 250 kip and through a system of gears, travels upward at constant velocity along the A-36 steel column, which has a height of 200 ft. The column has an outer diameter of 3 ft and is made from steel plate having a thickness of 0.25 in. Neglect the weight of the column, and determine the average normal stress in the column at its base, B, as a function of the cage's position y. Also, determine the displacement of end A as a function of y.

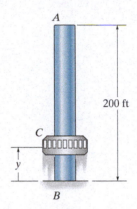

Prob. 4–28

4–29. Determine the elongation of the aluminum strap when it is subjected to an axial force of 30 kN. $E_{al} = 70$ GPa.

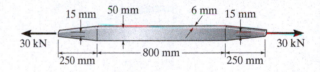

Prob. 4–29

4–30. The ball is truncated at its ends and is used to support the bearing load **P**. If the modulus of elasticity for the material is E, determine the decrease in the ball's height when the load is applied.

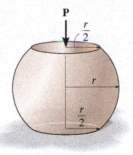

Prob. 4–30

4.3 PRINCIPLE OF SUPERPOSITION

The principle of superposition is often used to determine the stress or displacement at a point in a member when the member is subjected to a complicated loading. By subdividing the loading into components, this principle states that the resultant stress or displacement at the point can be determined by algebraically summing the stress or displacement caused by each load component applied separately to the member.

The following two conditions must be satisfied if the principle of superposition is to be applied.

1. **The loading N must be linearly related to the stress σ or displacement δ that is to be determined.** For example, the equations $\sigma = N/A$ and $\delta = NL/AE$ involve a linear relationship between σ and N, and δ and N.

2. **The loading must not significantly change the original geometry or configuration of the member.** If significant changes do occur, the direction and location of the applied forces and their moment arms will change. For example, consider the slender rod shown in Fig. 4–9a, which is subjected to the load **P**. In Fig. 4–9b, **P** is replaced by two of its components, $\mathbf{P} = \mathbf{P_1} + \mathbf{P_2}$. If **P** causes the rod to deflect a large amount, as shown, the moment of this load about its support, Pd, will not equal the sum of the moments of its component loads, $Pd \neq P_1 d_1 + P_2 d_2$, because $d_1 \neq d_2 \neq d$.

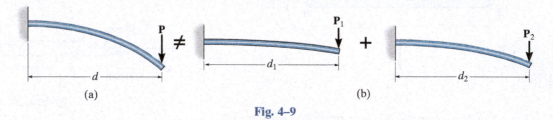

(a)   (b)

Fig. 4–9

4.4 STATICALLY INDETERMINATE AXIALLY LOADED MEMBERS

Consider the bar shown in Fig. 4–10a, which is fixed supported at both of its ends. From its free-body diagram, Fig. 4–10b, there are two unknown support reactions. Equilibrium requires

$$+\uparrow \Sigma F = 0; \qquad\qquad F_B + F_A - 500 \text{ N} = 0$$

This type of problem is called **statically indeterminate**, since the equilibrium equation is *not sufficient* to determine both reactions on the bar.

In order to establish an additional equation needed for solution, it is necessary to consider how points on the bar are displaced. Specifically, an equation that specifies the conditions for displacement is referred to as a **compatibility** or **kinematic condition**. In this case, a suitable compatibility condition would require the displacement of end A of the bar with respect to end B to equal zero, since the end supports are *fixed*, and so no relative movement can occur between them. Hence, the compatibility condition becomes

$$\delta_{A/B} = 0$$

This equation can be expressed in terms of the internal loads by using a **load–displacement relationship,** which depends on the material behavior. For example, if linear elastic behavior occurs, then $\delta = NL/AE$ can be used. Realizing that the internal force in segment AC is $+F_A$, and in segment CB it is $-F_B$, Fig. 4–10c, then the compatibility equation can be written as

$$\frac{F_A(2 \text{ m})}{AE} - \frac{F_B(3 \text{ m})}{AE} = 0$$

Since AE is constant, then $F_A = 1.5F_B$. Finally, using the equilibrium equation, the reactions are therefore

$$F_A = 300 \text{ N} \quad \text{and} \quad F_B = 200 \text{ N}$$

Since both of these results are positive, the directions of the reactions are shown correctly on the free-body diagram.

To solve for the reactions on any statically indeterminate problem, we must therefore satisfy both the equilibrium and compatibility equations, and relate the displacements to the loads using the load–displacement relations.

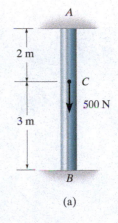

(a)

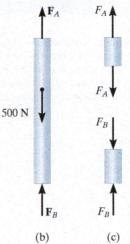

(b)   (c)

Fig. 4–10

IMPORTANT POINTS

- The *principle of superposition* is sometimes used to simplify stress and displacement problems having complicated loadings. This is done by subdividing the loading into components, then algebraically adding the results.

- Superposition requires that the loading be linearly related to the stress or displacement, and the loading must not significantly change the original geometry of the member.

- A problem is *statically indeterminate* if the equations of equilibrium are not sufficient to determine all the reactions on a member.

- *Compatibility conditions* specify the displacement constraints that occur at the supports or other points on a member.

Most concrete columns are reinforced with steel rods; and since these two materials work together in supporting the applied load, the force in each material must be determined using a statically indeterminate analysis.

PROCEDURE FOR ANALYSIS

The support reactions for statically indeterminate problems are determined by satisfying equilibrium, compatibility, and load–displacement requirements for the member.

Equilibrium.

- Draw a free-body diagram of the member in order to identify all the forces that act on it.

- The problem can be classified as statically indeterminate if the number of unknown reactions on the free-body diagram is greater than the number of available equations of equilibrium.

- Write the equations of equilibrium for the member.

Compatibility.

- Consider drawing a displacement diagram in order to investigate the way the member will elongate or contract when subjected to the external loads.

- Express the compatibility conditions in terms of the displacements caused by the loading.

Load–Displacement.

- Use a load–displacement relation, such as $\delta = NL/AE$, to relate the unknown displacements in the compatibility equation to the reactions.

- Solve all the equations for the reactions. If any of the results has a negative numerical value, it indicates that this force acts in the opposite sense of direction to that indicated on the free-body diagram.

EXAMPLE 4.5

The steel rod shown in Fig. 4–11a has a diameter of 10 mm. It is fixed to the wall at A, and before it is loaded, there is a gap of 0.2 mm between the wall at B' and the rod. Determine the reactions on the rod if it is subjected to an axial force of $P = 20$ kN. Neglect the size of the collar at C. Take $E_{st} = 200$ GPa.

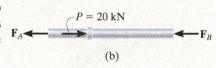

(a)

SOLUTION

Equilibrium. As shown on the free-body diagram, Fig. 4–11b, we will *assume* that force P is large enough to cause the rod's end B to contact the wall at B'. When this occurs, the problem becomes statically indeterminate since there are two unknowns and only one equation of equilibrium.

$$\xrightarrow{+} \Sigma F_x = 0; \qquad -F_A - F_B + 20(10^3) \text{ N} = 0 \qquad (1)$$

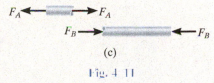

(b)

Compatibility. The force P causes point B to move to B', with no further displacement. Therefore the compatibility condition for the rod is

$$\delta_{B/A} = 0.0002 \text{ m}$$

Load–Displacement. This displacement can be expressed in terms of the unknown reactions using the load–displacement relationship, Eq. 4–2, applied to segments AC and CB, Fig. 4–11c. Working in units of newtons and meters, we have

$$\delta_{B/A} = \frac{F_A L_{AC}}{AE} - \frac{F_B L_{CB}}{AE} = 0.0002 \text{ m}$$

(c)

Fig. 4–11

$$\frac{F_A(0.4 \text{ m})}{\pi(0.005 \text{ m})^2 [200(10^9) \text{ N/m}^2]}$$

$$-\frac{F_B (0.8 \text{ m})}{\pi(0.005 \text{ m})^2 [200(10^9) \text{ N/m}^2]} = 0.0002 \text{ m}$$

or

$$F_A (0.4 \text{ m}) - F_B (0.8 \text{ m}) = 3141.59 \text{ N} \cdot \text{m} \qquad (2)$$

Solving Eqs. 1 and 2 yields

$$F_A = 16.0 \text{ kN} \qquad F_B = 4.05 \text{ kN} \qquad \textit{Ans.}$$

Since the answer for F_B is *positive*, indeed end B contacts the wall at B' as originally assumed.

NOTE: If F_B were a negative quantity, the problem would be statically determinate, so that $F_B = 0$ and $F_A = 20$ kN.

EXAMPLE 4.6

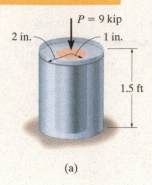

$P = 9$ kip

2 in.   1 in.

1.5 ft

(a)

$P = 9$ kip

F_{br}

F_{al}

(b)

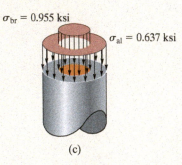

$\sigma_{br} = 0.955$ ksi

$\sigma_{al} = 0.637$ ksi

(c)

Fig. 4–12

The aluminum post shown in Fig. 4–12a is reinforced with a brass core. If this assembly supports an axial compressive load of $P = 9$ kip, applied to the rigid cap, determine the average normal stress in the aluminum and the brass. Take $E_{al} = 10(10^3)$ ksi and $E_{br} = 15(10^3)$ ksi.

SOLUTION

Equilibrium.   The free-body diagram of the post is shown in Fig. 4–12b. Here the resultant axial force at the base is represented by the unknown components carried by the aluminum, $\mathbf{F}_{al}$, and brass, $\mathbf{F}_{br}$. The problem is statically indeterminate.

Vertical force equilibrium requires

$$+\uparrow \Sigma F_y = 0; \qquad -9 \text{ kip} + F_{al} + F_{br} = 0 \qquad (1)$$

Compatibility.   The rigid cap at the top of the post causes both the aluminum and brass to be displaced the *same amount*. Therefore,

$$\delta_{al} = \delta_{br}$$

Load–Displacement.   Using the load–displacement relationships,

$$\frac{F_{al} L}{A_{al} E_{al}} = \frac{F_{br} L}{A_{br} E_{br}}$$

$$F_{al} = F_{br} \left(\frac{A_{al}}{A_{br}} \right) \left(\frac{E_{al}}{E_{br}} \right)$$

$$F_{al} = F_{br} \left[\frac{\pi[(2 \text{ in.})^2 - (1 \text{ in.})^2]}{\pi(1 \text{ in.})^2} \right] \left[\frac{10(10^3) \text{ ksi}}{15(10^3) \text{ ksi}} \right]$$

$$F_{al} = 2F_{br} \qquad (2)$$

Solving Eqs. 1 and 2 simultaneously yields

$$F_{al} = 6 \text{ kip} \qquad F_{br} = 3 \text{ kip}$$

The average normal stress in the aluminum and brass is therefore

$$\sigma_{al} = \frac{6 \text{ kip}}{\pi[(2 \text{ in.})^2 - (1 \text{ in.})^2]} = 0.637 \text{ ksi} \qquad \textit{Ans.}$$

$$\sigma_{br} = \frac{3 \text{ kip}}{\pi(1 \text{ in.})^2} = 0.955 \text{ ksi} \qquad \textit{Ans.}$$

NOTE: Using these results, the stress distributions within the materials are shown in Fig. 4–12c. Here the stiffer material, brass, is subjected to the larger stress.

EXAMPLE 4.7

The three A992 steel bars shown in Fig. 4–13a are pin connected to a *rigid* member. If the applied load on the member is 15 kN, determine the force developed in each bar. Bars *AB* and *EF* each have a cross-sectional area of 50 mm², and bar *CD* has a cross-sectional area of 30 mm².

SOLUTION

Equilibrium.   The free-body diagram of the rigid member is shown in Fig. 4–13b. This problem is statically indeterminate since there are three unknowns and only two available equilibrium equations.

$$+\uparrow \Sigma F_y = 0; \qquad F_A + F_C + F_E - 15 \text{ kN} = 0 \qquad (1)$$

$$\zeta + \Sigma M_C = 0; \quad -F_A(0.4 \text{ m}) + 15 \text{ kN}(0.2 \text{ m}) + F_E(0.4 \text{ m}) = 0 \quad (2)$$

Compatibility.   The applied load will cause the horizontal line *ACE* shown in Fig. 4–13c to move to the inclined position *A′C′E′*. The red displacements δ_A, δ_C, δ_E can be related by similar triangles. Thus the compatibility equation that relates these displacements is

$$\frac{\delta_A - \delta_E}{0.8 \text{ m}} = \frac{\delta_C - \delta_E}{0.4 \text{ m}}$$

$$\delta_C = \frac{1}{2}\delta_A + \frac{1}{2}\delta_E$$

Load–Displacement.   Using the load–displacement relationship, Eq. 4–2, we have

$$\frac{F_C L}{(30 \text{ mm}^2)E_{st}} = \frac{1}{2}\left[\frac{F_A L}{(50 \text{ mm}^2)E_{st}}\right] + \frac{1}{2}\left[\frac{F_E L}{(50 \text{ mm}^2)E_{st}}\right]$$

$$F_C = 0.3F_A + 0.3F_E \qquad (3)$$

Solving Eqs. 1–3 simultaneously yields

$$F_A = 9.52 \text{ kN} \qquad \qquad Ans.$$

$$F_C = 3.46 \text{ kN} \qquad \qquad Ans.$$

$$F_E = 2.02 \text{ kN} \qquad \qquad Ans.$$

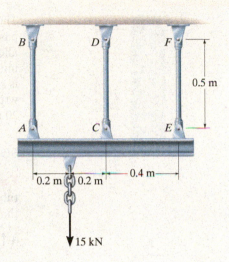

(a)

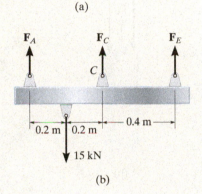

(b)

(c)

Fig. 4–13

EXAMPLE 4.8

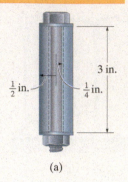

(a)

(b)

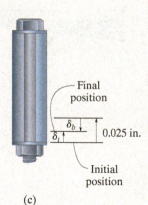

(c)

Fig. 4–14

The bolt shown in Fig. 4–14a is made of 2014-T6 aluminum alloy, and it passes through the cylindrical tube made of Am 1004-T61 magnesium alloy. The tube has an outer radius of $\frac{1}{2}$ in., and it is assumed that both the inner radius of the tube and the radius of the bolt are $\frac{1}{4}$ in. When the bolt is snug against the tube it produces negligible force in the tube. Using a wrench, the nut is then further tightened one-half turn. If the bolt has 20 threads per inch, determine the stress in the bolt.

SOLUTION

Equilibrium. The free-body diagram of a section of the bolt and the tube, Fig. 4–14b, is considered in order to relate the force in the bolt F_b to that in the tube, F_t. Equilibrium requires

$$+\uparrow \Sigma F_y = 0; \qquad\qquad F_b - F_t = 0 \qquad\qquad (1)$$

Compatibility. As noted in Fig. 4–14c, when the nut is tightened one-half turn on the bolt, it advances a distance of $\left(\frac{1}{2}\right)\left(\frac{1}{20}\text{ in.}\right) = 0.025$ in. This will cause the tube to shorten δ_t and the bolt to elongate δ_b. Thus, the compatibility of these displacements requires

$$(+\uparrow) \qquad\qquad \delta_t + \delta_b = 0.025 \text{ in.}$$

Load–Displacement. Taking the moduli of elasticity from the table on the inside back cover, and applying the load–displacement relationship, Eq. 4–2, yields

$$\frac{F_t \,(3 \text{ in.})}{\pi[(0.5 \text{ in.})^2 - (0.25 \text{ in.})^2]\,[6.48(10^3)\text{ ksi}]}$$

$$+ \frac{F_b \,(3 \text{ in.})}{\pi(0.25 \text{ in.})^2 \,[10.6(10^3)\text{ ksi}]} = 0.025 \text{ in.}$$

$$0.78595 F_t + 1.4414 F_b = 25 \qquad\qquad (2)$$

Solving Eqs. 1 and 2 simultaneously, we get

$$F_b = F_t = 11.22 \text{ kip}$$

The stresses in the bolt and tube are therefore

$$\sigma_b = \frac{F_b}{A_b} = \frac{11.22 \text{ kip}}{\pi(0.25 \text{ in.})^2} = 57.2 \text{ ksi} \qquad\qquad \textit{Ans.}$$

$$\sigma_t = \frac{F_t}{A_t} = \frac{11.22 \text{ kip}}{\pi[(0.5 \text{ in.})^2 - (0.25 \text{ in.})^2]} = 19.1 \text{ ksi}$$

These stresses are less than the reported yield stress for each material, $(\sigma_Y)_{al} = 60$ ksi and $(\sigma_Y)_{mg} = 22$ ksi (see the inside back cover), and therefore this "elastic" analysis is valid.

4.5   THE FORCE METHOD OF ANALYSIS FOR AXIALLY LOADED MEMBERS

It is also possible to solve statically indeterminate problems by writing the compatibility equation using the principle of superposition. This method of solution is often referred to as the *flexibility or force method of analysis*. To show how it is applied, consider again the bar in Fig. 4–15*a*. If we choose the support at *B* as "redundant" and *temporarily* remove it from the bar, then the bar will become statically determinate, as in Fig. 4–15*b*. Using the principle of superposition, however, we must add back the unknown redundant load $\mathbf{F}_B$, as shown in Fig. 4–15*c*.

Since the load $\mathbf{P}$ causes *B* to be displaced *downward* by an amount δ_P, the reaction $\mathbf{F}_B$ must displace end *B* of the bar *upward* by an amount δ_B, so that no displacement occurs at *B* when the two loadings are superimposed. Assuming displacements are positive downward, we have

$$(+\downarrow) \qquad\qquad 0 = \delta_P - \delta_B$$

This condition of $\delta_P = \delta_B$ represents the compatibility equation for displacements at point *B*.

Applying the load–displacement relationship to each bar, we have $\delta_P = 500\,\text{N}(2\,\text{m})/AE$ and $\delta_B = F_B(5\,\text{m})/AE$. Consequently,

$$0 = \frac{500\,\text{N}(2\,\text{m})}{AE} - \frac{F_B(5\,\text{m})}{AE}$$

$$F_B = 200\,\text{N}$$

From the free-body diagram of the bar, Fig. 4–15*d*, equilibrium requires

$$+\uparrow \Sigma F_y = 0; \qquad 200\,\text{N} + F_A - 500\,\text{N} = 0$$

Then

$$F_A = 300\,\text{N}$$

These results are the same as those obtained in Sec. 4.4.

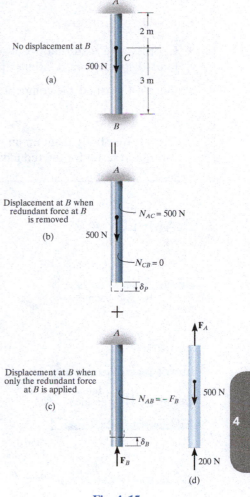

No displacement at *B*

(a)

||

Displacement at *B* when redundant force at *B* is removed

(b)

$N_{AC} = 500\,\text{N}$

$N_{CB} = 0$

δ_P

+

Displacement at *B* when only the redundant force at *B* is applied

(c)

$N_{AB} = -F_B$

δ_B

F_B

F_A

500 N

$N_{AB} = -F_B$

200 N

(d)

Fig. 4–15

▶ PROCEDURE FOR ANALYSIS

The force method of analysis requires the following steps.

Compatibility.

- Choose one of the supports as redundant and write the equation of compatibility. To do this, the known displacement at the redundant support, which is usually zero, is equated to the displacement at the support caused *only* by the external loads acting on the member *plus* (vectorially) the displacement at this support caused *only* by the redundant reaction acting on the member.

Load–Displacement.

- Express the external load and redundant displacements in terms of the loadings by using a load–displacement relationship, such as $\delta = NL/AE$.

- Once established, the compatibility equation can then be solved for the magnitude of the redundant force.

Equilibrium.

- Draw a free-body diagram and write the appropriate equations of equilibrium for the member using the calculated result for the redundant. Solve these equations for the other reactions.

EXAMPLE 4.9

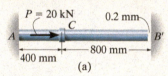

(a)

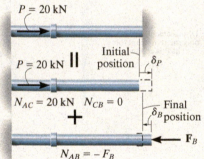

(b)

The A-36 steel rod shown in Fig. 4–16a has a diameter of 10 mm. It is fixed to the wall at A, and before it is loaded there is a gap between the wall and the rod of 0.2 mm. Determine the reactions at A and B'. Neglect the size of the collar at C. Take $E_{st} = 200$ GPa.

SOLUTION

Compatibility.   Here we will consider the support at B' as redundant. Using the principle of superposition, Fig. 4–16b, we have

$$(\overset{+}{\rightarrow}) \qquad\qquad 0.0002 \text{ m} = \delta_P - \delta_B \qquad\qquad (1)$$

Load–Displacement.   The deflections δ_P and δ_B are determined from Eq. 4–2.

$$\delta_P = \frac{N_{AC} L_{AC}}{AE} = \frac{[20(10^3) \text{ N}](0.4 \text{ m})}{\pi (0.005 \text{ m})^2 [200(10^9) \text{ N/m}^2]} = 0.5093(10^{-3}) \text{ m}$$

$$\delta_B = \frac{N_{AB} L_{AB}}{AE} = \frac{F_B (1.20 \text{ m})}{\pi (0.005 \text{ m})^2 [200(10^9) \text{ N/m}^2]} = 76.3944(10^{-9}) F_B$$

Substituting into Eq. 1, we get

$$0.0002 \text{ m} = 0.5093(10^{-3}) \text{ m} - 76.3944(10^{-9}) F_B$$

$$F_B = 4.05(10^3) \text{ N} = 4.05 \text{ kN} \qquad\qquad Ans.$$

Equilibrium.   From the free-body diagram, Fig. 4–16c,

$$\overset{+}{\rightarrow} \Sigma F_x = 0; \qquad -F_A + 20 \text{ kN} - 4.05 \text{ kN} = 0 \quad F_A = 16.0 \text{ kN} \qquad Ans.$$

(c)

Fig. 4–16

PROBLEMS

4–31. The column is constructed from high-strength concrete and eight A992 steel reinforcing rods. If the column is subjected to an axial force of 200 kip, determine the average normal stress in the concrete and in each rod. Each rod has a diameter of 1 in.

*4–32. The column is constructed from high-strength concrete and eight A992 steel reinforcing rods. If the column is subjected to an axial force of 200 kip, determine the required diameter of each rod so that 60% of the axial force is carried by the concrete.

4–34. If column AB is made from high strength precast concrete and reinforced with four $\frac{3}{4}$ in. diameter A-36 steel rods, determine the average normal stress developed in the concrete and in each rod. Set $P = 75$ kip.

4–35. If column AB is made from high strength precast concrete and reinforced with four $\frac{3}{4}$ in. diameter A-36 steel rods, determine the maximum allowable floor loadings P. The allowable normal stresses for the concrete and the steel are $(\sigma_{\text{allow}})_{\text{con}} = 2.5$ ksi and $(\sigma_{\text{allow}})_{\text{st}} = 24$ ksi, respectively.

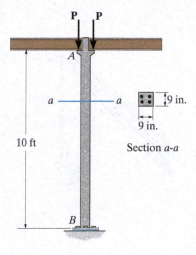

Probs. 4–34/35

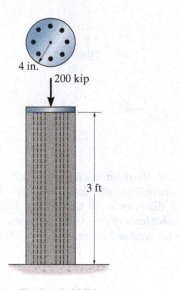

Probs. 4–31/32

4–33. The A-36 steel pipe has a 6061-T6 aluminum core. It is subjected to a tensile force of 200 kN. Determine the average normal stress in the aluminum and the steel due to this loading. The pipe has an outer diameter of 80 mm and an inner diameter of 70 mm.

*4–36. Determine the support reactions at the rigid supports A and C. The material has a modulus of elasticity of E.

4–37. If the supports at A and C are flexible and have a stiffness k, determine the support reactions at A and C. The material has a modulus of elasticity of E.

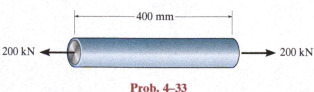

Prob. 4–33

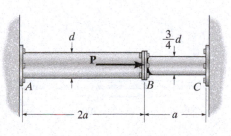

Probs. 4–36/37

4–38. The load of 2000 lb is to be supported by the two vertical steel wires for which $\sigma_Y = 70$ ksi. Originally wire AB is 60 in. long and wire AC is 60.04 in. long. Determine the force developed in each wire after the load is suspended. Each wire has a cross-sectional area of 0.02 in². $E_{st} = 29.0(10^3)$ ksi.

4–39. The load of 2000 lb is to be supported by the two vertical steel wires for which $\sigma_Y = 70$ ksi. Originally wire AB is 60 in. long and wire AC is 60.04 in. long. Determine the cross-sectional area of AB if the load is to be shared equally between both wires. Wire AC has a cross-sectional area of 0.02 in². $E_{st} = 29.0(10^3)$ ksi.

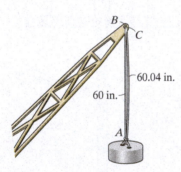

Probs. 4–38/39

*4–40.** The A-36 steel pipe has an outer radius of 20 mm and an inner radius of 15 mm. If it fits snugly between the fixed walls before it is loaded, determine the reaction at the walls when it is subjected to the load shown.

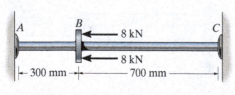

Prob. 4–40

4–41. The 10-mm-diameter steel bolt is surrounded by a bronze sleeve. The outer diameter of this sleeve is 20 mm, and its inner diameter is 10 mm. If the yield stress for the steel is $(\sigma_Y)_{st} = 640$ MPa, and for the bronze $(\sigma_Y)_{br} = 520$ MPa, determine the magnitude of the largest elastic load P that can be applied to the assembly. $E_{st} = 200$ GPa, $E_{br} = 100$ GPa.

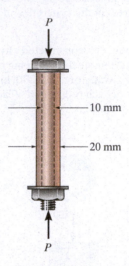

Prob. 4–41

4–42. The 10-mm-diameter steel bolt is surrounded by a bronze sleeve. The outer diameter of this sleeve is 20 mm, and its inner diameter is 10 mm. If the bolt is subjected to a compressive force of $P = 20$ kN, determine the average normal stress in the steel and the bronze. $E_{st} = 200$ GPa, $E_{br} = 100$ GPa.

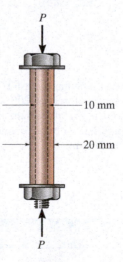

Prob. 4–42

4-43. The assembly consists of two red brass C83400 copper rods AB and CD of diameter 30 mm, a stainless 304 steel alloy rod EF of diameter 40 mm, and a rigid cap G. If the supports at A, C, and F are rigid, determine the average normal stress developed in the rods.

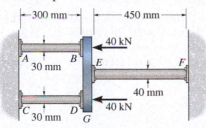

Prob. 4–43

*** 4-44.** The rigid beam is supported by the three suspender bars. Bars AB and EF are made of aluminum and bar CD is made of steel. If each bar has a cross-sectional area of 450 mm², determine the maximum value of P if the allowable stress is $(\sigma_{allow})_{st} = 200$ MPa for the steel and $(\sigma_{allow})_{al} = 150$ MPa for the aluminum. $E_{st} = 200$ GPa, $E_{al} = 70$ GPa.

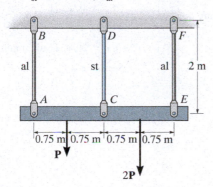

Prob. 4–44

4-45. The bolt AB has a diameter of 20 mm and passes through a sleeve that has an inner diameter of 40 mm and an outer diameter of 50 mm. The bolt and sleeve are made of A-36 steel and are secured to the rigid brackets as shown. If the bolt length is 220 mm and the sleeve length is 200 mm, determine the tension in the bolt when a force of 50 kN is applied to the brackets.

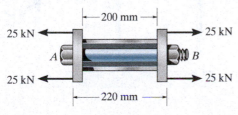

Prob. 4–45

4-46. If the gap between C and the rigid wall at D is initially 0.15 mm, determine the support reactions at A and D when the force $P = 200$ kN is applied. The assembly is made of solid A-36 steel cylinders.

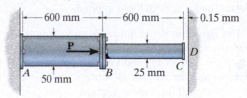

Prob. 4–46

4-47. The support consists of a solid red brass C83400 copper post surrounded by a 304 stainless steel tube. Before the load is applied the gap between these two parts is 1 mm. Given the dimensions shown, determine the greatest axial load that can be applied to the rigid cap A without causing yielding of any one of the materials.

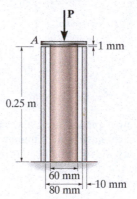

Prob. 4–47

*** 4-48.** The specimen represents a filament-reinforced matrix system made from plastic (matrix) and glass (fiber). If there are n fibers, each having a cross-sectional area of A_f and modulus of E_f, embedded in a matrix having a cross-sectional area of A_m and modulus of E_m, determine the stress in the matrix and in each fiber when the force P is applied on the specimen.

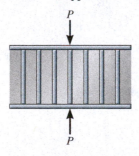

Prob. 4–48

4–49. The rigid bar is pinned at A and supported by two aluminum rods, each having a diameter of 1 in., a modulus of elasticity $E_{al} = 10(10^3)$ ksi, and yield stress of $(\sigma_Y)_{al} = 40$ ksi. If the bar is initially vertical, determine the displacement of the end B when the force of 20 kip is applied.

4–50. The rigid bar is pinned at A and supported by two aluminum rods, each having a diameter of 1 in. a modulus of elasticity $E_{al} = 10(10^3)$ ksi, and yield stress of $(\sigma_Y)_{al} = 40$ ksi. If the bar is initially vertical, determine the angle of tilt of the bar when the 20-kip load is applied.

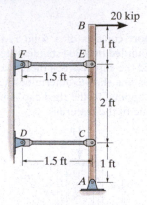

Probs. 4–49/50

4–51. The rigid bar is pinned at A and supported by two aluminum rods, each having a diameter of 1 in. and a modulus of elasticity $E_{al} = 10(10^3)$ ksi. If the bar is initially vertical, determine the displacement of the end B when the force of 2 kip is applied.

*4–52. The rigid bar is pinned at A and supported by two aluminum rods, each having a diameter of 1 in. and a modulus of elasticity $E_{al} = 10(10^3)$ ksi. If the bar is initially vertical, determine the force in each rod when the 2-kip load is applied.

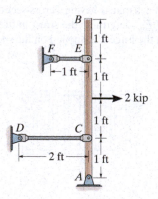

Probs. 4–51/52

4–53. The 2014-T6 aluminum rod AC is reinforced with the firmly bonded A992 steel tube BC. If the assembly fits snugly between the rigid supports so that there is no gap at C, determine the support reactions when the axial force of 400 kN is applied. The assembly is attached at D.

4–54. The 2014-T6 aluminum rod AC is reinforced with the firmly bonded A992 steel tube BC. When no load is applied to the assembly, the gap between end C and the rigid support is 0.5 mm. Determine the support reactions when the axial force of 400 kN is applied.

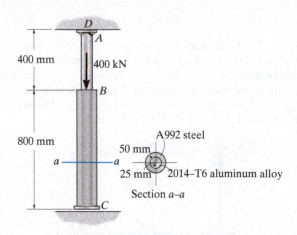

Probs. 4–53/54

4–55. The three suspender bars are made of A992 steel and have equal cross-sectional areas of 450 mm². Determine the average normal stress in each bar if the rigid beam is subjected to the loading shown.

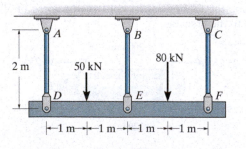

Prob. 4–55

*4–56.   The three A-36 steel wires each have a diameter of 2 mm and unloaded lengths of L_{AC} = 1.60 m and L_{AB} = L_{AD} = 2.00 m. Determine the force in each wire after the 150-kg mass is suspended from the ring at A.

4–57.   The A-36 steel wires AB and AD each have a diameter of 2 mm and the unloaded lengths of each wire are L_{AC} = 1.60 m and L_{AB} = L_{AD} = 2.00 m. Determine the required diameter of wire AC so that each wire is subjected to the same force when the 150-kg mass is suspended from the ring at A.

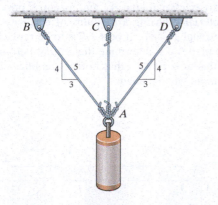

Probs. 4–56/57

4–58.   The post is made from 6061-T6 aluminum and has a diameter of 50 mm. It is fixed supported at A and B, and at its center C there is a coiled spring attached to the rigid collar. If the spring is originally uncompressed, determine the reactions at A and B when the force P = 40 kN is applied to the collar.

4–59.   The post is made from 6061-T6 aluminum and has a diameter of 50 mm. It is fixed supported at A and B, and at its center C there is a coiled spring attached to the rigid collar. If the spring is originally uncompressed, determine the compression in the spring when the load of P = 50 kN is applied to the collar.

Probs. 4–58/59

*4–60.   The bracket is held to the wall using three A-36 steel bolts at B, C, and D. Each bolt has a diameter of 0.5 in. and an unstretched length of 2 in. If a force of 800 lb is placed on the bracket as shown, determine the force developed in each bolt. For the calculation, assume that the bolts carry no shear; rather, the vertical force of 800 lb is supported by the toe at A. Also, assume that the wall and bracket are rigid. A greatly exaggerated deformation of the bolts is shown.

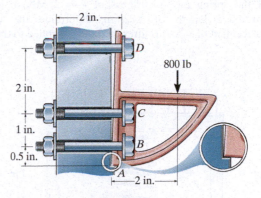

Prob. 4–60

4–61.   The bracket is held to the wall using three A-36 steel bolts at B, C, and D. Each bolt has a diameter of 0.5 in. and an unstretched length of 2 in. If a force of 800 lb is placed on the bracket as shown, determine how far, s, the top bracket at bolt D moves away from the wall. For the calculation, assume that the bolts carry no shear; rather, the vertical force of 800 lb is supported by the toe at A. Also, assume that the wall and bracket are rigid. A greatly exaggerated deformation of the bolts is shown.

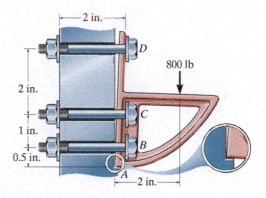

Prob. 4–61

4–62. The rigid bar is supported by the two short white spruce wooden posts and a spring. If each of the posts has an unloaded length of 1 m and a cross-sectional area of 600 mm², and the spring has a stiffness of $k = 2$ MN/m and an unstretched length of 1.02 m, determine the force in each post after the load is applied to the bar.

4–63. The rigid bar is supported by the two short white spruce wooden posts and a spring. If each of the posts has an unloaded length of 1 m and a cross-sectional area of 600 mm², and the spring has a stiffness of $k = 2$ MN/m and an unstretched length of 1.02 m, determine the vertical displacement of A and B after the load is applied to the bar.

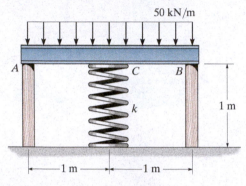

Probs. 4–62/63

*4–64. The assembly consists of two posts AB and CD each made from material 1 having a modulus of elasticity of E_1 and a cross-sectional area A_1, and a central post made from material 2 having a modulus of elasticity E_2 and cross-sectional area A_2. If a load **P** is applied to the rigid cap, determine the force in each material.

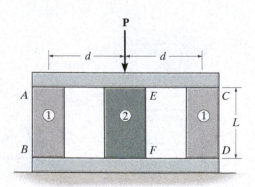

Prob. 4–64

4–65. The assembly consists of two posts AB and CD each made from material 1 having a modulus of elasticity of E_1 and a cross-sectional area A_1, and a central post EF made from material 2 having a modulus of elasticity E_2 and a cross-sectional area A_2. If posts AB and CD are to be replaced by those having a material 2, determine the required cross-sectional area of these new posts so that both assemblies deform the same amount when loaded.

4–66. The assembly consists of two posts AB and CD each made from material 1 having a modulus of elasticity of E_1 and a cross-sectional area A_1, and a central post EF made from material 2 having a modulus of elasticity E_2 and a cross-sectional area A_2. If post EF is to be replaced by one having a material 1, determine the required cross-sectional area of this new post so that both assemblies deform the same amount when loaded.

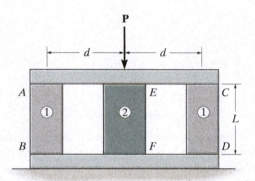

Probs. 4–65/66

4–67. The wheel is subjected to a force of 18 kN from the axle. Determine the force in each of the three spokes. Assume the rim is rigid and the spokes are made of the same material, and each has the same cross-sectional area.

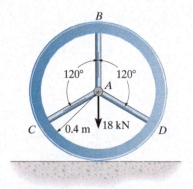

Prob. 4–67

4.6 THERMAL STRESS

A change in temperature can cause a body to change its dimensions. Generally, if the temperature increases, the body will expand, whereas if the temperature decreases, it will contract.* Ordinarily this expansion or contraction is *linearly* related to the temperature increase or decrease that occurs. If this is the case, and the material is homogeneous and isotropic, it has been found from experiment that the displacement of the end of a member having a length L can be calculated using the formula

$$\delta_T = \alpha \Delta T L \qquad (4\text{–}4)$$

Here

Most traffic bridges are designed with expansion joints to accommodate the thermal movement of the deck and thus avoid any thermal stress.

$\alpha =$ a property of the material, referred to as the ***linear coefficient of thermal expansion***. The units measure strain per degree of temperature. They are $1/°F$ (Fahrenheit) in the FPS system, and $1/°C$ (Celsius) or $1/K$ (Kelvin) in the SI system. Typical values are given on the inside back cover.

$\Delta T =$ the algebraic change in temperature of the member

$L =$ the original length of the member

$\delta_T =$ the algebraic change in the length of the member

The change in length of a *statically determinate* member can easily be calculated using Eq. 4–4, since the member is free to expand or contract when it undergoes a temperature change. However, for a *statically indeterminate* member, these thermal displacements will be constrained by the supports, thereby producing ***thermal stresses*** that must be considered in design. Using the methods outlined in the previous sections, it is possible to determine these thermal stresses, as illustrated in the following examples.

Long extensions of ducts and pipes that carry fluids are subjected to variations in temperature that will cause them to expand and contract. Expansion joints, such as the one shown, are used to mitigate thermal stress in the material.

---

*There are some materials, like Invar, an iron-nickel alloy, and scandium trifluoride, that behave in the opposite way, but we will not consider these here.

EXAMPLE 4.10

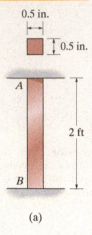

0.5 in.

0.5 in.

A

2 ft

B

(a)

F

F

(b)

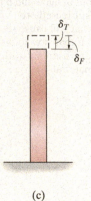

δ_T

δ_F

(c)

Fig. 4–17

The A-36 steel bar shown in Fig. 4–17a is constrained to just fit between two fixed supports when $T_1 = 60°F$. If the temperature is raised to $T_2 = 120°F$, determine the average normal thermal stress developed in the bar.

SOLUTION

Equilibrium.   The free-body diagram of the bar is shown in Fig. 4–17b. Since there is no external load, the force at A is equal but opposite to the force at B; that is,

$$+\uparrow\Sigma F_y = 0; \qquad\qquad F_A = F_B = F$$

The problem is statically indeterminate since this force cannot be determined from equilibrium.

Compatibility.   Since $\delta_{A/B} = 0$, the thermal displacement δ_T at A that occurs, Fig. 4–17c, is counteracted by the force $\mathbf{F}$ that is required to push the bar δ_F back to its original position. The compatibility condition at A becomes

$$(+\uparrow) \qquad\qquad \delta_{A/B} = 0 = \delta_T - \delta_F$$

Load–Displacement.   Applying the thermal and load–displacement relationships, we have

$$0 = \alpha\Delta TL - \frac{FL}{AE}$$

Using the value of α on the inside back cover yields

$$
\begin{aligned}
F &= \alpha\Delta TAE \\
&= [6.60(10^{-6})/°F](120°F - 60°F)(0.5 \text{ in.})^2 [29(10^3) \text{ kip/in}^2] \\
&= 2.871 \text{ kip}
\end{aligned}
$$

Since $\mathbf{F}$ also represents the internal axial force within the bar, the average normal compressive stress is thus

$$\sigma = \frac{F}{A} = \frac{2.871 \text{ kip}}{(0.5 \text{ in.})^2} = 11.5 \text{ ksi} \qquad\qquad Ans.$$

NOTE: The magnitude of $\mathbf{F}$ indicates that changes in temperature can cause large reaction forces in statically indeterminate members.

EXAMPLE 4.11

The rigid beam shown in Fig. 4–18a is fixed to the top of the three posts made of A992 steel and 2014-T6 aluminum. The posts each have a length of 250 mm when no load is applied to the beam, and the temperature is $T_1 = 20°C$. Determine the force supported by each post if the bar is subjected to a uniform distributed load of 150 kN/m and the temperature is raised to $T_2 = 80°C$.

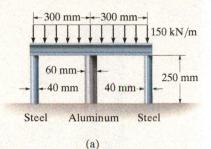

(a)

SOLUTION

Equilibrium. The free-body diagram of the beam is shown in Fig. 4–18b. Moment equilibrium about the beam's center requires the forces in the steel posts to be equal. Summing forces on the free-body diagram, we have

$$+\uparrow \Sigma F_y = 0; \qquad 2F_{st} + F_{al} - 90(10^3)\ \text{N} = 0 \qquad (1)$$

Compatibility. Due to load, geometry, and material symmetry, the top of each post is displaced by an equal amount. Hence,

$$(+\downarrow) \qquad \delta_{st} = \delta_{al} \qquad (2)$$

The final position of the top of each post is equal to its displacement caused by the temperature increase, plus its displacement caused by the internal axial compressive force, Fig. 4–18c. Thus, for the steel and aluminum post, we have

$$(+\downarrow) \qquad \delta_{st} = -(\delta_{st})_T + (\delta_{st})_F$$
$$(+\downarrow) \qquad \delta_{al} = -(\delta_{al})_T + (\delta_{al})_F$$

Applying Eq. 2 gives

$$-(\delta_{st})_T + (\delta_{st})_F = -(\delta_{al})_T + (\delta_{al})_F$$

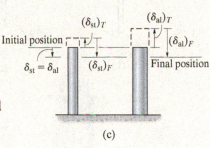

(c)

Fig. 4–18

Load–Displacement. Using Eqs. 4–2 and 4–4 and the material properties on the inside back cover, we get

$$-[12(10^{-6})/°C](80°C - 20°C)(0.250\ \text{m}) + \frac{F_{st}(0.250\ \text{m})}{\pi(0.020\ \text{m})^2[200(10^9)\ \text{N/m}^2]}$$

$$= -[23(10^{-6})/°C](80°C - 20°C)(0.250\ \text{m}) + \frac{F_{al}(0.250\ \text{m})}{\pi(0.030\ \text{m})^2[73.1(10^9)\ \text{N/m}^2]}$$

$$F_{st} = 1.216F_{al} - 165.9(10^3) \qquad (3)$$

To be *consistent*, all numerical data has been expressed in terms of newtons, meters, and degrees Celsius. Solving Eqs. 1 and 3 simultaneously yields

$$F_{st} = -16.4\ \text{kN} \quad F_{al} = 123\ \text{kN} \qquad \textit{Ans.}$$

The negative value for F_{st} indicates that this force acts opposite to that shown in Fig. 4–18b. In other words, the steel posts are in tension and the aluminum post is in compression.

EXAMPLE 4.12

(a)

F_s

F_b

(b)

Initial position

$(\delta_s)_T$

$(\delta_b)_T$

δ $(\delta_b)_F$ Final position

$(\delta_s)_F$

(c)

Fig. 4–19

A 2014-T6 aluminum tube having a cross-sectional area of 600 mm² is used as a sleeve for an A-36 steel bolt having a cross-sectional area of 400 mm², Fig. 4–19a. When the temperature is $T_1 = 15°C$, the nut holds the assembly in a snug position such that the axial force in the bolt is negligible. If the temperature increases to $T_2 = 80°C$, determine the force in the bolt and sleeve.

SOLUTION

Equilibrium.   The free-body diagram of a top segment of the assembly is shown in Fig. 4–19b. The forces F_b and F_s are produced since the sleeve has a higher coefficient of thermal expansion than the bolt, and therefore the sleeve will expand more when the temperature is increased. It is required that

$$+\uparrow \Sigma F_y = 0; \qquad\qquad F_s = F_b \qquad\qquad (1)$$

Compatibility.   The temperature increase causes the sleeve and bolt to expand $(\delta_s)_T$ and $(\delta_b)_T$, Fig. 4–19c. However, the redundant forces F_b and F_s elongate the bolt and shorten the sleeve. Consequently, the end of the assembly reaches a final position, which is not the same as its initial position. Hence, the compatibility condition becomes

$$(+\downarrow) \qquad\qquad \delta = (\delta_b)_T + (\delta_b)_F = (\delta_s)_T - (\delta_s)_F$$

Load–Displacement.   Applying Eqs. 4–2 and 4–4, and using the mechanical properties from the table on the inside back cover, we have

$$[12(10^{-6})/°C](80°C - 15°C)(0.150 \text{ m}) +$$

$$\frac{F_b (0.150 \text{ m})}{(400 \text{ mm}^2)(10^{-6} \text{ m}^2/\text{mm}^2)[200(10^9) \text{ N/m}^2]}$$

$$= [23(10^{-6})/°C](80°C - 15°C)(0.150 \text{ m})$$

$$- \frac{F_s (0.150 \text{ m})}{(600 \text{ mm}^2)(10^{-6} \text{ m}^2/\text{mm}^2)[73.1(10^9) \text{ N/m}^2]}$$

Using Eq. 1 and solving gives

$$F_s = F_b = 20.3 \text{ kN} \qquad\qquad Ans.$$

NOTE: Since linear elastic material behavior was assumed in this analysis, the average normal stresses should be checked to make sure that they do not exceed the proportional limits for the material.

PROBLEMS

*4–68. The C83400-red-brass rod *AB* and 2014-T6-aluminum rod *BC* are joined at the collar *B* and fixed connected at their ends. If there is no load in the members when $T_1 = 50°F$, determine the average normal stress in each member when $T_2 = 120°F$. Also, how far will the collar be displaced? The cross-sectional area of each member is 1.75 in².

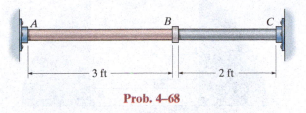

Prob. 4–68

4–69. The assembly has the diameters and material indicated. If it fits securely between its fixed supports when the temperature is $T_1 = 70°F$, determine the average normal stress in each material when the temperature reaches $T_2 = 110°F$.

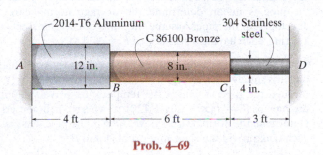

Prob. 4–69

4–70. The rod is made of A992 steel and has a diameter of 0.25 in. If the rod is 4 ft long when the springs are compressed 0.5 in. and the temperature of the rod is $T = 40°F$, determine the force in the rod when its temperature is $T = 160°F$.

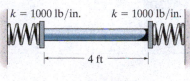

Prob. 4–70

4–71. The two cylindrical rod segments are fixed to the rigid walls such that there is a gap of 0.01 in. between them when $T_1 = 60°F$. What larger temperature T_2 is required in order to just close the gap? Each rod has a diameter of 1.25 in. Determine the average normal stress in each rod if $T_2 = 300°F$. Take $\alpha_{al} = 13(10^{-6})/°F$, $E_{al} = 10(10^3)$ ksi, $(\sigma_Y)_{al} = 40$ ksi, $\alpha_{cu} = 9.4(10^{-6})/°F$, $E_{cu} = 15(10^3)$ ksi, and $(\sigma_Y)_{cu} = 50$ ksi.

*4–72. The two cylindrical rod segments are fixed to the rigid walls such that there is a gap of 0.01 in. between them when $T_1 = 60°F$. Each rod has a diameter of 1.25 in. Determine the average normal stress in each rod if $T_2 = 400°F$, and also calculate the new length of the aluminum segment. Take $\alpha_{al} = 13(10^{-6})/°F$, $E_{al} = 10(10^3)$ ksi, $(\sigma_Y)_{al} = 40$ ksi, $\alpha_{cu} = 9.4(10^{-6})/°F$, $(\sigma_Y)_{cu} = 50$ ksi, and $E_{cu} = 15(10^3)$ ksi.

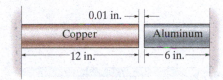

Probs. 4–71/72

4–73. The pipe is made of A992 steel and is connected to the collars at *A* and *B*. When the temperature is 60°F, there is no axial load in the pipe. If hot gas traveling through the pipe causes its temperature to rise by $\Delta T = (40 + 15x)°F$, where *x* is in feet, determine the average normal stress in the pipe. The inner diameter is 2 in., the wall thickness is 0.15 in.

4–74. The bronze C86100 pipe has an inner radius of 0.5 in. and a wall thickness of 0.2 in. If the gas flowing through it changes the temperature of the pipe uniformly from $T_A = 200°F$ at *A* to $T_B = 60°F$ at *B*, determine the axial force it exerts on the walls. The pipe was fitted between the walls when $T = 60°F$.

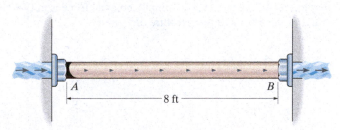

Probs. 4–73/74

4–75. The 40-ft-long A-36 steel rails on a train track are laid with a small gap between them to allow for thermal expansion. Determine the required gap δ so that the rails just touch one another when the temperature is increased from $T_1 = -20°F$ to $T_2 = 90°F$. Using this gap, what would be the axial force in the rails if the temperature rises to $T_3 = 110°F$? The cross-sectional area of each rail is 5.10 in².

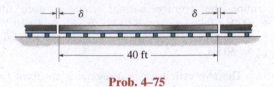

Prob. 4–75

*4–76. The device is used to measure a change in temperature. Bars AB and CD are made of A-36 steel and 2014-T6 aluminum alloy, respectively. When the temperature is at 75°F, ACE is in the horizontal position. Determine the vertical displacement of the pointer at E when the temperature rises to 150°F.

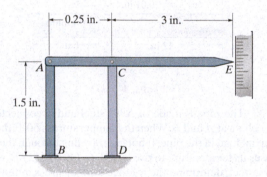

Prob. 4–76

4–77. The bar has a cross-sectional area A, length L, modulus of elasticity E, and coefficient of thermal expansion α. The temperature of the bar changes uniformly along its length from T_A at A to T_B at B so that at any point x along the bar $T = T_A + x(T_B - T_A)/L$. Determine the force the bar exerts on the rigid walls. Initially no axial force is in the bar and the bar has a temperature of T_A.

Prob. 4–77

4–78. When the temperature is at 30°C, the A-36 steel pipe fits snugly between the two fuel tanks. When fuel flows through the pipe, the temperatures at ends A and B rise to 130°C and 80°C, respectively. If the temperature drop along the pipe is linear, determine the average normal stress developed in the pipe. Assume each tank provides a rigid support at A and B.

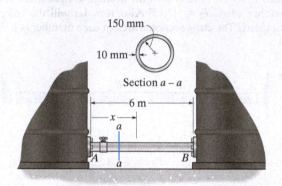

Prob. 4–78

4–79. When the temperature is at 30°C, the A-36 steel pipe fits snugly between the two fuel tanks. When fuel flows through the pipe, the temperatures at ends A and B rise to 130°C and 80°C, respectively. If the temperature drop along the pipe is linear, determine the average normal stress developed in the pipe. Assume the walls of each tank act as a spring, each having a stiffness of $k = 900$ MN/m.

*4–80. When the temperature is at 30°C, the A-36 steel pipe fits snugly between the two fuel tanks. When fuel flows through the pipe, it causes the temperature to vary along the pipe as $T = (\frac{5}{3}x^2 - 20x + 120)°C$, where x is in meters. Determine the normal stress developed in the pipe. Assume each tank provides a rigid support at A and B.

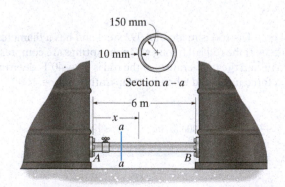

Probs. 4–79/80

4–81. The 50-mm-diameter cylinder is made from Am 1004-T61 magnesium and is placed in the clamp when the temperature is $T_1 = 20°$ C. If the 304-stainless-steel carriage bolts of the clamp each have a diameter of 10 mm, and they hold the cylinder snug with negligible force against the rigid jaws, determine the force in the cylinder when the temperature rises to $T_2 = 130°$C.

4–82. The 50-mm-diameter cylinder is made from Am 1004-T61 magnesium and is placed in the clamp when the temperature is $T_1 = 15°$C. If the two 304-stainless-steel carriage bolts of the clamp each have a diameter of 10 mm, and they hold the cylinder snug with negligible force against the rigid jaws, determine the temperature at which the average normal stress in either the magnesium or the steel first becomes 12 MPa.

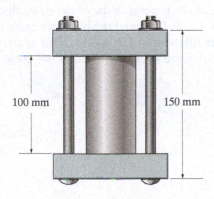

100 mm   150 mm

Probs. 4–81/82

4–83. The wires AB and AC are made of steel, and wire AD is made of copper. Before the 150-lb force is applied, AB and AC are each 60 in. long and AD is 40 in. long. If the temperature is increased by 80°F, determine the force in each wire needed to support the load. Each wire has a cross-sectional area of 0.0123 in². Take $E_{st} = 29(10^3)$ ksi, $E_{cu} = 17(10^3)$ ksi, $\alpha_{st} = 8(10^{-6})/°$F, $\alpha_{cu} = 9.60(10^{-6})/°$F.

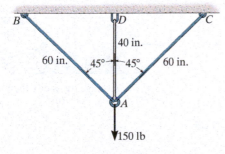

B   *D*   *C*

40 in.

60 in.   45° 45°   60 in.

A

150 lb

Prob. 4–83

*4–84. The cylinder CD of the assembly is heated from $T_1 = 30°$C to $T_2 = 180°$C using electrical resistance. At the lower temperature T_1 the gap between C and the rigid bar is 0.7 mm. Determine the force in rods AB and EF caused by the increase in temperature. Rods AB and EF are made of steel, and each has a cross-sectional area of 125 mm². CD is made of aluminum and has a cross-sectional area of 375 mm². $E_{st} = 200$ GPa, $E_{al} = 70$ GPa, and $\alpha_{al} = 23(10^{-6})/°$C.

4–85. The cylinder CD of the assembly is heated from $T_1 = 30°$C to $T_2 = 180°$C using electrical resistance. Also, the two end rods AB and EF are heated from $T_1 = 30°$C to $T_2 = 50°$C. At the lower temperature T_1 the gap between C and the rigid bar is 0.7 mm. Determine the force in rods AB and EF caused by the increase in temperature. Rods AB and EF are made of steel, and each has a cross-sectional area of 125 mm². CD is made of aluminum and has a cross-sectional area of 375 mm². $E_{st} = 200$ GPa, $E_{al} = 70$ GPa, $\alpha_{st} = 12(10^{-6})/°$C, and $\alpha_{al} = 23(10^{-6})/°$C.

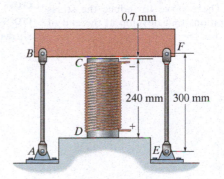

0.7 mm

B   *C*   *F*

240 mm   300 mm

D

A   *E*

Probs. 4–84/85

4–86. The metal strap has a thickness t and width w and is subjected to a temperature gradient T_1 to T_2 ($T_1 < T_2$). This causes the modulus of elasticity for the material to vary linearly from E_1 at the top to a smaller amount E_2 at the bottom. As a result, for any vertical position y, measured from the top surface, $E = [(E_2 - E_1)/w]y + E_1$. Determine the position d where the axial force P must be applied so that the bar stretches uniformly over its cross section.

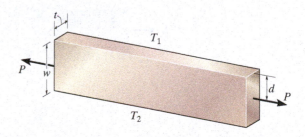

T_1

P w

d

P

T_2

Prob. 4–86

This saw blade has grooves cut into it in order to relieve both the dynamic stress that develops within it as it rotates and the thermal stress that develops as it heats up. Note the small circles at the end of each groove. These serve to reduce the stress concentrations that develop at the end of each groove.

4.7   STRESS CONCENTRATIONS

In Sec. 4.1, it was pointed out that when an axial force is applied to a bar, it creates a complex stress distribution within the localized region of the point of load application. However, complex stress distributions arise not only next to a concentrated loading; they can also arise at sections where the member's cross-sectional area changes. Consider, for example, the bar in Fig. 4–20a, which is subjected to an axial force N. Here the once horizontal and vertical grid lines deflect into an irregular pattern around the hole centered in the bar. The maximum normal stress in the bar occurs on section a–a, since it is located at the bar's *smallest* cross-sectional area. Provided the material behaves in a linear elastic manner, the stress distribution acting on this section can be determined either from a mathematical analysis, using the theory of elasticity, or experimentally by measuring the strain normal to section a–a and then calculating the stress using Hooke's law, $\sigma = E\epsilon$. Regardless of the method used, the general shape of the stress distribution will be like that shown in Fig. 4–20b. If instead the bar has a reduction in its cross section, using shoulder fillets as in Fig. 4–21a, then at the *smallest* cross-sectional area, section a–a, the stress distribution will look like that shown in Fig. 4–21b.

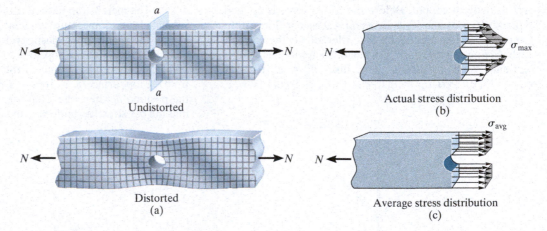

a

N N

a

Undistorted

N N

Distorted
(a)

N σ_{max}

Actual stress distribution
(b)

N σ_{avg}

Average stress distribution
(c)

Fig. 4–20

In both of these cases, *force equilibrium* requires the magnitude of the *resultant force* developed by the stress distribution at section *a–a* to be equal to *N*. In other words,

$$N = \int_A \sigma \, dA \qquad (4\text{--}5)$$

This integral *graphically* represents the total *volume* under each of the stress-distribution diagrams shown in Fig. 4–20b or Fig. 4–21b. Furthermore, the resultant *N* must act through the *centroid* of each of these *volumes*.

In engineering practice, the actual stress distributions in Fig. 4–20b and Fig. 4–21b do not have to be determined. Instead, for the purpose of design, only the *maximum stress* at these sections must be known. Specific values of this maximum normal stress have been determined for various dimensions of each bar, and the results have been reported in graphical form using a **stress concentration factor *K***, Figs. 4–23 and 4–24. We define *K* as a ratio of the maximum stress to the average normal stress acting at the cross section; i.e.,

$$K = \frac{\sigma_{max}}{\sigma_{avg}} \qquad (4\text{--}6)$$

Once *K* is determined from the graph, and the average normal stress has been calculated from $\sigma_{avg} = N/A$, where *A* is the *smallest* cross-sectional area, Figs. 4–20c and 4–21c, then the maximum normal stress at the cross section is determined from $\sigma_{max} = K(N/A)$.

Stress concentrations often arise at sharp corners on heavy machinery. Engineers can mitigate this effect by using stiffeners welded to the corners.

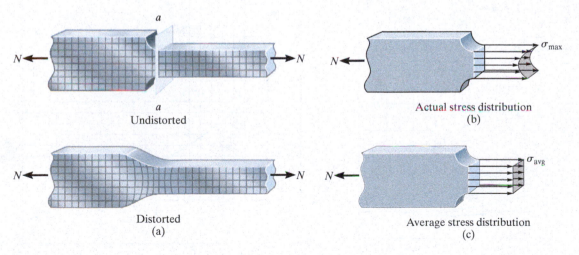

a

a

Undistorted

Distorted
(a)

Actual stress distribution
(b)

Average stress distribution
(c)

σ_{max}

σ_{avg}

Fig. 4–21

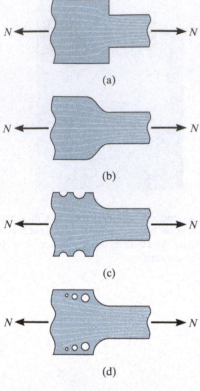

(a)

(b)

(c)

(d)

Fig. 4–22

Notice in Fig. 4–23 that, as the radius r of the shoulder fillet is *decreased*, the stress concentration is increased. For example, if a bar has a sharp corner, Fig. 4–22a, $r = 0$, and so the stress concentration factor will become greater than 3. In other words, the maximum normal stress will be more than three times greater than the average normal stress on the smallest cross section. Proper design can reduce this by introducing a rounded edge, Fig. 4–22b. A further reduction can be made by means of small grooves or holes placed at the transition, Fig. 4–22c and 4–22d. In all of these designs the rigidity of the material surrounding the corners is reduced, so that both the strain and the stress are more evenly spread throughout the bar.

Remember that the stress concentration factors given in Figs. 4–23 and 4–24 were determined on the basis of a static loading, with the assumption that the stress in the material does not exceed the proportional limit. If the material is *very brittle*, the proportional limit may be at the fracture stress, and so for this material, failure will begin at the point of stress concentration (σ_{max}). Essentially a crack begins to form at this point, and a higher stress concentration will develop at the *tip* of this crack. This, in turn, causes the crack to propagate over the cross section, resulting in sudden fracture. For this reason, it is very important to use stress concentration factors for the design of members made of brittle materials. On the other hand, if the material is ductile and subjected to a static load, it is often not necessary to use stress concentration factors since any stress that exceeds the proportional limit will not result in a crack. Instead, as will be shown in the next section, the material will have reserve strength due to yielding and strain hardening.

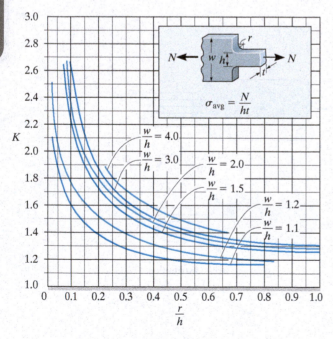

$$\sigma_{avg} = \frac{N}{ht}$$

Fig. 4–23

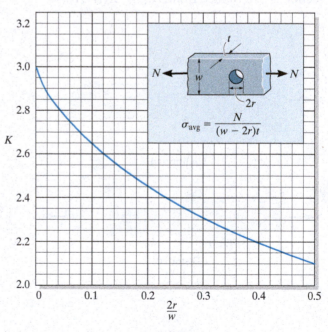

$$\sigma_{avg} = \frac{N}{(w - 2r)t}$$

Fig. 4–24

> ## IMPORTANT POINTS
>
> - *Stress concentrations* occur at sections where the cross-sectional area suddenly changes. The more severe the change, the larger the stress concentration.
>
> - For design or analysis, it is only necessary to determine the maximum stress acting on the smallest cross-sectional area. This is done using a *stress concentration factor, K*, that has been determined through experiment and is only a function of the geometry of the specimen.
>
> - Normally the stress concentration in a ductile specimen that is subjected to a static loading will *not* have to be considered in design; however, if the material is *brittle*, or subjected to *fatigue* loadings, then stress concentrations become important.

Failure of this steel pipe in tension occurred at its smallest cross-sectional area, which is through the hole. Notice how the material yielded around the fractured surface.

*4.8   INELASTIC AXIAL DEFORMATION

Up to this point we have only considered loadings that cause the material to behave elastically. Sometimes, however, a member may be designed so that the loading causes the material to yield and thereby permanently deform. Such members are often made of a highly ductile metal such as annealed low-carbon steel having a stress–strain diagram that is similar to that of Fig. 3–6, and for nonexcessive yielding can be *modeled* as shown in Fig. 4–25b. A material that exhibits this behavior is referred to as being **elastic perfectly plastic** or **elastoplastic.**

To illustrate physically how such a material behaves, consider the bar in Fig. 4–25a, which is subjected to the axial load N. If the load causes an *elastic stress* $\sigma = \sigma_1$ to be developed in the bar, then *equilibrium* requires $N = \int \sigma_1 \, dA = \sigma_1 A$. This stress causes the bar to strain ϵ_1 as indicated on the stress–strain diagram, Fig. 4–25b. If N is now increased such that it causes yielding of the material, then $\sigma = \sigma_Y$. This load N_p is called the **plastic load,** since it represents the maximum load that can be supported by an elastoplastic material. For this case, the strains are *not* uniquely defined. Instead, at the instant σ_Y is attained, the bar will be subjected to the yield strain ϵ_Y, Fig. 4–25b, then the bar will *continue to yield* (or elongate) producing the strains ϵ_2, then ϵ_3, etc. Since our "model" of the material exhibits perfectly plastic material behavior, this elongation is expected to continue indefinitely. However, the material will, after some yielding, begin to strain harden, so that the extra strength it attains will stop any further straining, thereby allowing the bar to support an additional load.

(a)

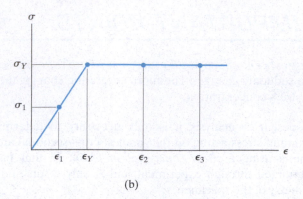

(b)

Fig. 4–25

To extend this discussion, now consider the case of a bar having a hole through it as shown in Fig. 4–26a. When N is applied, a stress concentration occurs in the material at the edge of the hole, on section a–a. The stress here will reach a maximum value of, say, $\sigma_{max} = \sigma_1$ and have a corresponding *elastic strain* of ϵ_1, Fig. 4–26b. The stresses and corresponding strains at other points on the cross section will be smaller, as indicated by the stress distribution shown in Fig. 4–26c. Equilibrium again requires $N = \int \sigma \, dA$, which is geometrically equivalent to the "volume" contained within the stress distribution. If the load is further increased to N', so that $\sigma_{max} = \sigma_Y$, then the material will begin to yield outward from the hole, until the equilibrium condition $N' = \int \sigma \, dA$ is satisfied, Fig. 4–26d. As shown, this produces a stress distribution that has a geometrically *greater* "volume" than that shown in Fig. 4–26c. A further increase in load will eventually cause the material over the *entire cross section* to yield, Fig. 4–26e. When this happens, *no greater load* can be sustained by the bar. This **plastic load** N_p is now

$$N_p = \int_A \sigma_Y \, dA = \sigma_Y A$$

where A is the bar's cross-sectional area at section a–a.

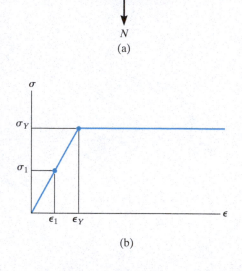

(a)

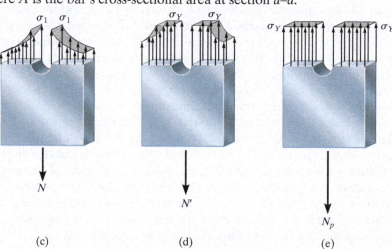

(c)   (d)   (e)

(b)

Fig. 4–26

*4.9   RESIDUAL STRESS

Consider a prismatic member made of an elastoplastic material having the stress–strain diagram shown in Fig. 4–27. If an axial load produces a stress σ_Y in the material and a corresponding strain ϵ_C, then when the load is *removed*, the material will respond elastically and follow the line CD in order to recover some of the strain. A recovery to zero stress at point O' will be possible if the member is statically determinate, since then the support reactions for the member will be zero when the load is removed. Under these circumstances the member will be permanently deformed so that the permanent set or strain in the member is $\epsilon_{O'}$.

   If the member is *statically indeterminate*, however, removal of the external load will cause the support forces to respond to the elastic recovery CD. Since these forces will constrain the member from full recovery, they will induce ***residual stresses*** in the member. To solve a problem of this kind, the complete cycle of loading and then unloading of the member can be considered as the *superposition* of a positive load (loading) on a negative load (unloading). The loading, O to C, results in a plastic stress distribution, whereas the unloading, along CD, results only in an elastic stress distribution. Superposition requires these loads to cancel; however, the stress distributions will not cancel, and so residual stresses will remain in the member. Examples 4.14 and 4.15 numerically illustrate this situation.

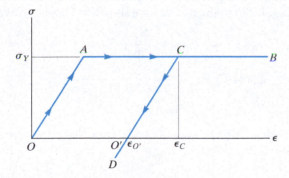

Fig. 4–27

EXAMPLE 4.13

The bar in Fig. 4–28a is made of steel that is assumed to be elastic perfectly plastic, with $\sigma_Y = 250$ MPa. Determine (a) the maximum value of the applied load N that can be applied without causing the steel to yield and (b) the maximum value of N that the bar can support. Sketch the stress distribution at the critical section for each case.

SOLUTION

Part (a). When the material behaves elastically, we must use a stress concentration factor determined from Fig. 4–23 that is unique for the bar's geometry. Here

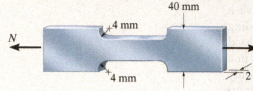

(a)

$$\frac{r}{h} = \frac{4 \text{ mm}}{(40 \text{ mm} - 8 \text{ mm})} = 0.125$$

$$\frac{w}{h} = \frac{40 \text{ mm}}{(40 \text{ mm} - 8 \text{ mm})} = 1.25$$

From the figure $K \approx 1.75$. The maximum load, without causing yielding, occurs when $\sigma_{max} = \sigma_Y$. The average normal stress is $\sigma_{avg} = N/A$. Using Eq. 4–6, we have

$$\sigma_{max} = K\sigma_{avg}; \qquad \sigma_Y = K\left(\frac{N_Y}{A}\right)$$

$$250(10^6) \text{ Pa} = 1.75\left[\frac{N_Y}{(0.002 \text{ m})(0.032 \text{ m})}\right]$$

$$N_Y = 9.14 \text{ kN} \qquad\qquad Ans.$$

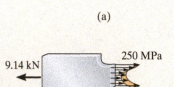

(b)

This load has been calculated using the *smallest* cross section. The resulting stress distribution is shown in Fig. 4–28b. For equilibrium, the "volume" contained within this distribution must equal 9.14 kN.

Part (b). The maximum load sustained by the bar will cause *all the material* at the smallest cross section to yield. Therefore, as N is increased to the *plastic load* N_p, it gradually changes the stress distribution from the elastic state shown in Fig. 4–28b to the plastic state shown in Fig. 4–28c. We require

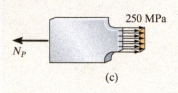

(c)

Fig. 4–28

$$\sigma_Y = \frac{N_p}{A}$$

$$250(10^6) \text{ Pa} = \frac{N_p}{(0.002 \text{ m})(0.032 \text{ m})}$$

$$N_p = 16.0 \text{ kN} \qquad\qquad Ans.$$

Here N_p equals the "volume" contained within the stress distribution, which in this case is $N_p = \sigma_Y A$.

EXAMPLE 4.14

Two steel wires are used to lift the weight of 3 kip, Fig. 4–29a. Wire AB has an unstretched length of 20.00 ft and wire AC has an unstretched length of 20.03 ft. If each wire has a cross-sectional area of 0.05 in^2, and the steel can be considered elastic perfectly plastic, as shown by the $\sigma-\epsilon$ graph in Fig. 4–29b, determine the force in each wire and its elongation.

(c)

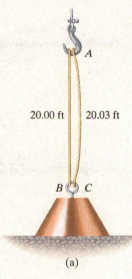

(a)

SOLUTION

Upon lifting the weight, the force (or stress) in each wire *depends* upon the strain in the wire. There are three possibilities: the strains in both wires are elastic, wire AB is plastically strained while wire AC is elastically strained, or both wires are plastically strained. We will assume that AC remains *elastic* and AB is plastically strained.

Investigation of the free-body diagram of the suspended weight, Fig. 4–29c, indicates that the problem is statically indeterminate. The equation of equilibrium is

$$+\uparrow \Sigma F_y = 0; \qquad T_{AB} + T_{AC} - 3 \text{ kip} = 0 \qquad (1)$$

Since AB becomes plastically strained then it must support its maximum load.

$$T_{AB} = \sigma_Y A_{AB} = 50 \text{ ksi } (0.05 \text{ in}^2) = 2.50 \text{ kip} \qquad Ans.$$

Therefore, from Eq. 1,

$$T_{AC} = 0.500 \text{ kip} \qquad Ans.$$

Note that wire AC remains elastic as assumed since the stress in the wire is $\sigma_{AC} = 0.500 \text{ kip}/0.05 \text{ in}^2 = 10 \text{ ksi} < 50 \text{ ksi}$. The corresponding elastic strain is determined by proportion, Fig. 4–29b; i.e.,

$$\frac{\epsilon_{AC}}{10 \text{ ksi}} = \frac{0.0017}{50 \text{ ksi}}$$

$$\epsilon_{AC} = 0.000340$$

The elongation of AC is thus

$$\delta_{AC} = (0.000340)(20.03 \text{ ft}) = 0.00681 \text{ ft} \qquad Ans.$$

And from Fig. 4–29d, the elongation of AB is then

$$\delta_{AB} = 0.03 \text{ ft} + 0.00681 \text{ ft} = 0.0368 \text{ ft} \qquad Ans.$$

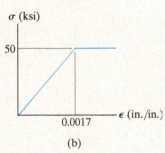

(b)

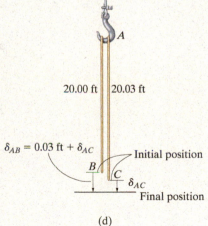

(d)

Fig. 4–29

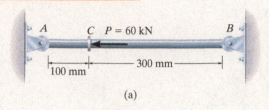

EXAMPLE   4.15

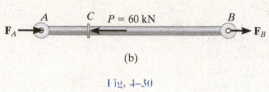

(a)

(b)

Fig. 4–30

The rod shown in Fig. 4–30a has a radius of 5 mm and is made of an elastic perfectly plastic material for which $\sigma_Y = 420$ MPa, $E = 70$ GPa, Fig. 4–30c. If a force of $P = 60$ kN is applied to the rod and then removed, determine the residual stress in the rod.

SOLUTION

The free-body diagram of the rod is shown in Fig. 4–30b. Application of the load **P** will cause one of three possibilities, namely, both segments AC and CB remain elastic, AC is plastic while CB is elastic, or both AC and CB are plastic.*

An el*astic analysis*, similar to that discussed in Sec. 4.4, will produce $F_A = 45$ kN and $F_B = 15$ kN at the supports. However, this results in a stress of

$$\sigma_{AC} = \frac{45 \text{ kN}}{\pi(0.005 \text{ m})^2} = 573 \text{ MPa (compression)} > \sigma_Y = 420 \text{ MPa}$$

$$\sigma_{CB} = \frac{15 \text{ kN}}{\pi(0.005 \text{ m})^2} = 191 \text{ MPa (tension)}$$

Since the material in segment AC will yield, we will assume that AC becomes plastic, while CB remains elastic.

For this case, the maximum possible force developed in AC is

$$(F_A)_Y = \sigma_Y A = 420(10^3) \text{ kN/m}^2 [\pi(0.005 \text{ m})^2] = 33.0 \text{ kN}$$

and from the equilibrium of the rod, Fig. 4–30b,

$$F_B = 60 \text{ kN} - 33.0 \text{ kN} = 27.0 \text{ kN}$$

The stress in each segment of the rod is therefore

$$\sigma_{AC} = \sigma_Y = 420 \text{ MPa (compression)}$$

$$\sigma_{CB} = \frac{27.0 \text{ kN}}{\pi(0.005 \text{ m})^2} = 344 \text{ MPa (tension)} < 420 \text{ MPa (OK)}$$

*The possibility of CB becoming plastic before AC will not occur because when point C moves, the *strain* in AC (since it is shorter) will always be larger than the strain in CB.

Residual Stress. In order to obtain the residual stress, it is also necessary to know the strain in each segment due to the loading. Since *CB* responds elastically,

$$\delta_C = \frac{F_B L_{CB}}{AE} = \frac{(27.0 \text{ kN})(0.300 \text{ m})}{\pi(0.005 \text{ m})^2 \, [70(10^6) \text{ kN/m}^2]} = 0.001474 \text{ m}$$

$$\epsilon_{CB} = \frac{\delta_C}{L_{CB}} = \frac{0.001474 \text{ m}}{0.300 \text{ m}} = +0.004913$$

$$\epsilon_{AC} = \frac{\delta_C}{L_{AC}} = -\frac{0.001474 \text{ m}}{0.100 \text{ m}} = -0.01474$$

Here the yield strain, Fig. 4–30c, is

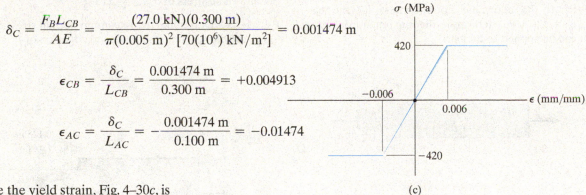

(c)

$$\epsilon_Y = \frac{\sigma_Y}{E} = \frac{420(10^6) \text{ N/m}^2}{70(10^9) \text{ N/m}^2} = 0.006$$

Therefore, when **P** is *applied*, the stress–strain behavior for the material in segment *CB* moves from *O* to *A'*, Fig. 4–30d, and the stress–strain behavior for the material in segment *AC* moves from *O* to *B'*. When the load **P** is applied in the reverse direction, in other words, the load is removed, then an elastic response occurs and a reverse force of $F_A = 45$ kN and $F_B = 15$ kN must be applied to each segment. As calculated previously, these forces now produce stresses $\sigma_{AC} = 573$ MPa (tension) and $\sigma_{CB} = 191$ MPa (compression), and as a result the residual stress in each member is

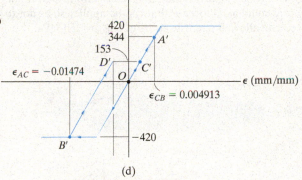

(d)

Fig. 4–30 (cont.)

$$(\sigma_{AC})_r = -420 \text{ MPa} + 573 \text{ MPa} = 153 \text{ MPa} \qquad \textit{Ans.}$$

$$(\sigma_{CB})_r = 344 \text{ MPa} - 191 \text{ MPa} = 153 \text{ MPa} \qquad \textit{Ans.}$$

This residual stress is the *same* for both segments, which is to be expected. Also note that the stress–strain behavior for segment *AC* moves from *B'* to *D'* in Fig. 4–30d, while the stress–strain behavior for the material in segment *CB* moves from *A'* to *C'* when the load is removed.

PROBLEMS

4–87. Determine the maximum normal stress developed in the bar when it is subjected to a tension of $P = 8$ kN.

*4–88. If the allowable normal stress for the bar is $\sigma_{allow} = 120$ MPa, determine the maximum axial force P that can be applied to the bar.

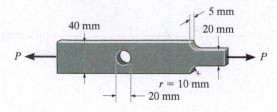

Probs. 4–87/88

4–89. The steel bar has the dimensions shown. Determine the maximum axial force P that can be applied so as not to exceed an allowable tensile stress of $\sigma_{allow} = 150$ MPa.

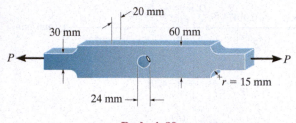

Prob. 4–89

4–90. The A-36 steel plate has a thickness of 12 mm. If $\sigma_{allow} = 150$ MPa, determine the maximum axial load P that it can support. Calculate its elongation, neglecting the effect of the fillets.

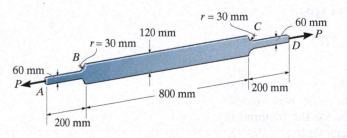

Prob. 4–90

4–91. Determine the maximum axial force P that can be applied to the bar. The bar is made from steel and has an allowable stress of $\sigma_{allow} = 21$ ksi.

*4–92. Determine the maximum normal stress developed in the bar when it is subjected to a tension of $P = 2$ kip.

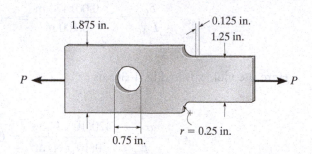

Probs. 4–91/92

4–93. The member is to be made from a steel plate that is 0.25 in. thick. If a 1-in. hole is drilled through its center, determine the approximate width w of the plate so that it can support an axial force of 3350 lb. The allowable stress is $\sigma_{allow} = 22$ ksi.

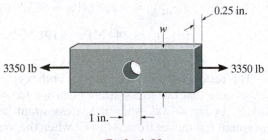

Prob. 4–93

4–94. The resulting stress distribution along section AB for the bar is shown. From this distribution, determine the approximate resultant axial force P applied to the bar. Also, what is the stress concentration factor?

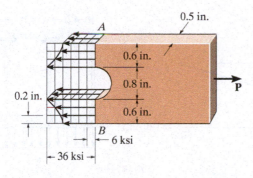

0.5 in.

0.6 in.

0.8 in.

P

0.2 in.

0.6 in.

A

B

6 ksi

36 ksi

Prob. 4–94

4–95. The resulting stress distribution along section AB for the bar is shown. From this distribution, determine the approximate resultant axial force P applied to the bar. Also, what is the stress concentration factor?

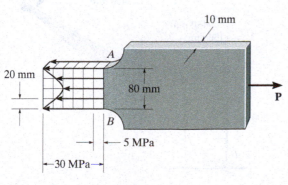

10 mm

20 mm

A

80 mm

P

B

5 MPa

30 MPa

Prob. 4–95

***4–96.** The weight is suspended from steel and aluminum wires, each having the same initial length of 3 m and cross-sectional area of 4 mm². If the materials can be assumed to be elastic perfectly plastic, with $(\sigma_Y)_{st} = 120$ MPa and $(\sigma_Y)_{al} = 70$ MPa, determine the force in each wire if the weight is (a) 600 N and (b) 720 N. $E_{al} = 70$ GPa, $E_{st} = 200$ GPa.

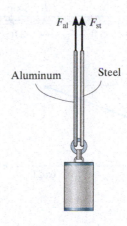

F_{al} F_{st}

Aluminum   Steel

Prob. 4–96

4–97. The weight is suspended from steel and aluminum wires, each having the same initial length of 3 m and cross-sectional area of 4 mm². If the materials can be assumed to be elastic perfectly plastic, with $(\sigma_Y)_{st} = 120$ MPa and $(\sigma_Y)_{al} = 70$ MPa, determine the force in each wire if the weight is (a) 600 N and (b) 720 N. $E_{al} = 70$ GPa, $E_{st} = 200$ GPa.

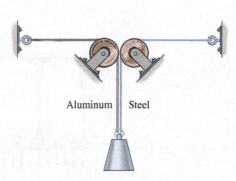

Aluminum   Steel

Prob. 4–97

4

4–98. The bar has a cross-sectional area of 0.5 in² and is made of a material that has a stress–strain diagram that can be approximated by the two line segments. Determine the elongation of the bar due to the applied loading.

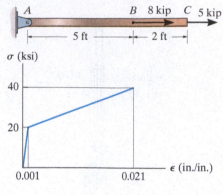

Prob. 4–98

4–99. The distributed loading is applied to the rigid beam, which is supported by the three bars. Each bar has a cross-sectional area of 1.25 in² and is made from a material having a stress–strain diagram that can be approximated by the two line segments. If a load of $w = 25$ kip/ft is applied to the beam, determine the stress in each bar and the vertical displacement of the beam.

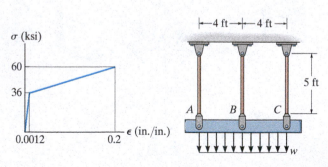

Prob. 4–99

*****4–100.** The distributed loading is applied to the rigid beam, which is supported by the three bars. Each bar has a cross-sectional area of 0.75 in² and is made from a material having a stress–strain diagram that can be approximated by the two line segments. Determine the intensity of the distributed loading w that will cause the beam to displace downward 1.5 in.

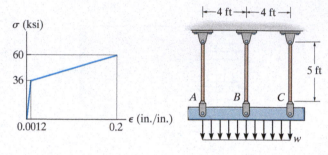

Prob. 4–100

4–101. The rigid lever arm is supported by two A-36 steel wires having the same diameter of 4 mm. If a force of $P = 3$ kN is applied to the handle, determine the force developed in both wires and their corresponding elongations. Consider A-36 steel as an elastic perfectly plastic material.

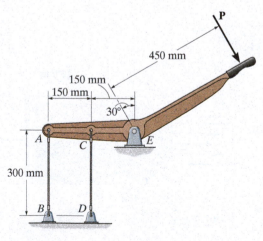

Prob. 4–101

4–102. The rigid lever arm is supported by two A-36 steel wires having the same diameter of 4 mm. Determine the smallest force **P** that will cause (a) only one of the wires to yield; (b) both wires to yield. Consider A-36 steel as an elastic perfectly plastic material.

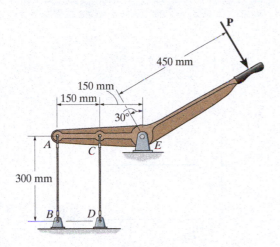

Prob. 4–102

4–103. The 300-kip weight is slowly set on the top of a post made of 2014-T6 aluminum with an A-36 steel core. If both materials can be considered elastic perfectly plastic, determine the average normal stress in each material.

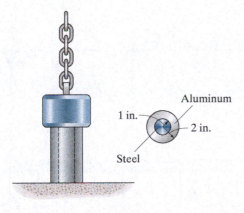

Prob. 4–103

*4–104. The rigid beam is supported by three 25-mm diameter A-36 steel rods. If the beam supports the force of $P = 230$ kN, determine the force developed in each rod. Consider the steel to be an elastic perfectly plastic material.

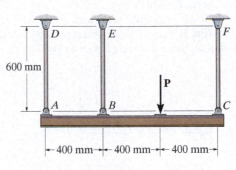

Prob. 4–104

4–105. The rigid beam is supported by three 25-mm diameter A-36 steel rods. If the force of $P = 230$ kN is applied on the beam and removed, determine the residual stresses in each rod. Consider the steel to be an elastic perfectly plastic material.

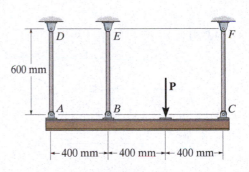

Prob. 4–105

4–106. The rigid beam is supported by the three posts A, B, and C of equal length. Posts A and C have a diameter of 75 mm and are made of a material for which $E = 70$ GPa and $\sigma_Y = 20$ MPa. Post B has a diameter of 20 mm and is made of a material for which $E' = 100$ GPa and $\sigma_{Y'} = 590$ MPa. Determine the smallest magnitude of **P** so that (a) only rods A and C yield and (b) all the posts yield.

4–107. The rigid beam is supported by the three posts A, B, and C. Posts A and C have a diameter of 60 mm and are made of a material for which $E = 70$ GPa and $\sigma_Y = 20$ MPa. Post B is made of a material for which $E' = 100$ GPa and $\sigma_{Y'} = 590$ MPa. If $P = 130$ kN, determine the diameter of post B so that all three posts are about to yield.

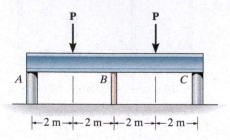

Probs. 4–106/107

*4–108. The bar having a diameter of 2 in. is fixed connected at its ends and supports the axial load **P**. If the material is elastic perfectly plastic as shown by the stress–strain diagram, determine the smallest load P needed to cause segment CB to yield. If this load is released, determine the permanent displacement of point C.

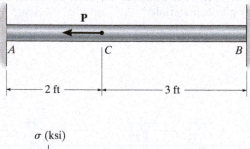

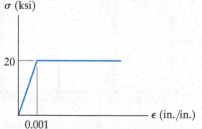

Prob. 4–108

4–109. Determine the elongation of the bar in Prob. 4–108 when both the load **P** and the supports are removed.

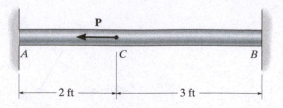

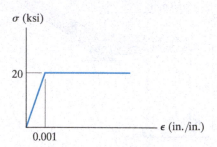

Prob. 4–109

4–110. The rigid beam is supported by three A-36 steel wires, each having a length of 4 ft. The cross-sectional area of AB and EF is 0.015 in², and the cross-sectional area of CD is 0.006 in². Determine the largest distributed load w that can be supported by the beam before any of the wires begin to yield. If the steel is assumed to be elastic perfectly plastic, determine how far the beam is displaced downward just before all the wires begin to yield.

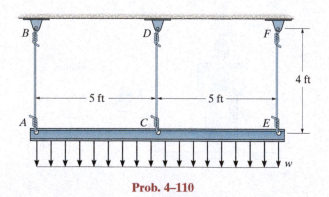

Prob. 4–110

CHAPTER REVIEW

When a loading is applied at a point on a body, it tends to create a stress distribution within the body that becomes more uniformly distributed at regions removed from the point of application of the load. This is called Saint-Venant's principle.	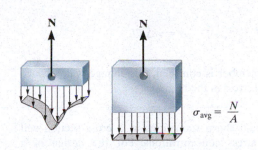 $$\sigma_{avg} = \frac{N}{A}$$
The relative displacement at the end of an axially loaded member relative to the other end is determined from $$\delta = \int_0^L \frac{N(x)dx}{A(x)E(x)}$$	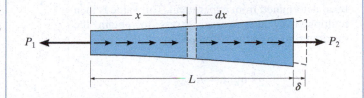
If a series of concentrated external axial forces are applied to a member and AE is also piecewise constant, then $$\delta = \sum \frac{NL}{AE}$$ For application, it is necessary to use a sign convention for the internal load N and displacement δ. We consider tension and elongation as positive values. Also, the material must not yield, but rather it must remain linear elastic.	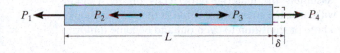
Superposition of load and displacement is possible provided the material remains linear elastic and no significant changes in the geometry of the member occur after loading.	
The reactions on a statically indeterminate bar can be determined using the equilibrium equations and compatibility conditions that relate the displacements at the supports. These displacements are then related to the loads using a load–displacement relationship such as $\delta = NL/AE$.	

A change in temperature can cause a member made of homogeneous isotropic material to change its length by

$$\delta = \alpha \Delta TL$$

If the member is confined, this change will produce thermal stress in the member.

Holes and sharp transitions at a cross section will create stress concentrations. For the design of a member made of brittle material one obtains the stress concentration factor K from a graph, which has been determined from experiment. This value is then multiplied by the average stress to obtain the maximum stress at the cross section.

$$\sigma_{\max} = K\sigma_{\text{avg}}$$

If the loading on a bar made of ductile material causes the material to yield, then the stress distribution that is produced can be determined from the strain distribution and the stress–strain diagram. Assuming the material is perfectly plastic, yielding will cause the stress distribution at the cross section of a hole or transition to even out and become uniform.

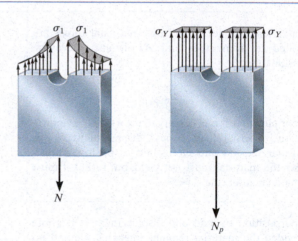

If a member is constrained and an external loading causes yielding, then when the load is released, it will cause residual stress in the member.

CONCEPTUAL PROBLEMS

C4–1. In each photo the concrete footings *A* were poured when the column was already in place. Later the concrete slab was poured. Explain why the 45° cracks formed in the slab at each corner of the square footing and not for the circular footing.

C4–1

C4–2. The row of bricks, along with mortar and an internal steel reinforcing rod, was intended to serve as a lintel beam to support the bricks above this ventilation opening on an exterior wall of a building. Explain what may have caused the bricks to fail in the manner shown.

C4–2

REVIEW PROBLEMS

R4–1. The assembly consists of two A992 steel bolts *AB* and *EF* and an 6061-T6 aluminum rod *CD*. When the temperature is at 30° C, the gap between the rod and rigid member *AE* is 0.1 mm. Determine the normal stress developed in the bolts and the rod if the temperature rises to 130° C. Assume *BF* is also rigid.

R4–2. The assembly shown consists of two A992 steel bolts *AB* and *EF* and an 6061-T6 aluminum rod *CD*. When the temperature is at 30° C, the gap between the rod and rigid member *AE* is 0.1 mm. Determine the highest temperature to which the assembly can be raised without causing yielding either in the rod or the bolts. Assume *BF* is also rigid.

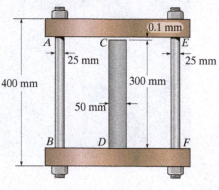

R4–1/2

R4–3. The rods each have the same 25-mm diameter and 600-mm length. If they are made of A992 steel, determine the forces developed in each rod when the temperature increases by 50° C.

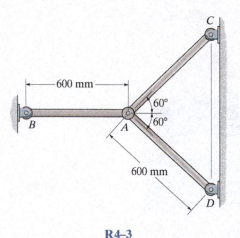

R4–3

*R4–4. Two A992 steel pipes, each having a cross-sectional area of 0.32 in², are screwed together using a union at *B*. Originally the assembly is adjusted so that no load is on the pipe. If the union is then tightened so that its screw, having a lead of 0.15 in., undergoes two full turns, determine the average normal stress developed in the pipe. Assume that the union and couplings at *A* and *C* are rigid. Neglect the size of the union. *Note:* The lead would cause the pipe, when *unloaded*, to shorten 0.15 in. when the union is rotated one revolution.

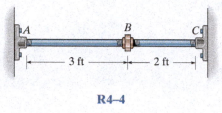

R4–4

R4–5. The force **P** is applied to the bar, which is made from an elastic perfectly plastic material. Construct a graph to show how the force in each section *AB* and *BC* (vertical axis) varies as *P* (horizontal axis) is increased. The bar has cross-sectional areas of 1 in² in region *AB* and 4 in² in region *BC*. Take $\sigma_Y = 30$ ksi.

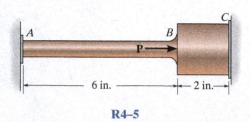

R4–5

R4–6. The 2014-T6 aluminum rod has a diameter of 0.5 in. and is lightly attached to the rigid supports at A and B when $T_1 = 70°F$. If the temperature becomes $T_2 = -10°F$, and an axial force of $P = 16$ lb is applied to the rigid collar as shown, determine the reactions at the rigid supports A and B.

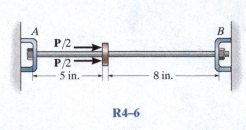

R4–6

R4–7. The 2014-T6 aluminum rod has a diameter of 0.5 in. and is lightly attached to the rigid supports at A and B when $T_1 = 70°F$. Determine the force P that must be applied to the collar so that, when $T = 0°F$, the reaction at B is zero.

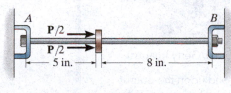

R4–7

*R4–8. The rigid link is supported by a pin at A and two A-36 steel wires, each having an unstretched length of 12 in. and cross-sectional area of 0.0125 in². Determine the force developed in the wires when the link supports the vertical load of 350 lb.

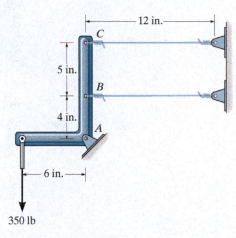

R4–8

R4–9. The joint is made from three A992 steel plates that are bonded together at their seams. Determine the displacement of end A with respect to end B when the joint is subjected to the axial loads. Each plate has a thickness of 5 mm.

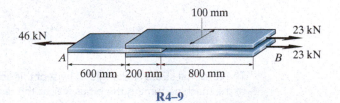

R4–9

CHAPTER 5

(© Jill Fromer/Getty Images)

The torsional stress and angle of twist of this soil auger depend upon the output of the machine turning the bit as well as the resistance of the soil in contact with the shaft.

TORSION

CHAPTER OBJECTIVES

■ In this chapter we will discuss the effects of applying a torsional loading to a long straight member such as a shaft or tube. Initially we will consider the member to have a circular cross section. We will show how to determine both the stress distribution within the member and the angle of twist. The statically indeterminate analysis of shafts and tubes will also be discussed, along with special topics that include those members having noncircular cross sections. Lastly, stress concentrations and residual stress caused by torsional loadings will be given special consideration.

5.1 TORSIONAL DEFORMATION OF A CIRCULAR SHAFT

Torque is a moment that tends to twist a member about its longitudinal axis. Its effect is of primary concern in the design of drive shafts used in vehicles and machinery, and for this reason it is important to be able to determine the stress and the deformation that occur in a shaft when it is subjected to torsional loads.

We can physically illustrate what happens when a torque is applied to a circular shaft by considering the shaft to be made of a highly deformable material such as rubber. When the torque is applied, the longitudinal grid lines originally marked on the shaft, Fig. 5–1a, tend to distort into a helix, Fig. 5–1b, that intersects the circles at equal angles. Also, all the cross sections of the shaft will remain flat—that is, they do not warp or bulge in or out—and radial lines remain straight and rotate during this deformation. Provided the angle of twist is *small*, then the length of the shaft and its radius will remain practically unchanged.

If the shaft is fixed at one end and a torque is applied to its other end, then the dark green shaded plane in Fig. 5–2a will distort into a skewed form as shown. Here a radial line located on the cross section at a distance x from the fixed end of the shaft will rotate through an angle $\phi(x)$. This angle is called the **angle of twist**. It depends on the position x and will vary along the shaft as shown.

In order to understand how this distortion strains the material, we will now isolate a small disk element located at x from the end of the shaft, Fig. 5–2b. Due to the deformation, the front and rear faces of the element will undergo rotation—the back face by $\phi(x)$, and the front face by $\phi(x) + d\phi$. As a result, the *difference* in these rotations, $d\phi$, causes the element to be subjected to a *shear strain*, γ (see Fig. 3–25b).

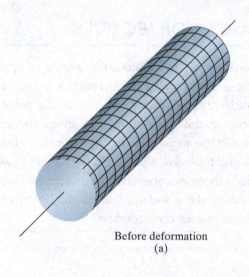

Before deformation
(a)

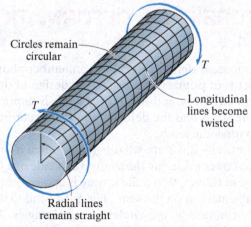

Circles remain circular

T

T

Longitudinal lines become twisted

Radial lines remain straight

After deformation
(b)

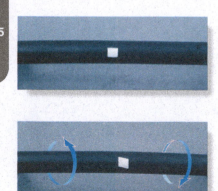

Notice the deformation of the rectangular element when this rubber bar is subjected to a torque.

Fig. 5–1

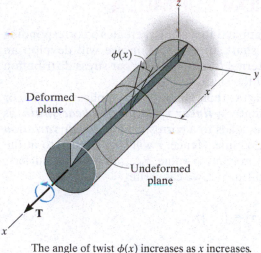

Deformed plane

Undeformed plane

T

The angle of twist $\phi(x)$ increases as x increases.

(a)

The shear strain at points on the cross section increases linearly with ρ, i.e., $\gamma = (\rho/c)\gamma_{max}$.

(b)

Fig. 5–2

This angle (or shear strain) can be related to the angle $d\phi$ by noting that the length of the red arc in Fig. 5–2b is

$$\rho\, d\phi = dx\, \gamma$$

or

$$\gamma = \rho\, \frac{d\phi}{dx} \qquad\qquad (5\text{--}1)$$

Since dx and $d\phi$ are the same for all elements, then $d\phi/dx$ is constant over the cross section, and Eq. 5–1 states that the magnitude of the shear strain varies only with its radial distance ρ from the axis of the shaft. Since $d\phi/dx = \gamma/\rho = \gamma_{max}/c$, then

$$\gamma = \left(\frac{\rho}{c}\right)\gamma_{max} \qquad\qquad (5\text{--}2)$$

In other words, the shear strain within the shaft *varies linearly* along any radial line, from zero at the axis of the shaft to a maximum γ_{max} at its outer boundary, Fig. 5–2b.

5

5.2   THE TORSION FORMULA

When an external torque is applied to a shaft, it creates a corresponding internal torque within the shaft. In this section, we will develop an equation that relates this internal torque to the shear stress distribution acting on the cross section of the shaft.

If the material is linear elastic, then Hooke's law applies, $\tau = G\gamma$, or $\tau_{max} = G\gamma_{max}$, and consequently a ***linear variation in shear strain***, as noted in the previous section, leads to a corresponding ***linear variation in shear stress*** along any radial line. Hence, τ will vary from zero at the shaft's longitudinal axis to a maximum value, τ_{max}, at its outer surface, Fig. 5–3. Therefore, similar to Eq. 5–2, we can write

$$\tau = \left(\frac{\rho}{c}\right)\tau_{max} \tag{5–3}$$

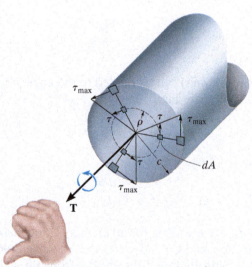

Shear stress varies linearly along each radial line of the cross section.

Fig. 5–3

Since each element of area dA, located at ρ, is subjected to a force of $dF = \tau\, dA$, Fig. 5–3, the torque produced by this force is then $dT = \rho(\tau\, dA)$. For the entire cross section we have

$$T = \int_A \rho(\tau\, dA) = \int_A \rho\left(\frac{\rho}{c}\right)\tau_{max}\, dA \qquad (5\text{–}4)$$

However, τ_{max}/c is constant, and so

$$T = \frac{\tau_{max}}{c} \int_A \rho^2\, dA \qquad (5\text{–}5)$$

The integral represents the **polar moment of inertia** of the shaft's cross-sectional area about the shaft's longitudinal axis. On the next page we will calculate its value, but here we will symbolize its value as J. As a result, the above equation can be rearranged and written in a more compact form, namely,

$$\boxed{\tau_{max} = \frac{Tc}{J}} \qquad (5\text{–}6)$$

Here

τ_{max} = the maximum shear stress in the shaft, which occurs at its outer surface

T = the resultant *internal torque* acting at the cross section. Its value is determined from the method of sections and the equation of moment equilibrium applied about the shaft's longitudinal axis

J = the polar moment of inertia of the cross-sectional area

c = the outer radius of the shaft

If Eq. 5–6 is substituted into Eq. 5–3, the shear stress at the intermediate distance ρ on the cross section can be determined.

$$\boxed{\tau = \frac{T\rho}{J}} \qquad (5\text{–}7)$$

Either of the above two equations is often referred to as the **torsion formula**. Recall that it is used only if the shaft has a circular cross section and the material is homogeneous and behaves in a linear elastic manner, since the derivation of Eq. 5–3 is based on Hooke's law.

The shaft attached to the center of this wheel is subjected to a torque, and the maximum stress it creates must be resisted by the shaft to prevent failure.

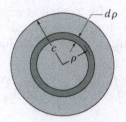

Fig. 5–4

Polar Moment of Inertia. If the shaft has a *solid* circular cross section, the polar moment of inertia J can be determined using an area element in the form of a *differential ring* or annulus having a thickness $d\rho$ and circumference $2\pi\rho$, Fig. 5–4. For this ring, $dA = 2\pi\rho\, d\rho$, and so

$$J = \int_A \rho^2\, dA = \int_0^c \rho^2 (2\pi\rho\, d\rho)$$

$$= 2\pi \int_0^c \rho^3\, d\rho = 2\pi\left(\frac{1}{4}\right)\rho^4 \Big|_0^c$$

$$\boxed{J = \frac{\pi}{2} c^4} \tag{5–8}$$

Solid Section

Note that J is always positive. Common units used for its measurement are mm^4 or in^4.

If a shaft has a tubular cross section, with inner radius c_i and outer radius c_o, Fig. 5–5, then from Eq. 5–8 we can determine its polar moment of inertia by subtracting J for a shaft of radius c_i from that determined for a shaft of radius c_o. The result is

$$\boxed{J = \frac{\pi}{2}(c_o^4 - c_i^4)} \tag{5–9}$$

Tube

Fig. 5–5

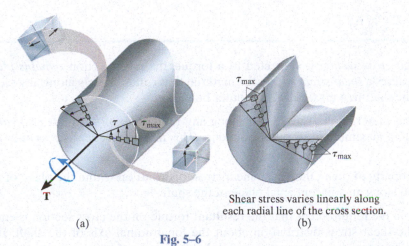

Shear stress varies linearly along
each radial line of the cross section.

(a)   (b)

Fig. 5–6

Shear Stress Distribution.

If an element of material on the cross section of the shaft or tube is isolated, then due to the complementary property of shear, equal shear stresses must also act on four of its adjacent faces, as shown in Fig. 5–6a. As a result, *the internal torque T develops a linear distribution of shear stress along each radial line in the plane of the cross-sectional area, and also an associated shear-stress distribution is developed along an axial plane*, Fig. 5–6b. It is interesting to note that because of this axial distribution of shear stress, shafts made of wood tend to *split* along the axial plane when subjected to excessive torque, Fig. 5–7. This is because wood is an anisotropic material, whereby its shear resistance parallel to its grains or fibers, directed along the axis of the shaft, is much less than its resistance perpendicular to the fibers within the plane of the cross section.

The tubular drive shaft for this truck was subjected to an excessive torque, resulting in failure caused by yielding of the material. Engineers deliberately design drive shafts to fail before torsional damage can occur to parts of the engine or transmission.

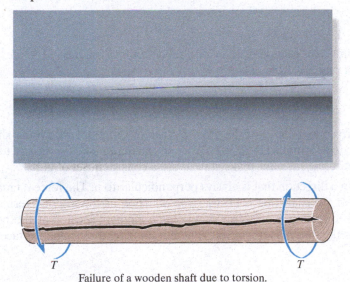

Failure of a wooden shaft due to torsion.

Fig. 5–7

IMPORTANT POINTS

- When a shaft having a *circular cross section* is subjected to a torque, the cross section *remains plane* while radial lines rotate. This causes a *shear strain* within the material that *varies linearly* along any radial line, from zero at the axis of the shaft to a maximum at its outer boundary.

- For linear elastic homogeneous material, the *shear stress* along any radial line of the shaft also *varies linearly*, from zero at its axis to a maximum at its outer boundary. This maximum shear stress *must not* exceed the proportional limit.

- Due to the complementary property of shear, the linear shear stress distribution within the plane of the cross section is also distributed along an adjacent axial plane of the shaft.

- The torsion formula is based on the requirement that the resultant torque on the cross section is equal to the torque produced by the shear stress distribution about the longitudinal axis of the shaft. It is required that the shaft or tube have a *circular* cross section and that it is made of *homogeneous* material which has *linear elastic* behavior.

PROCEDURE FOR ANALYSIS

The torsion formula can be applied using the following procedure.

Internal Torque.

- Section the shaft perpendicular to its axis at the point where the shear stress is to be determined, and use the necessary free-body diagram and equations of equilibrium to obtain the internal torque at the section.

Section Property.

- Calculate the polar moment of inertia of the cross-sectional area. For a solid section of radius c, $J = \pi c^4/2$, and for a tube of outer radius c_o and inner radius c_i, $J = \pi(c_o^4 - c_i^4)/2$.

Shear Stress.

- Specify the radial distance ρ, measured from the center of the cross section to the point where the shear stress is to be found. Then apply the torsion formula $\tau = T\rho/J$, or if the maximum shear stress is to be determined use $\tau_{\max} = Tc/J$. When substituting the data, make sure to use a consistent set of units.

- The shear stress acts on the cross section in a direction that is always perpendicular to ρ. The force it creates must contribute a torque about the axis of the shaft that is in the *same direction* as the internal resultant torque **T** acting on the section. Once this direction is established, a volume element located at the point where τ is determined can be isolated, and the direction of τ acting on the remaining three adjacent faces of the element can be shown.

EXAMPLE 5.1

The solid shaft and tube shown in Fig. 5–8 are made of a material having an allowable shear stress of 75 MPa. Determine the maximum torque that can be applied to each cross section, and show the stress acting on a small element of material at point A of the shaft, and points B and C of the tube.

SOLUTION

Section Properties.   The polar moments of inertia for the solid and tubular shafts are

$$J_s = \frac{\pi}{2} c^4 = \frac{\pi}{2} (0.1 \text{ m})^4 = 0.1571(10^{-3}) \text{ m}^4$$

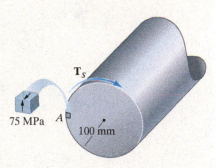

$$J_t = \frac{\pi}{2} (c_o^4 - c_i^4) = \frac{\pi}{2} \left[(0.1 \text{ m})^4 - (0.075 \text{ m})^4 \right] = 0.1074(10^{-3}) \text{ m}^4$$

Shear Stress.   The maximum torque in each case is

$$(\tau_{\max})_s = \frac{Tc}{J}; \qquad 75(10^6) \text{ N/m}^2 = \frac{T_s(0.1 \text{ m})}{0.1571(10^{-3}) \text{ m}^4}$$

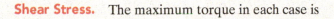

$$T_s = 118 \text{ kN} \cdot \text{m} \qquad\qquad Ans.$$

$$(\tau_{\max})_t = \frac{Tc}{J}; \qquad 75(10^6) \text{ N/m}^2 = \frac{T_t(0.1 \text{ m})}{0.1074(10^{-3}) \text{ m}^4}$$

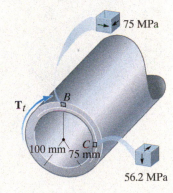

$$T_t = 80.5 \text{ kN} \cdot \text{m} \qquad\qquad Ans.$$

Also, the shear stress at the inner radius of the tube is

$$(\tau_i)_t = \frac{80.5(10^3) \text{ N} \cdot \text{m} (0.075 \text{ m})}{0.1074(10^{-3}) \text{ m}^4} = 56.2 \text{ MPa}$$

Fig. 5–8

These results are shown acting on small elements in Fig. 5–8. Notice how the shear stress on the front (shaded) face of the element contributes to the torque. As a consequence, shear stress components act on the other three faces. No shear stress acts on the outer surface of the shaft or tube or on the inner surface of the tube because it must be stress free.

5

EXAMPLE 5.2

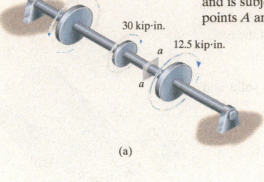

The 1.5-in.-diameter shaft shown in Fig. 5–9a is supported by two bearings and is subjected to three torques. Determine the shear stress developed at points A and B, located at section a–a of the shaft, Fig. 5–9c.

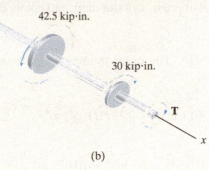

(a)

(b)

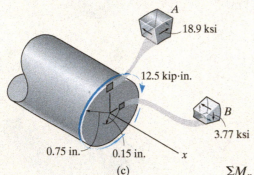

(c)

Fig. 5–9

SOLUTION

Internal Torque. Since the bearing reactions do not offer resistance to shaft rotation, the applied torques satisfy moment equilibrium about the shaft's axis.

The internal torque at section a–a will be determined from the free-body diagram of the left segment, Fig. 5–9b. We have

$$\Sigma M_x = 0; \quad 42.5 \text{ kip} \cdot \text{in.} - 30 \text{ kip} \cdot \text{in.} - T = 0 \quad T = 12.5 \text{ kip} \cdot \text{in.}$$

Section Property. The polar moment of inertia for the shaft is

$$J = \frac{\pi}{2}(0.75 \text{ in.})^4 = 0.497 \text{ in.}^4$$

Shear Stress. Since point A is at $\rho = c = 0.75$ in.,

$$\tau_A = \frac{Tc}{J} = \frac{(12.5 \text{ kip} \cdot \text{in.})(0.75 \text{ in.})}{(0.497 \text{ in.}^4)} = 18.9 \text{ ksi} \qquad \textit{Ans.}$$

Likewise for point B, at $\rho = 0.15$ in., we have

$$\tau_B = \frac{T\rho}{J} = \frac{(12.5 \text{ kip} \cdot \text{in.})(0.15 \text{ in.})}{(0.497 \text{ in.}^4)} = 3.77 \text{ ksi} \qquad \textit{Ans.}$$

NOTE: The directions of these stresses on each element at A and B, Fig. 5–9c, are established on the planes of each of these elements, so that they match the required clockwise torque.

EXAMPLE 5.3

The pipe shown in Fig. 5–10a has an inner radius of 40 mm and an outer radius of 50 mm. If its end is tightened against the support at A using the torque wrench, determine the shear stress developed in the material at the inner and outer walls along the central portion of the pipe.

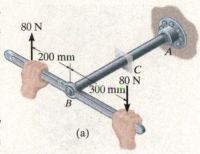

(a)

SOLUTION

Internal Torque. A section is taken at the intermediate location C along the pipe's axis, Fig. 5–10b. The only unknown at the section is the internal torque **T**. We require

$$\Sigma M_x = 0; \qquad 80 \text{ N}(0.3 \text{ m}) + 80 \text{ N}(0.2 \text{ m}) - T = 0$$

$$T = 40 \text{ N} \cdot \text{m}$$

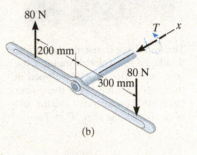

(b)

Section Property. The polar moment of inertia for the pipe's cross-sectional area is

$$J = \frac{\pi}{2} \left[(0.05 \text{ m})^4 - (0.04 \text{ m})^4 \right] = 5.796(10^{-6}) \text{ m}^4$$

Shear Stress. For any point lying on the outside surface of the pipe, $\rho = c_o = 0.05$ m, we have

$$\tau_o = \frac{Tc_o}{J} = \frac{40 \text{ N} \cdot \text{m}(0.05 \text{ m})}{5.796(10^{-6}) \text{ m}^4} = 0.345 \text{ MPa} \qquad \textit{Ans.}$$

And for any point located on the inside surface, $\rho = c_i = 0.04$ m, and so

$$\tau_i = \frac{Tc_i}{J} = \frac{40 \text{ N} \cdot \text{m}(0.04 \text{ m})}{5.796(10^{-6}) \text{ m}^4} = 0.276 \text{ MPa} \qquad \textit{Ans.}$$

These results are shown on two small elements in Fig. 5–10c.

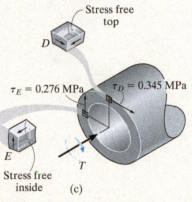

(c)

Fig. 5–10

NOTE: Since the top face of D and the inner face of E are in stress-free regions, no shear stress can exist on these faces or on the other corresponding faces of the elements.

5.3　POWER TRANSMISSION

Shafts and tubes having circular cross sections are often used to transmit power developed by a machine. When used for this purpose, they are subjected to a torque that depends on both the power generated by the machine and the angular speed of the shaft. *Power* is defined as the work performed per unit of time. Also, the work transmitted by a rotating shaft equals the torque applied times the angle of rotation. Therefore, if during an instant of time dt an applied torque $\mathbf{T}$ causes the shaft to rotate $d\theta$, then the work done is $Td\theta$ and the instantaneous power is

$$P = \frac{T\,d\theta}{dt}$$

Since the shaft's angular velocity is $\omega = d\theta/dt$, then the power is

$$P = T\omega \tag{5–10}$$

In the SI system, power is expressed in *watts* when torque is measured in newton-meters $(\text{N} \cdot \text{m})$ and ω is in radians per second (rad/s) $(1\text{ W} = 1\text{ N} \cdot \text{m/s})$. In the FPS system, the basic units of power are foot-pounds per second $(\text{ft} \cdot \text{lb/s})$; however, *horsepower* (hp) is often used in engineering practice, where

$$1\text{ hp} = 550\text{ ft} \cdot \text{lb/s}$$

For machinery, the *frequency* of a shaft's rotation, f, is often reported. This is a measure of the number of revolutions or "cycles" the shaft makes per second and is expressed in hertz $(1\text{ Hz} = 1\text{ cycle/s})$. Since 1 cycle $= 2\pi$ rad, then $\omega = 2\pi f$, and so the above equation for power can also be written as

$$P = 2\pi f T \tag{5–11}$$

Shaft Design.　When the power transmitted by a shaft and its frequency of rotation are known, the torque developed in the shaft can be determined from Eq. 5–11, that is, $T = P/2\pi f$. Knowing T and the allowable shear stress for the material, τ_{allow}, we can then determine the size of the shaft's cross section using the torsion formula. Specifically, the design or geometric parameter J/c becomes

$$\frac{J}{c} = \frac{T}{\tau_{\text{allow}}} \tag{5–12}$$

For a *solid shaft*, $J = (\pi/2)c^4$, and thus, upon substitution, a *unique value* for the shaft's radius c can be determined. If the shaft is *tubular*, so that $J = (\pi/2)(c_o^4 - c_i^4)$, design permits a wide range of possibilities for the solution. This is because an *arbitrary choice* can be made for either c_o or c_i and the other radius can then be determined from Eq. 5–12.

The belt drive transmits the torque developed by an electric motor to the shaft at A. The stress developed in the shaft depends upon the power transmitted by the motor and the rate of rotation of the shaft. $P = T\omega$.

EXAMPLE   5.4

A solid steel shaft AB, shown in Fig. 5–11, is to be used to transmit 5 hp from the motor M to which it is attached. If the shaft rotates at $\omega = 175$ rpm and the steel has an allowable shear stress of $\tau_{allow} = 14.5$ ksi, determine the required diameter of the shaft to the nearest $\frac{1}{8}$ in.

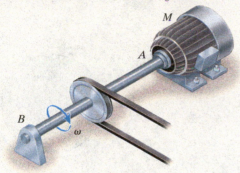

Fig. 5–11

SOLUTION

The torque on the shaft is determined from Eq. 5–10, that is, $P = T\omega$. Expressing P in foot-pounds per second and ω in radians/second, we have

$$P = 5 \text{ hp} \left(\frac{550 \text{ ft} \cdot \text{lb/s}}{1 \text{ hp}} \right) = 2750 \text{ ft} \cdot \text{lb/s}$$

$$\omega = \frac{175 \text{ rev}}{\text{min}} \left(\frac{2\pi \text{ rad}}{1 \text{ rev}} \right) \left(\frac{1 \text{ min}}{60 \text{ s}} \right) = 18.33 \text{ rad/s}$$

Thus,

$$P = T\omega; \qquad 2750 \text{ ft} \cdot \text{lb/s} = T(18.33 \text{ rad/s})$$

$$T = 150.1 \text{ ft} \cdot \text{lb}$$

Applying Eq. 5–12,

$$\frac{J}{c} = \frac{\pi}{2} \frac{c^4}{c} = \frac{T}{\tau_{allow}}$$

$$c = \left(\frac{2T}{\pi \tau_{allow}} \right)^{1/3} = \left(\frac{2(150.1 \text{ ft} \cdot \text{lb})(12 \text{ in./ft})}{\pi (14\ 500 \text{ lb/in}^2)} \right)^{1/3}$$

$$c = 0.429 \text{ in.}$$

Since $2c = 0.858$ in., select a shaft having a diameter of

$$d = \frac{7}{8} \text{ in.} = 0.875 \text{ in.} \qquad\qquad Ans.$$

5

PRELIMINARY PROBLEMS

P5–1.   Determine the internal torque at each section and show the shear stress on differential volume elements located at $A, B, C,$ and D.

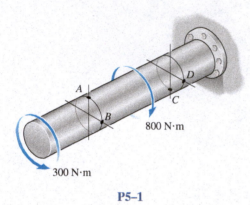

P5–1

P5–3.   The solid and hollow shafts are each subjected to the torque T. In each case, sketch the shear stress distribution along the two radial lines.

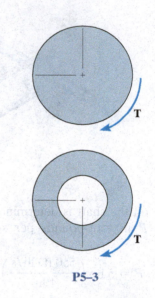

P5–3

P5–2.   Determine the internal torque at each section and show the shear stress on differential volume elements located at $A, B, C,$ and D.

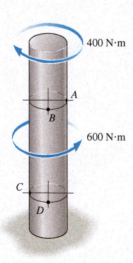

P5–2

P5–4.   The motor delivers 10 hp to the shaft. If it rotates at 1200 rpm, determine the torque produced by the motor.

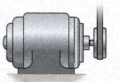

P5–4

FUNDAMENTAL PROBLEMS

F5–1. The solid circular shaft is subjected to an internal torque of $T = 5$ kN·m. Determine the shear stress at points A and B. Represent each state of stress on a volume element.

F5–3. The shaft is hollow from A to B and solid from B to C. Determine the maximum shear stress in the shaft. The shaft has an outer diameter of 80 mm, and the thickness of the wall of the hollow segment is 10 mm.

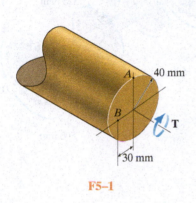

F5–1

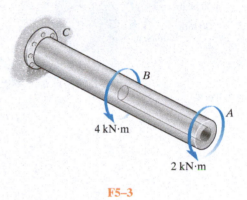

F5–3

F5–2. The hollow circular shaft is subjected to an internal torque of $T = 10$ kN·m. Determine the shear stress at points A and B. Represent each state of stress on a volume element.

F5–4. Determine the maximum shear stress in the 40-mm-diameter shaft.

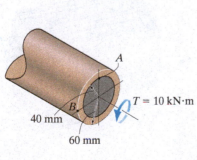

F5–2

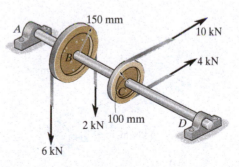

F5–4

5

F5–5. Determine the maximum shear stress in the shaft at section $a-a$.

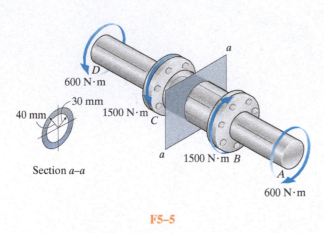

F5–5

F5–6. Determine the shear stress at point A on the surface of the shaft. Represent the state of stress on a volume element at this point. The shaft has a radius of 40 mm.

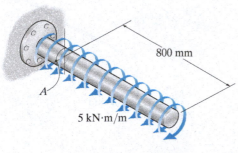

F5–6

F5–7. The solid 50-mm-diameter shaft is subjected to the torques applied to the gears. Determine the absolute maximum shear stress in the shaft.

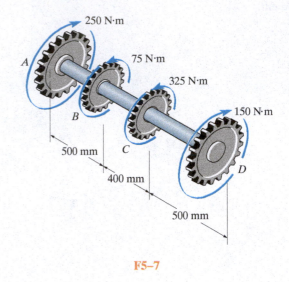

F5–7

F5–8. The gear motor can develop 3 hp when it turns at 150 rev/min. If the allowable shear stress for the shaft is $\tau_{allow} = 12$ ksi, determine the smallest diameter of the shaft to the nearest $\frac{1}{8}$ in. that can be used.

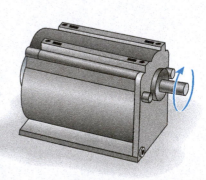

F5–8

PROBLEMS

5–1. The solid shaft of radius r is subjected to a torque **T**. Determine the radius r' of the inner core of the shaft that resists one-half of the applied torque $(T/2)$. Solve the problem two ways: (a) by using the torsion formula, (b) by finding the resultant of the shear-stress distribution.

5–2. The solid shaft of radius r is subjected to a torque **T**. Determine the radius r' of the inner core of the shaft that resists one-quarter of the applied torque $(T/4)$. Solve the problem two ways: (a) by using the torsion formula, (b) by finding the resultant of the shear-stress distribution.

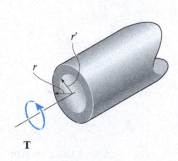

Probs. 5–1/2

5–3. A shaft is made of an aluminum alloy having an allowable shear stress of $\tau_{\text{allow}} = 100$ MPa. If the diameter of the shaft is 100 mm, determine the maximum torque **T** that can be transmitted. What would be the maximum torque **T'** if a 75-mm-diameter hole were bored through the shaft? Sketch the shear-stress distribution along a radial line in each case.

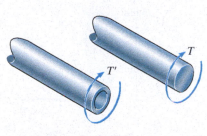

Prob. 5–3

*5–4. The copper pipe has an outer diameter of 40 mm and an inner diameter of 37 mm. If it is tightly secured to the wall and three torques are applied to it, determine the absolute maximum shear stress developed in the pipe.

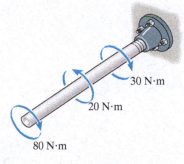

Prob. 5–4

5–5. The copper pipe has an outer diameter of 2.50 in. and an inner diameter of 2.30 in. If it is tightly secured to the wall and three torques are applied to it, determine the shear stress developed at points A and B. These points lie on the pipe's outer surface. Sketch the shear stress on volume elements located at A and B.

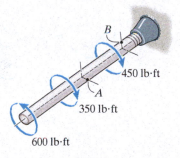

Prob. 5–5

5

5–6. The solid aluminum shaft has a diameter of 50 mm and an allowable shear stress of $\tau_{allow} = 60$ MPa. Determine the largest torque T_1 that can be applied to the shaft if it is also subjected to the other torsional loadings. It is required that $\mathbf{T_1}$ act in the direction shown. Also, determine the maximum shear stress within regions CD and DE.

5–7. The solid aluminum shaft has a diameter of 50 mm. Determine the absolute maximum shear stress in the shaft and sketch the shear-stress distribution along a radial line of the shaft where the shear stress is maximum. Set $T_1 = 2000$ N · m.

5–9. The solid shaft is fixed to the support at C and subjected to the torsional loadings. Determine the shear stress at points A and B on the surface, and sketch the shear stress on volume elements located at these points.

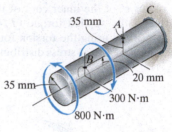

Prob. 5–9

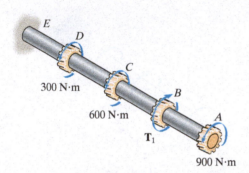

Probs. 5–6/7

*5–8. The solid 30-mm-diameter shaft is used to transmit the torques applied to the gears. Determine the absolute maximum shear stress in the shaft.

5–10. The link acts as part of the elevator control for a small airplane. If the attached aluminum tube has an inner diameter of 25 mm and a wall thickness of 5 mm, determine the maximum shear stress in the tube when the cable force of 600 N is applied to the cables. Also, sketch the shear-stress distribution over the cross section.

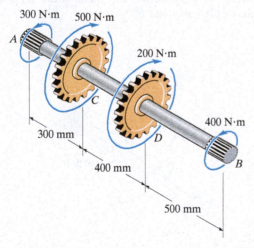

Prob. 5–8

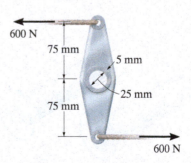

Prob. 5–10

5–11. The assembly consists of two sections of galvanized steel pipe connected together using a reducing coupling at *B*. The smaller pipe has an outer diameter of 0.75 in. and an inner diameter of 0.68 in., whereas the larger pipe has an outer diameter of 1 in. and an inner diameter of 0.86 in. If the pipe is tightly secured to the wall at *C*, determine the maximum shear stress in each section of the pipe when the couple is applied to the handles of the wrench.

5–14. A steel tube having an outer diameter of 2.5 in. is used to transmit 9 hp when turning at 27 rev/min. Determine the inner diameter d of the tube to the nearest $\frac{1}{8}$ in. if the allowable shear stress is $\tau_{allow} = 10$ ksi.

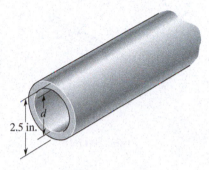

Prob. 5–14

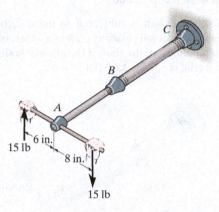

Prob. 5–11

*****5–12.** The shaft has an outer diameter of 100 mm and an inner diameter of 80 mm. If it is subjected to the three torques, determine the absolute maximum shear stress in the shaft. The smooth bearings *A* and *B* do not resist torque.

5–13. The shaft has an outer diameter of 100 mm and an inner diameter of 80 mm. If it is subjected to the three torques, plot the shear stress distribution along a radial line for the cross section within region *CD* of the shaft. The smooth bearings at *A* and *B* do not resist torque.

5–15. If the gears are subjected to the torques shown, determine the maximum shear stress in the segments *AB* and *BC* of the A-36 steel shaft. The shaft has a diameter of 40 mm.

*****5–16.** If the gears are subjected to the torques shown, determine the required diameter of the A-36 steel shaft to the nearest mm if $\tau_{allow} = 60$ MPa.

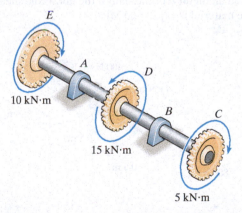

Probs. 5–12/13

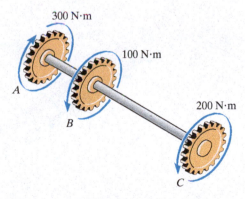

Probs. 5–15/16

5–17. The rod has a diameter of 1 in. and a weight of 10 lb/ft. Determine the maximum torsional stress in the rod at a section located at *A* due to the rod's weight.

5–18. The rod has a diameter of 1 in. and a weight of 15 lb/ft. Determine the maximum torsional stress in the rod at a section located at *B* due to the rod's weight.

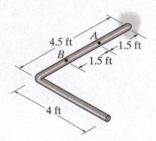

Probs. 5–17/18

5–19. The copper pipe has an outer diameter of 3 in. and an inner diameter of 2.5 in. If it is tightly secured to the wall at *C* and a uniformly distributed torque is applied to it as shown, determine the shear stress at points *A* and *B*. These points lie on the pipe's outer surface. Sketch the shear stress on volume elements located at *A* and *B*.

*5–20. The copper pipe has an outer diameter of 3 in. and an inner diameter of 2.50 in. If it is tightly secured to the wall at *C* and it is subjected to the uniformly distributed torque along its entire length, determine the absolute maximum shear stress in the pipe. Discuss the validity of this result.

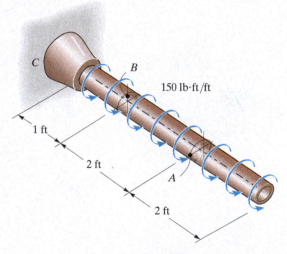

Probs. 5–19/20

5–21. The 60-mm-diameter solid shaft is subjected to the distributed and concentrated torsional loadings shown. Determine the shear stress at points *A* and *B*, and sketch the shear stress on volume elements located at these points.

5–22. The 60-mm-diameter solid shaft is subjected to the distributed and concentrated torsional loadings shown. Determine the absolute maximum and minimum shear stresses on the shaft's surface, and specify their locations, measured from the fixed end *C*.

5–23. The solid shaft is subjected to the distributed and concentrated torsional loadings shown. Determine the required diameter *d* of the shaft if the allowable shear stress for the material is $\tau_{allow} = 1.6$ MPa.

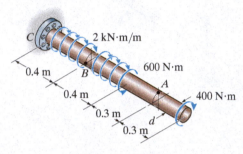

Probs. 5–21/22/23

*5–24. The 60-mm-diameter solid shaft is subjected to the distributed and concentrated torsional loadings shown. Determine the absolute maximum and minimum shear stresses on the shaft's surface and specify their locations, measured from the free end.

5–25. The solid shaft is subjected to the distributed and concentrated torsional loadings shown. Determine the required diameter *d* of the shaft if the allowable shear stress for the material is $\tau_{allow} = 60$ MPa.

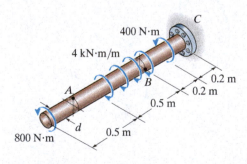

Probs. 5–24/25

5–26. When drilling a well at constant angular velocity, the bottom end of the drill pipe encounters a torsional resistance T_A. Also, soil along the sides of the pipe creates a distributed frictional torque along its length, varying uniformly from zero at the surface B to t_A at A. Determine the minimum torque T_B that must be supplied by the drive unit to overcome the resisting torques, and calculate the maximum shear stress in the pipe. The pipe has an outer radius r_o and an inner radius r_i.

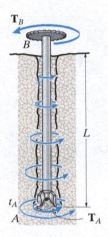

Prob. 5–26

5–27. The drive shaft AB of an automobile is made of a steel having an allowable shear stress of $\tau_{allow} = 8$ ksi. If the outer diameter of the shaft is 2.5 in. and the engine delivers 200 hp to the shaft when it is turning at 1140 rev/min, determine the minimum required thickness of the shaft's wall.

*5–28. The drive shaft AB of an automobile is to be designed as a thin-walled tube. The engine delivers 150 hp when the shaft is turning at 1000 rev/min. Determine the minimum thickness of the shaft's wall if the shaft's outer diameter is 2.5 in. The material has an allowable shear stress of $\tau_{allow} = 7$ ksi.

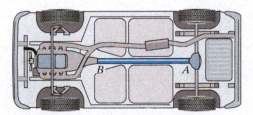

Probs. 5–27/28

5–29. The shaft is subjected to a distributed torque along its length of $t = (10x^2)$ N·m/m, where x is in meters. If the maximum stress in the shaft is to remain constant at 80 MPa, determine the required variation of the radius c of the shaft for $0 \le x \le 3$ m.

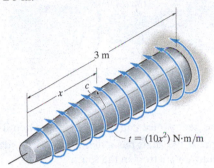

Prob. 5–29

5–30. The motor delivers 50 hp while turning at a constant rate of 1350 rpm at A. Using the belt and pulley system this loading is delivered to the steel blower shaft BC. Determine to the nearest $\frac{1}{8}$ in. the smallest diameter of this shaft if the allowable shear stress for steel is $\tau_{allow} = 12$ ksi.

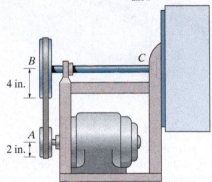

Prob. 5–30

5–31. The solid steel shaft AC has a diameter of 25 mm and is supported by smooth bearings at D and E. It is coupled to a motor at C, which delivers 3 kW of power to the shaft while it is turning at 50 rev/s. If gears A and B remove 1 kW and 2 kW, respectively, determine the maximum shear stress in the shaft within regions AB and BC. The shaft is free to turn in its support bearings D and E.

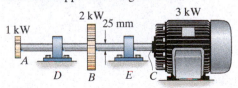

Prob. 5–31

*5–32   The pump operates using the motor that has a power of 85 W. If the impeller at B is turning at 150 rev/min, determine the maximum shear stress in the 20-mm-diameter transmission shaft at A.

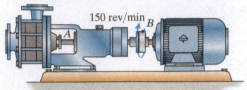

150 rev/min

Prob. 5–32

5–33.   The gear motor can develop $\frac{1}{10}$ hp when it turns at 300 rev/min. If the shaft has a diameter of $\frac{1}{2}$ in., determine the maximum shear stress in the shaft.

5–34.   The gear motor can develop $\frac{1}{10}$ hp when it turns at 80 rev/min. If the allowable shear stress for the shaft is $\tau_{allow} = 4$ ksi, determine the smallest diameter of the shaft to the nearest $\frac{1}{8}$ in. that can be used.

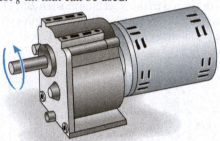

Probs. 5–33/34

5–35.   The gear motor can develop $\frac{1}{4}$ hp when it turns at 600 rev/min. If the shaft has a diameter of $\frac{1}{2}$ in., determine the maximum shear stress in the shaft.

*5–36.   The gear motor can develop 2 hp when it turns at 150 rev/min. If the allowable shear stress for the shaft is $\tau_{allow} = 8$ ksi, determine the smallest diameter of the shaft to the nearest $\frac{1}{8}$ in. that can be used.

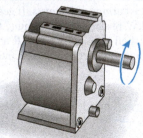

Probs. 5–35/36

5–37.   The 6-hp reducer motor can turn at 1200 rev/min. If the allowable shear stress for the shaft is $\tau_{allow} = 6$ ksi, determine the smallest diameter of the shaft to the nearest $\frac{1}{16}$ in. that can be used.

5–38.   The 6-hp reducer motor can turn at 1200 rev/min. If the shaft has a diameter of $\frac{5}{8}$ in., determine the maximum shear stress in the shaft.

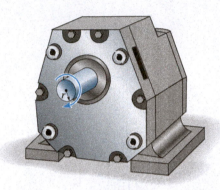

Probs. 5–37/38

5–39.   The solid steel shaft DF has a diameter of 25 mm and is supported by smooth bearings at D and E. It is coupled to a motor at F, which delivers 12 kW of power to the shaft while it is turning at 50 rev/s. If gears A, B, and C remove 3 kW, 4 kW, and 5 kW respectively, determine the maximum shear stress in the shaft within regions CF and BC. The shaft is free to turn in its support bearings D and E.

*5–40.   The solid steel shaft DF has a diameter of 25 mm and is supported by smooth bearings at D and E. It is coupled to a motor at F, which delivers 12 kW of power to the shaft while it is turning at 50 rev/s. If gears A, B, and C remove 3 kW, 4 kW, and 5 kW respectively, determine the absolute maximum shear stress in the shaft.

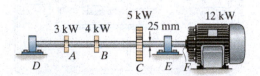

Probs. 5–39/40

5–41. The A-36 steel tubular shaft is 2 m long and has an outer diameter of 50 mm. When it is rotating at 40 rad/s, it transmits 25 kW of power from the motor M to the pump P. Determine the smallest thickness of the tube if the allowable shear stress is $\tau_{allow} = 80$ MPa.

5–42. The A-36 solid steel shaft is 2 m long and has a diameter of 60 mm. It is required to transmit 60 kW of power from the motor M to the pump P. Determine the smallest angular velocity the shaft if the allowable shear stress is $\tau_{allow} = 80$ MPa.

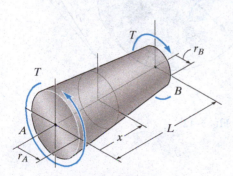

Probs. 5–41/42

5–43. The solid shaft has a linear taper from r_A at one end to r_B at the other. Derive an equation that gives the maximum shear stress in the shaft at a location x along the shaft's axis.

*5–44. The 1-in.-diameter bent rod is subjected to the load shown. Determine the maximum torsional stress in the rod at a section located through A.

5–45. The 1-in.-diameter bent rod is subjected to the load shown. Determine the maximum torsional stress in the rod at B.

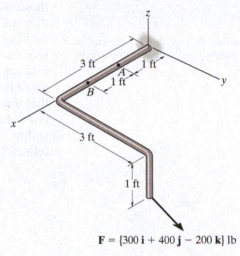

$\mathbf{F} = \{300\ \mathbf{i} + 400\ \mathbf{j} - 200\ \mathbf{k}\}$ lb

Probs. 5–44/45

5–46. A motor delivers 500 hp to the shaft, which is tubular and has an outer diameter of 2 in. If it is rotating at 200 rad/s, determine its largest inner diameter to the nearest $\frac{1}{8}$ in. if the allowable shear stress for the material is $\tau_{allow} = 25$ ksi.

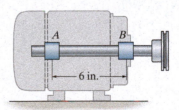

Prob. 5–46

Prob. 5–43

5.4   ANGLE OF TWIST

In this section we will develop a formula for determining the *angle of twist* ϕ (phi) of one end of a shaft with respect to its other end. To generalize this development, we will assume the shaft has a circular cross section that can gradually vary along its length, Fig. 5–12a. Also, the material is assumed to be homogeneous and to behave in a linear elastic manner when the torque is applied. As in the case of an axially loaded bar, we will neglect the localized deformations that occur at points of application of the torques and where the cross section changes abruptly. By Saint-Venant's principle, these effects occur within small regions of the shaft's length, and generally they will have only a slight effect on the final result.

Using the method of sections, a differential disk of thickness dx, located at position x, is isolated from the shaft, Fig. 5–12b. At this location, the internal torque is $T(x)$, since the external loading may cause it to change along the shaft. Due to $T(x)$, the disk will twist, such that the *relative rotation* of one of its faces with respect to the other face is $d\phi$. As a result an element of material located at an arbitrary radius ρ within the disk will undergo a shear strain γ. The values of γ and $d\phi$ are related by Eq. 5–1, namely,

$$d\phi = \gamma \frac{dx}{\rho} \qquad (5\text{–}13)$$

Long shafts subjected to torsion can, in some cases, have a noticeable elastic twist.

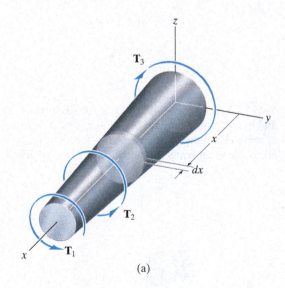

(a)

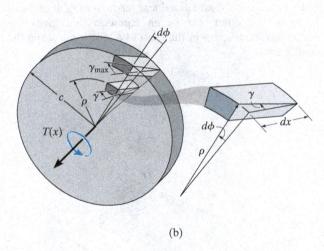

(b)

Fig. 5–12

Since Hooke's law, $\gamma = \tau/G$, applies and the shear stress can be expressed in terms of the applied torque using the torsion formula $\tau = T(x)\rho/J(x)$, then $\gamma = T(x)\rho/J(x)G(x)$. Substituting this into Eq. 5–13, the angle of twist for the disk is therefore

$$d\phi = \frac{T(x)}{J(x)G(x)}\,dx$$

Integrating over the entire length L of the shaft, we can obtain the angle of twist for the entire shaft, namely,

$$\phi = \int_0^L \frac{T(x)\,dx}{J(x)G(x)} \qquad (5\text{–}14)$$

When calculating both the stress and the angle of twist of this soil auger, it is necessary to consider the variable torsional loading which acts along its length.

Here

ϕ = the angle of twist of one end of the shaft with respect to the other end, measured in radians

$T(x)$ = the *internal torque* at the arbitrary position x, found from the method of sections and the equation of moment equilibrium applied about the shaft's axis

$J(x)$ = the shaft's polar moment of inertia expressed as a function of x

$G(x)$ = the shear modulus of elasticity for the material expressed as a function of x

Constant Torque and Cross-Sectional Area. Usually in engineering practice the material is homogeneous so that G is constant. Also, the cross-sectional area and the external torque are constant along the length of the shaft, Fig. 5–13. When this is the case, the internal torque $T(x) = T$, the polar moment of inertia $J(x) = J$, and Eq. 5–14 can be integrated, which gives

$$\phi = \frac{TL}{JG} \qquad (5\text{–}15)$$

Note the similarities between the above two equations and those for an axially loaded bar.

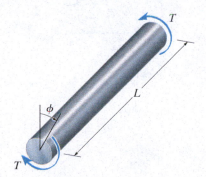

Fig. 5–13

Fig. 5–14

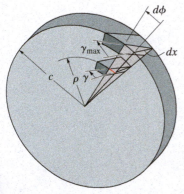

The shear strain at points on the cross section increases linearly with ρ, i.e., $\gamma = (\rho/c)\gamma_{max}$.

(b)

Fig. 5–12 (Repeated)

Equation 5–15 is often used to determine the shear modulus of elasticity, G, of a material. To do so, a specimen of known length and diameter is placed in a torsion testing machine like the one shown in Fig. 5–14. The applied torque T and angle of twist ϕ are then measured along the length L. From Eq. 5–15, we get $G = TL/J\phi$. To obtain a more reliable value of G, several of these tests are performed and the average value is used.

Multiple Torques.   If the shaft is subjected to several different torques, or the cross-sectional area or shear modulus changes abruptly from one region of the shaft to the next, as in Fig. 5–12, then Eq. 5–15 should be applied to each segment of the shaft where these quantities are all constant. The angle of twist of one end of the shaft with respect to the other is found from the algebraic addition of the angles of twist of each segment. For this case,

$$\phi = \sum \frac{TL}{JG} \qquad (5\text{–}16)$$

Sign Convention. The best way to apply this equation is to use a sign convention for both the internal torque and the angle of twist of one end of the shaft with respect to the other end. To do this, we will apply the right-hand rule, whereby both the torque and angle will be *positive*, provided the *thumb* is directed *outward* from the shaft while the fingers curl in the direction of the torque, Fig. 5–15.

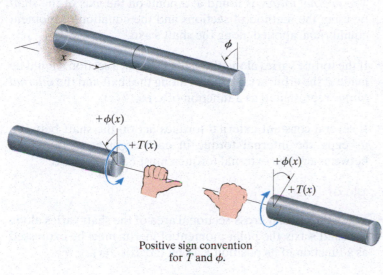

Positive sign convention for T and ϕ.

Fig. 5–15

> ## IMPORTANT POINT

- When applying Eq. 5–14 to determine the angle of twist, it is important that the applied torques do not cause yielding of the material, and that the material is homogeneous and behaves in a linear elastic manner.

5

▶ PROCEDURE FOR ANALYSIS

The angle of twist of one end of a shaft or tube with respect to the other end can be determined using the following procedure.

Internal Torque.

- The *internal torque* is found at a point on the axis of the shaft by using the method of sections and the equation of moment equilibrium, applied along the shaft's axis.

- If the torque varies along the shaft's length, a section should be made at the arbitrary position x along the shaft and the *internal torque* represented as a function of x, i.e., $T(x)$.

- If several constant external torques act on the shaft between its ends, the internal torque in each *segment* of the shaft, between any two external torques, must be determined.

Angle of Twist.

- When the circular cross-sectional area of the shaft varies along the shaft's axis, the polar moment of inertia must be expressed as a function of its position x along the axis, $J(x)$.

- If the polar moment of inertia or the internal torque *suddenly changes* between the ends of the shaft, then $\phi = \int (T(x)/J(x)G(x))\, dx$ or $\phi = TL/JG$ must be applied to *each segment* for which J, G, and T are continuous or constant.

- When the internal torque in each segment is determined, be sure to use a consistent sign convention for the shaft or its segments, such as the one shown in Fig. 5–15. Also make sure that a consistent set of units is used when substituting numerical data into the equations.

EXAMPLE 5.5

Determine the angle of twist of the end A of the A-36 steel shaft shown in Fig. 5–16a. Also, what is the angle of twist of A relative to C? The shaft has a diameter of 200 mm.

SOLUTION

Internal Torque. Using the method of sections, the *internal torques* are found in each segment as shown in Fig. 5–16b. By the right-hand rule, with positive torques directed away from the *sectioned end* of the shaft, we have $T_{AB} = +80$ kN·m, $T_{BC} = -70$ kN·m, and $T_{CD} = -10$ kN·m. These results are also shown on the **torque diagram**, which indicates how the internal torque varies along the axis of the shaft, Fig. 5–16c.

Angle of Twist. The polar moment of inertia for the shaft is

$$J = \frac{\pi}{2}(0.1 \text{ m})^4 = 0.1571(10^{-3}) \text{ m}^4$$

For A-36 steel, the table on the back cover gives $G = 75$ GPa. Therefore, the end A of the shaft has a rotation of

$$\phi_A = \Sigma \frac{TL}{JG} = \frac{80(10^3) \text{ N} \cdot \text{m} (3 \text{ m})}{(0.1571(10^{-3}) \text{ m}^4)(75(10^9) \text{ N/m}^2)}$$

$$+ \frac{-70(10^3) \text{ N} \cdot \text{m} (2 \text{ m})}{(0.1571(10^{-3}) \text{ m}^4)(75(10^9) \text{ N/m}^2)} + \frac{-10(10^3) \text{ N} \cdot \text{m} (1.5 \text{ m})}{(0.1571(10^{-3}) \text{ m}^4)(75(10^9) \text{ N/m}^2)}$$

$$\phi_A = 7.22(10^{-3}) \text{ rad} \qquad \textit{Ans.}$$

The relative angle of twist of A with respect to C involves only two segments of the shaft.

$$\phi_{A/C} = \Sigma \frac{TL}{JG} = \frac{80(10^3) \text{ N} \cdot \text{m} (3 \text{ m})}{(0.1571(10^{-3}) \text{ m}^4)(75(10^9) \text{ N/m}^2)}$$

$$+ \frac{-70(10^3) \text{ N} \cdot \text{m} (2 \text{ m})}{(0.1571(10^{-3}) \text{ m}^4)(75(10^9) \text{ N/m}^2)}$$

$$\phi_{A/C} = 8.49(10^{-3}) \text{ rad} \qquad \textit{Ans.}$$

Both results are positive, which means that end A will rotate as indicated by the curl of the right-hand fingers when the thumb is directed away from the shaft.

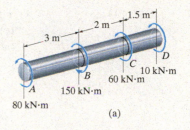

(a)

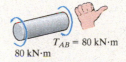

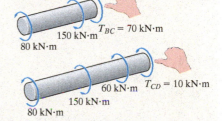

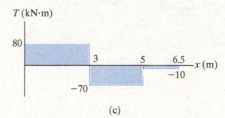

(b)

(c)

Fig. 5–16

EXAMPLE 5.6

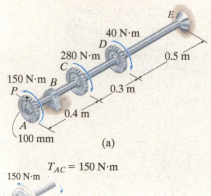

(a)

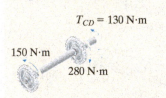

$T_{AC} = 150$ N·m

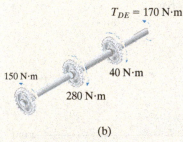

$T_{CD} = 130$ N·m

$T_{DE} = 170$ N·m

(b)

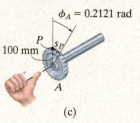

$\phi_A = 0.2121$ rad

(c)

Fig. 5-17

The gears attached to the fixed-end steel shaft are subjected to the torques shown in Fig. 5–17a. If the shaft has a diameter of 14 mm, determine the displacement of the tooth P on gear A. $G = 80$ GPa.

SOLUTION

Internal Torque. By inspection, the torques in segments AC, CD, and DE are different yet *constant* throughout each segment. Free-body diagrams of these segments along with the calculated internal torques are shown in Fig. 5–17b. Using the right-hand rule and the established sign convention that positive torque is directed away from the sectioned end of the shaft, we have

$$T_{AC} = +150 \text{ N·m} \qquad T_{CD} = -130 \text{ N·m} \qquad T_{DE} = -170 \text{ N·m}$$

Angle of Twist. The polar moment of inertia for the shaft is

$$J = \frac{\pi}{2}(0.007 \text{ m})^4 = 3.771(10^{-9}) \text{ m}^4$$

Applying Eq. 5–16 to each segment and adding the results algebraically, we have

$$\phi_A = \sum \frac{TL}{JG} = \frac{(+150 \text{ N·m})(0.4 \text{ m})}{3.771(10^{-9})\text{m}^4 [80(10^9)\text{N/m}^2]}$$

$$+ \frac{(-130 \text{ N·m})(0.3 \text{ m})}{3.771(10^{-9})\text{m}^4 [80(10^9)\text{N/m}^2]} + \frac{(-170 \text{ N·m})(0.5 \text{ m})}{3.771(10^{-9})\text{m}^4 [80(10^9) \text{ N/m}^2]}$$

$$\phi_A = -0.2121 \text{ rad}$$

Since the answer is negative, by the right-hand rule the thumb is directed *toward* the support E of the shaft, and therefore gear A will rotate as shown in Fig. 5–17c.

The displacement of tooth P on gear A is

$$s_P = \phi_A r = (0.2121 \text{ rad})(100 \text{ mm}) = 21.2 \text{ mm} \qquad \textit{Ans.}$$

EXAMPLE 5.7

The two solid steel shafts shown in Fig. 5–18a are coupled together using the meshed gears. Determine the angle of twist of end A of shaft AB when the torque $T = 45$ N·m is applied. Shaft DC is fixed at D. Each shaft has a diameter of 20 mm. $G = 80$ GPa.

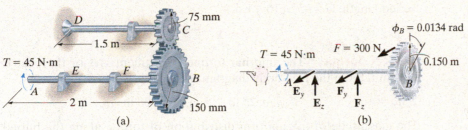

(a)

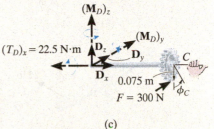

(b)

SOLUTION

Internal Torque. Free-body diagrams for each shaft are shown in Figs. 5–18b and 5–18c. Summing moments along the axis of shaft AB yields the tangential reaction between the gears of $F = 45$ N·m/0.15 m = 300 N. Summing moments about the axis of shaft DC, this force then creates a torque of $(T_D)_x = 300$ N (0.075 m) = 22.5 N·m in shaft DC.

Angle of Twist. To solve the problem, we will first calculate the rotation of gear C due to the torque of 22.5 N·m in shaft DC, Fig. 5–18c. This angle of twist is

$$\phi_C = \frac{TL_{DC}}{JG} = \frac{(+22.5 \text{ N·m})(1.5 \text{ m})}{(\pi/2)(0.010 \text{ m})^4 [80(10^9)\text{N/m}^2]} = +0.0269 \text{ rad}$$

Since the gears at the end of the shafts are in mesh, the rotation ϕ_C of gear C causes gear B to rotate ϕ_B, Fig. 5–18d, where

$$\phi_B(0.15 \text{ m}) = (0.0269 \text{ rad})(0.075 \text{ m})$$
$$\phi_B = 0.0134 \text{ rad}$$

We will now determine the angle of twist of end A with respect to end B of shaft AB caused by the 45 N·m torque, Fig. 5–18b. We have

$$\phi_{A/B} = \frac{T_{AB}L_{AB}}{JG} = \frac{(+45 \text{ N·m})(2 \text{ m})}{(\pi/2)(0.010 \text{ m})^4 [80(10^9) \text{ N/m}^2]} = +0.0716 \text{ rad}$$

The rotation of end A is therefore determined by adding ϕ_B and $\phi_{A/B}$, since both angles are in the *same direction*, Fig. 5–18b. We have

$$\phi_A = \phi_B + \phi_{A/B} = 0.0134 \text{ rad} + 0.0716 \text{ rad} = +0.0850 \text{ rad} \quad \textit{Ans.}$$

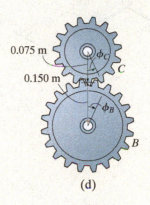

(c)

(d)

Fig. 5–18

EXAMPLE 5.8

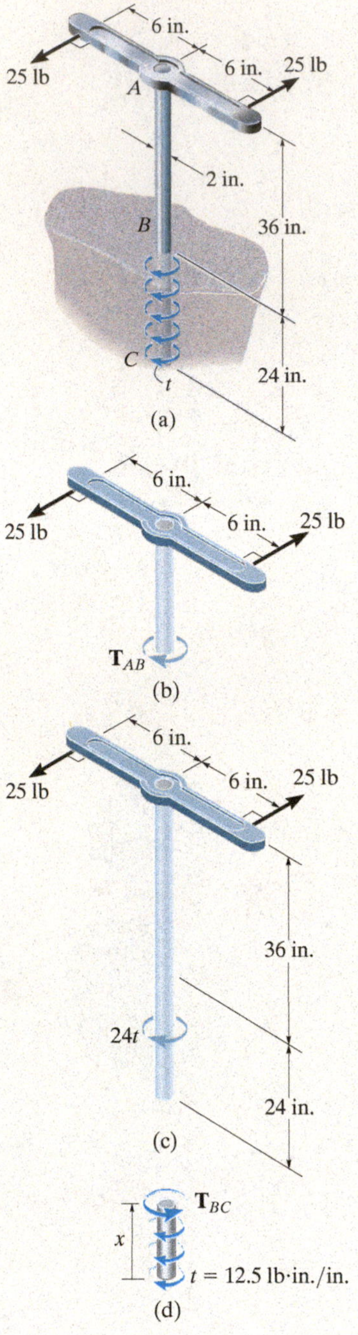

(a)

$\mathbf{T}_{AB}$
(b)

$24t$
(c)

$t = 12.5$ lb·in./in.
(d)

Fig. 5–19

The 2-in.-diameter solid cast-iron post shown in Fig. 5–19a is buried 24 in. in soil. If a torque is applied to its top using a rigid wrench, determine the maximum shear stress in the post and the angle of twist of the wrench. Assume that the torque is about to turn the bottom of the post, and the soil exerts a uniform torsional resistance of t lb·in./in. along its 24-in. buried length. $G = 5.5(10^3)$ ksi.

SOLUTION

Internal Torque. The internal torque in segment AB of the post is constant. From the free-body diagram, Fig. 5–19b, we have

$$\Sigma M_z = 0; \qquad T_{AB} = 25 \text{ lb } (12 \text{ in.}) = 300 \text{ lb·in.}$$

The magnitude of the uniform distribution of torque along the buried segment BC can be determined from equilibrium of the entire post, Fig. 5–19c. Here

$$\Sigma M_z = 0 \qquad 25 \text{ lb } (12 \text{ in.}) - t(24 \text{ in.}) = 0$$

$$t = 12.5 \text{ lb·in./in.}$$

Hence, from a free-body diagram of the bottom segment of the post, located at the position x, Fig. 5–19d, we have

$$\Sigma M_z = 0; \qquad T_{BC} - 12.5x = 0$$

$$T_{BC} = 12.5x$$

Maximum Shear Stress. The largest shear stress occurs in region AB, since the torque is largest there and J is constant for the post. Applying the torsion formula, we have

$$\tau_{max} = \frac{T_{AB}\, c}{J} = \frac{(300 \text{ lb·in.})(1 \text{ in.})}{(\pi/2)(1 \text{ in.})^4} = 191 \text{ psi} \qquad \textit{Ans.}$$

Angle of Twist. The angle of twist at the top can be determined relative to the bottom of the post, since it is fixed and yet is about to turn. Both segments AB and BC twist, and so in this case we have

$$\phi_A = \frac{T_{AB}\, L_{AB}}{JG} + \int_0^{L_{BC}} \frac{T_{BC}\, dx}{JG}$$

$$= \frac{(300 \text{ lb·in.})\, 36 \text{ in.}}{JG} + \int_0^{24 \text{ in.}} \frac{12.5x\, dx}{JG}$$

$$= \frac{10\,800 \text{ lb·in}^2}{JG} + \frac{12.5[(24)^2/2] \text{ lb·in}^2}{JG}$$

$$= \frac{14\,400 \text{ lb·in}^2}{(\pi/2)(1 \text{ in.})^4\, 5500(10^3) \text{ lb/in}^2} = 0.00167 \text{ rad} \qquad \textit{Ans.}$$

FUNDAMENTAL PROBLEMS

F5–9. The 60-mm-diameter steel shaft is subjected to the torques shown. Determine the angle of twist of end A with respect to C. Take $G = 75$ GPa.

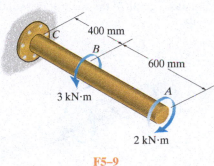

400 mm

600 mm

3 kN·m

2 kN·m

F5–9

F5–10. Determine the angle of twist of wheel B with respect to wheel A. The shaft has a diameter of 40 mm and is made of steel for which $G = 75$ GPa.

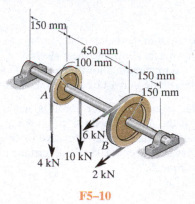

150 mm

450 mm
100 mm

150 mm
150 mm

6 kN

4 kN   10 kN

2 kN

F5–10

F5–11. The hollow 6061-T6 aluminum shaft has an outer and inner radius of $c_o = 40$ mm and $c_i = 30$ mm, respectively. Determine the angle of twist of end A. The support at B is flexible like a torsional spring, so that $T_B = k_B\,\phi_B$, where the torsional stiffness is $k_B = 90$ kN·m/rad.

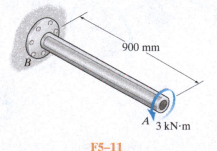

900 mm

3 kN·m

F5–11

F5–12. A series of gears are mounted on the 40-mm-diameter steel shaft. Determine the angle of twist of gear E relative to gear A. Take $G = 75$ GPa.

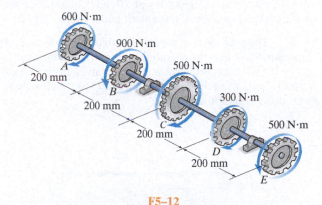

600 N·m

900 N·m

500 N·m

200 mm

200 mm

300 N·m

200 mm

500 N·m

200 mm

F5–12

F5–13. The 80-mm-diameter shaft is made of steel. If it is subjected to the uniform distributed torque, determine the angle of twist of end A. Take $G = 75$ GPa.

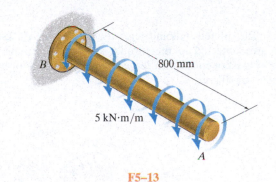

800 mm

5 kN·m/m

F5–13

F5–14. The 80-mm-diameter shaft is made of steel. If it is subjected to the triangular distributed load, determine the angle of twist of end A. Take $G = 75$ GPa.

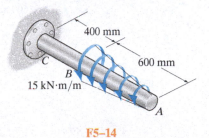

400 mm

600 mm

15 kN·m/m

F5–14

PROBLEMS

5–47. The propellers of a ship are connected to an A-36 steel shaft that is 60 m long and has an outer diameter of 340 mm and inner diameter of 260 mm. If the power output is 4.5 MW when the shaft rotates at 20 rad/s, determine the maximum torsional stress in the shaft and its angle of twist.

*5–48. The solid shaft of radius c is subjected to a torque **T** at its ends. Show that the maximum shear strain in the shaft is $\gamma_{max} = Tc/JG$. What is the shear strain on an element located at point A, $c/2$ from the center of the shaft? Sketch the shear strain distortion of this element.

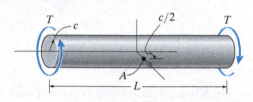

Prob. 5–48

5–49. The splined ends and gears attached to the A992 steel shaft are subjected to the torques shown. Determine the angle of twist of end B with respect to end A. The shaft has a diameter of 40 mm.

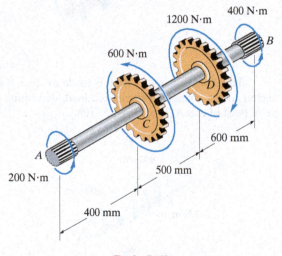

Prob. 5–49

5–50. The A-36 steel shaft has a diameter of 50 mm and is subjected to the distributed and concentrated loadings shown. Determine the absolute maximum shear stress in the shaft and plot a graph of the angle of twist of the shaft in radians versus x.

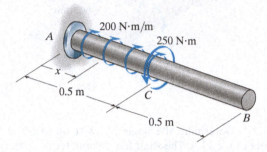

Prob. 5–50

5–51. The 60-mm-diameter shaft is made of 6061-T6 aluminum having an allowable shear stress of $\tau_{allow} = 80$ MPa. Determine the maximum allowable torque **T**. Also, find the corresponding angle of twist of disk A relative to disk C.

*5–52. The 60-mm-diameter shaft is made of 6061-T6 aluminum. If the allowable shear stress is $\tau_{allow} = 80$ MPa, and the angle of twist of disk A relative to disk C is limited so that it does not exceed 0.06 rad, determine the maximum allowable torque **T**.

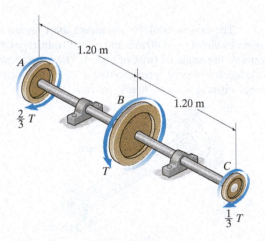

Probs. 5–51/52

5–53. The 50-mm-diameter A992 steel shaft is subjected to the torques shown. Determine the angle of twist of the end *A*.

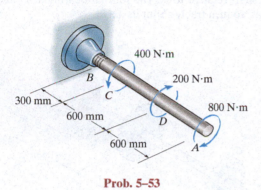

400 N·m

200 N·m

B

C

300 mm

600 mm

D

800 N·m

600 mm

A

Prob. 5–53

5–54. The shaft is made of A992 steel with the allowable shear stress of $\tau_{allow} = 75$ MPa. If gear *B* supplies 15 kW of power, while gears *A*, *C* and *D* withdraw 6 kW, 4 kW and 5 kW, respectively, determine the required minimum diameter *d* of the shaft to the nearest millimeter. Also, find the corresponding angle of twist of gear *A* relative to gear *D*. The shaft is rotating at 600 rpm.

5–55. Gear *B* supplies 15 kW of power, while gears *A*, *C* and *D* withdraw 6 kW, 4 kW and 5 kW, respectively. If the shaft is made of steel with the allowable shear stress of $\tau_{allow} = 75$ MPa, and the relative angle of twist between any two gears cannot exceed 0.05 rad, determine the required minimum diameter *d* of the shaft to the nearest millimeter. The shaft is rotating at 600 rpm.

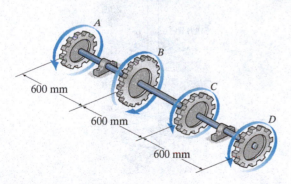

A

B

600 mm

600 mm

C

D

600 mm

Probs. 5–54/55

*5–56. The rotating flywheel-and-shaft, when brought to a sudden stop at *D*, begins to oscillate clockwise-counter-clockwise such that a point *A* on the outer edge of the flywheel is displaced through a 6-mm arc. Determine the maximum shear stress developed in the tubular A-36 steel shaft due to this oscillation. The shaft has an inner diameter of 24 mm and an outer diameter of 32 mm. The bearings at *B* and *C* allow the shaft to rotate freely, whereas the support at *D* holds the shaft fixed.

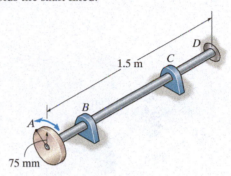

D

1.5 m

C

B

A

75 mm

Prob. 5–56

5–57. The turbine develops 150 kW of power, which is transmitted to the gears such that *C* receives 70% and *D* receives 30%. If the rotation of the 100-mm-diameter A-36 steel shaft is $\omega = 800$ rev/min., determine the absolute maximum shear stress in the shaft and the angle of twist of end *E* of the shaft relative to *B*. The journal bearing at *E* allows the shaft to turn freely about its axis.

5–58. The turbine develops 150 kW of power, which is transmitted to the gears such that both *C* and *D* receive an equal amount. If the rotation of the 100-mm-diameter A-36 steel shaft is $\omega = 500$ rev/min., determine the absolute maximum shear stress in the shaft and the rotation of end *B* of the shaft relative to *E*. The journal bearing at *E* allows the shaft to turn freely about its axis.

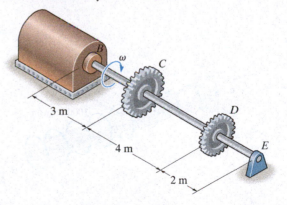

B

ω

C

3 m

D

4 m

E

2 m

Probs. 5–57/58

5–59. The shaft is made of A992 steel. It has a diameter of 1 in. and is supported by bearings at A and D, which allow free rotation. Determine the angle of twist of B with respect to D.

*5–60. The shaft is made of A-36 steel. It has a diameter of 1 in. and is supported by bearings at A and D, which allow free rotation. Determine the angle of twist of gear C with respect to B.

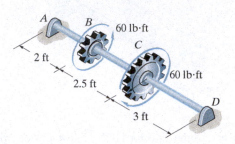

Probs. 5–59/60

5–61. The A992 steel shaft has a diameter of 50 mm and is subjected to the distributed loadings shown. Determine the absolute maximum shear stress in the shaft and plot a graph of the angle of twist of the shaft in radians versus x.

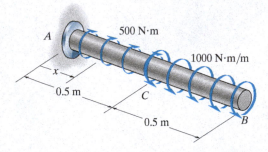

Prob. 5–61

5–62. The turbine develops 300 kW of power, which is transmitted to the gears such that both B and C receive an equal amount. If the rotation of the 100-mm-diameter A992 steel shaft is $\omega = 600$ rev/min., determine the absolute maximum shear stress in the shaft and the rotation of end D of the shaft relative to A. The journal bearing at D allows the shaft to turn freely about its axis.

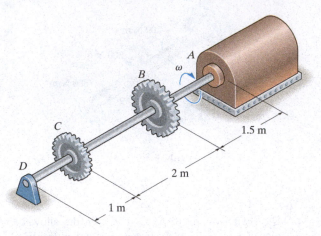

Prob. 5–62

5–63. The device shown is used to mix soils in order to provide in-situ stabilization. If the mixer is connected to an A-36 steel tubular shaft that has an inner diameter of 3 in. and an outer diameter of 4.5 in., determine the angle of twist of the shaft at A relative to C if each mixing blade is subjected to the torques shown.

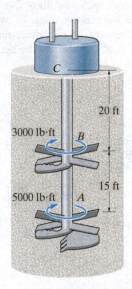

Prob. 5–63

*5–64. The device shown is used to mix soils in order to provide in-situ stabilization. If the mixer is connected to an A-36 steel tubular shaft that has an inner diameter of 3 in. and an outer diameter of 4.5 in, determine the angle of twist of the shaft at A relative to B and the absolute maximum shear stress in the shaft if each mixing blade is subjected to the torques shown.

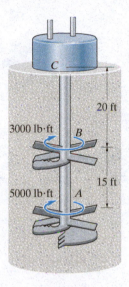

20 ft

3000 lb·ft B

15 ft

5000 lb·ft A

Prob. 5–64

5–65. The 6-in.-diameter L-2 steel shaft on the turbine is supported on journal bearings at A and B. If C is held fixed and the turbine blades create a torque on the shaft that increases linearly from zero at C to 2000 lb·ft at D, determine the angle of twist of the shaft at D relative to C. Also, calculate the absolute maximum shear stress in the shaft. Neglect the size of the blades.

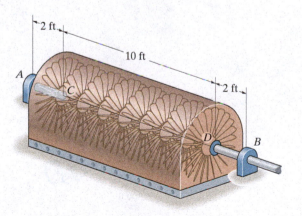

2 ft

10 ft

2 ft

A

C

D B

Prob. 5–65

5–66. The A-36 hollow steel shaft is 2 m long and has an outer diameter of 40 mm. When it is rotating at 80 rad/s, it transmits 32 kW of power from the engine E to the generator G. Determine the smallest thickness of the shaft if the allowable shear stress is $\tau_{allow} = 140$ MPa and the shaft is restricted not to twist more than 0.05 rad.

5–67. The A-36 solid steel shaft is 3 m long and has a diameter of 50 mm. It is required to transmit 35 kW of power from the engine E to the generator G. Determine the smallest angular velocity of the shaft if it is restricted not to twist more than 1°.

E G

Probs. 5–66/67

*5–68. The shaft of radius c is subjected to a distributed torque t, measured as torque/length of shaft. Determine the angle of twist at end A. The shear modulus is G.

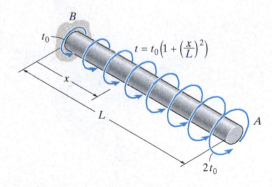

B

t_0

$t = t_0\left(1 + \left(\frac{x}{L}\right)^2\right)$

x

L

A

$2t_0$

Prob. 5–68

5–69. The A-36 steel bolt is tightened within a hole so that the reactive torque on the shank AB can be expressed by the equation $t = (kx^2)$ N·m/m, where x is in meters. If a torque of $T = 50$ N·m is applied to the bolt head, determine the constant k and the amount of twist in the 50-mm length of the shank. Assume the shank has a constant radius of 4 mm.

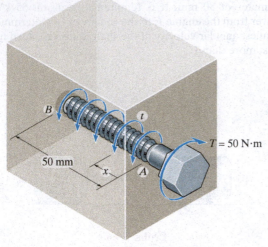

Prob. 5–69

5–70. The A-36 steel bolt is tightened within a hole so that the reactive torque on the shank AB can be expressed by the equation $t = (kx^{2/3})$ N·m/m, where x is in meters. If a torque of $T = 50$ N·m is applied to the bolt head, determine the constant k and the amount of twist in the 50-mm length of the shank. Assume the shank has a constant radius of 4 mm.

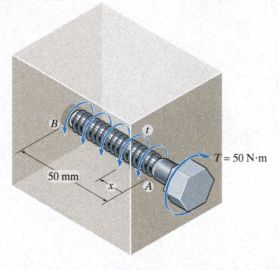

Prob. 5–70

5–71. The tubular drive shaft for the propeller of a hovercraft is 6 m long. If the motor delivers 4 MW of power to the shaft when the propellers rotate at 25 rad/s, determine the required inner diameter of the shaft if the outer diameter is 250 mm. What is the angle of twist of the shaft when it is operating? Take $\tau_{allow} = 90$ MPa and $G = 75$ GPa.

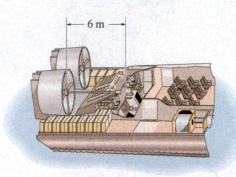

Prob. 5–71

*5–72.** The 60-mm-diameter solid shaft is made of 2014-T6 aluminum and is subjected to the distributed and concentrated torsional loadings shown. Determine the angle of twist at the free end A of the shaft.

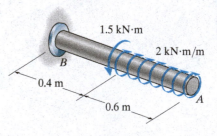

Prob. 5–72

5–73. The motor produces a torque of $T = 20\,\text{N}\cdot\text{m}$ on gear A. If gear C is suddenly locked so it does not turn, yet B can freely turn, determine the angle of twist of F with respect to E and F with respect to D of the L2-steel shaft, which has an inner diameter of 30 mm and an outer diameter of 50 mm. Also, calculate the absolute maximum shear stress in the shaft. The shaft is supported on journal bearings at G at H.

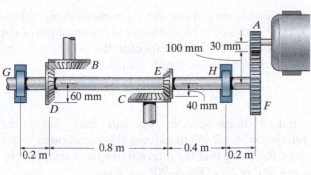

Prob. 5–73

5–74. The shaft has a radius c and is subjected to a torque per unit length of t_0, which is distributed uniformly over the shaft's entire length L. If it is fixed at its far end A, determine the angle of twist ϕ of end B. The shear modulus is G.

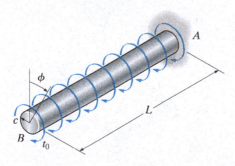

Prob. 5–74

5–75. The 60-mm-diameter solid shaft is made of A-36 steel and is subjected to the distributed and concentrated torsional loadings shown. Determine the angle of twist at the free end A of the shaft due to these loadings.

Prob. 5–75

*****5–76.** The contour of the surface of the shaft is defined by the equation $y = e^{ax}$, where a is a constant. If the shaft is subjected to a torque T at its ends, determine the angle of twist of end A with respect to end B. The shear modulus is G.

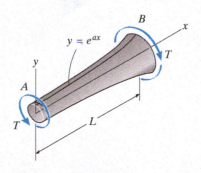

Prob. 5–76

5

5.5   STATICALLY INDETERMINATE TORQUE-LOADED MEMBERS

A torsionally loaded shaft will be statically indeterminate if the moment equation of equilibrium, applied about the axis of the shaft, is not adequate to determine the unknown torques acting on the shaft. An example of this situation is shown in Fig. 5–20a. As shown on the free-body diagram, Fig. 5–20b, the reactive torques at the supports A and B are unknown. Along the axis of the shaft, we require

$$\Sigma M = 0; \qquad\qquad 500\ \text{N} \cdot \text{m} - T_A - T_B = 0$$

In order to obtain a solution, we will use the same method of analysis discussed in Sec. 4.4. The necessary compatibility condition requires the angle of twist of one end of the shaft with respect to the other end to be equal to zero, since the end supports are fixed. Therefore,

$$\phi_{A/B} = 0$$

Provided the material is linear elastic, we can then apply the load–displacement relation $\phi = TL/JG$ to express this equation in terms of the unknown torques. Realizing that the internal torque in segment AC is $+T_A$ and in segment CB it is $-T_B$, Fig. 5–20c, we have

$$\frac{T_A(3\ \text{m})}{JG} - \frac{T_B(2\ \text{m})}{JG} = 0$$

Solving the above two equations for the reactions, we get

$$T_A = 200\ \text{N} \cdot \text{m} \quad \text{and} \quad T_B = 300\ \text{N} \cdot \text{m}$$

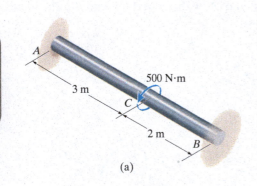

(a)

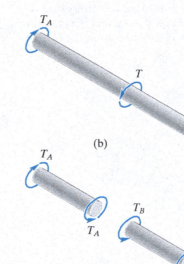

(b)

(c)

Fig. 5–20

PROCEDURE FOR ANALYSIS

The unknown torques in statically indeterminate shafts are determined by satisfying equilibrium, compatibility, and load–displacement requirements for the shaft.

Equilibrium.

- Draw a free-body diagram of the shaft in order to identify all the external torques that act on it. Then write the equation of moment equilibrium about the axis of the shaft.

Compatibility.

- Write the compatibility equation. Give consideration as to how the supports constrain the shaft when it is twisted.

Load–Displacement.

- Express the angles of twist in the compatibility condition in terms of the torques, using a load–displacement relation, such as $\phi = TL/JG$.

- Solve the equations for the unknown reactive torques. If any of the magnitudes have a negative numerical value, it indicates that this torque acts in the opposite sense of direction to that shown on the free-body diagram.

The shaft of this cutting machine is fixed at its ends and subjected to a torque at its center, allowing it to act as a torsional spring.

EXAMPLE 5.9

The solid steel shaft shown in Fig. 5–21a has a diameter of 20 mm. If it is subjected to the two torques, determine the reactions at the fixed supports A and B.

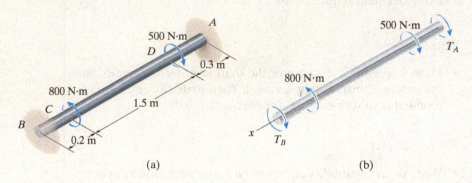

(a)   (b)

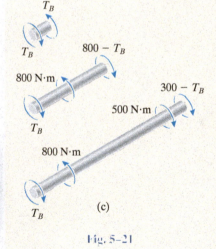

(c)

Fig. 5–21

SOLUTION

Equilibrium.   By inspection of the free-body diagram, Fig. 5–21b, it is seen that the problem is statically indeterminate, since there is only *one* available equation of equilibrium and there are two unknowns. We require

$$\Sigma M_x = 0; \qquad -T_B + 800 \text{ N} \cdot \text{m} - 500 \text{ N} \cdot \text{m} - T_A = 0 \qquad (1)$$

Compatibility.   Since the ends of the shaft are fixed, the angle of twist of one end of the shaft with respect to the other must be zero. Hence, the compatibility equation becomes

$$\phi_{A/B} = 0$$

Load–Displacement.   This condition can be expressed in terms of the unknown torques by using the load–displacement relationship, $\phi = TL/JG$. Here there are three regions of the shaft where the internal torque is constant. On the free-body diagrams in Fig. 5–21c we have shown the internal torques acting on the left segments of the shaft. This way the internal torque is only a function of T_B. Using the sign convention established in Sec. 5.4, we have

$$\frac{-T_B(0.2 \text{ m})}{JG} + \frac{(800 - T_B)(1.5 \text{ m})}{JG} + \frac{(300 - T_B)(0.3 \text{ m})}{JG} = 0$$

so that

$$T_B = 645 \text{ N} \cdot \text{m} \qquad \qquad \textit{Ans.}$$

Using Eq. 1,

$$T_A = -345 \text{ N} \cdot \text{m} \qquad \qquad \textit{Ans.}$$

The negative sign indicates that $\mathbf{T}_A$ acts in the opposite direction of that shown in Fig. 5–21b.

EXAMPLE 5.10

The shaft shown in Fig. 5–22a is made from a steel tube, which is bonded to a brass core. If a torque of $T = 250$ lb·ft is applied at its end, plot the shear-stress distribution along a radial line on its cross section. Take $G_{st} = 11.4(10^3)$ ksi, $G_{br} = 5.20(10^3)$ ksi.

SOLUTION

Equilibrium.   A free-body diagram of the shaft is shown in Fig. 5–22b. The reaction at the wall has been represented by the amount of torque resisted by the steel, T_{st}, and by the brass, T_{br}. Working in units of pounds and inches, equilibrium requires

$$-T_{st} - T_{br} + (250 \text{ lb} \cdot \text{ft})(12 \text{ in./ft}) = 0 \qquad (1)$$

Compatibility.   We require the angle of twist of end A to be the same for both the steel and brass since they are bonded together. Thus,

$$\phi = \phi_{st} = \phi_{br}$$

Load–Displacement.   Applying the load–displacement relationship, $\phi = TL/JG$,

$$\frac{T_{st}L}{(\pi/2)[(1 \text{ in.})^4 - (0.5 \text{ in.})^4] \, 11.4(10^3) \text{ kip/in}^2}$$
$$= \frac{T_{br}L}{(\pi/2)(0.5 \text{ in.})^4 \, 5.20(10^3) \text{ kip/in}^2}$$
$$T_{st} = 32.88 \, T_{br} \qquad (2)$$

Solving Eqs. 1 and 2, we get

$$T_{st} = 2911.5 \text{ lb} \cdot \text{in.} = 242.6 \text{ lb} \cdot \text{ft}$$
$$T_{br} = 88.5 \text{ lb} \cdot \text{in.} = 7.38 \text{ lb} \cdot \text{ft}$$

The shear stress in the brass core varies from zero at its center to a maximum at the interface where it contacts the steel tube. Using the torsion formula,

$$(\tau_{br})_{max} = \frac{(88.5 \text{ lb} \cdot \text{in.})(0.5 \text{ in.})}{(\pi/2)(0.5 \text{ in.})^4} = 451 \text{ psi}$$

For the steel, the minimum and maximum shear stresses are

$$(\tau_{st})_{min} = \frac{(2911.5 \text{ lb} \cdot \text{in.})(0.5 \text{ in.})}{(\pi/2)[(1 \text{ in.})^4 - (0.5 \text{ in.})^4]} = 989 \text{ psi}$$

$$(\tau_{st})_{max} = \frac{(2911.5 \text{ lb} \cdot \text{in.})(1 \text{ in.})}{(\pi/2)[(1 \text{ in.})^4 - (0.5 \text{ in.})^4]} = 1977 \text{ psi}$$

The results are plotted in Fig. 5–22c. Note the discontinuity of *shear stress* at the brass and steel interface. This is to be expected, since the materials have different moduli of rigidity; i.e., steel is stiffer than brass $(G_{st} > G_{br})$, and thus it carries more shear stress at the interface. Although the shear stress is discontinuous here, the *shear strain* is not. Rather, it is the *same* on either side of the brass–steel interface, Fig. 5–22d.

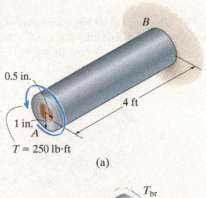

0.5 in.

4 ft

1 in.

$T = 250$ lb·ft

(a)

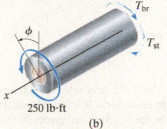

T_{br}

ϕ

T_{st}

x

250 lb·ft

(b)

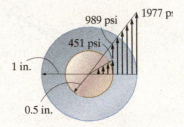

989 psi   1977 ps

451 psi

1 in.

0.5 in.

Shear–stress distribution

(c)

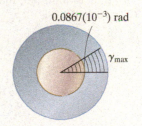

$0.0867(10^{-3})$ rad

γ_{max}

Shear–strain distribution

(d)

Fig. 5–22

PROBLEMS

5–77. The steel shaft has a diameter of 40 mm and is fixed at its ends A and B. If it is subjected to the couple, determine the maximum shear stress in regions AC and CB of the shaft. $G_{st} = 75$ GPa.

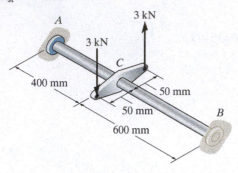

Prob. 5–77

5–78. The A992 steel shaft has a diameter of 60 mm and is fixed at its ends A and B. If it is subjected to the torques shown, determine the absolute maximum shear stress in the shaft.

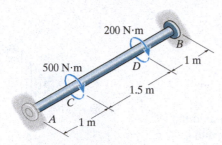

Prob. 5–78

5–79. The steel shaft is made from two segments: AC has a diameter of 0.5 in, and CB has a diameter of 1 in. If the shaft is fixed at its ends A and B and subjected to a torque of 500 lb·ft, determine the maximum shear stress in the shaft. $G_{st} = 10.8(10^3)$ ksi.

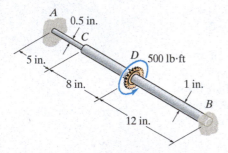

Prob. 5–79

*****5–80.** The bronze C86100 pipe has an outer diameter of 1.5 in. and a thickness of 0.125 in. The coupling on it at C is being tightened using a wrench. If the torque developed at A is 125 lb·in., determine the magnitude F of the couple forces. The pipe is fixed supported at end B.

5–81. The bronze C86100 pipe has an outer diameter of 1.5 in. and a thickness of 0.125 in. The coupling on it at C is being tightened using a wrench. If the applied force is $F = 20$ lb, determine the maximum shear stress in the pipe.

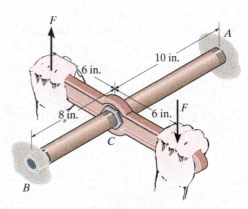

Probs. 5–80/81

5–82. The shaft is made of L2 tool steel, has a diameter of 40 mm, and is fixed at its ends A and B. If it is subjected to the torque, determine the maximum shear stress in regions AC and CB.

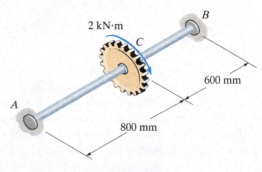

Prob. 5–82

5–83. The shaft is made of L2 tool steel, has a diameter of 40 mm, and is fixed at its ends A and B. If it is subjected to the couple, determine the maximum shear stress in regions AC and CB.

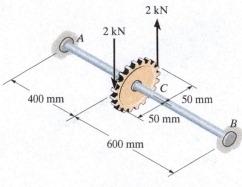

Prob. 5–83

***5–84.** The Am1004-T61 magnesium tube is bonded to the A-36 steel rod. If the allowable shear stresses for the magnesium and steel are $(\tau_{allow})_{mg} = 45$ MPa and $(\tau_{allow})_{st} = 75$ MPa, respectively, determine the maximum allowable torque that can be applied at A. Also, find the corresponding angle of twist of end A.

5–85. The Am1004-T61 magnesium tube is bonded to the A-36 steel rod. If a torque of $T = 5$ kN·m is applied to end A, determine the maximum shear stress in each material. Sketch the shear stress distribution.

Probs. 5–84/85

5–86. The two shafts are made of A-36 steel. Each has a diameter of 25 mm and they are connected using the gears fixed to their ends. Their other ends are attached to fixed supports at A and B. They are also supported by journal bearings at C and D, which allow free rotation of the shafts along their axes. If a torque of 500 N·m is applied to the gear at E, determine the reactions at A and B.

5–87. The two shafts are made of A-36 steel. Each has a diameter of 25 mm and they are connected using the gears fixed to their ends. Their other ends are attached to fixed supports at A and B. They are also supported by journal bearings at C and D, which allow free rotation of the shafts along their axes. If a torque of 500 N·m is applied to the gear at E, determine the rotation of this gear.

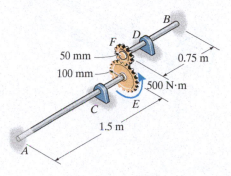

Probs. 5–86/87

***5–88.** A rod is made from two segments: AB is steel and BC is brass. It is fixed at its ends and subjected to a torque of $T = 680$ N·m. If the steel portion has a diameter of 30 mm, determine the required diameter of the brass portion so the reactions at the walls will be the same. $G_{st} = 75$ GPa, $G_{br} = 39$ GPa.

5–89. Determine the absolute maximum shear stress in the shaft of Prob. 5–88.

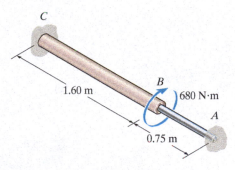

Probs. 5–88/89

5–90. The two 3-ft-long shafts are made of 2014-16 aluminum. Each has a diameter of 1.5 in. and they are connected using the gears fixed to their ends. Their other ends are attached to fixed supports at A and B. They are also supported by bearings at C and D, which allow free rotation of the shafts along their axes. If a torque of 600 lb·ft is applied to the top gear as shown, determine the maximum shear stress in each shaft.

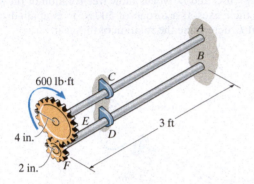

Prob. 5–90

5–91. The composite shaft consists of a mid-section that includes the 1-in.-diameter solid shaft and a tube that is welded to the rigid flanges at A and B. Neglect the thickness of the flanges and determine the angle of twist of end C of the shaft relative to end D. The shaft is subjected to a torque of 800 lb·ft. The material is A-36 steel.

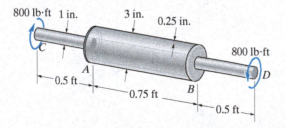

Prob. 5–91

*5–92. If the shaft is subjected to a uniform distributed torque of $t = 20$ kN·m/m, determine the maximum shear stress developed in the shaft. The shaft is made of 2014-T6 aluminum alloy and is fixed at A and C.

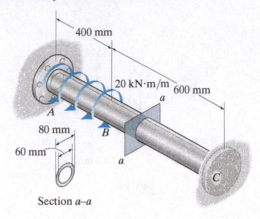

Section a–a

Prob. 5–92

5–93. The tapered shaft is confined by the fixed supports at A and B. If a torque T is applied at its mid-point, determine the reactions at the supports.

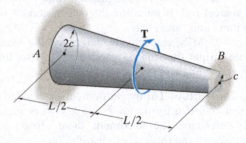

Prob. 5–93

5–94. The shaft of radius c is subjected to a distributed torque t, measured as torque/length of shaft. Determine the reactions at the fixed supports A and B.

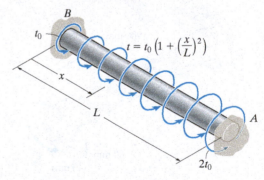

Prob. 5–94

*5.6   SOLID NONCIRCULAR SHAFTS

It was demonstrated in Sec. 5.1 that when a torque is applied to a shaft having a circular cross section—that is, one that is axisymmetric—the shear strains vary linearly from zero at its center to a maximum at its outer surface. Furthermore, due to uniformity, the cross sections do not deform, but rather remain plane after the shaft has twisted. Shafts that have a noncircular cross section, however, are *not* axisymmetric, and so their cross sections will **bulge** or **warp** when the shaft is twisted. Evidence of this can be seen from the way grid lines deform on a shaft having a square cross section, Fig. 5–23. Because of this deformation, the torsional analysis of *noncircular* shafts becomes considerably more complicated and will not be discussed in this text.

Using a mathematical analysis based on the theory of elasticity, however, the shear-stress distribution within a shaft of square cross section has been determined. Examples of how the shear stress varies along two radial lines of the shaft are shown in Fig. 5–24a, and because these shear-stress distributions are different, the shear strains they create will *warp* the cross section, as shown in Fig. 5–24b. In particular, notice that the corner points of the shaft must be subjected to zero shear stress and therefore zero shear strain. The reason for this can be shown by considering an element of material located at one of these corner points, Fig. 5–24c. One would expect the top face of this element to be subjected to a shear stress in order to contribute to the applied torque **T**. However, this cannot occur, since the complementary shear stresses τ and τ', acting on the *outer surface* of the shaft, must be *zero*.

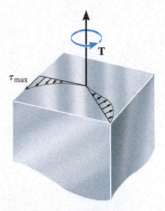

Shear stress distribution
along two radial lines

(a)

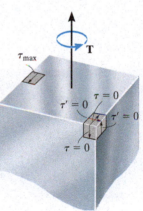

Warping of
cross-sectional area

(b)

Fig. 5–24

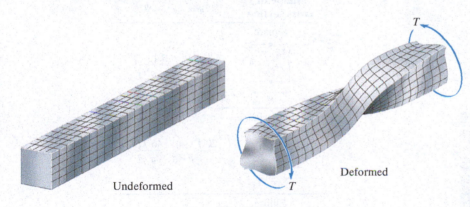

Undeformed   Deformed

Fig. 5–23

(c)

Notice the deformation of the square element when this rubber bar is subjected to a torque.

Using the theory of elasticity, Table 5–1 provides the results of the analysis for square cross sections, along with those for shafts having triangular and elliptical cross sections. In all cases, the *maximum shear stress* occurs at a point on the edge of the cross section that is closest to the center axis of the shaft. Also given are formulas for the angle of twist of each shaft. By extending these results, it can be shown that the *most efficient shaft* has a *circular cross section*, since it is subjected to both a *smaller* maximum shear stress and a *smaller* angle of twist than one having the same cross-sectional area, but not circular, and subjected to the same torque.

The shaft connected to the soil auger has a square cross section.

TABLE 5–1		
Shape of cross section	τ_{max}	ϕ
Square	$\dfrac{4.81\,T}{a^3}$	$\dfrac{7.10\,TL}{a^4G}$
Equilateral triangle	$\dfrac{20\,T}{a^3}$	$\dfrac{46\,TL}{a^4G}$
Ellipse	$\dfrac{2\,T}{\pi ab^2}$	$\dfrac{(a^2+b^2)TL}{\pi a^3b^3G}$

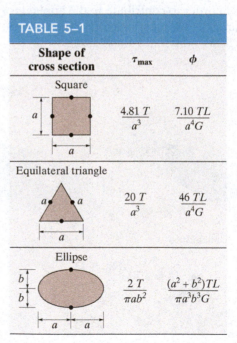

EXAMPLE   5.11

The 6061-T6 aluminum shaft shown in Fig. 5–25 has a cross-sectional area in the shape of an equilateral triangle. Determine the largest torque **T** that can be applied to the end of the shaft if the allowable shear stress is $\tau_{allow} = 8$ ksi and the angle of twist at its end is restricted to $\phi_{allow} = 0.02$ rad. How much torque can be applied to a shaft of circular cross section made from the same amount of material?

SOLUTION

By inspection, the resultant internal torque at any cross section along the shaft's axis is also **T**. Using the formulas for τ_{max} and ϕ in Table 5–1, we require

$$\tau_{allow} = \frac{20T}{a^3}; \qquad 8(10^3)\ \text{lb/in}^2 = \frac{20T}{(1.5\ \text{in.})^3}$$

$$T = 1350\ \text{lb} \cdot \text{in.}$$

Also,

$$\phi_{allow} = \frac{46TL}{a^4 G_{al}}; \qquad 0.02\ \text{rad} = \frac{46T(4\ \text{ft})(12\ \text{in./ft})}{(1.5\ \text{in.})^4\,[3.7(10^6)\ \text{lb/in}^2]}$$

$$T = 170\ \text{lb} \cdot \text{in.} \qquad\qquad \textit{Ans.}$$

By comparison, the torque is limited due to the angle of twist.

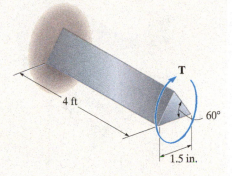

Fig. 5–25

Circular Cross Section.   If the same amount of aluminum is to be used in making the same length of shaft having a circular cross section, then the radius of the cross section can be calculated. We have

$$A_{circle} = A_{triangle}; \qquad \pi c^2 = \frac{1}{2}(1.5\ \text{in.})(1.5\sin 60°)$$

$$c = 0.557\ \text{in.}$$

The limitations of stress and angle of twist then require

$$\tau_{allow} = \frac{Tc}{J}; \qquad 8(10^3)\ \text{lb/in}^2 = \frac{T(0.557\ \text{in.})}{(\pi/2)(0.557\ \text{in.})^4}$$

$$T = 2170\ \text{lb} \cdot \text{in.}$$

$$\phi_{allow} = \frac{TL}{JG_{al}}; \qquad 0.02\ \text{rad} = \frac{T(4\ \text{ft})(12\ \text{in./ft})}{(\pi/2)(0.557\ \text{in.})^4\,[3.7(10^6)\ \text{lb/in}^2]}$$

$$T = 233\ \text{lb} \cdot \text{in.} \qquad\qquad \textit{Ans.}$$

Again, the angle of twist limits the applied torque.

NOTE: Comparing this result (233 lb·in.) with that given above (170 lb·in.), it is seen that a shaft of circular cross section can support 37% more torque than the one having a triangular cross section.

*5.7   THIN-WALLED TUBES HAVING CLOSED CROSS SECTIONS

Thin-walled tubes of noncircular cross section are often used to construct light-weight frameworks such as those used in aircraft. In some applications, they may be subjected to a torsional loading, and so in this section we will analyze the effects of twisting these members. Here we will consider a tube having a *closed* cross section, that is, one that does not have any breaks or slits along its length, Fig. 5–26a. Since the walls are thin, we will obtain the *average shear stress* by assuming that this stress is *uniformly distributed* across the thickness of the tube at any given location.

Shear Flow.   Shown in Figs. 5–26a and 5–26b is a small element of the tube having a finite length s and differential width dx. At one end, the element has a thickness t_A, and at the other end the thickness is t_B. Due to the torque **T**, shear stress is developed on the front face of the element. Specifically, at end A the shear stress is τ_A, and at end B it is τ_B. These stresses can be related by noting that equivalent shear stresses τ_A and τ_B must also act on the longitudinal sides of the element. These sides have a *constant* width dx, and so the forces acting on them are $dF_A = \tau_A(t_A\,dx)$ and $dF_B = \tau_B(t_B\,dx)$. Since equilibrium requires these forces to be of equal magnitude but opposite directions, we have

$$\tau_A\,t_A = \tau_B\,t_B$$

This important result shows that *the product of the average shear stress and the thickness of the tube is the same at each location on the cross section*. This product q is called *shear flow*,* and in general terms we can express it as

$$q = \tau_{\text{avg}}t \qquad (5\text{–}17)$$

Since q is constant over the cross section, the *largest* average shear stress must occur where the tube's thickness is the *smallest*.

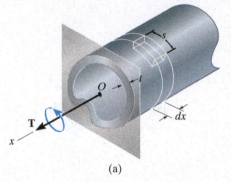

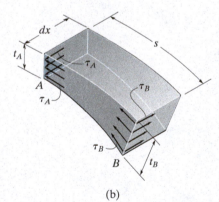

Fig. 5–26

5

* The terminology "flow" is used since q is analogous to water flowing through a channel of rectangular cross section having a constant depth and variable width.

If a differential element having a thickness t, length ds, and width dx is isolated from the tube, Fig. 5–26c, then the front face over which the average shear stress acts is $dA = t\,ds$, so that $dF = \tau_{avg}\,(t\,ds) = q\,ds$, or $q = dF/ds$. In other words, ***the shear flow measures the force per unit length along the cross section***.

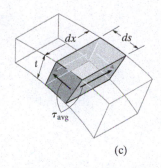

(c)

Average Shear Stress.

The average shear stress can be related to the torque T by considering the torque produced by this shear stress about a selected point O within the tube's boundary, Fig. 5–26d. As shown, the shear stress develops a force $dF = \tau_{avg}\,dA = \tau_{avg}\,(t\,ds)$ on an element of the tube. This force acts tangent to the centerline of the tube's wall, and if the moment arm is h, the torque is

$$dT = h(dF) = h(\tau_{avg}\,t\,ds)$$

For the entire cross section, we require

$$T = \oint h\tau_{avg}\,t\,ds$$

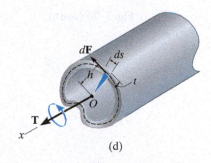

(d)

Fig. 5–26 (cont.)

Here the "line integral" indicates that integration must be performed *around* the entire boundary. Since the shear flow $q = \tau_{avg}\,t$ is *constant*, it can be factored out of the integral, so that

$$T = \tau_{avg}t \oint h\,ds$$

A graphical simplification can be made for evaluating the integral by noting that the *mean area*, shown by the blue colored triangle in Fig. 5–26d, is $dA_m = (1/2)h\,ds$. Thus,

$$T = 2\tau_{avg}\,t \int dA_m = 2\tau_{avg}\,tA_m$$

Solving for τ_{avg}, we have

$$\tau_{avg} = \frac{T}{2tA_m} \qquad (5\text{–}18)$$

Here

τ_{avg} = the average shear stress acting over a particular thickness of the tube

T = the resultant internal torque at the cross section

t = the thickness of the tube where τ_{avg} is to be determined

A_m = the mean area enclosed within the boundary of the centerline of the tube's thickness, shown shaded in Fig. 5–26e

Finally, since $q = \tau_{avg}\,t$, then the shear flow throughout the cross section becomes

$$q = \frac{T}{2A_m} \qquad (5\text{–}19)$$

Angle of Twist.

The angle of twist of a thin-walled tube of length L can be determined using energy methods, and the development of the necessary equation is given as a problem later in the book.* If the material behaves in a linear elastic manner and G is the shear modulus, then this angle ϕ, given in radians, can be expressed as

$$\phi = \frac{TL}{4A_m^2\,G} \oint \frac{ds}{t} \qquad (5\text{–}20)$$

Here again the integration must be performed around the entire boundary of the tube's cross-sectional area.

(e)

Fig. 5–26 (cont.)

▶ IMPORTANT POINTS

- Shear flow q is the product of the tube's thickness and the average shear stress. This value is the same at all points along the tube's cross section. As a result, the *largest* average shear stress on the cross section occurs where the thickness is *smallest*.

- Both shear flow and the average shear stress act *tangent* to the wall of the tube at all points and in a direction so as to contribute to the resultant internal torque.

*See Prob. 14-14.

EXAMPLE 5.12

Calculate the average shear stress in a thin-walled tube having a circular cross section of mean radius r_m and thickness t, which is subjected to a torque T, Fig. 5–27a. Also, what is the relative angle of twist if the tube has a length L?

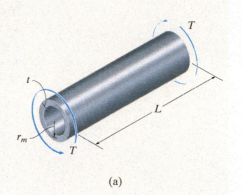

(a)

SOLUTION

Average Shear Stress.   The mean area for the tube is $A_m = \pi r_m^2$. Applying Eq. 5–18 gives

$$\tau_{avg} = \frac{T}{2tA_m} = \frac{T}{2\pi t r_m^2} \qquad \textit{Ans.}$$

We can check the validity of this result by applying the torsion formula. Here

$$J = \frac{\pi}{2}(r_o^4 - r_i^4)$$

$$= \frac{\pi}{2}(r_o^2 + r_i^2)(r_o^2 - r_i^2)$$

$$= \frac{\pi}{2}(r_o^2 + r_i^2)(r_o + r_i)(r_o - r_i)$$

Since $r_m \approx r_o \approx r_i$ and $t = r_o - r_i$, $J = \frac{\pi}{2}(2r_m^2)(2r_m)t = 2\pi r_m^3 t$

$$\tau_{avg} = \frac{T r_m}{J} = \frac{T r_m}{2\pi r_m^3 t} = \frac{T}{2\pi t r_m^2} \qquad \textit{Ans.}$$

which agrees with the previous result.

The average shear-stress distribution acting throughout the tube's cross section is shown in Fig. 5–27b. Also shown is the shear-stress distribution acting on a radial line as calculated using the torsion formula. Note that as the tube's thickness decreases, the shear stress throughout the tube approaches the average shear stress.

Actual shear-stress distribution (torsion formula)

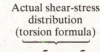

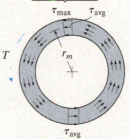

Average shear-stress distribution (thin-wall approximation)

(b)

Fig. 5–27

Angle of Twist.   Applying Eq. 5–20, we have

$$\phi = \frac{TL}{4A_m^2 G}\oint\frac{ds}{t} = \frac{TL}{4(\pi r_m^2)^2 Gt}\oint ds$$

The integral represents the length around the centerline boundary, which is $2\pi r_m$. Substituting, the final result is

$$\phi = \frac{TL}{2\pi r_m^3 Gt} \qquad \textit{Ans.}$$

Show that one obtains this same result using Eq. 5–15.

EXAMPLE   5.13

The tube is made of C86100 bronze and has a rectangular cross section as shown in Fig. 5–28a. If it is subjected to the two torques, determine the average shear stress in the tube at points A and B. Also, what is the angle of twist of end C? The tube is fixed at E.

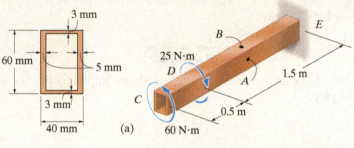

Fig. 5–28

SOLUTION

Average Shear Stress.   If the tube is sectioned through points A and B, the resulting free-body diagram is shown in Fig. 5–28b. The internal torque is 35 N · m. As shown in Fig. 5–28d, the mean area is

$$A_m = (0.035 \text{ m})(0.057 \text{ m}) = 0.00200 \text{ m}^2$$

Applying Eq. 5–18 for point A, $t = 5$ mm, and so

$$\tau_A = \frac{T}{2tA_m} = \frac{35 \text{ N} \cdot \text{m}}{2(0.005 \text{ m})(0.00200 \text{ m}^2)} = 1.75 \text{ MPa} \qquad Ans.$$

And for point B, $t = 3$ mm, and so

$$\tau_B = \frac{T}{2tA_m} = \frac{35 \text{ N} \cdot \text{m}}{2(0.003 \text{ m})(0.00200 \text{ m}^2)} = 2.92 \text{ MPa} \qquad Ans.$$

These results are shown on elements of material located at points A and B, Fig. 5–28e. Note carefully how the 35-N·m torque in Fig. 5–28b creates these stresses on the back sides of each element.

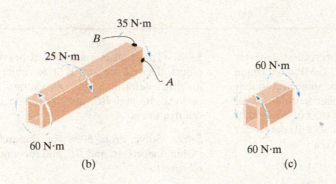

(b) (c)

Angle of Twist. From the free-body diagrams in Fig. 5–28b and 5–28c, the internal torques in regions *DE* and *CD* are 35 N · m and 60 N · m, respectively. Following the sign convention outlined in Sec. 5.4, these torques are both positive. Thus, Eq. 5–20 becomes

$$\phi = \Sigma \frac{TL}{4A_m^2 \, G} \oint \frac{ds}{t}$$

$$= \frac{60 \text{ N} \cdot \text{m} \, (0.5 \text{ m})}{4(0.00200 \text{ m}^2)^2 \, (38(10^9) \text{ N/m}^2)} \left[2\left(\frac{57 \text{ mm}}{5 \text{ mm}}\right) + 2\left(\frac{35 \text{ mm}}{3 \text{ mm}}\right) \right]$$

$$+ \frac{35 \text{ N} \cdot \text{m} \, (1.5 \text{ m})}{4(0.00200 \text{ m}^2)^2 \, (38(10^9) \text{ N/m}^2)} \left[2\left(\frac{57 \text{ mm}}{5 \text{ mm}}\right) + 2\left(\frac{35 \text{ mm}}{3 \text{ mm}}\right) \right]$$

$$= 6.29(10^{-3}) \text{ rad} = 0.360° \qquad\qquad \textit{Ans.}$$

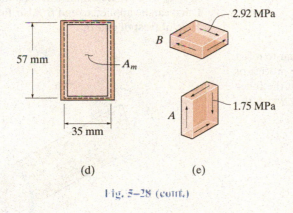

(d) (e)

Fig. 5–28 (cont.)

PROBLEMS

5–95. The brass wire has a triangular cross section, 2 mm on a side. If the yield stress for brass is $\tau_Y = 205$ MPa, determine the maximum torque T to which it can be subjected so that the wire will not yield. If this torque is applied to the 4-m-long segment, determine the greatest angle of twist of one end of the wire relative to the other end that will not cause permanent damage to the wire. $G_{br} = 37$ GPa.

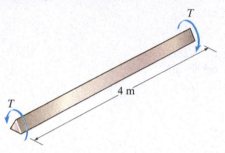

Prob. 5–95

*5–96. If $a = 25$ mm and $b = 15$ mm, determine the maximum shear stress in the circular and elliptical shafts when the applied torque is $T = 80$ N · m. By what percentage is the shaft of circular cross section more efficient at withstanding the torque than the shaft of elliptical cross section?

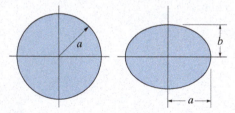

Prob. 5–96

5–97. It is intended to manufacture a circular bar to resist torque; however, the bar is made elliptical in the process of manufacturing, with one dimension smaller than the other by a factor k as shown. Determine the factor by which the maximum shear stress is increased.

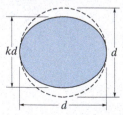

Prob. 5–97

5–98. The shaft is made of red brass C83400 and has an elliptical cross section. If it is subjected to the torsional loading, determine the maximum shear stress within regions AC and BC, and the angle of twist ϕ of end B relative to end A.

5–99. Solve Prob. 5–98 for the maximum shear stress within regions AC and BC, and the angle of twist ϕ of end B relative to C.

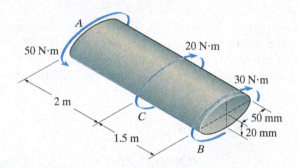

Probs. 5–98/99

*5–100. If end B of the shaft, which has an equilateral triangular cross section, is subjected to a torque of $T = 900$ lb · ft, determine the maximum shear stress in the shaft. Also, find the angle of twist of end B. The shaft is made from 6061-T1 aluminum.

5–101. If the shaft has an equilateral triangle cross section and is made from an alloy that has an allowable shear stress of $\tau_{allow} = 12$ ksi, determine the maximum allowable torque T that can be applied to end B. Also, find the corresponding angle of twist of end B.

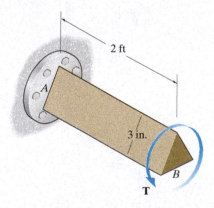

Probs. 5–100/101

5–102. The 2014-T6 aluminum strut is fixed between the two walls at A and B. If it has a 2 in. by 2 in. square cross section, and it is subjected to the torsional loading shown, determine the reactions at the fixed supports. Also, what is the angle of twist at C?

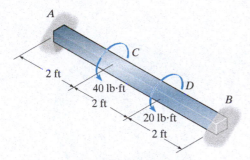

Prob. 5–102

5–103. A torque of 2 kip · in. is applied to the tube. If the wall thickness is 0.1 in., determine the average shear stress in the tube.

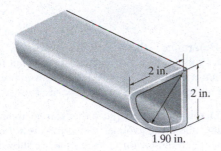

Prob. 5–103

*5–104. The 6061-T6 aluminum bar has a square cross section of 25 mm by 25 mm. If it is 2 m long, determine the maximum shear stress in the bar and the rotation of one end relative to the other end.

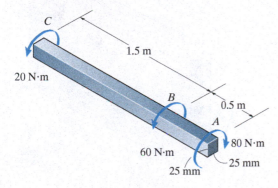

Prob. 5–104

5–105. If the shaft is subjected to the torque of 3 kN · m, determine the maximum shear stress developed in the shaft. Also, find the angle of twist of end B. The shaft is made from A-36 steel. Set $a = 50$ mm.

5–106. If the shaft is made from A-36 steel having an allowable shear stress of $\tau_{allow} = 75$ MPa, determine the minimum dimension a for the cross section to the nearest millimeter. Also, find the corresponding angle of twist at end B.

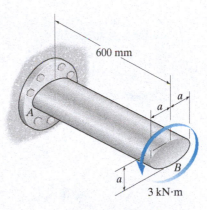

Probs. 5–105/106

5–107. If the solid shaft is made from red brass C83400 having an allowable shear stress of $\tau_{allow} = 4$ ksi, determine the maximum allowable torque **T** that can be applied at B.

*5–108. If the solid shaft is made from red brass C83400 and it is subjected to a torque $T = 6$ kip · ft at B, determine the maximum shear stress developed in segments AB and BC.

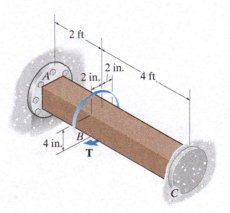

Probs. 5–107/108

5–109. For a given maximum average shear stress, determine the factor by which the torque-carrying capacity is increased if the half-circular section is reversed from the dashed-line position to the section shown. The tube is 0.1 in. thick.

*5–112. Due to a fabrication error the inner circle of the tube is eccentric with respect to the outer circle. By what percentage is the torsional strength reduced when the eccentricity e is one-fourth of the difference in the radii?

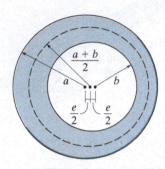

Prob. 5–112

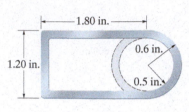

Prob. 5–109

5–113. Determine the constant thickness of the rectangular tube if the average shear stress is not to exceed 12 ksi when a torque of $T = 100$ kip·in. is applied to the tube. The mean dimensions of the tube are shown.

5–110. The plastic tube is subjected to a torque of 150 N·m. Determine the mean dimension a of its sides if the allowable shear stress is $\tau_{allow} = 60$ MPa. Each side has a thickness of $t = 3$ mm.

5–114. Determine the torque T that can be applied to the rectangular tube if the average shear stress is not to exceed 12 ksi. The mean dimensions of the tube are shown and the tube has a thickness of 0.25 in.

5–111. The plastic tube is subjected to a torque of 150 N·m. Determine the average shear stress in the tube if the mean dimension $a = 200$ mm. Each side has a thickness of $t = 3$ mm.

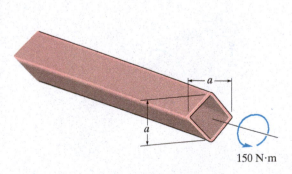

Probs. 5–110/111

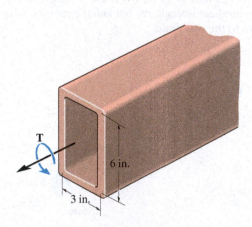

Probs. 5–113/114

5–115. The steel tube has an elliptical cross section of mean dimensions shown and a constant thickness of $t = 0.2$ in. If the allowable shear stress is $\tau_{\text{allow}} = 8$ ksi, and the tube is to resist a torque of $T = 250$ lb·ft, determine the necessary dimension b. The mean area for the ellipse is $A_m = \pi b(0.5b)$.

5–117. The 304 stainless steel tube has a thickness of 10 mm. If the allowable shear stress is $\tau_{\text{allow}} = 80$ MPa, determine the maximum torque T that it can transmit. Also, what is the angle of twist of one end of the tube with respect to the other if the tube is 4 m long? The mean dimensions are shown.

5–118. The 304 stainless steel tube has a thickness of 10 mm. If the applied torque is $T = 50$ N·m, determine the average shear stress in the tube. The mean dimensions are shown.

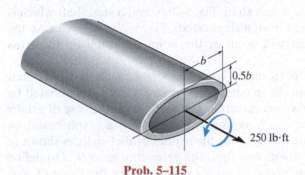

Prob. 5–115

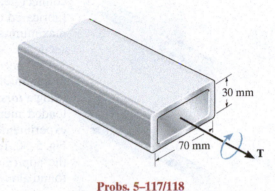

Probs. 5–117/118

*5–116. The plastic hexagonal tube is subjected to a torque of 150 N·m. Determine the mean dimension a of its sides if the allowable shear stress is $\tau_{\text{allow}} = 60$ MPa. Each side has a thickness of $t = 3$ mm.

5–119. The symmetric tube is made from a high-strength steel, having the mean dimensions shown and a thickness of 5 mm. If it is subjected to a torque of $T = 40$ N·m, determine the average shear stress developed at points A and B. Indicate the shear stress on volume elements located at these points.

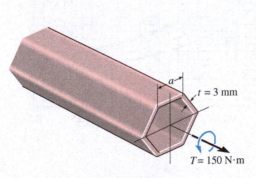

Prob. 5–116

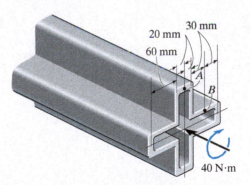

Prob. 5–119

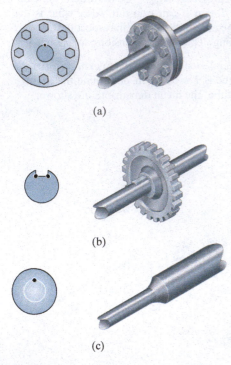

(a)

(b)

(c)

Fig. 5–29

5.8   STRESS CONCENTRATION

The torsion formula, $\tau_{max} = Tc/J$, cannot be applied to regions of a shaft having a sudden change in the cross section, because the shear-stress and shear-strain distributions in the shaft become complex. Results can be obtained, however, by using experimental methods or possibly by a mathematical analysis based on the theory of elasticity.

Three common discontinuities of the cross section that occur in practice are shown in Fig. 5–29. They are at *couplings*, which are used to connect two collinear shafts together, Fig. 5–29a, *keyways*, used to connect gears or pulleys to a shaft, Fig. 5–29b, and a step shaft which is fabricated or machined from a single shaft, Fig. 5–29c. In each case the maximum shear stress will occur at the point indicated on the cross section.

The necessity to perform a complex stress analysis at a shaft discontinuity to obtain the maximum shear stress can be eliminated by using a ***torsional stress concentration factor***, K. As in the case of axially loaded members, Sec. 4.7, K is usually taken from a graph based on experimental data. An example, for the shoulder-fillet shaft, is shown in Fig. 5–30. To use this graph, one finds the geometric ratio D/d to define the appropriate curve, and then after calculating r/d the value of K is found along the vertical axis.

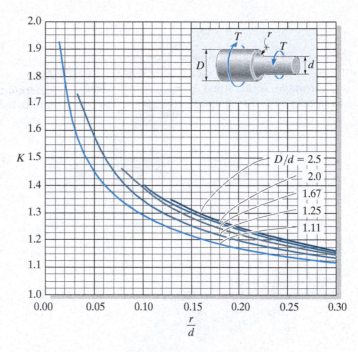

Fig. 5–30

The maximum shear stress is then determined from

$$\tau_{max} = K\frac{Tc}{J}$$ (5–21)

Stress concentrations can arise at the coupling of these shafts, and this must be taken into account when the shaft is designed.

Here the torsion formula is applied to the *smaller* of the two connected shafts, since τ_{max} occurs at the base of the fillet, Fig. 5–29c.

Note from the graph that an *increase* in fillet radius r causes a *decrease* in K. Hence the maximum shear stress in the shaft can be *reduced* by *increasing* the radius. Also, if the diameter of the larger section is reduced, the D/d ratio will be lower and so the value of K and therefore τ_{max} will be lower.

Like the case of axially loaded members, torsional stress concentration factors should *always* be used when designing shafts made of *brittle materials*, or when designing shafts that will be subjected to *fatigue or cyclic torsional loadings*. These conditions give rise to the formation of cracks at the stress concentration, and this can often lead to a sudden fracture. On the other hand, if large *static* torsional loadings are applied to a shaft made of *ductile material*, then *inelastic strains* will develop within the shaft. Yielding of the material will cause the stress distribution to become more *evenly distributed* throughout the shaft, so that the maximum stress will not be limited to the region of stress concentration. This effect is discussed further in the next section.

IMPORTANT POINTS

- *Stress concentrations* in shafts occur at points of sudden cross-sectional change, such as couplings, keyways, and step shafts. The more severe the change in geometry, the larger the stress concentration.

- For design or analysis, it is not necessary to know the exact shear-stress distribution on the cross section. Instead, it is possible to obtain the maximum shear stress using a stress concentration factor, K, that has been determined through experiment. Its value is only a function of the geometry of the shaft.

- Normally a stress concentration in a ductile shaft subjected to a static torque will *not* have to be considered in design; however, if the material is *brittle*, or subjected to *fatigue* loadings, then stress concentrations become important.

5

EXAMPLE 5.14

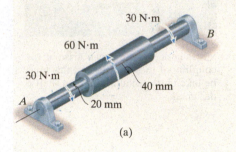

(a)

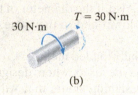

$T = 30 \text{ N·m}$

30 N·m

(b)

The stepped shaft shown in Fig. 5–31a is supported by bearings at A and B. Determine the maximum stress in the shaft due to the applied torques. The shoulder fillet at the junction of each shaft has a radius of $r = 6$ mm.

$\tau_{max} = 3.10$ MPa

Shear-stress distribution predicted by torsion formula

Actual shear-stress distribution caused by stress concentration

(c)

Fig. 5–31

SOLUTION

Internal Torque. By inspection, moment equilibrium about the axis of the shaft is satisfied. Since the maximum shear stress occurs at the rooted ends of the *smaller* diameter shafts, the internal torque (30 N·m) can be found there by applying the method of sections, Fig. 5–31b.

Maximum Shear Stress. The stress concentration factor can be determined by using Fig. 5–30. From the shaft geometry we have

$$\frac{D}{d} = \frac{2(40 \text{ mm})}{2(20 \text{ mm})} = 2$$

$$\frac{r}{d} = \frac{6 \text{ mm}}{2(20 \text{ mm})} = 0.15$$

Thus, the value of $K = 1.3$ is obtained.

Applying Eq. 5–21, we have

$$\tau_{max} = K\frac{Tc}{J}; \quad \tau_{max} = 1.3\left[\frac{30 \text{ N·m } (0.020 \text{ m})}{(\pi/2)(0.020 \text{ m})^4}\right] = 3.10 \text{ MPa} \qquad \textit{Ans.}$$

NOTE: From experimental evidence, the actual stress distribution along a radial line of the cross section at the critical section looks similar to that shown in Fig. 5–31c. Notice how this compares with the linear stress distribution found from the torsion formula.

*5.9   INELASTIC TORSION

If the torsional loadings applied to the shaft are excessive, then the material may yield, and, consequently, a "plastic analysis" must be used to determine the shear-stress distribution and the angle of twist.

It was shown in Sec. 5.1 that regardless of the material behavior, the shear strains that develop in a circular shaft will vary *linearly*, from zero at the center of the shaft to a maximum at its outer boundary, Fig. 5–32a. Also, the torque at the section must be equivalent to the torque caused by the entire shear-stress distribution acting on the cross section. Since the shear stress τ acting on an element of area dA, Fig. 5–32b, produces a force of $dF = \tau \, dA$, then the torque about the axis of the shaft is $dT = \rho \, dF = \rho(\tau \, dA)$. For the entire shaft we require

$$T = \int_A \rho\tau \, dA \qquad (5\text{–}22)$$

If the area dA over which τ acts is defined as a *differential ring* having an area of $dA = 2\pi\rho \, d\rho$, Fig. 5–32c, then the above equation can be written as

$$\boxed{T = 2\pi \int_0^c \tau\rho^2 \, d\rho} \qquad (5\text{–}23)$$

We will now apply this equation to a shaft that is subjected to two types of torque.

Severe twist of an aluminum specimen caused by the application of a plastic torque.

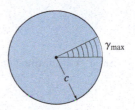

Linear shear–strain
distribution

(a)

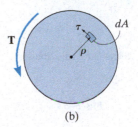

(b)

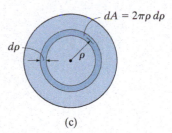

(c)

Fig. 5–32

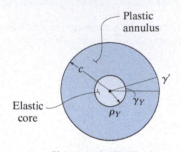

(a)

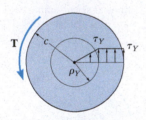

Shear–strain distribution

(b)

Shear–stress distribution

(c)

Fig. 5–33

Elastic-Plastic Torque.

Let us consider the material in the shaft to exhibit an elastic perfectly plastic behavior, as shown in Fig. 5–33a.

If the internal torque produces the maximum *elastic* shear strain γ_Y, at the outer boundary of the shaft, then the maximum elastic torque T_Y that produces this strain can be found from the torsion formula, $\tau_Y = T_Y c / [(\pi/2)c^4]$, so that

$$T_Y = \frac{\pi}{2}\tau_Y c^3 \qquad (5\text{–}24)$$

If the applied torque increases in magnitude above T_Y, it will begin to cause yielding, which will start at the outer boundary of the shaft, $\rho = c$. As the maximum shear strain increases to, say, γ', then if the material is *elastic perfectly plastic*, Fig. 5–33a, the yielding boundary will progress inward toward the shaft's center, Fig. 5–33b. As shown, this produces an *elastic core*, where, by proportion, the radius of the core is $\rho_Y = (\gamma_Y/\gamma')c$. The outer portion of the material forms a *plastic annulus* or ring, since the shear strains γ within this region are greater than γ_Y. The corresponding shear-stress distribution along a radial line of the shaft is shown in Fig. 5–33c. It is established by taking successive points on the shear-strain distribution in Fig. 5–33b and finding the corresponding value of shear stress from the τ–γ diagram, Fig. 5–33a. For example, at $\rho = c$, γ' gives τ_Y, and at $\rho = \rho_Y$, γ_Y also gives τ_Y, etc.

Since τ in Fig. 5–33c can now be expressed as a function of ρ, we can apply Eq. 5–23 to determine the torque. We have

$$
\begin{aligned}
T &= 2\pi \int_0^c \tau \rho^2 \, d\rho \\[2mm]
&= 2\pi \int_0^{\rho_Y} \left(\tau_Y \frac{\rho}{\rho_Y} \right) \rho^2 \, d\rho + 2\pi \int_{\rho_Y}^c \tau_Y \rho^2 \, d\rho \\[2mm]
&= \frac{2\pi}{\rho_Y} \tau_Y \int_0^{\rho_Y} \rho^3 \, d\rho + 2\pi \tau_Y \int_{\rho_Y}^c \rho^2 \, d\rho \\[2mm]
&= \frac{\pi}{2\rho_Y} \tau_Y \rho_Y^4 + \frac{2\pi}{3} \tau_Y (c^3 - \rho_Y^3) \\[2mm]
&= \frac{\pi \tau_Y}{6} (4c^3 - \rho_Y^3) \qquad (5\text{–}25)
\end{aligned}
$$

Plastic Torque.   Further increases in T tend to shrink the radius of the elastic core until all the material yields, i.e., $\rho_Y \to 0$, Fig. 5–33b. The material of the shaft will then be subjected to *perfectly plastic behavior* and the shear-stress distribution becomes uniform, so that $\tau = \tau_Y$. We can now apply Eq. 5–23 to determine the *plastic torque* T_p, which represents the largest possible torque the shaft will support.

$$T_p = 2\pi \int_0^c \tau_Y\, \rho^2 d\rho$$

$$= \frac{2\pi}{3}\, \tau_Y\, c^3 \qquad (5\text{–}26)$$

Compared with the maximum elastic torque T_Y, Eq. 5–24, it can be seen that

$$T_p = \frac{4}{3} T_Y$$

In other words, the plastic torque is 33% greater than the maximum elastic torque.

Unfortunately, the angle of twist ϕ for the shear-stress distribution in Fig. 5–33d *cannot* be uniquely defined. This is because $\tau = \tau_Y$ does not correspond to any unique value of shear strain $\gamma \geq \gamma_Y$. As a result, once $\mathbf{T}_p$ is applied, the shaft will continue to deform or twist with no corresponding increase in shear stress.

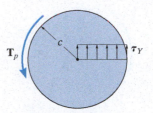

Fully plastic torque

(d)

Fig. 5–33 (cont.)

*5.10   RESIDUAL STRESS

When a shaft is subjected to plastic shear strains caused by torsion, removal of the torque will cause some shear stress to remain in the shaft. This stress is referred to as *residual stress*, and its distribution can be calculated using superposition.

For example, if T_p causes the material at the outer boundary of the shaft to be strained to γ_1, shown as point C on the $\tau–\gamma$ curve in Fig. 5–34, the release of T_p will cause a reverse shear stress, such that the material will recover some of the shear strain and follow the straight-lined segment CD. This will be an *elastic recovery*, and so this line is parallel to the initial straight-lined portion AB of the $\tau–\gamma$ diagram. In other words, both lines have the same slope G.

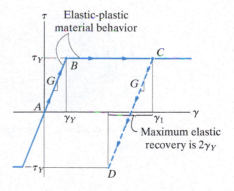

Fig. 5–34

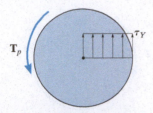

Plastic torque applied
causing plastic shear strains
throughout the shaft
(a)

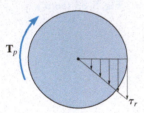

Plastic torque reversed
causing elastic shear strains
throughout the shaft
(b)

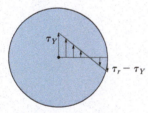

Residual shear–stress
distribution in shaft
(c)

Since elastic recovery occurs, we can superimpose on the plastic torque stress distribution in Fig. 5–35a a *linear stress distribution* caused by applying the plastic torque $\mathbf{T}_p$ in the *opposite* direction, Fig. 5–35b. Here the maximum shear stress τ_r for this stress distribution is called the **modulus of rupture** for torsion. It is determined from the torsion formula,* which gives

$$\tau_r = \frac{T_p c}{J} = \frac{T_p c}{(\pi/2)c^4}$$

Using Eq. 5–26,

$$\tau_r = \frac{\left[(2/3)\pi\tau_Y c^3\right]c}{(\pi/2)c^4} = \frac{4}{3}\tau_Y$$

Note that reversed application of $\mathbf{T}_p$ using the linear shear-stress distribution in Fig. 5–35b is possible here, since the maximum possible recovery for the elastic shear strain is $2\gamma_Y$, as noted in Fig. 5–34. This would correspond to a maximum applied shear stress of $2\tau_Y$, which is *greater* than the *required* shear stress of $\frac{4}{3}\tau_Y$ calculated above. Hence, by superimposing the stress distributions involving applications and then removal of the plastic torque, we obtain the residual shear-stress distribution in the shaft as shown in Fig. 5–35c. Actually the shear stress at the center of the shaft, shown as τ_Y, should be *zero*, since the material along the axis of the shaft is never strained. The reason it is not zero is that we assumed *all* the material of the shaft to have been strained beyond the yield point in order to determine the plastic torque, Fig. 5–35a. To be more realistic, however, an elastic-plastic torque should be considered when modeling the material behavior. Doing this leads to the superposition of the stress distribution shown in Fig. 5–35d.

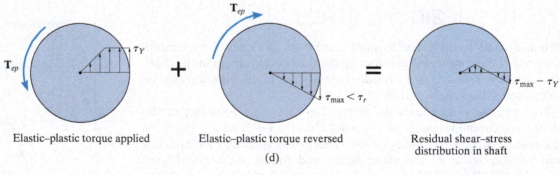

Elastic–plastic torque applied   Elastic–plastic torque reversed   Residual shear–stress
distribution in shaft
(d)

Fig. 5–35

*The torsion formula is valid only when the material behaves in a linear elastic manner; however, the modulus of rupture is so named because it assumes that the material behaves elastically and then suddenly ruptures at the proportional limit.

Ultimate Torque.

In the general case, most engineering materials will have a shear stress–strain diagram as shown in Fig. 5–36a. Consequently, if T is increased so that the maximum shear strain in the shaft becomes $\gamma = \gamma_u$, Fig. 5–36b, then, by proportion γ_Y occurs at $\rho_Y = (\gamma_Y/\gamma_u)c$. Likewise, the shear strains at, say, $\rho = \rho_1$ and $\rho = \rho_2$, can be found by proportion, i.e., $\gamma_1 = (\rho_1/c)\gamma_u$ and $\gamma_2 = (\rho_2/c)\gamma_u$. If the corresponding values of τ_1, τ_Y, τ_2, and τ_u are taken from the $\tau-\gamma$ diagram and plotted, we obtain the shear-stress distribution, which acts along a radial line on the cross section, Fig. 5–36c. The torque produced by this stress distribution is called the **ultimate torque**, T_u.

The magnitude of $\mathbf{T}_u$ can be determined by "graphically" integrating Eq. 5–23. To do this, the cross-sectional area of the shaft is segmented into a finite number of small rings, such as the one shown shaded in Fig. 5–36d. The area of this ring, $\Delta A = 2\pi\rho\,\Delta\rho$, is multiplied by the shear stress τ that acts on it, so that the force $\Delta F = \tau\,\Delta A$ can be determined. The torque created by this force is then $\Delta T = \rho\,\Delta F = \rho(\tau\,\Delta A)$. The addition of all the torques for the entire cross section, as determined in this manner, gives the ultimate torque T_u; that is, Eq. 5–23 becomes $T_u \approx 2\pi\Sigma\tau\rho^2\,\Delta\rho$. Of course, if the stress distribution can be expressed as an analytical function, $\tau = f(\rho)$, as in the elastic and plastic torque cases, then the integration of Eq. 5–23 can be carried out directly.

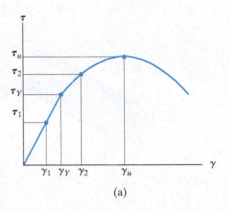

(a)

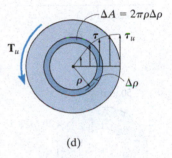

$\Delta A = 2\pi\rho\Delta\rho$

(d)

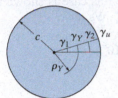

Ultimate shear-strain distribution

(b)

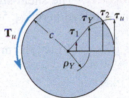

Ultimate shear-stress distribution

(c)

Fig. 5–36

IMPORTANT POINTS

- The *shear-strain distribution* along a radial line on the cross section of a shaft is based on geometric considerations, and it is found to *always* vary linearly along the radial line. Once it is established, the shear-stress distribution can then be determined using the shear stress–strain diagram.

- If the shear-stress distribution for the shaft is established, then the resultant torque it produces is equivalent to the resultant internal torque acting on the cross section.

- *Perfectly plastic behavior* assumes the shear-stress distribution is *constant* over each radial line. When it occurs, the shaft will continue to twist with no increase in torque. This torque is called the *plastic torque*.

EXAMPLE 5.15

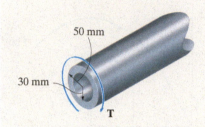

50 mm

30 mm

T

τ (MPa)

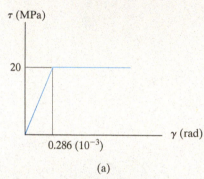

20

0.286 (10^{-3})

γ (rad)

(a)

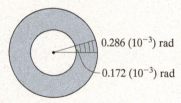

50 mm   12 MPa

20 MPa

30 mm

Elastic shear–stress distribution

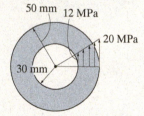

0.286 (10^{-3}) rad

0.172 (10^{-3}) rad

Elastic shear–strain distribution

(b)

Fig. 5–37

The tubular shaft in Fig. 5–37a is made of an aluminum alloy that is assumed to have an elastic perfectly plastic τ–γ diagram as shown. Determine the maximum torque that can be applied to the shaft without causing the material to yield, and the plastic torque that can be applied to the shaft. Also, what should the minimum shear strain at the outer wall be in order to develop a fully plastic torque?

SOLUTION

Maximum Elastic Torque. We require the shear stress at the outer fiber to be 20 MPa. Using the torsion formula, we have

$$\tau_Y = \frac{T_Y c}{J}; \qquad 20(10^6)\,\text{N/m}^2 = \frac{T_Y(0.05\,\text{m})}{(\pi/2)\big[(0.05\,\text{m})^4 - (0.03\,\text{m})^4\big]}$$

$$T_Y = 3.42\,\text{kN}\cdot\text{m} \qquad \textit{Ans.}$$

The shear-stress and shear-strain distributions for this case are shown in Fig. 5–37b. The values at the tube's inner wall have been obtained by proportion.

Plastic Torque. The shear-stress distribution in this case is shown in Fig. 5–37c. Application of Eq. 5–23 requires $\tau = \tau_Y$. We have

$$T_p = 2\pi \int_{0.03\,\text{m}}^{0.05\,\text{m}} \big[20(10^6)\,\text{N/m}^2\big]\rho^2\,d\rho = 125.66(10^6)\frac{1}{3}\rho^3 \,\bigg|_{0.03\,\text{m}}^{0.05\,\text{m}}$$

$$= 4.11\,\text{kN}\cdot\text{m} \qquad \textit{Ans.}$$

For this tube T_p represents a 20% increase in torque capacity compared with the elastic torque T_Y.

Outer Radius Shear Strain. The tube becomes fully plastic when the shear strain at the *inner wall* becomes $0.286(10^{-3})$ rad, as shown in Fig. 5–37c. Since the shear strain *remains linear* over the cross section, the plastic strain at the outer fibers of the tube in Fig. 5–37c is determined by proportion.

$$\frac{\gamma_o}{50\,\text{mm}} = \frac{0.286(10^{-3})\,\text{rad}}{30\,\text{mm}}$$

$$\gamma_o = 0.477(10^{-3})\,\text{rad} \qquad \textit{Ans.}$$

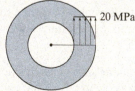

20 MPa

Plastic shear–stress distribution

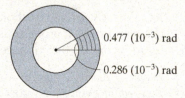

0.477 (10^{-3}) rad

0.286 (10^{-3}) rad

Initial plastic shear–strain distribution

(c)

EXAMPLE 5.16

A solid circular shaft has a radius of 20 mm and length of 1.5 m. The material has an elastic perfectly plastic τ–γ diagram as shown in Fig. 5–38a. Determine the torque needed to twist the shaft $\phi = 0.6$ rad.

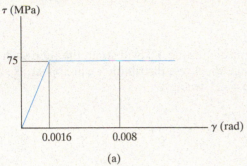

(a)

SOLUTION

We will first obtain the shear-strain distribution based on the required twist, then establish the shear-stress distribution. Once this is known, the applied torque can be determined.

The maximum shear strain occurs at the surface of the shaft, $\rho = c$. Since the angle of twist is $\phi = 0.6$ rad for the entire 1.5-m length of the shaft, then using Eq. 5–13, for the entire length we have

$$\phi = \gamma \frac{L}{\rho}; \qquad 0.6 = \frac{\gamma_{max}(1.5\text{ m})}{0.02\text{ m}}$$

$$\gamma_{max} = 0.008\text{ rad}$$

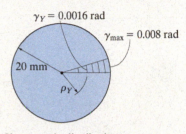

Shear–strain distribution

(b)

The shear-strain distribution is shown in Fig. 5–38b. Note that yielding of the material occurs since $\gamma_{max} > \gamma_Y = 0.0016$ rad in Fig. 5–38a. The radius of the elastic core, ρ_Y, can be obtained by proportion. From Fig. 5–38b,

$$\frac{\rho_Y}{0.0016} = \frac{0.02\text{ m}}{0.008}$$

$$\rho_Y = 0.004\text{ m} = 4\text{ mm}$$

Based on the shear-strain distribution, the shear-stress distribution, plotted over a radial line segment, is shown in Fig. 5–38c. The torque can now be obtained using Eq. 5–25. Substituting in the numerical data yields

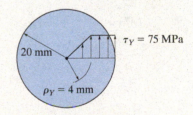

Shear–stress distribution

(c)

Fig. 5–38

$$T = \frac{\pi \tau_Y}{6}(4c^3 - \rho_Y^3)$$

$$= \frac{\pi\left[75(10^6)\text{ N/m}^2\right]}{6}[4(0.02\text{ m})^3 - (0.004\text{ m})^3]$$

$$= 1.25\text{ kN} \cdot \text{m} \qquad\qquad \textit{Ans.}$$

EXAMPLE 5.17

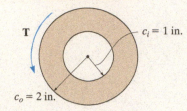

$c_i = 1$ in.

$c_o = 2$ in.

τ (ksi)

12

0.002 γ (rad)

(a)

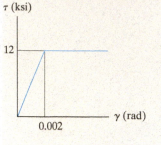

12 ksi

T_p

(b)

Plastic torque applied

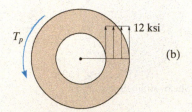

T_p

7.47 ksi

(c)

$\tau_r = 14.93$ ksi

Plastic torque reversed

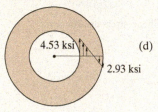

4.53 ksi

(d)

2.93 ksi

Residual shear–stress distribution

Fig. 5–39

The 5-ft-long tube and its elastic perfectly plastic τ–γ diagram are shown in Fig. 5–39a. Determine the plastic torque T_p. What is the residual shear-stress distribution in the tube if $\mathbf{T}_p$ is removed *just after* the tube becomes fully plastic?

SOLUTION

Plastic Torque.   The plastic torque $\mathbf{T}_p$ will strain the tube such that all the material yields. Hence the stress distribution will appear as shown in Fig. 5–39b. Applying Eq. 5–23, we have

$$T_p = 2\pi \int_{c_i}^{c_o} \tau_Y \rho^2 \, d\rho = \frac{2\pi}{3} \tau_Y (c_o^3 - c_i^3)$$

$$= \frac{2\pi}{3} (12(10^3) \text{ lb/in}^2)[(2 \text{ in.})^3 - (1 \text{ in.})^3] = 175.9 \text{ kip} \cdot \text{in.} \qquad Ans.$$

When the tube just becomes fully plastic, yielding has started at the inner wall, i.e., at $c_i = 1$ in., $\gamma_Y = 0.002$ rad, Fig. 5–39a. The angle of twist that occurs can be determined from Eq. 5–13, which for the entire tube becomes

$$\phi_p = \gamma_Y \frac{L}{c_i} = \frac{(0.002)(5 \text{ ft})(12 \text{ in./ft})}{(1 \text{ in.})} = 0.120 \text{ rad},$$

When $\mathbf{T}_p$ is *removed*, or in effect reapplied in the opposite direction, then the "fictitious" linear shear-stress distribution shown in Fig. 5–39c must be superimposed on the one shown in Fig. 5–39b. In Fig. 5–39c the maximum shear stress or the modulus of rupture is found from the torsion formula

$$\tau_r = \frac{T_p c_o}{J} = \frac{(175.9 \text{ kip} \cdot \text{in.})(2 \text{ in.})}{(\pi/2)[(2 \text{ in.})^4 - (1 \text{ in.})^4]} = 14.93 \text{ ksi}$$

Also, at the inner wall of the tube the shear stress is

$$\tau_i = (14.93 \text{ ksi})\left(\frac{1 \text{ in.}}{2 \text{ in.}}\right) = 7.47 \text{ ksi} \qquad Ans.$$

The resultant residual shear-stress distribution is shown in Fig. 5–39d.

PROBLEMS

*5–120.   The steel step shaft has an allowable shear stress of $\tau_{allow} = 8$ MPa. If the transition between the cross sections has a radius $r = 4$ mm, determine the maximum torque T that can be applied.

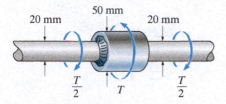

Prob. 5–120

5–121.   The built-up shaft is to be designed to rotate at 450 rpm while transmitting 230 kW of power. Is this possible? The allowable shear stress is $\tau_{allow} = 150$ MPa.

5–122.   The built-up shaft is designed to rotate at 450 rpm. If the radius of the fillet weld connecting the shafts is $r = 13.2$ mm, and the allowable shear stress for the material is $\tau_{allow} = 150$ MPa, determine the maximum power the shaft can transmit.

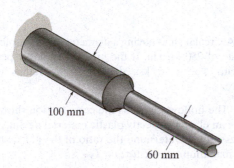

Probs. 5–121/122

5–123.   The shaft is used to transmit 30 hp while turning at 600 rpm. Determine the maximum shear stress in the shaft. The segments are connected together using a fillet weld having a radius of 0.18 in.

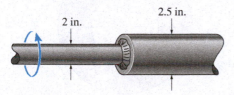

Prob. 5–123

*5–124.   The shaft is fixed to the wall at A and is subjected to the torques shown. Determine the maximum shear stress in the shaft. A fillet weld having a radius of 2.75 mm is used to connect the shafts at B.

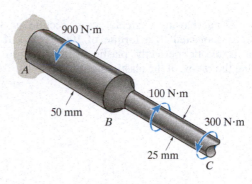

Prob. 5–124

5–125.   A solid shaft has a diameter of 40 mm and length of 1 m. It is made from an elastic-plastic material having a yield stress of $\tau_Y = 100$ MPa. Determine the maximum elastic torque T_Y and the corresponding angle of twist. What is the angle of twist if the torque is increased to $T = 1.2T_Y$? $G = 80$ GPa.

5–126.   The stepped shaft is subjected to a torque **T** that produces yielding on the surface of the larger diameter segment. Determine the radius of the elastic core produced in the smaller diameter segment. Neglect the stress concentration at the fillet.

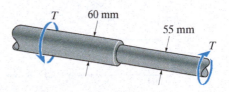

Prob. 5–126

5–127. Determine the torque needed to twist a short 2-mm-diameter steel wire through several revolutions if it is made from steel assumed to be elastic perfectly plastic and having a yield stress of $\tau_Y = 50$ MPa. Assume that the material becomes fully plastic.

*5–128. A bar having a circular cross section of 3 in. diameter is subjected to a torque of 100 in. · kip. If the material is elastic perfectly plastic, with $\tau_Y = 16$ ksi, determine the radius of the elastic core.

5–129. The solid shaft is made of an elastic perfectly plastic material. Determine the torque T needed to form an elastic core in the shaft having a radius of $\rho_Y = 20$ mm. If the shaft is 3 m long, through what angle does one end of the shaft twist with respect to the other end? When the torque is removed, determine the residual stress distribution in the shaft and the permanent angle of twist.

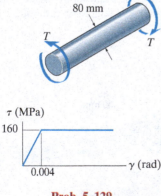

Prob. 5–129

5–130. The shaft is subjected to a maximum shear strain of 0.0048 rad. Determine the torque applied to the shaft.

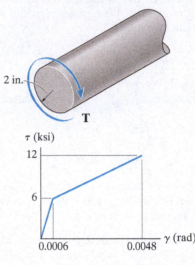

Prob. 5–130

5–131. A circular shaft having a diameter of 4 in. is subjected to a torque of 250 kip · in. If the material is elastic perfectly plastic, with $\tau_Y = 16$ ksi, determine the radius of the elastic core.

*5–132. The hollow shaft has the cross section shown and is made of an elastic perfectly plastic material having a yield shear stress of τ_Y. Determine the ratio of the plastic torque T_p to the maximum elastic torque T_Y.

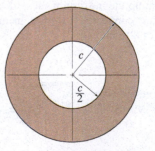

Prob. 5–132

5–133. The 2-m-long tube is made of an elastic perfectly plastic material as shown. Determine the applied torque T, which subjects the material at the tube's outer edge to a shear strain of $\gamma_{max} = 0.006$ rad. What would be the permanent angle of twist of the tube when this torque is removed? Sketch the residual stress distribution in the tube.

Prob. 5–133

5–134. The tube has a length of 2 m and is made of an elastic perfectly plastic material as shown. Determine the torque needed to just cause the material to become fully plastic. What is the permanent angle of twist of the tube when this torque is removed?

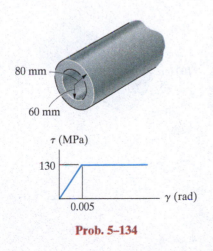

Prob. 5–134

5–135. The 3-in.-diameter shaft is made of an elastic perfectly plastic material as shown. Determine the radius of its elastic core if it is subjected to a torque of $T = 18$ kip · ft. If the shaft is 3 ft long, determine the angle of twist.

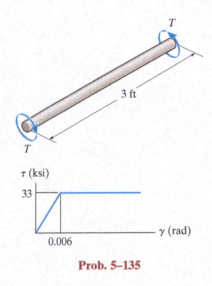

Prob. 5–135

*5–136. The shaft is made of an elastic perfectly plastic material as shown. Plot the shear-stress distribution acting along a radial line if it is subjected to a torque of $T = 20$ kN · m. What is the residual stress distribution in the shaft when the torque is removed?

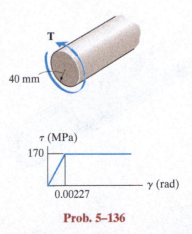

Prob. 5–136

5–137. A torque is applied to the shaft having a radius of 80 mm. If the material obeys a shear stress–strain relation of $\tau = 500\,\gamma^{1\!/\!4}$ MPa, determine the torque that must be applied to the shaft so that the maximum shear strain becomes 0.008 rad.

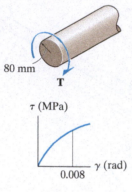

Prob. 5–137

5–138. A tubular shaft has an inner diameter of 60 mm, an outer diameter of 80 mm, and a length of 1 m. It is made of an elastic perfectly plastic material having a yield stress of $\tau_Y = 150$ MPa. Determine the maximum torque it can transmit. What is the angle of twist of one end with respect to the other end if the inner surface of the tube is about to yield? $G = 75$ GPa.

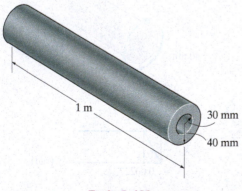

Prob. 5–138

5–139. The stepped shaft is subjected to a torque **T** that produces yielding on the surface of the larger diameter segment. Determine the radius of the elastic core produced in the smaller diameter segment. Neglect the stress concentration at the fillet.

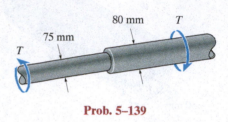

Prob. 5–139

*5–140. The tube has a length of 2 m and is made of an elastic perfectly plastic material as shown. Determine the torque needed to just cause the material to become fully plastic. What is the permanent angle of twist of the tube when this torque is removed?

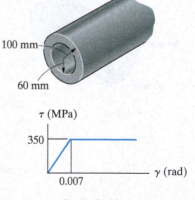

Prob. 5–140

5–141. A shaft of radius $c = 0.75$ in. is made from an elastic perfectly plastic material as shown. Determine the torque T that must be applied to its ends so that it has an elastic core of radius $\rho = 0.6$ in. If the shaft is 30 in. long, determine the angle of twist.

5–143. A steel alloy core is bonded firmly to the copper alloy tube to form the shaft shown. If the materials have the τ–γ diagrams shown, determine the torque resisted by the core and the tube.

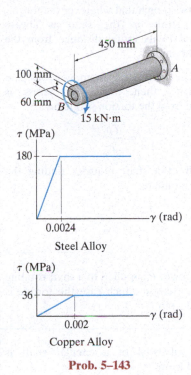

Prob. 5–143

5–142. The shaft consists of two sections that are rigidly connected. If the material is elastic perfectly plastic as shown, determine the largest torque T that can be applied to the shaft. Also, draw the shear-stress distribution over a radial line for each section. Neglect the effect of stress concentration.

5–144. The shear stress–strain diagram for a solid 50-mm-diameter shaft can be approximated as shown in the figure. Determine the torque required to cause a maximum shear stress in the shaft of 125 MPa. If the shaft is 3 m long, what is the corresponding angle of twist?

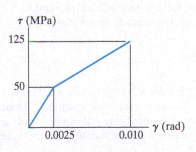

Prob. 5–144

Prob. 5–141

Prob. 5–142

CHAPTER REVIEW

Torque causes a shaft having a circular cross section to twist, such that whatever the torque, the shear strain in the shaft is always proportional to its radial distance from the center of the shaft.

Provided the material is homogeneous and linear elastic, then the shear stress is determined from the torsion formula,

$$\tau = \frac{T\rho}{J}$$

The design of a shaft requires finding the geometric parameter,

$$\frac{J}{c} = \frac{T}{\tau_{allow}}$$

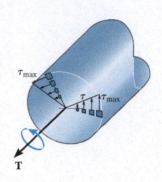
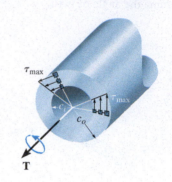

Often the power P supplied to a shaft rotating at ω is reported, in which case the torque is determined from $P = T\omega$.

The angle of twist of a circular shaft is determined from

$$\phi = \int_0^L \frac{T(x)\,dx}{J(x)G(x)}$$

If the internal torque and JG are constant within each segment of the shaft then

$$\phi = \sum \frac{TL}{JG}$$

For application, it is necessary to use a sign convention for the internal torque and to be sure the material remains linear elastic.

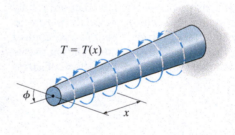
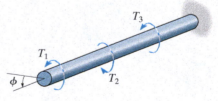

If the shaft is statically indeterminate, then the reactive torques are determined from equilibrium, compatibility of twist, and a load–displacement relationship, such as $\phi = TL/JG$.

Solid noncircular shafts tend to warp out of plane when subjected to a torque. Formulas are available to determine the maximum elastic shear stress and the twist for these cases.

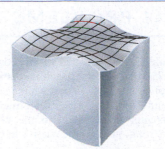

The average shear stress in thin-walled tubes is determined by assuming the shear flow across each thickness t of the tube is constant. The average shear stress value is determined from

$$\tau_{\text{avg}} = \frac{T}{2t\,A_m}$$

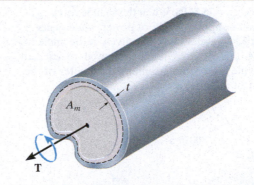

Stress concentrations occur in shafts when the cross section suddenly changes. The maximum shear stress is determined using a stress concentration factor K, which is determined from experiment and represented in graphical form. Once obtained, $\tau_{\text{max}} = K\left(\dfrac{Tc}{J}\right)$.

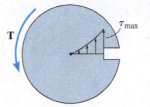

If the applied torque causes the material to exceed the elastic limit, then the stress distribution will not be proportional to the radial distance from the centerline of the shaft. Instead, the internal torque is related to the stress distribution using the shear stress–shear strain diagram and equilibrium.

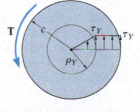

If a shaft is subjected to a plastic torque, which is then released, it will cause the material to respond elastically, thereby causing residual shear stress to be developed in the shaft.

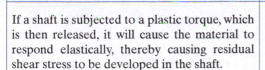

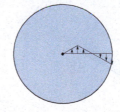

5

REVIEW PROBLEMS

R5–1.   The shaft is made of A992 steel and has an allowable shear stress of $\tau_{allow} = 75$ MPa. When the shaft is rotating at 300 rpm, the motor supplies 8 kW of power, while gears A and B withdraw 5 kW and 3 kW, respectively. Determine the required minimum diameter of the shaft to the nearest millimeter. Also, find the rotation of gear A relative to C.

R5–2.   The shaft is made of A992 steel and has an allowable shear stress of $\tau_{allow} = 75$ MPa. When the shaft is rotating at 300 rpm, the motor supplies 8 kW of power, while gears A and B withdraw 5 kW and 3 kW, respectively. If the angle of twist of gear A relative to C is not allowed to exceed 0.03 rad, determine the required minimum diameter of the shaft to the nearest millimeter.

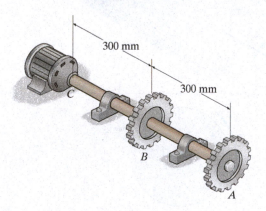

R5–1/2

R5–3.   The A-36 steel circular tube is subjected to a torque of 10 kN · m. Determine the shear stress at the mean radius $\rho = 60$ mm and calculate the angle of twist of the tube if it is 4 m long and fixed at its far end. Solve the problem using Eqs. 5–7 and 5–15 and by using Eqs. 5–18 and 5–20.

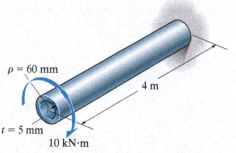

R5–3

*R5–4.   A portion of an airplane fuselage can be approximated by the cross section shown. If the thickness of its 2014-T6-aluminum skin is 10 mm, determine the maximum wing torque $\mathbf{T}$ that can be applied if $\tau_{allow} = 4$ MPa. Also, in a 4-m-long section, determine the angle of twist.

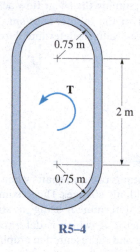

R5–4

R5–5.   The material of which each of three shafts is made has a yield stress of τ_Y and a shear modulus of G. Determine which shaft geometry will resist the largest torque without yielding. What percentage of this torque can be carried by the other two shafts? Assume that each shaft is made from the same amount of material and that it has the same cross-sectional area A.

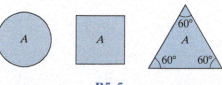

R5–5

R5–6. Segments AB and BC of the assembly are made from 6061-T6 aluminum and A992 steel, respectively. If couple forces $P = 3$ kip are applied to the lever arm, determine the maximum shear stress developed in each segment. The assembly is fixed at A and C.

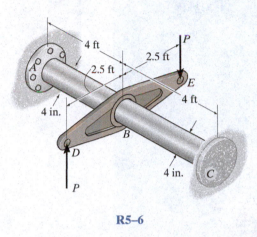

R5–6

R5–7. Segments AB and BC of the assembly are made from 6061-T6 aluminum and A992 steel, respectively. If the allowable shear stress for the aluminum is $(\tau_{allow})_{al} = 12$ ksi and for the steel $(\tau_{allow})_{st} = 10$ ksi, determine the maximum allowable couple forces P that can be applied to the lever arm. The assembly is fixed at A and C.

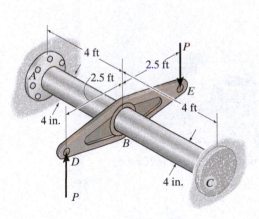

R5–7

*R5–8. The tapered shaft is made from 2014-T6 aluminum alloy, and has a radius which can be described by the equation $r = 0.02(1 + x^{3/2})$ m, where x is in meters. Determine the angle of twist of its end A if it is subjected to a torque of 450 N·m.

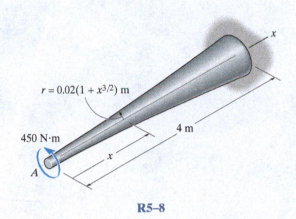

R5–8

R5–9. The 60-mm-diameter shaft rotates at 300 rev/min. This motion is caused by the unequal belt tensions on the pulley of 800 N and 450 N. Determine the power transmitted and the maximum shear stress developed in the shaft.

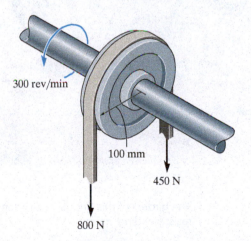

R5–9

CHAPTER 6

(© Construction Photography/Corbis)

The girders of this bridge have been designed on the basis of their ability to resist bending stress.

BENNDING

CHAPTER OBJECTIVES

■ In this chapter we will determine the stress in a beam or shaft caused by bending. The chapter begins with a discussion of how to find the variation of the shear and moment in these members. Then once the internal moment is determined, the maximum bending stress can be calculated. First we will consider members that are straight, have a symmetric cross section, and are made of homogeneous linear elastic material. Afterward we will discuss special cases involving unsymmetric bending and members made of composite materials. Consideration will also be given to curved members, stress concentrations, inelastic bending, and residual stresses.

6.1   SHEAR AND MOMENT DIAGRAMS

Members that are slender and support loadings that are applied perpendicular to their longitudinal axis are called *beams*. In general, beams are long, straight bars having a constant cross-sectional area. Often they are classified as to how they are supported. For example, a *simply supported beam* is pinned at one end and roller supported at the other, Fig. 6–1, a *cantilevered beam* is fixed at one end and free at the other, and an *overhanging beam* has one or both of its ends freely extended over

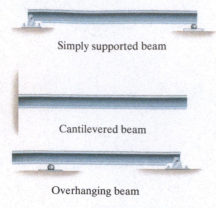

Simply supported beam

Cantilevered beam

Overhanging beam

Fig. 6–1

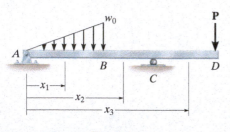

Fig. 6–2

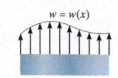

Positive external distributed load

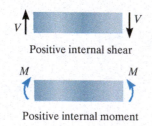

Positive internal shear

Positive internal moment

Beam sign convention

Fig. 6–3

the supports. Beams are considered among the most important of all structural elements. They are used to support the floor of a building, the deck of a bridge, or the wing of an aircraft. Also, the axle of an automobile, the boom of a crane, even many of the bones of the body act as beams.

Because of the applied loadings, beams develop an internal shear force and bending moment that, in general, vary from point to point along the axis of the beam. In order to properly design a beam it therefore becomes important to determine the *maximum* shear and moment in the beam. One way to do this is to express V and M as functions of their arbitrary position x along the beam's axis, and then plot these functions. They represent the ***shear and moment diagrams***, respectively. The maximum values of V and M can then be obtained directly from these graphs. Also, since the shear and moment diagrams provide detailed information about the *variation* of the shear and moment along the beam's axis, they are often used by engineers to decide where to place reinforcement materials within the beam or how to proportion the size of the beam at various points along its length.

In order to formulate V and M in terms of x we must choose the origin and the positive direction for x. Although the choice is arbitrary, most often the origin is located at the left end of the beam and the positive x direction is to the right.

Since beams can support portions of a distributed load and concentrated forces and couple moments, the internal shear and moment functions of x will be *discontinuous*, or their slopes will be discontinuous, at points where the loads are applied. Because of this, these functions must be determined for each region of the beam *between* any two discontinuities of loading. For example, coordinates x_1, x_2, and x_3 will have to be used to describe the variation of V and M throughout the length of the beam in Fig. 6–2. Here the coordinates are valid *only* within the regions from A to B for x_1, from B to C for x_2, and from C to D for x_3.

Beam Sign Convention.   Before presenting a method for determining the shear and moment as functions of x, and later plotting these functions (shear and moment diagrams), it is first necessary to establish a *sign convention* in order to define "positive" and "negative" values for V and M. Although the choice of a sign convention is arbitrary, here we will use the one often used in engineering practice. It is shown in Fig. 6–3. The *positive directions* are as follows: the *distributed load* acts *upward* on the beam, the internal *shear force* causes a *clockwise* rotation of the beam segment on which it acts, and the internal *moment* causes *compression* in the *top fibers* of the segment such that it bends the segment so that it "holds water". Loadings that are opposite to these are considered negative.

IMPORTANT POINTS

- *Beams* are long straight members that are subjected to loads perpendicular to their longitudinal axis. They are classified according to the way they are supported, e.g., simply supported, cantilevered, or overhanging.

- In order to properly design a beam, it is important to know the *variation* of the internal shear and moment along its axis in order to find the points where these values are a maximum.

- Using an established sign convention for positive shear and moment, the shear and moment in the beam can be determined as a function of their position x on the beam, and then these functions can be plotted to form the shear and moment diagrams.

PROCEDURE FOR ANALYSIS

The shear and moment diagrams for a beam can be constructed using the following procedure.

Support Reactions.

- Determine all the reactive forces and couple moments acting on the beam, and resolve all the forces into components acting perpendicular and parallel to the beam's axis.

Shear and Moment Functions.

- Specify separate coordinates x having an origin at the beam's *left end* and extending to regions of the beam between concentrated forces and/or couple moments, or where there is no discontinuity of distributed loading.

- Section the beam at each distance x, and draw the free-body diagram of one of the segments. Be sure **V** and **M** are shown acting in their positive sense, in accordance with the sign convention given in Fig. 6–3.

- The shear is obtained by summing forces perpendicular to the beam's axis.

- To eliminate V, the moment is obtained directly by summing moments about the sectioned end of the segment.

Shear and Moment Diagrams.

- Plot the shear diagram (V versus x) and the moment diagram (M versus x). If numerical values of the functions describing V and M are *positive*, the values are plotted above the x axis, whereas negative values are plotted below the axis.

- Generally it is convenient to show the shear and moment diagrams below the free-body diagram of the beam.

6

EXAMPLE 6.1

Draw the shear and moment diagrams for the beam shown in Fig. 6–4a.

SOLUTION

Support Reactions.   The support reactions are shown in Fig. 6–4c.

Shear and Moment Functions.   A free-body diagram of the left segment of the beam is shown in Fig. 6–4b. The distributed loading on this segment is represented by its resultant force $(3x)$ kN, which is found only *after* the segment is isolated as a free-body diagram. This force acts through the centroid of the area under the distributed loading, a distance of $x/2$ from the right end. Applying the two equations of equilibrium yields

$$+\uparrow \Sigma F_y = 0; \qquad\qquad 6\text{ kN} - (3x)\text{ kN} - V = 0$$

$$V = (6 - 3x)\text{ kN} \qquad (1)$$

$$\zeta + \Sigma M = 0; \qquad\qquad -6\text{ kN}(x) + (3x)\text{ kN}\left(\tfrac{1}{2}x\right) + M = 0$$

$$M = (6x - 1.5x^2)\text{ kN} \cdot \text{m} \qquad (2)$$

Shear and Moment Diagrams.   The shear and moment diagrams shown in Fig. 6–4c are obtained by plotting Eqs. 1 and 2. The point of *zero shear* can be found from Eq. 1:

$$V = (6 - 3x)\text{ kN} = 0$$

$$x = 2\text{ m}$$

NOTE: From the moment diagram, this value of x represents the point on the beam where the *maximum moment* occurs, since by Eq. 6–2 (see Sec. 6.2) the *slope* $V = dM/dx = 0$. From Eq. 2, we have

$$M_{\text{max}} = [6\,(2) - 1.5\,(2)^2]\text{ kN} \cdot \text{m}$$

$$= 6\text{ kN} \cdot \text{m}$$

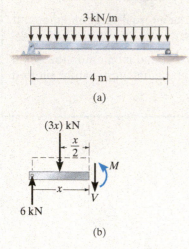

(a)

(b)

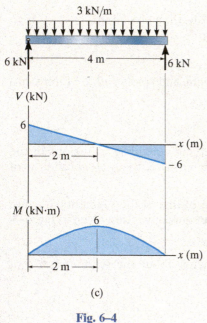

(c)

Fig. 6–4

EXAMPLE 6.2

Draw the shear and moment diagrams for the beam shown in Fig. 6–5a.

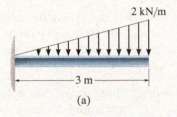

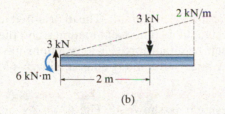

(a)   (b)

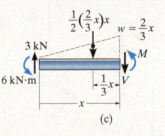

(c)

SOLUTION

Support Reactions. The distributed load is replaced by its resultant force, and the reactions have been determined, as shown in Fig. 6–5b.

Shear and Moment Functions. A free-body diagram of a beam segment of length x is shown in Fig. 6–5c. The intensity of the triangular load at the section is found by proportion, that is, $w/x = (2\text{ kN/m})/3\text{ m}$ or $w = \left(\frac{2}{3}x\right)\text{ kN/m}$. The resultant of the distributed loading is found from the area under the diagram. Thus,

$$+\uparrow \Sigma F_y = 0; \qquad 3\text{ kN} - \frac{1}{2}\left(\frac{2}{3}x\right)x - V = 0$$

$$V = \left(3 - \frac{1}{3}x^2\right)\text{ kN} \qquad (1)$$

$$\zeta + \Sigma M = 0; \qquad 6\text{ kN}\cdot\text{m} - (3\text{ kN})(x) + \frac{1}{2}\left(\frac{2}{3}x\right)x\left(\frac{1}{3}x\right) + M = 0$$

$$M = \left(-6 + 3x - \frac{1}{9}x^3\right)\text{ kN}\cdot\text{m} \qquad (2)$$

Shear and Moment Diagrams. The graphs of Eqs. 1 and 2 are shown in Fig. 6–5d.

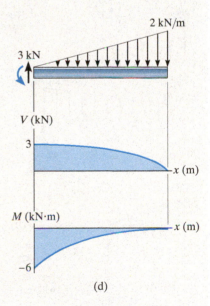

Fig. 6–5

EXAMPLE 6.3

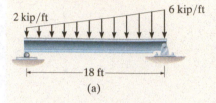

(a)

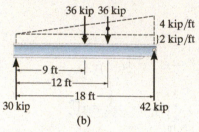

(b)

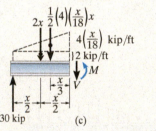

(c)

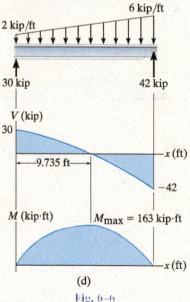

(d)

Fig. 6–6

Draw the shear and moment diagrams for the beam shown in Fig. 6–6a.

SOLUTION

Support Reactions. The distributed load is divided into triangular and rectangular component loadings, and these loadings are then replaced by their resultant forces. The reactions have been determined as shown on the beam's free-body diagram, Fig. 6–6b.

Shear and Moment Functions. A free-body diagram of the left segment is shown in Fig. 6–6c. As above, the trapezoidal loading is replaced by rectangular and triangular distributions. Here the intensity of the triangular load at the section is found by proportion. The resultant force and the location of each distributed loading are also shown. Applying the equilibrium equations, we have

$$+\uparrow \Sigma F_y = 0; \quad 30 \text{ kip} - (2 \text{ kip/ft})x - \frac{1}{2}(4 \text{ kip/ft})\left(\frac{x}{18 \text{ ft}}\right)x - V = 0$$

$$V = \left(30 - 2x - \frac{x^2}{9}\right) \text{kip} \tag{1}$$

$$\zeta + \Sigma M = 0;$$

$$-30 \text{ kip}(x) + (2 \text{ kip/ft})x\left(\frac{x}{2}\right) + \frac{1}{2}(4 \text{ kip/ft})\left(\frac{x}{18 \text{ ft}}\right)x\left(\frac{x}{3}\right) + M = 0$$

$$M = \left(30x - x^2 - \frac{x^3}{27}\right) \text{kip} \cdot \text{ft} \tag{2}$$

Shear and Moment Diagrams. Equations 1 and 2 are plotted in Fig. 6–6d. Since the point of maximum moment occurs when $dM/dx = V = 0$ (Eq. 6–2), then, from Eq. 1,

$$V = 0 = 30 - 2x - \frac{x^2}{9}$$

Choosing the positive root,

$$x = 9.735 \text{ ft}$$

Thus, from Eq. 2,

$$M_{max} = 30(9.735) - (9.735)^2 - \frac{(9.735)^3}{27} = 163 \text{ kip} \cdot \text{ft}$$

6

EXAMPLE 6.4

Draw the shear and moment diagrams for the beam shown in Fig. 6–7a.

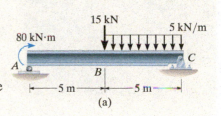

(a)

SOLUTION

Support Reactions. The reactions at the supports are shown on the free-body diagram of the beam, Fig. 6–7d.

Shear and Moment Functions. Since there is a discontinuity of distributed load and also a concentrated load at the beam's center, two regions of x must be considered in order to describe the shear and moment functions for the entire beam.

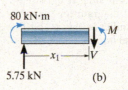

(b)

$0 \le x_1 < 5$ m, Fig. 6–7b:

$$+\uparrow \Sigma F_y = 0; \quad 5.75 \text{ kN} - V = 0$$

$$V = 5.75 \text{ kN} \tag{1}$$

$$\zeta + \Sigma M = 0; \quad -80 \text{ kN} \cdot \text{m} - 5.75 \text{ kN } x_1 + M = 0$$

$$M = (5.75x_1 + 80) \text{ kN} \cdot \text{m} \tag{2}$$

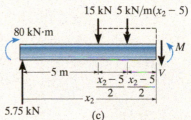

(c)

5 m $< x_2 \le 10$ m, Fig. 6–7c:

$$+\uparrow \Sigma F_y = 0; \quad 5.75 \text{ kN} - 15 \text{ kN} - 5 \text{ kN/m}(x_2 - 5 \text{ m}) - V = 0$$

$$V = (15.75 - 5x_2) \text{ kN} \tag{3}$$

$$\zeta + \Sigma M = 0; \quad -80 \text{ kN} \cdot \text{m} - 5.75 \text{ kN } x_2 + 15 \text{ kN}(x_2 - 5 \text{ m})$$

$$+ 5 \text{ kN/m}(x_2 - 5 \text{ m})\left(\frac{x_2 - 5 \text{ m}}{2}\right) + M = 0$$

$$M = (-2.5x_2^2 + 15.75x_2 + 92.5) \text{ kN} \cdot \text{m} \tag{4}$$

Shear and Moment Diagrams. Equations 1 through 4 are plotted in Fig. 6–7d.

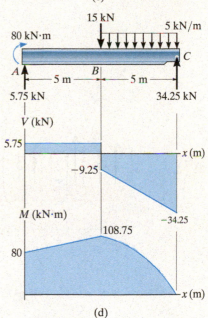

(d)

Fig. 6–7

Failure of this table occurred at the brace support on its right side. If drawn, the bending-moment diagram for the table loading would indicate this to be the point of maximum internal moment.

6.2 GRAPHICAL METHOD FOR CONSTRUCTING SHEAR AND MOMENT DIAGRAMS

In cases where a beam is subjected to *several* different loadings, determining V and M as functions of x and then plotting these equations can become quite tedious. In this section a simpler method for constructing the shear and moment diagrams is discussed—a method based on two differential relations, one that exists between the distributed load and shear, and the other between the shear and moment.

Regions of Distributed Load. For purposes of generality, consider the beam shown in Fig. 6–8a, which is subjected to an arbitrary loading. A free-body diagram for a very small segment Δx of the beam is shown in Fig. 6–8b. Since this segment has been chosen at a position x where there is no concentrated force or couple moment, the results to be obtained will *not* apply at these points.

Notice that all the loadings shown on the segment act in their positive directions according to the established sign convention, Fig. 6–3. Also, both the internal resultant shear and moment, acting on the right face of the segment, must be changed by a small amount in order to keep the segment in equilibrium. The distributed load, which is approximately constant over Δx, has been replaced by a resultant force $w\Delta x$ that acts at $\frac{1}{2}(\Delta x)$ from the right side. Applying the equations of equilibrium to the segment, we have

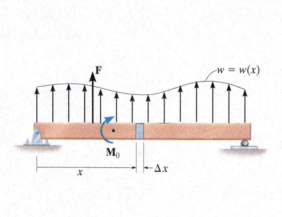

(a)

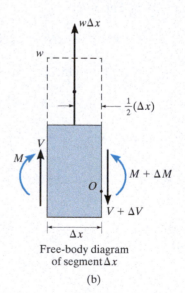

Free-body diagram
of segment Δx

(b)

Fig. 6–8

$$+\uparrow \Sigma F_y = 0; \qquad V + w\,\Delta x - (V + \Delta V) = 0$$

$$\Delta V = w\,\Delta x$$

$$\zeta + \Sigma M_O = 0; \qquad -V\,\Delta x - M - w\,\Delta x\left[\tfrac{1}{2}(\Delta x)\right] + (M + \Delta M) = 0$$

$$\Delta M = V\,\Delta x + w\,\tfrac{1}{2}(\Delta x)^2$$

Dividing by Δx and taking the limit as $\Delta x \to 0$, the above two equations become

$$\boxed{\frac{dV}{dx} = w} \qquad (6\text{–}1)$$

slope of	distributed
shear diagram	load intensity
at each point	at each point

$$\boxed{\frac{dM}{dx} = V} \qquad (6\text{–}2)$$

slope of	shear
moment diagram	at each
at each point	point

Equation 6–1 states that at any point the *slope* of the shear diagram equals the intensity of the distributed loading. For example, consider the beam in Fig. 6–9a. The distributed loading is negative and increases from zero to w_B. Knowing this provides a quick means for drawing the shape of the shear diagram. It must be a curve that has a *negative slope*, increasing from zero to $-w_B$. Specific slopes $w_A = 0$, $-w_C$, $-w_D$, and $-w_B$ are shown in Fig. 6–9b.

In a similar manner, Eq. 6–2 states that at any point the *slope* of the moment diagram is equal to the shear. Since the shear diagram in Fig. 6–9b starts at $+V_A$, decreases to zero, and then becomes negative and decreases to $-V_B$, the moment diagram (or curve) will then have an initial slope of $+V_A$ which decreases to zero, then the slope becomes negative and decreases to $-V_B$. Specific slopes V_A, V_C, V_D, 0, and $-V_B$ are shown in Fig. 6–9c.

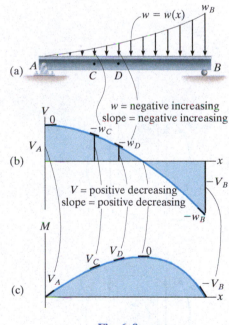

Fig. 6–9

6

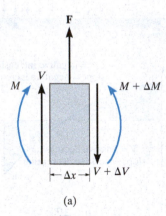

(d)

(e)

(f)

Fig. 6–9 (cont.)

Equations 6–1 and 6–2 may also be rewritten in the form $dV = w\,dx$ and $dM = V\,dx$. Since $w\,dx$ and $V\,dx$ represent differential areas under the distributed loading and the shear diagram, we can then integrate these areas between any two points C and D on the beam, Fig. 6–9d, and write

$$\Delta V = \int w\,dx \qquad (6\text{–}3)$$

change in   area under
shear   distributed loading

Equation 6–3 states that the *change in shear* between C and D is equal to the *area* under the distributed-loading curve between these two points, Fig. 6–9d. In this case the change is negative since the distributed load acts downward. Similarly, from Eq. 6–4, the change in moment between C and D, Fig. 6–9f, is equal to the area under the shear diagram within the region from C to D. Here the change is positive.

Regions of Concentrated Force and Moment.

A free-body diagram of a small segment of the beam in Fig. 6–8a taken from under the force is shown in Fig. 6–10a. Here force equilibrium requires

$$+\uparrow\Sigma F_y = 0; \qquad V + F - (V + \Delta V) = 0$$

$$\Delta V = F \qquad (6\text{–}5)$$

Thus, when $\mathbf{F}$ acts *upward* on the beam, then the change in shear, ΔV, is *positive* so the values of the shear on the shear diagram will "jump" *upward*. Likewise, if $\mathbf{F}$ acts *downward*, the jump (ΔV) will be *downward*.

When the beam segment includes the couple moment M_0, Fig. 6–10b, then moment equilibrium requires the change in moment to be

$$\zeta+\Sigma M_O = 0; \quad M + \Delta M - M_0 - V\,\Delta x - M = 0$$

Letting $\Delta x \approx 0$, we get

$$\Delta M = M_0 \qquad (6\text{–}6)$$

In this case, if $\mathbf{M_0}$ is applied *clockwise*, the change in moment, ΔM, is *positive* so the moment diagram will "jump" *upward*. Likewise, when $\mathbf{M_0}$ acts *counterclockwise*, the jump (ΔM) will be *downward*.

$$\Delta M = \int V\,dx \qquad (6\text{–}4)$$

change in   area under
moment   shear diagram

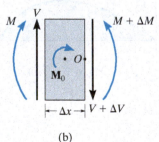

(a)

(b)

Fig. 6–10

▶ **PROCEDURE FOR ANALYSIS**

The following procedure provides a method for constructing the shear and moment diagrams for a beam based on the relations among distributed load, shear, and moment.

Support Reactions.

- Determine the support reactions and resolve the forces acting on the beam into components that are perpendicular and parallel to the beam's axis.

Shear Diagram.

- Establish the V and x axes and plot the known values of the shear at the two *ends* of the beam.

- Notice how the values of the distributed load vary along the beam, such as positive increasing, negative increasing, etc., and realize that each of these successive values indicates the way the shear diagram will slope ($dV/dx = w$). Here w is positive when it acts upward. Begin by sketching the slope at the end points.

- If a numerical value of the shear is to be determined at a point, one can find this value either by using the method of sections and the equation of force equilibrium, or by using $\Delta V = \int w\, dx$, which states that the *change in the shear* between any two points is equal to the *area under the load diagram* between the two points.

Moment Diagram.

- Establish the M and x axes and plot the known values of the moment at the *ends* of the beam.

- Notice how the values of the shear diagram vary along the beam, such as positive increasing, negative increasing, etc., and realize that each of these successive values indicates the way the moment diagram will slope ($dM/dx = V$). Begin by sketching the slope at the end points.

- At the point where the shear is zero, $dM/dx = 0$, and therefore this will be a point of maximum or minimum moment.

- If a numerical value of the moment is to be determined at the point, one can find this value either by using the method of sections and the equation of moment equilibrium, or by using $\Delta M = \int V\, dx$, which states that the *change in moment* between any two points is equal to the *area under the shear diagram* between the two points.

- Since w must be *integrated* to obtain ΔV, and V is integrated to obtain M, then if w is a curve of degree n, V will be a curve of degree $n + 1$ and M will be a curve of degree $n + 2$. For example, if w is uniform, V will be linear and M will be parabolic.

6

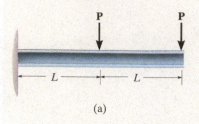

(a)

Draw the shear and moment diagrams for the beam shown in Fig. 6–11a.

SOLUTION

Support Reactions.   The reaction at the fixed support is shown on the free-body diagram, Fig. 6–11b.

Shear Diagram.   The shear at each end of the beam is plotted first, Fig. 6–11c. Since there is no distributed loading on the beam, the slope of the shear diagram is zero as indicated. Note how the force P at the center of the beam causes the shear diagram to jump downward an amount P, since this force acts downward.

Moment Diagram.   The moments at the ends of the beam are plotted, Fig. 6–11d. Here the moment diagram consists of two sloping lines, one with a slope of $+2P$ and the other with a slope of $+P$.

The value of the moment in the center of the beam can be determined by the method of sections, or from the area under the shear diagram. If we choose the left half of the shear diagram,

$$M|_{x=L} = M|_{x=0} + \Delta M$$
$$M|_{x=L} = -3PL + (2P)(L) = -PL$$

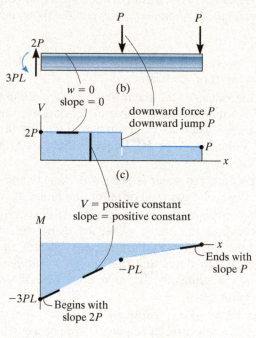

Fig. 6–11

EXAMPLE 6.6

Draw the shear and moment diagrams for the beam shown in Fig. 6–12a.

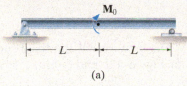

(a)

SOLUTION

Support Reactions.   The reactions are shown on the free-body diagram in Fig. 6–12b.

Shear Diagram.   The shear at each end is plotted first, Fig. 6–12c. Since there is no distributed load on the beam, the shear diagram has zero slope and is therefore a horizontal line.

Moment Diagram.   The moment is zero at each end, Fig. 6–12d. The moment diagram has a constant negative slope of $-M_0/2L$ since this is the shear in the beam at each point. However, here the couple moment M_0 causes a jump in the moment diagram at the beam's center.

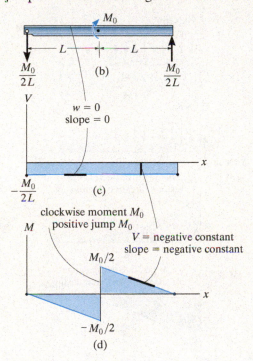

Fig. 6–12

EXAMPLE 6.7

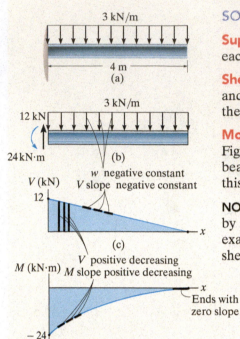

3 kN/m

4 m
(a)

3 kN/m

12 kN

24 kN·m   (b)

V (kN) w negative constant
12   V slope negative constant

x
(c)

M (kN·m) V positive decreasing
            M slope positive decreasing

x
Ends with
zero slope

−24
(d)

Fig. 6–13

Draw the shear and moment diagrams for each of the beams shown in Figs. 6–13a and 6–14a.

SOLUTION

Support Reactions.   The reactions at the fixed support are shown on each free-body diagram, Figs. 6–13b and 6–14b.

Shear Diagram.   The shear at each end point is plotted first, Figs. 6–13c and 6–14c. The distributed loading on each beam indicates the slope of the shear diagram and thus produces the shapes shown.

Moment Diagram.   The moment at each end point is plotted first, Figs. 6–13d and 6–14d. Various values of the shear at each point on the beam indicate the slope of the moment diagram at the point. Notice how this variation produces the curves shown.

NOTE: Observe how the degree of the curves from w to V to M increases by one due to the integration of $dV = w\,dx$ and $dM = V\,dx$. For example, in Fig. 6–14, the linear distributed load produces a parabolic shear diagram and cubic moment diagram.

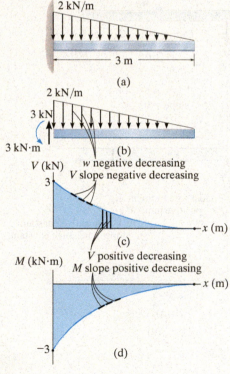

2 kN/m

3 m
(a)

2 kN/m

3 kN

3 kN·m   (b)

V (kN) w negative decreasing
3   V slope negative decreasing

x (m)
(c)

M (kN·m) V positive decreasing
             M slope positive decreasing

x (m)

−3   (d)

Fig. 6–14

EXAMPLE 6.8

Draw the shear and moment diagrams for the cantilever beam in Fig. 6–15a.

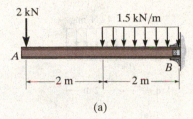

2 kN

1.5 kN/m

A

B

├── 2 m ──┼── 2 m ──┤

(a)

SOLUTION

Support Reactions. The support reactions at the fixed support B are shown in Fig. 6–15b.

Shear Diagram. The shear at the ends is plotted first, Fig. 6–15c. Notice how the shear diagram is constructed by following the slopes defined by the loading w.

Moment Diagram. The moments at the ends of the beam are plotted first, Fig. 6–15d. Notice how the moment diagram is constructed based on knowing its slope, which is equal to the shear at each point. The moment at $x = 2$ m can be found from the area under the shear diagram. We have

$$M|_{x=2\,m} = M|_{x=0} + \Delta M = 0 + [-2\text{ kN}(2\text{ m})] = -4\text{ kN} \cdot \text{m}$$

Of course, this same value can be determined from the method of sections, Fig. 6–15e.

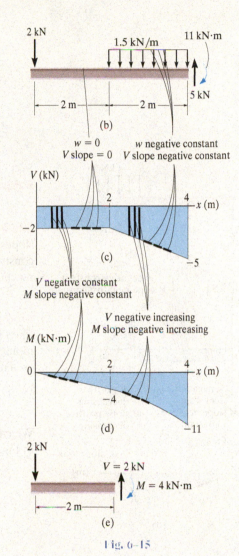

2 kN

1.5 kN/m   11 kN·m

├── 2 m ──┼── 2 m ──┤   5 kN

(b)

$w = 0$ w negative constant
V slope = 0 V slope negative constant

V (kN)

(c)

V negative constant
M slope negative constant

V negative increasing
M slope negative increasing

M (kN·m)

(d)

2 kN $V = 2$ kN $M = 4$ kN·m

├── 2 m ──┤

(e)

Fig. 6–15

EXAMPLE 6.9

Draw the shear and moment diagrams for the overhang beam in Fig. 6–16a.

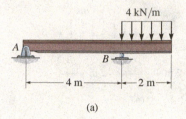

(a)

SOLUTION

Support Reactions.   The support reactions are shown in Fig. 6–16b.

Shear Diagram.   The shear at the ends is plotted first, Fig. 6–16c. The slopes are determined from the loading and from this the shear diagram is constructed. Notice the positive jump of 10 kN at $x = 4$ m due to the force reaction.

Moment Diagram.   The moments at the ends are plotted first, Fig. 6–16d. Then following the behavior of the slope found from the shear diagram, the moment diagram is constructed. The moment at $x = 4$ m is found from the area under the shear diagram.

$$M|_{x=4\,m} = M|_{x=0} + \Delta M = 0 + [-2 \text{ kN}(4 \text{ m})] = -8 \text{ kN} \cdot \text{m}$$

We can also obtain this value by using the method of sections, as shown in Fig. 6–16e.

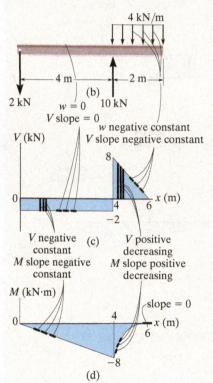

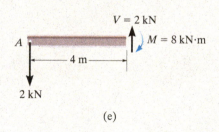

(e)

Fig. 6–16

EXAMPLE 6.10

The shaft in Fig. 6–17a is supported by a thrust bearing at A and a journal bearing at B. Draw the shear and moment diagrams.

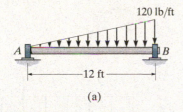

120 lb/ft

A   B

12 ft

(a)

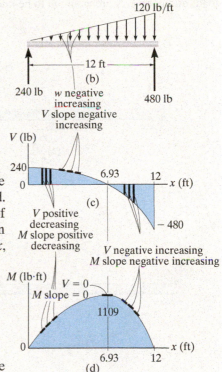

120 lb/ft

12 ft

(b)

240 lb   w negative
increasing
V slope negative
increasing

V (lb)

240   6.93   12
0   x (ft)

(c)

480 lb

V positive
decreasing
M slope positive
decreasing

V negative increasing
M slope negative increasing

– 480

M (lb·ft)   V = 0
M slope = 0

1109

0
6.93   12   x (ft)

(d)

SOLUTION

Support Reactions. The support reactions are shown in Fig. 6–17b.

Shear Diagram. As shown in Fig. 6–17c, the shears at the ends of the beam are +240 lb and –480 lb. The point where $V = 0$ must be located. To do this we will use the method of sections. The free-body diagram of the left segment of the shaft, sectioned at an arbitrary position x, is shown in Fig. 6–17e. Here the intensity of the distributed load at x is $w = 10x$, which has been found by proportional triangles, i.e., $120/12 = w/x$. Thus, for $V = 0$,

$$+\uparrow \Sigma F_y = 0; \qquad 240 \text{ lb} - \tfrac{1}{2}(10x)x = 0$$

$$x = 6.93 \text{ ft}$$

Moment Diagram. The moment diagram starts and ends at 0. The maximum moment occurs at $x = 6.93$ ft, where the shear is equal to zero, since $dM/dx = V = 0$, Fig. 6–17d. From Fig. 6–17e, we have

$$\zeta + \Sigma M = 0; \quad M_{max} + \tfrac{1}{2}[(10)(6.93)]\, 6.93\left(\tfrac{1}{3}(6.93)\right) - 240(6.93) = 0$$

$$M_{max} = 1109 \text{ lb} \cdot \text{ft}$$

Finally, notice how integration, first of the loading w, which is linear, produces a shear diagram which is parabolic, and then a moment diagram which is cubic.

NOTE: Now test yourself by covering over the shear and moment diagrams in Examples 6.1 through 6.4, and see if you can construct them based on the concepts discussed here.

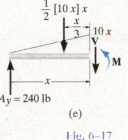

$\tfrac{1}{2}[10\,x]\,x$

$\dfrac{x}{3}$ $10\,x$

V

M

x

$A_y = 240$ lb

(e)

Fig. 6–17

PRELIMINARY PROBLEMS

P6–1. In each case, the beam is subjected to the loadings shown. Draw the free-body diagram of the beam, and sketch the general shape of the shear and moment diagrams. The loads and geometry are assumed to be known.

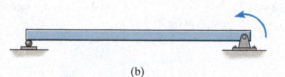

(a)

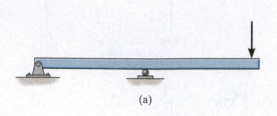

(e)

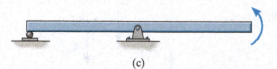

(b)

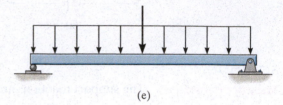

(f)

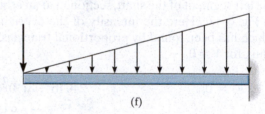

(c)

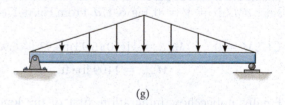

(g)

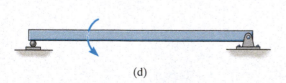

(d)

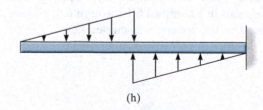

(h)

FUNDAMENTAL PROBLEMS

In each case, express the shear and moment functions in terms of x, and then draw the shear and moment diagrams for the beam.

In each case, draw the shear and moment diagrams for the beam.

F6–1.

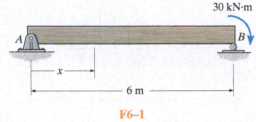

F6–1

F6–5.

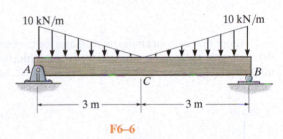

F6–5

F6–2.

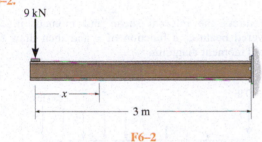

F6–2

F6–6.

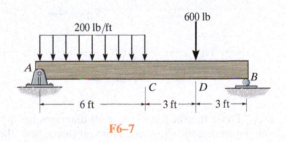

F6–6

F6–3.

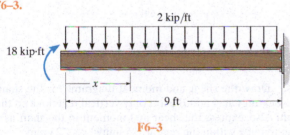

F6–3

F6–7.

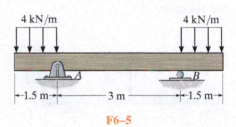

F6–7

F6–4.

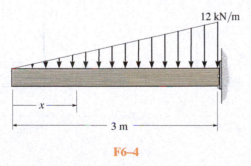

F6–4

F6–8.

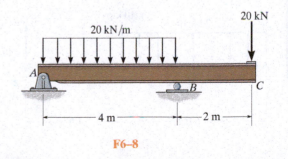

F6–8

PROBLEMS

6–1. Draw the shear and moment diagrams for the shaft and determine the shear and moment throughout the shaft as a function of x for $0 \le x < 3$ ft, 3 ft $< x < 5$ ft, and 5 ft $< x < 6$ ft. The bearings at A and B exert only vertical reactions on the shaft.

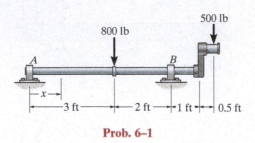

Prob. 6–1

6–2. Draw the shear and moment diagrams for the beam, and determine the shear and moment in the beam as functions of x for $0 \le x < 4$ ft, 4 ft $< x < 10$ ft, and 10 ft $< x < 14$ ft.

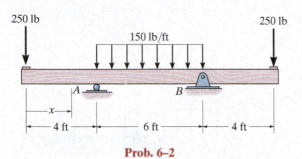

Prob. 6–2

6–3. Draw the shear and moment diagrams for the beam, and determine the shear and moment throughout the beam as functions of x for $0 \le x \le 6$ ft and 6 ft $\le x \le 10$ ft.

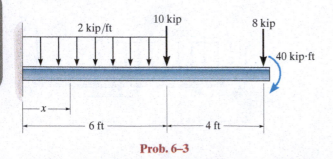

Prob. 6–3

*6–4. Express the shear and moment in terms of x for $0 < x < 3$ m and 3 m $< x < 4.5$ m, and then draw the shear and moment diagrams for the simply supported beam.

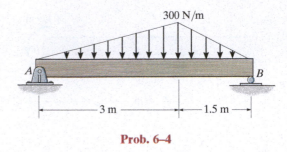

Prob. 6–4

6–5. Express the internal shear and moment in the cantilevered beam as a function of x and then draw the shear and moment diagrams.

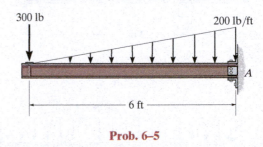

Prob. 6–5

6–6. Draw the shear and moment diagrams for the shaft. The bearings at A and B exert only vertical reactions on the shaft. Also, express the shear and moment in the shaft as a function of x within the region 125 mm $< x < 725$ mm.

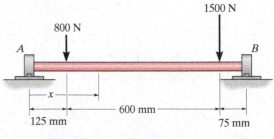

Prob. 6–6

6–7. Express the internal shear and moment in terms of x for $0 \le x < L/2$, and $L/2 < x < L$, and then draw the shear and moment diagrams.

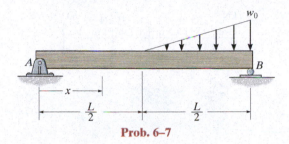

Prob. 6–7

*6–8. Draw the shear and moment diagrams for the beam, and determine the shear and moment throughout the beam as functions of x for $0 \le x \le 6$ ft and 6 ft $\le x \le 9$ ft.

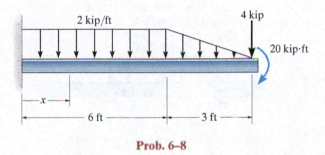

Prob. 6–8

6–9. If the force applied to the handle of the load binder is 50 lb, determine the tensions T_1 and T_2 in each end of the chain and then draw the shear and moment diagrams for the arm ABC.

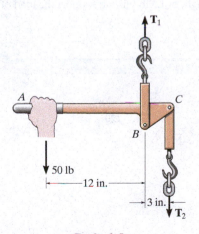

Prob. 6–9

6–10. Draw the shear and moment diagrams for the shaft. The bearings at A and D exert only vertical reactions on the shaft.

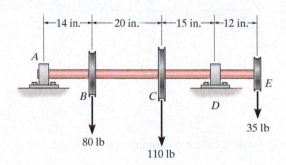

Prob. 6–10

6–11. The crane is used to support the engine, which has a weight of 1200 lb. Draw the shear and moment diagrams of the boom ABC when it is in the horizontal position.

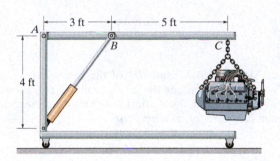

Prob. 6–11

*6–12. Draw the shear and moment diagrams for the beam.

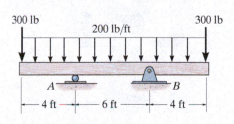

Prob. 6–12

6–13. Draw the shear and moment diagrams for the beam.

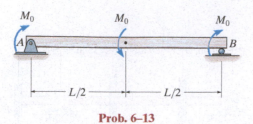

Prob. 6–13

6–14. Draw the shear and moment diagrams for the beam.

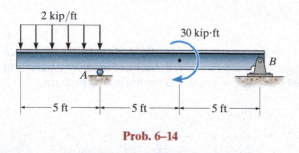

Prob. 6–14

6–15. Members ABC and BD of the counter chair are rigidly connected at B and the smooth collar at D is allowed to move freely along the vertical post. Draw the shear and moment diagrams for member ABC.

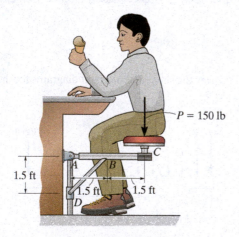

Prob. 6–15

*6–16. A reinforced concrete pier is used to support the stringers for a bridge deck. Draw the shear and moment diagrams for the pier. Assume the columns at A and B exert only vertical reactions on the pier.

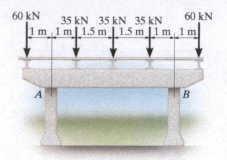

Prob. 6–16

6–17. Draw the shear and moment diagrams for the beam and determine the shear and moment in the beam as functions of x, where $4 \text{ ft} < x < 10 \text{ ft}$.

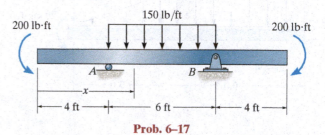

Prob. 6–17

6–18. The industrial robot is held in the stationary position shown. Draw the shear and moment diagrams of the arm ABC if it is pin connected at A and connected to a hydraulic cylinder (two-force member) BD. Assume the arm and grip have a uniform weight of 1.5 lb/in. and support the load of 40 lb at C.

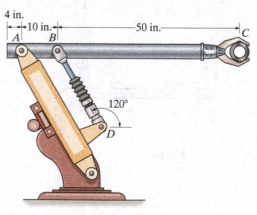

Prob. 6–18

6–19. Determine the placement distance a of the roller support so that the largest absolute value of the moment is a minimum. Draw the shear and moment diagrams for this condition.

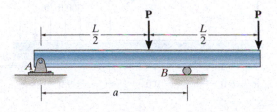

Prob. 6–19

*6–20. Draw the shear and moment diagrams for the beam.

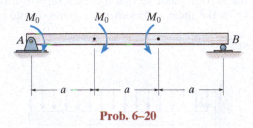

Prob. 6–20

6–21. Draw the shear and moment diagrams for the beam.

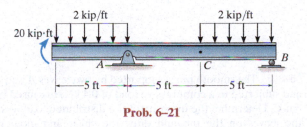

Prob. 6–21

6–22. Draw the shear and moment diagrams for the overhanging beam.

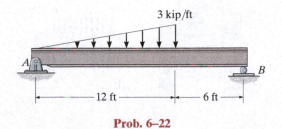

Prob. 6–22

6–23. The 150-lb man sits in the center of the boat, which has a uniform width and a weight per linear foot of 3 lb/ft. Determine the maximum internal bending moment. Assume that the water exerts a uniform distributed load upward on the bottom of the boat.

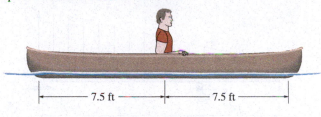

Prob. 6–23

*6–24. Draw the shear and moment diagrams for the beam.

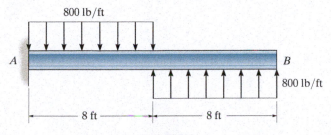

Prob. 6–24

6–25. The footing supports the load transmitted by the two columns. Draw the shear and moment diagrams for the footing if the soil pressure on the footing is assumed to be uniform.

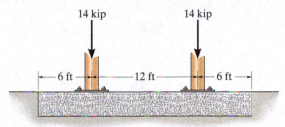

Prob. 6–25

6–26. Draw the shear and moment diagrams for the beam.

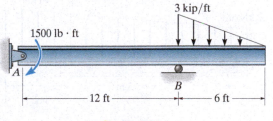

Prob. 6–26

6–27. Draw the shear and moment diagrams for the beam.

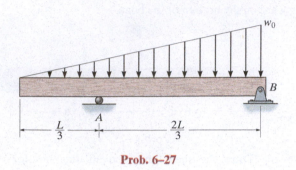

Prob. 6–27

6–30. Draw the shear and moment diagrams for the beam.

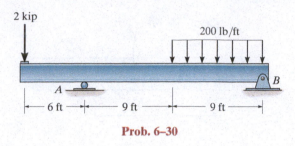

Prob. 6–30

6–31. The support at A allows the beam to slide freely along the vertical guide so that it cannot support a vertical force. Draw the shear and moment diagrams for the beam.

***6–28.** Draw the shear and moment diagrams for the beam.

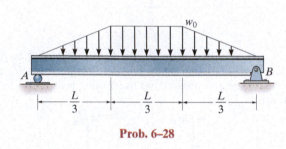

Prob. 6–28

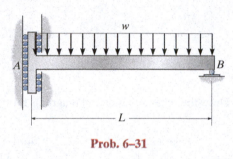

Prob. 6–31

***6–32.** The smooth pin is supported by two leaves A and B and subjected to a compressive load of 0.4 kN/m caused by bar C. Determine the intensity of the distributed load w_0 of the leaves on the pin and draw the shear and moment diagram for the pin.

6–29. Draw the shear and moment diagrams for the beam.

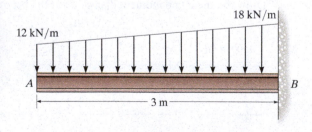

Prob. 6–29

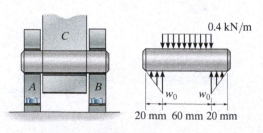

Prob. 6–32

6–33. The shaft is supported by a smooth thrust bearing at *A* and smooth journal bearing at *B*. Draw the shear and moment diagrams for the shaft.

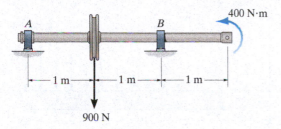

Prob. 6–33

6–34. Draw the shear and moment diagrams for the cantilever beam.

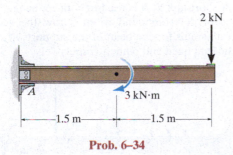

Prob. 6–34

6–35. Draw the shear and moment diagrams for the beam.

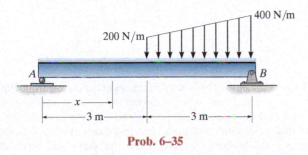

Prob. 6–35

*****6–36.** Draw the shear and moment diagrams for the rod. Only vertical reactions occur at its ends *A* and *B*.

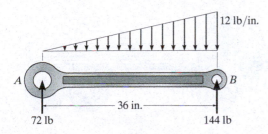

Prob. 6–36

6–37. Draw the shear and moment diagrams for the beam.

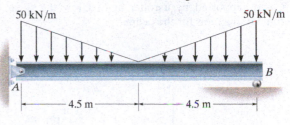

Prob. 6–37

6–38. The beam is used to support a uniform load along *CD* due to the 6-kN weight of the crate. Also, the reaction at the bearing support *B* can be assumed uniformly distributed along its width. Draw the shear and moment diagrams for the beam.

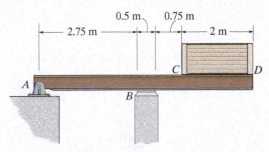

Prob. 6–38

6–39. Draw the shear and moment diagrams for the double overhanging beam.

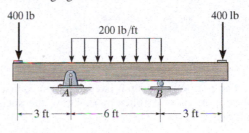

Prob. 6–39

*****6–40.** Draw the shear and moment diagrams for the simply supported beam.

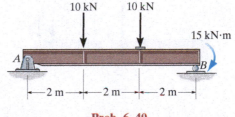

Prob. 6–40

6–41. The compound beam is fixed at A, pin connected at B, and supported by a roller at C. Draw the shear and moment diagrams for the beam.

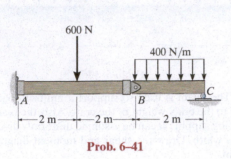

Prob. 6–41

6–42. Draw the shear and moment diagrams for the compound beam.

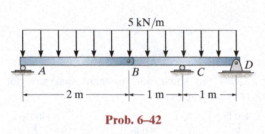

Prob. 6–42

6–43. The compound beam is fixed at A, pin connected at B, and supported by a roller at C. Draw the shear and moment diagrams for the beam.

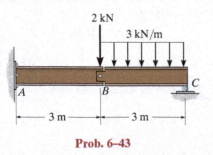

Prob. 6–43

*6–44. Draw the shear and moment diagrams for the beam.

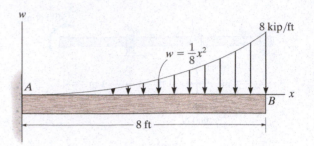

Prob. 6–44

6–45. A short link at B is used to connect beams AB and BC to form the compound beam. Draw the shear and moment diagrams for the beam if the supports at A and C are considered fixed and pinned, respectively.

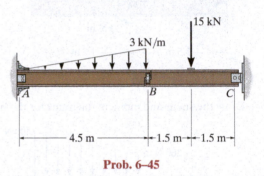

Prob. 6–45

6–46. The truck is to be used to transport the concrete column. If the column has a uniform weight of w (force/length), determine the equal placement a of the supports from the ends so that the absolute maximum bending moment in the column is as small as possible. Also, draw the shear and moment diagrams for the column.

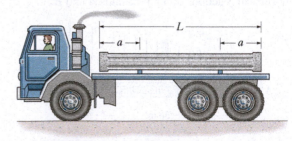

Prob. 6–46

6.3 BENDING DEFORMATION OF A STRAIGHT MEMBER

In this section, we will discuss the deformations that occur when a straight prismatic beam, made of homogeneous material, is subjected to bending. The discussion will be limited to beams having a cross-sectional area that is symmetrical with respect to an axis, and the bending moment is applied about an axis perpendicular to this axis of symmetry, as shown in Fig. 6–18. The behavior of members that have unsymmetrical cross sections, or are made of several different materials, is based on similar observations and will be discussed separately in later sections of this chapter.

Consider the undeformed bar in Fig. 6–19a, which has a square cross section and is marked with horizontal and vertical grid lines. When a bending moment is applied, it tends to distort these lines into the pattern shown in Fig. 6–19b. Here the horizontal lines become *curved*, while the vertical lines *remain straight* but undergo a *rotation*. The bending moment causes the material within the *bottom* portion of the bar to *stretch* and the material within the *top* portion to *compress*. Consequently, between these two regions there must be a surface, called the **neutral surface,** in which horizontal fibers of the material will not undergo a change in length, Fig. 6–18. As noted, we will refer to the z axis that lies along the neutral surface as the **neutral axis.**

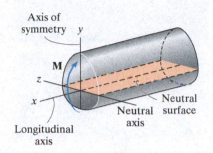

Fig. 6–18

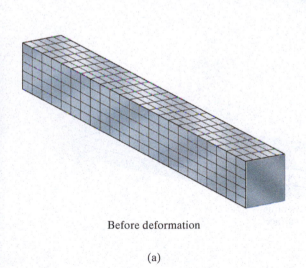

Before deformation

(a)

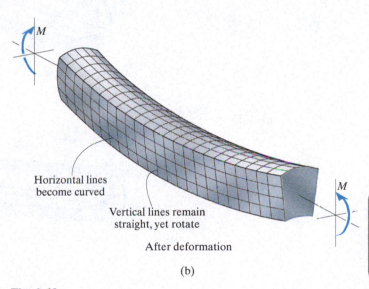

Horizontal lines become curved

Vertical lines remain straight, yet rotate

After deformation

(b)

Fig. 6–19

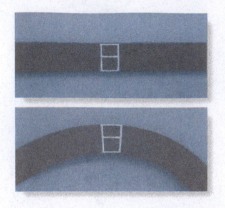

Note the distortion of the lines due to bending of this rubber bar. The top line stretches, the bottom line compresses, and the center line remains the same length. Furthermore the vertical lines rotate and yet remain straight.

From these observations we will make the following three assumptions regarding the way the moment deforms the material. First, the longitudinal axis, which lies within the neutral surface, Fig. 6–20a, does not experience any change in length. Rather the moment will tend to deform the beam so that this line becomes a curve that lies in the vertical plane of symmetry, Fig. 6–20b. Second, all cross sections of the beam remain plane and perpendicular to the longitudinal axis during the deformation. And third, the small lateral strains due to the Poisson effect discussed in Sec. 3.6 will be neglected. In other words, the cross section in Fig. 6–19 retains its shape.

With the above assumptions, we will now consider how the bending moment distorts a small element of the beam located a distance x along the beam's length, Fig. 6–20. This element is shown in profile view in the undeformed and deformed positions in Fig. 6–21. Here the line segment

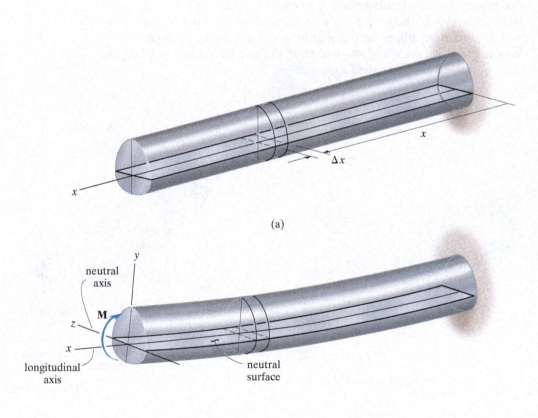

(a)

(b)

Fig. 6–20

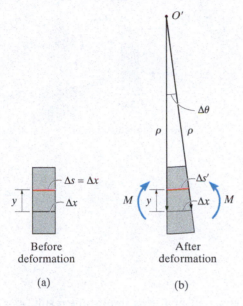

Before deformation

(a)

After deformation

(b)

Fig. 6–21

Δx, located on the neutral surface, does not change its length, whereas any line segment Δs, located at the arbitrary distance y above the neutral surface, will contract and become $\Delta s'$ after deformation. By definition, the normal strain along Δs is determined from Eq. 2–2, namely,

$$\epsilon = \lim_{\Delta s \to 0} \frac{\Delta s' - \Delta s}{\Delta s}$$

Now let's represent this strain in terms of the location y of the segment and the radius of curvature ρ of the longitudinal axis of the element. Before deformation, $\Delta s = \Delta x$, Fig. 6–21a. After deformation, Δx has a radius of curvature ρ, with center of curvature at point O', Fig. 6–21b, so that $\Delta x = \Delta s = \rho \Delta \theta$. Also, since $\Delta s'$ has a radius of curvature of $\rho - y$, then $\Delta s' = (\rho - y)\Delta \theta$. Substituting these results into the above equation, we get

$$\epsilon = \lim_{\Delta \theta \to 0} \frac{(\rho - y)\Delta \theta - \rho \Delta \theta}{\rho \Delta \theta}$$

or

$$\epsilon = -\frac{y}{\rho} \tag{6–7}$$

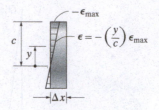

Normal strain distribution

Fig. 6–22

Since $1/\rho$ is constant at x, this important result, $\epsilon = -y/\rho$, indicates that the *longitudinal normal strain will vary linearly* with y measured from the neutral axis. A contraction $(-\epsilon)$ will occur in fibers located above the neutral axis $(+y)$, whereas elongation $(+\epsilon)$ will occur in fibers located below the axis $(-y)$. This variation in strain over the cross section is shown in Fig. 6–22. Here the maximum strain occurs at the outermost fiber, located a distance of $y = c$ from the neutral axis. Using Eq. 6–7, since $\epsilon_{max} = c/\rho$, then by division,

$$\frac{\epsilon}{\epsilon_{max}} = -\left(\frac{y/\rho}{c/\rho}\right)$$

So that

$$\epsilon = -\left(\frac{y}{c}\right)\epsilon_{max} \tag{6–8}$$

This normal strain depends only on the assumptions made with regard to the deformation.

6.4   THE FLEXURE FORMULA

In this section, we will develop an equation that relates the stress distribution within a straight beam to the bending moment acting on its cross section. To do this we will assume that the material behaves in a linear elastic manner, so that by Hooke's law, a linear variation of normal strain, Fig. 6–23a, must result in a linear variation in normal stress, Fig. 6–23b. Hence, like the normal strain variation, σ will vary from zero at the member's neutral axis to a maximum value, σ_{max}, a distance c farthest from the neutral axis. Because of the proportionality of triangles, Fig. 6–23b, or by using Hooke's law, $\sigma = E\epsilon$, and Eq. 6–8, we can write

$$\sigma = -\left(\frac{y}{c}\right)\sigma_{max} \qquad (6\text{–}9)$$

This equation describes the stress distribution over the cross-sectional area. The sign convention established here is significant. For positive **M**, which acts in the $+z$ direction, positive values of y give negative values for σ, that is, a compressive stress, since it acts in the negative x direction. Similarly, negative y values will give positive or tensile values for σ.

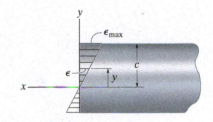

Normal strain variation
(profile view)

(a)

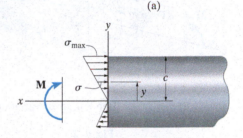

Bending stress variation
(profile view)

(b)

Fig. 6–23

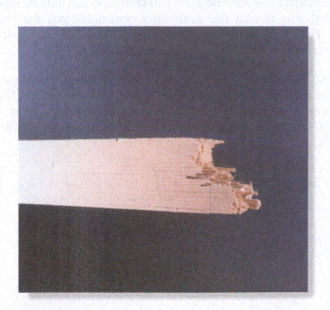

This wood specimen failed in bending due to its fibers being crushed at its top and torn apart at its bottom.

Location of Neutral Axis.   To locate the position of the neutral axis, we require the *resultant force* produced by the stress distribution acting over the cross-sectional area to be equal to *zero*. Noting that the force $dF = \sigma \, dA$ acts on the arbitrary element dA in Fig. 6–24, we have

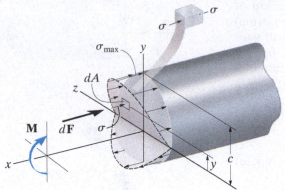

Bending stress variation

Fig. 6–24

$$F_R = \Sigma F_x; \qquad 0 = \int_A dF = \int_A \sigma \, dA$$

$$= \int_A -\left(\frac{y}{c}\right)\sigma_{max} \, dA$$

$$= \frac{-\sigma_{max}}{c} \int_A y \, dA$$

Since σ_{max}/c is not equal to zero, then

$$\int_A y \, dA = 0 \qquad (6\text{–}10)$$

In other words, the first moment of the member's cross-sectional area about the neutral axis must be zero. This condition can only be satisfied if the neutral axis is also the horizontal centroidal axis for the cross section.* Therefore, once the centroid for the member's cross-sectional area is determined, the location of the neutral axis is known.

Bending Moment.   We can determine the stress in the beam if we require the moment M to be equal to the moment produced by the stress distribution about the neutral axis. The moment of $d\mathbf{F}$ in Fig. 6–24 is $dM = y \, dF$. Since $dF = \sigma \, dA$, using Eq. 6–9, we have for the entire cross section,

$$(M_R)_z = \Sigma M_z; \qquad M = \int_A y \, dF = \int_A y \, (\sigma \, dA) = \int_A y \left(\frac{y}{c}\sigma_{max}\right) dA$$

or

$$M = \frac{\sigma_{max}}{c} \int_A y^2 \, dA \qquad (6\text{–}11)$$

*Recall that the location $\bar{y}$ for the centroid of an area is defined from the equation $\bar{y} = \int y \, dA / \int dA$. If $\int y \, dA = 0$, then $\bar{y} = 0$, and so the centroid lies on the reference (neutral) axis. See Appendix A.

The integral represents the ***moment of inertia*** of the cross-sectional area about the neutral axis.* We will symbolize its value as *I*. Hence, Eq. 6–11 can be solved for σ_{max} and written as

$$\sigma_{max} = \frac{Mc}{I}$$

(6–12)

Here

σ_{max} = the maximum normal stress in the member, which occurs at a point on the cross-sectional area *farthest away* from the neutral axis

M = the resultant internal moment, determined from the method of sections and the equations of equilibrium, and calculated about the neutral axis of the cross section

c = perpendicular distance from the neutral axis to a point farthest away from the neutral axis. This is where σ_{max} acts.

I = moment of inertia of the cross-sectional area about the neutral axis

Since $\sigma_{max}/c = -\sigma/y$, Eq. 6–9, the normal stress at any distance *y* can be determined from an equation similar to Eq. 6–12. We have

$$\sigma = -\frac{My}{I}$$

(6–13)

Either of the above two equations is often referred to as the ***flexure formula.*** Although we have assumed that the member is prismatic, we can conservatively also use the flexure formula to determine the normal stress in members that have a *slight taper.* For example, using a mathematical analysis based on the theory of elasticity, a member having a rectangular cross section and a length that is tapered 15° will have an actual maximum normal stress that is about 5.4% *less* than that calculated using the flexure formula.

*See Appendix A for a discussion on how to determine the moment of inertia for various shapes.

▶ IMPORTANT POINTS

- The cross section of a straight beam *remains plane* when the beam deforms due to bending. This causes tensile stress on one portion of the cross section and compressive stress on the other portion. In between these portions, there exists the *neutral axis* which is subjected to *zero stress*.

- Due to the deformation, the *longitudinal strain* varies *linearly* from zero at the neutral axis to a maximum at the outer fibers of the beam. Provided the material is homogeneous and linear elastic, then the *stress* also varies in a *linear* fashion over the cross section.

- Since there is no resultant normal force on the cross section, then the neutral axis must pass through the *centroid* of the cross-sectional area.

- The flexure formula is based on the requirement that the internal moment on the cross section is equal to the moment produced by the normal stress distribution about the neutral axis.

▶ PROCEDURE FOR ANALYSIS

In order to apply the flexure formula, the following procedure is suggested.

Internal Moment.

- Section the member at the point where the bending or normal stress is to be determined, and obtain the internal moment M at the section. The centroidal or neutral axis for the cross section must be known, since M *must* be calculated about this axis.

- If the absolute maximum bending stress is to be determined, then draw the moment diagram in order to determine the maximum moment in the member.

Section Property.

- Determine the moment of inertia of the cross-sectional area about the neutral axis. Methods used for its calculation are discussed in Appendix A, and a table listing values of I for several common shapes is given on the inside front cover.

Normal Stress.

- Specify the location y, measured perpendicular to the neutral axis to the point where the normal stress is to be determined. Then apply the equation $\sigma = -My/I$, or if the maximum bending stress is to be calculated, use $\sigma_{max} = Mc/I$. When substituting the data, make sure the units are consistent.

- The stress acts in a direction such that the force it creates at the point contributes a moment about the neutral axis that is in the same direction as the internal moment **M**. In this manner the stress distribution acting over the entire cross section can be sketched, or a volume element of the material can be isolated and used to graphically represent the normal stress acting at the point, see Fig. 6–24.

6

EXAMPLE   6.11

A beam has a rectangular cross section and is subjected to the stress distribution shown in Fig. 6–25a. Determine the internal moment **M** at the section caused by the stress distribution (a) using the flexure formula, (b) by finding the resultant of the stress distribution using basic principles.

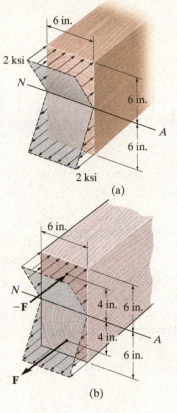

(a)

(b)

Fig. 6–25

SOLUTION

Part (a).   The flexure formula is $\sigma_{max} = Mc/I$. From Fig. 6–25a, $c = 6$ in. and $\sigma_{max} = 2$ ksi. The neutral axis is defined as line *NA*, because it passes through the centroid of the cross section, and the stress is zero along this line. Since the cross section has a rectangular shape, the moment of inertia for the area about *NA* is determined from the formula for a rectangle given on the inside front cover; i.e.,

$$I = \frac{1}{12}bh^3 = \frac{1}{12}(6\text{ in.})(12\text{ in.})^3 = 864\text{ in}^4$$

Therefore,

$$\sigma_{max} = \frac{Mc}{I}; \qquad 2\text{ kip/in}^2 = \frac{M(6\text{ in.})}{864\text{ in}^4}$$

$$M = 288\text{ kip} \cdot \text{in.} = 24\text{ kip} \cdot \text{ft} \qquad \textit{Ans.}$$

Part (b).   The resultant force for each of the two *triangular* stress distributions in Fig. 6–25b is graphically equivalent to the *volume* contained within each stress distribution. This volume is

$$F = \frac{1}{2}(6\text{ in.})(2\text{ kip/in}^2)(6\text{ in.}) = 36\text{ kip}$$

These forces, which form a couple, act in the same direction as the stresses within each distribution, Fig. 6–25b. Furthermore, they act through the *centroid* of each volume, i.e., $\frac{2}{3}(6\text{ in.}) = 4$ in. from the neutral axis of the beam. Hence the distance between them is 8 in. as shown. The moment of the couple is therefore

$$M = 36\text{ kip}(8\text{ in.}) = 288\text{ kip} \cdot \text{in.} = 24\text{ kip} \cdot \text{ft} \qquad \textit{Ans.}$$

6

EXAMPLE   6.12

The simply supported beam in Fig. 6–26a has the cross-sectional area shown in Fig. 6–26b. Determine the absolute maximum bending stress in the beam and draw the stress distribution over the cross section at this location. Also, what is the stress at point B?

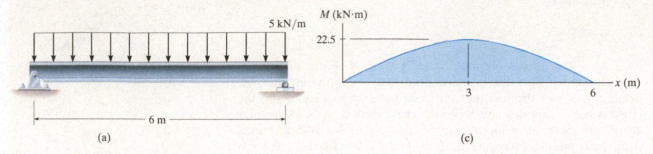

(a)

(c)

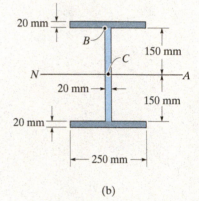

(b)

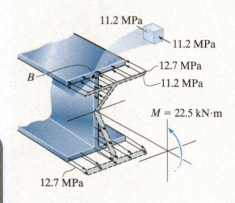

(d)

Fig. 6–26

SOLUTION

Maximum Internal Moment.   The maximum internal moment in the beam, $M = 22.5$ kN · m, occurs at the center, as indicated on the moment diagram, Fig. 6–26c.

Section Property.   By reasons of symmetry, the neutral axis passes through the centroid C at the midheight of the beam, Fig. 6–26b. The area is subdivided into the three parts shown, and the moment of inertia of each part is calculated about the neutral axis using the parallel-axis theorem. (See Eq. A–5 of Appendix A.) Choosing to work in meters, we have

$$I = \Sigma(\bar{I} + Ad^2)$$

$$= 2\left[\frac{1}{12}(0.25 \text{ m})(0.020 \text{ m})^3 + (0.25 \text{ m})(0.020 \text{ m})(0.160 \text{ m})^2\right]$$

$$+ \left[\frac{1}{12}(0.020 \text{ m})(0.300 \text{ m})^3\right]$$

$$= 301.3(10^{-6}) \text{ m}^4$$

$$\sigma_{max} = \frac{Mc}{I}; \quad \sigma_{max} = \frac{22.5(10^3) \text{ N} \cdot \text{m}(0.170 \text{ m})}{301.3(10^{-6}) \text{ m}^4} = 12.7 \text{ MPa}   \textit{Ans.}$$

A three-dimensional view of the stress distribution is shown in Fig. 6–26d. Specifically, at point B, $y_B = 150$ mm, and so as shown in Fig. 6–26d,

$$\sigma_B = -\frac{My_B}{I}; \quad \sigma_B = -\frac{22.5(10^3) \text{ N} \cdot \text{m}(0.150 \text{ m})}{301.3(10^{-6}) \text{ m}^4} = -11.2 \text{ MPa}   \textit{Ans.}$$

EXAMPLE 6.13

The beam shown in Fig. 6–27a has a cross-sectional area in the shape of a channel, Fig. 6–27b. Determine the maximum bending stress that occurs in the beam at section a–a.

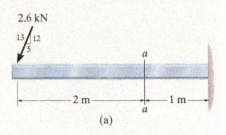

(a)

SOLUTION

Internal Moment. Here the beam's support reactions do not have to be determined. Instead, by the method of sections, the segment to the left of section a–a can be used, Fig. 6–27c. It is important that the resultant internal axial force **N** passes through the centroid of the cross section. Also, realize that the resultant internal moment must be calculated about the beam's neutral axis at section a–a.

To find the location of the neutral axis, the cross-sectional area is subdivided into three composite parts as shown in Fig. 6–27b. Using Eq. A–2 of Appendix A, we have

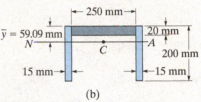

(b)

$$\bar{y} = \frac{\Sigma \tilde{y} A}{\Sigma A} = \frac{2[0.100 \text{ m}](0.200 \text{ m})(0.015 \text{ m}) + [0.010 \text{ m}](0.02 \text{ m})(0.250 \text{ m})}{2(0.200 \text{ m})(0.015 \text{ m}) + 0.020 \text{ m}(0.250 \text{ m})}$$

$$= 0.05909 \text{ m} = 59.09 \text{ mm}$$

This dimension is shown in Fig. 6–27c.

Applying the moment equation of equilibrium about the neutral axis, we have

$$\zeta + \Sigma M_{NA} = 0; \quad 2.4 \text{ kN}(2 \text{ m}) + 1.0 \text{ kN}(0.05909 \text{ m}) - M = 0$$
$$M = 4.859 \text{ kN} \cdot \text{m}$$

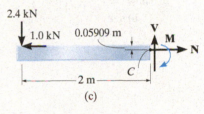

(c)

Fig. 6–27

Section Property. The moment of inertia of the cross-sectional area about the neutral axis is determined using $I = \Sigma (\bar{I} + Ad^2)$ applied to each of the three composite parts of the area. Working in meters, we have

$$I = \left[\frac{1}{12}(0.250 \text{ m})(0.020 \text{ m})^3 + (0.250 \text{ m})(0.020 \text{ m})(0.05909 \text{ m} - 0.010 \text{ m})^2 \right]$$

$$+ 2\left[\frac{1}{12}(0.015 \text{ m})(0.200 \text{ m})^3 + (0.015 \text{ m})(0.200 \text{ m})(0.100 \text{ m} - 0.05909 \text{ m})^2 \right]$$

$$= 42.26(10^{-6}) \text{ m}^4$$

Maximum Bending Stress. The maximum bending stress occurs at points farthest away from the neutral axis. This is at the bottom of the beam, $c = 0.200 \text{ m} - 0.05909 \text{ m} = 0.1409 \text{ m}$. Here the stress is compressive. Thus,

$$\sigma_{max} = \frac{Mc}{I} = \frac{4.859(10^3) \text{ N} \cdot \text{m}(0.1409 \text{ m})}{42.26(10^{-6}) \text{ m}^4} = 16.2 \text{ MPa (C)} \qquad Ans.$$

Show that at the top of the beam the bending stress is $\sigma' = 6.79$ MPa.

NOTE: The normal force of $N = 1$ kN and shear force $V = 2.4$ kN will also contribute additional stress on the cross section. The superposition of all these effects will be discussed in Chapter 8.

EXAMPLE 6.14

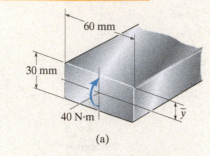

(a)

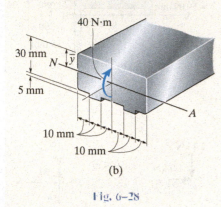

(b)

Fig. 6–28

The member having a rectangular cross section, Fig. 6–28a, is designed to resist a moment of 40 N·m. In order to increase its strength and rigidity, it is proposed that two small ribs be added at its bottom, Fig. 6–28b. Determine the maximum normal stress in the member for both cases.

SOLUTION

Without Ribs.   Clearly the neutral axis is at the center of the cross section, Fig. 6–28a, so $\bar{y} = c = 15$ mm $= 0.015$ m. Thus,

$$I = \frac{1}{12}bh^3 = \frac{1}{12}(0.060 \text{ m})(0.030 \text{ m})^3 = 0.135(10^{-6}) \text{ m}^4$$

Therefore the maximum normal stress is

$$\sigma_{max} = \frac{Mc}{I} = \frac{(40 \text{ N} \cdot \text{m})(0.015 \text{ m})}{0.135(10^{-6}) \text{ m}^4} = 4.44 \text{ MPa} \qquad \textit{Ans.}$$

With Ribs.   From Fig. 6–28b, segmenting the area into the large main rectangle and the bottom two rectangles (ribs), the location $\bar{y}$ of the centroid and the neutral axis is determined as follows:

$$\bar{y} = \frac{\Sigma \tilde{y}A}{\Sigma A}$$

$$= \frac{[0.015 \text{ m}](0.030 \text{ m})(0.060 \text{ m}) + 2[0.0325 \text{ m}](0.005 \text{ m})(0.010 \text{ m})}{(0.03 \text{ m})(0.060 \text{ m}) + 2(0.005 \text{ m})(0.010 \text{ m})}$$

$$= 0.01592 \text{ m}$$

This value does not represent c. Instead

$$c = 0.035 \text{ m} - 0.01592 \text{ m} = 0.01908 \text{ m}$$

Using the parallel-axis theorem, the moment of inertia about the neutral axis is

$$I = \left[\frac{1}{12}(0.060 \text{ m})(0.030 \text{ m})^3 + (0.060 \text{ m})(0.030 \text{ m})(0.01592 \text{ m} - 0.015 \text{ m})^2 \right]$$

$$+ 2\left[\frac{1}{12}(0.010 \text{ m})(0.005 \text{ m})^3 + (0.010 \text{ m})(0.005 \text{ m})(0.0325 \text{ m} - 0.01592 \text{ m})^2 \right]$$

$$= 0.1642(10^{-6}) \text{ m}^4$$

Therefore, the maximum normal stress is

$$\sigma_{max} = \frac{Mc}{I} = \frac{40 \text{ N} \cdot \text{m}(0.01908 \text{ m})}{0.1642(10^{-6}) \text{ m}^4} = 4.65 \text{ MPa} \qquad \textit{Ans.}$$

NOTE: This surprising result indicates that the addition of the ribs to the cross section will *increase* the maximum normal stress rather than decrease it, and for this reason, the ribs should be omitted.

PRELIMINARY PROBLEMS

P6–2. Determine the moment of inertia of the cross section about the neutral axis.

P6–4. In each case, show how the bending stress acts on a differential volume element located at point A and point B.

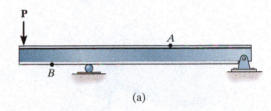

(a)

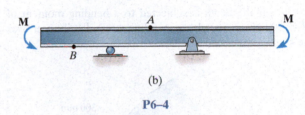

(b)

P6–4

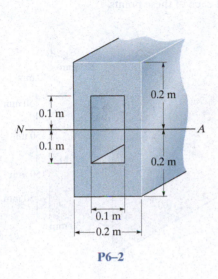

P6–2

P6–3. Determine the location of the centroid, y, and the moment of inertia of the cross section about the neutral axis.

P6–5. Sketch the bending stress distribution over each cross section.

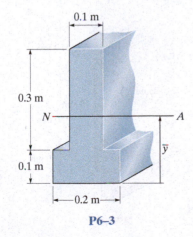

P6–3

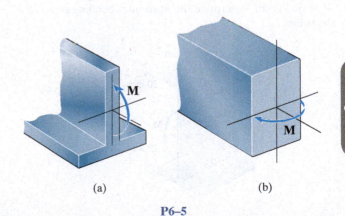

(a) (b)

P6–5

6

FUNDAMENTAL PROBLEMS

F6–9. If the beam is subjected to a bending moment of $M = 20$ kN·m, determine the maximum bending stress in the beam.

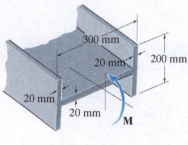

300 mm

20 mm   200 mm

20 mm

20 mm

M

F6–9

F6–10. If the beam is subjected to a bending moment of $M = 50$ kN·m, sketch the bending stress distribution over the beam's cross section.

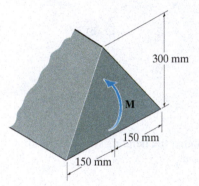

300 mm

M

150 mm

150 mm

F6–10

F6–11. If the beam is subjected to a bending moment of $M = 50$ kN·m, determine the maximum bending stress in the beam.

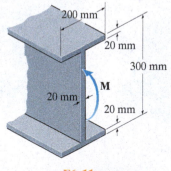

200 mm

20 mm

300 mm

20 mm

M

20 mm

F6–11

F6–12. If the beam is subjected to a bending moment of $M = 10$ kN·m, determine the bending stress in the beam at points A and B, and sketch the results on a differential element at each of these points.

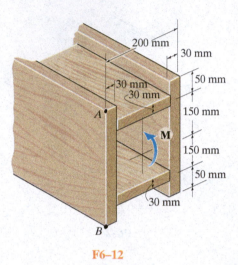

200 mm

30 mm

30 mm   50 mm

30 mm

150 mm

A

M

150 mm

50 mm

30 mm

B

F6–12

F6–13. If the beam is subjected to a bending moment of $M = 5$ kN·m, determine the bending stress developed at point A and sketch the result on a differential element at this point.

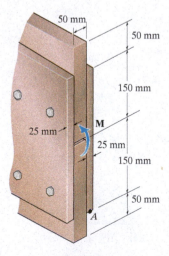

50 mm

50 mm

150 mm

25 mm   **M**

25 mm

150 mm

50 mm

A

F6–13

PROBLEMS

6–47. An A-36 steel strip has an allowable bending stress of 165 MPa. If it is rolled up, determine the smallest radius r of the spool if the strip has a width of 10 mm and a thickness of 1.5 mm. Also, find the corresponding maximum internal moment developed in the strip.

6–50. The beam is constructed from four pieces of wood, glued together as shown. If $M = 10$ kip · ft, determine the maximum bending stress in the beam. Sketch a three-dimensional view of the stress distribution acting over the cross section.

6–51. The beam is constructed from four pieces of wood, glued together as shown. If $M = 10$ kip · ft, determine the resultant force this moment exerts on the top and bottom boards of the beam.

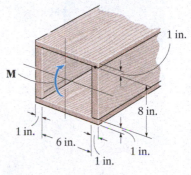

Probs. 6–50/51

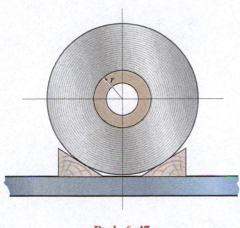

Prob. 6–47

*6–48. Determine the moment M that will produce a maximum stress of 10 ksi on the cross section.

6–49. Determine the maximum tensile and compressive bending stress in the beam if it is subjected to a moment of $M = 4$ kip · ft.

*6–52. The beam is made from three boards nailed together as shown. If the moment acting on the cross section is $M = 600$ N · m, determine the maximum bending stress in the beam. Sketch a three-dimensional view of the stress distribution and cover the cross section.

6–53. The beam is made from three boards nailed together as shown. If the moment acting on the cross section is $M = 600$ N · m, determine the resultant force the bending stress produces on the top board.

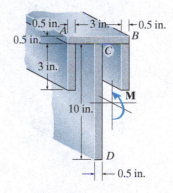

Probs. 6–48/49

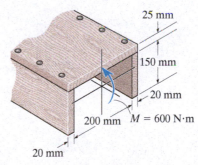

Probs. 6–52/53

6–54. If the built-up beam is subjected to an internal moment of $M = 75$ kN · m, determine the maximum tensile and compressive stress acting in the beam.

6–55. If the built-up beam is subjected to an internal moment of $M = 75$ kN · m, determine the amount of this internal moment resisted by plate A.

6–58. The beam is made from three boards nailed together as shown. If the moment acting on the cross section is $M = 1$ kip · ft, determine the maximum bending stress in the beam. Sketch a three-dimensional view of the stress distribution acting over the cross section.

6–59. If $M = 1$ kip · ft, determine the resultant force the bending stresses produce on the top board A of the beam.

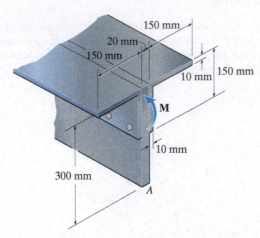

Probs. 6–54/55

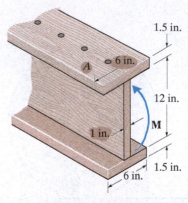

Probs. 6–58/59

*6–56. The beam is subjected to a moment M. Determine the percentage of this moment that is resisted by the stresses acting on both the top and bottom boards of the beam.

6–57. Determine the moment M that should be applied to the beam in order to create a compressive stress at point D of $\sigma_D = 10$ MPa. Also sketch the stress distribution acting over the cross section and calculate the maximum stress developed in the beam.

*6–60. The beam is subjected to a moment of 15 kip · ft. Determine the resultant force the bending stress produces on the top flange A and bottom flange B. Also calculate the maximum bending stress developed in the beam.

6–61. The beam is subjected to a moment of 15 kip · ft. Determine the percentage of this moment that is resisted by the web D of the beam.

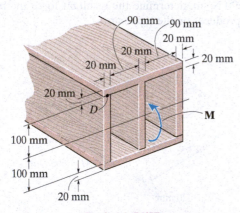

Probs. 6–56/57

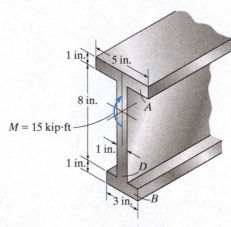

Probs. 6–60/61

6–62. The beam is subjected to a moment of $M = 40$ kN · m. Determine the bending stress at points A and B. Sketch the results on a volume element acting at each of these points.

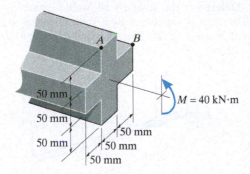

50 mm
50 mm
50 mm
50 mm
50 mm
50 mm

$M = 40$ kN·m

Prob. 6–62

6–63. The steel shaft has a diameter of 2 in. It is supported on smooth journal bearings A and B, which exert only vertical reactions on the shaft. Determine the absolute maximum bending stress in the shaft if it is subjected to the pulley loadings shown.

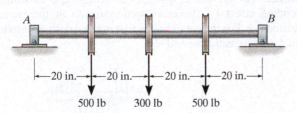

—20 in.——20 in.——20 in.——20 in.—

500 lb   300 lb   500 lb

Prob. 6–63

*6–64. The beam is made of steel that has an allowable stress of $\sigma_{allow} = 24$ ksi. Determine the largest internal moment the beam can resist if the moment is applied (a) about the z axis, (b) about the y axis.

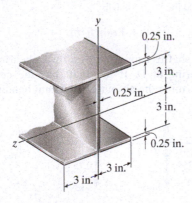

0.25 in.

3 in.

0.25 in.

3 in.

0.25 in.

3 in.

3 in.

3 in.

Prob. 6–64

6–65. A shaft is made of a polymer having an elliptical cross section. If it resists an internal moment of $M = 50$ N · m, determine the maximum bending stress in the material (a) using the flexure formula, where $I_z = \frac{1}{4}\pi(0.08 \text{ m})(0.04 \text{ m})^3$, (b) using integration. Sketch a three-dimensional view of the stress distribution acting over the cross-sectional area. Here $I_x = \frac{1}{4}\pi(0.08 \text{ m})(0.04 \text{ m})^3$.

6–66. Solve Prob. 6–65 if the moment $M = 50$ N · m is applied about the y axis instead of the x axis. Here $I_y = \frac{1}{4}\pi (0.04 \text{ m})(0.08 \text{ m})^3$.

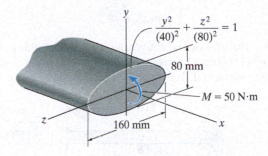

$\dfrac{y^2}{(40)^2} + \dfrac{z^2}{(80)^2} = 1$

80 mm

$M = 50$ N·m

160 mm

Probs. 6–65/66

6–67. The shaft is supported by smooth journal bearings at A and B that only exert vertical reactions on the shaft. If $d = 90$ mm, determine the absolute maximum bending stress in the beam, and sketch the stress distribution acting over the cross section.

*6–68. The shaft is supported by smooth journal bearings at A and B that only exert vertical reactions on the shaft. Determine its smallest diameter d if the allowable bending stress is $\sigma_{allow} = 180$ MPa.

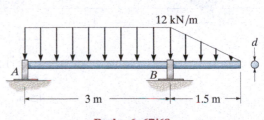

12 kN/m

d

A

B

3 m   1.5 m

Probs. 6–67/68

6

6–69. The axle of the freight car is subjected to a wheel loading of 20 kip. If it is supported by two journal bearings at C and D, determine the maximum bending stress developed at the center of the axle, where the diameter is 5.5 in.

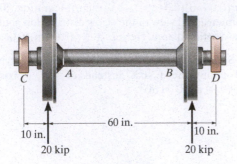

10 in.

60 in.

10 in.

20 kip

20 kip

Prob. 6–69

6–70. The strut on the utility pole supports the cable having a weight of 600 lb. Determine the absolute maximum bending stress in the strut if A, B, and C are assumed to be pinned.

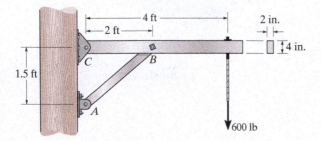

4 ft

2 ft

2 in.

4 in.

1.5 ft

C

B

A

600 lb

Prob. 6–70

6–71. The boat has a weight of 2300 lb and a center of gravity at G. If it rests on the trailer at the smooth contact A and can be considered pinned at B, determine the absolute maximum bending stress developed in the main strut of the trailer which is pinned at C. Consider the strut to be a box-beam having the dimensions shown.

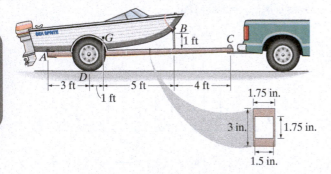

3 ft

1 ft

5 ft

4 ft

1 ft

B

C

G

A

D

1.75 in.

3 in.

1.75 in.

1.5 in.

Prob. 6–71

*6–72. Determine the absolute maximum bending stress in the 1.5-in.-diameter shaft. The shaft is supported by a thrust bearing at A and a journal bearing at B.

6–73. Determine the smallest allowable diameter of the shaft. The shaft is supported by a thrust bearing at A and a journal bearing at B. The allowable bending stress is $\sigma_{allow} = 22$ ksi.

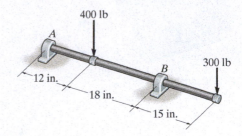

400 lb

A

B

300 lb

12 in.

18 in.

15 in.

Probs. 6–72/73

6–74. The pin is used to connect the three links together. Due to wear, the load is distributed over the top and bottom of the pin as shown on the free-body diagram. If the diameter of the pin is 0.40 in., determine the maximum bending stress on the cross-sectional area at the center section $a–a$. For the solution it is first necessary to determine the load intensities w_1 and w_2.

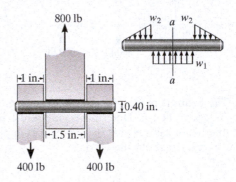

800 lb

w_2 a w_2

w_1

a

1 in.

1 in.

0.40 in.

1.5 in.

400 lb

400 lb

Prob. 6–74

6–75. The shaft is supported by a thrust bearing at A and journal bearing at D. If the shaft has the cross section shown, determine the absolute maximum bending stress in the shaft.

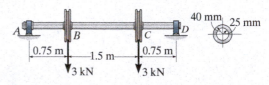

A

B

C

D

40 mm

25 mm

0.75 m

1.5 m

0.75 m

3 kN

3 kN

Prob. 6–75

*6–76. A timber beam has a cross section which is originally square. If it is oriented as shown, determine the dimension h' so that it can resist the maximum moment possible. By what factor is this moment greater than that of the beam without its top or bottom flattened?

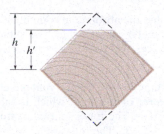

Prob. 6–76

6–77. If the beam is subjected to an internal moment of $M = 2$ kip · ft, determine the maximum tensile and compressive stress in the beam. Also, sketch the bending stress distribution on the cross section.

6–78. If the allowable tensile and compressive stress for the beam are $(\sigma_{\text{allow}})_t = 2$ ksi and $(\sigma_{\text{allow}})_c = 3$ ksi, respectively, determine the maximum moment **M** that can be applied on the cross section.

6–79. If the beam is subjected to an internal moment of $M = 2$ kip · ft, determine the resultant force of the bending stress distribution acting on the top board A.

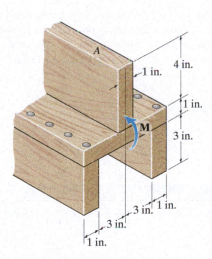

Probs. 6–77/78/79

*6–80. If the beam is subjected to a moment of $M = 100$ kN · m, determine the bending stress at points A, B, and C. Sketch the bending stress distribution on the cross section.

6–81. If the beam is made of material having an allowable tensile and compressive stress of $(\sigma_{\text{allow}})_t = 125$ MPa and $(\sigma_{\text{allow}})_c = 150$ MPa, respectively, determine the maximum moment **M** that can be applied to the beam.

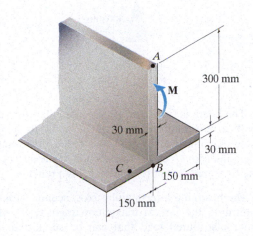

Probs. 6–80/81

6–82. The shaft is supported by a smooth thrust bearing at A and smooth journal bearing at C. If $d = 3$ in., determine the absolute maximum bending stress in the shaft.

6–83. The shaft is supported by a thrust bearing at A and journal bearing at C. If the material has an allowable bending stress of $\sigma_{\text{allow}} = 24$ ksi, determine the required minimum diameter d of the shaft to the nearest $\frac{1}{16}$ in.

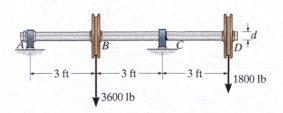

Probs. 6–82/83

*6–84. If the intensity of the load $w = 15$ kN/m, determine the absolute maximum tensile and compressive stress in the beam.

6–85. If the allowable bending stress is $\sigma_{allow} = 150$ MPa, determine the maximum intensity w of the uniform distributed load.

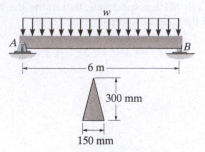

Probs. 6–84/85

6–86. The beam is subjected to the triangular distributed load with a maximum intensity of $w_0 = 300$ lb/ft. If the allowable bending stress is $\sigma_{allow} = 1.40$ ksi, determine the required dimension b of its cross section to the nearest $\frac{1}{8}$ in. Assume the support at A is a pin and B is a roller.

6–87. The beam has a rectangular cross section with $b = 4$ in. Determine the largest maximum intensity w_0 of the triangular distributed load that can be supported if the allowable bending stress is $\sigma_{allow} = 1.40$ ksi.

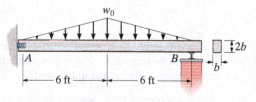

Probs. 6–86/87

*6–88. Determine the absolute maximum bending stress in the beam. Each segment has a rectangular cross section with a base of 4 in. and height of 12 in.

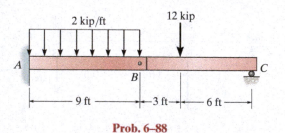

Prob. 6–88

6–89. If the compound beam in Prob. 6–42 has a square cross section of side length a, determine the minimum value of a if the allowable bending stress is $\sigma_{allow} = 150$ MPa.

6–90. If the beam in Prob. 6–28 has a rectangular cross section with a width b and a height h, determine the absolute maximum bending stress in the beam.

6–91. Determine the absolute maximum bending stress in the 80-mm-diameter shaft which is subjected to the concentrated forces. There is a journal bearing at A and a thrust bearing at B.

*6–92. Determine, to the nearest millimeter, the smallest allowable diameter of the shaft which is subjected to the concentrated forces. There is a journal bearing at A and a thrust bearing at B. The allowable bending stress is $\sigma_{allow} = 150$ MPa.

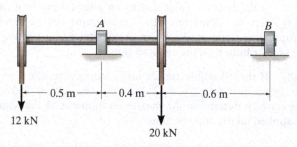

Probs. 6–91/92

6–93. Determine the absolute maximum bending stress in the beam, assuming that the support at B exerts a uniformly distributed reaction on the beam. The cross section is rectangular with a base of 3 in. and height of 6 in.

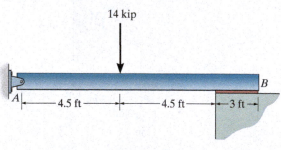

Prob. 6–93

6–94. Determine the absolute maximum bending stress in the 2-in.-diameter shaft. There is a journal bearing at A and a thrust bearing at B.

6–95. Determine the smallest diameter of the shaft to the nearest $\frac{1}{8}$ in. There is a journal bearing at A and a thrust bearing at B. The allowable bending stress is $\sigma_{allow} = 22$ ksi.

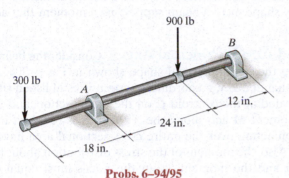

Probs. 6–94/95

*6–96. A log that is 2 ft in diameter is to be cut into a rectangular section for use as a simply supported beam. If the allowable bending stress is $\sigma_{allow} = 8$ ksi, determine the required width b and height h of the beam that will support the largest load possible. What is this load?

6–97. A log that is 2 ft in diameter is to be cut into a rectangular section for use as a simply supported beam. If the allowable bending stress is $\sigma_{allow} = 8$ ksi, determine the largest load P that can be supported if the width of the beam is $b = 8$ in.

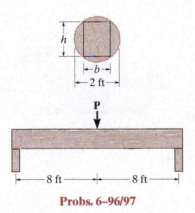

Probs. 6–96/97

6–98. If the beam in Prob. 6–3 has a rectangular cross section with a width of 8 in. and a height of 16 in., determine the absolute maximum bending stress in the beam.

6–99. The simply supported truss is subjected to the central distributed load. Neglect the effect of the diagonal lacing and determine the absolute maximum bending stress in the truss. The top member is a pipe having an outer diameter of 1 in. and thickness of $\frac{3}{16}$ in., and the bottom member is a solid rod having a diameter of $\frac{1}{2}$ in.

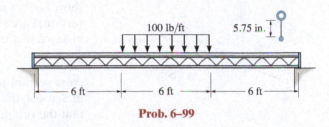

Prob. 6–99

*6–100. If $d = 450$ mm, determine the absolute maximum bending stress in the overhanging beam.

6–101. If the allowable bending stress is $\sigma_{allow} = 6$ MPa, determine the minimum dimension d of the beam's cross-sectional area to the nearest mm.

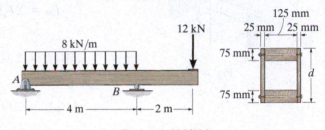

Probs. 6–100/101

6–102. The beam has a rectangular cross section as shown. Determine the largest intensity w of the uniform distributed load so that the bending stress in the beam does not exceed $\sigma_{max} = 10$ MPa.

6–103. The beam has the rectangular cross section shown. If $w = 1$ kN/m, determine the maximum bending stress in the beam. Sketch the stress distribution acting over the cross section.

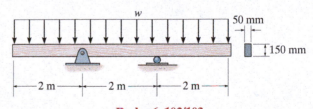

Probs. 6–102/103

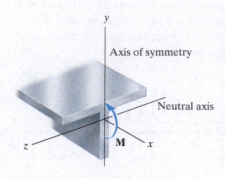

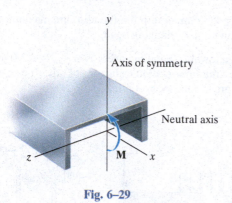

Fig. 6–29

6.5 UNSYMMETRIC BENDING

When developing the flexure formula, we required the cross-sectional area to be *symmetric* about an axis perpendicular to the neutral axis and the resultant moment **M** to act along the neutral axis. Such is the case for the "T" and channel sections shown in Fig. 6–29. In this section we will show how to apply the flexure formula either to a beam having a cross-sectional area of any shape or to a beam supporting a moment that acts in any direction.

Moment Applied About Principal Axis.   Consider the beam's cross section to have the unsymmetrical shape shown in Fig. 6–30*a*. As in Sec. 6.4, the right-handed *x, y, z* coordinate system is established such that the origin is located at the centroid *C* on the cross section, and the resultant internal moment **M** acts along the +*z* axis. It is required that the stress distribution acting over the entire cross-sectional area have a zero force resultant. Also, the moment of the stress distribution about the *y* axis must be zero, and the moment about the *z* axis must equal **M**. These three conditions can be expressed mathematically by considering the force acting on the differential element *dA* located at (0, *y*, *z*), Fig. 6–30*a*. Since this force is $dF = \sigma\, dA$, we have

$$F_R = \Sigma F_x; \qquad\qquad 0 = -\int_A \sigma\, dA \qquad\qquad (6\text{–}14)$$

$$(M_R)_y = \Sigma M_y; \qquad\qquad 0 = -\int_A z\sigma\, dA \qquad\qquad (6\text{–}15)$$

$$(M_R)_z = \Sigma M_z; \qquad\qquad M = \int_A y\sigma\, dA \qquad\qquad (6\text{–}16)$$

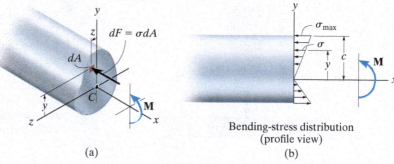

(a)

Bending-stress distribution
(profile view)

(b)

Fig. 6–30

As shown in Sec. 6.4, Eq. 6–14 is satisfied since the z axis passes through the *centroid* of the area. Also, since the z axis represents the *neutral axis* for the cross section, the normal stress will vary linearly from zero at the neutral axis to a maximum at $|y| = c$, Fig. 6–30b. Hence the stress distribution is defined by $\sigma = -(y/c)\sigma_{max}$. When this equation is substituted into Eq. 6–16 and integrated, it leads to the flexure formula $\sigma_{max} = Mc/I$. When it is substituted into Eq. 6–15, we get

$$0 = \frac{-\sigma_{max}}{c} \int_A yz\, dA$$

which requires

$$\int_A yz\, dA = 0$$

Z-sectioned members are often used in light-gage metal building construction to support roofs. To design them to support bending loads, it is necessary to determine their principal axes of inertia.

This integral is called the **product of inertia** for the area. As indicated in Appendix A, it will indeed be zero provided the y and z axes are chosen as **principal axes of inertia** for the area. For an arbitrarily shaped area, such as the one in Fig. 6–30a, the orientation of the principal axes can always be determined, using the inertia transformation equations as explained in Appendix A, Sec. A.4. If the area has an axis of symmetry, however, the **principal axes** can easily be established **since they will always be oriented along the axis of symmetry and perpendicular to it.**

For example, consider the members shown in Fig. 6–31. In each of these cases, y and z represent the principal axes of inertia for the cross section. In Fig. 6–31a the principal axes are located by symmetry, and in Figs. 6–31b and 6–31c their orientation is determined using the methods of Appendix A. Since **M** is applied only about one of the principal axes (the z axis), the stress distribution has a linear variation, and is determined from the flexure formula, $\sigma = -My/I_z$, as shown for each case.

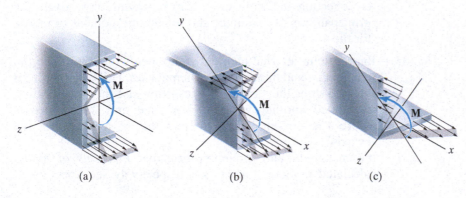

(a)   (b)   (c)

Fig. 6–31

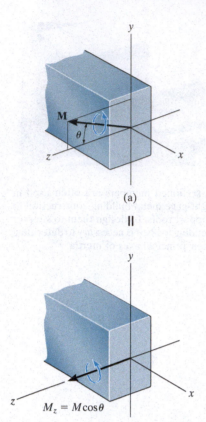

(a)

‖

$M_z = M\cos\theta$

(b)

+

$M_y = M\sin\theta$

(c)

Fig. 6–32

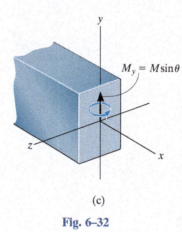

Moment Arbitrarily Applied.   Sometimes a member may be loaded such that M does not act about one of the principal axes of the cross section. When this occurs, the moment should first be resolved into components directed along the principal axes, then the flexure formula can be used to determine the normal stress caused by *each* moment component. Finally, using the principle of superposition, the resultant normal stress at the point can be determined.

To formalize this procedure, consider the beam to have a rectangular cross section and to be subjected to the moment M, Fig. 6–32a, where M makes an angle θ with the maximum principal z axis, i.e., the axis of maximum moment of inertia for the cross section. We will assume θ is positive when it is directed from the $+z$ axis towards the $+y$ axis. Resolving M into components, we have $M_z = M \cos \theta$ and $M_y = M \sin \theta$, Figs. 6–32b and 6–32c. The normal-stress distributions that produce M and its components M_z and M_y are shown in Figs. 6–32d, 6–32e, and 6–32f, where it is assumed that $(\sigma_x)_{max} > (\sigma'_x)_{max}$. By inspection, the maximum tensile and compressive stresses $[(\sigma_x)_{max} + (\sigma'_x)_{max}]$ occur at two opposite corners of the cross section, Fig. 6–32d.

Applying the flexure formula to each moment component in Figs. 6–32b and 6–32c, and adding the results algebraically, the resultant normal stress at any point on the cross section, Fig. 6–32d, is therefore

$$\sigma = -\frac{M_z y}{I_z} + \frac{M_y z}{I_y} \tag{6–17}$$

Here,

σ = the normal stress at the point. Tensile stress is positive and compressive stress is negative.

y, z = the coordinates of the point measured from a *right-handed coordinate system*, x, y, z, having their origin at the centroid of the cross-sectional area. The x axis is directed outward from the cross section and the y and z axes represent, respectively, the principal axes of minimum and maximum moment of inertia for the area.

M_z, M_y = the resultant internal moment components directed along the maximum z and minimum y principal axes. They are positive if directed along the $+z$ and $+y$ axes, otherwise they are negative. Or, stated another way, $M_y = M \sin \theta$ and $M_z = M \cos \theta$, where θ is measured positive from the $+z$ axis towards the $+y$ axis.

I_z, I_y = the maximum and minimum *principal moments of inertia* calculated about the z and y axes, respectively. See Appendix A.

Orientation of the Neutral Axis.

The equation defining the neutral axis, and its inclination α, Fig. 6–32d, can be determined by applying Eq. 6–17 to a point y, z where $\sigma = 0$, since by definition no normal stress acts on the neutral axis. We have

$$y = \frac{M_y I_z}{M_z I_y} z$$

Since $M_z = M \cos \theta$ and $M_y = M \sin \theta$, then

$$y = \left(\frac{I_z}{I_y} \tan \theta \right) z \qquad (6–18)$$

Since the slope of this line is $\tan \alpha = y/z$, then

$$\boxed{\tan \alpha = \frac{I_z}{I_y} \tan \theta} \qquad (6–19)$$

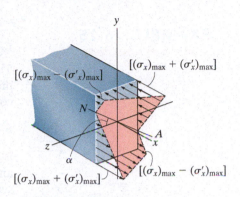

$[(\sigma_x)_{max} - (\sigma'_x)_{max}]$ $[(\sigma_x)_{max} + (\sigma'_x)_{max}]$

$[(\sigma_x)_{max} + (\sigma'_x)_{max}]$ $[(\sigma_x)_{max} - (\sigma'_x)_{max}]$

(d)

=

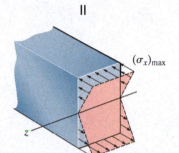

$(\sigma_x)_{max}$

$(\sigma_x)_{max}$

(e)

+

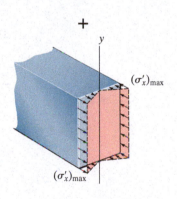

$(\sigma'_x)_{max}$

$(\sigma'_x)_{max}$

(f)

Fig. 6–32 (cont.)

▶ IMPORTANT POINTS

- The flexure formula can be applied only when bending occurs about axes that represent the *principal axes of inertia* for the cross section. These axes have their origin at the centroid and are oriented along an axis of symmetry, if there is one, and perpendicular to it.

- If the moment is applied about some arbitrary axis, then the moment must be resolved into components along each of the principal axes, and the stress at a point is determined by superposition of the stress caused by each of the moment components.

EXAMPLE 6.15

The rectangular cross section shown in Fig. 6–33a is subjected to a bending moment of $M = 12$ kN·m. Determine the normal stress developed at each corner of the section, and specify the orientation of the neutral axis.

SOLUTION

Internal Moment Components. By inspection it is seen that the y and z axes represent the principal axes of inertia since they are axes of symmetry for the cross section. As required we have established the z axis as the principal axis for *maximum* moment of inertia. The moment is resolved into its y and z components, where

$$M_y = -\frac{4}{5}(12 \text{ kN·m}) = -9.60 \text{ kN·m}$$

$$M_z = \frac{3}{5}(12 \text{ kN·m}) = 7.20 \text{ kN·m}$$

Section Properties. The moments of inertia about the y and z axes are

$$I_y = \frac{1}{12}(0.4 \text{ m})(0.2 \text{ m})^3 = 0.2667(10^{-3}) \text{ m}^4$$

$$I_z = \frac{1}{12}(0.2 \text{ m})(0.4 \text{ m})^3 = 1.067(10^{-3}) \text{ m}^4$$

Bending Stress. Thus,

$$\sigma = -\frac{M_z y}{I_z} + \frac{M_y z}{I_y}$$

$$\sigma_B = -\frac{7.20(10^3) \text{ N·m}(0.2 \text{ m})}{1.067(10^{-3}) \text{ m}^4} + \frac{-9.60(10^3) \text{ N·m}(-0.1 \text{ m})}{0.2667(10^{-3}) \text{ m}^4} = 2.25 \text{ MPa} \qquad \textit{Ans.}$$

$$\sigma_C = -\frac{7.20(10^3) \text{ N·m}(0.2 \text{ m})}{1.067(10^{-3}) \text{ m}^4} + \frac{-9.60(10^3) \text{ N·m}(0.1 \text{ m})}{0.2667(10^{-3}) \text{ m}^4} = -4.95 \text{ MPa} \qquad \textit{Ans.}$$

$$\sigma_D = -\frac{7.20(10^3) \text{ N·m}(-0.2 \text{ m})}{1.067(10^{-3}) \text{ m}^4} + \frac{-9.60(10^3) \text{ N·m}(0.1 \text{ m})}{0.2667(10^{-3}) \text{ m}^4} = -2.25 \text{ MPa} \qquad \textit{Ans.}$$

$$\sigma_E = -\frac{7.20(10^3) \text{ N·m}(-0.2 \text{ m})}{1.067(10^{-3}) \text{ m}^4} + \frac{-9.60(10^3) \text{ N·m}(-0.1 \text{ m})}{0.2667(10^{-3}) \text{ m}^4} = 4.95 \text{ MPa} \qquad \textit{Ans.}$$

The resultant normal-stress distribution has been sketched using these values, Fig. 6–33b. Since superposition applies, the distribution is linear as shown.

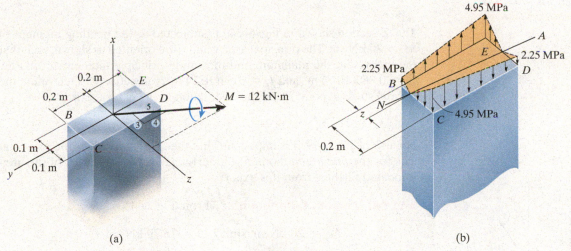

(a)

(b)

Fig. 6-33

Orientation of Neutral Axis. The location z of the neutral axis (NA), Fig. 6–33b, can be established by proportion. Along the edge BC, we require

$$\frac{2.25 \text{ MPa}}{z} = \frac{4.95 \text{ MPa}}{(0.2 \text{ m} - z)}$$

$$0.450 - 2.25z = 4.95z$$

$$z = 0.0625 \text{ m}$$

In the same manner this is also the distance from D to the neutral axis.

We can also establish the orientation of the NA using Eq. 6–19, which is used to specify the angle α that the axis makes with the z or *maximum* principal axis. According to our sign convention, θ must be measured from the $+z$ axis toward the $+y$ axis. By comparison, in Fig. 6–33c, $\theta = -\tan^{-1}\frac{4}{3} = -53.1°$ (or $\theta = +306.9°$). Thus,

$$\tan \alpha = \frac{I_z}{I_y}\tan \theta$$

$$\tan \alpha = \frac{1.067(10^{-3}) \text{ m}^4}{0.2667(10^{-3}) \text{ m}^4}\tan(-53.1°)$$

$$\alpha = -79.4° \qquad \qquad \textit{Ans.}$$

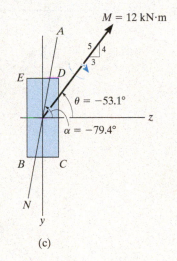

(c)

This result is shown in Fig. 6–33c. Using the value of z calculated above, verify, using the geometry of the cross section, that one obtains the same answer.

EXAMPLE 6.16

The Z-section shown in Fig. 6–34a is subjected to the bending moment of $M = 20$ kN · m. The principal axes y and z are oriented as shown, such that they represent the minimum and maximum principal moments of inertia, $I_y = 0.960(10^{-3})$ m^4 and $I_z = 7.54(10^{-3})$ m^4, respectively.* Determine the normal stress at point P and the orientation of the neutral axis.

SOLUTION

For use of Eq. 6–19, it is important that the z axis represent the principal axis for the *maximum* moment of inertia. (For this case most of the area is located farthest from this axis.)

Internal Moment Components.   From Fig. 6–34a,

$$M_y = 20 \text{ kN} \cdot \text{m} \sin 57.1° = 16.79 \text{ kN} \cdot \text{m}$$

$$M_z = 20 \text{ kN} \cdot \text{m} \cos 57.1° = 10.86 \text{ kN} \cdot \text{m}$$

Bending Stress.   The y and z coordinates of point P must be determined first. Note that the y', z' coordinates of P are $(-0.2$ m, 0.35 m$)$. Using the colored triangles from the construction shown in Fig. 6–34b, we have

$$y_P = -0.35 \sin 32.9° - 0.2 \cos 32.9° = -0.3580 \text{ m}$$

$$z_P = 0.35 \cos 32.9° - 0.2 \sin 32.9° = 0.1852 \text{ m}$$

Applying Eq. 6–17,

$$\sigma_P = -\frac{M_z y_P}{I_z} + \frac{M_y z_P}{I_y}$$

$$= -\frac{(10.86(10^3) \text{ N} \cdot \text{m})(-0.3580 \text{ m})}{7.54(10^{-3}) \text{ m}^4} + \frac{(16.79(10^3) \text{ N} \cdot \text{m})(0.1852 \text{ m})}{0.960(10^{-3}) \text{ m}^4}$$

$$= 3.76 \text{ MPa} \qquad\qquad\qquad\qquad\qquad\qquad\qquad Ans.$$

Orientation of Neutral Axis.   Using the angle $\theta = 57.1°$ between **M** and the z axis, Fig. 6–34a, we have

$$\tan \alpha = \left[\frac{7.54(10^{-3}) \text{ m}^4}{0.960(10^{-3}) \text{ m}^4}\right] \tan 57.1°$$

$$\alpha = 85.3° \qquad\qquad\qquad\qquad\qquad\qquad\qquad Ans.$$

The neutral axis is oriented as shown in Fig. 6–34b.

* These values are obtained using the methods of Appendix A. (See Example A.4 or A.5.)

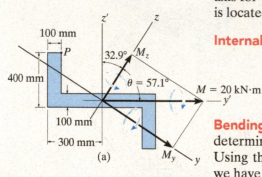

(a)

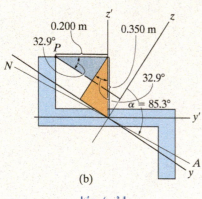

(b)

Fig. 6–34

FUNDAMENTAL PROBLEMS

F6–14. Determine the bending stress at corners A and B. What is the orientation of the neutral axis?

F6–15. Determine the maximum bending stress in the beam's cross section.

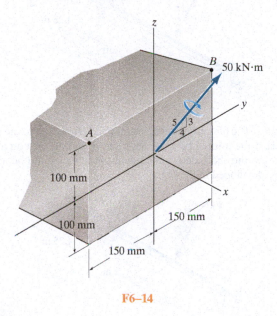

F6–14

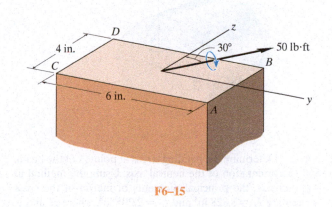

F6–15

PROBLEMS

*6–104. The member has a square cross section and is subjected to the moment $M = 850 \ \text{N} \cdot \text{m}$. Determine the stress at each corner and sketch the stress distribution. Set $\theta = 45°$.

6–105. The member has a square cross section and is subjected to the moment $M = 850 \ \text{N} \cdot \text{m}$ as shown. Determine the stress at each corner and sketch the stress distribution. Set $\theta = 30°$.

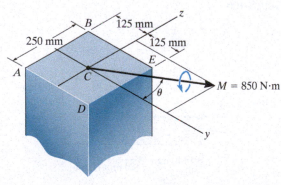

Prob. 6–104

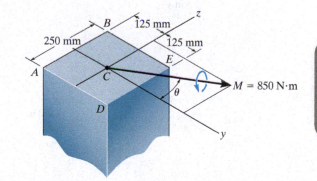

Prob. 6–105

6–106. Consider the general case of a prismatic beam subjected to bending-moment components M_y and M_z when the x, y, z axes pass through the centroid of the cross section. If the material is linear elastic, the normal stress in the beam is a linear function of position such that $\sigma = a + by + cz$. Using the equilibrium conditions $0 = \int_A \sigma \, dA$, $M_y = \int_A z\sigma \, dA$, $M_z = \int_A -y\sigma \, dA$, determine the constants a, b, and c, and show that the normal stress can be determined from the equation $\sigma = [-(M_zI_y + M_yI_{yz})y + (M_yI_z + M_zI_{yz})z]/(I_yI_z - I_{yz}^2)$, where the moments and products of inertia are defined in Appendix A.

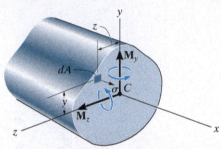

Prob. 6–106

6–107. Determine the bending stress at point A of the beam, and the orientation of the neutral axis. Using the method in Appendix A, the principal moments of inertia of the cross section are $I_{z'} = 8.828$ in^4 and $I_{y'} = 2.295$ in^4, where z' and y' are the principal axes. Solve the problem using Eq. 6–17.

*6–108. Determine the bending stress at point A of the beam using the result obtained in Prob. 6–106. The moments of inertia of the cross-sectional area about the z and y axes are $I_z = I_y = 5.561$ in^4 and the product of inertia of the cross sectional area with respect to the z and y axes is $I_{yz} = -3.267$ in^4. (See Appendix A.)

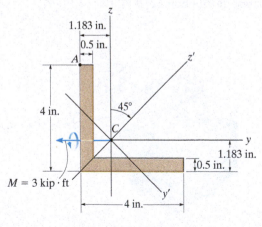

Probs. 6–107/108

6–109. The steel shaft is subjected to the two loads. If journal bearings at A and B do not exert an axial force on the shaft, determine the required diameter of the shaft if the allowable bending stress is $\sigma_{\text{allow}} = 180$ MPa.

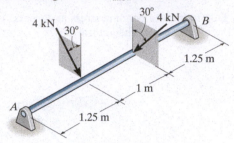

Prob. 6–109

6–110. The 65-mm-diameter steel shaft is subjected to the two loads. If the journal bearings at A and B do not exert an axial force on the shaft, determine the absolute maximum bending stress developed in the shaft.

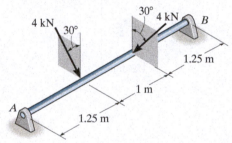

Prob. 6–110

6–111. For the section, $I_{z'} = 31.7(10^{-6})$ m^4, $I_{y'} = 114(10^{-6})$ m^4, $I_{y'z'} = -15.8(10^{-6})$ m^4. Using the techniques outlined in Appendix A, the member's cross-sectional area has principal moments of inertia of $I_z = 28.8(10^{-6})$ m^4 and $I_y = 117(10^{-6})$ m^4, calculated about the principal axes of inertia y and z, respectively. If the section is subjected to the moment $M = 15$ kN $\cdot$ m, determine the stress at point A using Eq. 6–17.

*6–112. Solve Prob. 6–111 using the equation developed in Prob. 6–106.

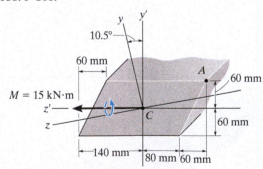

Probs. 6–111/112

6–113. The box beam is subjected to a moment of $M = 15$ kip · ft. Determine the maximum bending stress in the beam and the orientation of the neutral axis.

6–114. Determine the maximum magnitude of the bending moment **M** so that the bending stress in the member does not exceed 15 ksi.

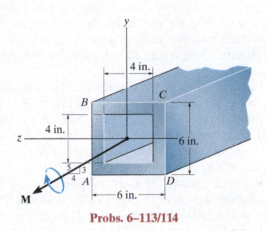

Probs. 6–113/114

6–115. The shaft is subjected to the vertical and horizontal loadings of two pulleys D and E as shown. It is supported on two journal bearings at A and B which offer no resistance to axial loading. Furthermore, the coupling to the motor at C can be assumed not to offer any support to the shaft. Determine the required diameter d of the shaft if the allowable bending stress is $\sigma_{allow} = 180$ MPa.

*6–116. For the section, $I_{y'} = 31.7(10^{-6})$ m^4, $I_{z'} = 114(10^{-6})$ m^4, $I_{y'z'} = 15.8(10^{-6})$ m^4. Using the techniques outlined in Appendix A, the member's cross-sectional area has principal moments of inertia of $I_y = 28.8(10^{-6})$ m^4 and $I_z = 117(10^{-6})$ m^4, calculated about the principal axes of inertia y and z, respectively. If the section is subjected to a moment of $M = 2500$ N · m, determine the stress produced at point A, using Eq. 6–17.

6–117. Solve Prob. 6–116 using the equation developed in Prob. 6–106.

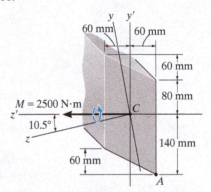

Probs. 6–116/117

6–118. If the applied distributed loading of $w = 4$ kN/m can be assumed to pass through the centroid of the beam's cross-sectional area, determine the absolute maximum bending stress in the joist and the orientation of the neutral axis. The beam can be considered simply supported at A and B.

6–119. Determine the maximum allowable intensity w of the uniform distributed load that can be applied to the beam. Assume w passes through the centroid of the beam's cross-sectional area, and the beam is simply supported at A and B. The allowable bending stress is $\sigma_{allow} = 165$ MPa.

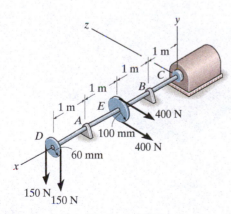

Prob. 6–115

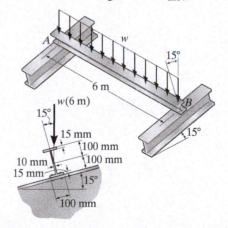

Probs. 6–118/119

*6.6   COMPOSITE BEAMS

Beams constructed of two or more different materials are referred to as *composite beams.* An example is a beam made of wood with straps of steel at its top and bottom, Fig. 6–35. Engineers purposely design beams in this manner in order to develop a more efficient means for supporting loads.

Since the flexure formula was developed only for beams made of homogeneous material, this formula cannot be applied to directly determine the normal stress in a composite beam. In this section, however, we will develop a method for modifying or "transforming" a composite beam's cross section into one made of a single material. Once this has been done, the flexure formula can then be used to determine the bending stress in the beam.

To explain how to do this we will consider a composite beam made of two materials, 1 and 2, bonded together as shown in Fig. 6–36a. If a bending moment is applied to this beam, then, like one that is homogeneous, the total cross-sectional area will *remain plane* after bending, and hence the normal strains will vary linearly from zero at the neutral axis to a maximum farthest from this axis, Fig. 6–36b. Provided the material is linear elastic, then at any point the normal stress in material 1 is determined from $\sigma = E_1\epsilon$, and for material 2 the stress is found from $\sigma = E_2\epsilon$. Assuming material 1 is stiffer than material 2, then $E_1 > E_2$ and so the stress distribution will look like that shown in Fig. 6–36c or 6–36d. In particular, notice the jump in stress that occurs at the juncture of the two materials. Here the *strain* is the *same,* but since the modulus of elasticity for the materials suddenly changes, so does the stress.

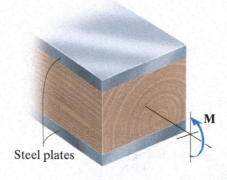

Steel plates

Fig. 6–35

Rather than using this complicated stress distribution, it is simpler to transform the beam into one made of a single material. For example, if the beam is thought to consist entirely of the less stiff material 2, then the cross section will look like that shown in Fig. 6–36e. Here the height h of the beam remains the *same*, since the strain distribution in Fig. 6–36b must be the same. However, the upper portion of the beam must be widened in order to carry a load *equivalent* to that carried by the stiffer material 1 in Fig. 6–36d. This necessary width can be determined by considering the force $d\mathbf{F}$ acting on an area $dA = dz\,dy$ of the beam in Fig. 6–36a. It is $dF = \sigma\,dA = E_1\epsilon\,(dz\,dy)$. Assuming the width of a *corresponding* element of height dy in Fig. 6–36e is $n\,dz$, then $dF' = \sigma'dA' = E_2\epsilon\,(n\,dz\,dy)$. Equating these forces, so that they produce the same moment about the z (neutral) axis, we have

$$E_1\epsilon\,(dz\,dy) = E_2\epsilon\,(n\,dz\,dy)$$

or

$$n = \frac{E_1}{E_2} \tag{6–20}$$

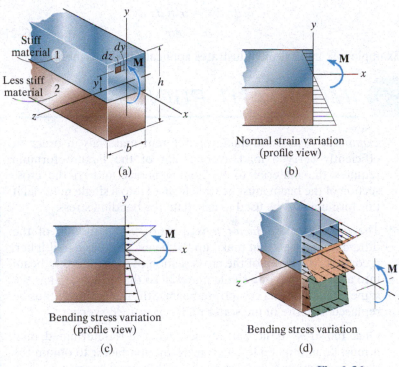

Stiff material 1

Less stiff material 2

(a)

Normal strain variation
(profile view)

(b)

Bending stress variation
(profile view)

(c)

Bending stress variation

(d)

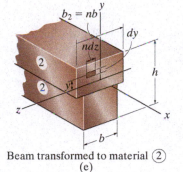

$b_2 = nb$

Beam transformed to material ②

(e)

Fig. 6–36

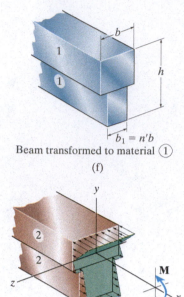

Beam transformed to material ①

(f)

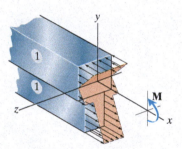

Bending stress variation for
beam transformed to material ②

(g)

Bending stress variation for
beam transformed to material ①

(h)

Fig. 6–36 (cont.)

This dimensionless number n is called the ***transformation factor.*** It indicates that the cross section, having a width b on the original beam, Fig. 6–36a, must be increased in width to $b_2 = nb$ in the region where material 1 is being transformed into material 2, Fig. 6–36e.

In a similar manner, if the less stiff material 2 is transformed into the stiffer material 1, the cross section will look like that shown in Fig. 6–36f. Here the width of material 2 has been changed to $b_1 = n'b$, where $n' = E_2/E_1$. In this case the transformation factor n' will be *less than one* since $E_1 > E_2$. In other words, we need less of the stiffer material to support the moment.

Once the beam has been transformed into one having a single material, the normal-stress distribution over the transformed cross section will be linear as shown in Fig. 6–36g or Fig. 6–36h. Consequently, the flexure formula can now be applied in the usual manner to determine the stress at each point on the transformed beam. Of course, the stress in the transformed beam will be equivalent to the stress in the same material of the actual beam; however, the stress in the transformed material has to be multiplied by the transformation factor n (or n') to obtain the stress in any other actual material that was transformed. This is because the area of the transformed material, $dA' = n\,dz\,dy$, is n times the area of actual material $dA = dz\,dy$. That is,

$$dF = \sigma\,dA = \sigma'dA'$$
$$\sigma\,dz\,dy = \sigma'n\,dz\,dy$$
$$\sigma = n\sigma' \qquad (6\text{–}21)$$

Example 6.17 numerically illustrates application of this method.

IMPORTANT POINTS

- *Composite beams* are made of different materials in order to efficiently carry a load. Application of the flexure formula requires the material to be homogeneous, and so the cross section of the beam must be transformed into a single material if this formula is to be used to calculate the bending stress.

- The *transformation factor* n is a ratio of the moduli of the different materials that make up the beam. Used as a multiplier, it converts the width of the cross section of the composite beam into a beam made of a single material so that this beam has the same strength as the composite beam. Stiff material will thus be replaced by more of the softer material and vice versa.

- Once the stress in the transformed material is determined, then it must be multiplied by the transformation factor to obtain the stress in any transformed material of the actual beam.

*6.7 REINFORCED CONCRETE BEAMS

All beams subjected to pure bending must resist both tensile and compressive stresses. Concrete, however, is very susceptible to cracking when it is in tension, and therefore by itself it will not be suitable for resisting a bending moment.* In order to circumvent this shortcoming, engineers place steel reinforcing rods within a concrete beam at a location where the concrete is in tension, Fig. 6–37a. To be most effective, these rods are located farthest from the beam's neutral axis, so that the moment created by the forces developed in them is greatest about the neutral axis. Furthermore, the rods are required to have some concrete coverage to protect them from corrosion or loss of strength in the event of a fire. Codes used for actual reinforced concrete design assume the concrete will not be able to support any tensile loading, since the possible cracking of concrete is unpredictable. As a result, the normal-stress distribution acting on the cross-sectional area of a reinforced concrete beam is assumed to look like that shown in Fig. 6–37b.

The stress analysis requires locating the neutral axis and determining the maximum stress in the steel and concrete. To do this, the area of steel A_{st} is first transformed into an equivalent area of concrete using the transformation factor $n = E_{st}/E_{conc}$, as discussed in Sec. 6.6. This ratio, which gives $n > 1$, requires a "greater" amount of concrete to replace the steel. The transformed area is nA_{st} and the transformed section looks like that shown in Fig. 6–37c. Here d represents the distance from the top of the beam to the thin strip of (transformed) steel, b is the beam's width, and h' is the yet unknown distance from the top of the beam to the neutral axis. To obtain h', we require the neutral axis to pass through the centroid C of the cross-sectional area of the transformed section, Fig. 6–37c. With reference to the neutral axis, therefore, the moment of the two areas together, $\Sigma \tilde{y} A$, must be zero, since $\bar{y} = \Sigma \tilde{y} A / \Sigma A = 0$. Thus,

$$bh'\left(\frac{h'}{2}\right) - nA_{st}(d - h') = 0$$

$$\frac{b}{2}h'^2 + nA_{st}h' - nA_{st}d = 0$$

Once h' is obtained from this quadratic equation, the solution proceeds in the usual manner for obtaining the stress in the beam. Example 6.18 numerically illustrates application of this method.

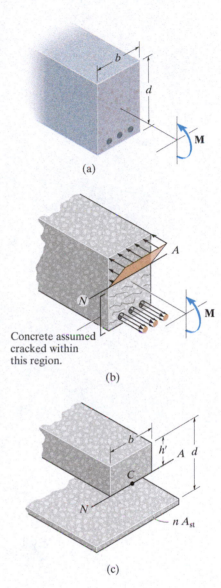

(a)

Concrete assumed cracked within this region.

(b)

(c)

Fig. 6–37

---

*Inspection of its stress–strain diagram in Fig. 3–12 reveals that concrete can be 12.5 times stronger in compression than in tension.

EXAMPLE 6.17

The composite beam in Fig. 6–38a is made of wood and reinforced with a steel strap located on its bottom side. If the beam is subjected to a bending moment of $M = 2$ kN·m, determine the normal stress at points B and C. Take $E_w = 12$ GPa and $E_{st} = 200$ GPa.

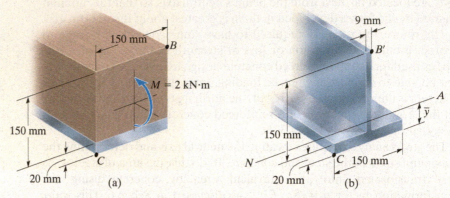

Fig. 6–38

SOLUTION

Section Properties. Although the choice is arbitrary, here we will transform the section into one made entirely of steel. Since steel has a greater stiffness than wood ($E_{st} > E_w$), the width of the wood is *reduced* to an equivalent width for steel. For this to be the case, $n = E_w/E_{st}$, so that

$$b_{st} = nb_w = \frac{12 \text{ GPa}}{200 \text{ GPa}}(150 \text{ mm}) = 9 \text{ mm}$$

The transformed section is shown in Fig. 6–38b.

The location of the centroid (neutral axis), calculated from the *bottom* of the section, is

$$\bar{y} = \frac{\Sigma \bar{y}A}{\Sigma A} = \frac{[0.01 \text{ m}](0.02 \text{ m})(0.150 \text{ m}) + [0.095 \text{ m}](0.009 \text{ m})(0.150 \text{ m})}{0.02 \text{ m}(0.150 \text{ m}) + 0.009 \text{ m}(0.150 \text{ m})} = 0.03638 \text{ m}$$

The moment of inertia about the neutral axis is therefore

$$I_{NA} = \left[\frac{1}{12}(0.150 \text{ m})(0.02 \text{ m})^3 + (0.150 \text{ m})(0.02 \text{ m})(0.03638 \text{ m} - 0.01 \text{ m})^2 \right]$$

$$+ \left[\frac{1}{12}(0.009 \text{ m})(0.150 \text{ m})^3 + (0.009 \text{ m})(0.150 \text{ m})(0.095 \text{ m} - 0.03638 \text{ m})^2 \right]$$

$$= 9.358(10^{-6}) \text{ m}^4$$

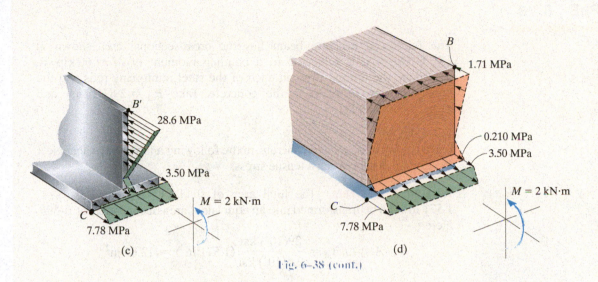

Fig. 6-38 (cont.)

Normal Stress.   Applying the flexure formula, the normal stress at B' and C is

$$\sigma_{B'} = \frac{2(10^3)\ \text{N}\cdot\text{m}(0.170\ \text{m}\ -\ 0.03638\ \text{m})}{9.358(10^{-6})\ \text{m}^4} = 28.6\ \text{MPa}$$

$$\sigma_C = \frac{2(10^3)\ \text{N}\cdot\text{m}(0.03638\ \text{m})}{9.358(10^{-6})\ \text{m}^4} = 7.78\ \text{MPa} \qquad \textit{Ans.}$$

The normal-stress distribution on the transformed (all steel) section is shown in Fig. 6–38c.

The normal stress in the wood at B in Fig. 6–38a is determined from Eq. 6–21; that is,

$$\sigma_B = n\sigma_{B'} = \frac{12\ \text{GPa}}{200\ \text{GPa}}(28.56\ \text{MPa}) = 1.71\ \text{MPa} \qquad \textit{Ans.}$$

Using these concepts, show that the normal stress in the steel and the wood at the point where they are in contact is $\sigma_{st} = 3.50\ \text{MPa}$ and $\sigma_w = 0.210\ \text{MPa}$, as shown in Fig. 6–38d.

EXAMPLE 6.18

The reinforced concrete beam has the cross-sectional area shown in Fig. 6–39a. If it is subjected to a bending moment of $M = 60$ kip·ft, determine the normal stress in each of the steel reinforcing rods and the maximum normal stress in the concrete. Take $E_{st} = 29(10^3)$ ksi and $E_{conc} = 3.6(10^3)$ ksi.

SOLUTION

Since the beam is made of concrete, in the following analysis we will neglect its strength in supporting a tensile stress.

Section Properties. The total area of steel, $A_{st} = 2[\pi(0.5 \text{ in.})^2] = 1.571 \text{ in}^2$, will be transformed into an equivalent area of concrete, Fig. 6–39b. Here

$$A' = nA_{st} = \frac{29(10^3) \text{ ksi}}{3.6(10^3) \text{ ksi}}(1.571 \text{ in}^2) = 12.65 \text{ in}^2$$

We require the centroid to lie on the neutral axis. Thus $\Sigma \tilde{y} A = 0$, or

$$12 \text{ in. } (h')\frac{h'}{2} - 12.65 \text{ in}^2(16 \text{ in. } - h') = 0$$

$$h'^2 + 2.11h' - 33.7 = 0$$

Solving for the positive root,

$$h' = 4.85 \text{ in.}$$

Using this value for h', the moment of inertia of the transformed section about the neutral axis is

$$I = \left[\frac{1}{12}(12 \text{ in.})(4.85 \text{ in.})^3 + 12 \text{ in. } (4.85 \text{ in.})\left(\frac{4.85 \text{ in.}}{2}\right)^2\right] +$$

$$12.65 \text{ in}^2(16 \text{ in. } - 4.85 \text{ in.})^2 = 2029 \text{ in}^4$$

Normal Stress. Applying the flexure formula to the transformed section, the maximum normal stress in the concrete is

$$(\sigma_{conc})_{max} = \frac{[60 \text{ kip} \cdot \text{ft}(12 \text{ in./ft})](4.85 \text{ in.})}{2029 \text{ in}^4} = 1.72 \text{ ksi} \qquad Ans.$$

The normal stress resisted by the "concrete" strip that replaced the steel is

$$\sigma'_{conc} = \frac{[60 \text{ kip} \cdot \text{ft}(12 \text{ in./ft})](16 \text{ in. } - 4.85 \text{ in.})}{2029 \text{ in}^4} = 3.96 \text{ ksi}$$

The normal stress in each of the two reinforcing rods is therefore

$$\sigma_{st} = n\sigma'_{conc} = \left(\frac{29(10^3) \text{ ksi}}{3.6(10^3) \text{ ksi}}\right)3.96 \text{ ksi} = 31.9 \text{ ksi} \qquad Ans.$$

The normal-stress distribution is shown graphically in Fig. 6–39c.

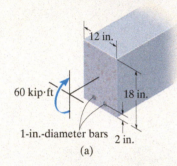

12 in.

60 kip·ft

18 in.

1-in.-diameter bars   2 in.

(a)

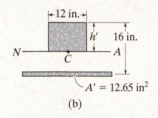

|←12 in.→|

h' 16 in.

N ———— A

C

$A' = 12.65 \text{ in}^2$

(b)

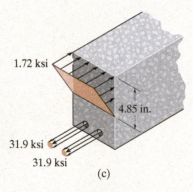

1.72 ksi

4.85 in.

31.9 ksi

31.9 ksi

(c)

Fig. 6–39

*6.8   CURVED BEAMS

The flexure formula applies to a straight member, because the normal strain within the member varies linearly from the neutral axis. If the member is *curved,* however, the strain will not be linear, and so we must develop another method to describe the stress distribution. In this section we will consider the analysis of a **curved beam,** that is, a member that has a curved axis and is subjected to bending. Typical examples include hooks and rings. In all cases, the members are not slender, but rather have a sharp curve, and their cross-sectional dimensions will be large compared with their radius of curvature.

The following analysis assumes that the cross section is constant and has an axis of symmetry that is perpendicular to the direction of the applied moment **M,** Fig. 6–40a. This moment is *positive* if it tends to straighten out the member. Also, the material is homogeneous and isotropic, and it behaves in a linear elastic manner when the load is applied. Like the case of a straight beam, we will also assume that the cross sections of the member remain plane after the moment is applied. Furthermore, any distortion of the cross section within its own plane, as caused by Poisson's effect, will be neglected.

To perform the analysis, three radii, extending from the center of curvature O' of the member, are identified in Fig. 6–40a. Here $\bar{r}$ references the known location of the centroid for the cross-sectional area, R references the yet unspecified location of the neutral axis, and r locates the arbitrary point or area element dA on the cross section.

This crane hook represents a typical example of a curved beam.

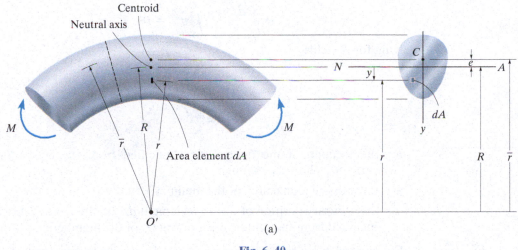

(a)

Fig. 6–40

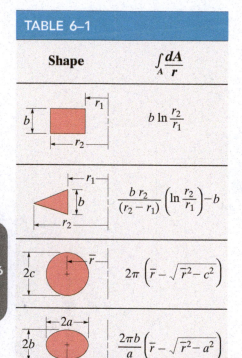

(b)

Fig. 6–40 (cont.)

If we isolate a differential segment of the beam, Fig. 6–40b, the stress tends to deform the material such that each cross section will rotate through an angle $\delta\theta/2$. The normal strain ϵ in the strip (or line) of material located at r will now be determined. This strip has an original length $r\,d\theta$; however, due to the rotations $\delta\theta/2$ the strip's total change in length is $\delta\theta(R - r)$. Consequently, $\epsilon = \delta\theta(R - r)/r\,d\theta$. If we let $k = \delta\theta/d\theta$, a constant, since it is the *same* for any particular strip, we have $\epsilon = k(R - r)/r$. Unlike the case of straight beams, here it can be seen that the *normal strain* is a nonlinear function of r, in fact it varies in a *hyperbolic fashion*. This occurs even though the cross section of the beam remains plane after deformation. Since the material is linear elastic, then $\sigma = E\epsilon$, and so

$$\sigma = Ek\left(\frac{R - r}{r}\right) \tag{6–22}$$

This variation is also hyperbolic, and now that it has been established, we can determine the location of the neutral axis, and relate the stress distribution to the internal moment M.

Location of Neutral Axis. To obtain the location R of the neutral axis, we require the resultant internal force caused by the stress distribution acting over the cross section to be equal to zero; i.e.,

$$F_R = \Sigma F_x; \qquad \int_A \sigma\,dA = 0$$

$$\int_A Ek\left(\frac{R - r}{r}\right)dA = 0$$

Since Ek and R are constants, we have

$$R\int_A \frac{dA}{r} - \int_A dA = 0$$

Solving for R yields

$$R = \frac{A}{\displaystyle\int_A \frac{dA}{r}} \tag{6–23}$$

Here

$R =$ the location of the neutral axis, specified from the center of curvature O' of the member

$A =$ the cross-sectional area of the member

$r =$ the arbitrary position of the area element dA on the cross section, specified from the center of curvature O' of the member

The integral in Eq. 6–23 has been evaluated for various cross-sectional geometries, and the results for some common cross sections are listed in Table 6–1.

TABLE 6–1

Shape	$\displaystyle\int_A \frac{dA}{r}$
rectangle, b, r_1, r_2	$b\ln\dfrac{r_2}{r_1}$
triangle, r_1, b, r_2	$\dfrac{b\,r_2}{(r_2 - r_1)}\left(\ln\dfrac{r_2}{r_1}\right) - b$
circle, $2c$, $\bar{r}$	$2\pi\left(\bar{r} - \sqrt{\bar{r}^2 - c^2}\right)$
ellipse, $2a$, $2b$, $\bar{r}$	$\dfrac{2\pi b}{a}\left(\bar{r} - \sqrt{\bar{r}^2 - a^2}\right)$

Bending Moment.   In order to relate the stress distribution to the resultant bending moment, we require the resultant internal moment to be equal to the moment of the stress distribution calculated about the neutral axis. From Fig. 6–40c, the stress σ, acting on the area element dA and located a distance y from the neutral axis, creates a moment about the neutral axis of $dM = y(\sigma\, dA)$. For the entire cross section, we require $M = \int y\sigma\, dA$. Since $y = R - r$, and σ is defined by Eq. 6–22, we have

$$M = \int_A (R - r)Ek\left(\frac{R - r}{r}\right) dA$$

Expanding, realizing that Ek and R are constants, then

$$M = Ek\left(R^2 \int_A \frac{dA}{r} - 2R \int_A dA + \int_A r\, dA\right)$$

The first integral is equivalent to A/R as determined from Eq. 6–23, and the second integral is simply the cross-sectional area A. Realizing that the location of the centroid of the cross section is determined from $\bar{r} = \int r\, dA / A$, the third integral can be replaced by $\bar{r}A$. Thus,

$$M = EkA(\bar{r} - R)$$

Finally, solving for Ek in Eq. 6–22, substituting into the above equation, and solving for σ, we have

$$\boxed{\sigma = \frac{M(R - r)}{Ar(\bar{r} - R)}} \qquad (6\text{–}24)$$

Here

σ = the normal stress in the member

M = the internal moment, determined from the method of sections and the equations of equilibrium, and calculated about the neutral axis for the cross section. This moment is *positive* if it tends to *increase* the member's radius of curvature, i.e., it tends to straighten out the member.

A = the cross-sectional area of the member

R = the distance measured from the center of curvature to the neutral axis, determined from Eq. 6–23

$\bar{r}$ = the distance measured from the center of curvature to the centroid of the cross section

r = the distance measured from the center of curvature to the point where the stress σ is to be determined

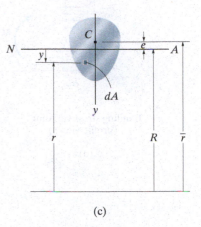

(c)

Fig. 6–40 (cont.)

6

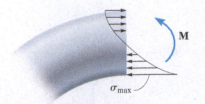

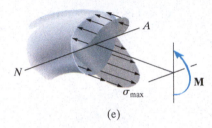

Bending stress variation
(profile view)

(d)

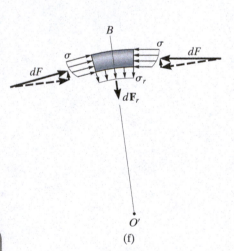

(e)

(f)

Fig. 6–40 (cont.)

From Fig. 6–40a, $r = R - y$. Also, the constant and usually very small distance between the neutral axis and the centroid is $e = \bar{r} - R$. When these results are substituted into Eq. 6–24, we can also write

$$\sigma = \frac{My}{Ae(R - y)}$$

(6–25)

These two equations represent two forms of the **curved-beam formula,** which like the flexure formula can be used to determine the normal-stress distribution in a curved member. This distribution is, as previously stated, hyperbolic. An example is shown in Figs. 6–40d and 6–40e. Since the stress acts along the circumference of the beam, it is sometimes called **circumferential stress**, Fig. 6–40f.

Radial Stress.   Due to the curvature of the beam, the circumferential stress will create a corresponding component of **radial stress**, so called since this component acts in the radial direction. To show how it is developed, consider the free-body diagram of the segment shown in Fig. 6–40f. Here the radial stress σ_r is necessary, since it creates the force dF_r that is required to balance the two components of circumferential forces dF which act along the radial line $O'B$.

Limitations.   Sometimes the radial stresses within curved members may become significant, especially if the member is constructed from thin plates and has, for example, the shape of an I or T section. In this case the radial stress can become as large as the circumferential stress, and consequently the member must be designed to resist both stresses. For most cases, however, these stresses can be neglected, especially if the member has a *solid section*. Here the curved-beam formula gives results that are in very close agreement with those determined either by experiment or by a mathematical analysis based on the theory of elasticity.

The curved-beam formula is normally used when the curvature of the member is very pronounced, as in the case of hooks or rings. However, if the radius of curvature is greater than five times the depth of the member, the *flexure formula* can normally be used to determine the stress. For example, for rectangular sections for which this ratio equals 5, the maximum normal stress, when determined by the flexure formula, will be about 7% *less* than its value when determined by the more accurate curved-beam formula. This error is further reduced when the radius of curvature-to-depth ratio is more than 5.*

---

*See, for example, Boresi, A. P. and Schmidt, R. J. *Advanced Mechanics of Materials*, John Wiley & Sons, New York.

IMPORTANT POINTS

- Due to the curvature of the beam, the normal strain in the beam *does not* vary linearly with depth as in the case of a straight beam. As a result, the neutral axis generally does not pass through the centroid of the cross section.

- The radial stress component caused by bending can generally be neglected, especially if the cross section is a solid section and not made from thin plates.

- The *curved-beam formula* should be used to determine the circumferential stress in a beam when the radius of curvature is less than five times the depth of the beam.

PROCEDURE FOR ANALYSIS

In order to apply the curved-beam formula the following procedure is suggested.

Section Properties.

- Determine the cross-sectional area A and the location of the centroid, $\bar{r}$, measured from the center of curvature.

- Find the location of the neutral axis, R, using Eq. 6–23 or Table 6–1. If the cross-sectional area consists of n "composite" parts, determine $\int dA/r$ for *each part*. Then, from Eq. 6–23, for the entire section, $R = \Sigma A / \Sigma(\int dA/r)$.

Normal Stress.

- The normal stress located at a point r away from the center of curvature is determined from Eq. 6–24. If the distance y to the point is measured from the neutral axis, then find $e = \bar{r} - R$ and use Eq. 6–25.

- Since $\bar{r} - R$ generally produces a very *small number*, it is best to calculate $\bar{r}$ and R with sufficient accuracy so that the subtraction leads to a number e having at least four significant figures.

- According to the established sign convention, positive M tends to straighten out the member, and so if the stress is positive it will be tensile, whereas if it is negative it will be compressive.

- The stress distribution over the entire cross section can be graphed, or a volume element of the material can be isolated and used to represent the stress acting at the point on the cross section where it has been calculated.

6

EXAMPLE   6.19

The curved bar has a cross-sectional area shown in Fig. 6–41a. If it is subjected to bending moments of 4 kN · m, determine the maximum normal stress developed in the bar.

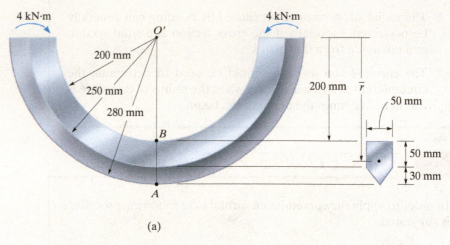

(a)

Fig. 6–41

SOLUTION

Internal Moment.   Each section of the bar is subjected to the same resultant internal moment of 4 kN · m. Since this moment tends to decrease the bar's radius of curvature, it is negative. Thus, $M = -4$ kN · m.

Section Properties.   Here we will consider the cross section to be composed of a rectangle and triangle. The total cross-sectional area is

$$\Sigma A = (0.05 \text{ m})^2 + \frac{1}{2}(0.05 \text{ m})(0.03 \text{ m}) = 3.250(10^{-3}) \text{ m}^2$$

The location of the centroid is determined with reference to the center of curvature, point O', Fig. 6–41a.

$$\bar{r} = \frac{\Sigma \tilde{r} A}{\Sigma A}$$

$$= \frac{[0.225 \text{ m}](0.05 \text{ m})(0.05 \text{ m}) + [0.260 \text{ m}]\frac{1}{2}(0.050 \text{ m})(0.030 \text{ m})}{3.250(10^{-3}) \text{ m}^2}$$

$$= 0.233077 \text{ m}$$

We can find $\int_A dA/r$ for each part using Table 6–1. For the rectangle,

$$\int_A \frac{dA}{r} = b \ln \frac{r_2}{r_1} = 0.05 \text{ m} \left(\ln \frac{0.250 \text{ m}}{0.200 \text{ m}} \right) = 0.0111572 \text{ m}$$

And for the triangle,

$$\int_A \frac{dA}{r} = \frac{br_2}{(r_2 - r_1)} \left(\ln \frac{r_2}{r_1} \right) - b = \frac{(0.05 \text{ m})(0.280 \text{ m})}{(0.280 \text{ m} - 0.250 \text{ m})} \left(\ln \frac{0.280 \text{ m}}{0.250 \text{ m}} \right) - 0.05 \text{ m} = 0.00288672 \text{ m}$$

Thus the location of the neutral axis is determined from

$$R = \frac{\Sigma A}{\Sigma \int_A dA/r} = \frac{3.250(10^{-3}) \text{ m}^2}{0.0111572 \text{ m} + 0.00288672 \text{ m}} = 0.231417 \text{ m}$$

The calculations were performed with sufficient accuracy so that $(\bar{r} - R) = 0.233077 \text{ m} - 0.231417 \text{ m} = 0.001660 \text{ m}$ is now accurate to four significant figures.

Normal Stress. The maximum normal stress occurs either at A or B. Applying the curved-beam formula to calculate the normal stress at B, $r_B = 0.200$ m, we have

$$\sigma_B = \frac{M(R - r_B)}{Ar_B(\bar{r} - R)} = \frac{(-4 \text{ kN} \cdot \text{m})(0.231417 \text{ m} - 0.200 \text{ m})}{3.250(10^{-3}) \text{ m}^2(0.200 \text{ m})(0.001660 \text{ m})}$$

$$= -116 \text{ MPa}$$

At point A, $r_A = 0.280$ m and the normal stress is

$$\sigma_A = \frac{M(R - r_A)}{Ar_A(\bar{r} - R)} = \frac{(-4 \text{ kN} \cdot \text{m})(0.231417 \text{ m} - 0.280 \text{ m})}{3.250(10^{-3}) \text{ m}^2(0.280 \text{ m})(0.001660 \text{ m})}$$

$$= 129 \text{ MPa} \qquad \qquad \textit{Ans.}$$

By comparison, the maximum normal stress is at A. A two-dimensional representation of the stress distribution is shown in Fig. 6–41b.

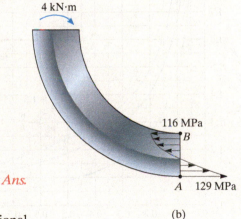

Fig. 6–41 (cont.)

(b)

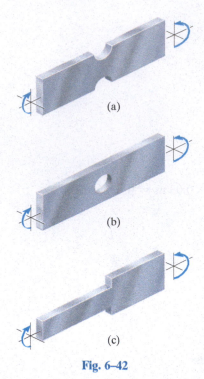

(a)

(b)

(c)

Fig. 6–42

6.9 STRESS CONCENTRATIONS

The flexure formula cannot be used to determine the stress distribution within regions of a member where the cross-sectional area suddenly changes, since the normal stress and strain distributions at this location are *nonlinear.* The results can only be obtained through experiment or, in some cases, by using the theory of elasticity. Common discontinuities include members having notches on their surfaces, Fig. 6–42*a*, holes for passage of fasteners or other items, Fig. 6–42*b*, or abrupt changes in the outer dimensions of the member's cross section, Fig. 6–42*c*. The *maximum* normal stress at each of these discontinuities occurs at the section taken through the *smallest* cross-sectional area.

For design, it is generally important to only know the maximum normal stress developed at these sections, not the actual stress distribution. As in the previous cases of axially loaded bars and torsionally loaded shafts, we can obtain this stress due to bending using a stress concentration factor K. For example, Fig. 6–43 gives values of K for a flat bar that has a change in cross section using shoulder fillets. To use this graph, simply find the

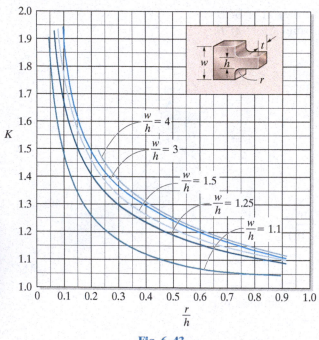

Fig. 6–43

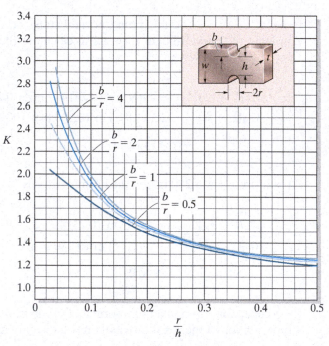

Fig. 6–44

geometric ratios w/h and r/h, and then find the corresponding value of K. Once K is obtained, the maximum bending stress shown in Fig. 6–45 is determined using

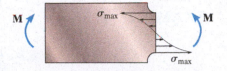

Fig. 6–45

$$\sigma_{max} = K \frac{Mc}{I} \qquad (6\text{–}26)$$

In the same manner, Fig. 6–44 can be used if the discontinuity consists of circular grooves or notches.

Like axial load and torsion, stress concentration for bending should always be considered when designing members made of brittle materials, or those that are subjected to fatigue or cyclic loadings. The stress concentration factors apply *only* when the material is subjected to *elastic behavior*. If the applied moment causes yielding of the material, as in the case of ductile materials, the stress becomes redistributed throughout the member, and the maximum stress that results will be *lower* than that determined using stress-concentration factors. This effect is discussed further in the next section.

Stress concentrations caused by bending occur at the sharp corners of this window lintel and are responsible for the crack at the corner.

> ## IMPORTANT POINTS

- Stress concentrations occur at points of sudden cross-sectional change, caused by notches and holes, because here the stress and strain become nonlinear. The more severe the change, the larger the stress concentration.

- For design or analysis, the maximum normal stress occurs on the *smallest* cross-sectional area. This stress can be obtained by using a stress concentration factor, K, that has been determined through experiment and is only a function of the geometry of the member.

- Normally, the stress concentration in a ductile material subjected to a static moment will not have to be considered in design; however, if the material is *brittle*, or subjected to *fatigue* loading, then stress concentrations become important.

EXAMPLE 6.20

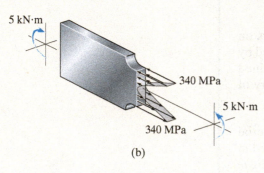

(a)

The transition in the cross-sectional area of the steel bar is achieved using shoulder fillets, as shown in Fig. 6–46a. If the bar is subjected to a bending moment of 5 kN · m, determine the maximum normal stress developed in the steel. The yield stress is $\sigma_Y = 500$ MPa.

SOLUTION

The moment creates the largest stress in the bar at the base of the fillet, where the cross-sectional area is smallest, and a stress concentration arises. The stress concentration factor can be determined by using Fig. 6–43. From the geometry of the bar, we have $r = 16$ mm, $h = 80$ mm, $w = 120$ mm. Thus,

$$\frac{r}{h} = \frac{16 \text{ mm}}{80 \text{ mm}} = 0.2 \qquad \frac{w}{h} = \frac{120 \text{ mm}}{80 \text{ mm}} = 1.5$$

These values give $K = 1.45$. Applying Eq. 6–26,

$$\sigma_{max} = K\frac{Mc}{I} = (1.45)\frac{(5(10^3) \text{ N} \cdot \text{m})(0.04 \text{ m})}{\left[\frac{1}{12}(0.020 \text{ m})(0.08 \text{ m})^3\right]} = 340 \text{ MPa} \qquad \textit{Ans.}$$

This result indicates that the steel remains elastic since the stress is below the yield stress (500 MPa).

NOTE: The normal-stress distribution is nonlinear and is shown in Fig. 6–46b. Realize, however, that by Saint-Venant's principle, Sec. 4.1, these localized stresses smooth out and become linear when one moves (approximately) a distance of 80 mm or more to the right of the transition. In this case, the flexure formula gives $\sigma_{max} = 234$ MPa, Fig. 6–46c.

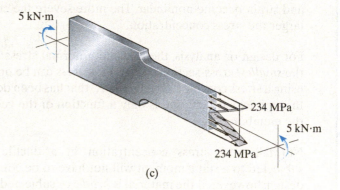

(b)

(c)

Fig. 6–46

PROBLEMS

*6–120.   The composite beam is made of steel (A) bonded to brass (B) and has the cross section shown. If it is subjected to a moment of $M = 6.5$ kN $\cdot$ m, determine the maximum bending stress in the brass and steel. Also, what is the stress in each material at the seam where they are bonded together? $E_{br} = 100$ GPa, $E_{st} = 200$ GPa.

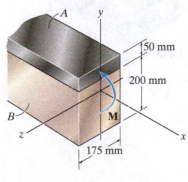

Prob. 6–120

6–121.   The composite beam is made of steel (A) bonded to brass (B) and has the cross section shown. If the allowable bending stress for the steel is $(\sigma_{allow})_{st} = 180$ MPa, and for the brass $(\sigma_{allow})_{br} = 60$ MPa, determine the maximum moment M that can be applied to the beam. $E_{br} = 100$ GPa, $E_{st} = 200$ GPa.

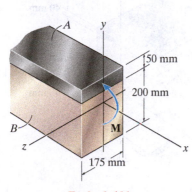

Prob. 6–121

6–122.   Segment A of the composite beam is made from 2014-T6 aluminum alloy and segment B is A-36 steel. If $w = 0.9$ kip/ft, determine the absolute maximum bending stress in the aluminum and steel. Sketch the stress distribution on the cross section.

6–123.   Segment A of the composite beam is made from 2014-T6 aluminum alloy and segment B is A-36 steel. The allowable bending stress for the aluminum and steel are $(\sigma_{allow})_{al} = 15$ ksi and $(\sigma_{allow})_{st} = 22$ ksi. Determine the maximum allowable intensity w of the uniform distributed load.

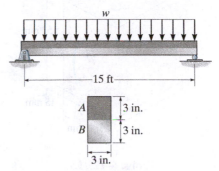

Probs. 6–122/123

*6–124.   The white spruce beam is reinforced with A-992 steel straps at its center and sides. Determine the maximum stress developed in the wood and steel if the beam is subjected to a bending moment of $M_z = 10$ kip $\cdot$ ft. Sketch the stress distribution acting over the cross section.

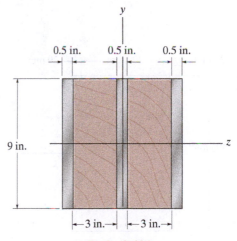

Prob. 6–124

6–125. The wooden section of the beam is reinforced with two steel plates as shown. Determine the maximum moment M that the beam can support if the allowable stresses for the wood and steel are $(\sigma_{allow})_w = 6$ MPa, and $(\sigma_{allow})_{st} = 150$ MPa, respectively. Take $E_w = 10$ GPa and $E_{st} = 200$ GPa.

6–126. The wooden section of the beam is reinforced with two steel plates as shown. If the beam is subjected to a moment of $M = 30$ kN · m, determine the maximum bending stresses in the steel and wood. Sketch the stress distribution over the cross section. Take $E_w = 10$ GPa and $E_{st} = 200$ GPa.

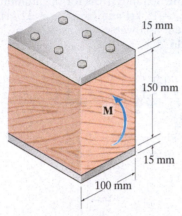

15 mm

150 mm

M

15 mm

100 mm

Probs. 6–125/126

6–127. The Douglas Fir beam is reinforced with A-992 steel straps at its sides. Determine the maximum stress in the wood and steel if the beam is subjected to a moment of $M_z = 80$ kN · m. Sketch the stress distribution acting over the cross section.

*6–128.** The steel channel is used to reinforce the wood beam. Determine the maximum stress in the steel and in the wood if the beam is subjected to a moment of $M = 850$ lb · ft. $E_{st} = 29(10^3)$ ksi, $E_w = 1600$ ksi.

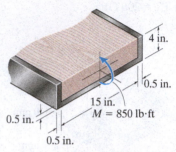

4 in.

0.5 in.

15 in.

$M = 850$ lb·ft

0.5 in.

0.5 in.

Prob. 6–128

6–129. A wood beam is reinforced with steel straps at its top and bottom as shown. Determine the maximum bending stress developed in the wood and steel if the beam is subjected to a moment of $M = 150$ kN · m. Sketch the stress distribution acting over the cross section. Take $E_w = 10$ GPa, $E_{st} = 200$ GPa.

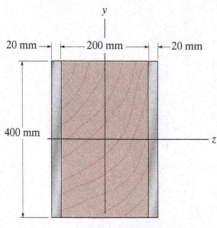

y

20 mm — |← 200 mm →| ←20 mm

400 mm

z

Prob. 6–127

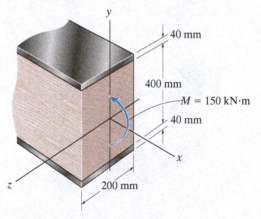

y

40 mm

400 mm

$M = 150$ kN·m

40 mm

x

z

200 mm

Prob. 6–129

6–130. A bimetallic strip is made from pieces of 2014-T6 aluminum and C83400 red brass, having the cross section shown. A temperature increase causes its neutral surface to be bent into a circular arc having a radius of 16 in. Determine the moment that must be acting on its cross section due to the thermal stress.

*6–132. The composite beam is made of A-36 steel (A) bonded to C83400 red brass (B) and has the cross section shown. If it is subjected to a moment of $M = 6.5$ kN · m, determine the maximum stress in the brass and steel. Also, what is the stress in each material at the seam where they are bonded together?

6–133. The composite beam is made of A-36 steel (A) bonded to C83400 red brass (B) and has the cross section shown. If the allowable bending stress for the steel is $(\sigma_{allow})_{st} = 180$ MPa and for the brass $(\sigma_{allow})_{br} = 60$ MPa, determine the maximum moment M that can be applied to the beam.

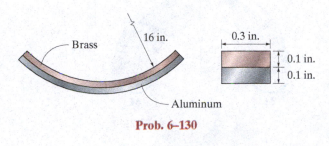

Prob. 6–130

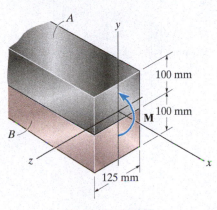

Probs. 6–132/133

6–131. Determine the maximum uniform distributed load w_0 that can be supported by the reinforced concrete beam if the allowable tensile stress for the steel is $(\sigma_{st})_{allow} = 28$ ksi and the allowable compressive stress for the concrete is $(\sigma_{conc})_{allow} = 3$ ksi. Assume the concrete cannot support a tensile stress. Take $E_{st} = 29(10^3)$ ksi, $E_{conc} = 3.6(10^3)$ ksi.

6–134. If the beam is subjected to a moment of $M = 45$ kN · m, determine the maximum bending stress in the A-36 steel section A and the 2014-T6 aluminum alloy section B.

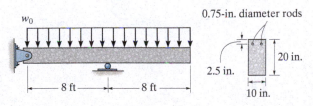

Prob. 6–131

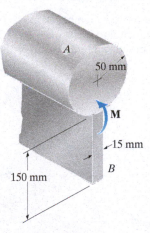

Prob. 6–134

6–135. The Douglas Fir beam is reinforced with A-36 steel straps at its sides. Determine the maximum stress in the wood and steel if the beam is subjected to a bending moment of $M_z = 4$ kN · m. Sketch the stress distribution acting over the cross section.

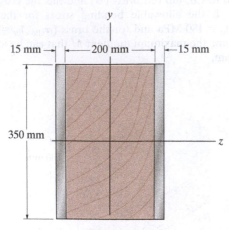

Prob. 6–135

*6–136. For the curved beam in Fig. 6–40a, show that when the radius of curvature approaches infinity, the curved-beam formula, Eq. 6–24, reduces to the flexure formula, Eq. 6–13.

6–137. The curved member is subjected to the moment of $M = 50$ kN · m. Determine the percentage error introduced in the calculation of maximum bending stress using the flexure formula for straight members.

6–138. The curved member is made from material having an allowable bending stress of $\sigma_{allow} = 100$ MPa. Determine the maximum allowable moment M that can be applied to the member.

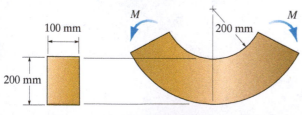

Probs. 6–137/138

6–139. The curved beam is subjected to a moment of $M = 40$ lb · ft. Determine the maximum bending stress in the beam. Also, sketch a two-dimensional view of the stress distribution acting on section a–a.

*6–140. The curved beam is made from material having an allowable bending stress of $\sigma_{allow} = 24$ ksi. Determine the maximum moment M that can be applied to the beam.

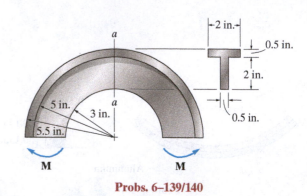

Probs. 6–139/140

6–141. If $P = 3$ kN, determine the bending stress at points A, B, and C of the cross section at section a–a. Using these results, sketch the stress distribution on section a–a.

6–142. If the maximum bending stress at section a–a is not allowed to exceed $\sigma_{allow} = 150$ MPa, determine the maximum allowable force **P** that can be applied to the end E.

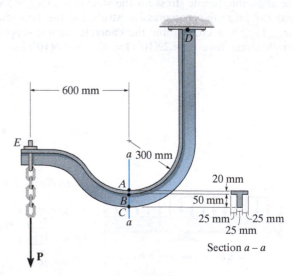

Probs. 6–141/142

6–143. The elbow of the pipe has an outer radius of 0.75 in. and an inner radius of 0.63 in. If the assembly is subjected to the moments of $M = 25$ lb·in., determine the maximum stress at section a–a.

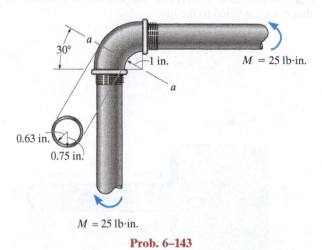

Prob. 6–143

6–145. The curved bar used on a machine has a rectangular cross section. If the bar is subjected to a couple as shown, determine the maximum tensile and compressive stresses acting at section a–a. Sketch the stress distribution on the section in three dimensions.

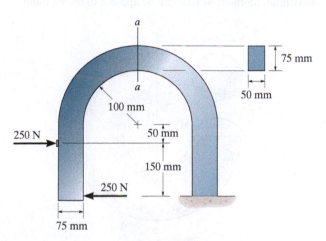

Prob. 6–145

***6–144.** The curved bar used on a machine has a rectangular cross section. If the bar is subjected to a couple as shown, determine the maximum tensile and compressive stresses acting at section a–a. Sketch the stress distribution on the section in three dimensions.

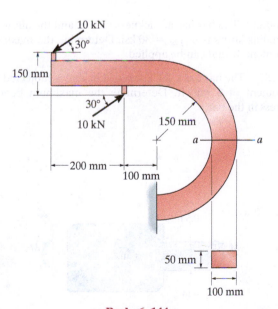

Prob. 6–144

6–146. The steel rod has a circular cross section. If it is gripped at its ends and a couple moment of $M = 12$ lb·in is developed at each grip, determine the stress acting at points A and B and at the centroid C.

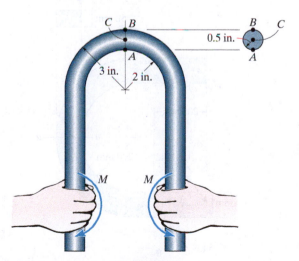

Prob. 6–146

6–147. The member has a circular cross section. If it is subjected to a moment of $M = 5$ kN $\cdot$ m, determine the stress at points A and B. Is the stress at point A', which is located on the member near the wall, the same as that at A? Explain.

*6–148. The member has a circular cross section. If the allowable bending stress is $\sigma_{allow} = 100$ MPa, determine the maximum moment M that can be applied to the member.

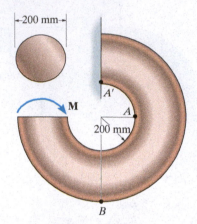

Probs. 6–147/148

6–149. The curved bar used on a machine has a rectangular cross section. If the bar is subjected to a couple as shown, determine the maximum tensile and compressive stress acting at section a–a. Sketch the stress distribution on the section in three dimensions.

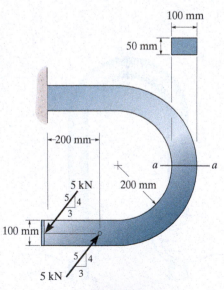

Prob. 6–149

6–150. The bar is subjected to a moment of $M = 100$ N $\cdot$ m. Determine the maximum bending stress in the bar and sketch, approximately, how the stress varies over the critical section.

6–151. The allowable bending stress for the bar is $\sigma_{allow} = 200$ MPa. Determine the maximum moment M that can be applied to the bar.

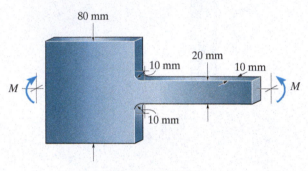

Probs. 6–150/151

*6–152. The bar has a thickness of 1 in. and the allowable bending stress is $\sigma_{allow} = 30$ ksi. Determine the maximum moment M that can be applied.

6–153. The bar has a thickness of 1 in. and is subjected to a moment of 3 kip $\cdot$ ft. Determine the maximum bending stress in the bar.

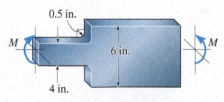

Probs. 6–152/153

6–154. The bar has a thickness of 0.5 in. and the allowable bending stress is σ_{allow} = 20 ksi. Determine the maximum moment M that can be applied.

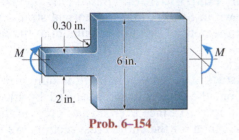

Prob. 6–154

6–155. If the radius of each notch on the plate is r = 10 mm, determine the largest moment M that can be applied. The allowable bending stress is σ_{allow} = 180 MPa.

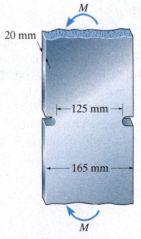

Prob. 6–155

***6–156.** The stepped bar has a thickness of 10 mm. Determine the maximum moment that can be applied to its ends if the allowable bending stress is σ_{allow} = 150 MPa.

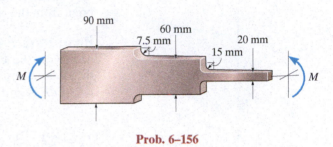

Prob. 6–156

6–157. The bar has a thickness of 0.5 in. and is subjected to a moment of 600 lb · ft. Determine the maximum bending stress in the bar.

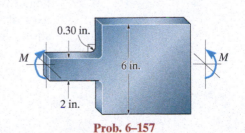

Prob. 6–157

6

*6.10   INELASTIC BENDING

The previous equations for determining the normal stress due to bending are valid only if the material behaves in a linear elastic manner. If the applied moment causes the material to *yield*, a plastic analysis must then be used to determine the stress distribution. For bending of straight members three conditions must be met.

Linear Normal-Strain Distribution.   Based only on geometric considerations, it was shown in Sec. 6.3 that the normal strains always vary *linearly* from zero at the neutral axis to a maximum at the farthest point from the neutral axis.

Resultant Force Equals Zero.   Since there is only a moment acting on the cross section, the resultant force caused by the stress distribution must be equal to zero. Since σ creates a force on the area dA of $dF = \sigma\, dA$, Fig. 6–47, then for the entire cross section, we have

$$F_R = \Sigma F_x; \qquad\qquad \int_A \sigma\, dA = 0 \qquad\qquad (6\text{–}27)$$

This equation provides the means for obtaining the *location of the neutral axis*.

Resultant Moment.   The moment at the section must be equivalent to the moment caused by the stress distribution about the neutral axis. The moment of the force $dF = \sigma\, dA$ about the neutral axis is $dM = y(\sigma\, dA)$, Fig. 6–47, and so for the entire cross section, we have

$$(M_R)_z = \Sigma M_z; \qquad\qquad M = \int_A y(\sigma\, dA) \qquad\qquad (6\text{–}28)$$

These conditions of geometry and loading will now be used to show how to determine the stress distribution in a beam when it is subjected to an internal moment that causes yielding of the material. Throughout the discussion we will assume that the material has a stress–strain diagram that is the *same* in tension as it is in compression. For simplicity, we will begin by considering the beam to have a cross-sectional area with two axes of symmetry; in this case, a rectangle of height h and width b, as shown in Fig. 6–48a.

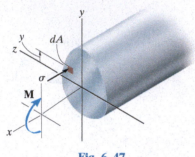

Fig. 6–47

Plastic Moment. Some materials, such as steel, tend to exhibit *elastic perfectly plastic behavior* when the stress in the material reaches σ_Y. If the moment $M = M_Y$ is just sufficient to produce yielding in the top and bottom fibers of the beam, then we can determine M_Y using the flexure formula $\sigma_Y = M_Y(h/2)/[bh^3/12]$ or

$$M_Y = \frac{1}{6}bh^2\sigma_Y \qquad (6\text{–}29)$$

If the moment $M > M_Y$, the material at the top and bottom of the beam will begin to yield, causing a redistribution of stress over the cross section until the required moment M is developed. For example, if M causes the normal-strain distribution shown in Fig. 6–48*b*, then the corresponding normal-stress distribution must be determined from the stress–strain diagram, Fig. 6–48*c*. If the strains ϵ_1, ϵ_Y, ϵ_2 correspond to stresses σ_1, σ_Y, σ_Y, respectively, then these and others like them produce the stress distribution shown in Fig. 6–48*d* or 6–48*e*. The resultant forces of the component rectangular and triangular stress blocks are equivalent to their volumes.

$$T_1 = C_1 = \frac{1}{2}y_Y\sigma_Y b \qquad T_2 = C_2 = \left(\frac{h}{2} - y_Y\right)\sigma_Y b$$

Because of the symmetry, Eq. 6–27 is satisfied, and the neutral axis passes through the centroid of the cross section as shown. The moment M can be related to the yield stress σ_Y using Eq. 6–28. From Fig. 6–48*e*, we require

$$
\begin{aligned}
M &= T_1\left(\frac{2}{3}y_Y\right) + C_1\left(\frac{2}{3}y_Y\right) + T_2\left[y_Y + \frac{1}{2}\left(\frac{h}{2} - y_Y\right)\right] \\
&\quad + C_2\left[y_Y + \frac{1}{2}\left(\frac{h}{2} - y_Y\right)\right] \\
&= 2\left(\frac{1}{2}y_Y\sigma_Y b\right)\left(\frac{2}{3}y_Y\right) + 2\left[\left(\frac{h}{2} - y_Y\right)\sigma_Y b\right]\left[\frac{1}{2}\left(\frac{h}{2} + y_Y\right)\right] \\
&= \frac{1}{4}bh^2\sigma_Y\left(1 - \frac{4}{3}\frac{y_Y^2}{h^2}\right)
\end{aligned}
$$

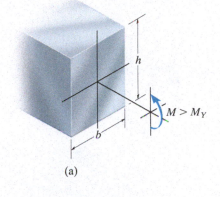

(a)

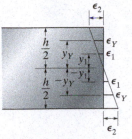

Strain distribution
(profile view)

(b)

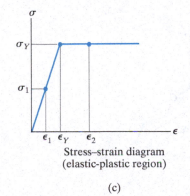

Stress–strain diagram
(elastic-plastic region)

(c)

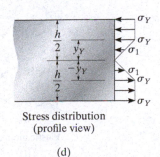

Stress distribution
(profile view)

(d)

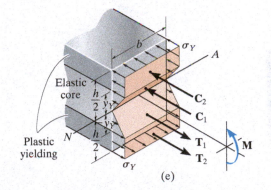

(e)

Fig. 6–48

Or using Eq. 6–29,

$$M = \frac{3}{2}M_Y\left(1 - \frac{4}{3}\frac{y_Y^2}{h^2}\right) \qquad (6\text{--}30)$$

As **M** increases in magnitude, the distance y_Y in Fig. 6–48e approaches zero, and the material becomes entirely plastic, resulting in a stress distribution that looks like that shown in Fig. 6–48f. Finding the moments of the stress "blocks" around the neutral axis, we can express this limiting value as

$$M_p = \frac{1}{4}bh^2\sigma_Y \qquad (6\text{--}31)$$

Using Eq. 6–29, or Eq. 6–30, with $y_Y = 0$, we also have

$$M_p = \frac{3}{2}M_Y \qquad (6\text{--}32)$$

This moment is referred to as the ***plastic moment***. Its value applies only for a rectangular section, since the analysis here depends on the geometry of the cross section.

　Beams used for steel building frames are sometimes designed to resist a plastic moment. When this is the case, codes usually list a design property for a beam called the shape factor. The ***shape factor*** is defined as the ratio

$$k = \frac{M_p}{M_Y} \qquad (6\text{--}33)$$

By definition, this value specifies the additional moment capacity that a beam can support beyond its maximum elastic moment. For example, from Eq. 6–32, a beam having a rectangular cross section has a shape factor of $k = 1.5$. Therefore this section will support 50% more bending moment than its maximum elastic moment when it becomes fully plastic.

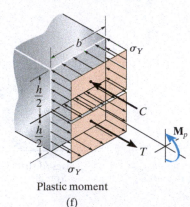

Plastic moment

(f)

Fig. 6–48 (cont.)

Residual Stress.   When the plastic moment in Fig. 6–48*f* is removed, it will cause **residual stress** in the beam. For example, let's say M_p causes the material at the top and bottom of the beam to be strained to $\epsilon_1 (\gg \epsilon_Y)$, as shown by point *B* on the $\sigma-\epsilon$ curve in Fig. 6–49*a*. A release of this moment will cause the material to recover some of this strain elastically by following the dashed path *BC*. Since this recovery is elastic, we can then superimpose on the stress distribution in Fig. 6–49*b* a linear stress distribution caused by applying the plastic moment in the opposite direction, Fig. 6–49*c*. Here the maximum stress for this distribution, which is called the **modulus of rupture** for bending, σ_r, can be determined from the flexure formula when the beam is loaded with the plastic moment. We have

$$\sigma_{max} = \frac{Mc}{I} = \frac{M_p\left(\frac{1}{2}h\right)}{\left(\frac{1}{12}bh^3\right)} = \frac{\left(\frac{1}{4}bh^2\sigma_Y\right)\left(\frac{1}{2}h\right)}{\left(\frac{1}{12}bh^3\right)} = 1.5\sigma_Y$$

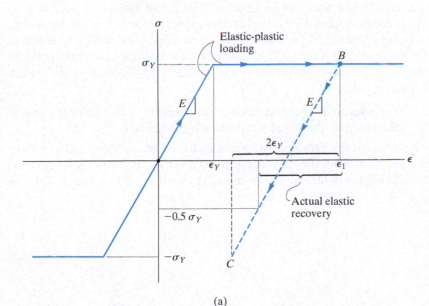

(a)

Fig. 6–49

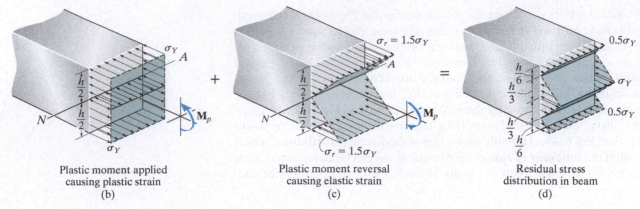

| Plastic moment applied causing plastic strain (b) | Plastic moment reversal causing elastic strain (c) | Residual stress distribution in beam (d) |

Fig. 6–49 (cont.)

Fortunately this value is less than $2\sigma_Y$, which is caused by the greatest possible strain recovery, $2\epsilon_Y$, Fig. 6–49a.

The superposition of the plastic moment, Fig. 6–49b, and its removal, Fig. 6–49c, gives the residual-stress distribution shown in Fig. 6–49d. As an exercise, use the component triangular "blocks" that represent this stress distribution and show that it results in a zero-force and zero-moment resultant on the member.

Ultimate Moment.

Consider now the more general case of a beam having a cross section that is symmetrical only with respect to the vertical axis, while the moment is applied about the horizontal axis, Fig. 6–50a. Here we will assume that the material exhibits strain hardening and that its stress–strain diagrams for tension and compression are different, Fig. 6–50b.

If the moment **M** produces yielding of the beam, difficulty arises in finding *both* the location of the neutral axis and the maximum strain in the beam. To solve this problem, a trial-and-error procedure requires the following steps:

1. For a given moment **M**, *assume* the location of the neutral axis and the slope of the linear strain distribution, Fig. 6–50c.

2. Graphically establish the stress distribution on the member's cross section using the $\sigma-\epsilon$ curve to plot values of stress corresponding to values of strain. The resulting stress distribution, Fig. 6–50d, will then have the same shape as the $\sigma-\epsilon$ curve.

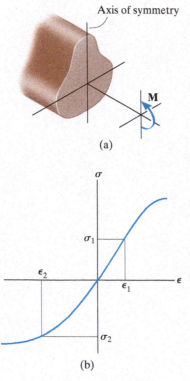

Fig. 6–50

3. Determine the volumes enclosed by the tensile and compressive stress "blocks." (As an approximation, this may require dividing each block into composite regions.) Equation 6–27 requires the volumes of these blocks to be *equal*, since they represent the resultant tensile force **T** and resultant compressive force **C** on the section, Fig. 6–50e. If these forces are unequal, an adjustment as to the *location* of the neutral axis must be made (point of *zero strain*), and the process repeated until Eq. 6–27 ($T = C$) is satisfied.

4. Once $T = C$, the moments produced by **T** and **C** can be calculated about the neutral axis. Here the moment arms for **T** and **C** are measured from the neutral axis to the *centroids of the volumes* defined by the stress distributions, Fig. 6–50e. Equation 6–28 requires $M = Ty' + Cy''$. If this equation is not satisfied, the *slope* of the strain distribution must be adjusted, and the computations for T and C and the moment must be repeated until close agreement is obtained.

   This trial-and-error procedure is obviously very tedious, and fortunately it does not occur very often in engineering practice. Most beams are symmetric about two axes, and they are constructed from materials that are assumed to have similar tension-and-compression stress–strain diagrams. Fortunately, when this occurs, the neutral axis will pass through the centroid of the cross section, and the process of relating the stress distribution to the resultant moment will be simplified.

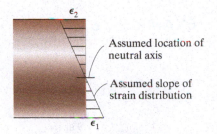

Assumed location of neutral axis

Assumed slope of strain distribution

Strain distribution (profile view)

(c)

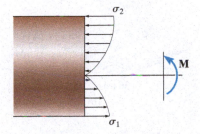

Stress distribution (profile view)

(d)

> ## ▶ IMPORTANT POINTS
>
> - The *normal-strain distribution* over the cross section of a beam is based only on geometric considerations and has been found to always remain *linear*, regardless of the applied load. The normal-stress distribution, however, must be determined from the material behavior, or stress–strain diagram, once the strain distribution is established.
>
> - The *location of the neutral axis* is determined from the condition that the *resultant force* on the cross section is *zero*.
>
> - The internal moment on the cross section must be equal to the moment of the stress distribution about the neutral axis.
>
> - Perfectly plastic behavior assumes the normal stress is *constant* over the cross section, and the beam will continue to bend, with no increase in moment. This moment is called the *plastic moment*.

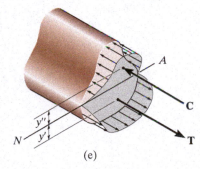

(e)

Fig. 6–50 (cont.)

6

EXAMPLE 6.21

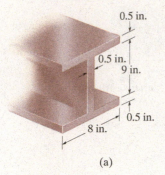

0.5 in.

0.5 in.

9 in.

0.5 in.

8 in.

(a)

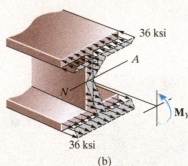

36 ksi

A

N

M_Y

36 ksi

(b)

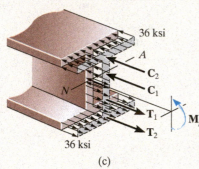

36 ksi

A

C_2

C_1

N

T_1

T_2 M_p

36 ksi

(c)

Fig. 6–51

The steel wide-flange beam has the dimensions shown in Fig. 6–51a. If it is made of an elastic perfectly plastic material having a tensile and compressive yield stress of $\sigma_Y = 36$ ksi, determine the shape factor for the beam.

SOLUTION

In order to determine the shape factor, it is first necessary to calculate the maximum elastic moment M_Y and the plastic moment M_p.

Maximum Elastic Moment. The normal-stress distribution for the maximum elastic moment is shown in Fig. 6–51b. The moment of inertia about the neutral axis is

$$I = \left[\frac{1}{12}(0.5\text{ in.})(9\text{ in.})^3\right] + 2\left[\frac{1}{12}(8\text{ in.})(0.5\text{ in.})^3\right.$$
$$\left. + 8\text{ in. }(0.5\text{ in.})(4.75\text{ in.})^2\right] = 211.0\text{ in}^4$$

Applying the flexure formula, we have

$$\sigma_{\max} = \frac{Mc}{I}; \quad 36\text{ kip/in}^2 = \frac{M_Y(5\text{ in.})}{211.0\text{ in}^4} \quad M_Y = 1519.5\text{ kip}\cdot\text{in.}$$

Plastic Moment. The plastic moment causes the steel over the entire cross section of the beam to yield, so that the normal-stress distribution looks like that shown in Fig. 6–51c. Due to symmetry, and since the tension and compression stress–strain diagrams are the same, the neutral axis passes through the centroid of the cross section. In order to determine the plastic moment, the stress distribution is divided into four composite rectangular "blocks," and the force produced by each "block" is equal to the volume of the block. Therefore, we have

$$C_1 = T_1 = 36\text{ kip/in}^2(0.5\text{ in.})(4.5\text{ in.}) = 81\text{ kip}$$
$$C_2 = T_2 = 36\text{ kip/in}^2(0.5\text{ in.})(8\text{ in.}) = 144\text{ kip}$$

These forces act through the *centroid* of each volume. Calculating the moments of these forces about the neutral axis, we obtain the plastic moment.

$$M_p = 2[(2.25\text{ in.})(81\text{ kip})] + 2[(4.75\text{ in.})(144\text{ kip})] = 1732.5\text{ kip}\cdot\text{in.}$$

Shape Factor. Applying Eq. 6–33 gives

$$k = \frac{M_p}{M_Y} = \frac{1732.5\text{ kip}\cdot\text{in.}}{1519.5\text{ kip}\cdot\text{in.}} = 1.14 \qquad \qquad \textit{Ans.}$$

NOTE: This value indicates that a wide-flange beam provides a very efficient section for resisting an *elastic moment*. Most of the moment is developed in the flanges, i.e., in the top and bottom segments, whereas the web or vertical segment contributes very little. In this particular case, only 14% additional moment can be supported by the beam beyond that which can be supported elastically.

6

EXAMPLE 6.22

A T-beam has the dimensions shown in Fig. 6–52a. If it is made of an elastic perfectly plastic material having a tensile and compressive yield stress of $\sigma_Y = 250$ MPa, determine the plastic moment that can be resisted by the beam.

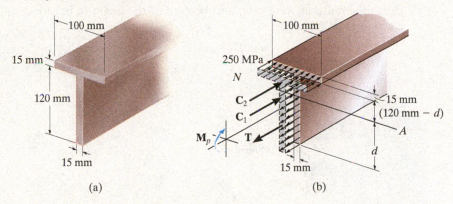

(a)   (b)

Fig. 6–52

SOLUTION

The "plastic" stress distribution acting over the beam's cross-sectional area is shown in Fig. 6–52b. In this case the cross section is not symmetric with respect to a horizontal axis, and consequently, the neutral axis will *not* pass through the centroid of the cross section. To determine the *location* of the neutral axis, d, we require the stress distribution to produce a zero resultant force on the cross section. Assuming that $d \le 120$ mm, we have

$$\int_A \sigma \, dA = 0; \qquad\qquad T - C_1 - C_2 = 0$$

$$250 \text{ MPa } (0.015 \text{ m})(d) - 250 \text{ MPa } (0.015 \text{ m})(0.120 \text{ m} - d)$$
$$- 250 \text{ MPa } (0.015 \text{ m})(0.100 \text{ m}) = 0$$

$$d = 0.110 \text{ m} < 0.120 \text{ m} \quad \text{OK}$$

Using this result, the forces acting on each segment are

$$T = 250 \text{ MN/m}^2 (0.015 \text{ m})(0.110 \text{ m}) = 412.5 \text{ kN}$$
$$C_1 = 250 \text{ MN/m}^2 (0.015 \text{ m})(0.010 \text{ m}) = 37.5 \text{ kN}$$
$$C_2 = 250 \text{ MN/m}^2 (0.015 \text{ m})(0.100 \text{ m}) = 375 \text{ kN}$$

Hence the resultant plastic moment about the neutral axis is

$$M_p = 412.5 \text{ kN}\left(\frac{0.110 \text{ m}}{2}\right) + 37.5 \text{ kN}\left(\frac{0.01 \text{ m}}{2}\right) + 375 \text{ kN}\left(0.01 \text{ m} + \frac{0.015 \text{ m}}{2}\right)$$

$$M_p = 29.4 \text{ kN} \cdot \text{m} \qquad\qquad\qquad\qquad\qquad\qquad\qquad \textit{Ans.}$$

EXAMPLE 6.23

The beam in Fig. 6–53a is made of a titanium alloy that has a stress–strain diagram that can in part be approximated by two straight lines. If the material behavior is the *same* in both tension and compression, determine the bending moment that can be applied to the beam that will cause the material at the top and bottom of the beam to be subjected to a strain of 0.050 in./in.

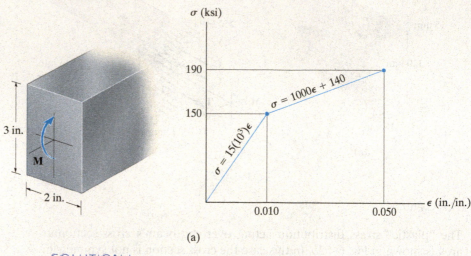

(a)

SOLUTION I

By inspection of the stress–strain diagram, the material is said to exhibit "elastic-plastic behavior with strain hardening." Since the cross section is symmetric and the tension–compression $\sigma-\epsilon$ diagrams are the same, the neutral axis must pass through the centroid of the cross section. The strain distribution, which is always linear, is shown in Fig. 6–53b. In particular, the point where maximum elastic strain (0.010 in./in.) occurs has been determined by proportion, that is, 0.05/1.5 in. = 0.010/y or $y = 0.3$ in.

The corresponding normal-stress distribution acting over the cross section is shown in Fig. 6–53c. The moment produced by this distribution can be calculated by finding the "volume" of the stress blocks. To do so we will subdivide this distribution into two triangular blocks and a rectangular block in both the tension and compression regions, Fig. 6–53d. Since the beam is 2 in. wide, the resultants and their locations are determined as follows:

$$T_1 = C_1 = \frac{1}{2}(1.2 \text{ in.})(40 \text{ kip/in}^2)(2 \text{ in.}) = 48 \text{ kip}$$

$$y_1 = 0.3 \text{ in.} + \frac{2}{3}(1.2 \text{ in.}) = 1.10 \text{ in.}$$

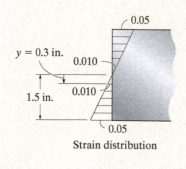

Strain distribution

(b)

Fig. 6–53

$$T_2 = C_2 = (1.2 \text{ in.})(150 \text{ kip/in}^2)(2 \text{ in.}) = 360 \text{ kip}$$

$$y_2 = 0.3 \text{ in.} + \frac{1}{2}(1.2 \text{ in.}) = 0.90 \text{ in.}$$

$$T_3 = C_3 = \frac{1}{2}(0.3 \text{ in.})(150 \text{ kip/in}^2)(2 \text{ in.}) = 45 \text{ kip}$$

$$y_3 = \frac{2}{3}(0.3 \text{ in.}) = 0.2 \text{ in.}$$

The moment produced by this normal-stress distribution about the neutral axis is therefore

$$M = 2[48 \text{ kip } (1.10 \text{ in.}) + 360 \text{ kip } (0.90 \text{ in.}) + 45 \text{ kip } (0.2 \text{ in.})]$$
$$= 772 \text{ kip} \cdot \text{in.} \qquad \qquad \qquad Ans.$$

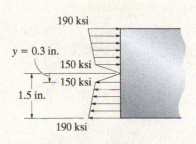

Stress distribution

(c)

SOLUTION II

Rather than using the above semigraphical technique, it is also possible to find the moment analytically. To do this we must express the stress distribution in Fig. 6–53c as a function of position y along the beam. Here $\sigma = f(\epsilon)$ has been given in Fig. 6–53a, and from Fig. 6–53b, the normal strain can be determined as a function of position y by proportional triangles; i.e.,

$$\epsilon = \frac{0.05}{1.5}y \qquad 0 \le y \le 1.5 \text{ in.}$$

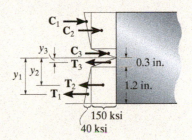

(d)

Substituting this into the $\sigma - \epsilon$ functions shown in Fig. 6–53a gives

$$\sigma = 500y \qquad \qquad 0 \le y \le 0.3 \text{ in.} \qquad (1)$$
$$\sigma = 33.33y + 140 \qquad 0.3 \text{ in.} \le y \le 1.5 \text{ in.} \qquad (2)$$

From Fig. 6–53e, the moment caused by σ acting on the area strip $dA = 2\,dy$ is

$$dM = y(\sigma\, dA) = y\sigma(2\, dy)$$

Using Eqs. 1 and 2, the moment for the entire cross section is therefore

$$M = 2\left[2\int_{0}^{0.3 \text{ in.}} 500y^2\, dy + 2\int_{0.3 \text{ in.}}^{1.5 \text{ in.}} (33.3y^2 + 140y)\, dy\right]$$
$$= 772 \text{ kip} \cdot \text{in.} \qquad \qquad \qquad Ans.$$

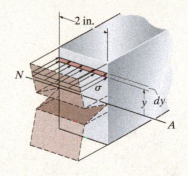

(e)

Fig. 6–53 (cont.)

6

EXAMPLE 6.24

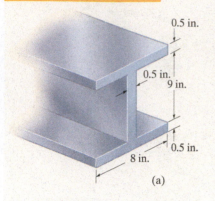

0.5 in.

0.5 in.

9 in.

0.5 in.

8 in.

(a)

The steel wide-flange beam shown in Fig. 6–54a is subjected to a fully plastic moment of $\mathbf{M}_p$. If this moment is removed, determine the residual stress distribution in the beam. The material is elastic perfectly plastic and has a yield stress of $\sigma_Y = 36$ ksi.

SOLUTION

The normal-stress distribution in the beam caused by $\mathbf{M}_p$ is shown in Fig. 6–54b. When $\mathbf{M}_p$ is removed, the material responds elastically, and therefore the stress distribution is as shown in Fig. 6–54c. The modulus of rupture σ_r is calculated from the flexure formula. Using $M_p = 1732.5$ kip·in. and $I = 211.0$ in^4 from Example 6.21, we have

$$\sigma_{max} = \frac{Mc}{I};$$

$$\sigma_r = \frac{1732.5 \text{ kip} \cdot \text{in. (5 in.)}}{211.0 \text{ in}^4} = 41.1 \text{ ksi}$$

As expected, $\sigma_r < 2\sigma_Y$.

Superposition of the stresses gives the residual stress distribution shown in Fig. 6–54d. The point of zero normal stress was determined by proportion; i.e., from Figs. 6–54b and 6–54c, we require that

$$\frac{41.1 \text{ ksi}}{5 \text{ in.}} = \frac{36 \text{ ksi}}{y}$$

$$y = 4.38 \text{ in.}$$

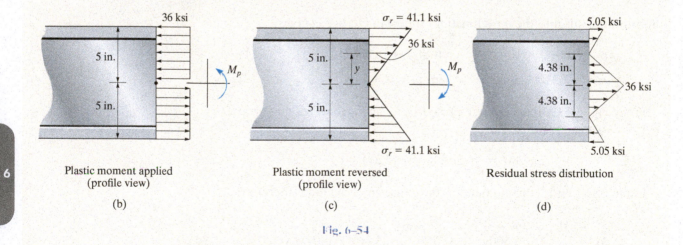

36 ksi

5 in.

5 in.

M_p

Plastic moment applied
(profile view)

(b)

$\sigma_r = 41.1$ ksi

36 ksi

5 in.

y

5 in.

M_p

$\sigma_r = 41.1$ ksi

Plastic moment reversed
(profile view)

(c)

5.05 ksi

4.38 in.

36 ksi

4.38 in.

5.05 ksi

Residual stress distribution

(d)

Fig. 6–54

PROBLEMS

6–158. Determine the shape factor for the wide-flange beam.

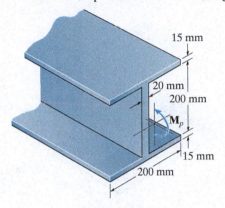

15 mm

20 mm
200 mm

$\mathbf{M}_p$

15 mm

200 mm

Prob. 6–158

6–159. The wide-flange member is made from an elastic perfectly plastic material. Determine the shape factor for the beam.

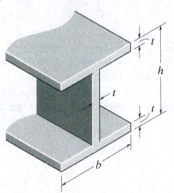

t

h

t

t

b

Prob. 6–159

*6–160. The rod has a circular cross section. If it is made of an elastic perfectly plastic material, determine the shape factor.

100 mm

Prob. 6–160

6–161. The rod has a circular cross section. If it is made of an elastic perfectly plastic material where $\sigma_Y = 345$ MPa, determine the maximum elastic moment and plastic moment that can be applied to the cross section.

100 mm

Prob. 6–161

6–162. The beam is made of an elastic perfectly plastic material. Determine the plastic moment $\mathbf{M}_p$ that can be supported by a beam having the cross section shown. $\sigma_Y = 30$ ksi.

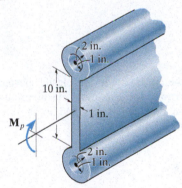

2 in.
1 in.

10 in.

1 in.

$\mathbf{M}_p$

2 in.
1 in.

Prob. 6–162

6–163. Determine the plastic moment $\mathbf{M}_p$ that can be supported by a beam having the cross section shown. $\sigma_Y = 30$ ksi.

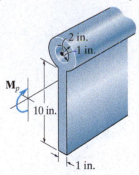

2 in.
1 in.

$\mathbf{M}_p$

10 in.

1 in.

Prob. 6–163

6

*6–164. Determine the shape factor for the beam.

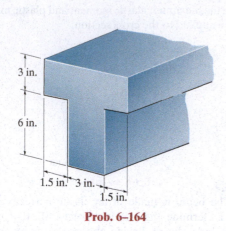

1.5 in. 3 in.

1.5 in.

Prob. 6–164

6–165. The beam is made of elastic perfectly plastic material. Determine the maximum elastic moment and the plastic moment that can be applied to the cross section. Take $\sigma_Y = 36$ ksi.

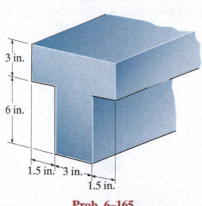

1.5 in. 3 in.

1.5 in.

Prob. 6–165

6–166. Determine the shape factor for the beam.

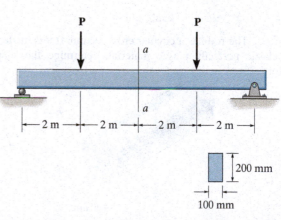

10 mm

15 mm

200 mm

$\mathbf{M}_p$

10 mm

200 mm

Prob. 6–166

6–167. The beam is made of an elastic perfectly plastic material for which $\sigma_Y = 200$ MPa. If the largest moment in the beam occurs within the center section a–a, determine the magnitude of each force **P** that causes this moment to be (a) the largest elastic moment and (b) the largest plastic moment.

P   P

a

a

2 m 2 m 2 m 2 m

200 mm

100 mm

Prob. 6–167

*6–168. The beam is made of elastic perfectly plastic material for which $\sigma_Y = 345$ MPa. Determine the maximum elastic moment and the plastic moment that can be applied to the cross section.

6–169. Determine the shape factor of the cross section.

*6–172. Determine the shape factor of the cross section.

6–173. The beam is made of elastic perfectly plastic material. Determine the maximum elastic moment and the plastic moment that can be applied to the cross section. Take $a = 50$ mm and $\sigma_Y = 230$ MPa.

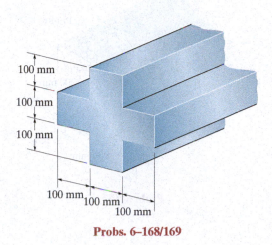

Probs. 6–168/169

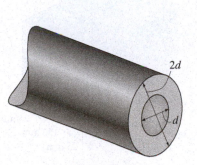

Probs. 6–172/173

6–170. The rod has a circular cross section. If it is made of an elastic perfectly plastic material, determine the shape factor for the rod.

6–171. The rod has a circular cross section. If it is made of an elastic perfectly plastic material, determine the maximum elastic moment and plastic moment that can be applied to the cross section. Take $r = 3$ in., $\sigma_Y = 36$ ksi.

6–174. Determine the shape factor for the member having the tubular cross section.

Probs. 6–170/171

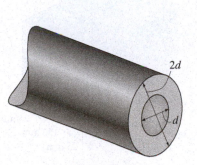

Prob. 6–174

6–175. Determine the shape factor of the cross section.

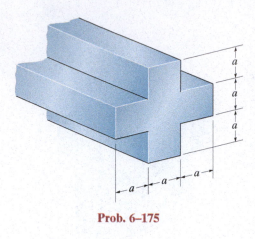

Prob. 6–175

6–177. The beam is made of an elastic perfectly plastic material for which $\sigma_Y = 250$ MPa. Determine the residual stress in the beam at its top and bottom after the plastic moment $\mathbf{M}_p$ is applied and then released.

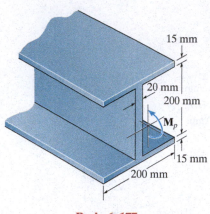

Prob. 6–177

*6–176. The box beam is made of an elastic perfectly plastic material for which $\sigma_Y = 250$ MPa. Determine the residual stress in the top and bottom of the beam after the plastic moment $\mathbf{M}_p$ is applied and then released.

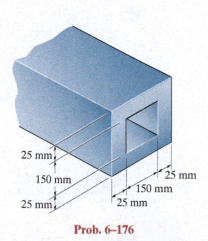

Prob. 6–176

6–178. The plexiglass bar has a stress–strain curve that can be approximated by the straight-line segments shown. Determine the largest moment M that can be applied to the bar before it fails.

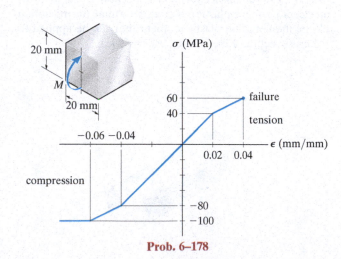

Prob. 6–178

6–179. The stress–strain diagram for a titanium alloy can be approximated by the two straight lines. If a strut made of this material is subjected to bending, determine the moment resisted by the strut if the maximum stress reaches a value of (a) σ_A and (b) σ_B.

6–181. The bar is made of an aluminum alloy having a stress–strain diagram that can be approximated by the straight line segments shown. Assuming that this diagram is the same for both tension and compression, determine the moment the bar will support if the maximum strain at the top and bottom fibers of the beam is $\epsilon_{max} = 0.05$.

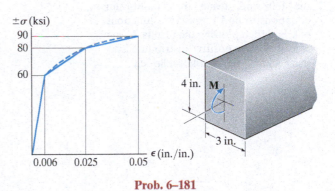

Prob. 6–181

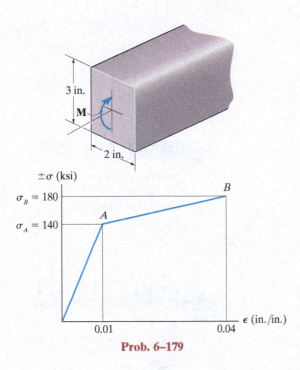

Prob. 6–179

6–182. The beam is made of phenolic, a structural plastic, that has the stress–strain curve shown. If a portion of the curve can be represented by the equation $\sigma = (5(10^6)\epsilon)^{1/2}$ MPa, determine the magnitude w of the distributed load that can be applied to the beam without causing the maximum strain in its fibers at the critical section to exceed $\epsilon_{max} = 0.005$ mm/mm.

*6–180. A beam is made from polypropylene plastic and has a stress–strain diagram that can be approximated by the curve shown. If the beam is subjected to a maximum tensile and compressive strain of $\epsilon = 0.02$ mm/mm, determine the moment M.

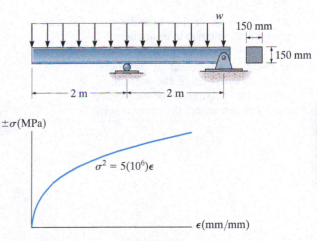

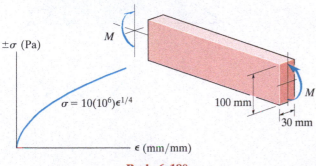

Prob. 6–180

Prob. 6–182

CHAPTER REVIEW

Shear and moment diagrams are graphical representations of the internal shear and moment within a beam. They can be constructed by sectioning the beam an arbitrary distance x from the left end, using the equilibrium equations to find V and M as functions of x, and then plotting the results. A sign convention for positive distributed load, shear, and moment must be followed.

Positive external distributed load

Positive internal shear

Positive internal moment

Beam sign convention

It is also possible to plot the shear and moment diagrams by realizing that at each point the slope of the shear diagram is equal to the intensity of the distributed loading at the point.

$$w = \frac{dV}{dx}$$

Likewise, the slope of the moment diagram is equal to the shear at the point.

$$V = \frac{dM}{dx}$$

The area under the distributed-loading diagram between the points represents the change in shear.

$$\Delta V = \int w\, dx$$

And the area under the shear diagram represents the change in moment.

$$\Delta M = \int V\, dx$$

The shear and moment at any point can be obtained using the method of sections. The maximum (or minimum) moment occurs where the shear is zero.

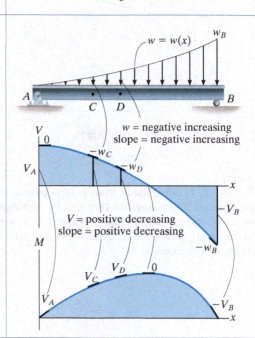

A bending moment tends to produce a linear variation of normal strain within a straight beam. Provided the material is homogeneous and linear elastic, then equilibrium can be used to relate the internal moment in the beam to the stress distribution. The result is the flexure formula,

$$\sigma_{max} = \frac{Mc}{I}$$

where I and c are determined from the neutral axis that passes through the centroid of the cross section.

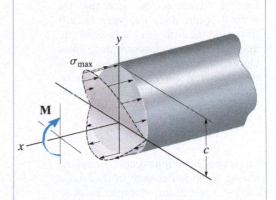

If the cross-sectional area of the beam is not symmetric about an axis that is perpendicular to the neutral axis, then unsymmetrical bending will occur. The maximum stress can be determined from formulas, or the problem can be solved by considering the superposition of bending caused by the moment components $\mathbf{M}_y$ and $\mathbf{M}_z$ about the principal axes of inertia for the area.

$$\sigma = -\frac{M_z y}{I_z} + \frac{M_y z}{I_y}$$

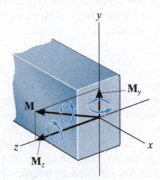

Beams made of composite materials can be "transformed" so their cross section is considered as if it were made of a single material. To do this, the transformation factor n, which is a ratio of the moduli of elasticity of the materials, is used to change the width b of the beam.

   Once the cross section is transformed, then the stress in the material that was transformed is determined using the flexure formula multiplied by n.

$$n = \frac{E_1}{E_2}$$

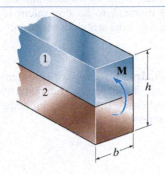

6

Curved beams deform such that the normal strain does not vary linearly from the neutral axis. Provided the material is homogeneous and linear elastic, and the cross section has an axis of symmetry, then the curved-beam formula can be used to determine the bending stress.	$$\sigma = \frac{M(R - r)}{Ar(\bar{r} - R)}$$ or $$\sigma = \frac{My}{Ae(R - y)}$$	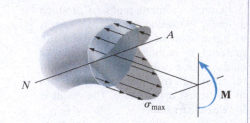
Stress concentrations occur in members having a sudden change in their cross section, caused, for example, by holes and notches. The maximum bending stress at these locations is determined using a stress concentration factor K that is found from graphs determined from experiment.	$$\sigma_{max} = K\frac{Mc}{I}$$	
If the bending moment causes the stress in the material to exceed its elastic limit, then the normal strain will remain linear; however, the stress distribution will vary in accordance with the stress–strain diagram. The plastic and ultimate moments supported by the beam can be determined by requiring the resultant force to be zero and the resultant moment to be equivalent to the moment of the stress distribution.		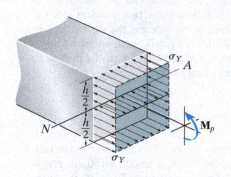
If an applied plastic or ultimate moment is released, it will cause the material to respond elastically, thereby inducing residual stresses in the beam.		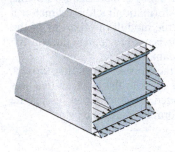

6

CONCEPTUAL PROBLEMS

C6–1. The steel saw blade passes over the drive wheel of the band saw. Using appropriate measurements and data, explain how to determine the bending stress in the blade.

C6–1

C6–2. The crane boom has a noticeable taper along its length. Explain why. To do so, assume the boom is in the horizontal position and in the process of hoisting a load at its end, so that the reaction on the support *A* becomes zero. Use realistic dimensions and a load, to justify your reasoning.

C6–2

C6–3. Use reasonable dimensions for this hammer and a loading to show through an analysis why this hammer failed in the manner shown.

C6–3

C6–4. These garden shears were manufactured using an inferior material. Using a loading of 50 lb applied normal to the blades, and appropriate dimensions for the shears, determine the absolute maximum bending stress in the material and show why the failure occurred at the critical location on the handle.

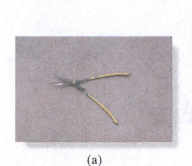

(a)   (b)

C6–4

REVIEW PROBLEMS

R6–1. Determine the shape factor for the wide-flange beam.

R6–3. The composite beam consists of a wood core and two plates of steel. If the allowable bending stress for the wood is $(\sigma_{allow})_w = 20$ MPa, and for the steel $(\sigma_{allow})_{st} = 130$ MPa, determine the maximum moment that can be applied to the beam. $E_w = 11$ GPa, $E_{st} = 200$ GPa.

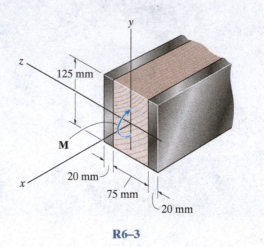

R6–3

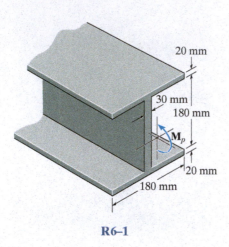

R6–1

*R6–4. A shaft is made of a polymer having a parabolic upper and lower cross section. If it resists a moment of $M = 125$ N·m, determine the maximum bending stress in the material (a) using the flexure formula and (b) using integration. Sketch a three-dimensional view of the stress distribution acting over the cross-sectional area. *Hint:* The moment of inertia is determined using Eq. A–3 of Appendix A.

R6–2. The compound beam consists of two segments that are pinned together at B. Draw the shear and moment diagrams if it supports the distributed loading shown.

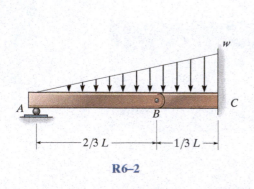

R6–2

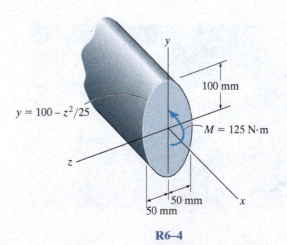

R6–4

R6–5. Determine the maximum bending stress in the handle of the cable cutter at section *a–a*. A force of 45 lb is applied to the handles.

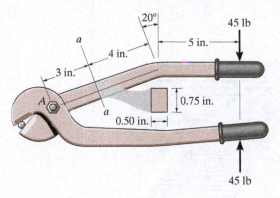

R6–5

R6–6. The curved beam is subjected to a bending moment of $M = 85$ N·m as shown. Determine the stress at points A and B and show the stress on a volume element located at these points.

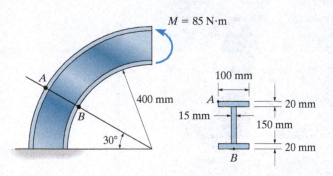

R6–6

R6–7. Determine the shear and moment in the beam as functions of x, where $0 \le x < 6$ ft, then draw the shear and moment diagrams for the beam.

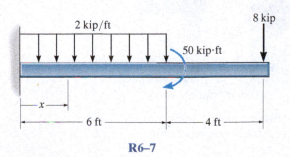

R6–7

*R6–8. A wooden beam has a square cross section as shown. Determine which orientation of the beam provides the greatest strength at resisting the moment **M**. What is the difference in the resulting maximum stress in both cases?

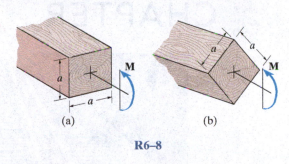

R6–8

R6–9. Draw the shear and moment diagrams for the shaft if it is subjected to the vertical loadings. The bearings at A and B exert only vertical reactions on the shaft.

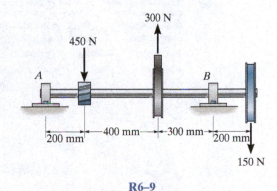

R6–9

R6–10. The strut has a square cross section a by a and is subjected to the bending moment **M** applied at an angle θ as shown. Determine the maximum bending stress in terms of a, M, and θ. What angle θ will give the largest bending stress in the strut? Specify the orientation of the neutral axis for this case.

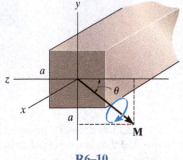

R6–10

6

CHAPTER 7

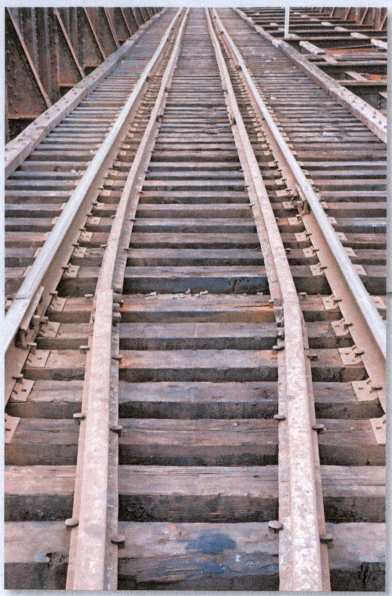

(© Bert Folsom/Alamy)

Railroad ties act as beams that support very large transverse shear loadings. As a result, if they are made of wood they will tend to split at their ends, where the shear loads are the largest.

TRANSVERSE SHEAR

CHAPTER OBJECTIVES

■ In this chapter we will develop a method for finding the shear stress in a beam and discuss a way to find the spacing of fasteners along the beam's length. The concept of shear flow will be presented, and used to find the average stress within thin-walled members. The chapter ends with a discussion of how to prevent twisting of a beam when it supports a load.

7.1 SHEAR IN STRAIGHT MEMBERS

In general, a beam will support both an internal shear and a moment. The shear **V** is the result of a *transverse* shear-stress distribution that acts over the beam's cross section, Fig. 7–1. Due to the complementary property of shear, this stress will also create corresponding *longitudinal* shear stress that acts along the length of the beam.

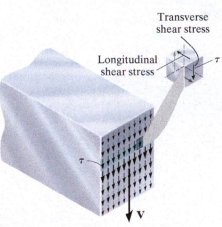

Fig. 7–1

7

Boards not bonded together
(a)

Boards bonded together
(b)

Fig. 7–2

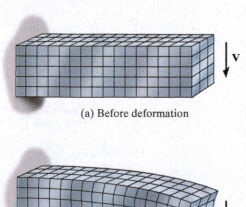

Shear connectors are "tack welded" to this corrugated metal floor liner so that when the concrete floor is poured, the connectors will prevent the concrete slab from slipping on the liner surface. The two materials will thus act as a composite slab.

To illustrate the effect caused by the longitudinal shear stress, consider the beam made from three boards shown in Fig. 7–2a. If the top and bottom surfaces of each board are smooth, and the boards are *not* bonded together, then application of the load **P** will cause the boards to *slide* relative to one another when the beam deflects. However, if the boards are bonded together, then the longitudinal shear stress acting between the boards will prevent their relative sliding, and consequently the beam will act as a single unit, Fig. 7–2b.

As a result of the shear stress, shear strains will be developed and these will tend to distort the cross section in a rather complex manner. For example, consider the short bar in Fig. 7–3a made of a highly deformable material and marked with horizontal and vertical grid lines. When the shear force **V** is applied, it tends to deform these lines into the pattern shown in Fig. 7–3b. This nonuniform shear-strain distribution will cause the cross section to *warp*; and as a result, when a beam is subjected to *both* bending and shear, the cross section will not remain plane as assumed in the development of the flexure formula.

(a) Before deformation

(b) After deformation

Fig. 7–3

7.2   THE SHEAR FORMULA

Because the strain distribution for shear is not easily defined, as in the case of axial load, torsion, and bending, we will obtain the shear-stress distribution in an indirect manner. To do this we will consider the horizontal force equilibrium of a portion of an element taken from the beam in Fig. 7–4a. A free-body diagram of the entire element is shown in Fig. 7–4b. The normal-stress distribution acting on it is caused by the bending moments M and $M + dM$. Here we have excluded the effects of V, $V + dV$, and $w(x)$, since these loadings are vertical and will therefore not be involved in a horizontal force summation. Notice that $\Sigma F_x = 0$ is satisfied since the stress distribution on each side of the element forms only a couple moment, and therefore a zero force resultant.

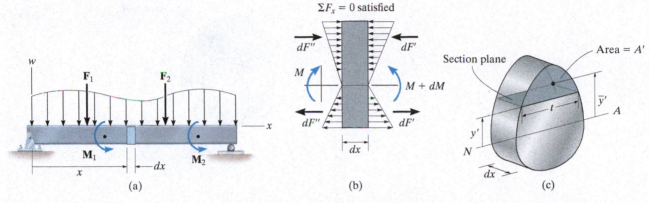

(a)

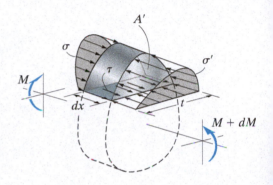

Fig. 7–4

Now let's consider the shaded *top portion* of the element that has been sectioned at y' from the neutral axis, Fig. 7–4c. It is on this sectioned plane that we want to find the shear stress. This top segment has a width t at the section, and the two cross-sectional sides each have an area A'. The segment's free-body diagram is shown in Fig. 7–4d. The resultant moments on each side of the element differ by dM, so that $\Sigma F_x = 0$ will not be satisfied unless a longitudinal shear stress τ acts over the bottom sectioned plane. To simplify the analysis, we will assume that this shear stress is *constant* across the width t of the bottom face. To find the horizontal force created by the bending moments, we will assume that the effect of warping due to shear is small, so that it can generally be *neglected*. This assumption is particularly true for the most common case of a *slender beam*, that is, one that has a small depth compared to its length. Therefore, using the flexure formula, Eq. 6–13, we have

$$\xrightarrow{+}\ \Sigma F_x = 0; \qquad \int_{A'} \sigma'\,dA' - \int_{A'} \sigma\,dA' - \tau(t\,dx) = 0$$

$$\int_{A'} \left(\frac{M + dM}{I}\right)y\,dA' - \int_{A'}\left(\frac{M}{I}\right)y\,dA' - \tau(t\,dx) = 0$$

$$\left(\frac{dM}{I}\right)\int_{A'} y\,dA' = \tau(t\,dx) \qquad (7\text{–}1)$$

Solving for τ, we get

$$\tau = \frac{1}{It}\left(\frac{dM}{dx}\right)\int_{A'} y\,dA'$$

Here $V = dM/dx$ (Eq. 6–2). Also, the integral represents the *moment of the area A' about the neutral axis*, which we will denote by the symbol Q. Since the location of the centroid of A' is determined from $\bar{y}' = \int_{A'} y\,dA'/A'$, we can also write

$$Q = \int_{A'} y\,dA' = \bar{y}'A' \qquad (7\text{–}2)$$

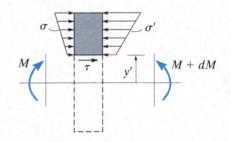

Three-dimensional view

Profile view

(d)

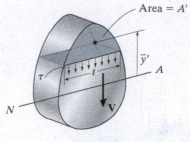

Fig. 7–5

The final result is called the **shear formula**, namely

$$\tau = \frac{VQ}{It} \tag{7–3}$$

With reference to Fig. 7–5,

$\tau =$ the shear stress in the member at the point located a distance y from the neutral axis. This stress is assumed to be constant and therefore *averaged* across the width t of the member

$V =$ the shear force, determined from the method of sections and the equations of equilibrium

$I =$ the moment of inertia of the *entire* cross-sectional area calculated about the neutral axis

$t =$ the width of the member's cross section, measured at the point where τ is to be determined

$Q = \bar{y}'A'$, where A' is the area of the top (or bottom) portion of the member's cross section, above (or below) the section plane where t is measured, and $\bar{y}'$ is the distance from the neutral axis to the centroid of A'

Although for the derivation we considered only the shear stress acting on the beam's longitudinal plane, the formula applies as well for finding the transverse shear stress on the beam's cross section, because these stresses are complementary and numerically equal.

Calculating Q.   Of all the variables in the shear formula, Q is usually the most difficult to define properly. Try to remember that it represents *the moment of the cross-sectional area that is above or below the point where the shear stress is to be determined*. It is this area A' that is "held onto" the rest of the beam by the longitudinal shear stress as the beam undergoes bending, Fig. 7–4d. The examples shown in Fig. 7–6 will help to illustrate this point. Here the stress at point P is to be determined, and so A' represents the dark shaded region. The value of Q for each case is reported under each figure. These same results can *also* be obtained for Q by considering A' to be the light shaded area below P, although here y' is a negative quantity when a portion of A' is below the neutral axis.

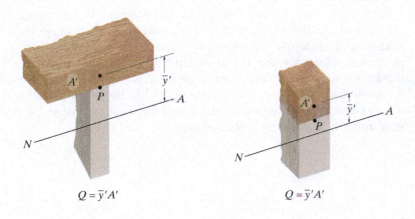

$$Q = \bar{y}'A' \qquad\qquad Q = \bar{y}'A'$$

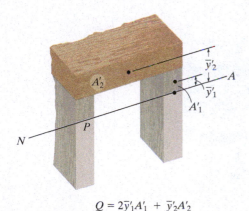

$$Q = 2\bar{y}'_1 A'_1 + \bar{y}'_2 A'_2$$

Fig. 7–6

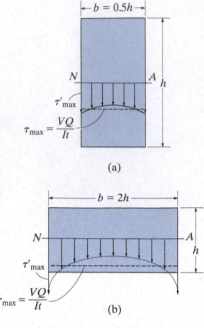

(a)

$$\tau_{max} = \frac{VQ}{It}$$

$$\tau_{max} = \frac{VQ}{It}$$

(b)

Fig. 7–7

Limitations on the Use of the Shear Formula.

One of the major assumptions used in the development of the shear formula is that the shear stress is *uniformly* distributed over the *width t* at the section. In other words, the *average shear stress* is calculated across the width. We can test the accuracy of this assumption by comparing it with a more exact mathematical analysis based on the theory of elasticity. For example, if the beam's cross section is rectangular, the shear-stress distribution across the neutral axis actually varies as shown in Fig. 7–7. The maximum value, τ'_{max}, occurs at the *sides* of the cross section, and its magnitude depends on the ratio b/h (width/depth). For sections having a $b/h = 0.5$, τ'_{max} is only about 3% greater than the shear stress calculated from the shear formula, Fig. 7–7a. However, for *flat sections*, say $b/h = 2$, τ'_{max} is about 40% greater than τ_{max}, Fig. 7–7b. The error becomes even greater as the section becomes flatter, that is, as the b/h ratio increases. Errors of this magnitude are certainly intolerable if one attempts to use the shear formula to determine the shear stress in the *flange* of the wide-flange beam shown in Fig. 7–8.

It should also be noted that the shear formula will not give accurate results when used to determine the shear stress at the flange–web junction of this beam, since this is a point of sudden cross-sectional change and therefore a *stress concentration* occurs here. Fortunately, engineers must only use the shear formula to calculate the average maximum shear stress in a beam, and for a wide-flange section this occurs at the neutral axis, where the b/h (width/depth) ratio for the web is *very small*, and therefore the calculated result is very close to the *actual* maximum shear stress as explained above.

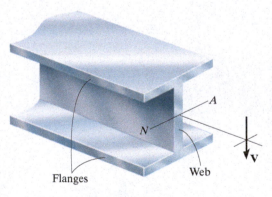

Fig. 7–8

Another important limitation on the use of the shear formula can be illustrated with reference to Fig. 7–9a, which shows a member having a cross section with an irregular boundary. If we apply the shear formula to determine the (average) shear stress τ along the line AB, it will be directed downward across this line as shown in Fig. 7–9b. However, an element of material taken from the boundary point B, Fig. 7–9c, must not have any shear stress on its outer surface. In other words, the shear stress acting on this element must *be directed tangent to the boundary*, and so the shear-stress distribution across line AB is actually directed as shown in Fig. 7–9d. As a result, the shear formula can only be applied at sections shown by the blue lines in Fig. 7–9a, because these lines intersect the tangents to the boundary at *right angles*, Fig. 7–9e.

To summarize the above points, the shear formula does not give accurate results when applied to members having cross sections that are *short or flat*, or at points where the cross section suddenly changes. Nor should it be applied across a section that intersects the boundary of the member at an angle other than 90°.

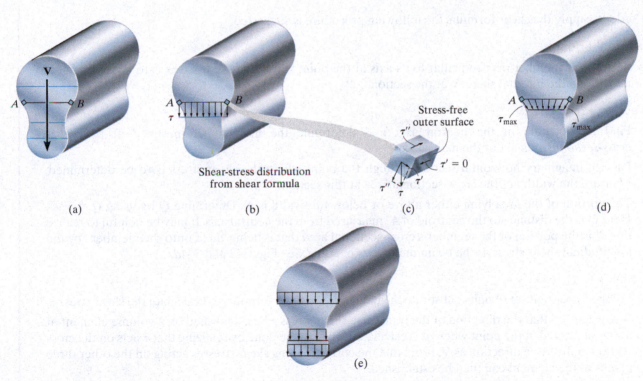

Shear-stress distribution from shear formula

(a)   (b)   (c)   (d)

(e)

Fig. 7–9

IMPORTANT POINTS

- Shear forces in beams cause *nonlinear shear-strain* distributions over the cross section, causing it to *warp*.

- Due to the complementary property of shear, the shear stress developed in a beam acts over the cross section of the beam and along its longitudinal planes.

- The *shear formula* was derived by considering horizontal force equilibrium of a portion of a differential segment of the beam.

- The shear formula is to be used on straight prismatic members made of homogeneous material that has linear elastic behavior. Also, the internal resultant shear force must be directed along an axis of symmetry for the cross section.

- The shear formula should not be used to determine the shear stress on cross sections that are short or flat, at points of sudden cross-sectional changes, or across a section that intersects the boundary of the member at an angle other than 90°.

PROCEDURE FOR ANALYSIS

In order to apply the shear formula, the following procedure is suggested.

Internal Shear.

- Section the member perpendicular to its axis at the point where the shear stress is to be determined, and obtain the internal shear V at the section.

Section Properties.

- Find the location of the neutral axis, and determine the moment of inertia I of the *entire cross-sectional area* about the neutral axis.
- Pass an imaginary horizontal section through the point where the shear stress is to be determined. Measure the width t of the cross-sectional area at this section.
- The portion of the area lying either above or below this width is A'. Determine Q by using $Q = \bar{y}'A'$. Here $\bar{y}'$ is the distance to the centroid of A', measured from the neutral axis. It may be helpful to realize that A' is the portion of the member's cross-sectional area that is being "held onto the member" by the longitudinal shear stress as the beam undergoes bending. See Figs. 7–2 and 7–4d.

Shear Stress.

- Using a consistent set of units, substitute the data into the shear formula and calculate the shear stress τ.
- It is suggested that the direction of the transverse shear stress τ be established on a volume element of material located at the point where it is calculated. This can be done by realizing that τ acts on the cross section in the same direction as V. From this, the corresponding shear stresses acting on the other three planes of the element can then be established.

EXAMPLE 7.1

The beam shown in Fig. 7–10a is made from two boards. Determine the maximum shear stress in the glue necessary to hold the boards together along the seam where they are joined.

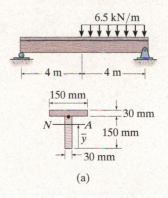

(a)

SOLUTION

Internal Shear. The support reactions and the shear diagram for the beam are shown in Fig. 7–10b. It is seen that the maximum shear in the beam is 19.5 kN.

Section Properties. The centroid and therefore the neutral axis will be determined from the reference axis placed at the bottom of the cross-sectional area, Fig. 7–10a. Working in units of meters, we have

$$\bar{y} = \frac{\Sigma \tilde{y} A}{\Sigma A}$$

$$= \frac{[0.075 \text{ m}](0.150 \text{ m})(0.030 \text{ m}) + [0.165 \text{ m}](0.030 \text{ m})(0.150 \text{ m})}{(0.150 \text{ m})(0.030 \text{ m}) + (0.030 \text{ m})(0.150 \text{ m})} = 0.120 \text{ m}$$

The moment of inertia about the neutral axis, Fig. 7–10a, is therefore

$$I = \left[\frac{1}{12}(0.030 \text{ m})(0.150 \text{ m})^3 + (0.150 \text{ m})(0.030 \text{ m})(0.120 \text{ m} - 0.075 \text{ m})^2 \right]$$

$$+ \left[\frac{1}{12}(0.150 \text{ m})(0.030 \text{ m})^3 + (0.030 \text{ m})(0.150 \text{ m})(0.165 \text{ m} - 0.120 \text{ m})^2 \right]$$

$$= 27.0(10^{-6}) \text{ m}^4$$

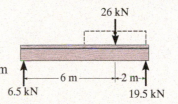

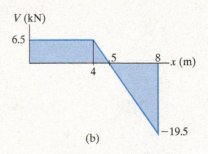

(b)

The top board (flange) is *held* onto the bottom board (web) by the glue, which is applied over the thickness $t = 0.03$ m. Consequently Q is taken from the area of the top board, Fig. 7–10a. We have

$$Q = \bar{y}'A' = [0.180 \text{ m} - 0.015 \text{ m} - 0.120 \text{ m}](0.03 \text{ m})(0.150 \text{ m})$$

$$= 0.2025(10^{-3}) \text{ m}^3$$

Shear Stress. Applying the shear formula,

$$\tau_{max} = \frac{VQ}{It} = \frac{19.5(10^3) \text{ N}(0.2025(10^{-3}) \text{ m}^3)}{27.0(10^{-6}) \text{ m}^4(0.030 \text{ m})} = 4.88 \text{ MPa} \qquad Ans.$$

The shear stress acting at the top of the bottom board is shown in Fig. 7–10c.

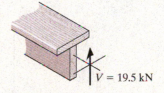

(c)

Fig. 7–10

NOTE: It is the glue's resistance to this *longitudinal shear* stress that holds the boards from slipping at the right support.

EXAMPLE 7.2

Determine the distribution of the shear stress over the cross section of the beam shown in Fig. 7–11a.

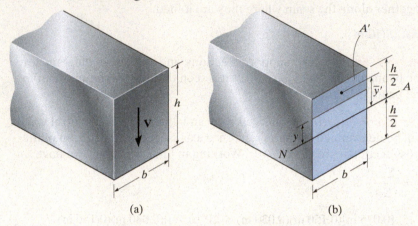

(a)   (b)

SOLUTION

The distribution can be determined by finding the shear stress at an *arbitrary height y* from the neutral axis, Fig. 7–11b, and then plotting this function. Here, the dark colored area A' will be used for Q.* Hence

$$Q = \bar{y}'A' = \left[y + \frac{1}{2}\left(\frac{h}{2} - y\right)\right]\left(\frac{h}{2} - y\right)b = \frac{1}{2}\left(\frac{h^2}{4} - y^2\right)b$$

Applying the shear formula, we have

$$\tau = \frac{VQ}{It} = \frac{V\left(\frac{1}{2}\right)\left[(h^2/4) - y^2\right]b}{\left(\frac{1}{12}bh^3\right)b} = \frac{6V}{bh^3}\left(\frac{h^2}{4} - y^2\right) \tag{1}$$

This result indicates that the shear-stress distribution over the cross section is *parabolic*. As shown in Fig. 7–11c, the intensity varies from zero at the top and bottom, $y = \pm h/2$, to a maximum value at the neutral axis, $y = 0$. Specifically, since the area of the cross section is $A = bh$, then at $y = 0$ Eq. 1 becomes

$$\boxed{\tau_{max} = 1.5\frac{V}{A}} \tag{2}$$

Rectangular cross section

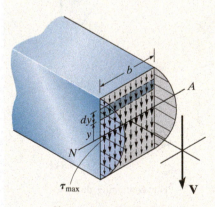

Shear–stress distribution

(c)

Fig. 7–11

*The area below y can also be used $[A' = b(h/2 + y)]$, but doing so involves a bit more algebraic manipulation.

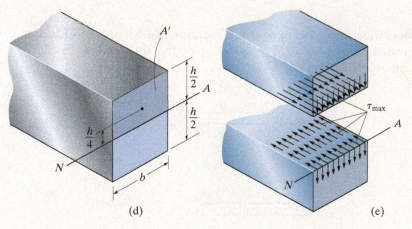

(d)   (e)

Typical shear failure of this wooden beam occurred at the support and through the approximate center of its cross section.

Fig. 7–11 (cont.)

This same value for τ_{max} can be obtained directly from the shear formula, $\tau = VQ/It$, by realizing that τ_{max} occurs where Q is *largest*, since V, I, and t are *constant*. By inspection, Q will be a maximum when the entire area above (or below) the neutral axis is considered; that is, $A' = bh/2$ and $\bar{y}' = h/4$, Fig. 7–11d. Thus,

$$\tau_{max} = \frac{VQ}{It} = \frac{V(h/4)(bh/2)}{\left[\frac{1}{12}bh^3\right]b} = 1.5\frac{V}{A}$$

By comparison, τ_{max} is 50% greater than the *average* shear stress determined from Eq. 1–7; that is, $\tau_{avg} = V/A$.

It is important to realize that τ_{max} also acts in the longitudinal direction of the beam, Fig. 7–11e. It is this stress that can cause a timber beam to fail at its supports, as shown Fig. 7–11f. Here horizontal splitting of the wood starts to occur through the neutral axis at the beam's ends, since there the vertical reactions subject the beam to large shear stress, and wood has a low resistance to shear along its grains, which are oriented in the longitudinal direction.

It is instructive to show that when the shear-stress distribution, Eq. 1, is integrated over the cross section it produces the resultant shear V. To do this, a differential strip of area $dA = b \, dy$ is chosen, Fig. 7–11c, and since τ acts uniformly over this strip, we have

$$\int_A \tau \, dA = \int_{-h/2}^{h/2} \frac{6V}{bh^3}\left(\frac{h^2}{4} - y^2\right)b \, dy$$

$$= \frac{6V}{h^3}\left[\frac{h^2}{4}y - \frac{1}{3}y^3\right]_{-h/2}^{h/2}$$

$$= \frac{6V}{h^3}\left[\frac{h^2}{4}\left(\frac{h}{2} + \frac{h}{2}\right) - \frac{1}{3}\left(\frac{h^3}{8} + \frac{h^3}{8}\right)\right] = V$$

(f)

EXAMPLE 7.3

A steel wide-flange beam has the dimensions shown in Fig. 7–12a. If it is subjected to a shear of $V = 80$ kN, plot the shear-stress distribution acting over the beam's cross section.

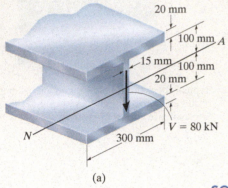

(a)

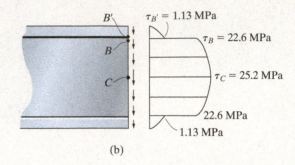

(b)

SOLUTION

Since the flange and web are rectangular elements, then like the previous example, the shear-stress distribution will be parabolic and in this case it will vary in the manner shown in Fig. 7–12b. Due to symmetry, only the shear stresses at points B', B, and C have to be determined. To show how these values are obtained, we must first determine the moment of inertia of the cross-sectional area about the neutral axis. Working in meters, we have

$$
I = \left[\frac{1}{12}(0.015 \text{ m})(0.200 \text{ m})^3 \right]
$$

$$
+ 2\left[\frac{1}{12}(0.300 \text{ m})(0.02 \text{ m})^3 + (0.300 \text{ m})(0.02 \text{ m})(0.110 \text{ m})^2 \right]
$$

$$
= 155.6(10^{-6}) \text{ m}^4
$$

For point B', $t_{B'} = 0.300$ m, and A' is the dark shaded area shown in Fig. 7–12c. Thus,

$$
Q_{B'} = \bar{y}'A' = [0.110 \text{ m}](0.300 \text{ m})(0.02 \text{ m}) = 0.660(10^{-3}) \text{ m}^3
$$

so that

$$
\tau_{B'} = \frac{VQ_{B'}}{It_{B'}} = \frac{80(10^3) \text{ N}(0.660(10^{-3}) \text{ m}^3)}{155.6(10^{-6}) \text{ m}^4(0.300 \text{ m})} = 1.13 \text{ MPa}
$$

For point B, $t_B = 0.015$ m and $Q_B = Q_{B'}$, Fig. 7–12c. Hence

$$
\tau_B = \frac{VQ_B}{It_B} = \frac{80(10^3) \text{ N}(0.660(10^{-3}) \text{ m}^3)}{155.6(10^{-6}) \text{ m}^4(0.015 \text{ m})} = 22.6 \text{ MPa}
$$

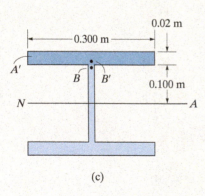

(c)

Fig. 7–12

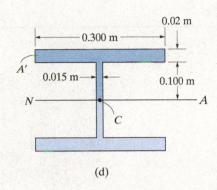

(d)

Fig. 7–12 (cont.)

Note from our discussion of the "Limitations on the Use of the Shear Formula" that the calculated values for both $\tau_{B'}$ and τ_B are actually very misleading. Why?

For point C, $t_C = 0.015$ m and A' is the dark shaded area shown in Fig. 7–12d. Considering this area to be composed of two rectangles, we have

$$Q_C = \Sigma \bar{y}'A' = [0.110 \text{ m}](0.300 \text{ m})(0.02 \text{ m})$$

$$+ [0.05 \text{ m}](0.015 \text{ m})(0.100 \text{ m})$$

$$= 0.735(10^{-3}) \text{ m}^3$$

Thus,

$$\tau_C = \tau_{max} = \frac{VQ_C}{It_C} = \frac{80(10^3) \text{ N}[0.735(10^{-3}) \text{ m}^3]}{155.6(10^{-6}) \text{ m}^4(0.015 \text{ m})} = 25.2 \text{ MPa}$$

From Fig. 7–12b, note that the largest shear stress occurs in the web and is almost uniform throughout its depth, varying from 22.6 MPa to 25.2 MPa. It is for this reason that for design, some codes permit the use of calculating the *average* shear stress on the cross section of the web, rather than using the shear formula; that is,

$$\tau_{avg} = \frac{V}{A_w} = \frac{80(10^3) \text{ N}}{(0.015 \text{ m})(0.2 \text{ m})} = 26.7 \text{ MPa}$$

This will be discussed further in Chapter 11.

PRELIMINARY PROBLEMS

P7–1. In each case, calculate the value of Q and t that are used in the shear formula for finding the shear stress at A. Also, show how the shear stress acts on a differential volume element located at point A.

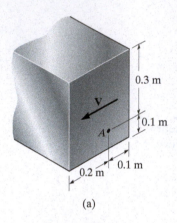

(a)

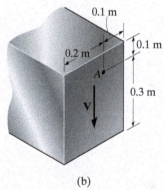

(b)

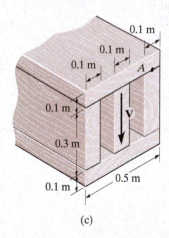

(c)

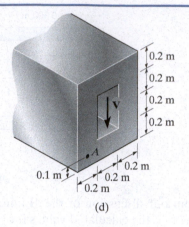

(d)

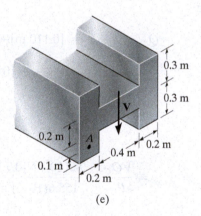

(e)

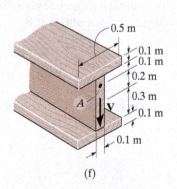

(f)

P7–1

FUNDAMENTAL PROBLEMS

F7–1. If the beam is subjected to a shear force of $V = 100$ kN, determine the shear stress at point A. Represent the state of stress on a volume element at this point.

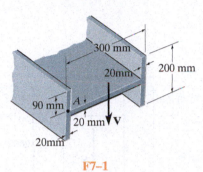

F7–1

F7–2. Determine the shear stress at points A and B if the beam is subjected to a shear force of $V = 600$ kN. Represent the state of stress on a volume element of these points.

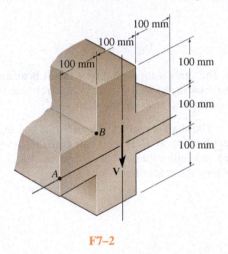

F7–2

F7–3. Determine the absolute maximum shear stress in the beam.

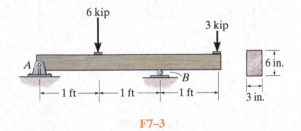

F7–3

F7–4. If the beam is subjected to a shear force of $V = 20$ kN, determine the maximum shear stress in the beam.

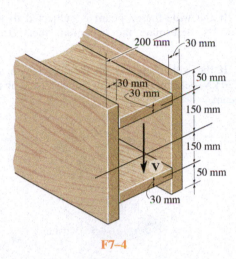

F7–4

F7–5. If the beam is made from four plates and subjected to a shear force of $V = 20$ kN, determine the shear stress at point A. Represent the state of stress on a volume element at this point.

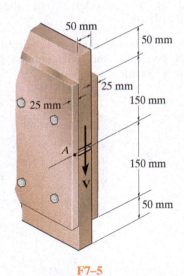

F7–5

PROBLEMS

7–1. If the wide-flange beam is subjected to a shear of $V = 20$ kN, determine the shear stress on the web at A. Indicate the shear-stress components on a volume element located at this point.

7–2. If the wide-flange beam is subjected to a shear of $V = 20$ kN, determine the maximum shear stress in the beam.

7–3. If the wide-flange beam is subjected to a shear of $V = 20$ kN, determine the shear force resisted by the web of the beam.

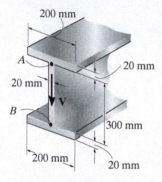

Probs. 7–1/2/3

*7–4. If the beam is subjected to a shear of $V = 30$ kN, determine the web's shear stress at A and B. Indicate the shear-stress components on a volume element located at these points. Set $w = 200$ mm. Show that the neutral axis is located at $\bar{y} = 0.2433$ m from the bottom and $I = 0.5382(10^{-3})$ m^4.

7–5. If the wide-flange beam is subjected to a shear of $V = 30$ kN, determine the maximum shear stress in the beam. Set $w = 300$ mm.

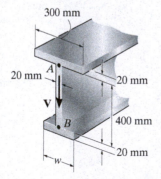

Probs. 7–4/5

7–6. The wood beam has an allowable shear stress of $\tau_{allow} = 7$ MPa. Determine the maximum shear force V that can be applied to the cross section.

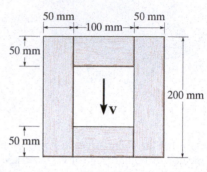

Prob. 7–6

7–7. The shaft is supported by a thrust bearing at A and a journal bearing at B. If $P = 20$ kN, determine the absolute maximum shear stress in the shaft.

*7–8. The shaft is supported by a thrust bearing at A and a journal bearing at B. If the shaft is made from a material having an allowable shear stress of $\tau_{allow} = 75$ MPa, determine the maximum value for P.

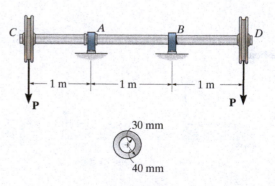

Probs. 7–7/8

7–9. Determine the largest shear force V that the member can sustain if the allowable shear stress is $\tau_{allow} = 8$ ksi.

7–10. If the applied shear force $V = 18$ kip, determine the maximum shear stress in the member.

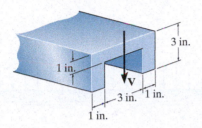

Probs. 7–9/10

7–11. The overhang beam is subjected to the uniform distributed load having an intensity of $w = 50$ kN/m. Determine the maximum shear stress in the beam.

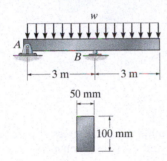

Prob. 7–11

*7–12. The beam is made from a polymer and is subjected to a shear of $V = 7$ kip. Determine the maximum shear stress in the beam and plot the shear-stress distribution over the cross section. Report the values of the shear stress every 0.5 in. of beam depth.

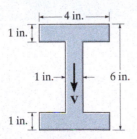

Prob. 7–12

7–13. Determine the maximum shear stress in the strut if it is subjected to a shear force of $V = 20$ kN.

7–14. Determine the maximum shear force V that the strut can support if the allowable shear stress for the material is $\tau_{allow} = 40$ MPa.

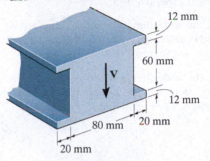

Probs. 7–13/14

7–15. Sketch the intensity of the shear-stress distribution acting over the beam's cross-sectional area, and determine the resultant shear force acting on the segment AB. The shear force acting at the section is $V = 35$ kip. Show that $I_{NA} = 872.49$ in^4.

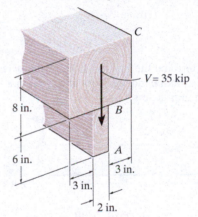

Prob. 7–15

*7–16. Plot the shear-stress distribution over the cross section of a rod that has a radius c. By what factor is the maximum shear stress greater than the average shear stress acting over the cross section?

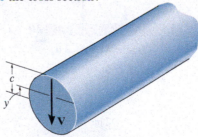

Prob. 7–16

7–17. If the beam is subjected to a shear of $V = 15$ kN, determine the web's shear stress at A and B. Indicate the shear-stress components on a volume element located at these points. Set $w = 125$ mm. Show that the neutral axis is located at $\bar{y} = 0.1747$ m from the bottom and $I_{NA} = 0.2182(10^{-3})$ m^4.

7–18. If the wide-flange beam is subjected to a shear of $V = 30$ kN, determine the maximum shear stress in the beam. Set $w = 200$ mm.

7–19. If the wide-flange beam is subjected to a shear of $V = 30$ kN, determine the shear force resisted by the web of the beam. Set $w = 200$ mm.

Probs. 7–17/18/19

*7–20. Determine the length of the cantilevered beam so that the maximum bending stress in the beam is equivalent to the maximum shear stress.

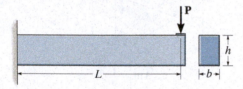

Prob. 7–20

7–21. If the beam is made from wood having an allowable shear stress $\tau_{allow} = 400$ psi, determine the maximum magnitude of **P**. Set $d = 4$ in.

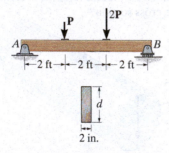

Prob. 7–21

7–22. Determine the largest intensity w of the distributed load that the member can support if the allowable shear stress is $\tau_{allow} = 800$ psi. The supports at A and B are smooth.

7–23. If $w = 800$ lb/ft, determine the absolute maximum shear stress in the beam. The supports at A and B are smooth.

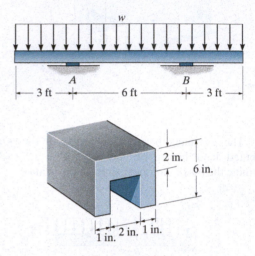

Probs. 7–22/23

*7–24. Determine the shear stress at point B on the web of the cantilevered strut at section a–a.

7–25. Determine the maximum shear stress acting at section a–a of the cantilevered strut.

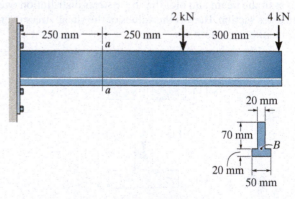

Probs. 7–24/25

7–26. Railroad ties must be designed to resist large shear loadings. If the tie is subjected to the 34-kip rail loadings and an assumed uniformly distributed ground reaction, determine the intensity w for equilibrium, and calculate the maximum shear stress in the tie at section a–a, which is located just to the left of the rail.

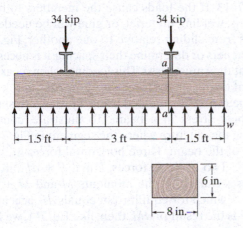

Prob. 7–26

7–27. The beam is slit longitudinally along both sides. If it is subjected to a shear of $V = 250$ kN, compare the maximum shear stress in the beam before and after the cuts were made.

*7–28. The beam is to be cut longitudinally along both sides as shown. If it is made from a material having an allowable shear stress of $\tau_{\text{allow}} = 75$ MPa, determine the maximum allowable shear force $\mathbf{V}$ that can be applied before and after the cut is made.

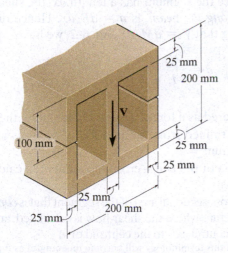

Probs. 7–27/28

7–29. The composite beam is constructed from wood and reinforced with a steel strap. Use the method of Sec. 6.6 and calculate the maximum shear stress in the beam when it is subjected to a shear of $V = 50$ kN. Take $E_{\text{st}} = 200$ GPa, $E_{\text{w}} = 15$ GPa.

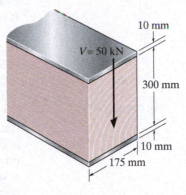

Prob. 7–29

7–30. The beam has a rectangular cross section and is subjected to a load P that is just large enough to develop a fully plastic moment $M_p = PL$ at the fixed support. If the material is elastic perfectly plastic, then at a distance $x < L$ the moment $M = Px$ creates a region of plastic yielding with an associated elastic core having a height $2y'$. This situation has been described by Eq. 6–30 and the moment $\mathbf{M}$ is distributed over the cross section as shown in Fig. 6–48e. Prove that the maximum shear stress in the beam is given by $\tau_{\text{max}} = \frac{3}{2}(P/A')$, where $A' = 2y'b$, the cross-sectional area of the elastic core.

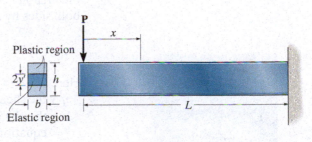

Prob. 7–30

7–31. The beam in Fig. 6–48f is subjected to a fully plastic moment $\mathbf{M}_p$. Prove that the longitudinal and transverse shear stresses in the beam are zero. *Hint:* Consider an element of the beam shown in Fig. 7–4d.

7.3 SHEAR FLOW IN BUILT-UP MEMBERS

Fig. 7–13

Occasionally in engineering practice, members are "built up" from several composite parts in order to achieve a greater resistance to loads. An example is shown in Fig. 7–13. If the loads cause the members to bend, fasteners such as nails, bolts, welding material, or glue will be needed to keep the component parts from sliding relative to one another, Fig. 7–2. In order to design these fasteners or determine their spacing, it is necessary to know the shear force that they must resist. This loading, when measured as a force per unit length of beam, is referred to as *shear flow, q.**

The magnitude of the shear flow is obtained using a procedure similar to that for finding the shear stress in a beam. To illustrate, consider finding the shear flow along the juncture where the segment in Fig. 7–14*a* is connected to the flange of the beam. Three horizontal forces must act on this segment, Fig. 7–14*b*. Two of these forces, *F* and *F* + *dF*, are the result of the normal stresses caused by the moments *M* and *M* + *dM*, respectively. The third force, which for equilibrium equals *dF*, acts at the juncture. Realizing that *dF* is the result of *dM*, then, like Eq. 7–1, we have

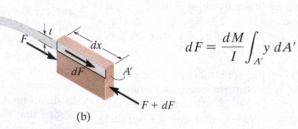

$$dF = \frac{dM}{I}\int_{A'} y\, dA'$$

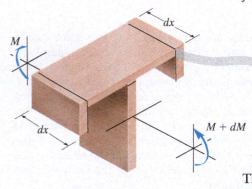

(a)

(b)

The integral represents *Q*, that is, the moment of the segment's area *A'* about the neutral axis. Since the segment has a length *dx*, the shear flow, or *force per unit length along the beam*, is *q* = *dF/dx*. Hence dividing both sides by *dx* and noting that *V* = *dM/dx*, Eq. 6–2, we have

$$q = \frac{VQ}{I} \qquad (7\text{–}4)$$

Here

  q = the shear flow, measured as a force per unit length along the beam

  V = the shear force, determined from the method of sections and the equations of equilibrium

  I = the moment of inertia of the *entire* cross-sectional area calculated about the neutral axis

  Q = $\bar{y}'A'$, where *A'* is the cross-sectional area of the segment that is *connected to the beam* at the juncture where the shear flow is calculated, and $\bar{y}'$ is the distance from the neutral axis to the centroid of *A'*

Fig. 7–14

*The use of the word "flow" in this terminology will become meaningful as it pertains to the discussion in Sec. 7.4.

Fastener Spacing.   When segments of a beam are connected by fasteners, such as nails or bolts, their spacing s along the beam can be determined. For example, let's say that a fastener, such as a nail, can support a maximum shear force of F (N) before it fails, Fig. 7–15a. If these nails are used to construct the beam made from two boards, as shown in Fig. 7–15b, then the nails must resist the shear flow q (N/m) between the boards. In other words, the nails are used to "hold" the top board to the bottom board so that no slipping occurs during bending. (See Fig. 7–2a.) As shown in Fig. 7–15c, the nail spacing is therefore determined from

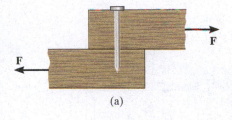
(a)

$$F \text{ (N)} = q \text{ (N/m)} \, s \text{ (m)}$$

The examples that follow illustrate application of this equation.

Other examples of shaded segments connected to built-up beams by fasteners are shown in Fig. 7–16. The shear flow here must be found at the thick black line, and is determined by using a value of Q calculated from A' and $\bar{y}'$ indicated in each figure. This value of q will be resisted by a *single* fastener in Fig. 7–16a, by *two* fasteners in Fig. 7–16b, and by *three* fasteners in Fig. 7–16c. In other words, the fastener in Fig. 7–16a supports the calculated value of q, and in Figs. 7–16b and 7–16c each fastener supports $q/2$ and $q/3$, respectively.

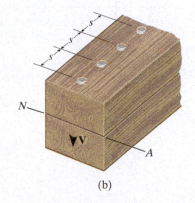
(b)

(c)

Fig. 7–15

> ▶ **IMPORTANT POINT**
>
> - *Shear flow* is a measure of the force per unit length along the axis of a beam. This value is found from the shear formula and is used to determine the shear force developed in fasteners and glue that holds the various segments of a composite beam together.

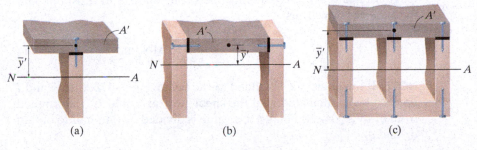

Fig. 7–16

EXAMPLE 7.4

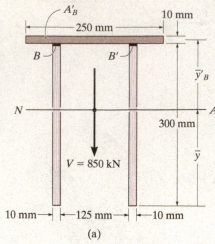

(a)

The beam is constructed from three boards glued together as shown in Fig. 7–17a. If it is subjected to a shear of $V = 850$ kN, determine the shear flow at B and B' that must be resisted by the glue.

SOLUTION

Section Properties.   The neutral axis (centroid) will be located from the bottom of the beam, Fig. 7–17a. Working in units of meters, we have

$$\bar{y} = \frac{\Sigma \tilde{y} A}{\Sigma A} = \frac{2[0.15 \text{ m}](0.3 \text{ m})(0.01 \text{ m}) + [0.305 \text{ m}](0.250 \text{ m})(0.01 \text{ m})}{2(0.3 \text{ m})(0.01 \text{ m}) + 0.250 \text{ m}(0.01 \text{ m})}$$

$$= 0.1956 \text{ m}$$

The moment of inertia of the cross section about the neutral axis is thus

$$I = 2\left[\frac{1}{12}(0.01 \text{ m})(0.3 \text{ m})^3 + (0.01 \text{ m})(0.3 \text{ m})(0.1956 \text{ m} - 0.150 \text{ m})^2\right]$$

$$+ \left[\frac{1}{12}(0.250 \text{ m})(0.01 \text{ m})^3 + (0.250 \text{ m})(0.01 \text{ m})(0.305 \text{ m} - 0.1956 \text{ m})^2\right]$$

$$= 87.42(10^{-6}) \text{ m}^4$$

The glue at both B and B' in Fig. 7–17a "holds" the top board to the beam. Here

$$Q_B = \bar{y}_B' A_B' = [0.305 \text{ m} - 0.1956 \text{ m}](0.250 \text{ m})(0.01 \text{ m})$$

$$= 0.2735(10^{-3}) \text{ m}^3$$

Shear Flow.

$$q = \frac{VQ_B}{I} = \frac{850(10^3) \text{ N}(0.2735(10^{-3}) \text{ m}^3)}{87.42(10^{-6}) \text{ m}^4} = 2.66 \text{ MN/m}$$

Since *two seams* are used to secure the board, the glue per meter length of beam at each seam must be strong enough to resist *one-half* of this shear flow. Thus,

$$q_B = q_{B'} = \frac{q}{2} = 1.33 \text{ MN/m} \qquad \qquad Ans.$$

NOTE: If the board CC' is added to the beam, Fig. 7–17b, then $\bar{y}$ and I have to be recalculated, and the shear flow at C and C' determined from $q = V y_C' A_C'/I$. Finally, this value is divided by one-half to obtain q_C and $q_{C'}$.

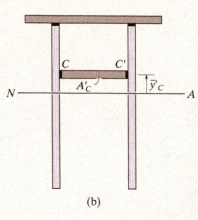

(b)

Fig. 7–17

EXAMPLE 7.5

A box beam is constructed from four boards nailed together as shown in Fig. 7–18a. If each nail can support a maximum shear force of 30 lb, determine the maximum spacing s of the nails at B and at C so that the beam can support the force of 80 lb.

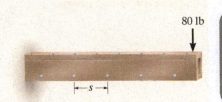

SOLUTION

Internal Shear. If the beam is sectioned at an *arbitrary point* along its length, the internal shear required for equilibrium is always $V = 80$ lb.

Section Properties. The moment of inertia of the cross-sectional area about the neutral axis can be determined by considering a 7.5 in. $\times$ 7.5 in. square minus a 4.5 in. $\times$ 4.5 in. square.

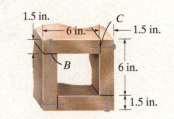

(a)

$$I = \frac{1}{12}(7.5 \text{ in.})(7.5 \text{ in.})^3 - \frac{1}{12}(4.5 \text{ in.})(4.5 \text{ in.})^3 = 229.5 \text{ in}^4$$

The shear flow at B is determined using Q_B found from the darker shaded area shown in Fig. 7–18b. It is this "symmetric" portion of the beam that is to be "held onto" the rest of the beam by nails on the left side and by the fibers of the board on the right side, B'. Thus,

$$Q_B = \bar{y}'A' = [3 \text{ in.}](7.5 \text{ in.})(1.5 \text{ in.}) = 33.75 \text{ in}^3$$

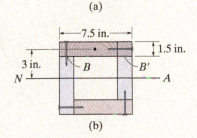

(b)

Likewise, the shear flow at C can be determined using the "symmetric" shaded area shown in Fig. 7–18c. We have

$$Q_C = \bar{y}'A' = [3 \text{ in.}](4.5 \text{ in.})(1.5 \text{ in.}) = 20.25 \text{ in}^3$$

Shear Flow.

$$q_B = \frac{VQ_B}{I} = \frac{80 \text{ lb}(33.75 \text{ in}^3)}{229.5 \text{ in}^4} = 11.76 \text{ lb/in.}$$

$$q_C = \frac{VQ_C}{I} = \frac{80 \text{ lb}(20.25 \text{ in}^3)}{229.5 \text{ in}^4} = 7.059 \text{ lb/in.}$$

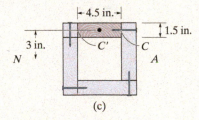

(c)

Fig. 7–18

These values represent the shear force per unit length of the beam that must be resisted by the nails at B and the fibers at B', Fig. 7–18b, and the nails at C and the fibers at C', Fig. 7–18c, respectively. Since in each case the shear flow is resisted at *two* surfaces and each nail can resist 30 lb, for B the spacing is

$$s_B = \frac{30 \text{ lb}}{(11.76/2) \text{ lb/in.}} = 5.10 \text{ in.} \quad \text{Use } s_B = 5 \text{ in.} \qquad Ans.$$

And for C,

$$s_C = \frac{30 \text{ lb}}{(7.059/2) \text{ lb/in.}} = 8.50 \text{ in.} \quad \text{Use } s_C = 8.5 \text{ in.} \qquad Ans.$$

EXAMPLE 7.6

Nails, each having a total shear strength of 40 lb, are used in a beam that can be constructed either as in Case I or as in Case II, Fig. 7–19. If the nails are spaced at 9 in., determine the largest vertical shear that can be supported in each case so that the fasteners will not fail.

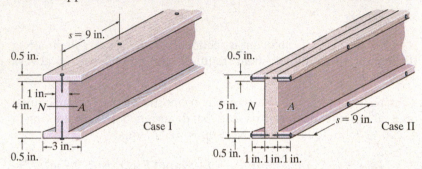

Fig. 7–19

SOLUTION

Since the cross section is the same in both cases, the moment of inertia about the neutral axis is calculated using one large rectangle and two smaller side rectangles.

$$I = \frac{1}{12}(3 \text{ in.})(5 \text{ in.})^3 - 2\left[\frac{1}{12}(1 \text{ in.})(4 \text{ in.})^3\right] = 20.58 \text{ in}^4$$

Case I.   For this design a single row of nails holds the top or bottom flange onto the web. For one of these flanges,

$$Q = \bar{y}'A' = [2.25 \text{ in.}](3 \text{ in.}(0.5 \text{ in.})) = 3.375 \text{ in}^3$$

so that

$$q = \frac{VQ}{I}; \qquad \frac{40 \text{ lb}}{9 \text{ in.}} = \frac{V(3.375 \text{ in}^3)}{20.58 \text{ in}^4}$$

$$V = 27.1 \text{ lb} \qquad\qquad Ans.$$

Case II.   Here a single row of nails holds *one* of the side boards onto the web. Thus,

$$Q = \bar{y}'A' = [2.25 \text{ in.}](1 \text{ in.}(0.5 \text{ in.})) = 1.125 \text{ in}^3$$

$$q = \frac{F}{s} = \frac{VQ}{I}; \qquad \frac{40 \text{ lb}}{9 \text{ in.}} = \frac{V(1.125 \text{ in}^3)}{20.58 \text{ in}^4}$$

$$V = 81.3 \text{ lb} \qquad\qquad Ans.$$

Or, we can also say two rows of nails hold two side boards onto the web, so that

$$q = \frac{F}{s} = \frac{VQ}{It}; \qquad \frac{2(40 \text{ lb})}{9 \text{ in.}} = \frac{V[2(1.125 \text{ in}^3)]}{20.58 \text{ in}^4}$$

$$V = 81.3 \text{ lb} \qquad\qquad Ans.$$

FUNDAMENTAL PROBLEMS

F7–6. The two identical boards are bolted together to form the beam. Determine the maximum spacing s of the bolts to the nearest mm if each bolt has a shear strength of 15 kN. The beam is subjected to a shear force of $V = 50$ kN.

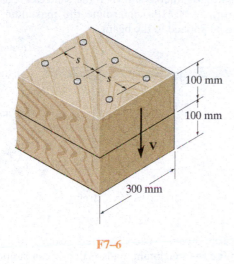

100 mm

100 mm

300 mm

F7–6

F7–8. The boards are bolted together to form the built-up beam. If the beam is subjected to a shear force of $V = 20$ kN, determine the maximum spacing s of the bolts to the nearest mm if each bolt has a shear strength of 8 kN.

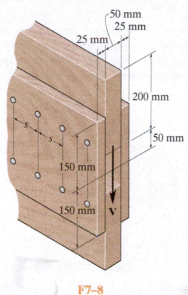

50 mm
25 mm
25 mm
200 mm
50 mm
150 mm
150 mm

F7–8

F7–7. Two identical 20-mm-thick plates are bolted to the top and bottom flange to form the built-up beam. If the beam is subjected to a shear force of $V = 300$ kN, determine the maximum spacing s of the bolts to the nearest mm if each bolt has a shear strength of 30 kN.

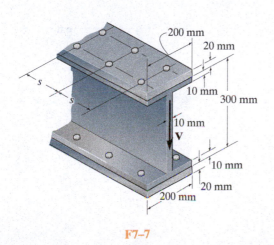

200 mm
20 mm
10 mm
300 mm
10 mm
10 mm
20 mm
200 mm

F7–7

F7–9. The boards are bolted together to form the built-up beam. If the beam is subjected to a shear force of $V = 15$ kip, determine the maximum spacing s of the bolts to the nearest $\frac{1}{8}$ in. if each bolt has a shear strength of 6 kip.

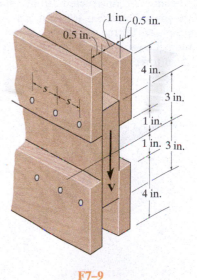

1 in.   0.5 in.
0.5 in.
4 in.
3 in.
1 in.
1 in. 3 in.
4 in.

F7–9

PROBLEMS

*7–32. The beam is constructed from two boards fastened together at the top and bottom with three rows of nails spaced every 4 in. If each nail can support a 400-lb shear force, determine the maximum shear force V that can be applied to the beam.

7–33. The beam is constructed from two boards fastened together at the top and bottom with three rows of nails spaced every 4 in. If a shear force of $V = 900$ lb is applied to the boards, determine the shear force resisted by each nail.

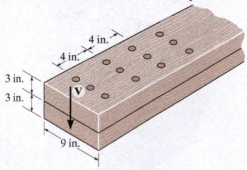

Probs. 7–32/33

7–34. The beam is constructed from three boards. If it is subjected to a shear of $V = 5$ kip, determine the maximum allowable spacing s of the nails used to hold the top and bottom flanges to the web. Each nail can support a shear force of 500 lb.

7–35. The beam is constructed from three boards. Determine the maximum shear V that it can support if the allowable shear stress for the wood is $\tau_{allow} = 400$ psi. What is the maximum allowable spacing s of the nails if each nail can resist a shear force of 400 lb?

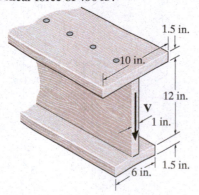

Probs. 7–34/35

*7–36. The double T-beam is fabricated by welding the three plates together as shown. Determine the shear stress in the weld necessary to support a shear force of $V = 80$ kN.

7–37. The double T-beam is fabricated by welding the three plates together as shown. If the weld can resist a shear stress $\tau_{allow} = 90$ MPa, determine the maximum shear V that can be applied to the beam.

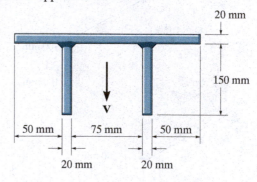

Probs. 7–36/37

7–38. The beam is constructed from three boards. Determine the maximum loads P that it can support if the allowable shear stress for the wood is $\tau_{allow} = 400$ psi. What is the maximum allowable spacing s of the nails used to hold the top and bottom flanges to the web if each nail can resist a shear force of 400 lb?

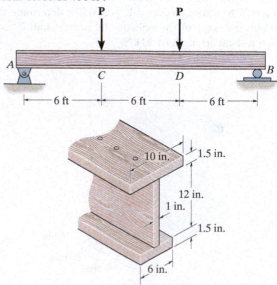

Prob. 7–38

7–39. A beam is constructed from three boards bolted together as shown. Determine the shear force in each bolt if the bolts are spaced $s = 250$ mm apart and the shear is $V = 35$ kN.

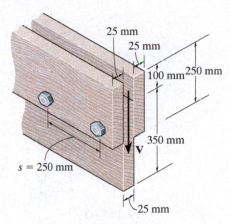

25 mm

25 mm

100 mm 250 mm

350 mm

V

$s = 250$ mm

25 mm

Prob. 7–39

*7–40. The simply supported beam is built up from three boards by nailing them together as shown. The wood has an allowable shear stress of $\tau_{\text{allow}} = 1.5$ MPa, and an allowable bending stress of $\sigma_{\text{allow}} = 9$ MPa. The nails are spaced at $s = 75$ mm, and each has a shear strength of 1.5 kN. Determine the maximum allowable force **P** that can be applied to the beam.

7–41. The simply supported beam is built up from three boards by nailing them together as shown. If $P = 12$ kN, determine the maximum allowable spacing s of the nails to support that load, if each nail can resist a shear force of 1.5 kN.

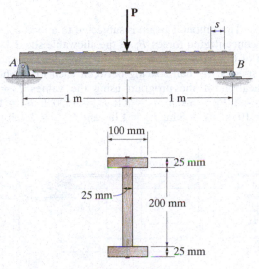

P

A   B

1 m   1 m

100 mm

25 mm

25 mm

200 mm

25 mm

Probs. 7–40/41

7–42. The T-beam is constructed as shown. If each nail can support a shear force of 950 lb, determine the maximum shear force V that the beam can support and the corresponding maximum nail spacing s to the nearest $\frac{1}{8}$ in. The allowable shear stress for the wood is $\tau_{\text{allow}} = 450$ psi.

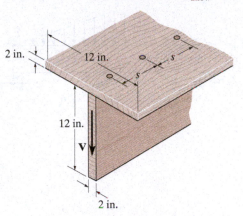

2 in.   12 in. s

s

12 in.

V

2 in.

Prob. 7–42

7–43. The box beam is constructed from four boards that are fastened together using nails spaced along the beam every 2 in. If each nail can resist a shear force of 50 lb, determine the largest force P that can be applied to the beam without causing failure of the nails.

*7–44. The box beam is constructed from four boards that are fastened together using nails spaced along the beam every 2 in. If a force $P = 2$ kip is applied to the beam, determine the shear force resisted by each nail at A and B.

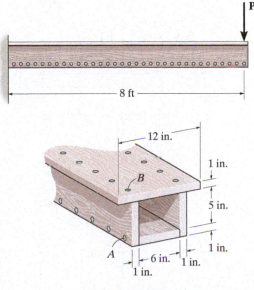

P

8 ft

12 in.

1 in.

B

5 in.

A

6 in.   1 in.

1 in.   1 in.

Probs. 7–43/44

7–45. The member consists of two plastic channel strips 0.5 in. thick, glued together at A and B. If the distributed load has a maximum intensity of $w_0 = 3$ kip/ft, determine the maximum shear stress resisted by the glue.

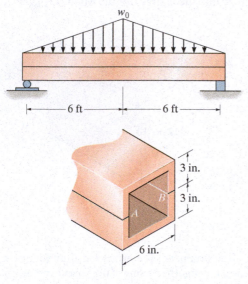

Prob. 7–45

7–46. The member consists of two plastic channel strips 0.5 in. thick, glued together at A and B. If the glue can support an allowable shear stress of $\tau_{allow} = 600$ psi, determine the maximum intensity w_0 of the triangular distributed loading that can be applied to the member based on the strength of the glue.

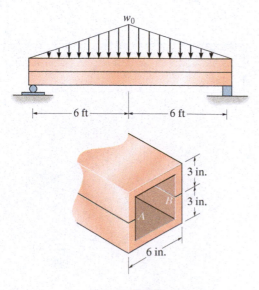

Prob. 7–46

7–47. The beam is made from four boards nailed together as shown. If the nails can each support a shear force of 100 lb., determine their required spacing s' and s if the beam is subjected to a shear of $V = 700$ lb.

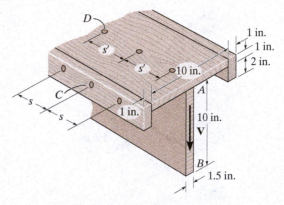

Prob. 7–47

*7–48. The beam is made from three polystyrene strips that are glued together as shown. If the glue has a shear strength of 80 kPa, determine the maximum load P that can be applied without causing the glue to lose its bond.

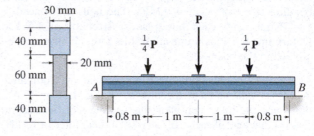

Prob. 7–48

7–49. The timber T-beam is subjected to a load consisting of n concentrated forces, P_n. If the allowable shear V_{nail} for each of the nails is known, write a computer program that will specify the nail spacing between each load. Show an application of the program using the values $L = 15$ ft, $a_1 = 4$ ft, $P_1 = 600$ lb, $a_2 = 8$ ft, $P_2 = 1500$ lb, $b_1 = 1.5$ in., $h_1 = 10$ in., $b_2 = 8$ in., $h_2 = 1$ in., and $V_{nail} = 200$ lb.

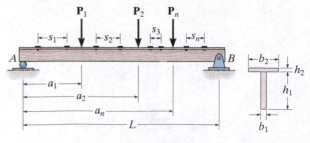

Prob. 7–49

7.4 SHEAR FLOW IN THIN-WALLED MEMBERS

In this section we will show how to describe the shear-flow *distribution* throughout a member's cross-sectional area. As with most structural members, we will assume that the member has *thin walls*, that is, the wall thickness is small compared to its height or width.

Before we determine the shear-flow distribution, we will first show how to establish its direction. To begin, consider the beam in Fig. 7–20*a*, and the free-body diagram of segment *B* taken from the top flange, Fig. 7–20*b*. The force *dF* must act on the longitudinal section in order to balance the normal forces *F* and *F* + *dF* created by the moments *M* and *M* + *dM*, respectively. Because *q* (and *τ*) are complementary, *transverse components* of *q* must act on the cross section as shown on the corner element in Fig. 7–20*b*.

Although it is also true that *V* + *dV* will create a *vertical* shear-flow component on this element, Fig. 7–20*c*, here we will neglect its effects. This is because the flange is thin, and the top and bottom surfaces of the flange are free of stress. To summarize then, only the shear flow component that acts *parallel* to the sides of the flange will be considered.

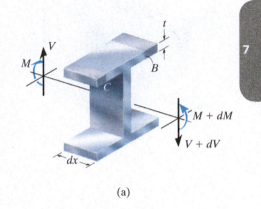

(a)

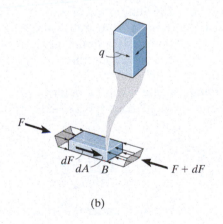

(b)

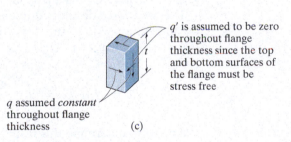

q assumed *constant* throughout flange thickness

q' is assumed to be zero throughout flange thickness since the top and bottom surfaces of the flange must be stress free

(c)

Fig. 7–20

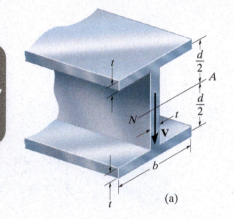

(a)

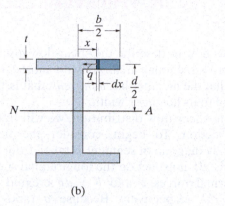

(b)

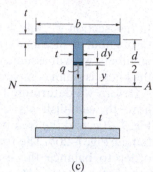

(c)

Fig. 7–21

Shear Flow in Flanges.

The shear-flow distribution along the top flange of the beam in Fig. 7–21a can be found by considering the shear flow q, acting on the dark blue element dx, located an arbitrary distance x from the centerline of the cross section, Fig. 7–21b. Here $Q = \bar{y}'A' = [d/2](b/2 - x)t$, so that

$$q = \frac{VQ}{I} = \frac{V[d/2](b/2 - x)t}{I} = \frac{Vtd}{2I}\left(\frac{b}{2} - x\right) \qquad (7\text{–}5)$$

By inspection, this distribution varies in a *linear manner* from $q = 0$ at $x = b/2$ to $(q_{max})_f = Vtdb/4I$ at $x = 0$. (The limitation of $x = 0$ is possible here since the member is assumed to have "thin walls" and so the thickness of the web is neglected.) Due to symmetry, a similar analysis yields the same distribution of shear flow for the other three flange segments. These results are as shown in Fig. 7–21d.

The total *force* developed in each flange segment can be determined by integration. Since the force on the element dx in Fig. 7–21b is $dF = q\,dx$, then

$$F_f = \int q\,dx = \int_0^{b/2} \frac{Vt\,d}{2I}\left(\frac{b}{2} - x\right)dx = \frac{Vt\,db^2}{16I}$$

We can also determine this result by finding the area under the triangle in Fig. 7–21d. Hence,

$$F_f = \frac{1}{2}(q_{max})_f\left(\frac{b}{2}\right) = \frac{Vt\,db^2}{16I}$$

All four of these forces are shown in Fig. 7–21e, and we can see from their direction that horizontal force equilibrium on the cross section is maintained.

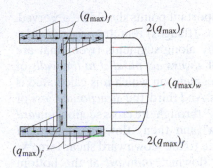

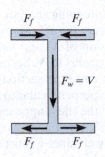

Shear-flow distribution

(d)   (e)

Fig. 7–21 (cont.)

Shear Flow in Web.

A similar analysis can be performed for the web, Fig. 7–21c. Here q must act downward, and at element dy we have $Q = \Sigma \bar{y}'A' = [d/2](bt) + [y + (1/2)(d/2 - y)]t(d/2 - y) = bt\,d/2 + (t/2)(d^2/4 - y^2)$, so that

$$q = \frac{VQ}{I} = \frac{Vt}{I}\left[\frac{db}{2} + \frac{1}{2}\left(\frac{d^2}{4} - y^2\right)\right] \qquad (7\text{–}6)$$

For the web, the shear flow varies in a *parabolic manner*, from $q = 2(q_{max})_f = Vt\,db/2I$ at $y = d/2$ to $(q_{max})_w = (Vt\,d/I)(b/2 + d/8)$ at $y = 0$, Fig. 7–21d.

Integrating to determine the force in the web, F_w, we have

$$\begin{aligned}
F_w &= \int q\,dy = \int_{-d/2}^{d/2} \frac{Vt}{I}\left[\frac{db}{2} + \frac{1}{2}\left(\frac{d^2}{4} - y^2\right)\right] dy \\
&= \frac{Vt}{I}\left[\frac{db}{2}y + \frac{1}{2}\left(\frac{d^2}{4}y - \frac{1}{3}y^3\right)\right]\Bigg|_{-d/2}^{d/2} \\
&= \frac{Vt\,d^2}{4I}\left(2b + \frac{1}{3}d\right)
\end{aligned}$$

Simplification is possible by noting that the moment of inertia for the cross-sectional area is

$$I = 2\left[\frac{1}{12}bt^3 + bt\left(\frac{d}{2}\right)^2\right] + \frac{1}{12}td^3$$

Neglecting the first term, since the thickness of each flange is small, then

$$I = \frac{td^2}{4}\left(2b + \frac{1}{3}d\right)$$

Substituting this into the above equation, we see that $F_w = V$, which is to be expected, Fig. 7–21e.

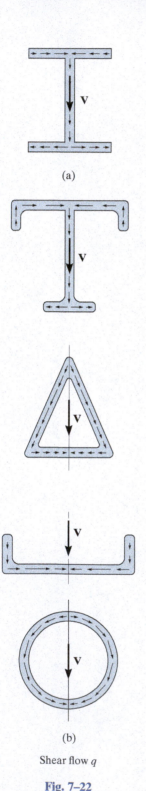

(a)

(b)

Shear flow q

Fig. 7–22

From the foregoing analysis, three important points should be observed. First, q will vary *linearly* along segments (flanges) that are *perpendicular* to the direction of **V**, and *parabolically* along segments (web) that are *inclined or parallel* to **V**. Second, q will *always act parallel to the walls* of the member, since the section of the segment on which q is calculated is always taken perpendicular to the walls. And third, the *directional sense* of q is such that the shear appears to "*flow*" through the cross section, *inward* at the beam's top flange, "combining" and then "flowing" *downward* through the web, since it must contribute to the downward shear force **V**, Fig. 7–22a, and then separating and "flowing" *outward* at the bottom flange. If one is able to "visualize" this "flow" it will provide an easy means for establishing not only the direction of q, but *also* the corresponding direction of τ. Other examples of how q is directed along the segments of thin-walled members are shown in Fig. 7–22b. In all cases, symmetry prevails about an axis that is collinear with **V**, and so q "flows" in a direction such that it will provide the vertical force **V** and yet also satisfy horizontal force equilibrium for the cross section.

> ## IMPORTANT POINTS
>
> - The shear flow formula $q = VQ/I$ can be used to determine the *distribution* of the shear flow throughout a thin-walled member, provided the shear **V** acts along an axis of symmetry or principal centroidal axis of inertia for the cross section.
>
> - If a member is made from segments having thin walls, then only the shear flow *parallel* to the walls of the member is important.
>
> - The shear flow varies *linearly* along segments that are *perpendicular* to the direction of the shear **V**.
>
> - The shear flow varies *parabolically* along segments that are *inclined* or *parallel* to the direction of the shear **V**.
>
> - On the cross section, the shear "flows" along the segments so that it results in the vertical shear force **V** and yet satisfies horizontal force equilibrium.

EXAMPLE 7.7

The thin-walled box beam in Fig. 7–23a is subjected to a shear of 10 kip. Determine the variation of the shear flow throughout the cross section.

SOLUTION

By symmetry, the neutral axis passes through the center of the cross section. For thin-walled members we use centerline dimensions for calculating the moment of inertia.

$$I = \frac{1}{12}(2 \text{ in.})(7 \text{ in.})^3 + 2\,[(5 \text{ in.})(1 \text{ in.})(3.5 \text{ in.})^2] = 179.7 \text{ in.}^4$$

Only the shear flow at points B, C, and D has to be determined. For point B, the area $A' \approx 0$, Fig. 7–23b, since it can be thought of as being located entirely at point B. Alternatively, A' can also represent the *entire* cross-sectional area, in which case $Q_B = \bar{y}'A' = 0$ since $\bar{y}' = 0$. Because $Q_B = 0$, then

$$q_B = 0$$

For point C, the area A' is shown dark shaded in Fig. 7–23c. Here, we have used the mean dimensions since point C is on the centerline of each segment. We have

$$Q_C = \bar{y}'A' = (3.5 \text{ in.})(5 \text{ in.})(1 \text{ in.}) = 17.5 \text{ in}^3$$

Since there are two points of attachment,

$$q_C = \frac{1}{2}\left(\frac{VQ_C}{I}\right) = \frac{1}{2}\left(\frac{10 \text{ kip}(17.5 \text{ in}^3))}{179.7 \text{ in}^4}\right) = 0.487 \text{ kip/in.}$$

The shear flow at D is determined using the three dark-shaded rectangles shown in Fig. 7–23d. Again, using centerline dimensions

$$Q_D = \Sigma\bar{y}'A' = 2\left[\frac{3.5 \text{ in.}}{2}\right](1 \text{ in.})(3.5 \text{ in.}) + [3.5 \text{ in.}](5 \text{ in.})(1 \text{ in.}) = 29.75 \text{ in}^3$$

Because there are two points of attachment,

$$q_D = \frac{1}{2}\left(\frac{VQ_D}{I}\right) = \frac{1}{2}\left(\frac{10 \text{ kip}(29.75 \text{ in}^3)}{179.7 \text{ in}^4}\right) = 0.828 \text{ kip/in.}$$

Using these results, and the symmetry of the cross section, the shear-flow distribution is plotted in Fig. 7–23e. The distribution is linear along the horizontal segments (perpendicular to **V**) and parabolic along the vertical segments (parallel to **V**).

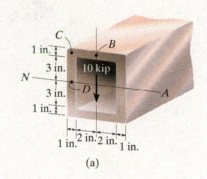

(a)

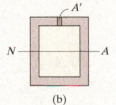

(b)

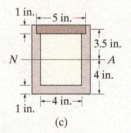

(c)

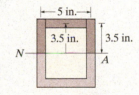

(d)

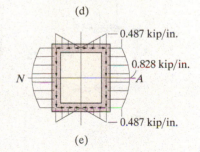

0.487 kip/in.

0.828 kip/in.

0.487 kip/in.

(e)

Fig. 7-23

*7.5   SHEAR CENTER FOR OPEN THIN-WALLED MEMBERS

In the previous section, the internal shear $\mathbf{V}$ was applied along a principal centroidal axis of inertia that *also* represents an *axis of symmetry* for the cross section. In this section we will consider the effect of applying the shear along a principal centroidal axis that is *not an axis of symmetry*. As before, only open thin-walled members will be analyzed, where the dimensions to the centerline of the walls of the members will be used.

A typical example of this case is the channel shown in Fig. 7–24a. Here it is cantilevered from a fixed support and subjected to the force $\mathbf{P}$. If this force is applied through the *centroid C* of the cross section, the channel will not only bend downward, but *it will also twist* clockwise as shown.

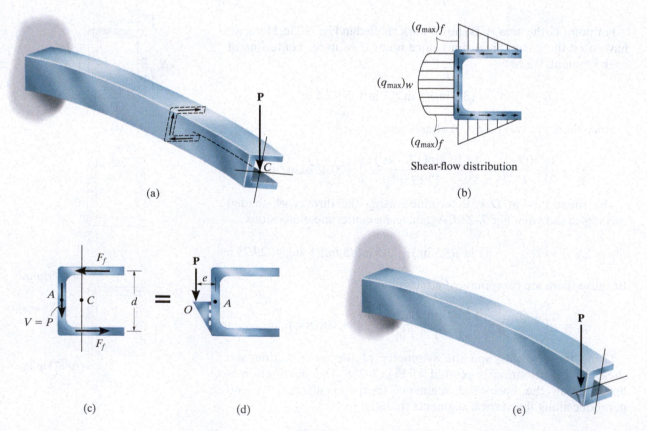

(a)

$(q_{max})_f$

$(q_{max})_w$

$(q_{max})_f$

Shear-flow distribution

(b)

F_f

A

$\cdot C$

$V = P$

d

F_f

(c)

$\mathbf{P}$

e

O

$\cdot A$

(d)

$\mathbf{P}$

(e)

Fig. 7–24

The reason the member twists has to do with the shear-flow distribution along the channel's flanges and web, Fig. 7–24b. When this distribution is integrated over the flange and web areas, it will give resultant forces of F_f in each flange and a force of $V = P$ in the web, Fig. 7–24c. If the moments of these three forces are summed about point A, the unbalanced couple or torque created by the flange forces is seen to be responsible for twisting the member. The actual twist is clockwise when viewed from the front of the beam, as shown in Fig. 7–24a, because *reactive* internal "equilibrium" forces F_f cause the twisting. In order to *prevent* this twisting and therefore cancel the unbalanced moment, it is necessary to apply **P** at a point O located an eccentric distance e from the web, as shown in Fig. 7–24d. We require $\Sigma M_A = F_f d = Pe$, or

$$e = \frac{F_f d}{P}$$

The point O so located is called the **shear center** or **flexural center**. When **P** is applied at this point, the **beam will bend without twisting**, Fig. 7–24e. Design handbooks often list the location of the shear center for a variety of thin-walled beam cross sections that are commonly used in practice.

From this analysis, it should be noted that **the shear center will always lie on an axis of symmetry** of a member's cross-sectional area. For example, if the channel is rotated 90° and **P** is applied at A, Fig. 7–25a, no twisting will occur since the shear flow in the web and flanges for this case is *symmetrical*, and therefore the force resultants in these elements will create zero moments about A, Fig. 7–25b. Obviously, if a member has a cross section with *two* axes of symmetry, as in the case of a wide-flange beam, the shear center will coincide with the intersection of these axes (the centroid).

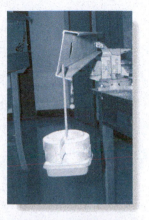

Notice how this cantilever beam deflects when loaded through the centroid (above) and through the shear center (below).

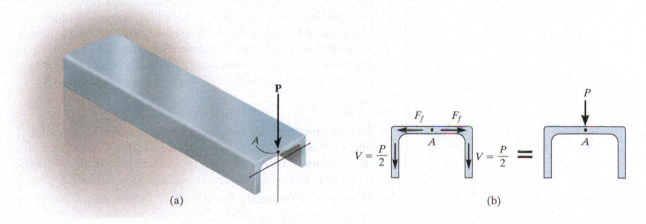

(a)   (b)

Fig. 7–25

▶ **IMPORTANT POINTS**

- The *shear center* is the point through which a force can be applied which will cause a beam to bend and yet not twist.

- The shear center will always lie on an axis of symmetry of the cross section.

- The location of the shear center is only a function of the geometry of the cross section, and does not depend upon the applied loading.

▶ **PROCEDURE FOR ANALYSIS**

The location of the shear center for an open thin-walled member for which the internal shear is in the *same direction* as a principal centroidal axis for the cross section may be determined by using the following procedure.

Shear-Flow Resultants.

- By observation, determine the direction of the shear flow through the various segments of the cross section, and sketch the force resultants on each segment of the cross section. (For example, see Fig. 7–24c.) Since the shear center is determined by taking the moments of these force resultants about a point, A, choose this point at a location that eliminates the moments of as many force resultants as possible.

- The magnitudes of the force resultants that create a moment about A must be calculated. For any segment this is done by determining the shear flow q at an arbitrary point on the segment and then integrating q along the segment's length. Realize that $\mathbf{V}$ will create a *linear* variation of shear flow in segments that are *perpendicular* to $\mathbf{V}$, and a *parabolic* variation of shear flow in segments that are *parallel or inclined* to $\mathbf{V}$.

Shear Center.

- Sum the moments of the shear-flow resultants about point A and set this moment equal to the moment of $\mathbf{V}$ about A. Solve this equation to determine the moment-arm or eccentric distance e, which locates the line of action of $\mathbf{V}$ from A.

- If an *axis of symmetry* for the cross section exists, the shear center lies at a point on this axis.

EXAMPLE 7.8

Determine the location of the shear center for the thin-walled channel having the dimensions shown in Fig. 7–26a.

SOLUTION

Shear-Flow Resultants.   A vertical downward shear V applied to the section causes the shear to flow through the flanges and web as shown in Fig. 7–26b. This in turn creates force resultants F_f and V in the flanges and web as shown in Fig. 7–26c. We will take moments about point A so that only the force F_f on the lower flange has to be determined.

The cross-sectional area can be divided into three component rectangles—a web and two flanges. Since each component is assumed to be thin, the moment of inertia of the area about the neutral axis is

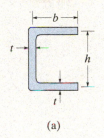

(a)

$$I = \frac{1}{12}th^3 + 2\left[bt\left(\frac{h}{2}\right)^2 \right] = \frac{th^2}{2}\left(\frac{h}{6} + b\right)$$

From Fig. 7–26d, q at the arbitrary position x is

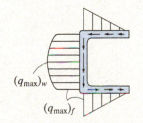

$(q_{max})_w$

$(q_{max})_f$

Shear flow distribution

(b)

$$q = \frac{VQ}{I} = \frac{V(h/2)[b - x]t}{(th^2/2)[(h/6) + b]} = \frac{V(b - x)}{h[(h/6) + b]}$$

Hence, the force F_f in the flange is

$$F_f = \int_0^b q \, dx = \frac{V}{h[(h/6) + b]}\int_0^b (b - x)\, dx = \frac{Vb^2}{2h[(h/6) + b]}$$

This same result can also be determined without integration by first finding $(q_{max})_f$, Fig. 7–26b, then determining the triangular area $\frac{1}{2}b(q_{max})_f = F_f$.

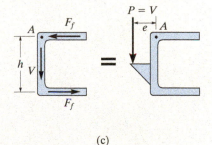

(c)

Shear Center.   Summing moments about point A, Fig. 7–26c, we require

$$Ve = F_f h = \frac{Vb^2 h}{2h[(h/6) + b]}$$

Thus,

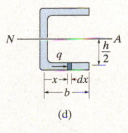

(d)

Fig. 7–26

$$e = \frac{b^2}{[(h/3) + 2b]} \qquad\qquad Ans.$$

EXAMPLE 7.9

Determine the location of the shear center for the angle having equal legs, Fig. 7–27a. Also, find the internal shear-force resultant in each leg.

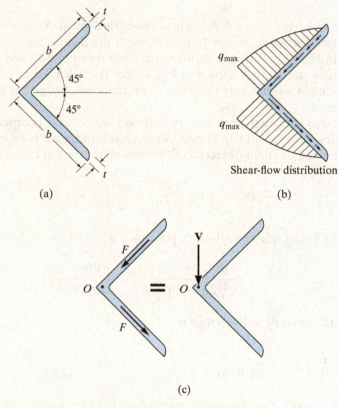

Shear-flow distribution

(a)   (b)

(c)

Fig. 7–27

SOLUTION

When a vertical downward shear **V** is applied at the section, the shear flow and shear-flow resultants are directed as shown in Figs. 7–27b and 7–27c, respectively. Note that the force F in each leg must be equal, since for equilibrium the sum of their horizontal components must be equal to zero. Also, the lines of action of both forces intersect point O; therefore, this point *must be the shear center*, since the sum of the moments of these forces and **V** about O is zero, Fig. 7–27c.

The magnitude of **F** can be determined by first finding the shear flow at the arbitrary location s along the top leg, Fig. 7–27d. Here

$$Q = \bar{y}'A' = \frac{1}{\sqrt{2}}\left((b-s)+\frac{s}{2}\right)ts = \frac{1}{\sqrt{2}}\left(b-\frac{s}{2}\right)st$$

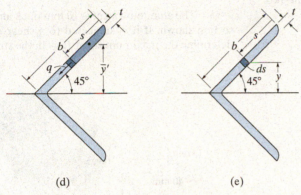

(d)   (e)

Fig. 7–27 (cont.)

The moment of inertia of the angle about the neutral axis must be determined from "first principles," since the legs are inclined with respect to the neutral axis. For the area element $dA = t\,ds$, Fig. 7–27e, we have

$$I = \int_A y^2\,dA = 2\int_0^b\left[\frac{1}{\sqrt{2}}(b-s)\right]^2 t\,ds = t\left(b^2 s - bs^2 + \frac{1}{3}s^3\right)\Big|_0^b = \frac{tb^3}{3}$$

Thus, the shear flow is

$$q = \frac{VQ}{I} = \frac{V}{(tb^3/3)}\left[\frac{1}{\sqrt{2}}\left(b-\frac{s}{2}\right)st\right]$$

$$= \frac{3V}{\sqrt{2}b^3}s\left(b-\frac{s}{2}\right)$$

The variation of q is parabolic, and it reaches a maximum value when $s = b$, as shown in Fig. 7–27b. The force F is therefore

$$F = \int_0^b q\,ds = \frac{3V}{\sqrt{2}b^3}\int_0^b s\left(b-\frac{s}{2}\right)ds$$

$$= \frac{3V}{\sqrt{2}b^3}\left(b\frac{s^2}{2}-\frac{1}{6}s^3\right)\Big|_0^b$$

$$= \frac{1}{\sqrt{2}}V \qquad\qquad \textit{Ans.}$$

NOTE: This result can be easily verified, since the sum of the vertical components of the force F in each leg must equal V and, as stated previously, the sum of the horizontal components equals zero.

PROBLEMS

7–50. A shear force of $V = 300$ kN is applied to the box girder. Determine the shear flow at points A and B.

7–51. A shear force of $V = 450$ kN is applied to the box girder. Determine the shear flow at points C and D.

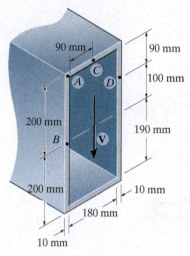

Probs. 7–50/51

*7–52. A shear force of $V = 18$ kN is applied to the box girder. Determine the shear flow at points A and B.

7–53. A shear force of $V = 18$ kN is applied to the box girder. Determine the shear flow at point C.

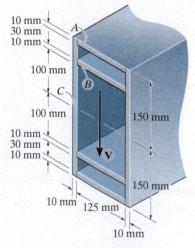

Probs. 7–52/53

7–54. The aluminum strut is 10 mm thick and has the cross section shown. If it is subjected to a shear of $V = 150$ N, determine the shear flow at points A and B.

7–55. The aluminum strut is 10 mm thick and has the cross section shown. If it is subjected to a shear of $V = 150$ N, determine the maximum shear flow in the strut.

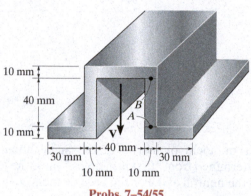

Probs. 7–54/55

*7–56. The beam is subjected to a shear force of $V = 50$ kip. Determine the shear flow at points A and B.

7–57. The beam is subjected to a shear force of $V = 50$ kip. Determine the maximum shear flow in the cross section.

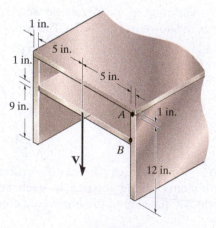

Probs. 7–56/57

7–58. The H-beam is subjected to a shear of $V = 80$ kN. Determine the shear flow at point A.

7–59. The H-beam is subjected to a shear of $V = 80$ kN. Sketch the shear-stress distribution acting along one of its side segments. Indicate all peak values.

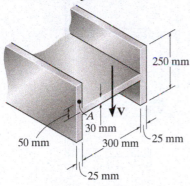

Probs. 7–58/59

*7–60. The built-up beam is formed by welding together the thin plates of thickness 5 mm. Determine the location of the shear center O.

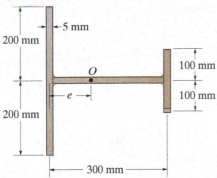

Prob. 7–60

7–61. The assembly is subjected to a vertical shear of $V = 7$ kip. Determine the shear flow at points A and B and the maximum shear flow in the cross section.

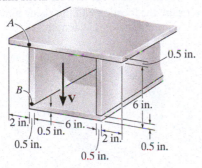

Prob. 7–61

7–62. The box girder is subjected to a shear of $V = 15$ kN. Determine the shear flow at point B and the maximum shear flow in the girder's web AB.

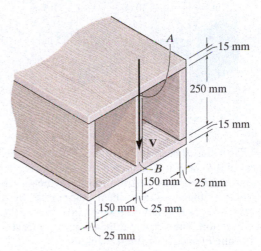

Prob. 7–62

7–63. Determine the location e of the shear center, point O, for the thin-walled member having a slit along its section.

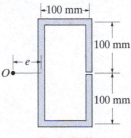

Prob. 7–63

*7–64. Determine the location e of the shear center, point O, for the thin-walled member. The member segments have the same thickness t.

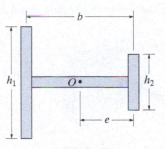

Prob. 7–64

7–65. The beam supports a vertical shear of $V = 7$ kip. Determine the resultant force in segment AB of the beam.

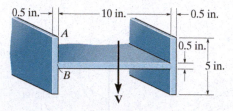

0.5 in. ┤├ 10 in. ┤├ 0.5 in.

0.5 in.

5 in.

Prob. 7–65

7–66. The stiffened beam is constructed from plates having a thickness of 0.25 in. If it is subjected to a shear of $V = 8$ kip, determine the shear-flow distribution in segments AB and CD. What is the resultant shear supported by these segments? Also, sketch how the shear flow passes through the cross section. The vertical dimensions are measured to the centerline of each horizontal segment.

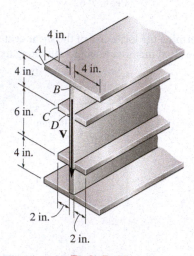

4 in.

4 in. 4 in.

6 in.

4 in.

2 in.

2 in.

Prob. 7–66

7–67. The pipe is subjected to a shear force of $V = 8$ kip. Determine the shear flow in the pipe at points A and B.

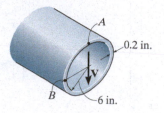

0.2 in.

6 in.

Prob. 7–67

*7–68. Determine the location e of the shear center, point O, for the thin-walled member. The member segments have the same thickness t.

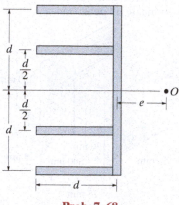

d

$\dfrac{d}{2}$

$\dfrac{d}{2}$

d

e

d

O

Prob. 7–68

7–69. A thin plate of thickness t is bent to form the beam having the cross section shown. Determine the location of the shear center O.

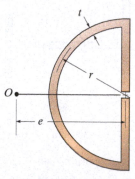

t

r

O

e

Prob. 7–69

7–70. Determine the location e of the shear center, point O, for the tube having a slit along its length.

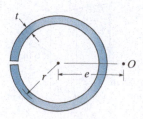

t

r e O

Prob. 7–70

CHAPTER REVIEW

Transverse shear stress in beams is determined indirectly by using the flexure formula and the relationship between moment and shear ($V = dM/dx$). The result is the shear formula

$$\tau = \frac{VQ}{It}$$

In particular, the value for Q is the moment of the area A' about the neutral axis, $Q = \bar{y}'A'$. This area is the portion of the cross-sectional area that is "held onto" the beam above (or below) the thickness t where τ is to be determined.

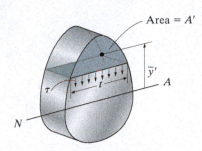

If the beam has a rectangular cross section, then the shear-stress distribution will be parabolic, having a maximum value at the neutral axis. For this special case, the maximum shear stress can be determined using $\tau_{max} = 1.5 \dfrac{V}{A}$.

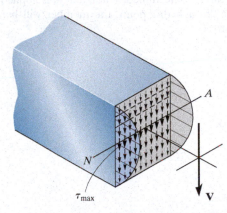

Shear–stress distribution

Fasteners, such as nails, bolts, glue, or weld, are used to connect the composite parts of a "built-up" section. The shear force resisted by these fasteners is determined from the shear flow, q, or force per unit length, that must be supported by the beam. The shear flow is

$$q = \frac{VQ}{I}$$

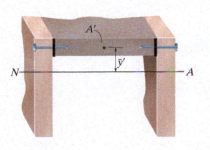

If the beam is made from thin-walled segments, then the shear-flow distribution along each segment can be determined. This distribution will vary linearly along horizontal segments, and parabolically along inclined or vertical segments.

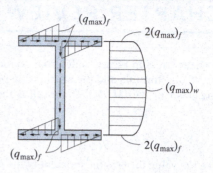

Shear-flow distribution

Provided the shear-flow distribution in each segment of an open thin-walled section is known, then using a balance of moments, the location O of the shear center for the cross section can be determined. When a load is applied to the member through this point, the member will bend, and not twist.

REVIEW PROBLEMS

R7–1. The beam is fabricated from four boards nailed together as shown. Determine the shear force each nail along the sides C and the top D must resist if the nails are uniformly spaced at $s = 3$ in. The beam is subjected to a shear of $V = 4.5$ kip.

R7–3. The member is subject to a shear force of $V = 2$ kN. Determine the shear flow at points A, B, and C. The thickness of each thin-walled segment is 15 mm.

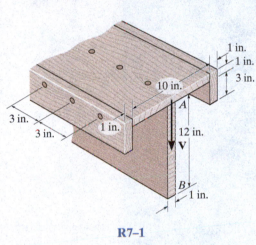

R7–1

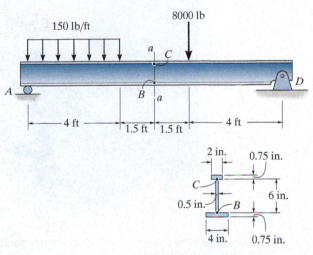

200 mm

B

100 mm

A C

300 mm

$V = 2$ kN

R7–3

*R7–4. Determine the shear stress at points B and C on the web of the beam located at section a–a.

R7–5. Determine the maximum shear stress acting at section a–a in the beam.

R7–2. The T-beam is subjected to a shear of $V = 150$ kN. Determine the amount of this force that is supported by the web B.

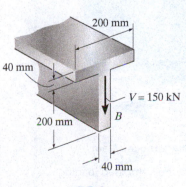

200 mm

40 mm

$V = 150$ kN

200 mm

B

40 mm

R7–2

8000 lb

150 lb/ft

a C

A

B a

D

4 ft

1.5 ft   1.5 ft

4 ft

2 in.   0.75 in.

C

6 in.

0.5 in.   B

4 in.   0.75 in.

R7–4/5

CHAPTER 8

(© ImageBroker/Alamy)

The offset hanger supporting this ski gondola is subjected to the combined loadings of axial force and bending moment.

COMBINED LOADINGS

- This chapter begins with an analysis of stress developed in thin-walled pressure vessels. Then we will use the formulas for axial load, torsion, bending, and shear to determine the stress in a member subjected to several loadings.

8.1 THIN-WALLED PRESSURE VESSELS

Cylindrical or spherical pressure vessels are commonly used in industry to serve as boilers or storage tanks. The stresses acting in the wall of these vessels can be analyzed in a simple manner provided it has a *thin wall*, that is, the inner-radius-to-wall-thickness ratio is 10 or more ($r/t \geq 10$). Specifically, when $r/t = 10$ the results of a thin-wall analysis will predict a stress that is approximately 4% *less* than the actual maximum stress in the vessel. For larger r/t ratios this error will be even smaller.

In the following analysis, we will assume the gas pressure in the vessel is the *gage pressure*, that is, it is the pressure *above* atmospheric pressure, since atmospheric pressure is assumed to exist both inside and outside the vessel's wall before the vessel is pressurized.

Cylindrical pressure vessels, such as this gas tank, have semispherical end caps rather than flat ones in order to reduce the stress in the tank.

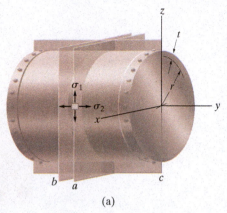

σ_1

σ_2

(a)

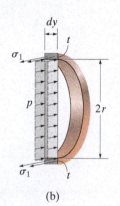

(b)

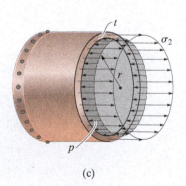

Fig. 8–1

Cylindrical Vessels.

The cylindrical vessel in Fig. 8–1a has a wall thickness t, inner radius r, and is subjected to an internal gas pressure p. To find the **circumferential** or **hoop stress**, we can section the vessel by planes a, b, and c. A free-body diagram of the back segment along with its contained gas is then shown in Fig. 8–1b. Here only the loadings in the x direction are shown. They are caused by the uniform hoop stress σ_1, acting on the vessel's wall, and the pressure acting on the vertical face of the gas. For equilibrium in the x direction, we require

$$\Sigma F_x = 0; \qquad 2[\sigma_1(t\, dy)] - p(2r\, dy) = 0$$

$$\boxed{\sigma_1 = \frac{pr}{t}} \qquad (8\text{–}1)$$

The longitudinal stress can be determined by considering the left portion of section b, Fig. 8–1a. As shown on its free-body diagram, Fig. 8–1c, σ_2 acts uniformly throughout the wall, and p acts on the section of the contained gas. Since the mean radius is approximately equal to the vessel's inner radius, equilibrium in the y direction requires

$$\Sigma F_y = 0; \qquad \sigma_2(2\pi rt) - p(\pi r^2) = 0$$

$$\boxed{\sigma_2 = \frac{pr}{2t}} \qquad (8\text{–}2)$$

For these two equations,

$\sigma_1, \sigma_2 =$ the normal stress in the hoop and longitudinal directions, respectively. Each is assumed to be *constant* throughout the wall of the cylinder, and each subjects the material to tension.

$p =$ the internal gage pressure developed by the contained gas

$r =$ the inner radius of the cylinder

$t =$ the thickness of the wall ($r/t \geq 10$)

By comparison, note that the hoop or circumferential stress is *twice as large* as the longitudinal or axial stress. Consequently, when fabricating cylindrical pressure vessels from rolled-formed plates, it is important that the longitudinal joints be designed to carry twice as much stress as the circumferential joints.

Spherical Vessels.

We can analyze a spherical pressure vessel in a similar manner. If the vessel in Fig. 8–2a is sectioned in half, the resulting free-body diagram is shown in Fig. 8–2b. Like the cylinder, equilibrium in the y direction requires

This thin-walled pipe was subjected to an excessive gas pressure that caused it to rupture in the circumferential or hoop direction. The stress in this direction is twice that in the axial direction as noted by Eqs. 8–1 and 8–2.

$$\Sigma F_y = 0; \qquad\qquad \sigma_2(2\pi r t) - p(\pi r^2) = 0$$

$$\boxed{\sigma_2 = \frac{pr}{2t}} \qquad\qquad (8\text{–}3)$$

This is the same result as that obtained for the longitudinal stress in the cylindrical pressure vessel, although this stress will be the same regardless of the orientation of the hemispheric free-body diagram.

Limitations.

The above analysis indicates that an element of material taken from either a cylindrical or a spherical pressure vessel is subjected to **biaxial stress**, i.e., normal stress existing in only two directions. Actually, however, the pressure also subjects the material to a **radial stress**, σ_3, which acts along a radial line. This stress has a maximum value equal to the pressure p at the interior wall and it decreases through the wall to zero at the exterior surface of the vessel, since the pressure there is zero. For thin-walled vessels, however, we will *ignore* this stress component, since our limiting assumption of $r/t = 10$ results in σ_2 and σ_1 being, respectively, 5 and 10 times *higher* than the maximum radial stress, $(\sigma_3)_{max} = p$. Finally, note that if the vessel is subjected to an *external pressure,* the resulting compressive stresses within the wall may cause the wall to suddenly collapse inward or buckle rather than causing the material to fracture.

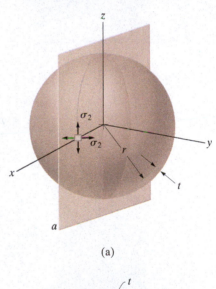

(a)

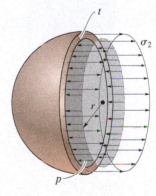

(b)

Fig. 8–2

EXAMPLE 8.1

A cylindrical pressure vessel has an inner diameter of 4 ft and a thickness of $\frac{1}{2}$ in. Determine the maximum internal pressure it can sustain so that neither its circumferential nor its longitudinal stress component exceeds 20 ksi. Under the same conditions, what is the maximum internal pressure that a similar-size spherical vessel can sustain?

SOLUTION

Cylindrical Pressure Vessel. The maximum stress occurs in the circumferential direction. From Eq. 8–1 we have

$$\sigma_1 = \frac{pr}{t}; \qquad\qquad 20 \text{ kip/in}^2 = \frac{p(24 \text{ in.})}{\frac{1}{2} \text{ in.}}$$

$$p = 417 \text{ psi} \qquad\qquad\qquad Ans.$$

Note that when this pressure is reached, from Eq. 8–2, the stress in the longitudinal direction will be $\sigma_2 = \frac{1}{2}(20 \text{ ksi}) = 10 \text{ ksi}$, and the *maximum stress* in the *radial direction* is at the inner wall of the vessel, $(\sigma_3)_{\text{max}} = p = 417 \text{ psi}$. This value is 48 times smaller than the circumferential stress (20 ksi), and as stated earlier, its effect will be neglected.

Spherical Vessel. Here the maximum stress occurs in any two perpendicular directions on an element of the vessel, Fig. 8–2a. From Eq. 8–3, we have

$$\sigma_2 = \frac{pr}{2t}; \qquad\qquad 20 \text{ kip/in}^2 = \frac{p(24 \text{ in.})}{2\left(\frac{1}{2} \text{ in.}\right)}$$

$$p = 833 \text{ psi} \qquad\qquad\qquad Ans.$$

NOTE: Although it is more difficult to fabricate, the spherical pressure vessel will carry twice as much internal pressure as a cylindrical vessel.

PROBLEMS

8–1. A spherical gas tank has an inner radius of $r = 1.5$ m. If it is subjected to an internal pressure of $p = 300$ kPa, determine its required thickness if the maximum normal stress is not to exceed 12 MPa.

8–2. A pressurized spherical tank is made of 0.5-in.-thick steel. If it is subjected to an internal pressure of $p = 200$ psi, determine its outer radius if the maximum normal stress is not to exceed 15 ksi.

8–3. The thin-walled cylinder can be supported in one of two ways as shown. Determine the state of stress in the wall of the cylinder for both cases if the piston P causes the internal pressure to be 65 psi. The wall has a thickness of 0.25 in. and the inner diameter of the cylinder is 8 in.

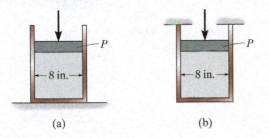

(a)   (b)

Prob. 8–3

*8–4. The tank of the air compressor is subjected to an internal pressure of 90 psi. If the inner diameter of the tank is 22 in., and the wall thickness is 0.25 in., determine the stress components acting at point A. Draw a volume element of the material at this point, and show the results on the element.

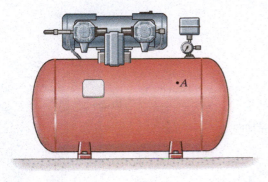

Prob. 8–4

8–5. Air pressure in the cylinder is increased by exerting forces $P = 2$ kN on the two pistons, each having a radius of 45 mm. If the cylinder has a wall thickness of 2 mm, determine the state of stress in the wall of the cylinder.

8–6. Determine the maximum force P that can be exerted on each of the two pistons so that the circumferential stress in the cylinder does not exceed 3 MPa. Each piston has a radius of 45 mm and the cylinder has a wall thickness of 2 mm.

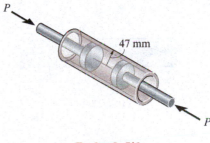

Probs. 8–5/6

8–7. A boiler is constructed of 8-mm-thick steel plates that are fastened together at their ends using a butt joint consisting of two 8-mm cover plates and rivets having a diameter of 10 mm and spaced 50 mm apart as shown. If the steam pressure in the boiler is 1.35 MPa, determine (a) the circumferential stress in the boiler's plate away from the seam, (b) the circumferential stress in the outer cover plate along the rivet line a–a, and (c) the shear stress in the rivets.

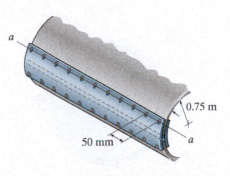

Prob. 8–7

8

*8–8. The steel water pipe has an inner diameter of 12 in. and a wall thickness of 0.25 in. If the valve A is opened and the flowing water has a pressure of 250 psi as it passes point B, determine the longitudinal and hoop stress developed in the wall of the pipe at point B.

8–9. The steel water pipe has an inner diameter of 12 in. and a wall thickness of 0.25 in. If the valve A is closed and the water pressure is 300 psi, determine the longitudinal and hoop stress developed in the wall of the pipe at point B. Draw the state of stress on a volume element located on the wall.

8–11. The gas pipe line is supported every 20 ft by concrete piers and also lays on the ground. If there are rigid retainers at the piers that hold the pipe fixed, determine the longitudinal and hoop stress in the pipe if the temperature rises 60° F from the temperature at which it was installed. The gas within the pipe is at a pressure of 600 lb/in². The pipe has an inner diameter of 20 in. and thickness of 0.25 in. The material is A-36 steel.

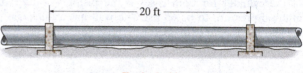

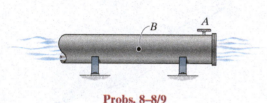

Probs. 8–8/9

Prob. 8–11

8–10. The A-36-steel band is 2 in. wide and is secured around the smooth rigid cylinder. If the bolts are tightened so that the tension in them is 400 lb, determine the normal stress in the band, the pressure exerted on the cylinder, and the distance half the band stretches.

*8–12. A pressure-vessel head is fabricated by welding the circular plate to the end of the vessel as shown. If the vessel sustains an internal pressure of 450 kPa, determine the average shear stress in the weld and the state of stress in the wall of the vessel.

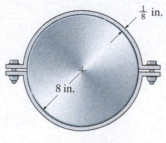

Prob. 8–10

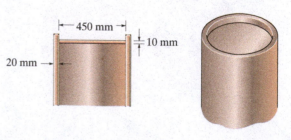

Prob. 8–12

8–13. An A-36-steel hoop has an inner diameter of 23.99 in., thickness of 0.25 in., and width of 1 in. If it and the 24-in.-diameter rigid cylinder have a temperature of 65° F, determine the temperature to which the hoop should be heated in order for it to just slip over the cylinder. What is the pressure the hoop exerts on the cylinder, and the tensile stress in the ring when it cools back down to 65° F?

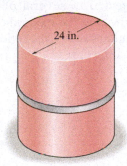

Prob. 8–13

8–14. The ring, having the dimensions shown, is placed over a flexible membrane which is pumped up with a pressure p. Determine the change in the inner radius of the ring after this pressure is applied. The modulus of elasticity for the ring is E.

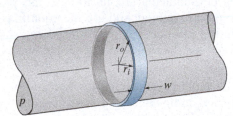

Prob. 8–14

8–15. The inner ring A has an inner radius r_1 and outer radius r_2. The outer ring B has an inner radius r_3 and an outer radius r_4, and $r_2 > r_3$. If the outer ring is heated and then fitted over the inner ring, determine the pressure between the two rings when ring B reaches the temperature of the inner ring. The material has a modulus of elasticity of E and a coefficient of thermal expansion of α.

Prob. 8–15

*__8–16.__ Two hemispheres having an inner radius of 2 ft and wall thickness of 0.25 in. are fitted together, and the inside pressure is reduced to −10 psi. If the coefficient of static friction is $\mu_s = 0.5$ between the hemispheres, determine (a) the torque T needed to initiate the rotation of the top hemisphere relative to the bottom one, (b) the vertical force needed to pull the top hemisphere off the bottom one, and (c) the horizontal force needed to slide the top hemisphere off the bottom one.

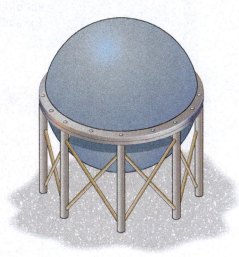

Prob. 8–16

8–17. In order to increase the strength of the pressure vessel, filament winding of the same material is wrapped around the circumference of the vessel as shown. If the pretension in the filament is T and the vessel is subjected to an internal pressure p, determine the hoop stresses in the filament and in the wall of the vessel. Use the free-body diagram shown, and assume the filament winding has a thickness t' and width w for a corresponding length L of the vessel.

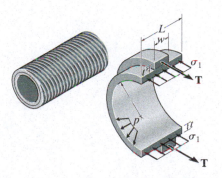

Prob. 8–17

This chimney is subjected to the combined internal loading caused by the wind and the chimney's weight.

8.2   STATE OF STRESS CAUSED BY COMBINED LOADINGS

In the previous chapters we showed how to determine the stress in a member subjected to either an internal axial force, a shear force, a bending moment, or a torsional moment. Most often, however, the cross section of a member will be subjected to several of these loadings simultaneously, and when this occurs, then the method of superposition should be used to determine the resultant stress. The following procedure for analysis provides a method for doing this.

PROCEDURE FOR ANALYSIS

Here it is required that the material be homogeneous and behave in a linear elastic manner. Also, Saint-Venant's principle requires that the stress be determined at a point far removed from any discontinuities in the cross section or points of applied load.

Internal Loading.

- Section the member perpendicular to its axis at the point where the stress is to be determined; and use the equations of equilibrium to obtain the resultant internal normal and shear force components, and the bending and torsional moment components.

- The force components should act through the *centroid* of the cross section, and the moment components should be calculated about *centroidal axes*, which represent the principal axes of inertia for the cross section.

Stress Components.

- Determine the stress component associated with *each* internal loading.

   Normal Force.

   - The normal force is related to a uniform normal-stress distribution determined from $\sigma = N/A$.

Shear Force.

- The shear force is related to a shear-stress distribution determined from the shear formula, $\tau = VQ/It$.

Bending Moment.

- For *straight members* the bending moment is related to a normal-stress distribution that varies linearly from zero at the neutral axis to a maximum at the outer boundary of the member. This stress distribution is determined from the flexure formula, $\sigma = -My/I$. If the member is *curved*, the stress distribution is nonlinear and is determined from $\sigma = My/[Ae(R - y)]$.

Torsional Moment.

- For circular shafts and tubes the torsional moment is related to a shear-stress distribution that varies linearly from zero at the center of the shaft to a maximum at the shaft's outer boundary. This stress distribution is determined from the torsion formula, $\tau = T\rho/J$.

Thin-Walled Pressure Vessels.

- If the vessel is a thin-walled cylinder, the internal pressure p will cause a biaxial state of stress in the material such that the hoop or circumferential stress component is $\sigma_1 = pr/t$, and the longitudinal stress component is $\sigma_2 = pr/2t$. If the vessel is a thin-walled sphere, then the biaxial state of stress is represented by two equivalent components, each having a magnitude of $\sigma_2 = pr/2t$.

Superposition.

- Once the normal and shear stress components for each loading have been calculated, use the principle of superposition and determine the resultant normal and shear stress components.
- Represent the results on an element of material located at a point, or show the results as a distribution of stress acting over the member's cross-sectional area.

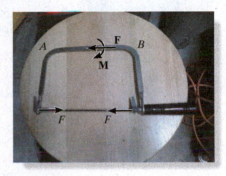

When a pretension force F is developed in the blade of this coping saw, it will produce both a compressive force F and bending moment M at the section AB of the frame. The material must therefore resist the normal stress produced by both of these loadings.

Problems in this section, which involve combined loadings, serve as a basic *review* of the application of the stress equations mentioned above. A thorough understanding of how these equations are applied, as indicated in the previous chapters, is necessary if one is to successfully solve the problems at the end of this section. The following examples should be carefully studied before proceeding to solve the problems.

EXAMPLE 8.2

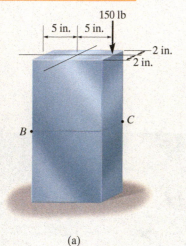

(a)

Fig. 8–3

(b)

A force of 150 lb is applied to the edge of the member shown in Fig. 8–3a. Neglect the weight of the member and determine the state of stress at points B and C.

SOLUTION

Internal Loadings. The member is sectioned through B and C, Fig. 8–3b. For equilibrium at the section there must be an axial force of 150 lb acting through the *centroid* and a bending moment of 750 lb·in. about the centroidal principal axis, Fig. 8–3b.

Stress Components.

Normal Force. The uniform normal-stress distribution due to the normal force is shown in Fig. 8–3c. Here

$$\sigma = \frac{N}{A} = \frac{150 \text{ lb}}{(10 \text{ in.})(4 \text{ in})} = 3.75 \text{ psi}$$

Bending Moment. The normal-stress distribution due to the bending moment is shown in Fig. 8–3d. The maximum stress is

$$\sigma_{\max} = \frac{Mc}{I} = \frac{750 \text{ lb} \cdot \text{in. (5 in.)}}{\frac{1}{12} (4 \text{ in.}) (10 \text{ in.})^3} = 11.25 \text{ psi}$$

Superposition. Algebraically adding the stresses at B and C, we get

$$\sigma_B = -\frac{N}{A} + \frac{Mc}{I} = -3.75 \text{ psi} + 11.25 \text{ psi} = 7.5 \text{ psi} \quad \text{(tension)} \qquad \textit{Ans.}$$

$$\sigma_C = -\frac{N}{A} - \frac{Mc}{I} = -3.75 \text{ psi} - 11.25 \text{ psi} = -15 \text{ psi (compression)}$$

$$\textit{Ans.}$$

NOTE: The resultant stress distribution over the cross section is shown in Fig. 8–3e, where the location of the line of zero stress can be determined by proportional triangles; i.e.,

$$\frac{7.5 \text{ psi}}{x} = \frac{15 \text{ psi}}{(10 \text{ in.} - x)}; \quad x = 3.33 \text{ in.}$$

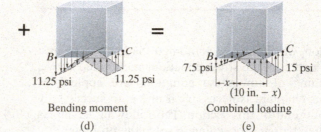

Normal force
(c)

Bending moment
(d)

Combined loading
(e)

(f)   (g)

EXAMPLE 8.3

The gas tank in Fig. 8–4a has an inner radius of 24 in. and a thickness of 0.5 in. If it supports the 1500-lb load at its top, and the gas pressure within it is 2 lb/in², determine the state of stress at point A.

SOLUTION

Internal Loadings. The free-body diagram of the section of the tank above point A is shown in Fig. 8–4b.

Stress Components.

Circumferential Stress. Since $r/t = 24 \text{ in.}/0.5 \text{ in.} = 48 > 10$, the tank is a thin-walled vessel. Applying Eq. 8–1, using the inner radius $r = 24$ in., we have

$$\sigma_1 = \frac{pr}{t} = \frac{2 \text{ lb/in}^2 (24 \text{ in.})}{0.5 \text{ in.}} = 96 \text{ psi} \qquad \textit{Ans.}$$

Longitudinal Stress. Here the wall of the tank uniformly supports the load of 1500 lb (compression) and the pressure stress (tensile). Thus, we have

$$\sigma_2 = -\frac{N}{A} + \frac{pr}{2t} = -\frac{1500 \text{ lb}}{\pi[(24.5 \text{ in.})^2 - (24 \text{ in.})^2]} + \frac{2 \text{ lb/in}^2 (24 \text{ in.})}{2 (0.5 \text{ in.})}$$

$$= 28.3 \text{ psi} \qquad \textit{Ans.}$$

Point A is therefore subjected to the biaxial stress shown in Fig. 8–4c.

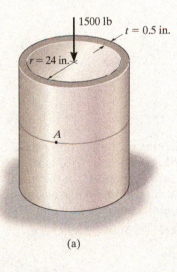

(a)

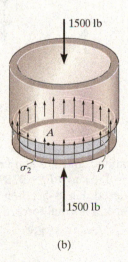

(b)

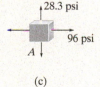

(c)

Fig. 8–4

EXAMPLE   8.4

The member shown in Fig. 8–5a has a rectangular cross section. Determine the state of stress that the loading produces at point C and point D.

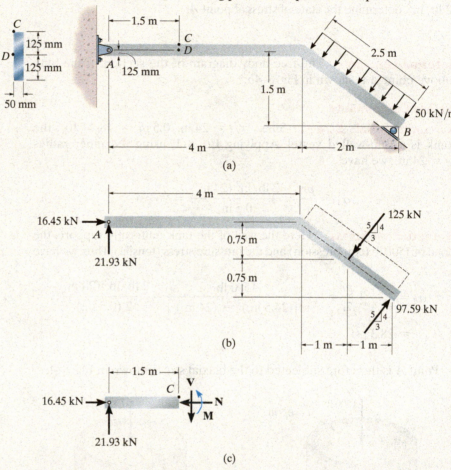

(a)

(b)

(c)

Fig. 8–5

SOLUTION

Internal Loadings. The support reactions on the member have been determined and are shown in Fig. 8–5b. (As a review of statics, apply $\Sigma M_A = 0$ to show $F_B = 97.59$ kN.) If the left segment AC of the member is considered, Fig. 8–5c, then the resultant internal loadings at the section consist of a normal force, a shear force, and a bending moment. They are

$$N = 16.45 \text{ kN} \qquad V = 21.93 \text{ kN} \qquad M = 32.89 \text{ kN} \cdot \text{m}$$

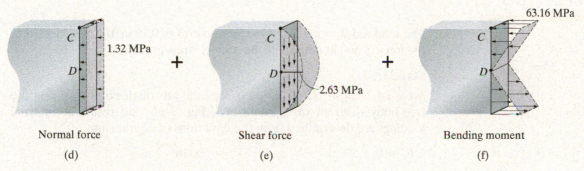

Normal force   Shear force   Bending moment

(d)   (e)   (f)

Fig. 8–5 (cont.)

Stress Components at C.

Normal Force.   The uniform normal-stress distribution acting over the cross section is produced by the normal force, Fig. 8–5d. At point C,

$$\sigma_C = \frac{N}{A} = \frac{16.45(10^3) \text{ N}}{(0.050 \text{ m})(0.250 \text{ m})} = 1.32 \text{ MPa}$$

Shear Force.   Here the area $A' = 0$, since point C is located at the top of the member. Thus $Q = \bar{y}'A' = 0$, Fig. 8–5e. The shear stress is therefore

$$\tau_C = 0$$

Bending Moment.   Point C is located at $y = c = 0.125$ m from the neutral axis, so the bending stress at C, Fig. 8–5f, is

$$\sigma_C = \frac{Mc}{I} = \frac{(32.89(10^3) \text{ N} \cdot \text{m})(0.125 \text{ m})}{\left[\frac{1}{12}(0.050 \text{ m})(0.250 \text{ m})^3\right]} = 63.16 \text{ MPa}$$

Superposition.   There is no shear-stress component. Adding the normal stresses gives a compressive stress at C having a value of

$$\sigma_C = 1.32 \text{ MPa} + 63.16 \text{ MPa} = 64.5 \text{ MPa} \qquad \textit{Ans.}$$

This result, acting on an element at C, is shown in Fig. 8–5g.

64.5 MPa

(g)

Stress Components at D.

Normal Force.   This is the same as at C, $\sigma_D = 1.32$ MPa, Fig. 8–5d.

Shear Force.   Since D is at the neutral axis, and the cross section is rectangular, we can use the special form of the shear formula, Fig. 8–5e.

$$\tau_D = 1.5\frac{V}{A} = 1.5\frac{21.93(10^3) \text{ N}}{(0.25 \text{ m})(0.05 \text{ m})} = 2.63 \text{ MPa} \qquad \textit{Ans.}$$

Bending Moment.   Here D is on the neutral axis and so $\sigma_D = 0$.

2.63 MPa
1.32 MPa

Superposition.   The resultant stress on the element is shown in Fig. 8–5h.

(h)

8

EXAMPLE 8.5

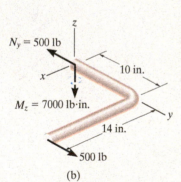

(a)

(b)

The solid rod shown in Fig. 8–6a has a radius of 0.75 in. If it is subjected to the force of 500 lb, determine the state of stress at point A.

SOLUTION

Internal Loadings.   The rod is sectioned through point A. Using the free-body diagram of segment AB, Fig. 8–6b, the resultant internal loadings are determined from the equations of equilibrium.

$$\Sigma F_y = 0;\quad 500\ \text{lb} - N_y = 0;\quad N_y = 500\ \text{lb}$$

$$\Sigma M_z = 0;\quad 500\ \text{lb}(14\ \text{in.}) - M_z = 0;\quad M_z = 7000\ \text{lb} \cdot \text{in.}$$

In order to better "visualize" the stress distributions due to these loadings, we can consider the *equal but opposite resultants* acting on segment AC, Fig. 8–6c.

Stress Components.

Normal Force.   The normal-stress distribution is shown in Fig. 8–6d. For point A, we have

$$(\sigma_A)_y = \frac{N}{A} = \frac{500\ \text{lb}}{\pi(0.75\ \text{in.})^2} = 283\ \text{psi} = 0.283\ \text{ksi}$$

Bending Moment.   For the moment, $c = 0.75$ in., so the bending stress at point A, Fig. 8–6e, is

$$(\sigma_A)_y = \frac{Mc}{I} = \frac{7000\ \text{lb} \cdot \text{in.}(0.75\ \text{in.})}{\left[\frac{1}{4}\pi(0.75\ \text{in.})^4\right]}$$

$$= 21\,126\ \text{psi} = 21.13\ \text{ksi}$$

Superposition.   When the above results are superimposed, it is seen that an element at A, Fig. 8–6f, is subjected to the normal stress

$$(\sigma_A)_y = 0.283\ \text{ksi} + 21.13\ \text{ksi} = 21.4\ \text{ksi}\qquad \textit{Ans.}$$

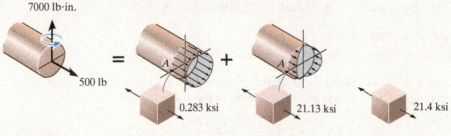

Normal force   Bending moment

(c)   (d)   (e)   (f)

Fig. 8–6

EXAMPLE 8.6

The solid rod shown in Fig. 8–7a has a radius of 0.75 in. If it is subjected to the force of 800 lb, determine the state of stress at point A.

SOLUTION

Internal Loadings. The rod is sectioned through point A. Using the free-body diagram of segment AB, Fig. 8–7b, the resultant internal loadings are determined from the equations of equilibrium. Take a moment to verify these results. The *equal but opposite resultants* are shown acting on segment AC, Fig. 8–7c.

$$\Sigma F_z = 0; \quad V_z - 800 \text{ lb} = 0; \quad V_z = 800 \text{ lb}$$
$$\Sigma M_x = 0; \quad M_x - 800 \text{ lb}(10 \text{ in.}) = 0; \quad M_x = 8000 \text{ lb} \cdot \text{in.}$$
$$\Sigma M_y = 0; \quad -M_y + 800 \text{ lb}(14 \text{ in.}) = 0; \quad M_y = 11\,200 \text{ lb} \cdot \text{in.}$$

Stress Components.
Shear Force. The shear-stress distribution is shown in Fig. 8–7d. For point A, Q is determined from the grey shaded *semicircular* area. Using the table on the inside front cover, we have

$$Q = \bar{y}'A' = \frac{4(0.75 \text{ in.})}{3\pi}\left[\frac{1}{2}\pi(0.75 \text{ in.})^2\right] = 0.28125 \text{ in}^3$$

so that

$$(\tau_{yz})_A = \frac{VQ}{It} = \frac{800 \text{ lb}(0.28125 \text{ in}^3)}{\left[\frac{1}{4}\pi(0.75 \text{ in.})^4\right]2(0.75 \text{ in.})}$$
$$= 604 \text{ psi} = 0.604 \text{ ksi}$$

Bending Moment. Since point A lies on the neutral axis, Fig. 8–7e, the bending stress is

$$\sigma_A = 0$$

Torque. At point A, $\rho_A = c = 0.75$ in., Fig. 8–7f. Thus the shear stress is

$$(\tau_{yz})_A = \frac{Tc}{J} = \frac{11\,200 \text{ lb} \cdot \text{in.}(0.75 \text{ in.})}{\left[\frac{1}{2}\pi(0.75 \text{ in.})^4\right]} = 16\,901 \text{ psi} = 16.90 \text{ ksi}$$

Superposition. Here the element of material at A is subjected only to a shear stress component, Fig. 8–7g, where

$$(\tau_{yz})_A = 0.604 \text{ ksi} + 16.90 \text{ ksi} = 17.5 \text{ ksi} \qquad Ans.$$

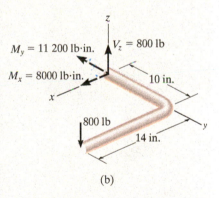

(a)

(b)

Fig. 8–7

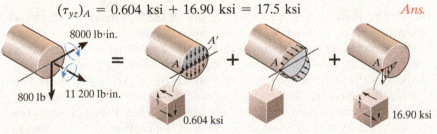

(c)

Shear force   Bending moment   Torsional moment

(d)   (e)   (f)   (g)

EXAMPLE 8.7

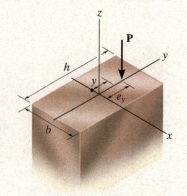

(a)

||

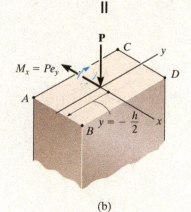

(b)

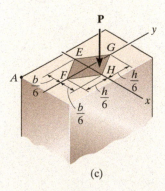

(c)

Fig. 8–8

A rectangular block has a negligible weight and is subjected to a vertical force **P**, Fig. 8–8a. (a) Determine the range of values for the eccentricity e_y of the load along the y axis, so that it does not cause any tensile stress in the block. (b) Specify the region on the cross section where **P** may be applied without causing tensile stress.

SOLUTION

Part (a).   When **P** is moved to the centroid of the cross section, Fig. 8–8b, it is necessary to add a bending moment $M_x = Pe_y$ in order to maintain a statically equivalent loading. The combined normal stress at any coordinate location y on the cross section caused by these two loadings is therefore

$$\sigma = -\frac{P}{A} - \frac{(Pe_y)y}{I_x} = -\frac{P}{A}\left(1 + \frac{Ae_yy}{I_x}\right)$$

Here the negative sign indicates compressive stress. For positive e_y, Fig. 8–8a, the *smallest* compressive stress will occur along edge AB, where $y = -h/2$, Fig. 8–8b. (By inspection, **P** causes compression there, but $\mathbf{M}_x$ causes tension.) Hence,

$$\sigma_{\min} = -\frac{P}{A}\left(1 - \frac{Ae_yh}{2I_x}\right)$$

This stress will remain negative, i.e., compressive, provided the term in parentheses is positive; i.e.,

$$1 > \frac{Ae_yh}{2I_x}$$

Since $A = bh$ and $I_x = \frac{1}{12}bh^3$, then

$$1 > \frac{6e_y}{h} \quad \text{or} \quad e_y < \frac{1}{6}h \qquad \textit{Ans.}$$

In other words, if $-\frac{1}{6}h \le e_y \le \frac{1}{6}h$, the stress in the block along edge AB or CD will be zero or remain *compressive*.

NOTE: This is sometimes referred to as the "***middle-third rule.***" It is very important to keep this rule in mind when loading columns or arches having a rectangular cross section and made of material such as stone or concrete, which can support little or no tensile stress. We can extend this analysis in the same way by placing **P** along the x axis in Fig. 8–8b. The result will produce a shaded parallelogram, shown in Fig. 8–8c. This region is referred to as the core or ***kern*** of the section. When **P** is applied within the kern, the normal stress at the corners of the cross section will always be compressive.

PRELIMINARY PROBLEMS

P8–1. In each case, determine the internal loadings that act on the indicated section. Show the results on the left segment.

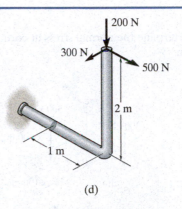

(d)   **P8–1**

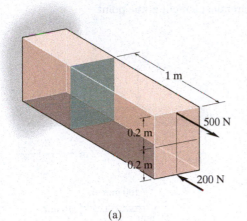

(a)

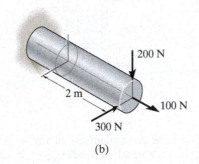

(b)

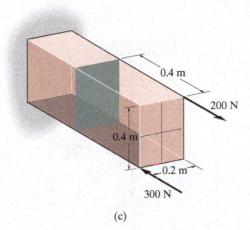

(c)

P8–2. The internal loadings act on the section. Show the stress that each of these loads produce on differential elements located at point *A* and point *B*.

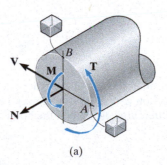

(a)

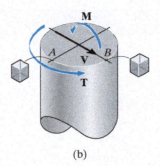

(b)   **P8–2**

FUNDAMENTAL PROBLEMS

F8–1. Determine the normal stress at corners A and B of the column.

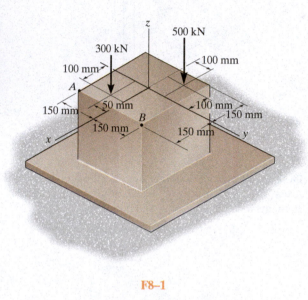

F8–1

F8–2. Determine the state of stress at point A on the cross section at section a–a of the cantilever beam. Show the results in a differential element at the point.

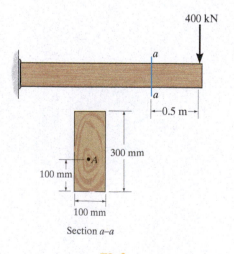

Section a–a

F8–2

F8–3. Determine the state of stress at point A on the cross section of the beam at section a–a. Show the results in a differential element at the point.

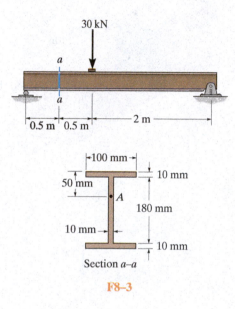

Section a–a

F8–3

F8–4. Determine the magnitude of the load P that will cause a maximum normal stress of $\sigma_{max} = 30$ ksi in the link along section a–a.

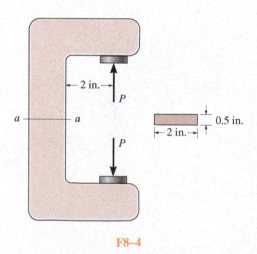

F8–4

F8–5. The beam has a rectangular cross section and is subjected to the loading shown. Determine the state of stress at point B. Show the results in a differential element at the point.

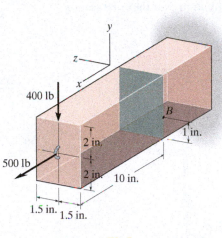

F8–5

F8–6. Determine the state of stress at point A on the cross section of the pipe assembly at section a–a. Show the results in a differential element at the point.

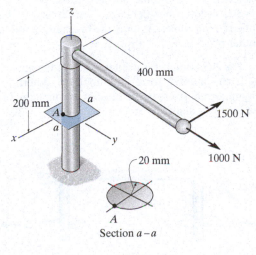

Section a–a

F8–6

F8–7. Determine the state of stress at point A on the cross section of the pipe at section a–a. Show the results in a differential element at the point.

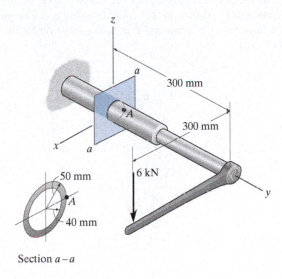

Section a–a

F8–7

F8–8. Determine the state of stress at point A on the cross section of the shaft at section a–a. Show the results in a differential element at the point.

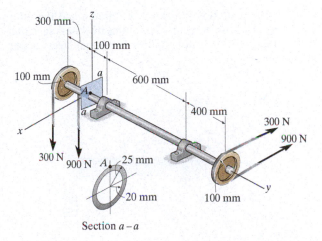

Section a–a

F8–8

PROBLEMS

8–18. Determine the shortest distance d to the edge of the plate at which the force **P** can be applied so that it produces no compressive stresses in the plate at section a–a. The plate has a thickness of 10 mm and **P** acts along the centerline of this thickness.

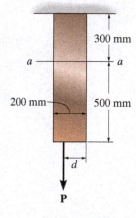

Prob. 8–18

8–19. Determine the maximum distance d to the edge of the plate at which the force **P** can be applied so that it produces no compressive stresses on the plate at section a–a. The plate has a thickness of 20 mm and **P** acts along the centerline of this thickness.

*8–20. The plate has a thickness of 20 mm and the force $P = 3$ kN acts along the centerline of this thickness such that $d = 150$ mm. Plot the distribution of normal stress acting along section a–a.

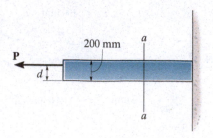

Probs. 8–19/20

8–21. If the load has a weight of 600 lb, determine the maximum normal stress on the cross section of the supporting member at section a–a. Also, plot the normal-stress distribution over the cross section.

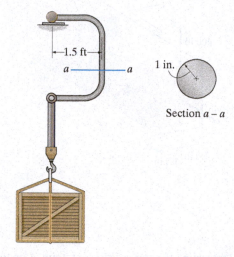

Prob. 8–21

8–22. The steel bracket is used to connect the ends of two cables. If the allowable normal stress for the steel is $\sigma_{allow} = 30$ ksi, determine the largest tensile force P that can be applied to the cables. Assume the bracket is a rod having a diameter of 1.5 in.

8–23. The steel bracket is used to connect the ends of two cables. If the applied force $P = 1.50$ kip, determine the maximum normal stress in the bracket. Assume the bracket is a rod having a diameter of 1.5 in.

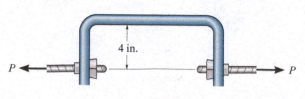

Probs. 8–22/23

*8–24. The column is built up by gluing the two boards together. Determine the maximum normal stress on the cross section when the eccentric force of $P = 50$ kN is applied.

8–25. The column is built up by gluing the two boards together. If the wood has an allowable normal stress of $\sigma_{allow} = 6$ MPa, determine the maximum allowable eccentric force **P** that can be applied to the column.

*8–28. The joint is subjected to the force system shown. Sketch the normal-stress distribution acting over section a–a if the member has a rectangular cross section of width 0.5 in. and thickness 1 in.

8–29. The joint is subjected to the force system shown. Determine the state of stress at points A and B, and sketch the results on differential elements located at these points. The member has a rectangular cross-sectional area of width 0.5 in. and thickness 1 in.

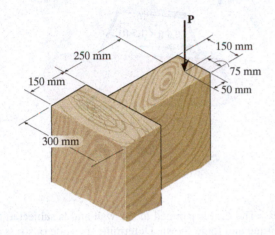

Probs. 8–24/25

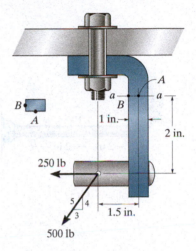

Probs. 8–28/29

8–26. The screw of the clamp exerts a compressive force of 500 lb on the wood blocks. Determine the maximum normal stress along section a–a. The cross section is rectangular, 0.75 in. by 0.50 in.

8–27. The screw of the clamp exerts a compressive force of 500 lb on the wood blocks. Sketch the stress distribution along section a–a of the clamp. The cross section is rectangular, 0.75 in. by 0.50 in.

8–30. The rib-joint pliers are used to grip the smooth pipe C. If the force of 100 N is applied to the handles, determine the state of stress at points A and B on the cross section of the jaw at section a–a. Indicate the results on an element at each point.

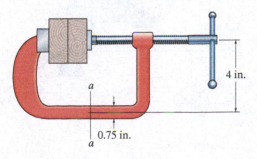

Probs. 8–26/27

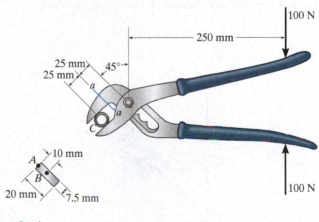

Prob. 8–30

8–31. The $\frac{1}{2}$-in.-diameter bolt hook is subjected to the load of $F = 150$ lb. Determine the stress components at point A on the shank. Show the result on a volume element located at this point.

*8–32. The $\frac{1}{2}$-in.-diameter bolt hook is subjected to the load of $F = 150$ lb. Determine the stress components at point B on the shank. Show the result on a volume element located at this point.

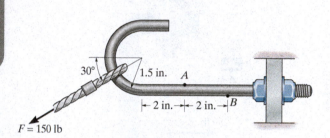

Probs. 8–31/32

8–33. The block is subjected to the eccentric load shown. Determine the normal stress developed at points A and B. Neglect the weight of the block.

8–34. The block is subjected to the eccentric load shown. Sketch the normal-stress distribution acting over the cross section at section a–a. Neglect the weight of the block.

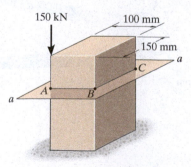

Probs. 8–33/34

8–35. The spreader bar is used to lift the 2000-lb tank. Determine the state of stress at points A and B, and indicate the results on a differential volume element.

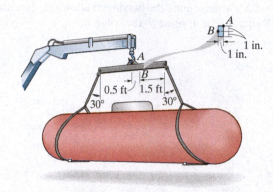

Prob. 8–35

*8–36. The drill is jammed in the wall and is subjected to the torque and force shown. Determine the state of stress at point A on the cross section of the drill bit at section a–a.

8–37. The drill is jammed in the wall and is subjected to the torque and force shown. Determine the state of stress at point B on the cross section of the drill bit at section a–a.

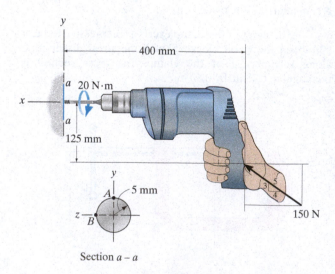

Section a – a

Probs. 8–36/37

8–38. The frame supports the distributed load shown. Determine the state of stress acting at point D. Show the results on a differential element at this point.

8–39. The frame supports the distributed load shown. Determine the state of stress acting at point E. Show the results on a differential element at this point.

8–42. The beveled gear is subjected to the loads shown. Determine the stress components acting on the shaft at point A, and show the results on a volume element located at this point. The shaft has a diameter of 1 in. and is fixed to the wall at C.

8–43. The beveled gear is subjected to the loads shown. Determine the stress components acting on the shaft at point B, and show the results on a volume element located at this point. The shaft has a diameter of 1 in. and is fixed to the wall at C.

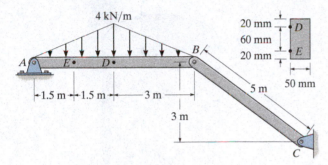

Probs. 8–38/39

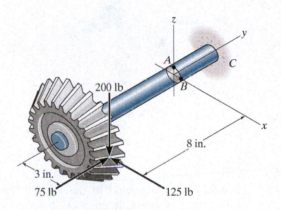

Probs. 8–42/43

*****8–40.** The rod has a diameter of 40 mm. If it is subjected to the force system shown, determine the stress components that act at point A, and show the results on a volume element located at this point.

8–41. Solve Prob. 8–40 for point B.

*****8–44.** Determine the normal stress developed at points A and B. Neglect the weight of the block.

8–45. Sketch the normal-stress distribution acting over the cross section at section a–a. Neglect the weight of the block.

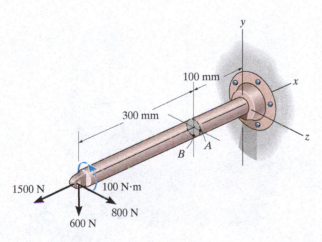

Probs. 8–40/41

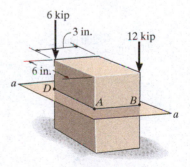

Probs. 8–44/45

8–46. The man has a mass of 100 kg and center of mass at G. If he holds himself in the position shown, determine the maximum tensile and compressive stress developed in the curved bar at section a–a. He is supported uniformly by two bars, each having a diameter of 25 mm. Assume the floor is smooth. Use the curved-beam formula to calculate the bending stress.

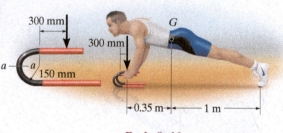

Prob. 8–46

8–47. The solid rod is subjected to the loading shown. Determine the state of stress at point A, and show the results on a differential volume element located at this point.

*8–48. The solid rod is subjected to the loading shown. Determine the state of stress at point B, and show the results on a differential volume element at this point.

8–49. The solid rod is subjected to the loading shown. Determine the state of stress at point C, and show the results on a differential volume element at this point.

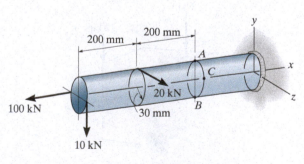

Probs. 8–47/48/49

8–50. The post has a circular cross section of radius c. Determine the maximum radius e at which the load $\mathbf{P}$ can be applied so that no part of the post experiences a tensile stress. Neglect the weight of the post.

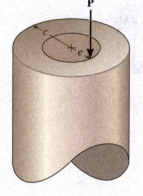

Prob. 8–50

8–51. The post having the dimensions shown is subjected to the load $\mathbf{P}$. Specify the region to which this load can be applied without causing tensile stress at points $A, B, C,$ and D.

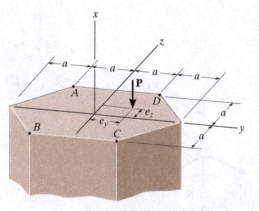

Prob. 8–51

*8–52. The vertebra of the spinal column can support a maximum compressive stress of σ_{max}, before undergoing a compression fracture. Determine the smallest force P that can be applied to a vertebra, if we assume this load is applied at an eccentric distance e from the centerline of the bone, and the bone remains elastic. Model the vertebra as a hollow cylinder with an inner radius r_i and outer radius r_o.

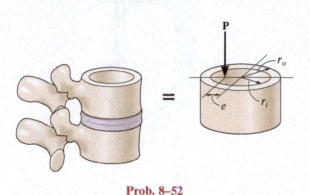

Prob. 8–52

8–53. The coiled spring is subjected to a force P. If we assume the shear stress caused by the shear force at any vertical section of the coil wire to be uniform, show that the maximum shear stress in the coil is $\tau_{max} = P/A + PRr/J$, where J is the polar moment of inertia of the coil wire and A is its cross-sectional area.

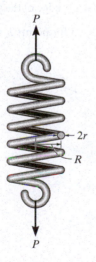

Prob. 8–53

8–54. The frame supports a centrally applied distributed load of 1.8 kip/ft. Determine the state of stress at points A and B on member CD and indicate the results on a volume element located at each of these points. The pins at C and D are at the same location as the neutral axis for the cross section.

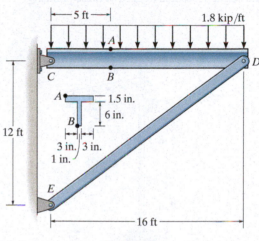

Prob. 8–54

8–55. The 1-in.-diameter rod is subjected to the loads shown. Determine the state of stress at point A, and show the results on a differential volume element located at this point.

*8–56. The 1-in.-diameter rod is subjected to the loads shown. Determine the state of stress at point B, and show the results on a differential volume element located at this point.

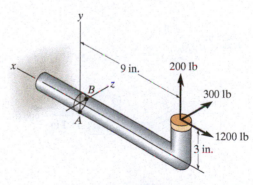

Probs. 8–55/56

8–57. The sign is subjected to the uniform wind loading. Determine the stress components at points A and B on the 100-mm-diameter supporting post. Show the results on a volume element located at each of these points.

8–58. The sign is subjected to the uniform wind loading. Determine the stress components at points C and D on the 100-mm-diameter supporting post. Show the results on a volume element located at each of these points.

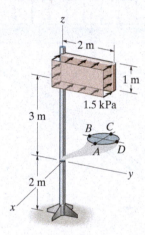

Probs. 8–57/58

8–59. The bearing pin supports the load of 900 lb. Determine the stress components in the support member at point A. Represent the state of stress at point A with a differential element.

*8–60. The bearing pin supports the load of 900 lb. Determine the stress components in the support member at point B. Represent the state of stress at point B with a differential element.

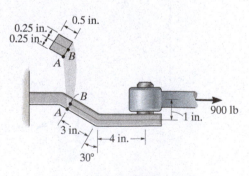

Probs. 8–59/60

8–61. The C-frame is used in a riveting machine. If the force at the ram on the clamp at D is $P = 8$ kN, sketch the stress distribution acting over the section a–a.

8–62. Determine the maximum ram force P that can be applied to the clamp at D if the allowable normal stress for the material is $\sigma_{allow} = 180$ MPa.

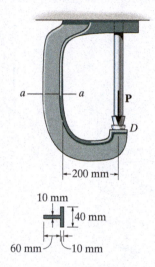

Probs. 8–61/62

8–63. The uniform sign has a weight of 1500 lb and is supported by the pipe AB, which has an inner radius of 2.75 in. and an outer radius of 3.00 in. If the face of the sign is subjected to a uniform wind pressure of $p = 150$ lb/ft^2, determine the state of stress at points C and D. Show the results on a differential volume element located at each of these points. Neglect the thickness of the sign, and assume that it is supported along the outside edge of the pipe.

*8–64. Solve Prob. 8–63 for points E and F.

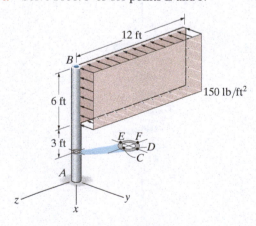

Probs. 8–63/64

8–65. The pin support is made from a steel rod and has a diameter of 20 mm. Determine the stress components at points A and B and represent the results on a volume element located at each of these points.

8–66. Solve Prob. 8–65 for points C and D.

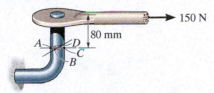

Probs. 8–65/66

8–67. The handle of the press is subjected to a force of 20 lb. Due to internal gearing, this causes the block to be subjected to a compressive force of 80 lb. Determine the normal-stress acting in the frame at points along the outside flanges A and B. Use the curved-beam formula to calculate the bending stress.

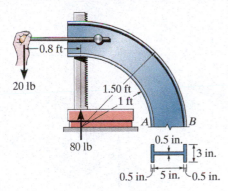

Prob. 8–67

*8–68. The bar has a diameter of 40 mm. Determine the state of stress at point A and show the results on a differential volume element located at this point.

8–69. Solve Prob. 8–68 for point B.

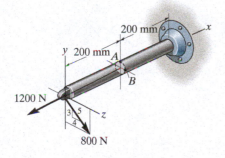

Probs. 8–68/69

8–70. The $\frac{3}{4}$-in.-diameter shaft is subjected to the loading shown. Determine the stress components at point A. Sketch the results on a volume element located at this point. The journal bearing at C can exert only force components $\mathbf{C}_y$ and $\mathbf{C}_z$ on the shaft, and the thrust bearing at D can exert force components $\mathbf{D}_x$, $\mathbf{D}_y$, and $\mathbf{D}_z$ on the shaft.

8–71. Solve Prob. 8–70 for the stress components at point B.

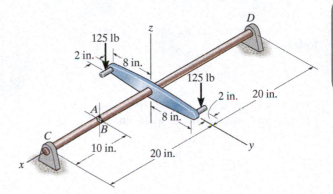

Probs. 8–70/71

*8–72. The hook is subjected to the force of 80 lb. Determine the state of stress at point A at section a–a. The cross section is circular and has a diameter of 0.5 in. Use the curved-beam formula to calculate the bending stress.

8–73. The hook is subjected to the force of 80 lb. Determine the state of stress at point B at section a–a. The cross section has a diameter of 0.5 in. Use the curved-beam formula to calculate the bending stress.

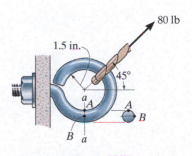

Probs. 8–72/73

CHAPTER REVIEW

A pressure vessel is considered to have a thin wall provided $r/t \geq 10$. If the vessel contains gas having a gage pressure p, then for a cylindrical vessel, the circumferential or hoop stress is

$$\sigma_1 = \frac{pr}{t}$$

This stress is twice as great as the longitudinal stress,

$$\sigma_2 = \frac{pr}{2t}$$

Thin-walled spherical vessels have the same stress within their walls in all directions. It is

$$\sigma_1 = \sigma_2 = \frac{pr}{2t}$$

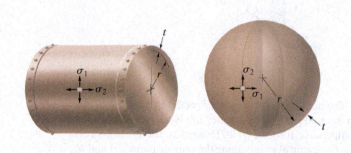

Superposition of stress components can be used to determine the normal and shear stress at a point in a member subjected to a combined loading. To do this, it is first necessary to determine the resultant axial and shear forces and the resultant torsional and bending moments at the section where the point is located. Then the normal and shear stress resultant components at the point are determined by algebraically adding the normal and shear stress components of each loading.

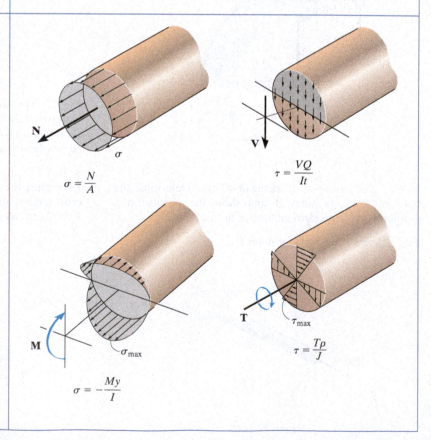

$$\sigma = \frac{N}{A}$$

$$\tau = \frac{VQ}{It}$$

$$\sigma = -\frac{My}{I}$$

$$\tau = \frac{T\rho}{J}$$

CONCEPTUAL PROBLEMS

C8–1. Explain why failure of this garden hose occurred near its end and why the tear occurred along its length. Use numerical values to explain your result. Assume the water pressure is 30 psi.

C8–1

C8–2. This open-ended silo contains granular material. It is constructed from wood slats and held together with steel bands. Explain, using numerical values, why the bands are not spaced evenly along the height of the cylinder. Also, how would you find this spacing if each band is to be subjected to the same stress?

C8–2

C8–3. Unlike the turnbuckle at B, which is connected along the axis of the rod, the one at A has been welded to the edges of the rod, and so it will be subjected to additional stress. Use the same numerical values for the tensile load in each rod and the rod's diameter, and compare the stress in each rod.

C8–3

C8–4. A constant wind blowing against the side of this chimney has caused creeping strains in the mortar joints, such that the chimney has a noticeable deformation. Explain how to obtain the stress distribution over a section at the base of the chimney, and sketch this distribution over the section.

C8–4

REVIEW PROBLEMS

R8–1. The eye hook has the dimensions shown. If it supports a cable loading of 800 lb, determine the maximum normal stress at section a–a and sketch the stress distribution acting over the cross section. Use the curved-beam formula to calculate the bending stress.

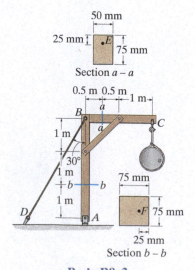

Prob. R8–1

R8–2. The 20-kg drum is suspended from the hook mounted on the wooden frame. Determine the state of stress at point E on the cross section of the frame at section a–a. Indicate the results on an element.

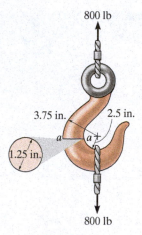

Prob. R8–2

R8–3. The 20-kg drum is suspended from the hook mounted on the wooden frame. Determine the state of stress at point F on the cross section of the frame at section b–b. Indicate the results on an element.

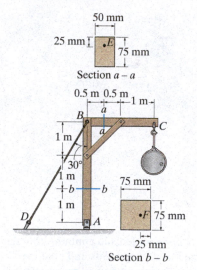

Prob. R8–3

*R8–4. The gondola and passengers have a weight of 1500 lb and center of gravity at G. The suspender arm AE has a square cross-sectional area of 1.5 in. by 1.5 in., and is pin connected at its ends A and E. Determine the largest tensile stress developed in regions AB and DC of the arm.

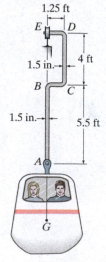

Prob. R8–4

R8–5. If the cross section of the femur at section *a–a* can be approximated as a circular tube as shown, determine the maximum normal stress developed on the cross section at section *a–a* due to the load of 75 lb.

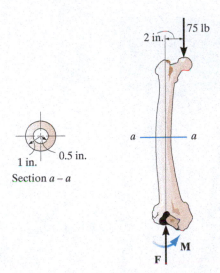

Section *a – a*

Prob. R8–5

R8–6. A bar having a square cross section of 30 mm by 30 mm is 2 m long and is held upward. If it has a mass of 5 kg/m, determine the largest angle θ, measured from the vertical, at which it can be supported before it is subjected to a tensile stress along its axis near the grip.

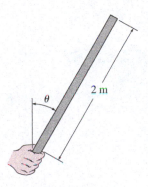

2 m

Prob. R8–6

R8–7. The wall hanger has a thickness of 0.25 in. and is used to support the vertical reactions of the beam that is loaded as shown. If the load is transferred uniformly to each strap of the hanger, determine the state of stress at points *C* and *D* on the strap at *A*. Assume the vertical reaction **F** at this end acts in the center and on the edge of the bracket as shown.

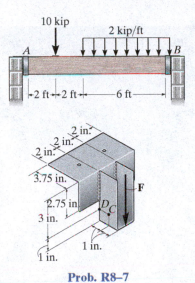

Prob. R8–7

*R8–8. The wall hanger has a thickness of 0.25 in. and is used to support the vertical reactions of the beam that is loaded as shown. If the load is transferred uniformly to each strap of the hanger, determine the state of stress at points *C* and *D* on the strap at *B*. Assume the vertical reaction **F** at this end acts in the center and on the edge of the bracket as shown.

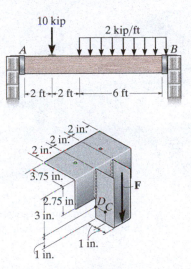

Prob. R8–8

CHAPTER 9

(© R.G. Henry/Fotolia)

These turbine blades are subjected to a complex pattern of stress. For design it is necessary to determine where and in what direction the maximum stress occurs.

STRESS TRANSFORMATION

CHAPTER OBJECTIVES

- In this chapter, we will show how to transform the stress components acting on an element at a point into components acting on a corresponding element having a different orientation. Once the method for doing this is established, we will then be able to find the maximum normal and maximum shear stress at the point, and find the orientation of the elements upon which they act.

9.1 PLANE-STRESS TRANSFORMATION

It was shown in Sec. 1.3 that the general state of stress at a point is characterized by *six* normal and shear-stress components, shown in Fig. 9–1a. This state of stress, however, is not often encountered in engineering practice. Instead, most loadings are coplanar, and so the stress these loadings produce can be analyzed in a *single plane*. When this is the case, the material is then said to be subjected to **plane stress**.

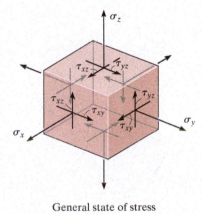

General state of stress

(a)

Fig. 9–1

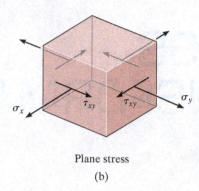

Plane stress

(b)

Fig. 9–1 (cont.)

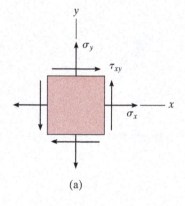

(a)

||

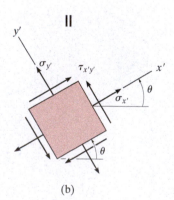

(b)

Fig. 9–2

The general state of plane stress at a point, shown in Fig. 9–1b, is therefore represented by a combination of two normal-stress components, σ_x, σ_y, and one shear-stress component, τ_{xy}, which act on only four faces of the element. For convenience, in this text we will view this state of stress in the x–y plane, as shown in Fig. 9–2a. Realize, however, that if this state of stress is produced on an element having a *different orientation* θ, as in Fig. 9–2b, then it will be subjected to three *different* stress components, $\sigma_{x'}$, $\sigma_{y'}$, $\tau_{x'y'}$, measured relative to the x', y' axes. In other words, *the state of plane stress at the point is uniquely represented by two normal-stress components and one shear-stress component acting on an element. To be equivalent, these three components will be different for each specific orientation θ of the element at the point.*

If these three stress components act on the element in Fig. 9–2a, we will now show what their values will have to be when they act on the element in Fig. 9–2b. This is similar to knowing the two force components $\mathbf{F}_x$ and $\mathbf{F}_y$ directed along the x, y axes, and then finding the force components $\mathbf{F}_{x'}$ and $\mathbf{F}_{y'}$ directed along the x', y' axes, so they produce the *same* resultant force. The transformation of force must only account for the force component's magnitude and direction. The transformation of stress components, however, is more difficult since it must account for the magnitude and direction of each stress *and* the orientation of the area upon which it acts.

PROCEDURE FOR ANALYSIS

If the state of stress at a point is known for a given orientation of an element, Fig. 9–3a, then the state of stress on an element having some other orientation θ, Fig. 9–3b, can be determined as follows.

- The normal and shear stress components $\sigma_{x'}$, $\tau_{x'y'}$ acting on the $+x'$ face of the element, Fig. 9–3b, can be determined from an arbitrary section of the element in Fig. 9–3a as shown in Fig. 9–3c. If the sectioned area is ΔA, then the adjacent areas of the segment will be $\Delta A \sin \theta$ and $\Delta A \cos \theta$.

- Draw the *free-body diagram* of the segment, which requires showing the *forces* that act on the segment, Fig. 9–3d. This is done by multiplying the stress components on each face by the area upon which they act.

- When $\Sigma F_{x'} = 0$ is applied to the free-body diagram, the area ΔA will cancel out of each term and a *direct* solution for $\sigma_{x'}$ will be possible. Likewise, $\Sigma F_{y'} = 0$ will yield $\tau_{x'y'}$.

- If $\sigma_{y'}$, acting on the $+y'$ face of the element in Fig. 9–3b, is to be determined, then it is necessary to consider an arbitrary segment of the element as shown in Fig. 9–3e. Applying $\Sigma F_{y'} = 0$ to its free-body diagram will give $\sigma_{y'}$.

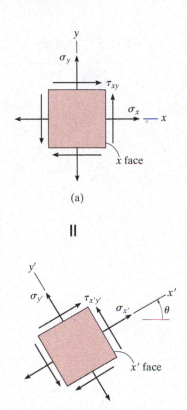

(a)

$\|$

(b)

9

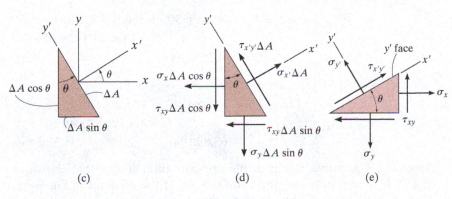

(c)   (d)   (e)

Fig. 9–3

EXAMPLE 9.1

The state of plane stress at a point on the surface of the airplane fuselage is represented on the element oriented as shown in Fig. 9–4a. Represent the state of stress at the point on an element that is oriented 30° clockwise from this position.

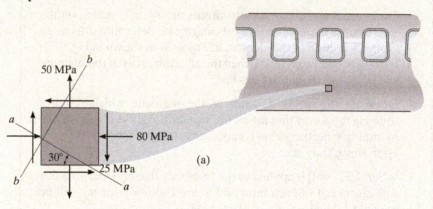

(a)

SOLUTION

The rotated element is shown in Fig. 9–4d. To obtain the stress components on this element we will first section the element in Fig. 9–4a by the line a–a. The bottom segment is removed, and assuming the sectioned (inclined) plane has an area ΔA, the horizontal and vertical planes have the areas shown in Fig. 9–4b. The free-body diagram of this segment is shown in Fig. 9–4c. Notice that the sectioned x' face is defined by the *outward normal x' axis*, and the y' axis is *along* the face.

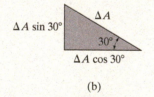

(b)

Equilibrium. If we apply the equations of force equilibrium in the x' and y' directions, not the x and y directions, we will be able to obtain *direct solutions* for $\sigma_{x'}$ and $\tau_{x'y'}$.

$$+\nearrow\Sigma F_{x'} = 0;\qquad \sigma_{x'}\Delta A - (50\ \Delta A\cos 30°)\cos 30°$$
$$+ (25\ \Delta A\cos 30°)\sin 30° + (80\ \Delta A\sin 30°)\sin 30°$$
$$+ (25\ \Delta A\sin 30°)\cos 30° = 0$$
$$\sigma_{x'} = -4.15\ \text{MPa}\qquad\qquad Ans.$$

$$+\nwarrow\Sigma F_{y'} = 0;\qquad \tau_{x'y'}\Delta A - (50\ \Delta A\cos 30°)\sin 30°$$
$$- (25\ \Delta A\cos 30°)\cos 30° - (80\ \Delta A\sin 30°)\cos 30°$$
$$+ (25\ \Delta A\sin 30°)\sin 30° = 0$$
$$\tau_{x'y'} = 68.8\ \text{MPa}\qquad\qquad Ans.$$

Since $\sigma_{x'}$ is negative, it acts in the opposite direction of that shown in Fig. 9–4c. The results are shown on the *top* of the element in Fig. 9–4d, since this surface is the one considered in Fig. 9–4c.

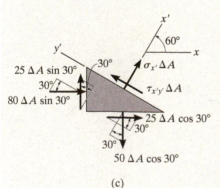

(c)

Fig. 9–4

We must now repeat the procedure to obtain the stress on the *perpendicular* plane b–b. Sectioning the element in Fig. 9–4a along b–b results in a segment having sides with areas shown in Fig. 9–4e. Orienting the $+x'$ axis outward, perpendicular to the sectioned face, the associated free-body diagram is shown in Fig. 9–4f. Thus,

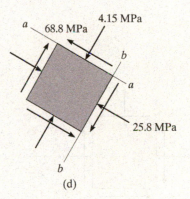

(d)

$$+\searrow\Sigma F_{x'} = 0; \quad \sigma_{x'}\Delta A - (25\ \Delta A\cos 30°)\sin 30°$$
$$+ (80\ \Delta A\cos 30°)\cos 30° - (25\ \Delta A\sin 30°)\cos 30°$$
$$- (50\ \Delta A\sin 30°)\sin 30° = 0$$
$$\sigma_{x'} = -25.8\ \text{MPa} \qquad\qquad \textit{Ans.}$$

$$+\nearrow\Sigma F_{y'} = 0; \quad \tau_{x'y'}\ \Delta A + (25\ \Delta A\cos 30°)\cos 30°$$
$$+ (80\ \Delta A\cos 30°)\sin 30° - (25\ \Delta A\sin 30°)\sin 30°$$
$$+ (50\ \Delta A\sin 30°)\cos 30° = 0$$
$$\tau_{x'y'} = -68.8\ \text{MPa} \qquad\qquad \textit{Ans.}$$

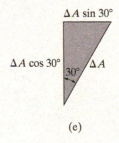

(e)

Since both $\sigma_{x'}$ and $\tau_{x'y'}$ are negative quantities, they act opposite to their direction shown in Fig. 9–4f. The stress components are shown acting on the *right side* of the element in Fig. 9–4d.

From this analysis we may therefore conclude that the state of stress at the point can be represented by a stress component acting on an element removed from the fuselage and oriented as shown in Fig. 9–4a, or by choosing one removed and oriented as shown in Fig. 9–4d. In other words, these states of stress are equivalent.

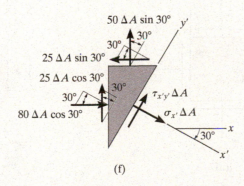

(f)

9.2 GENERAL EQUATIONS OF PLANE-STRESS TRANSFORMATION

The method of transforming the normal and shear stress components from the x, y to the x', y' coordinate axes, as discussed in the previous section, can be developed in a general manner and expressed as a set of stress-transformation equations.

Sign Convention.
To apply these equations we must first establish a sign convention for the stress components. As shown in Fig. 9–5, the $+x$ and $+x'$ axes are used to define the outward normal on the right-hand face of the element, so that σ_x and $\sigma_{x'}$ are positive when they act in the positive x and x' directions, and τ_{xy} and $\tau_{x'y'}$ are positive when they act in the positive y and y' directions.

The orientation of the face upon which the normal and shear stress components are to be determined will be defined by the angle θ, which is measured from the $+x$ axis to the $+x'$ axis, Fig. 9–5b. Notice that the unprimed and primed sets of axes in this figure both form right-handed coordinate systems; that is, the positive z (or z') axis always points out of the page. The *angle* θ will be *positive* when it follows the curl of the right-hand fingers, i.e., counterclockwise as shown in Fig. 9–5b.

Normal and Shear Stress Components.
Using this established sign convention, the element in Fig. 9–6a is sectioned along the inclined plane and the segment shown in Fig. 9–6b is isolated. Assuming the sectioned area is ΔA, then the horizontal and vertical faces of the segment have an area of $\Delta A \sin \theta$ and $\Delta A \cos \theta$, respectively.

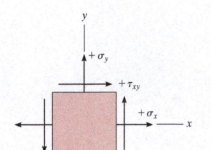

(a)

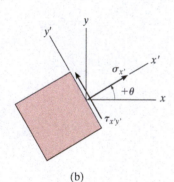

(b)

Positive sign convention

Fig. 9–5

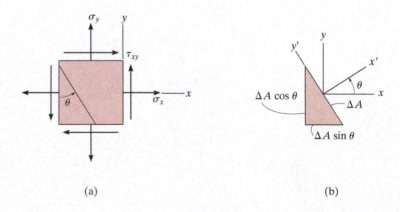

(a)   (b)

Fig. 9–6

The resulting *free-body diagram* of the segment is shown in Fig. 9–6c. If we apply the equations of equilibrium along the x' and y' axes, we can obtain a direct solution for $\sigma_{x'}$ and $\tau_{x'y'}$. We have

$$+\nearrow \Sigma F_{x'} = 0; \quad \sigma_{x'}\Delta A - (\tau_{xy}\Delta A \sin\theta)\cos\theta - (\sigma_y\Delta A \sin\theta)\sin\theta$$
$$- (\tau_{xy}\Delta A \cos\theta)\sin\theta - (\sigma_x\Delta A \cos\theta)\cos\theta = 0$$
$$\sigma_{x'} = \sigma_x\cos^2\theta + \sigma_y\sin^2\theta + \tau_{xy}(2\sin\theta\cos\theta)$$

$$+\nwarrow \Sigma F_{y'} = 0; \quad \tau_{x'y'}\Delta A + (\tau_{xy}\Delta A \sin\theta)\sin\theta - (\sigma_y\Delta A \sin\theta)\cos\theta$$
$$- (\tau_{xy}\Delta A \cos\theta)\cos\theta + (\sigma_x\Delta A \cos\theta)\sin\theta = 0$$
$$\tau_{x'y'} = (\sigma_y - \sigma_x)\sin\theta\cos\theta + \tau_{xy}(\cos^2\theta - \sin^2\theta)$$

To simplify these two equations, use the trigonometric identities $\sin 2\theta = 2\sin\theta\cos\theta$, $\sin^2\theta = (1 - \cos 2\theta)/2$, and $\cos^2\theta = (1 + \cos 2\theta)/2$. Therefore,

$$\sigma_{x'} = \frac{\sigma_x + \sigma_y}{2} + \frac{\sigma_x - \sigma_y}{2}\cos 2\theta + \tau_{xy}\sin 2\theta \tag{9–1}$$

$$\tau_{x'y'} = -\frac{\sigma_x - \sigma_y}{2}\sin 2\theta + \tau_{xy}\cos 2\theta \tag{9–2}$$

Stress Components Acting along x', y' Axes

If the normal stress acting in the y' direction is needed, it can be obtained by simply substituting $\theta + 90°$ for θ into Eq. 9–1, Fig. 9–6d. This yields

$$\sigma_{y'} = \frac{\sigma_x + \sigma_y}{2} - \frac{\sigma_x - \sigma_y}{2}\cos 2\theta - \tau_{xy}\sin 2\theta \tag{9–3}$$

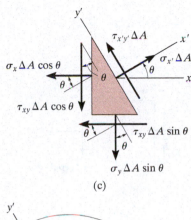

(c)

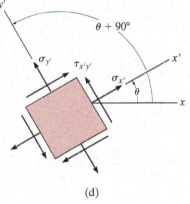

(d)

Fig. 9–6

9

PROCEDURE FOR ANALYSIS

To apply the stress transformation Eqs. 9–1 and 9–2, it is simply necessary to substitute in the known data for σ_x, σ_y, τ_{xy}, and θ in accordance with the established sign convention, Fig. 9–5. Remember that the x' axis is *always* directed positive outward from the plane upon which the normal stress is to be determined. The angle θ is *positive counterclockwise*, from the x to the x' axis. If $\sigma_{x'}$ and $\tau_{x'y'}$ are calculated as positive quantities, then these stresses act in the positive direction of the x' and y' axes.

For convenience, these equations can easily be programmed on a pocket calculator.

EXAMPLE 9.2

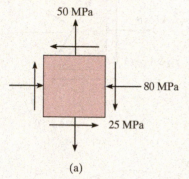

(a)

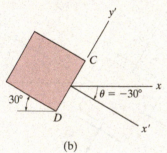

(b)

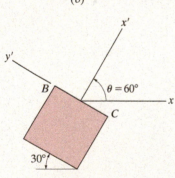

(c)

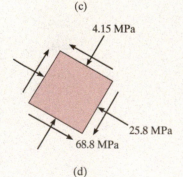

(d)

Fig. 9–7

The state of plane stress at a point is represented on the element shown in Fig. 9–7a. Determine the state of stress at this point on another element oriented 30° clockwise from the position shown.

SOLUTION

This problem was solved in Example 9.1 using basic principles. Here we will apply Eqs. 9–1 and 9–2. From the established sign convention, Fig. 9–5, it is seen that

$$\sigma_x = -80 \text{ MPa} \qquad \sigma_y = 50 \text{ MPa} \qquad \tau_{xy} = -25 \text{ MPa}$$

Plane CD. To obtain the stress components on plane CD, Fig. 9–7b, the positive x' axis must be directed outward, perpendicular to CD, and the associated y' axis is directed along CD. The angle measured from the x to the x' axis is $\theta = -30°$ (clockwise). Applying Eqs. 9–1 and 9–2 yields

$$\sigma_{x'} = \frac{\sigma_x + \sigma_y}{2} + \frac{\sigma_x - \sigma_y}{2} \cos 2\theta + \tau_{xy} \sin 2\theta$$

$$= \frac{-80 + 50}{2} + \frac{-80 - 50}{2} \cos 2(-30°) + (-25) \sin 2(-30°)$$

$$= -25.8 \text{ MPa} \qquad\qquad\qquad\qquad Ans.$$

$$\tau_{x'y'} = -\frac{\sigma_x - \sigma_y}{2} \sin 2\theta + \tau_{xy} \cos 2\theta$$

$$= -\frac{-80 - 50}{2} \sin 2(-30°) + (-25) \cos 2(-30°)$$

$$= -68.8 \text{ MPa} \qquad\qquad\qquad\qquad Ans.$$

The negative signs indicate that $\sigma_{x'}$ and $\tau_{x'y'}$ act in the negative x' and y' directions, respectively. The results are shown acting on the element in Fig. 9–7d.

Plane BC. Establishing the x' axis outward from plane BC, Fig. 9–7c, then between the x and x' axes, $\theta = 60°$ (counterclockwise). Applying Eqs. 9–1 and 9–2,* we get

$$\sigma_{x'} = \frac{-80 + 50}{2} + \frac{-80 - 50}{2} \cos 2(60°) + (-25) \sin 2(60°)$$

$$= -4.15 \text{ MPa} \qquad\qquad\qquad\qquad Ans.$$

$$\tau_{x'y'} = -\frac{-80 - 50}{2} \sin 2(60°) + (-25) \cos 2(60°)$$

$$= 68.8 \text{ MPa} \qquad\qquad\qquad\qquad Ans.$$

Here $\tau_{x'y'}$ has been calculated twice in order to provide a check. The negative sign for $\sigma_{x'}$ indicates that this stress acts in the negative x' direction, Fig. 9–7c. The results are shown on the element in Fig. 9–7d.

*Alternatively, we could apply Eq. 9–3 with $\theta = -30°$ rather than Eq. 9–1.

9.3   PRINCIPAL STRESSES AND MAXIMUM IN-PLANE SHEAR STRESS

Since $\sigma_x, \sigma_y, \tau_{xy}$ are all constant, then from Eqs. 9–1 and 9–2 it can be seen that the magnitudes of $\sigma_{x'}$ and $\tau_{x'y'}$ only depend on the angle of inclination θ of the planes on which these stresses act. In engineering practice it is often important to determine the orientation that causes the normal stress to be a maximum, and the orientation that causes the shear stress to be a maximum. We will now consider each of these cases.

In-Plane Principal Stresses. To determine the maximum and minimum *normal stress*, we must differentiate Eq. 9–1 with respect to θ and set the result equal to zero. This gives

$$\frac{d\sigma_{x'}}{d\theta} = -\frac{\sigma_x - \sigma_y}{2}(2\sin 2\theta) + 2\tau_{xy}\cos 2\theta = 0$$

Solving we obtain the orientation $\theta = \theta_p$ of the planes of maximum and minimum normal stress.

$$\tan 2\theta_p = \frac{\tau_{xy}}{(\sigma_x - \sigma_y)/2} \qquad (9\text{–}4)$$

Orientation of Principal Planes

The solution has two roots, θ_{p_1} and θ_{p_2}. Specifically, the values of $2\theta_{p_1}$ and $2\theta_{p_2}$ are 180° apart, so θ_{p_1} and θ_{p_2} will be 90° apart.

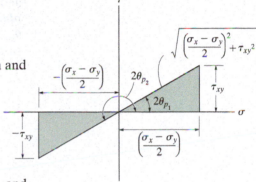

Fig. 9–8

The cracks in this concrete beam were caused by tension stress, even though the beam was subjected to both an internal moment and shear. The stress transformation equations can be used to predict the direction of the cracks, and the principal normal stresses that caused them.

To obtain the maximum and minimum normal stress, we must substitute these angles into Eq. 9–1. Here the necessary sine and cosine of $2\theta_{p_1}$ and $2\theta_{p_2}$ can be found from the shaded triangles shown in Fig. 9–8, which are constructed based on Eq. 9–4, assuming that τ_{xy} and $(\sigma_x - \sigma_y)$ are both positive or both negative quantities.

After substituting and simplifying, we obtain two roots, σ_1 and σ_2. They are

$$\sigma_{1,2} = \frac{\sigma_x + \sigma_y}{2} \pm \sqrt{\left(\frac{\sigma_x - \sigma_y}{2}\right)^2 + \tau_{xy}^2} \qquad (9\text{–}5)$$

Principal Stresses

These two values, with $\sigma_1 \geq \sigma_2$, are called the in-plane **principal stresses**, and the corresponding planes on which they act are called the **principal planes** of stress, Fig. 9–9. Finally, if the trigonometric relations for θ_{p_1} or θ_{p_2} are substituted into Eq. 9–2, it will be seen that $\tau_{x'y'} = 0$; in other words, **no shear stress acts on the principal planes**, Fig. 9–9.

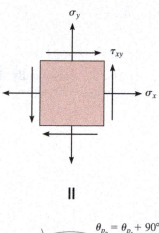

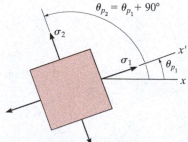

In-plane principal stresses

Fig. 9–9

Maximum In-Plane Shear Stress. The orientation of the element that is subjected to maximum shear stress can be determined by taking the derivative of Eq. 9–2 with respect to θ, and setting the result equal to zero. This gives

$$\tan 2\theta_s = \frac{-(\sigma_x - \sigma_y)/2}{\tau_{xy}} \qquad (9\text{–}6)$$

Orientation of Maximum In-Plane Shear Stress

The two roots of this equation, θ_{s_1} and θ_{s_2}, can be determined from the shaded triangles shown in Fig. 9–10a. Since $\tan 2\theta_s$, Eq. 9–6, is the negative reciprocal of $\tan 2\theta_p$, Eq. 9–4, then each root $2\theta_s$ is 90° from $2\theta_p$, and the roots θ_s and θ_p are 45° apart. Therefore, an element subjected to *maximum shear stress must be oriented 45° from the position of an element that is subjected to the principal stress*.

The maximum shear stress can be found by taking the trigonometric values of $\sin 2\theta_s$ and $\cos 2\theta_s$ from Fig. 9–10 and substituting them into Eq. 9–2. The result is

$$\tau_{\substack{\text{max} \\ \text{in-plane}}} = \sqrt{\left(\frac{\sigma_x - \sigma_y}{2}\right)^2 + \tau_{xy}^2} \qquad (9\text{–}7)$$

Maximum In-Plane Shear Stress

Here $\tau_{\substack{\text{max} \\ \text{in-plane}}}$ is referred to as the *maximum in-plane shear stress*, because it acts on the element in the *x–y* plane.

Finally, when the values for $\sin 2\theta_s$ and $\cos 2\theta_s$ are substituted into Eq. 9–1, we see that there is *also an average normal stress* on the planes of maximum in-plane shear stress. It is

$$\sigma_{\text{avg}} = \frac{\sigma_x + \sigma_y}{2} \qquad (9\text{–}8)$$

Average Normal Stress

For numerical applications, it is suggested that Eqs. 9–1 through 9–8 be programmed for use on a pocket calculator.

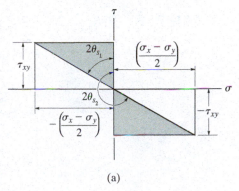

(a)

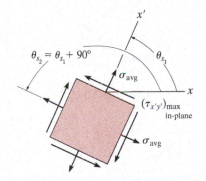

Maximum in-plane shear stresses

(b)

Fig. 9–10

> ## IMPORTANT POINTS

- The *principal stresses* represent the maximum and minimum normal stress at the point.
- When the state of stress is represented by the principal stresses, *no shear stress* will act on the element.
- The state of stress at the point can also be represented in terms of the *maximum in-plane shear stress*. In this case an *average normal stress* will also act on the element.
- The element representing the maximum in-plane shear stress with the associated average normal stresses is oriented 45° from the element representing the principal stresses.

EXAMPLE 9.3

The state of stress at a point just before failure of this shaft is shown in Fig. 9–11a. Represent this state of stress in terms of its principal stresses.

SOLUTION

From the established sign convention,

$$\sigma_x = -20 \text{ MPa} \qquad \sigma_y = 90 \text{ MPa} \qquad \tau_{xy} = 60 \text{ MPa}$$

Orientation of Element.   Applying Eq. 9–4,

$$\tan 2\theta_p = \frac{\tau_{xy}}{(\sigma_x - \sigma_y)/2} = \frac{60}{(-20 - 90)/2}$$

Solving, and referring to this first angle as θ_{p_2}, we have

$$2\theta_{p_2} = -47.49° \qquad \theta_{p_2} = -23.7°$$

Since the difference between $2\theta_{p_1}$ and $2\theta_{p_2}$ is 180°, the second angle is

$$2\theta_{p_1} = 180° + 2\theta_{p_2} = 132.51° \qquad \theta_{p_1} = 66.3°$$

In both cases, θ must be measured positive *counterclockwise* from the x axis to the outward normal (x' axis) on the face of the element, and so the element showing the principal stresses will be oriented as shown in Fig. 9–11b.

Principal Stress.   We have

$$\sigma_{1,2} = \frac{\sigma_x + \sigma_y}{2} \pm \sqrt{\left(\frac{\sigma_x - \sigma_y}{2}\right)^2 + \tau_{xy}^2}$$

$$= \frac{-20 + 90}{2} \pm \sqrt{\left(\frac{-20 - 90}{2}\right)^2 + (60)^2}$$

$$= 35.0 \pm 81.4$$

$$\sigma_1 = 116 \text{ MPa} \hspace{4cm} \textit{Ans.}$$

$$\sigma_2 = -46.4 \text{ MPa} \hspace{3.5cm} \textit{Ans.}$$

The principal plane on which each normal stress acts can be determined by applying Eq. 9–1 with, say, $\theta = \theta_{p_2} = -23.7°$. We have

$$\sigma_{x'} = \frac{\sigma_x + \sigma_y}{2} + \frac{\sigma_x - \sigma_y}{2} \cos 2\theta + \tau_{xy} \sin 2\theta$$

$$= \frac{-20 + 90}{2} + \frac{-20 - 90}{2} \cos 2(-23.7°) + 60 \sin 2(-23.7°)$$

$$= -46.4 \text{ MPa}$$

Hence, $\sigma_2 = -46.4$ MPa acts on the plane defined by $\theta_{p_2} = -23.7°$, whereas $\sigma_1 = 116$ MPa acts on the plane defined by $\theta_{p_1} = 66.3°$, Fig. 9–11c. Recall that no shear stress acts on this element.

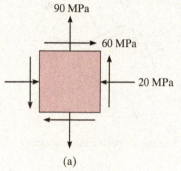

(a)

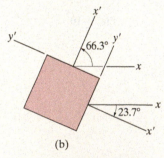

(b)

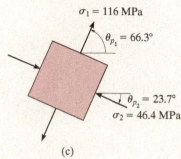

(c)

Fig. 9–11

EXAMPLE 9.4

The state of plane stress at a point on a body is represented on the element shown in Fig. 9–12a. Represent this state of stress in terms of its maximum in-plane shear stress and associated average normal stress.

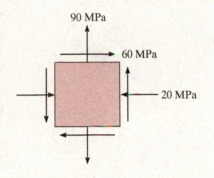

(a)

SOLUTION

Orientation of Element. Since $\sigma_x = -20$ MPa, $\sigma_y = 90$ MPa, and $\tau_{xy} = 60$ MPa, applying Eq. 9–6, the two angles are

$$\tan 2\theta_s = \frac{-(\sigma_x - \sigma_y)/2}{\tau_{xy}} = \frac{-(-20 - 90)/2}{60}$$

$$2\theta_{s_2} = 42.5° \qquad\qquad \theta_{s_2} = 21.3°$$

$$2\theta_{s_1} = 180° + 2\theta_{s_2} \qquad\qquad \theta_{s_1} = 111.3°$$

Note how these angles are formed between the x and x' axes, Fig. 9–12b. They happen to be 45° away from the principal planes of stress, which were determined in Example 9.3.

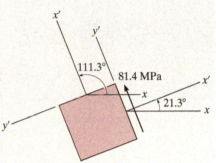

(b)

Maximum In-Plane Shear Stress. Applying Eq. 9–7,

$$\tau_{\text{in-plane}}^{\max} = \sqrt{\left(\frac{\sigma_x - \sigma_y}{2}\right)^2 + \tau_{xy}^2} = \sqrt{\left(\frac{-20 - 90}{2}\right)^2 + (60)^2}$$

$$= \pm 81.4 \text{ MPa} \qquad\qquad\qquad Ans.$$

The proper direction of $\tau_{\text{in-plane}}^{\max}$ on the element can be determined by substituting $\theta = \theta_{s_2} = 21.3°$ into Eq. 9–2. We have

$$\tau_{x'y'} = -\left(\frac{\sigma_x - \sigma_y}{2}\right)\sin 2\theta + \tau_{xy}\cos 2\theta$$

$$= -\left(\frac{-20 - 90}{2}\right)\sin 2(21.3°) + 60\cos 2(21.3°)$$

$$= 81.4 \text{ MPa}$$

This positive result indicates that $\tau_{\text{in-plane}}^{\max} = \tau_{x'y'}$ acts in the *positive y'* direction on this face ($\theta = 21.3°$), Fig. 9–12b. The shear stresses on the other three faces are directed as shown in Fig. 9–12c.

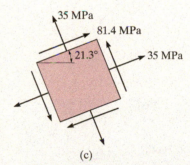

(c)

Fig. 9–12

Average Normal Stress. Besides the maximum shear stress, the element is also subjected to an average normal stress determined from Eq. 9–8; that is,

$$\sigma_{\text{avg}} = \frac{\sigma_x + \sigma_y}{2} = \frac{-20 + 90}{2} = 35 \text{ MPa} \qquad\qquad Ans.$$

This is a tensile stress. The results are shown in Fig. 9–12c.

EXAMPLE 9.5

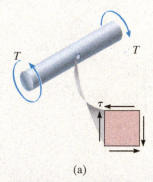

(a)

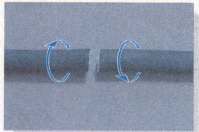

Torsion failure of mild steel.

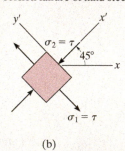

(b)

Fig. 9–13

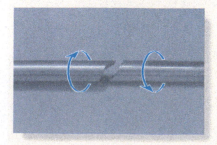

Torsion failure of cast iron.

When the torsional loading T is applied to the bar in Fig. 9–13a, it produces a state of pure shear stress in the material. Determine (a) the maximum in-plane shear stress and the associated average normal stress, and (b) the principal stress.

SOLUTION
From the established sign convention,

$$\sigma_x = 0 \qquad \sigma_y = 0 \qquad \tau_{xy} = -\tau$$

Maximum In-Plane Shear Stress.   Applying Eqs. 9–7 and 9–8, we have

$$\tau_{\substack{max \\ in\text{-}plane}} = \sqrt{\left(\frac{\sigma_x - \sigma_y}{2}\right)^2 + \tau_{xy}^2} = \sqrt{(0)^2 + (-\tau)^2} = \pm\tau \quad Ans.$$

$$\sigma_{avg} = \frac{\sigma_x + \sigma_y}{2} = \frac{0 + 0}{2} = 0 \qquad\qquad Ans.$$

Thus, as expected, the maximum in-plane shear stress is represented by the element in Fig. 9–13a.

NOTE: Through experiment it has been found that materials that are *ductile* actually fail due to *shear stress*. As a result, if the bar in Fig. 9–13a is made of mild steel, the maximum in-plane shear stress will cause it to fail as shown in the adjacent photo.

Principal Stress.   Applying Eqs. 9–4 and 9–5 yields

$$\tan 2\theta_p = \frac{\tau_{xy}}{(\sigma_x - \sigma_y)/2} = \frac{-\tau}{(0 - 0)/2}, \theta_{p_2} = 45°, \theta_{p_1} = -45°$$

$$\sigma_{1,2} = \frac{\sigma_x + \sigma_y}{2} \pm \sqrt{\left(\frac{\sigma_x - \sigma_y}{2}\right)^2 + \tau_{xy}^2} = 0 \pm \sqrt{(0)^2 + \tau^2} = \pm\tau \quad Ans.$$

If we now apply Eq. 9–1 with $\theta_{p_2} = 45°$, then

$$\sigma_{x'} = \frac{\sigma_x + \sigma_y}{2} + \frac{\sigma_x - \sigma_y}{2}\cos 2\theta + \tau_{xy}\sin 2\theta$$

$$= 0 + 0 + (-\tau)\sin 90° = -\tau$$

Thus, $\sigma_2 = -\tau$ acts at $\theta_{p_2} = 45°$ as shown in Fig. 9–13b, and $\sigma_1 = \tau$ acts on the other face, $\theta_{p_1} = -45°$.

NOTE: Materials that are *brittle* fail due to *normal stress*. Therefore, if the bar in Fig. 9–13a is made of cast iron it will fail in tension at a 45° inclination as seen in the adjacent photo.

EXAMPLE 9.6

When the axial loading P is applied to the bar in Fig. 9–14a, it produces a tensile stress in the material. Determine (a) the principal stress and (b) the maximum in-plane shear stress and associated average normal stress.

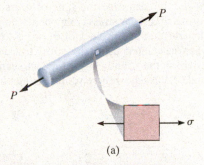

(a)

SOLUTION
From the established sign convention,

$$\sigma_x = \sigma \qquad \sigma_y = 0 \qquad \tau_{xy} = 0$$

Principal Stress.   By observation, the element oriented as shown in Fig. 9–14a illustrates a condition of principal stress since no shear stress acts on this element. This can also be shown by direct substitution of the above values into Eqs. 9–4 and 9–5. Thus,

$$\sigma_1 = \sigma \qquad \sigma_2 = 0 \qquad \textit{Ans.}$$

NOTE: *Brittle materials* will fail due to normal stress, and therefore, if the bar in Fig. 9–14a is made of cast iron, it will fail as shown in the adjacent photo.

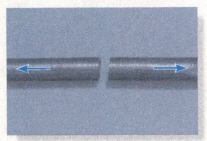

Axial failure of cast iron.

Maximum In-Plane Shear Stress.   Applying Eqs. 9–6, 9–7, and 9–8, we have

$$\tan 2\theta_s = \frac{-(\sigma_x - \sigma_y)/2}{\tau_{xy}} = \frac{-(\sigma - 0)/2}{0} ; \theta_{s_1} = 45°, \theta_{s_2} = -45°$$

$$\tau_{\substack{max \\ in\text{-}plane}} = \sqrt{\left(\frac{\sigma_x - \sigma_y}{2}\right)^2 + \tau_{xy}{}^2} = \sqrt{\left(\frac{\sigma - 0}{2}\right)^2 + (0)^2} = \pm\frac{\sigma}{2} \quad \textit{Ans.}$$

$$\sigma_{avg} = \frac{\sigma_x + \sigma_y}{2} = \frac{\sigma + 0}{2} = \frac{\sigma}{2} \qquad \textit{Ans.}$$

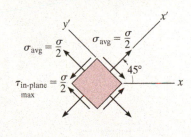

(b)

To determine the proper orientation of the element, apply Eq. 9–2.

$$\tau_{x'y'} = -\frac{\sigma_x - \sigma_y}{2}\sin 2\theta + \tau_{xy}\cos 2\theta = -\frac{\sigma - 0}{2}\sin 90° + 0 = -\frac{\sigma}{2}$$

This negative shear stress acts on the x' face in the negative y' direction, as shown in Fig. 9–14b.

NOTE: If the bar in Fig. 9–14a is made of a *ductile material* such as mild steel then *shear stress* will cause it to fail. This can be noted in the adjacent photo, where within the region of necking, shear stress has caused "slipping" along the steel's crystalline boundaries, resulting in a plane of failure that has formed a *cone* around the bar oriented at approximately 45° as calculated above.

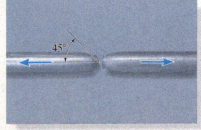

Axial failure of mild steel.

Fig. 9–14

PRELIMINARY PROBLEMS

P9–1. In each case, the state of stress σ_x, σ_y, τ_{xy} produces normal and shear stress components along section AB of the element that have values of $\sigma_{x'} = -5$ kPa and $\tau_{x'y'} = 8$ kPa when calculated using the stress transformation equations. Establish the x' and y' axes for each segment and specify the angle θ, then show these results acting on each segment.

P9–2. Given the state of stress shown on the element, find σ_{avg} and $\tau_{\text{in-plane}}^{\max}$ and show the results on a properly oriented element.

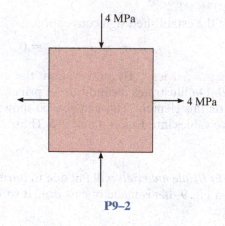

P9–2

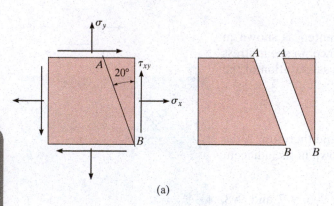

(a)

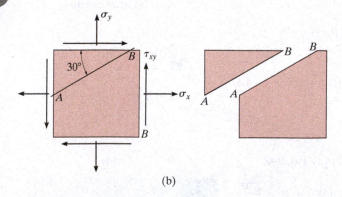

(b)

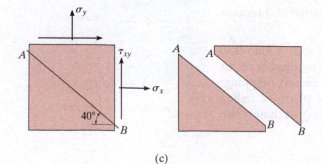

(c)

P9–1

FUNDAMENTAL PROBLEMS

F9–1. Determine the normal stress and shear stress acting on the inclined plane AB. Sketch the result on the sectioned element.

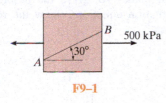

F9–1

F9–2. Determine the equivalent state of stress on an element at the same point oriented 45° clockwise with respect to the element shown.

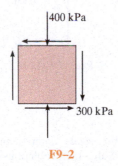

F9–2

F9–3. Determine the equivalent state of stress on an element at the same point that represents the principal stresses at the point. Also, find the corresponding orientation of the element with respect to the element shown.

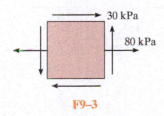

F9–3

F9–4. Determine the equivalent state of stress on an element at the same point that represents the maximum in-plane shear stress at the point.

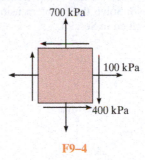

F9–4

F9–5. The beam is subjected to the load at its end. Determine the maximum principal stress at point B.

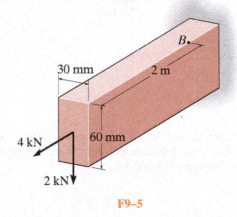

F9–5

F9–6. The beam is subjected to the loading shown. Determine the principal stress at point C.

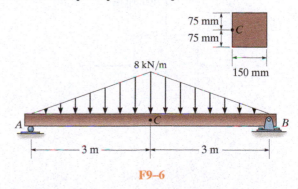

F9–6

PROBLEMS

9–1. Prove that the sum of the normal stresses $\sigma_x + \sigma_y = \sigma_{x'} + \sigma_{y'}$ is constant. See Figs. 9–2a and 9–2b.

9–2. Determine the stress components acting on the inclined plane AB. Solve the problem using the method of equilibrium described in Sec. 9.1.

*9–4. Determine the normal stress and shear stress acting on the inclined plane AB. Solve the problem using the method of equilibrium described in Sec. 9.1.

9–5. Determine the normal stress and shear stress acting on the inclined plane AB. Solve the problem using the stress transformation equations. Show the results on the sectional element.

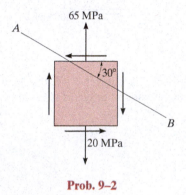

Prob. 9–2

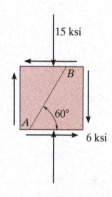

Probs. 9–4/5

9–3. Determine the stress components acting on the inclined plane AB. Solve the problem using the method of equilibrium described in Sec. 9.1.

9–6. Determine the stress components acting on the inclined plane AB. Solve the problem using the method of equilibrium described in Sec. 9.1.

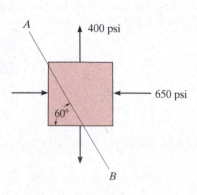

Prob. 9–3

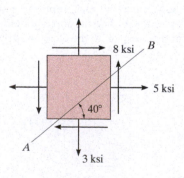

Prob. 9–6

9–7. Determine the stress components acting on the inclined plane AB. Solve the problem using the method of equilibrium described in Sec. 9.1.

*9–8. Solve Prob. 9–7 using the stress transformation equations developed in Sec. 9.2.

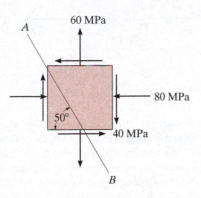

60 MPa

A

80 MPa

50°

40 MPa

B

Probs. 9–7/8

9–11. Determine the equivalent state of stress on an element at the same point oriented 60° clockwise with respect to the element shown. Sketch the results on the element.

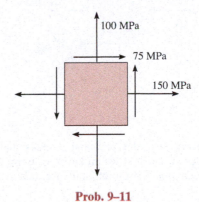

100 MPa

75 MPa

150 MPa

Prob. 9–11

9–9. Determine the stress components acting on the plane AB. Solve the problem using the method of equilibrium described in Sec. 9.1.

9–10. Solve Prob. 9–9 using the stress transformation equation developed in Sec. 9.2.

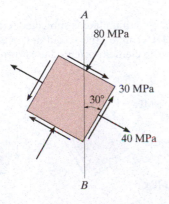

A

80 MPa

30 MPa

30°

40 MPa

B

Probs. 9–9/10

*9–12. Determine the equivalent state of stress on an element at the same point oriented 60° counterclockwise with respect to the element shown. Sketch the results on the element.

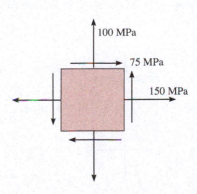

100 MPa

75 MPa

150 MPa

Prob. 9–12

9–13. Determine the stress components acting on the inclined plane AB. Solve the problem using the method of equilibrium described in Sec. 9.1.

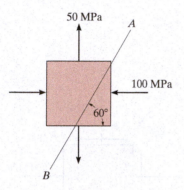

50 MPa

A

100 MPa

60°

B

Prob. 9–13

9–14. Determine (a) the principal stresses and (b) the maximum in-plane shear stress and average normal stress at the point. Specify the orientation of the element in each case.

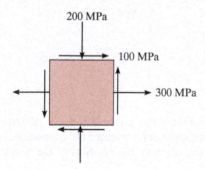

200 MPa

100 MPa

300 MPa

Prob. 9–14

9–15. The state of stress at a point is shown on the element. Determine (a) the principal stresses and (b) the maximum in-plane shear stress and average normal stress at the point. Specify the orientation of the element in each case.

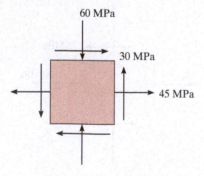

60 MPa

30 MPa

45 MPa

Prob. 9–15

***9–16.** Determine the equivalent state of stress on an element at the point which represents (a) the principal stresses and (b) the maximum in-plane shear stress and the associated average normal stress. Also, for each case, determine the corresponding orientation of the element with respect to the element shown and sketch the results on the element.

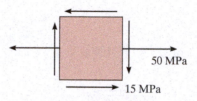

50 MPa

15 MPa

Prob. 9–16

9–17. Determine the equivalent state of stress on an element at the same point which represents (a) the principal stress, and (b) the maximum in-plane shear stress and the associated average normal stress. Also, for each case, determine the corresponding orientation of the element with respect to the element shown and sketch the results on the element.

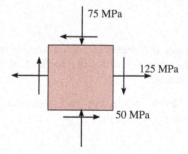

75 MPa

125 MPa

50 MPa

Prob. 9–17

9–18. A point on a thin plate is subjected to the two stress components. Determine the resultant state of stress represented on the element oriented as shown on the right.

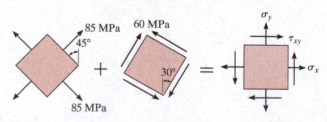

85 MPa

45°

60 MPa

30°

σ_y

τ_{xy}

σ_x

85 MPa

Prob. 9–18

9–19. Determine the equivalent state of stress on an element at the same point which represents (a) the principal stress, and (b) the maximum in-plane shear stress and the associated average normal stress. Also, for each case, determine the corresponding orientation of the element with respect to the element shown and sketch the results on the element.

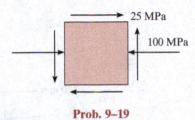

25 MPa

100 MPa

Prob. 9–19

*9–20. The stress along two planes at a point is indicated. Determine the normal stresses on plane *b–b* and the principal stresses.

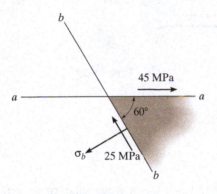

b

45 MPa

a   a

60°

σ_b 25 MPa

b

Prob. 9–20

9–21. The stress acting on two planes at a point is indicated. Determine the shear stress on plane *a–a* and the principal stresses at the point.

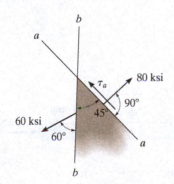

b

a

τ_a 80 ksi

90°

45°

60 ksi

60°

a

b

Prob. 9–21

9–22. The state of stress at a point in a member is shown on the element. Determine the stress components acting on the plane *AB*.

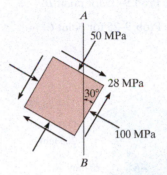

A

50 MPa

28 MPa

30°

100 MPa

B

Prob. 9–22

The following problems involve material covered in Chapter 8.

9–23. The grains of wood in the board make an angle of 20° with the horizontal as shown. Determine the normal and shear stress that act perpendicular and parallel to the grains if the board is subjected to an axial load of 250 N.

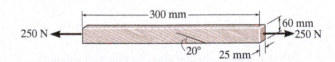

300 mm

250 N   60 mm
                                           250 N
20°   25 mm

Prob. 9–23

9–24. The wood beam is subjected to a load of 12 kN. If grains of wood in the beam at point *A* make an angle of 25° with the horizontal as shown, determine the normal and shear stress that act perpendicular to the grains due to the loading.

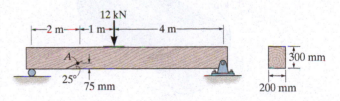

12 kN

2 m   1 m   4 m

A

300 mm

25°   75 mm

200 mm

Prob. 9–24

9–25. The internal loadings at a section of the beam are shown. Determine the in-plane principal stresses at point *A*. Also compute the maximum in-plane shear stress at this point.

9–26. Solve Prob. 9–25 for point *B*.

9–27. Solve Prob. 9–25 for point *C*.

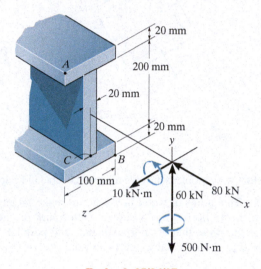

Probs. 9–25/26/27

9–29. The bell crank is pinned at *A* and supported by a short link *BC*. If it is subjected to the force of 80 N, determine the principal stresses at (a) point *D* and (b) point *E*. The crank is constructed from an aluminum plate having a thickness of 20 mm.

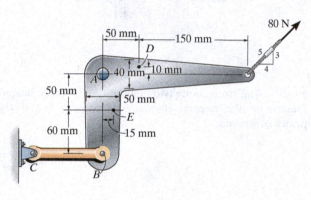

Prob. 9–29

*9–28 A rod has a circular cross section with a diameter of 2 in. It is subjected to a torque of 12 kip · in. and a bending moment **M**. The greater principal stress at the point of maximum flexural stress is 15 ksi. Determine the magnitude of the bending moment.

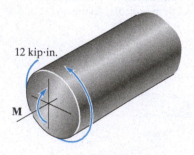

Prob. 9–28

9–30. The beam has a rectangular cross section and is subjected to the loadings shown. Determine the principal stresses at point *A* and point *B*, which are located just to the left of the 20-kN load. Show the results on elements located at these points.

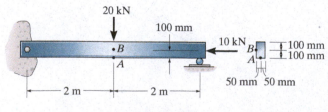

Prob. 9–30

9–31. A paper tube is formed by rolling a cardboard strip in a spiral and then gluing the edges together as shown. Determine the shear stress acting along the seam, which is at 50° from the horizontal, when the tube is subjected to an axial compressive force of 200 N. The paper is 2 mm thick and the tube has an outer diameter of 100 mm.

*9–32.** Solve Prob. 9–31 for the normal stress acting perpendicular to the seam.

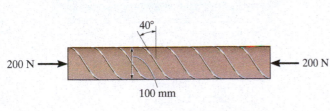

Probs. 9–31/32

9–33. The 2-in.-diameter drive shaft *AB* on the helicopter is subjected to an axial tension of 10 000 lb and a torque of 300 lb·ft. Determine the principal stresses and the maximum in-plane shear stress that act at a point on the surface of the shaft.

Prob. 9–33

9–34. Determine the principal stresses in the cantilevered beam at points *A* and *B*.

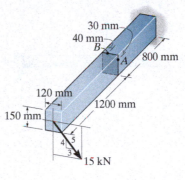

Prob. 9–34

9–35. The internal loadings at a cross section through the 6-in.-diameter drive shaft of a turbine consist of an axial force of 2500 lb, a bending moment of 800 lb·ft, and a torsional moment of 1500 lb·ft. Determine the principal stresses at point *A*. Also calculate the maximum in-plane shear stress at this point.

*9–36.** The internal loadings at a cross section through the 6-in.-diameter drive shaft of a turbine consist of an axial force of 2500 lb, a bending moment of 800 lb·ft, and a torsional moment of 1500 lb·ft. Determine the principal stresses at point *B*. Also calculate the maximum in-plane shear stress at this point.

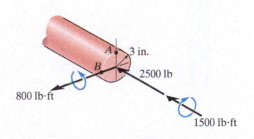

Probs. 9–35/36

9–37. The shaft has a diameter d and is subjected to the loadings shown. Determine the principal stresses and the maximum in-plane shear stress at point A. The bearings only support vertical reactions.

*9–40. The wide-flange beam is subjected to the 50-kN force. Determine the principal stresses in the beam at point A located on the *web* at the bottom of the upper flange. Although it is not very accurate, use the shear formula to calculate the shear stress.

9–41. Solve Prob. 9–40 for point B located on the *web* at the top of the bottom flange.

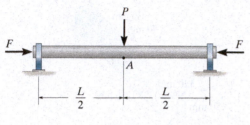

Prob. 9–37

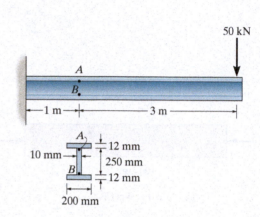

Probs. 9–40/41

9–38. The steel pipe has an inner diameter of 2.75 in. and an outer diameter of 3 in. If it is fixed at C and subjected to the horizontal 60-lb force acting on the handle of the pipe wrench at its end, determine the principal stresses in the pipe at point A, which is located on the outer surface of the pipe.

9–39. Solve Prob. 9–38 for point B, which is located on the outer surface of the pipe.

9–42. The box beam is subjected to the 26-kN force that is applied at the center of its width, 75 mm from each side. Determine the principal stresses at point A and show the results in an element located at this point. Use the shear formula to calculate the shear stress.

9–43. Solve Prob. 9–42 for point B.

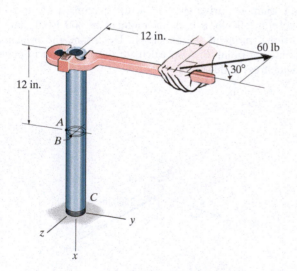

Probs. 9–38/39

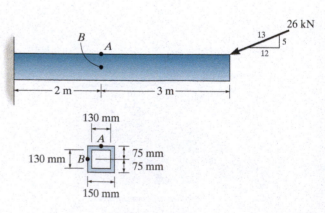

Probs. 9–42/43

9.4 MOHR'S CIRCLE—PLANE STRESS

In this section, we will show how to apply the equations for plane-stress transformation using a *graphical* procedure that is often convenient to use and easy to remember. Furthermore, this approach will allow us to "visualize" how the normal and shear stress components $\sigma_{x'}$ and $\tau_{x'y'}$ vary as the plane on which they act changes its direction, Fig. 9–15a.

If we write Eqs. 9–1 and 9–2 in the form

$$\sigma_{x'} - \left(\frac{\sigma_x + \sigma_y}{2}\right) = \left(\frac{\sigma_x - \sigma_y}{2}\right)\cos 2\theta + \tau_{xy}\sin 2\theta \qquad (9\text{–}9)$$

$$\tau_{x'y'} = -\left(\frac{\sigma_x - \sigma_y}{2}\right)\sin 2\theta + \tau_{xy}\cos 2\theta \qquad (9\text{–}10)$$

then the parameter θ can be eliminated by squaring each equation and adding them together. The result is

$$\left[\sigma_{x'} - \left(\frac{\sigma_x + \sigma_y}{2}\right)\right]^2 + \tau_{x'y'}^2 = \left(\frac{\sigma_x - \sigma_y}{2}\right)^2 + \tau_{xy}^2$$

Finally, since σ_x, σ_y, τ_{xy} are *known constants*, then the above equation can be written in a more compact form as

$$(\sigma_{x'} - \sigma_{avg})^2 + \tau_{x'y'}^2 = R^2 \qquad (9\text{–}11)$$

where

$$\sigma_{avg} = \frac{\sigma_x + \sigma_y}{2}$$

$$R = \sqrt{\left(\frac{\sigma_x - \sigma_y}{2}\right)^2 + \tau_{xy}^2} \qquad (9\text{–}12)$$

If we establish coordinate axes, σ *positive to the right* and τ *positive downward*, and then plot Eq. 9–11, it will be seen that this equation represents a *circle* having a radius R and center on the σ axis at point $C(\sigma_{avg}, 0)$, Fig. 9–15b. This circle is called **Mohr's circle**, because it was developed by the German engineer Otto Mohr.

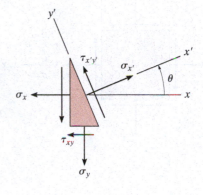

(a)

Fig. 9–15

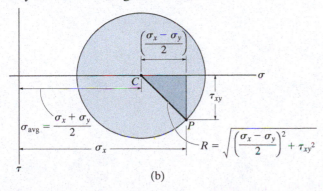

(b)

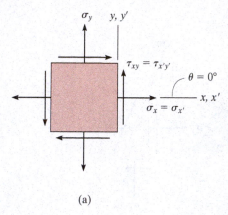

(a)

Each point on Mohr's circle represents the two stress components $\sigma_{x'}$ and $\tau_{x'y'}$ acting on the side of the element defined by the outward x' axis, when this axis is in a specific direction θ. For example, when x' is coincident with the x axis as shown in Fig. 9–16a, then $\theta = 0°$ and $\sigma_{x'} = \sigma_x, \tau_{x'y'} = \tau_{xy}$. We will refer to this as the "reference point" A and plot its coordinates $A(\sigma_x, \tau_{xy})$, Fig. 9–16c.

Now consider rotating the x' axis 90° counterclockwise, Fig. 9–16b. Then $\sigma_{x'} = \sigma_y, \tau_{x'y'} = -\tau_{xy}$. These values are the coordinates of point $G(\sigma_y, -\tau_{xy})$ on the circle, Fig. 9–16c. Hence, the radial line CG is 180° counterclockwise from the radial "reference line" CA. In other words, a rotation θ of the x' axis on the element will correspond to a rotation 2θ on the circle in the *same direction*.

As discussed in the following procedure, Mohr's circle can be used to determine the principal stresses, the maximum in-plane shear stress, or the stress on any arbitrary plane.

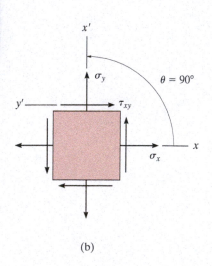

(b)

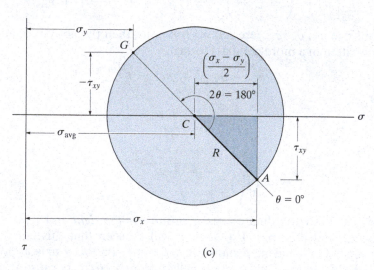

(c)

Fig. 9–16

PROCEDURE FOR ANALYSIS

The following steps are required to draw and use Mohr's circle.

Construction of the Circle.

- Establish a coordinate system such that the horizontal axis represents the normal stress σ, with *positive to the right*, and the vertical axis represents the shear stress τ, with *positive downwards*, Fig. 9–17a.*

- Using the positive sign convention for σ_x, σ_y, τ_{xy}, Fig. 9–17a, plot the center of the circle C, which is located on the σ axis at a distance $\sigma_{avg} = (\sigma_x + \sigma_y)/2$ from the origin, Fig. 9–17a.

- Plot the "reference point" A having coordinates $A(\sigma_x, \tau_{xy})$. This point represents the normal and shear stress components on the element's right-hand vertical face, and since the x' axis coincides with the x axis, this represents $\theta = 0°$, Fig. 9–17a.

- Connect point A with the center C of the circle and determine CA by trigonometry. This represents the radius R of the circle, Fig. 9–17a.

- Once R has been determined, sketch the circle.

Principal Stress.

- The principal stresses σ_1 and σ_2 ($\sigma_1 \geq \sigma_2$) are the coordinates of points B and D, where the circle intersects the σ axis, i.e., where $\tau = 0$, Fig. 9–17a.

- These stresses act on planes defined by angles θ_{p_1} and θ_{p_2}, Fig. 9–17b. One of these angles is represented on the circle as $2\theta_{p_1}$. It is measured *from* the radial reference line CA to line CB.

- Using trigonometry, determine θ_{p_1} from the circle. Remember that the direction of rotation $2\theta_p$ on the circle (here it happens to be counterclockwise) represents the *same* direction of rotation θ_p from the reference axis ($+x$) to the principal plane ($+x'$), Fig. 9–17b.*

Maximum In-Plane Shear Stress.

- The average normal stress and maximum in-plane shear stress components are determined from the circle as the coordinates of either point E or F, Fig. 9–17a.

- In this case the angles θ_{s_1} and θ_{s_2} give the orientation of the planes that contain these components, Fig. 9–17c. The angle $2\theta_{s_1}$ is shown in Fig. 9–17a and can be determined using trigonometry. Here the rotation happens to be clockwise, from CA to CE, and so θ_{s_1} must be clockwise on the element, Fig. 9–17c.*

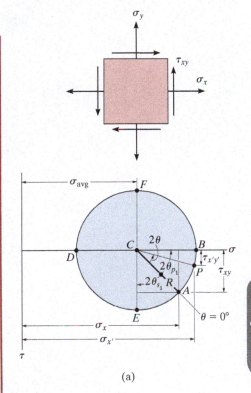

(a)

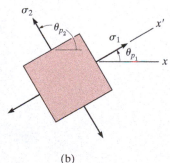

(b)

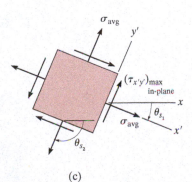

(c)

Fig. 9–17

9

Stresses on Arbitrary Plane.

- The normal and shear stress components $\sigma_{x'}$ and $\tau_{x'y'}$ acting on a specified plane or x' axis, defined by the angle θ, Fig. 9–17d, can be obtained by finding the coordinates of point P on the circle using trigonometry, Fig. 9–17a.

- To locate P, the known angle θ (in this case counterclockwise), Fig. 9–17d, must be measured on the circle in the *same direction* 2θ (counterclockwise) *from* the radial reference line CA *to* the radial line CP, Fig. 9–17a.*

*If the τ axis were constructed *positive upwards*, then the angle 2θ on the circle would be measured in the *opposite direction* to the orientation θ of the x' axis.

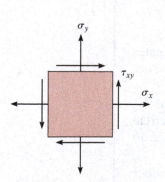

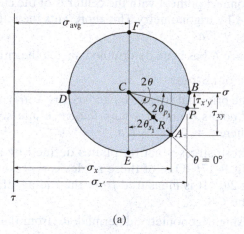

(a)

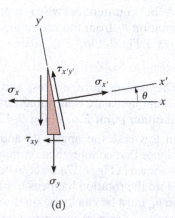

(d)

Fig. 9–17 (cont.)

EXAMPLE 9.7

Due to the applied loading, the element at point A on the solid shaft in Fig. 9–18a is subjected to the state of stress shown. Determine the principal stresses acting at this point.

SOLUTION

Construction of the Circle. From Fig. 9–18a,

$$\sigma_x = -12 \text{ ksi} \qquad \sigma_y = 0 \qquad \tau_{xy} = -6 \text{ ksi}$$

The center of the circle is located on the σ axis at the point

$$\sigma_{avg} = \frac{-12 + 0}{2} = -6 \text{ ksi}$$

The reference point $A(-12, -6)$ and the center $C(-6, 0)$ are plotted in Fig. 9–18b. From the shaded triangle, the circle is constructed having a radius of

$$R = \sqrt{(12 - 6)^2 + (6)^2} = 8.49 \text{ ksi}$$

Principal Stress. The principal stresses are indicated by the coordinates of points B and D. We have, for $\sigma_1 > \sigma_2$,

$$\sigma_1 = 8.49 - 6 = 2.49 \text{ ksi} \qquad\qquad \textit{Ans.}$$

$$\sigma_2 = -6 - 8.49 = -14.5 \text{ ksi} \qquad\qquad \textit{Ans.}$$

The orientation of the element can be determined by calculating the angle $2\theta_{p_2}$ in Fig. 9–18b, which here is measured *counterclockwise* from CA to CD. It defines the direction θ_{p_2} of σ_2 and its associated principal plane. We have

$$2\theta_{p_2} = \tan^{-1} \frac{6}{12 - 6} = 45.0°$$

$$\theta_{p_2} = 22.5°$$

The element is oriented such that the x' axis or σ_2 is directed $22.5°$ *counterclockwise* from the horizontal (x axis), as shown in Fig. 9–18c.

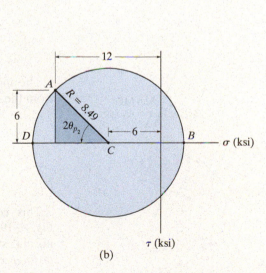

(a)

(b)

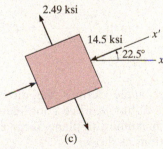

(c)

Fig. 9–18

EXAMPLE 9.8

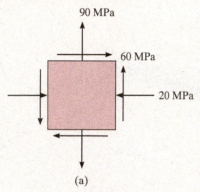

90 MPa

60 MPa

20 MPa

(a)

The state of plane stress at a point is shown on the element in Fig. 9–19a. Determine the maximum in-plane shear stress at this point.

SOLUTION

Construction of the Circle.   From the problem data,

$$\sigma_x = -20 \text{ MPa} \qquad \sigma_y = 90 \text{ MPa} \qquad \tau_{xy} = 60 \text{ MPa}$$

The σ, τ axes are established in Fig. 9–19b. The center of the circle C is located on the σ axis, at the point

$$\sigma_{\text{avg}} = \frac{-20 + 90}{2} = 35 \text{ MPa}$$

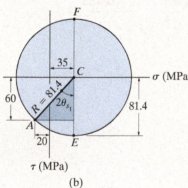

τ (MPa)

(b)

Point C and the reference point $A(-20, 60)$ are plotted. Applying the Pythagorean theorem to the shaded triangle to determine the circle's radius CA, we have

$$R = \sqrt{(60)^2 + (55)^2} = 81.4 \text{ MPa}$$

Maximum In-Plane Shear Stress.   The maximum in-plane shear stress and the average normal stress are identified by point E (or F) on the circle. The coordinates of point $E(35, 81.4)$ give

$$\sigma_{\text{avg}} = 35 \text{ MPa} \qquad\qquad Ans.$$

$$\tau_{\substack{\text{max} \\ \text{in-plane}}} = 81.4 \text{ MPa} \qquad\qquad Ans.$$

The angle θ_{s_1}, measured *counterclockwise* from CA to CE, can be found from the circle, identified as $2\theta_{s_1}$. We have

$$2\theta_{s_1} = \tan^{-1}\left(\frac{20 + 35}{60}\right) = 42.5°$$

$$\theta_{s_1} = 21.3° \qquad\qquad Ans.$$

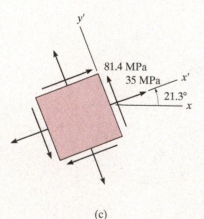

81.4 MPa

35 MPa

21.3°

(c)

Fig. 9–19

This *counterclockwise* angle defines the direction of the x' axis, Fig. 9–19c. Since point E has *positive* coordinates, then the average normal stress and the maximum in-plane shear stress both act in the *positive x' and y'* directions as shown.

EXAMPLE 9.9

The state of plane stress at a point is shown on the element in Fig. 9–20a. Represent this state of stress on an element oriented 30° counterclockwise from the position shown.

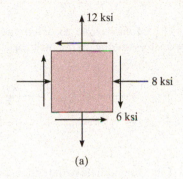

(a)

SOLUTION

Construction of the Circle. From the problem data,

$$\sigma_x = -8 \text{ ksi} \qquad \sigma_y = 12 \text{ ksi} \qquad \tau_{xy} = -6 \text{ ksi}$$

The σ and τ axes are established in Fig. 9–20b. The center of the circle C is on the σ axis at the point

$$\sigma_{\text{avg}} = \frac{-8 + 12}{2} = 2 \text{ ksi}$$

The reference point for $\theta = 0°$ has coordinates $A(-8, -6)$. Hence from the shaded triangle the radius CA is

$$R = \sqrt{(10)^2 + (6)^2} = 11.66$$

Stresses on 30° Element. Since the element is to be rotated 30° *counterclockwise*, we must construct a radial line CP, $2(30°) = 60°$ *counterclockwise*, measured from CA ($\theta = 0°$), Fig. 9–20b. The coordinates of point $P(\sigma_{x'}, \tau_{x'y'})$ must then be obtained. From the geometry of the circle,

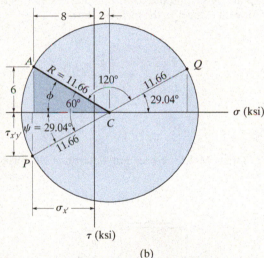

(b)

$$\phi = \tan^{-1}\frac{6}{10} = 30.96° \qquad \psi = 60° - 30.96° = 29.04°$$

$$\sigma_{x'} = 2 - 11.66 \cos 29.04° = -8.20 \text{ ksi} \qquad \textit{Ans.}$$

$$\tau_{x'y'} = 11.66 \sin 29.04° = 5.66 \text{ ksi} \qquad \textit{Ans.}$$

These two stress components act on face BD of the element shown in Fig. 9–20c, since the x' axis for this face is oriented 30° *counterclockwise* from the x axis.

The stress components acting on the adjacent face DE of the element, which is 60° *clockwise* from the positive x axis, Fig. 9–20c, are represented by the coordinates of point Q on the circle. This point lies on the radial line CQ, which is 180° from CP, or 120° *clockwise* from CA. The coordinates of point Q are

$$\sigma_{x'} = 2 + 11.66 \cos 29.04° = 12.2 \text{ ksi} \qquad \textit{Ans.}$$

$$\tau_{x'y'} = -(11.66 \sin 29.04) = -5.66 \text{ ksi} \quad \text{(check)} \qquad \textit{Ans.}$$

NOTE: Here $\tau_{x'y'}$ acts in the $-y'$ direction, Fig. 9–20c.

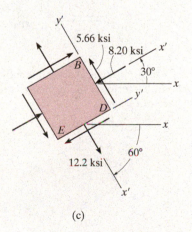

(c)

Fig. 9–20

FUNDAMENTAL PROBLEMS

F9–7. Use Mohr's circle to determine the normal stress and shear stress acting on the inclined plane *AB*.

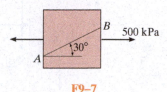

F9–7

F9–8. Use Mohr's circle to determine the principal stresses at the point. Also, find the corresponding orientation of the element with respect to the element shown.

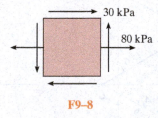

F9–8

F9–9. Draw Mohr's circle and determine the principal stresses.

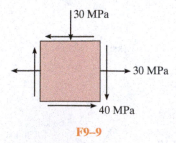

F9–9

F9–10. The hollow circular shaft is subjected to the torque of 4 kN · m. Determine the principal stresses at a point on the surface of the shaft.

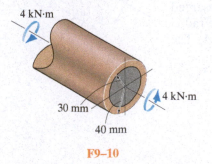

F9–10

F9–11. Determine the principal stresses at point *A* on the cross section of the beam at section *a–a*.

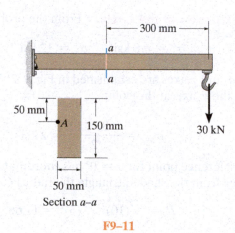

Section *a–a*

F9–11

F9–12. Determine the maximum in-plane shear stress at point *A* on the cross section of the beam at section *a–a*, which is located just to the left of the 60-kN force. Point *A* is just below the flange.

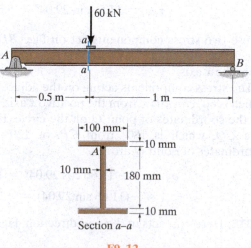

Section *a–a*

F9–12

PROBLEMS

*9–44. Solve Prob. 9–2 using Mohr's circle.

9–45. Solve Prob. 9–3 using Mohr's circle.

9–46. Solve Prob. 9–6 using Mohr's circle.

9–47. Solve Prob. 9–11 using Mohr's circle.

*9–48. Solve Prob. 9–15 using Mohr's circle.

9–49. Solve Prob. 9–16 using Mohr's circle.

9–50. Mohr's circle for the state of stress is shown in Fig. 9–17a. Show that finding the coordinates of point $P(\sigma_{x'}, \tau_{x'y'})$ on the circle gives the same value as the stress transformation Eqs. 9–1 and 9–2.

9–51. Determine (a) the principal stresses and (b) the maximum in-plane shear stress and average normal stress. Specify the orientation of the element in each case.

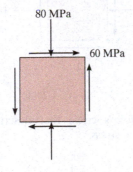

Prob. 9–51

*9–52. Determine (a) the principal stresses and (b) the maximum in-plane shear stress and average normal stress. Specify the orientation of the element in each case.

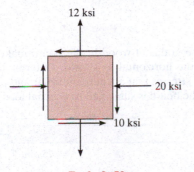

Prob. 9–52

9–53. Determine the equivalent state of stress if an element is oriented 60° clockwise from the element shown.

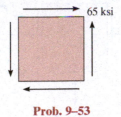

Prob. 9–53

9–54. Draw Mohr's circle that describes each of the following states of stress.

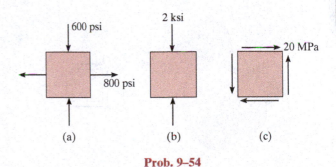

Prob. 9–54

9–55. Draw Mohr's circle that describes each of the following states of stress.

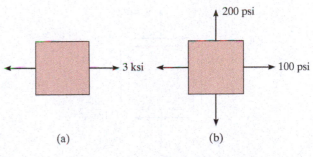

Prob. 9–55

*9–56. Determine (a) the principal stresses and (b) the maximum in-plane shear stress and average normal stress. Specify the orientation of the element in each case.

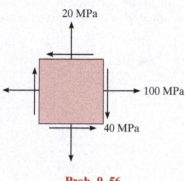

Prob. 9–56

9–57. Determine (a) the principal stresses and (b) the maximum in-plane shear stress and average normal stress. Specify the orientation of the element in each case.

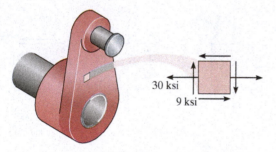

Prob. 9–57

9–58. Determine (a) the principal stresses and (b) the maximum in-plane shear stress and average normal stress. Specify the orientation of the element in each case.

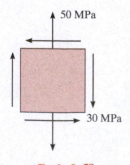

Prob. 9–58

9–59. Determine (a) the principal stresses and (b) the maximum in-plane shear stress and average normal stress. Specify the orientation of the element in each case.

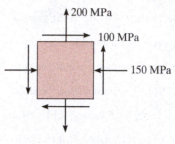

Prob. 9–59

*9–60. Determine (a) the principal stresses and (b) the maximum in-plane shear stress and average normal stress. Specify the orientation of the element in each case.

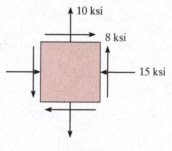

Prob. 9–60

9–61. Draw Mohr's circle that describes each of the following states of stress.

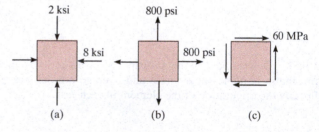

Prob. 9–61

9–62. The grains of wood in the board make an angle of 20° with the horizontal as shown. Determine the normal and shear stresses that act perpendicular and parallel to the grains if the board is subjected to an axial load of 250 N.

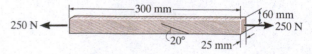

Prob. 9–62

9–63. The post is fixed supported at its base and a horizontal force is applied at its end as shown, determine (a) the maximum in-plane shear stress developed at A and (b) the principal stresses at A.

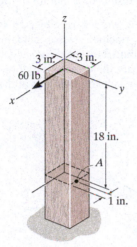

Prob. 9–63

*9–64. Determine the principal stresses, the maximum in-plane shear stress, and average normal stress. Specify the orientation of the element in each case.

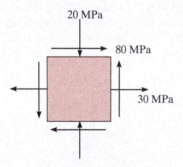

Prob. 9–64

9–65. The thin-walled pipe has an inner diameter of 0.5 in. and a thickness of 0.025 in. If it is subjected to an internal pressure of 500 psi and the axial tension and torsional loadings shown, determine the principal stress at a point on the surface of the pipe.

Prob. 9–65

9–66. The frame supports the triangular distributed load shown. Determine the normal and shear stresses at point D that act perpendicular and parallel, respectively, to the grains. The grains at this point make an angle of 35° with the horizontal as shown.

9–67. The frame supports the triangular distributed load shown. Determine the normal and shear stresses at point E that act perpendicular and parallel, respectively, to the grains. The grains at this point make an angle of 45° with the horizontal as shown.

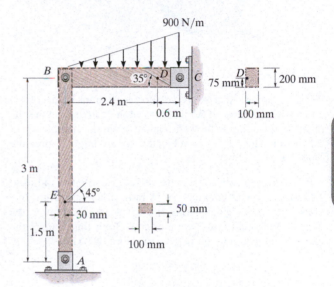

Probs. 9–66/67

*9–68. The rotor shaft of the helicopter is subjected to the tensile force and torque shown when the rotor blades provide the lifting force to suspend the helicopter at midair. If the shaft has a diameter of 6 in., determine the principal stresses and maximum in-plane shear stress at a point located on the surface of the shaft.

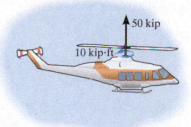

Prob. 9–68

9–69. The pedal crank for a bicycle has the cross section shown. If it is fixed to the gear at B and does not rotate while subjected to a force of 75 lb, determine the principal stresses on the cross section at point C.

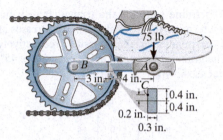

Prob. 9–69

9–70. A spherical pressure vessel has an inner radius of 5 ft and a wall thickness of 0.5 in. Draw Mohr's circle for the state of stress at a point on the vessel and explain the significance of the result. The vessel is subjected to an internal pressure of 80 psi.

9–71. The cylindrical pressure vessel has an inner radius of 1.25 m and a wall thickness of 15 mm. It is made from steel plates that are welded along the 45° seam. Determine the normal and shear stress components along this seam if the vessel is subjected to an internal pressure of 8 MPa.

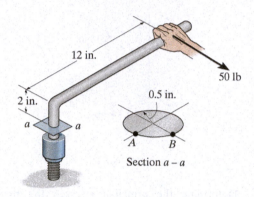

Prob. 9–71

***9–72.** Determine the normal and shear stresses at point D that act perpendicular and parallel, respectively, to the grains. The grains at this point make an angle of 30° with the horizontal as shown. Point D is located just to the left of the 10-kN force.

9–73. Determine the principal stress at point D, which is located just to the left of the 10-kN force.

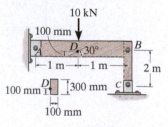

Probs. 9–72/73

9–74. If the box wrench is subjected to the 50 lb force, determine the principal stresses and maximum in-plane shear stress at point A on the cross section of the wrench at section a–a. Specify the orientation of these states of stress and indicate the results on elements at the point.

9–75. If the box wrench is subjected to the 50-lb force, determine the principal stresses and maximum in-plane shear stress at point B on the cross section of the wrench at section a–a. Specify the orientation of these states of stress and indicate the results on elements at the point.

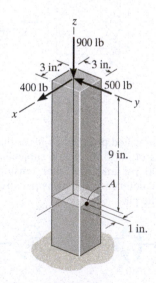

Probs. 9–74/75

***9–76.** The post is fixed supported at its base and the loadings are applied at its end as shown. Determine (a) the maximum in-plane shear stress developed at A and (b) the principal stresses at A.

Prob. 9–76

9.5 ABSOLUTE MAXIMUM SHEAR STRESS

Since the strength of a ductile material depends upon its ability to resist shear stress, it becomes important to find the **absolute maximum shear stress** in the material when it is subjected to a loading. To show how this can be done, we will confine our attention only to the most common case of plane stress,* as shown in Fig. 9–21a. Here *both* σ_1 and σ_2 are tensile. If we view the element in two dimensions at a time, that is, in the y–z, x–z, and x–y planes, Figs. 9–21b, 9–21c, and 9–21d, then we can use Mohr's circle to determine the maximum in-plane shear stress for each case. For example, Mohr's circle extends between 0 and σ_2 for the case shown in Fig. 9–21b. From this circle, Fig. 9–21e, the maximum in-plane shear stress is $\tau_{\substack{max\\in\text{-plane}}} = \sigma_2/2$. Mohr's circles for the other two cases are also shown in Fig. 9–21e. Comparing all three circles, the absolute maximum shear stress is

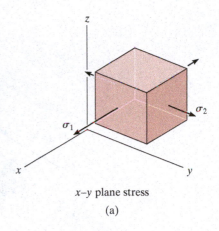

x–y plane stress

(a)

$$\tau_{\substack{abs\\max}} = \frac{\sigma_1}{2} \qquad (9\text{–}13)$$

σ_1 and σ_2 have
the same sign

It occurs on an element that is rotated 45° about the y axis from the element shown in Fig. 9–21a or Fig. 9–21c. It is this out of plane shear stress that will cause the material to fail, not $\tau_{\substack{max\\in\text{-plane}}}$.

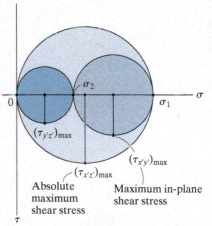

(e)

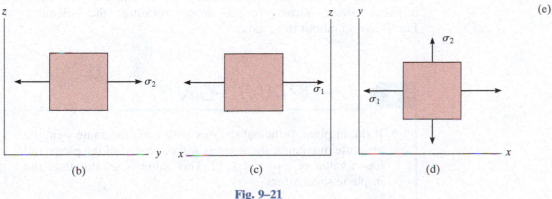

(b)   (c)   (d)

Fig. 9–21

*The case for three-dimensional stress is discussed in books related to advanced mechanics of materials and the theory of elasticity.

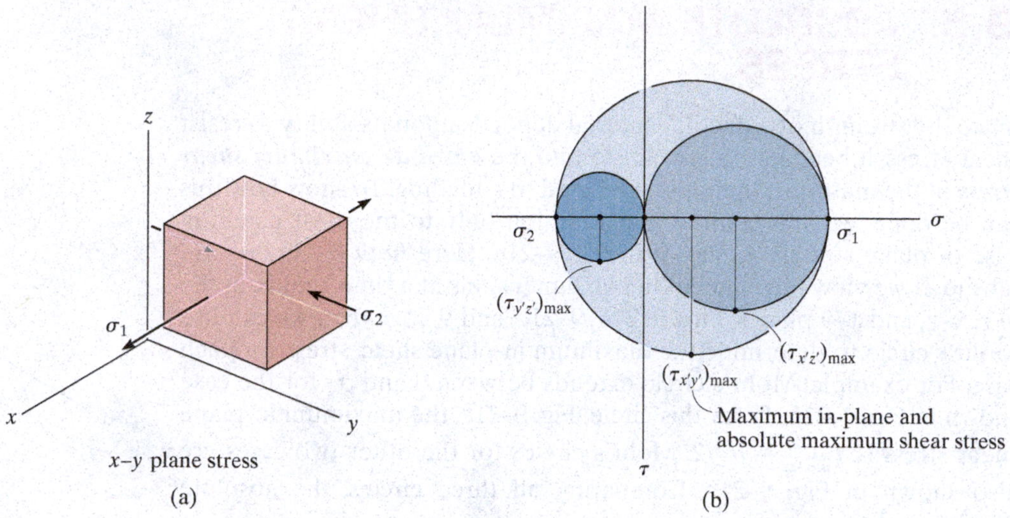

x–y plane stress

(a)

Maximum in-plane and
absolute maximum shear stress

(b)

Fig. 9–22

In a similar manner, if one of the in-plane principal stresses has the *opposite sign* of the other, Fig. 9–22a, then the three Mohr's circles that describe the state of stress for the element when viewed from each plane are shown in Fig. 9–22b. Clearly, in this case

$$\tau_{\substack{\text{abs} \\ \text{max}}} = \frac{\sigma_1 - \sigma_2}{2} \qquad (9\text{–}14)$$

σ_1 and σ_2 have
opposite signs

Here the absolute maximum shear stress is equal to the maximum in-plane shear stress found from rotating the element in Fig. 9–22a, 45° about the z axis.

> ### IMPORTANT POINTS
>
> - If the in-plane principal stresses both have the same sign, the absolute maximum shear stress will occur out of the plane and has a value of $\tau_{\substack{\text{abs} \\ \text{max}}} = \sigma_{max}/2$. This value is greater than the in-plane shear stress.
>
> - If the in-plane principal stresses are of opposite signs, then the absolute maximum shear stress will equal the maximum in-plane shear stress; that is, $\tau_{\substack{\text{abs} \\ \text{max}}} = (\sigma_{max} - \sigma_{min})/2$.

EXAMPLE 9.10

The point on the surface of the pressure vessel in Fig. 9–23a is subjected to the state of plane stress. Determine the absolute maximum shear stress at this point.

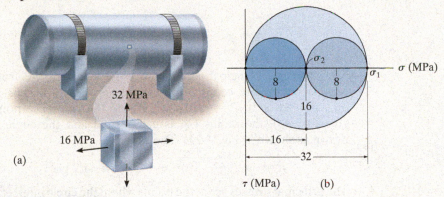

(a)

(b)

Fig. 9–23

SOLUTION

The principal stresses are $\sigma_1 = 32$ MPa, $\sigma_2 = 16$ MPa. If these stresses are plotted along the σ axis, the three Mohr's circles can be constructed that describe the state of stress viewed in each of the three perpendicular planes, Fig. 9–23b. The largest circle has a radius of 16 MPa and describes the state of stress in the plane only containing $\sigma_1 = 32$ MPa, shown shaded in Fig. 9–23a. An orientation of an element 45° within this plane yields the state of absolute maximum shear stress and the associated average normal stress, namely,

$$\tau_{\max}^{\text{abs}} = 16 \text{ MPa} \qquad\qquad Ans.$$

$$\sigma_{\text{avg}} = 16 \text{ MPa}$$

This same result for $\tau_{\max}^{\text{abs}}$ can be obtained from direct application of Eq. 9–13.

$$\tau_{\max}^{\text{abs}} = \frac{\sigma_1}{2} = \frac{32}{2} = 16 \text{ MPa} \qquad\qquad Ans.$$

$$\sigma_{\text{avg}} = \frac{32 + 0}{2} = 16 \text{ MPa}$$

By comparison, the maximum in-plane shear stress can be determined from the Mohr's circle drawn between $\sigma_1 = 32$ MPa and $\sigma_2 = 16$ MPa, Fig. 9–23b. This gives a value of

$$\tau_{\substack{\max \\ \text{in-plane}}} = \frac{32 - 16}{2} = 8 \text{ MPa}$$

$$\sigma_{\text{avg}} = \frac{32 + 16}{2} = 24 \text{ MPa}$$

EXAMPLE 9.11

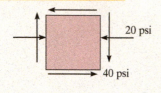

(a)

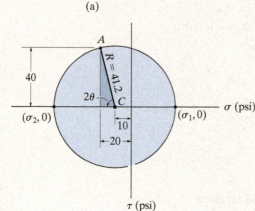

(b)

Due to an applied loading, an element at a point on a machine shaft is subjected to the state of plane stress shown in Fig. 9–24a. Determine the principal stresses and the absolute maximum shear stress at the point.

SOLUTION

Principal Stresses.
The in-plane principal stresses can be determined from Mohr's circle. The center of the circle is on the σ axis at $\sigma_{avg} = (-20 + 0)/2 = -10$ psi. Plotting the reference point $A(-20, -40)$, the radius CA is established and the circle is drawn as shown in Fig. 9–24b. The radius is

$$R = \sqrt{(20 - 10)^2 + (40)^2} = 41.2 \text{ psi}$$

The principal stresses are at the points where the circle intersects the σ axis; i.e.,

$$\sigma_1 = -10 + 41.2 = 31.2 \text{ psi}$$
$$\sigma_2 = -10 - 41.2 = -51.2 \text{ psi}$$

From the circle, the *counterclockwise* angle 2θ, measured from CA to the $-\sigma$ axis, is

$$2\theta = \tan^{-1}\left(\frac{40}{20 - 10}\right) = 76.0°$$

Thus,

$$\theta = 38.0°$$

This *counterclockwise* rotation defines the direction of the x' axis and σ_2, Fig. 9–24c. We have

$$\sigma_1 = 31.2 \text{ psi} \quad \sigma_2 = -51.2 \text{ psi} \qquad \textit{Ans.}$$

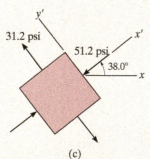

(c)

Absolute Maximum Shear Stress. Since these stresses have opposite signs, applying Eq. 9–14 we have

$$\tau_{\substack{abs \\ max}} = \frac{\sigma_1 - \sigma_2}{2} = \frac{31.2 - (-51.2)}{2} = 41.2 \text{ psi} \qquad \textit{Ans.}$$

$$\sigma_{avg} = \frac{31.2 - 51.2}{2} = -10 \text{ psi}$$

These same results can also be obtained by drawing Mohr's circle for each orientation of an element about the x, y, and z axes, Fig. 9–24d. Since σ_1 and σ_2 are of *opposite signs*, then the absolute maximum shear stress as noted equals the maximum in-plane shear stress.

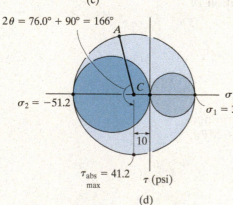

(d)

Fig. 9–24

PROBLEMS

9–77. Draw the three Mohr's circles that describe each of the following states of stress.

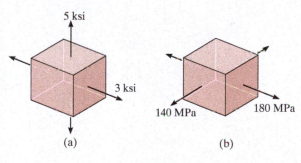

(a)   (b)

Prob. 9–77

9–78. Draw the three Mohr's circles that describe the following state of stress.

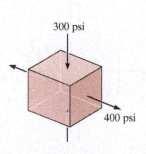

Prob. 9–78

9–79. Draw the three Mohr's circles that describe the following state of stress.

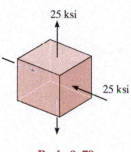

Prob. 9–79

*9–80. Determine the principal stresses and the absolute maximum shear stress.

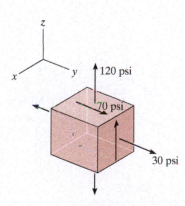

Prob. 9–80

9

9–81. Determine the principal stresses and the absolute maximum shear stress.

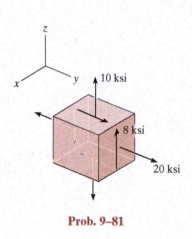

Prob. 9–81

9–82. Determine the principal stresses and the absolute maximum shear stress.

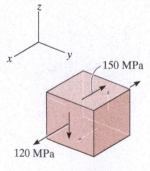

Prob. 9–82

9–83. Determine the principal stresses and the absolute maximum shear stress.

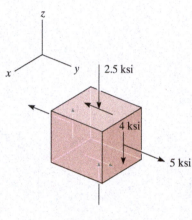

Prob. 9–83

***9–84.** Consider the general case of plane stress as shown. Write a computer program that will show a plot of the three Mohr's circles for the element, and will also determine the maximum in-plane shear stress and the absolute maximum shear stress.

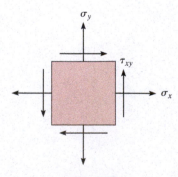

Prob. 9–84

9–85. The solid shaft is subjected to a torque, bending moment, and shear force. Determine the principal stresses at points A and B and the absolute maximum shear stress.

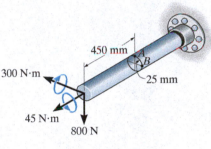

450 mm

A
B

300 N·m

25 mm

45 N·m

800 N

Prob. 9–85

9–86. The frame is subjected to a horizontal force and couple moment. Determine the principal stresses and the absolute maximum shear stress at point A. The cross-sectional area at this point is shown.

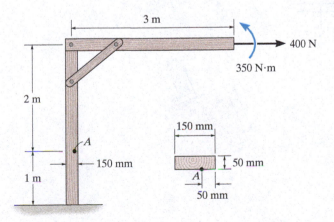

3 m

400 N

350 N·m

2 m

150 mm

A

150 mm

50 mm

A

1 m

50 mm

Prob. 9–86

9–87. The bolt is fixed to its support at C. If a force of 18 lb is applied to the wrench to tighten it, determine the principal stresses and the absolute maximum shear stress in the bolt shank at point A. Represent the results on an element located at this point. The shank has a diameter of 0.25 in.

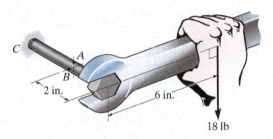

C

A

B

2 in.

6 in.

18 lb

Prob. 9–87

***9–88.** The bolt is fixed to its support at C. If a force of 18 lb is applied to the wrench to tighten it, determine the principal stresses and the absolute maximum shear stress developed in the bolt shank at point B. Represent the results on an element located at this point. The shank has a diameter of 0.25 in.

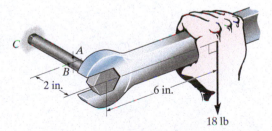

C

A

B

2 in.

6 in.

18 lb

Prob. 9–88

9

CHAPTER REVIEW

Plane stress occurs when the material at a point is subjected to two normal stress components σ_x and σ_y and a shear stress τ_{xy}. Provided these components are known, then the stress components acting on an element having a different orientation θ can be determined using the equations of stress transformation.

$$\sigma_{x'} = \frac{\sigma_x + \sigma_y}{2} + \frac{\sigma_x - \sigma_y}{2}\cos 2\theta + \tau_{xy}\sin 2\theta$$

$$\tau_{x'y'} = -\frac{\sigma_x - \sigma_y}{2}\sin 2\theta + \tau_{xy}\cos 2\theta$$

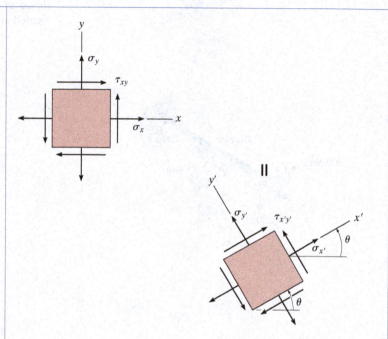

For design, it is important to determine either the maximum principal normal stress or the maximum in-plane shear stress at a point.

No shear stress acts on the planes of principal stress, where

$$\sigma_{1,2} = \frac{\sigma_x + \sigma_y}{2} \pm \sqrt{\left(\frac{\sigma_x - \sigma_y}{2}\right)^2 + \tau_{xy}^2}$$

On the planes of maximum in-plane shear stress there is an associated average normal stress, where

$$\tau_{\substack{\max \\ \text{in-plane}}} = \sqrt{\left(\frac{\sigma_x - \sigma_y}{2}\right)^2 + \tau_{xy}^2}$$

$$\sigma_{\text{avg}} = \frac{\sigma_x + \sigma_y}{2}$$

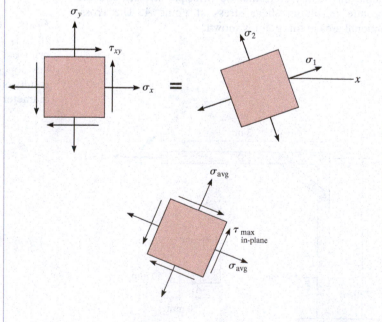

Mohr's circle provides a semigraphical method for finding either the stress on any plane, or the principal normal stresses, or the maximum in-plane shear stress. To draw the circle, first establish the σ and τ axes, then plot the center of the circle $C[(\sigma_x + \sigma_y)/2, 0]$ and the reference point $A(\sigma_x, \tau_{xy})$. The radius R of the circle extends between these two points and is determined from trigonometry.

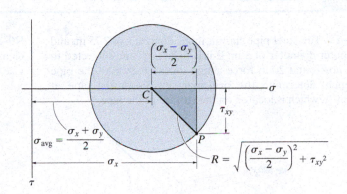

If σ_1 and σ_2 are of the same sign, then the absolute maximum shear stress at a point will lie out of plane.

$$\tau_{\substack{abs \\ max}} = \frac{\sigma_1}{2}$$

If σ_1 and σ_2 have the opposite sign, then the absolute maximum shear stress will be equal to the maximum in-plane shear stress.

$$\tau_{\substack{abs \\ max}} = \frac{\sigma_1 - \sigma_2}{2}$$

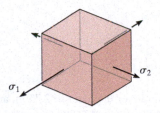

x–y plane stress

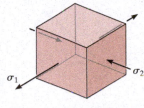

x–y plane stress

9

REVIEW PROBLEMS

R9–1. The steel pipe has an inner diameter of 2.75 in. and an outer diameter of 3 in. If it is fixed at C and subjected to the horizontal 20-lb force acting on the handle of the pipe wrench, determine the principal stresses in the pipe at point A, which is located on the surface of the pipe.

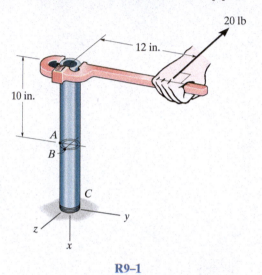

R9–1

R9–2. The steel pipe has an inner diameter of 2.75 in. and an outer diameter of 3 in. If it is fixed at C and subjected to the horizontal 20-lb force acting on the handle of the pipe wrench, determine the principal stresses in the pipe at point B, which is located on the surface of the pipe.

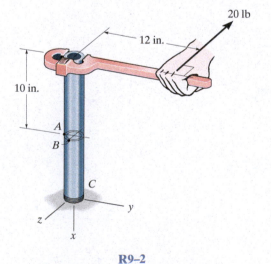

R9–2

R9–3. Determine the equivalent state of stress if an element is oriented 40° clockwise from the element shown. Use Mohr's circle.

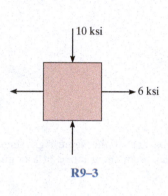

R9–3

*R9–4. The crane is used to support the 350-lb load. Determine the principal stresses acting in the boom at points A and B. The cross section is rectangular and has a width of 6 in. and a thickness of 3 in. Use Mohr's circle.

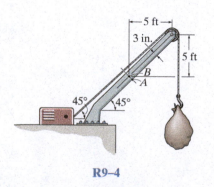

R9–4

R9–5. Determine the equivalent state of stress on an element which represents (a) the principal stresses, and (b) the maximum in-plane shear stress and the associated average normal stress. Also, for each case, determine the corresponding orientation of the element with respect to the element shown and sketch the results on the element.

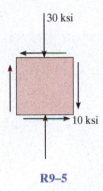

30 ksi

10 ksi

R9–5

R9–6. The propeller shaft of the tugboat is subjected to the compressive force and torque shown. If the shaft has an inner diameter of 100 mm and an outer diameter of 150 mm, determine the principal stresses at a point A located on the outer surface.

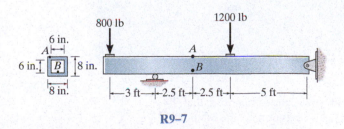

10 kN

2 kN·m

R9–6

R9–7. Determine the principal stresses in the box beam at points A and B.

6 in.

6 in.

8 in.

8 in.

800 lb

1200 lb

A

B

3 ft — 2.5 ft — 2.5 ft — 5 ft

R9–7

***R9–8.** Determine (a) the principal stresses and (b) the maximum in-plane shear stress and average normal stress at the point. Specify the orientation of the element in each case.

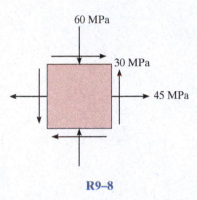

60 MPa

30 MPa

45 MPa

R9–8

R9–9. Determine the stress components acting on the inclined plane AB. Solve the problem using the method of equilibrium described in Sec. 9.1.

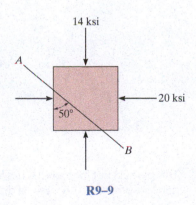

14 ksi

A

50°

20 ksi

B

R9–9

CHAPTER 10

(© Peter Steiner/Alamy)

This pin support for a bridge has been tested with strain gages to ensure that the principal strains in the material do not exceed a failure criterion for the material.

STRAIN TRANSFORMATION

CHAPTER OBJECTIVES

■ The transformation of strain at a point is similar to the transformation of stress, and as a result the methods of the previous chapter will be applied in this chapter. Here we will also discuss various ways for measuring strain and develop some important material-property relationships, including a generalized form of Hooke's law. At the end of the chapter, a few of the theories used to predict the failure of a material will be discussed.

10.1 PLANE STRAIN

As outlined in Sec. 2.2, the general state of strain at a point in a body is represented by a combination of three components of normal strain, ϵ_x, ϵ_y, ϵ_z, and three components of shear strain, γ_{xy}, γ_{xz}, γ_{yz}, Fig. 2–4c. The normal strains cause a change in the volume of the element, and the shear strains cause a change in its shape. Like stress, these six components depend upon the orientation of the element, and in many situations, engineers must transform the strains in order to obtain their values in other directions.

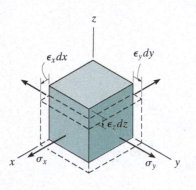

Plane stress, σ_x, σ_y, does not cause plane strain in the x–y plane since $\epsilon_z \neq 0$.

Fig. 10–1

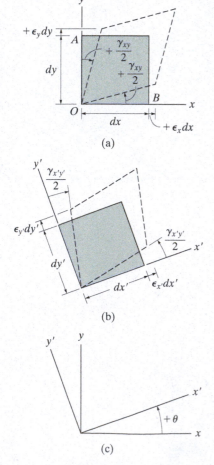

To understand how this is done, we will direct our attention to a study of **plane strain**, whereby the element is subjected to two components of normal strain, ϵ_x, ϵ_y, and one component of shear strain, γ_{xy}.* Although plane strain and plane stress each have three components lying in the same plane, realize that plane stress *does not* necessarily cause plane strain or vice versa. The reason for this has to do with the Poisson effect discussed in Sec. 3.6. For example, the element in Fig. 10–1 is subjected to *plane stress* caused by σ_x and σ_y. Not only are normal strains ϵ_x and ϵ_y produced, but there is *also* an associated normal strain, ϵ_z, and so this is *not* a case of plane strain.

Actually, a case of plane strain rarely occurs in practice, because many materials are never constrained between rigid surfaces so as not to permit any distortion in, say, the z direction (see photo). In spite of this, the analysis of plane strain, as outlined in the following section, is still of great importance, because it will allow us to convert strain-gage data, measured at a point on the surface of a body, into plane stress at the point.

10.2   GENERAL EQUATIONS OF PLANE-STRAIN TRANSFORMATION

For plane-strain analysis it is important to establish strain transformation equations that can be used to determine the components of normal and shear strain at a point, $\epsilon_{x'}$, $\epsilon_{y'}$, $\gamma_{x'y'}$, Fig. 10–2b, provided the components ϵ_x, ϵ_y, γ_{xy} are known, Fig. 10–2a. So in other words, if we know how the element of material in Fig. 10–2a deforms, we want to know how the tipped element of material in Fig. 10–2b will deform. To do this requires relating the deformations and rotations of line segments which represent the sides of differential elements that are parallel to the x, y and x', y' axes.

Sign Convention.   To begin, we must first establish a sign convention for strain. The *normal strains* ϵ_x and ϵ_y in Fig. 10–2a are *positive* if they cause *elongation* along the x and y axes, respectively, and the *shear strain* γ_{xy} is *positive* if the interior angle *AOB* becomes *smaller* than 90°. This sign convention also follows the corresponding one used for plane stress, Fig. 9–5a, that is, positive σ_x, σ_y, τ_{xy} will cause the element to *deform* in the positive ϵ_x, ϵ_y, γ_{xy} directions, respectively. Finally, if the angle between the x and x' axes is θ, then, like the case of plane stress, θ will be *positive* provided it follows the curl of the right-hand fingers, i.e., counterclockwise, as shown in Fig. 10–2c.

*Three-dimensional strain analysis is discussed in books related to advanced mechanics of materials or the theory of elasticity.

Fig. 10–2

Normal and Shear Strains.   To determine $\epsilon_{x'}$, we must find the elongation of a line segment dx' that lies along the x' axis and is subjected to strain components ϵ_x, ϵ_y, γ_{xy}. As shown in Fig. 10–3a, the components of line dx' along the x and y axes are

$$dx = dx' \cos \theta$$
$$dy = dx' \sin \theta \tag{10-1}$$

When the positive normal strain ϵ_x occurs, dx is elongated $\epsilon_x\, dx$, Fig. 10–3b, which causes dx' to elongate $\epsilon_x\, dx \cos \theta$. Likewise, when ϵ_y occurs, dy elongates $\epsilon_y\, dy$, Fig. 10–3c, which causes dx' to elongate $\epsilon_y\, dy \sin \theta$. Finally, assuming that dx remains fixed in position, the shear strain γ_{xy} in Fig. 10–3d, which is the change in angle between dx and dy, causes the top of line dy to be displaced $\gamma_{xy}\, dy$ to the right. This causes dx' to elongate $\gamma_{xy}\, dy \cos \theta$. If all three of these (red) elongations are added together, the resultant elongation of dx' is then

$$\delta x' = \epsilon_x\, dx \cos \theta + \epsilon_y\, dy \sin \theta + \gamma_{xy}\, dy \cos \theta$$

Since the normal strain along line dx' is $\epsilon_{x'} = \delta x'/dx'$, then using Eqs. 10–1, we have

$$\epsilon_{x'} = \epsilon_x \cos^2 \theta + \epsilon_y \sin^2 \theta + \gamma_{xy} \sin \theta \cos \theta \tag{10-2}$$

This normal strain is shown in Fig. 10–2b.

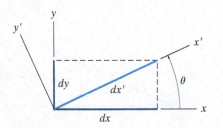

Before deformation

(a)

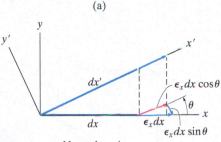

Normal strain ϵ_x

(b)

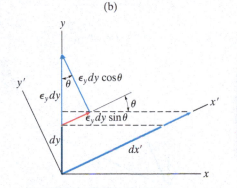

Normal strain ϵ_y

(c)

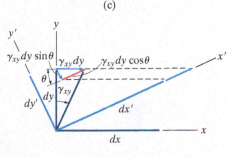

Shear strain γ_{xy}

(d)

Fig. 10–3

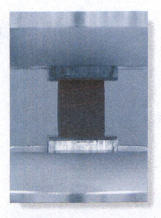

The rubber specimen is constrained between the two fixed supports, and so it will undergo plane strain when loads are applied to it in the horizontal plane.

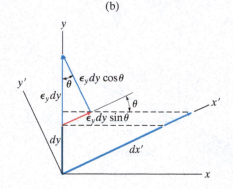

Normal strain ϵ_x

(b)

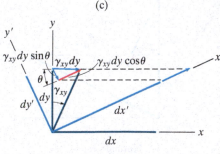

Normal strain ϵ_y

(c)

To determine $\gamma_{x'y'}$, we must find the rotation of each of the line segments dx' and dy' when they are subjected to the strain components ϵ_x, ϵ_y, γ_{xy}. First we will consider the counterclockwise rotation α of dx', Fig. 10–3e. Here $\alpha = \delta y'/dx'$. The displacement $\delta y'$ consists of three displacement components: one from ϵ_x, giving $-\epsilon_x\, dx \sin \theta$, Fig. 10–3b; another from ϵ_y, giving $\epsilon_y\, dy \cos \theta$, Fig. 10–3c; and the last from γ_{xy}, giving $-\gamma_{xy}\, dy \sin \theta$, Fig. 10–3d. Thus, $\delta y'$ is

$$\delta y' = -\epsilon_x\, dx \sin \theta + \epsilon_y\, dy \cos \theta - \gamma_{xy}\, dy \sin \theta$$

Using Eq. 10–1, we therefore have

$$\alpha = \frac{\delta y'}{dx'} = (-\epsilon_x + \epsilon_y) \sin \theta \cos \theta - \gamma_{xy} \sin^2 \theta \qquad (10\text{–}3)$$

Finally, line dy' rotates by an amount β, Fig. 10–3e. We can determine this angle by a similar analysis, or by simply substituting $\theta + 90°$ for θ into Eq. 10–3. Using the identities $\sin(\theta + 90°) = \cos \theta$, $\cos(\theta + 90°) = -\sin \theta$, we have

$$\beta = (-\epsilon_x + \epsilon_y) \sin(\theta + 90°) \cos(\theta + 90°) - \gamma_{xy} \sin^2(\theta + 90°)$$

$$= -(-\epsilon_x + \epsilon_y) \cos \theta \sin \theta - \gamma_{xy} \cos^2 \theta$$

Since α and β must represent the rotation of the sides dx' and dy' in the manner shown in Fig. 10–3c, then the element is subjected to a shear strain of

$$\gamma_{x'y'} = \alpha - \beta = -2(\epsilon_x - \epsilon_y) \sin \theta \cos \theta + \gamma_{xy}(\cos^2 \theta - \sin^2 \theta) \qquad (10\text{–}4)$$

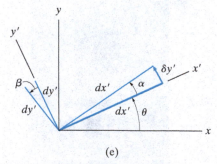

Shear strain γ_{xy}

(d)

(e)

Fig. 10–3 (cont.)

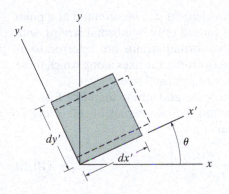

Positive normal strain, $\epsilon_{x'}$

(a)

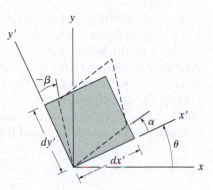

Positive shear strain, $\gamma_{x'y'}$

(b)

Fig. 10–4

Using the trigonometric identities $\sin 2\theta = 2\sin\theta\cos\theta$, $\cos^2\theta = (1 + \cos 2\theta)/2$, and $\sin^2\theta + \cos^2\theta = 1$, Eqs. 10–2 and 10–4 can be written in the final form

$$\epsilon_{x'} = \frac{\epsilon_x + \epsilon_y}{2} + \frac{\epsilon_x - \epsilon_y}{2}\cos 2\theta + \frac{\gamma_{xy}}{2}\sin 2\theta \qquad (10\text{–}5)$$

$$\frac{\gamma_{x'y'}}{2} = -\left(\frac{\epsilon_x - \epsilon_y}{2}\right)\sin 2\theta + \frac{\gamma_{xy}}{2}\cos 2\theta \qquad (10\text{–}6)$$

Normal and Shear Strain Components

According to our sign convention, if $\epsilon_{x'}$ is *positive*, the element *elongates* in the positive x' direction, Fig. 10–4a, and if $\gamma_{x'y'}$ is positive, the element deforms as shown in Fig. 10–4b.

If the normal strain in the y' direction is required, it can be obtained from Eq. 10–5 by simply substituting $(\theta + 90°)$ for θ. The result is

$$\epsilon_{y'} = \frac{\epsilon_x + \epsilon_y}{2} - \frac{\epsilon_x - \epsilon_y}{2}\cos 2\theta - \frac{\gamma_{xy}}{2}\sin 2\theta \qquad (10\text{–}7)$$

The similarity between the above three equations and those for plane-stress transformation, Eqs. 9–1, 9–2, and 9–3, should be noted. Making the comparison, $\sigma_x, \sigma_y, \sigma_{x'}, \sigma_{y'}$ correspond to $\epsilon_x, \epsilon_y, \epsilon_{x'}, \epsilon_{y'}$; and $\tau_{xy}, \tau_{x'y'}$ correspond to $\gamma_{xy}/2, \gamma_{x'y'}/2$.

Principal Strains. Like stress, an element can be oriented at a point so that the element's deformation is caused only by normal strains, with *no* shear strain. When this occurs the normal strains are referred to as *principal strains*, and if the material is isotropic, the axes along which these strains occur will coincide with the axes of principal stress.

From the correspondence between stress and strain, then like Eqs. 9–4 and 9–5, the direction of the x' axis and the two values of the principal strains ϵ_1 and ϵ_2 are determined from

$$\tan 2\theta_p = \frac{\gamma_{xy}}{\epsilon_x - \epsilon_y} \tag{10–8}$$

Orientation of principal planes

Complex stresses are often developed at the joints where the cylindrical and hemispherical vessels are joined together. The stresses are determined by making measurements of strain.

$$\epsilon_{1,2} = \frac{\epsilon_x + \epsilon_y}{2} \pm \sqrt{\left(\frac{\epsilon_x - \epsilon_y}{2}\right)^2 + \left(\frac{\gamma_{xy}}{2}\right)^2} \tag{10–9}$$

Principal strains

Maximum In-Plane Shear Strain. Similar to Eqs. 9–6, 9–7, and 9–8, the direction of the x' axis and the maximum in-plane shear strain and associated average normal strain are determined from the following equations:

$$\tan 2\theta_s = -\left(\frac{\epsilon_x - \epsilon_y}{\gamma_{xy}}\right) \tag{10–10}$$

Orientation of maximum in-plane shear strain

$$\frac{\gamma_{\substack{max \\ \text{in-plane}}}}{2} = \sqrt{\left(\frac{\epsilon_x - \epsilon_y}{2}\right)^2 + \left(\frac{\gamma_{xy}}{2}\right)^2} \tag{10–11}$$

Maximum in-plane shear strain

$$\epsilon_{avg} = \frac{\epsilon_x + \epsilon_y}{2} \tag{10–12}$$

Average normal strain

▶ IMPORTANT POINTS

- In the case of plane stress, plane-strain analysis may be used within the plane of the stresses to analyze the data from strain gages. Remember, though, there will be a normal strain that is perpendicular to the gages due to the Poisson effect.

- When the state of strain is represented by the principal strains, no shear strain will act on the element.

- When the state of strain is represented by the maximum in-plane shear strain, an associated average normal strain will also act on the element.

EXAMPLE 10.1

The state of plane strain at a point has components of $\epsilon_x = 500(10^{-6})$, $\epsilon_y = -300(10^{-6})$, $\gamma_{xy} = 200(10^{-6})$, which tends to distort the element as shown in Fig. 10–5a. Determine the equivalent strains acting on an element of the material oriented *clockwise* 30°.

SOLUTION

The strain transformation Eqs. 10–5 and 10–6 will be used to solve the problem. Since θ is *positive counterclockwise*, then for this problem $\theta = -30°$. Thus,

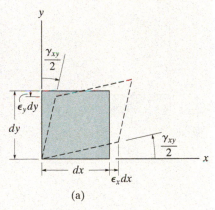

(a)

$$\epsilon_{x'} = \frac{\epsilon_x + \epsilon_y}{2} + \frac{\epsilon_x - \epsilon_y}{2}\cos 2\theta + \frac{\gamma_{xy}}{2}\sin 2\theta$$

$$= \left[\frac{500 + (-300)}{2}\right](10^{-6}) + \left[\frac{500 - (-300)}{2}\right](10^{-6})\cos(2(-30°))$$

$$+ \left[\frac{200(10^{-6})}{2}\right]\sin(2(-30°))$$

$$\epsilon_{x'} = 213(10^{-6}) \qquad \qquad \textit{Ans.}$$

$$\frac{\gamma_{x'y'}}{2} = -\left(\frac{\epsilon_x - \epsilon_y}{2}\right)\sin 2\theta + \frac{\gamma_{xy}}{2}\cos 2\theta$$

$$= -\left[\frac{500 - (-300)}{2}\right](10^{-6})\sin(2(-30°)) + \frac{200(10^{-6})}{2}\cos(2(-30°))$$

$$\gamma_{x'y'} = 793(10^{-6}) \qquad \qquad \textit{Ans.}$$

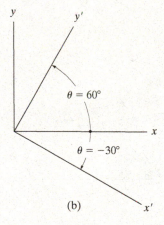

(b)

The strain in the y' direction can be obtained from Eq. 10–7 with $\theta = -30°$. However, we can also obtain $\epsilon_{y'}$ using Eq. 10–5 with $\theta = 60°(\theta = -30° + 90°)$, Fig. 10–5b. We have with $\epsilon_{y'}$ replacing $\epsilon_{x'}$,

$$\epsilon_{y'} = \frac{\epsilon_x + \epsilon_y}{2} + \frac{\epsilon_x - \epsilon_y}{2}\cos 2\theta + \frac{\gamma_{xy}}{2}\sin 2\theta$$

$$= \left[\frac{500 + (-300)}{2}\right](10^{-6}) + \left[\frac{500 - (-300)}{2}\right](10^{-6})\cos(2(60°))$$

$$+ \frac{200(10^{-6})}{2}\sin(2(60°))$$

$$\epsilon_{y'} = -13.4(10^{-6}) \qquad \qquad \textit{Ans.}$$

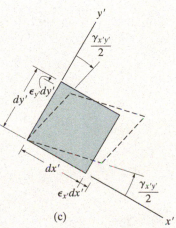

(c)

These results tend to distort the element as shown in Fig. 10–5c.

Fig. 10–5

EXAMPLE 10.2

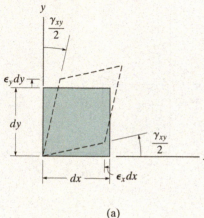

(a)

The state of plane strain at a point has components of $\epsilon_x = -350(10^{-6})$, $\epsilon_y = 200(10^{-6})$, $\gamma_{xy} = 80(10^{-6})$, Fig. 10–6a. Determine the principal strains at the point and the orientation of the element upon which they act.

SOLUTION

Orientation of the Element. From Eq. 10–8 we have

$$\tan 2\theta_p = \frac{\gamma_{xy}}{\epsilon_x - \epsilon_y}$$

$$= \frac{80(10^{-6})}{(-350 - 200)(10^{-6})}$$

Thus, $2\theta_p = -8.28°$ and $-8.28° + 180° = 171.72°$, so that

$$\theta_p = -4.14° \text{ and } 85.9° \qquad \textit{Ans.}$$

Each of these angles is measured *positive counterclockwise*, from the x axis to the outward normals on each face of the element. The angle of $-4.14°$ is shown in Fig. 10–6b.

Principal Strains. The principal strains are determined from Eq. 10–9. We have

$$\epsilon_{1,2} = \frac{\epsilon_x + \epsilon_y}{2} \pm \sqrt{\left(\frac{\epsilon_x - \epsilon_y}{2}\right)^2 + \left(\frac{\gamma_{xy}}{2}\right)^2}$$

$$= \frac{(-350 + 200)(10^{-6})}{2} \pm \left[\sqrt{\left(\frac{-350 - 200}{2}\right)^2 + \left(\frac{80}{2}\right)^2}\right](10^{-6})$$

$$= -75.0(10^{-6}) \pm 277.9(10^{-6})$$

$$\epsilon_1 = 203(10^{-6}) \qquad \epsilon_2 = -353(10^{-6}) \qquad \textit{Ans.}$$

To determine the direction of each of these strains we will apply Eq. 10–5 with $\theta = -4.14°$, Fig. 10–6b. Thus,

$$\epsilon_{x'} = \frac{\epsilon_x + \epsilon_y}{2} + \frac{\epsilon_x - \epsilon_y}{2} \cos 2\theta + \frac{\gamma_{xy}}{2} \sin 2\theta$$

$$= \left(\frac{-350 + 200}{2}\right)(10^{-6}) + \left(\frac{-350 - 200}{2}\right)(10^{-6}) \cos 2(-4.14°)$$

$$+ \frac{80(10^{-6})}{2} \sin 2(-4.14°)$$

$$\epsilon_{x'} = -353(10^{-6})$$

Hence $\epsilon_{x'} = \epsilon_2$. When subjected to the principal strains, the element is distorted as shown in Fig. 10–6b.

(b)

Fig. 10–6

10

EXAMPLE 10.3

The state of plane strain at a point has components of $\epsilon_x = -350(10^{-6})$, $\epsilon_y = 200(10^{-6})$, $\gamma_{xy} = 80(10^{-6})$, Fig. 10–7a. Determine the maximum in-plane shear strain at the point and the orientation of the element upon which it acts.

SOLUTION

Orientation of the Element.   From Eq. 10–10 we have

$$\tan 2\theta_s = -\left(\frac{\epsilon_x - \epsilon_y}{\gamma_{xy}}\right) = -\frac{(-350 - 200)(10^{-6})}{80(10^{-6})}$$

Thus, $2\theta_s = 81.72°$ and $81.72° + 180° = 261.72°$, so that

$$\theta_s = 40.9° \text{ and } 131°$$

Notice that this orientation is 45° from that shown in Fig. 10–6b.

Maximum In-Plane Shear Strain.   Applying Eq. 10–11 gives

$$\frac{\gamma_{\substack{max \\ in\text{-}plane}}}{2} = \sqrt{\left(\frac{\epsilon_x - \epsilon_y}{2}\right)^2 + \left(\frac{\gamma_{xy}}{2}\right)^2}$$

$$= \left[\sqrt{\left(\frac{-350 - 200}{2}\right)^2 + \left(\frac{80}{2}\right)^2}\right](10^{-6})$$

$$\gamma_{\substack{max \\ in\text{-}plane}} = 556(10^{-6}) \qquad \textit{Ans.}$$

The square root gives two signs for $\gamma_{\substack{max \\ in\text{-}plane}}$. The proper one for each angle can be obtained by applying Eq. 10–6. When $\theta_s = 40.9°$, we have

$$\frac{\gamma_{x'y'}}{2} = -\frac{\epsilon_x - \epsilon_y}{2} \sin 2\theta + \frac{\gamma_{xy}}{2} \cos 2\theta$$

$$= -\left(\frac{-350 - 200}{2}\right)(10^{-6}) \sin 2(40.9°) + \frac{80(10^{-6})}{2} \cos 2(40.9°)$$

$$\gamma_{x'y'} = 556(10^{-6})$$

This result is positive and so $\gamma_{\substack{max \\ in\text{-}plane}}$ tends to distort the element so that the right angle between dx' and dy' is *decreased* (positive sign convention), Fig. 10–7b.

Also, there are associated average normal strains imposed on the element that are determined from Eq. 10–12.

$$\epsilon_{avg} = \frac{\epsilon_x + \epsilon_y}{2} = \frac{-350 + 200}{2} (10^{-6}) = -75(10^{-6})$$

These strains tend to cause the element to contract, Fig. 10–7b.

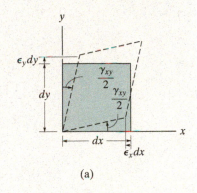

(a)

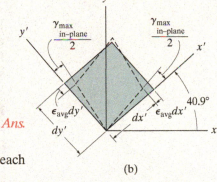

(b)

Fig. 10–7

*10.3   MOHR'S CIRCLE—PLANE STRAIN

Since the equations of plane-strain transformation are mathematically similar to the equations of plane-stress transformation, we can also solve problems involving the transformation of strain using Mohr's circle.

Like the case for stress, the parameter θ in Eqs. 10–5 and 10–6 can be eliminated and the result rewritten in the form

$$(\epsilon_{x'} - \epsilon_{avg})^2 + \left(\frac{\gamma_{x'y'}}{2}\right)^2 = R^2 \qquad (10\text{–}13)$$

where

$$\epsilon_{avg} = \frac{\epsilon_x + \epsilon_y}{2}$$

$$R = \sqrt{\left(\frac{\epsilon_x - \epsilon_y}{2}\right)^2 + \left(\frac{\gamma_{xy}}{2}\right)^2}$$

Equation 10–13 represents the equation of Mohr's circle for strain. It has a center on the ϵ axis at point $C(\epsilon_{avg}, 0)$ and a radius R. As described in the following procedure, Mohr's circle can be used to determine the principal strains, the maximum in-plane strain, or the strains on an arbitrary plane.

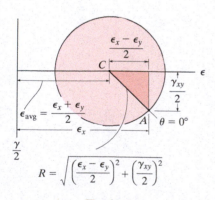

$$R = \sqrt{\left(\frac{\epsilon_x - \epsilon_y}{2}\right)^2 + \left(\frac{\gamma_{xy}}{2}\right)^2}$$

Fig. 10–8

▶ PROCEDURE FOR ANALYSIS

The procedure for drawing Mohr's circle for strain follows the same one established for stress.

Construction of the Circle.

- Establish a coordinate system such that the horizontal axis represents the normal strain ϵ, with *positive to the right*, and the vertical axis represents *half* the value of the shear strain, $\gamma/2$, with *positive downward*, Fig. 10–8.

- Using the positive sign convention for ϵ_x, ϵ_y, γ_{xy}, Fig. 10–2, determine the center of the circle C, located $\epsilon_{avg} = (\epsilon_x + \epsilon_y)/2$ from the origin, Fig. 10–8.

- Plot the reference point A having coordinates $A(\epsilon_x, \gamma_{xy}/2)$. This point represents the case when the x' axis coincides with the x axis. Hence $\theta = 0°$, Fig. 10–8.

- Connect point A with C and from the shaded triangle determine the radius R of the circle, Fig. 10–8.

- Once R has been determined, sketch the circle.

Principal Strains.

- The principal strains ϵ_1 and ϵ_2 are determined from the circle as the coordinates of points B and D, that is, where $\gamma/2 = 0$, Fig. 10–9a.

- The orientation of the plane on which ϵ_1 acts can be determined from the circle by calculating $2\theta_{p_1}$ using trigonometry. Here this angle happens to be counterclockwise, measured *from CA to CB*, Fig. 10–9a. Remember that the *rotation* of θ_{p_1} must be in this *same direction*, from the element's reference axis x to the x' axis, Fig. 10–9b.*

- When ϵ_1 and ϵ_2 are positive as in Fig. 10–9a, the element in Fig. 10–9b will elongate in the x' and y' directions as shown by the dashed outline.

Maximum In-Plane Shear Strain.

- The average normal strain and half the maximum in-plane shear strain are determined from the circle as the coordinates of point E or F, Fig. 10–9a.

- The orientation of the plane on which $\gamma_{\substack{max \\ in\text{-}plane}}$ and ϵ_{avg} act can be determined from the circle in Fig. 10–9a, by calculating $2\theta_{s_1}$ using trigonometry. Here this angle happens to be clockwise *from CA to CE*. Remember that the *rotation* of θ_{s_1} must be in this *same direction*, from the element's reference axis x to the x' axis, Fig. 10–9c.*

Strains on Arbitrary Plane.

- The normal and shear strain components $\epsilon_{x'}$ and $\gamma_{x'y'}$ for an element oriented at an angle θ, Fig. 10–9d, can be obtained from the circle using trigonometry to determine the coordinates of point P, Fig. 10–9a.

- To locate P, the known counterclockwise angle θ of the x' axis, Fig. 10–9d, is measured counterclockwise on the circle as 2θ. This measurement is made *from CA to CP*.

- If the value of $\epsilon_{y'}$ is required, it can be determined by calculating the ϵ coordinate of point Q in Fig. 10–9a. The line CQ lies 180° away from CP and thus represents a 90° rotation of the x' axis.

*If the $\gamma/2$ axis were constructed *positive upwards*, then the angle 2θ on the circle would be measured in the *opposite direction* to the orientation θ of the plane.

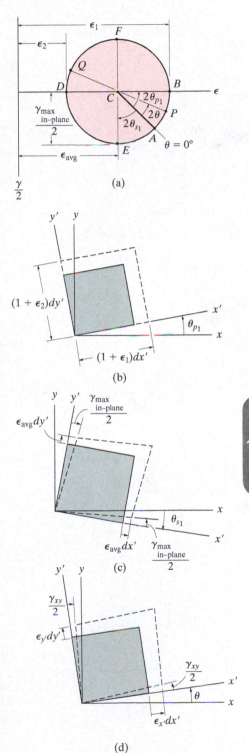

(a)

(b)

(c)

(d)

Fig. 10–9

EXAMPLE 10.4

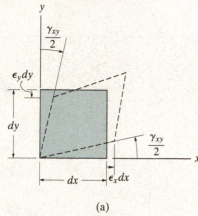

(a)

The state of plane strain at a point has components of $\epsilon_x = 250(10^{-6})$, $\epsilon_y = -150(10^{-6})$, $\gamma_{xy} = 120(10^{-6})$, Fig. 10–10a. Determine the principal strains and the orientation of the element upon which they act.

SOLUTION

Construction of the Circle. The ϵ and $\gamma/2$ axes are established in Fig. 10–10b. Remember that the *positive* $\gamma/2$ axis must be directed *downward* so that *counterclockwise* rotations of the element correspond to *counterclockwise* rotation around the circle, and vice versa. The center of the circle C is located at

$$\epsilon_{avg} = \frac{250 + (-150)}{2}(10^{-6}) = 50(10^{-6})$$

Since $\gamma_{xy}/2 = 60(10^{-6})$, the reference point A ($\theta = 0°$) has coordinates $A(250(10^{-6}), 60(10^{-6}))$. From the shaded triangle in Fig. 10–10b, the radius of the circle is

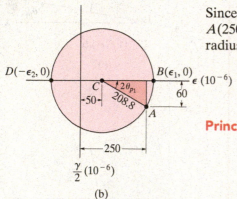

(b)

$$R = \left[\sqrt{(250 - 50)^2 + (60)^2}\,\right](10^{-6}) = 208.8(10^{-6})$$

Principal Strains. The ϵ coordinates of points B and D are therefore

$$\epsilon_1 = (50 + 208.8)(10^{-6}) = 259(10^{-6}) \qquad \textit{Ans.}$$

$$\epsilon_2 = (50 - 208.8)(10^{-6}) = -159(10^{-6}) \qquad \textit{Ans.}$$

The direction of the positive principal strain ϵ_1 in Fig. 10–10b is defined by the *counterclockwise* angle $2\theta_{p_1}$, measured from CA ($\theta = 0°$) to CB. We have

$$\tan 2\theta_{p_1} = \frac{60}{(250 - 50)}$$

$$\theta_{p_1} = 8.35° \qquad \textit{Ans.}$$

Hence, the side dx' of the element is inclined *counterclockwise* 8.35° as shown in Fig. 10–10c. This also defines the direction of ϵ_1. The deformation of the element is also shown in the figure.

(c)

Fig. 10–10

EXAMPLE 10.5

The state of plane strain at a point has components of $\epsilon_x = 250(10^{-6})$, $\epsilon_y = -150(10^{-6})$, $\gamma_{xy} = 120(10^{-6})$, Fig. 10–11a. Determine the maximum in-plane shear strains and the orientation of the element upon which they act.

SOLUTION
The circle has been established in the previous example and is shown in Fig. 10–11b.

Maximum In-Plane Shear Strain. Half the maximum in-plane shear strain and average normal strain are represented by the coordinates of point E or F on the circle. From the coordinates of point E,

$$\frac{(\gamma_{x'y'})_{\text{in-plane}}^{\text{max}}}{2} = 208.8(10^{-6})$$

$$(\gamma_{x'y'})_{\text{in-plane}}^{\text{max}} = 418(10^{-6}) \qquad Ans.$$

$$\epsilon_{\text{avg}} = 50(10^{-6})$$

To orient the element, we will determine the clockwise angle $2\theta_{s_1}$, measured from CA ($\theta = 0°$) to CE.

$$2\theta_{s_1} = 90° - 2(8.35°)$$

$$\theta_{s_1} = 36.7° \qquad Ans.$$

This angle is shown in Fig. 10–11c. Since the shear strain defined from point E on the circle has a positive value and the average normal strain is also positive, these strains deform the element into the dashed shape shown in the figure.

(a)

(b)

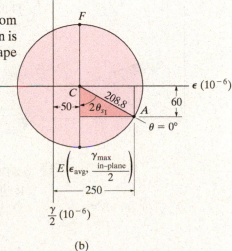

(c)

Fig. 10–11

EXAMPLE 10.6

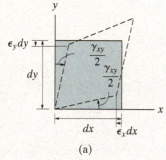

(a)

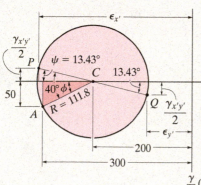

(b)

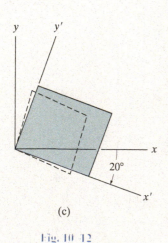

(c)

Fig. 10–12

The state of plane strain at a point has components of $\epsilon_x = -300(10^{-6})$, $\epsilon_y = -100(10^{-6})$, $\gamma_{xy} = 100(10^{-6})$, Fig. 10–12a. Determine the state of strain on an element oriented 20° clockwise from this position.

SOLUTION

Construction of the Circle. The ϵ and $\gamma/2$ axes are established in Fig. 10–12b. The center of the circle is at

$$\epsilon_{avg} = \left(\frac{-300 - 100}{2}\right)(10^{-6}) = -200(10^{-6})$$

The reference point A has coordinates $A(-300(10^{-6}), 50(10^{-6}))$, and so the radius CA, determined from the shaded triangle, is

$$R = \left[\sqrt{(300 - 200)^2 + (50)^2}\right](10^{-6}) = 111.8(10^{-6})$$

Strains on Inclined Element. Since the element is to be oriented 20° *clockwise*, we must consider the radial line CP, $2(20°) = 40°$ *clockwise*, measured from CA ($\theta = 0°$), Fig. 10–12b. The coordinates of point P are obtained from the geometry of the circle. Note that

$$\phi = \tan^{-1}\left(\frac{50}{(300 - 200)}\right) = 26.57°, \qquad \psi = 40° - 26.57° = 13.43°$$

Thus,

$$\epsilon_{x'} = -(200 + 111.8 \cos 13.43°)(10^{-6})$$
$$= -309(10^{-6}) \hspace{3cm} \textit{Ans.}$$

$$\frac{\gamma_{x'y'}}{2} = -(111.8 \sin 13.43°)(10^{-6})$$

$$\gamma_{x'y'} = -52.0(10^{-6}) \hspace{3cm} \textit{Ans.}$$

The normal strain $\epsilon_{y'}$ can be determined from the ϵ coordinate of point Q on the circle, Fig. 10–12b.

$$\epsilon_{y'} = -(200 - 111.8 \cos 13.43°)(10^{-6}) = -91.3(10^{-6}) \hspace{1cm} \textit{Ans.}$$

As a result of these strains, the element deforms relative to the x', y' axes as shown in Fig. 10–12c.

PROBLEMS

10–1. Prove that the sum of the normal strains in perpendicular directions is constant, i.e., $\epsilon_x + \epsilon_y = \epsilon_{x'} + \epsilon_{y'}$.

10–2. The state of strain at the point on the arm has components of $\epsilon_x = 200(10^{-6})$, $\epsilon_y = -300(10^{-6})$, and $\gamma_{xy} = 400(10^{-6})$. Use the strain transformation equations to determine the equivalent in-plane strains on an element oriented at an angle of 30° counterclockwise from the original position. Sketch the deformed element due to these strains within the x–y plane.

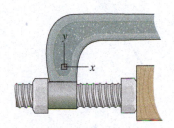

Prob. 10–2

10–3. The state of strain at the point on the pin leaf has components of $\epsilon_x = 200(10^{-6})$, $\epsilon_y = 180(10^{-6})$, and $\gamma_{xy} = -300(10^{-6})$. Use the strain transformation equations and determine the equivalent in-plane strains on an element oriented at an angle of $\theta = 60°$ counterclockwise from the original position. Sketch the deformed element due to these strains within the x–y plane.

*10–4. Solve Prob. 10–3 for an element oriented $\theta = 30°$ clockwise.

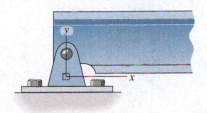

Probs. 10–3/4

10–5. The state of strain at the point on the leaf of the caster assembly has components of $\epsilon_x = -400(10^{-6})$, $\epsilon_y = 860(10^{-6})$, and $\gamma_{xy} = 375(10^{-6})$. Use the strain transformation equations to determine the equivalent in-plane strains on an element oriented at an angle of $\theta = 30°$ counterclockwise from the original position. Sketch the deformed element due to these strains within the x–y plane.

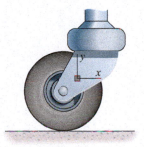

Prob. 10–5

10–6. The state of strain at a point on the bracket has components of $\epsilon_x = 150(10^{-6})$, $\epsilon_y = 200(10^{-6})$, $\gamma_{xy} = -700(10^{-6})$. Use the strain transformation equations and determine the equivalent in-plane strains on an element oriented at an angle of $\theta = 60°$ counterclockwise from the original position. Sketch the deformed element within the x–y plane due to these strains.

10–7. Solve Prob. 10–6 for an element oriented $\theta = 30°$ clockwise.

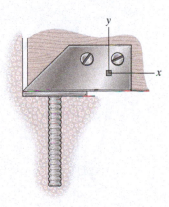

Probs. 10–6/7

10

*10–8.   The state of strain at the point on the spanner wrench has components of $\epsilon_x = 260(10^{-6})$, $\epsilon_y = 320(10^{-6})$, and $\gamma_{xy} = 180(10^{-6})$. Use the strain transformation equations to determine (a) the in-plane principal strains and (b) the maximum in-plane shear strain and average normal strain. In each case specify the orientation of the element and show how the strains deform the element within the x–y plane.

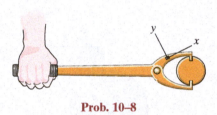

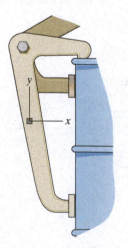

Prob. 10–8

10–9.   The state of strain at the point on the member has components of $\epsilon_x = 180(10^{-6})$, $\epsilon_y = -120(10^{-6})$, and $\gamma_{xy} = -100(10^{-6})$. Use the strain transformation equations to determine (a) the in-plane principal strains and (b) the maximum in-plane shear strain and average normal strain. In each case specify the orientation of the element and show how the strains deform the element within the x–y plane.

10–10.   The state of strain at the point on the support has components of $\epsilon_x = 350(10^{-6})$, $\epsilon_y = 400(10^{-6})$, $\gamma_{xy} = -675(10^{-6})$. Use the strain-transformation equations to determine (a) the in-plane principal strains and (b) the maximum in-plane shear strain and average normal strain. In each case specify the orientation of the element and show how the strains deform the element within the x–y plane.

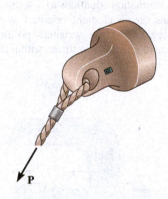

Prob. 10–10

10–11.   Due to the load **P**, the state of strain at the point on the bracket has components of $\epsilon_x = 500(10^{-6})$, $\epsilon_y = 350(10^{-6})$, and $\gamma_{xy} = -430(10^{-6})$. Use the strain transformation equations to determine the equivalent in-plane strains on an element oriented at an angle of $\theta = 30°$ clockwise from the original position. Sketch the deformed element due to these strains within the x–y plane.

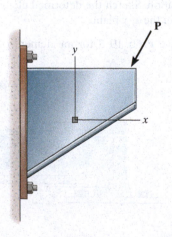

Prob. 10–9

Prob. 10–11

*10–12. The state of strain on an element has components $\epsilon_x = -400(10^{-6})$, $\epsilon_y = 0$, $\gamma_{xy} = 150(10^{-6})$. Determine the equivalent state of strain on an element at the same point oriented 30° clockwise with respect to the original element. Sketch the results on this element.

10–13. The state of plane strain on the element is $\epsilon_x = -300(10^{-6})$, $\epsilon_y = 0$, and $\gamma_{xy} = 150(10^{-6})$. Determine the equivalent state of strain which represents (a) the principal strains, and (b) the maximum in-plane shear strain and the associated average normal strain. Specify the orientation of the corresponding elements for these states of strain with respect to the original element.

10–15. Consider the general case of plane strain where ϵ_x, ϵ_y, and γ_{xy} are known. Write a computer program that can be used to determine the normal and shear strain, $\epsilon_{x'}$ and $\gamma_{x'y'}$, on the plane of an element oriented θ from the horizontal. Also, include the principal strains and the element's orientation, and the maximum in-plane shear strain, the average normal strain, and the element's orientation.

*10–16. The state of strain on the element has components $\epsilon_x = -300(10^{-6})$, $\epsilon_y = 100(10^{-6})$, $\gamma_{xy} = 150(10^{-6})$. Determine the equivalent state of strain, which represents (a) the principal strains, and (b) the maximum in-plane shear strain and the associated average normal strain. Specify the orientation of the corresponding elements for these states of strain with respect to the original element.

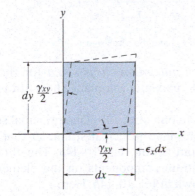

Probs. 10–12/13

10–14. The state of strain at the point on a boom of a shop crane has components of $\epsilon_x = 250(0^{-6})$, $\epsilon_y = 300(10^{-6})$, and $\gamma_{xy} = -180(10^{-6})$. Use the strain transformation equations to determine (a) the in-plane principal strains and (b) the maximum in-plane shear strain and average normal strain. In each case, specify the orientation of the element and show how the strains deform the element within the x–y plane.

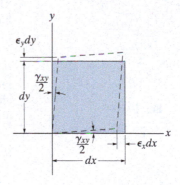

Prob. 10–16

10–17. Solve Prob. 10–3 using Mohr's circle.

10–18. Solve Prob. 10–4 using Mohr's circle.

10–19. Solve Prob. 10–5 using Mohr's circle.

Prob. 10–14

*10–20. Solve Prob. 10–8 using Mohr's circle.

10–21. Solve Prob. 10–7 using Mohr's circle.

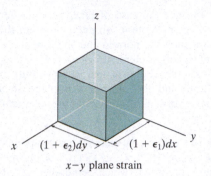

$x-y$ plane strain

(a)

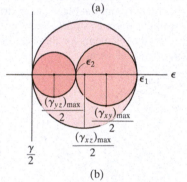

(b)

Fig. 10–13

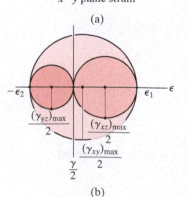

$x-y$ plane strain

(a)

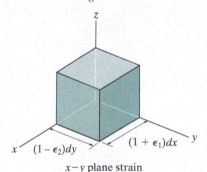

(b)

Fig. 10–14

*10.4   ABSOLUTE MAXIMUM SHEAR STRAIN

In Sec. 9.5 it was pointed out that in the case of plane stress, the absolute maximum shear stress in an element of material will occur *out of the plane* when the principal stresses have the *same sign*, i.e., both are tensile or both are compressive. A similar result occurs for plane strain. For example, if the principal in-plane strains cause elongations, Fig. 10–13a, then the three Mohr's circles describing the normal and shear strain components for the element rotations about the x, y, and z axes are shown in Fig. 10–13b. By inspection, the largest circle has a radius $R = (\gamma_{xz})_{max}/2$, and so

$$\gamma_{\substack{abs \\ max}} = (\gamma_{xz})_{max} = \epsilon_1 \qquad (10\text{–}14)$$

ϵ_1 and ϵ_2 have the same sign

This value gives the *absolute maximum shear strain* for the material. Note that it is *larger* than the maximum in-plane shear strain, which is $(\gamma_{xy})_{max} = \epsilon_1 - \epsilon_2$.

Now consider the case where one of the in-plane principal strains is of *opposite sign* to the other in-plane principal strain, so that ϵ_1 causes elongation and ϵ_2 causes contraction, Fig. 10–14a. The three Mohr's circles, which describe the strain components on the element rotated about the x, y, z axes, are shown in Fig. 10–14b. Here

$$\gamma_{\substack{abs \\ max}} = (\gamma_{xy})_{\substack{in\text{-}plane \\ max}} = \epsilon_1 - \epsilon_2 \qquad (10\text{–}15)$$

ϵ_1 and ϵ_2 have opposite signs

IMPORTANT POINTS

- If the in-plane principal strains both have the same sign, the absolute maximum shear strain will occur out of plane and has a value of $\gamma_{\substack{abs \\ max}} = \epsilon_{max}$. This value is greater than the maximum in-plane shear strain.

- If the in-plane principal strains are of opposite signs, then the absolute maximum shear strain equals the maximum in-plane shear strain, $\gamma_{\substack{abs \\ max}} = \epsilon_1 - \epsilon_2$.

EXAMPLE 10.7

The state of plane strain at a point has strain components of $\epsilon_x = -400(10^{-6})$, $\epsilon_y = 200(10^{-6})$, and $\gamma_{xy} = 150(10^{-6})$, Fig. 10–15a. Determine the maximum in-plane shear strain and the absolute maximum shear strain.

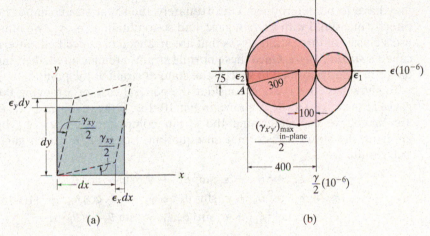

(a)   (b)

Fig. 10–15

SOLUTION

Maximum In-Plane Shear Strain.   We will solve this problem using Mohr's circle. The center of the circle is at

$$\epsilon_{\text{avg}} = \frac{-400 + 200}{2}(10^{-6}) = -100(10^{-6})$$

Since $\gamma_{xy}/2 = 75(10^{-6})$, the reference point A has coordinates $(-400(10^{-6}), 75(10^{-6}))$, Fig. 10–15b. The radius of the circle is therefore

$$R = \left[\sqrt{(400 - 100)^2 + (75)^2} \right](10^{-6}) = 309(10^{-6})$$

From the circle, the in-plane principal strains are

$$\epsilon_1 = (-100 + 309)(10^{-6}) = 209(10^{-6})$$
$$\epsilon_2 = (-100 - 309)(10^{-6}) = -409(10^{-6})$$

Also, the maximum in-plane shear strain is

$$\gamma_{\text{in-plane}}^{\text{max}} = \epsilon_1 - \epsilon_2 = [209 - (-409)](10^{-6}) = 618(10^{-6}) \qquad \textit{Ans.}$$

Absolute Maximum Shear Strain.   Since the *principal in-plane strains* have *opposite signs*, the maximum in-plane shear strain is *also* the absolute maximum shear strain; i.e.,

$$\gamma_{\substack{\text{abs} \\ \text{max}}} = 618(10^{-6}) \qquad \textit{Ans.}$$

The three Mohr's circles, plotted for element orientations about each of the x, y, z axes, are also shown in Fig. 10–15b.

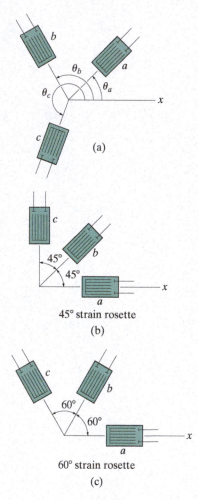

(a)

45° strain rosette

(b)

60° strain rosette

(c)

Fig. 10–16

10.5   STRAIN ROSETTES

The normal strain on the free surface of a body can be measured in a particular direction using an electrical resistance strain gage. For example, in Sec. 3.1 we showed how this type of gage is used to find the axial strain in a specimen when performing a tension test. When the body is subjected to several loads, however, then the strains $\epsilon_x, \epsilon_y, \gamma_{xy}$ at a point on its surface may have to be determined. Unfortunately, the shear strain cannot be directly measured with a strain gage, and so to obtain $\epsilon_x, \epsilon_y, \gamma_{xy}$, we must use a cluster of three strain gages that are arranged in a specified pattern called a **strain rosette**. Once these normal strains are measured, then the data can be transformed to specify the state of strain at the point.

To show how this is done, consider the general case of arranging the gages at the angles $\theta_a, \theta_b, \theta_c$ shown in Fig. 10–16a. If the readings $\epsilon_a, \epsilon_b, \epsilon_c$ are taken, we can determine the strain components $\epsilon_x, \epsilon_y, \gamma_{xy}$ by applying the strain transformation equation, Eq. 10–2, for each gage. The results are

$$\epsilon_a = \epsilon_x \cos^2 \theta_a + \epsilon_y \sin^2 \theta_a + \gamma_{xy} \sin \theta_a \cos \theta_a$$
$$\epsilon_b = \epsilon_x \cos^2 \theta_b + \epsilon_y \sin^2 \theta_b + \gamma_{xy} \sin \theta_b \cos \theta_b \qquad (10\text{–}16)$$
$$\epsilon_c = \epsilon_x \cos^2 \theta_c + \epsilon_y \sin^2 \theta_c + \gamma_{xy} \sin \theta_c \cos \theta_c$$

The values of $\epsilon_x, \epsilon_y, \gamma_{xy}$ are determined by solving these three equations simultaneously.

Normally, strain rosettes are arranged in 45° or 60° patterns. In the case of the 45° or "rectangular" strain rosette, Fig. 10–16b, $\theta_a = 0°, \theta_b = 45°$, $\theta_c = 90°$, so that Eq. 10–16 gives

$$\epsilon_x = \epsilon_a$$
$$\epsilon_y = \epsilon_c$$
$$\gamma_{xy} = 2\epsilon_b - (\epsilon_a + \epsilon_c)$$

And for the 60° strain rosette, Fig. 10–16c, $\theta_a = 0°, \theta_b = 60°, \theta_c = 120°$. Here Eq. 10–16 gives

$$\epsilon_x = \epsilon_a$$
$$\epsilon_y = \frac{1}{3}(2\epsilon_b + 2\epsilon_c - \epsilon_a) \qquad (10\text{–}17)$$
$$\gamma_{xy} = \frac{2}{\sqrt{3}}(\epsilon_b - \epsilon_c)$$

Once $\epsilon_x, \epsilon_y, \gamma_{xy}$ are determined, then the strain transformation equations or Mohr's circle can be used to determine the principal in-plane strains ϵ_1 and ϵ_2, or the maximum in-plane shear strain $\gamma_{\max}^{\text{in-plane}}$. The stress in the material that causes these strains can then be determined using Hooke's law, which is discussed in the next section.

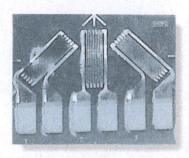

Typical electrical resistance 45° strain rosette.

EXAMPLE 10.8

The state of strain at point A on the bracket in Fig. 10–17a is measured using the strain rosette shown in Fig. 10–17b. The readings from the gages give $\epsilon_a = 60(10^{-6})$, $\epsilon_b = 135(10^{-6})$, and $\epsilon_c = 264(10^{-6})$. Determine the in-plane principal strains at the point and the directions in which they act.

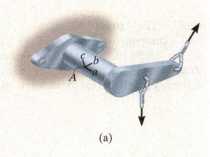

(a)

SOLUTION

We will use Eqs. 10–16 for the solution. Establishing an x axis, Fig. 10–17b, and measuring the angles counterclockwise from this axis to the centerlines of each gage, we have $\theta_a = 0°$, $\theta_b = 60°$, and $\theta_c = 120°$. Substituting these results, along with the problem data, into the equations gives

$$60(10^{-6}) = \epsilon_x \cos^2 0° + \epsilon_y \sin^2 0° + \gamma_{xy} \sin 0° \cos 0°$$
$$= \epsilon_x \tag{1}$$
$$135(10^{-6}) = \epsilon_x \cos^2 60° + \epsilon_y \sin^2 60° + \gamma_{xy} \sin 60° \cos 60°$$
$$= 0.25\epsilon_x + 0.75\epsilon_y + 0.433\gamma_{xy} \tag{2}$$
$$264(10^{-6}) = \epsilon_x \cos^2 120° + \epsilon_y \sin^2 120° + \gamma_{xy} \sin 120° \cos 120°$$
$$= 0.25\epsilon_x + 0.75\epsilon_y - 0.433\gamma_{xy} \tag{3}$$

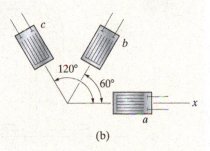

(b)

Using Eq. 1 and solving Eqs. 2 and 3 simultaneously, we get

$$\epsilon_x = 60(10^{-6}) \quad \epsilon_y = 246(10^{-6}) \quad \gamma_{xy} = -149(10^{-6})$$

These same results can also be obtained in a more direct manner from Eq. 10–17.

The in-plane principal strains will be determined using Mohr's circle. The center, C, is at $\epsilon_{avg} = 153(10^{-6})$, and the reference point on the circle is at $A[60(10^{-6}), -74.5(10^{-6})]$, Fig. 10–17c. From the shaded triangle, the radius is

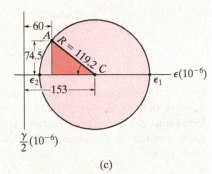

(c)

$$R = \left[\sqrt{(153 - 60)^2 + (74.5)^2} \right](10^{-6}) = 119.1(10^{-6})$$

The in-plane principal strains are therefore

$$\epsilon_1 = 153(10^{-6}) + 119.1(10^{-6}) = 272(10^{-6}) \qquad \textit{Ans.}$$
$$\epsilon_2 = 153(10^{-6}) - 119.1(10^{-6}) = 33.9(10^{-6}) \qquad \textit{Ans.}$$
$$2\theta_{p_2} = \tan^{-1} \frac{74.5}{(153 - 60)} = 38.7°$$
$$\theta_{p_2} = 19.3° \qquad \textit{Ans.}$$

NOTE: The deformed element is shown in the dashed position in Fig. 10–17d. Realize that, due to the Poisson effect, the element is *also* subjected to an out-of-plane strain, i.e., in the z direction, although this value will not influence the calculated results.

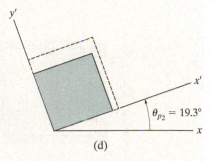

(d)

Fig. 10–17

10

PROBLEMS

10–22. The strain at point A on the bracket has components $\epsilon_x = 300(10^{-6})$, $\epsilon_y = 550(10^{-6})$, $\gamma_{xy} = -650(10^{-6})$, $\epsilon_z = 0$. Determine (a) the principal strains at A in the x–y plane, (b) the maximum shear strain in the x–y plane, and (c) the absolute maximum shear strain.

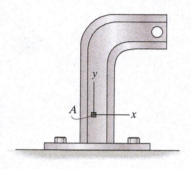

Prob. 10–22

10–23. The strain at point A on a beam has components $\epsilon_x = 450(10^{-6})$, $\epsilon_y = 825(10^{-6})$, $\gamma_{xy} = 275(10^{-6})$, $\epsilon_z = 0$. Determine (a) the principal strains at A, (b) the maximum shear strain in the x–y plane, and (c) the absolute maximum shear strain.

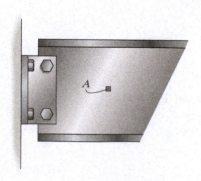

Prob. 10–23

*10–24.** The strain at point A on the pressure-vessel wall has components $\epsilon_x = 480(10^{-6})$, $\epsilon_y = 720(10^{-6})$, $\gamma_{xy} = 650(10^{-6})$. Determine (a) the principal strains at A, in the x–y plane, (b) the maximum shear strain in the x–y plane, and (c) the absolute maximum shear strain.

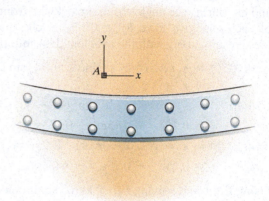

Prob. 10–24

10–25. The 45° strain rosette is mounted on the surface of a shell. The following readings are obtained for each gage: $\epsilon_a = -200(10^{-6})$, $\epsilon_b = 300(10^{-6})$, and $\epsilon_c = 250(10^{-6})$. Determine the in-plane principal strains.

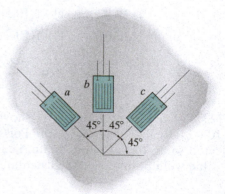

Prob. 10–25

10–26. The 45° strain rosette is mounted on the surface of a pressure vessel. The following readings are obtained for each gage: $\epsilon_a = 475(10^{-6})$, $\epsilon_b = 250(10^{-6})$, and $\epsilon_c = -360(10^{-6})$. Determine the in-plane principal strains.

*10–28. The 45° strain rosette is mounted on a steel shaft. The following readings are obtained from each gage: $\epsilon_a = 800(10^{-6})$, $\epsilon_b = 520(10^{-6})$, $\epsilon_c = -450(10^{-6})$. Determine the in-plane principal strains.

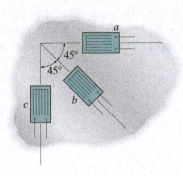

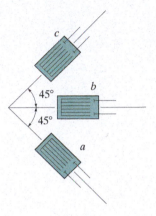

Prob. 10–28

Prob. 10–26

10–27. The 60° strain rosette is mounted on the surface of the bracket. The following readings are obtained for each gage: $\epsilon_a = -780(10^{-6})$, $\epsilon_b = 400(10^{-6})$, and $\epsilon_c = 500(10^{-6})$. Determine (a) the principal strains and (b) the maximum in-plane shear strain and associated average normal strain. In each case show the deformed element due to these strains.

10–29. Consider the general orientation of three strain gages at a point as shown. Write a computer program that can be used to determine the principal in-plane strains and the maximum in-plane shear strain at the point. Show an application of the program using the values $\theta_a = 40°$, $\epsilon_a = 160(10^{-6})$, $\theta_b = 125°$, $\epsilon_b = 100(10^{-6})$, $\theta_c = 220°$, $\epsilon_c = 80(10^{-6})$.

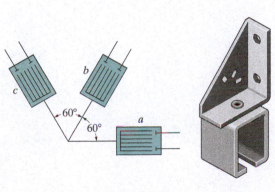

Prob. 10–27

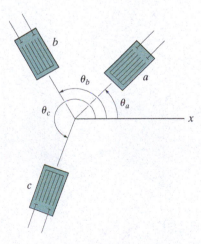

Prob. 10–29

10.6   MATERIAL PROPERTY RELATIONSHIPS

In this section we will present some important material property relationships that are used when the material is subjected to multiaxial stress and strain. In all cases, we will assume that the material is homogeneous and isotropic, and behaves in a linear elastic manner.

Generalized Hooke's Law.   When the material at a point is subjected to a state of ***triaxial stress***, σ_x, σ_y, σ_z, Fig. 10–18a, then these stresses can be related to the normal strains ϵ_x, ϵ_y, ϵ_z by using the principle of superposition, Poisson's ratio, $\epsilon_{\text{lat}} = -\nu\epsilon_{\text{long}}$, and Hooke's law as it applies in the uniaxial direction, $\epsilon = \sigma/E$. For example, consider the normal strain of the element in the x direction, caused by separate application of each normal stress. When σ_x is applied, Fig. 10–18b, the element elongates with a strain ϵ_x', where

$$\epsilon_x' = \frac{\sigma_x}{E}$$

Application of σ_y causes the element to contract with a strain ϵ_x'', Fig. 10–18c. Here

$$\epsilon_x'' = -\nu\frac{\sigma_y}{E}$$

Finally, application of σ_z, Fig. 10–18d, causes a contraction strain ϵ_x''', so that

$$\epsilon_x''' = -\nu\frac{\sigma_z}{E}$$

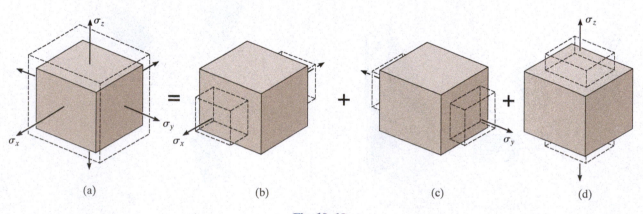

(a)   =   (b)   +   (c)   +   (d)

Fig. 10–18

We can obtain the resultant strain ϵ_x by adding these three strains algebraically. Similar equations can be developed for the normal strains in the y and z directions, and so the final results can be written as

$$\epsilon_x = \frac{1}{E}[\sigma_x - \nu(\sigma_y + \sigma_z)]$$

$$\epsilon_y = \frac{1}{E}[\sigma_y - \nu(\sigma_x + \sigma_z)] \qquad (10\text{–}18)$$

$$\epsilon_z = \frac{1}{E}[\sigma_z - \nu(\sigma_x + \sigma_y)]$$

These three equations represent the general form of Hooke's law for a triaxial state of stress. For application, tensile stress is considered a positive quantity, and a compressive stress is negative. If a resulting normal strain is *positive*, it indicates that the material *elongates*, whereas a *negative* normal strain indicates the material *contracts*.

If we only apply a shear stress τ_{xy} to the element, Fig. 10–19*a*, experimental observations indicate that the material will change its shape, but it will not change its volume. In other words, τ_{xy} will only cause the shear strain γ_{xy} in the material. Likewise, τ_{yz} and τ_{xz} will only cause shear strains γ_{yz} and γ_{xz}, Figs. 10–19*b* and 10–19*c*. Therefore, Hooke's law for shear stress and shear strain becomes

$$\gamma_{xy} = \frac{1}{G}\tau_{xy} \qquad \gamma_{yz} = \frac{1}{G}\tau_{yz} \qquad \gamma_{xz} = \frac{1}{G}\tau_{xz} \qquad (10\text{–}19)$$

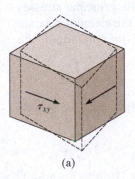

(a)

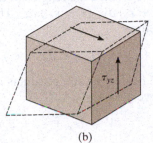

(b)

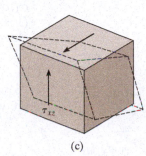

(c)

Fig. 10–19

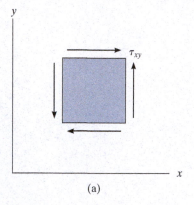

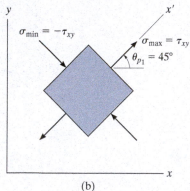

Fig. 10–20

Relationship Involving *E*, *ν*, and *G*.

In Sec. 3.7 it was stated that the modulus of elasticity *E* is related to the shear modulus *G* by Eq. 3–11, namely,

$$G = \frac{E}{2(1 + \nu)} \tag{10–20}$$

One way to derive this relationship is to consider an element of the material to be subjected only to shear, Fig. 10–20*a*. Applying Eq. 9–5 (see Example 9–5), the principal stresses at the point are $\sigma_{\max} = \tau_{xy}$ and $\sigma_{\min} = -\tau_{xy}$, where this element must be oriented $\theta_{p_1} = 45°$ counterclockwise from the *x* axis, as shown in Fig. 10–20*b*. If the three principal stresses $\sigma_{\max} = \tau_{xy}$, $\sigma_{\text{int}} = 0$, and $\sigma_{\min} = -\tau_{xy}$ are then substituted into the first of Eqs. 10–18, the principal strain $\epsilon_{\max}$ can be related to the shear stress τ_{xy}. The result is

$$\epsilon_{\max} = \frac{\tau_{xy}}{E}(1 + \nu) \tag{10–21}$$

This strain, which deforms the element along the *x'* axis, can also be related to the shear strain γ_{xy}. From Fig. 10–20*a*, $\sigma_x = \sigma_y = \sigma_z = 0$. Substituting these results into the first and second Eqs. 10–18 gives $\epsilon_x = \epsilon_y = 0$. Now apply the strain transformation Eq. 10–9, which gives

$$\epsilon_1 = \epsilon_{\max} = \frac{\gamma_{xy}}{2}$$

By Hooke's law, $\gamma_{xy} = \tau_{xy}/G$, so that $\epsilon_{\max} = \tau_{xy}/2G$. Finally, substituting this into Eq. 10–21 and rearranging the terms gives our result, namely, Eq. 10–20.

Dilatation.

When an elastic material is subjected to normal stress, the strains that are produced will cause its volume to change. For example, if the volume element in Fig. 10–21*a* is subjected to the principal stresses $\sigma_1, \sigma_2, \sigma_3$, Fig. 10–21*b*, then the lengths of the sides of the element become $(1 + \epsilon_x)\,dx$, $(1 + \epsilon_y)\,dy$, $(1 + \epsilon_z)\,dz$. The change in volume of the element is therefore

$$\delta V = (1 + \epsilon_x)(1 + \epsilon_y)(1 + \epsilon_z)\,dx\,dy\,dz - dx\,dy\,dz$$

Expanding, and neglecting the products of the strains, since the strains are very small, we get

$$\delta V = (\epsilon_x + \epsilon_y + \epsilon_z)\,dx\,dy\,dz$$

The change in volume per unit volume is called the "volumetric strain" or the ***dilatation*** *e*.

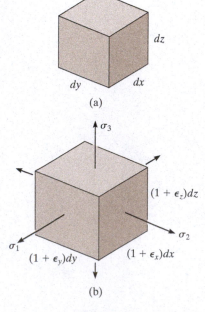

Fig. 10–21

$$e = \frac{\delta V}{dV} = \epsilon_x + \epsilon_y + \epsilon_z \tag{10–22}$$

If we use Hooke's law, Eq. 10–18, we can also express the dilatation in terms of the applied stress. We have

$$e = \frac{1 - 2\nu}{E}(\sigma_1 + \sigma_2 + \sigma_3) \qquad (10\text{–}23)$$

Bulk Modulus. According to Pascal's law, when a volume element of material is subjected to a uniform pressure p caused by a static fluid, the pressure will be the same in all directions. Shear stresses will not be present, since the fluid does not flow around the element. This state of "hydrostatic" loading therefore requires $\sigma_1 = \sigma_2 = \sigma_3 = -p$, Fig. 10–22. Substituting into Eq. 10–23 and rearranging terms yields

$$\frac{p}{e} = -\frac{E}{3(1 - 2\nu)} \qquad (10\text{–}24)$$

The term on the right is called the ***volume modulus of elasticity*** or the ***bulk modulus***, since this ratio, p/e, is *similar* to the ratio of one-dimensional linear elastic stress to strain, which defines E, i.e., $\sigma/\epsilon = E$. The bulk modulus has the same units as stress and is symbolized by the letter k, so that

$$k = \frac{E}{3(1 - 2\nu)} \qquad (10\text{–}25)$$

For most metals $\nu \approx \frac{1}{3}$ and so $k \approx E$. However, if we assume the material did not change its volume when loaded, then $\delta V = e = 0$, and k would be infinite. As a result, Eq. 10–25 would then indicate the theoretical *maximum* value for Poisson's ratio to be $\nu = 0.5$.

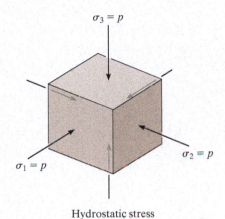

Hydrostatic stress

Fig. 10–22

> ## IMPORTANT POINTS

- When a homogeneous isotropic material is subjected to a state of triaxial stress, the strain in each direction is influenced by the strains produced by *all* the stresses. This is the result of the Poisson effect, and the stress is then related to the strain in the form of a generalized Hooke's law.

- When a shear stress is applied to homogeneous isotropic material, it will only produce shear strain in the same plane.

- The material constants E, G, and ν are all related by Eq. 10–20.

- *Dilatation*, or *volumetric strain*, is caused only by normal strain, not shear strain.

- The *bulk modulus* is a measure of the stiffness of a volume of material. This material property provides an upper limit to Poisson's ratio of $\nu = 0.5$.

EXAMPLE 10.9

The bracket in Example 10.8, Fig. 10–23a, is made of steel for which $E_{st} = 200$ GPa, $\nu_{st} = 0.3$. Determine the principal stresses at point A.

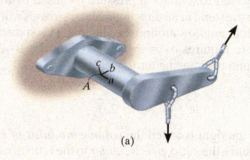

(a)

Fig. 10–23

SOLUTION I

From Example 10.8 the principal strains have been determined as

$$\epsilon_1 = 272(10^{-6})$$

$$\epsilon_2 = 33.9(10^{-6})$$

Since point A is on the *surface* of the bracket, for which there is no loading, the stress on the surface is zero, and so point A is subjected to plane stress (not plane strain). Applying Hooke's law with $\sigma_3 = 0$, we have

$$\epsilon_1 = \frac{\sigma_1}{E} - \frac{\nu}{E}\sigma_2; \quad 272(10^{-6}) = \frac{\sigma_1}{200(10^9)} - \frac{0.3}{200(10^9)}\sigma_2$$

$$54.4(10^6) = \sigma_1 - 0.3\sigma_2 \quad\quad (1)$$

$$\epsilon_2 = \frac{\sigma_2}{E} - \frac{\nu}{E}\sigma_1; \quad 33.9(10^{-6}) = \frac{\sigma_2}{200(10^9)} - \frac{0.3}{200(10^9)}\sigma_1$$

$$6.78(10^6) = \sigma_2 - 0.3\sigma_1 \quad\quad (2)$$

Solving Eqs. 1 and 2 simultaneously yields

$$\sigma_1 = 62.0 \text{ MPa} \quad\quad\quad\quad Ans.$$

$$\sigma_2 = 25.4 \text{ MPa} \quad\quad\quad\quad Ans.$$

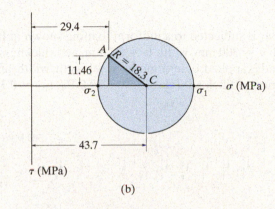

Fig. 10–23 (cont.)

SOLUTION II

It is also possible to solve this problem using the given state of strain as specified in Example 10.8.

$$\epsilon_x = 60(10^{-6}) \qquad \epsilon_y = 246(10^{-6}) \qquad \gamma_{xy} = -149(10^{-6})$$

Applying Hooke's law in the x–y plane, we have

$$\epsilon_x = \frac{\sigma_x}{E} - \frac{\nu}{E}\sigma_y; \quad 60(10^{-6}) = \frac{\sigma_x}{200(10^9)\text{ Pa}} - \frac{0.3\sigma_y}{200(10^9)\text{ Pa}}$$

$$\epsilon_y = \frac{\sigma_y}{E} - \frac{\nu}{E}\sigma_x; \quad 246(10^{-6}) = \frac{\sigma_y}{200(10^9)\text{ Pa}} - \frac{0.3\sigma_x}{200(10^9)\text{ Pa}}$$

$$\sigma_x = 29.4\text{ MPa} \qquad \sigma_y = 58.0\text{ MPa}$$

The shear stress is determined using Hooke's law for shear. First, however, we must calculate G.

$$G = \frac{E}{2(1 + \nu)} = \frac{200\text{ GPa}}{2(1 + 0.3)} = 76.9\text{ GPa}$$

Thus,

$$\tau_{xy} = G\gamma_{xy}; \qquad \tau_{xy} = 76.9(10^9)[-149(10^{-6})] = -11.46\text{ MPa}$$

The Mohr's circle for this state of plane stress has a center at $\sigma_{avg} = 43.7$ MPa and a reference point $A(29.4\text{ MPa}, -11.46\text{ MPa})$, Fig. 10–23b. The radius is determined from the shaded triangle.

$$R = \sqrt{(43.7 - 29.4)^2 + (11.46)^2} = 18.3\text{ MPa}$$

Therefore,

$$\sigma_1 = 43.7\text{ MPa} + 18.3\text{ MPa} = 62.0\text{ MPa} \qquad \textit{Ans.}$$

$$\sigma_2 = 43.7\text{ MPa} - 18.3\text{ MPa} = 25.4\text{ MPa} \qquad \textit{Ans.}$$

NOTE: Each of these solutions is valid provided the material is both linear elastic and isotropic, since only then will the directions of the principal stress and strain coincide.

EXAMPLE 10.10

The copper bar is subjected to a uniform loading shown in Fig. 10–24. If it has a length $a = 300$ mm, width $b = 50$ mm, and thickness $t = 20$ mm before the load is applied, determine its new length, width, and thickness after application of the load. Take $E_{cu} = 120$ GPa, $\nu_{cu} = 0.34$.

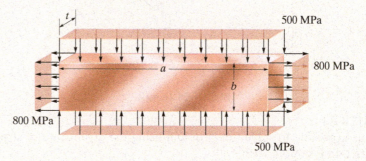

Fig. 10–24

SOLUTION

By inspection, the bar is subjected to a state of plane stress. From the loading we have

$$\sigma_x = 800 \text{ MPa} \qquad \sigma_y = -500 \text{ MPa} \qquad \tau_{xy} = 0 \qquad \sigma_z = 0$$

The associated normal strains are determined from Hooke's law, Eq. 10–18; that is,

$$\epsilon_x = \frac{\sigma_x}{E} - \frac{\nu}{E}(\sigma_y + \sigma_z)$$

$$= \frac{800 \text{ MPa}}{120(10^3) \text{ MPa}} - \frac{0.34}{120(10^3) \text{ MPa}}(-500 \text{ MPa} + 0) = 0.00808$$

$$\epsilon_y = \frac{\sigma_y}{E} - \frac{\nu}{E}(\sigma_x + \sigma_z)$$

$$= \frac{-500 \text{ MPa}}{120(10^3) \text{ MPa}} - \frac{0.34}{120(10^3) \text{ MPa}}(800 \text{ MPa} + 0) = -0.00643$$

$$\epsilon_z = \frac{\sigma_z}{E} - \frac{\nu}{E}(\sigma_x + \sigma_y)$$

$$= 0 - \frac{0.34}{120(10^3) \text{ MPa}}(800 \text{ MPa} - 500 \text{ MPa}) = -0.000850$$

The new bar length, width, and thickness are therefore

$$a' = 300 \text{ mm} + 0.00808(300 \text{ mm}) = 302.4 \text{ mm} \qquad \textit{Ans.}$$

$$b' = 50 \text{ mm} + (-0.00643)(50 \text{ mm}) = 49.68 \text{ mm} \qquad \textit{Ans.}$$

$$t' = 20 \text{ mm} + (-0.000850)(20 \text{ mm}) = 19.98 \text{ mm} \qquad \textit{Ans.}$$

EXAMPLE 10.11

If the rectangular block shown in Fig. 10–25 is subjected to a uniform pressure of $p = 20$ psi, determine the dilatation and the change in length of each side. Take $E = 600$ psi, $v = 0.45$.

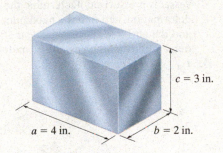

$c = 3$ in.

$a = 4$ in. $b = 2$ in.

Fig. 10–25

SOLUTION

Dilatation.   The dilatation can be determined using Eq. 10–23 with $\sigma_x = \sigma_y = \sigma_z = -20$ psi. We have

$$e = \frac{1 - 2v}{E}(\sigma_x + \sigma_y + \sigma_z)$$

$$= \frac{1 - 2(0.45)}{600 \text{ psi}}[3(-20 \text{ psi})]$$

$$= -0.01 \text{ in}^3/\text{in}^3 \qquad\qquad\qquad Ans.$$

Change in Length.   The normal strain on each side is determined from Hooke's law, Eq. 10–18; that is,

$$\epsilon = \frac{1}{E}[\sigma_x - v(\sigma_y + \sigma_z)]$$

$$= \frac{1}{600 \text{ psi}}[-20 \text{ psi} - (0.45)(-20 \text{ psi} - 20 \text{ psi})] = -0.00333 \text{ in./in.}$$

Thus, the change in length of each side is

$$\delta a = -0.00333(4 \text{ in.}) = -0.0133 \text{ in.} \qquad Ans.$$

$$\delta b = -0.00333(2 \text{ in.}) = -0.00667 \text{ in.} \qquad Ans.$$

$$\delta c = -0.00333(3 \text{ in.}) = -0.0100 \text{ in.} \qquad Ans.$$

The negative signs indicate that each dimension is decreased.

PROBLEMS

10–30. For the case of plane stress, show that Hooke's law can be written as

$$\sigma_x = \frac{E}{(1 - \nu^2)}(\epsilon_x + \nu\epsilon_y), \quad \sigma_y = \frac{E}{(1 - \nu^2)}(\epsilon_y + \nu\epsilon_x)$$

10–31. Use Hooke's law, Eq. 10–18, to develop the strain tranformation equations, Eqs. 10–5 and 10–6, from the stress tranformation equations, Eqs. 9–1 and 9–2.

*10–32. The principal plane stresses and associated strains in a plane at a point are $\sigma_1 = 36$ ksi, $\sigma_2 = 16$ ksi, $\epsilon_1 = 1.02(10^{-3})$, $\epsilon_2 = 0.180(10^{-3})$. Determine the modulus of elasticity and Poisson's ratio.

10–33. A rod has a radius of 10 mm. If it is subjected to an axial load of 15 N such that the axial strain in the rod is $\epsilon_x = 2.75(10^{-6})$, determine the modulus of elasticity E and the change in the rod's diameter. $\nu = 0.23$.

10–34. The polyvinyl chloride bar is subjected to an axial force of 900 lb. If it has the original dimensions shown, determine the change in the angle θ after the load is applied. $E_{pvc} = 800(10^3)$ psi, $\nu_{pvc} = 0.20$.

10–35. The polyvinyl chloride bar is subjected to an axial force of 900 lb. If it has the original dimensions shown, determine the value of Poisson's ratio if the angle θ decreases by $\Delta\theta = 0.01°$ after the load is applied. $E_{pvc} = 800(10^3)$ psi.

*10–36. The spherical pressure vessel has an inner diameter of 2 m and a thickness of 10 mm. A strain gage having a length of 20 mm is attached to it, and it is observed to increase in length by 0.012 mm when the vessel is pressurized. Determine the pressure causing this deformation, and find the maximum in-plane shear stress, and the absolute maximum shear stress at a point on the outer surface of the vessel. The material is steel, for which $E_{st} = 200$ GPa and $\nu_{st} = 0.3$.

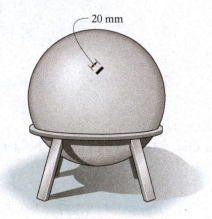

20 mm

Prob. 10–36

10–37. Determine the bulk modulus for each of the following materials: (a) rubber, $E_r = 0.4$ ksi, $\nu_r = 0.48$, and (b) glass, $E_g = 8(10^3)$ ksi, $\nu_g = 0.24$.

10–38. The strain gage is placed on the surface of the steel boiler as shown. If it is 0.5 in. long, determine the pressure in the boiler when the gage elongates $0.2(10^{-3})$ in. The boiler has a thickness of 0.5 in. and inner diameter of 60 in. Also, determine the maximum x, y in-plane shear strain in the material. $E_{st} = 29(10^3)$ ksi, $\nu_{st} = 0.3$.

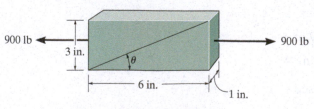

900 lb   900 lb

3 in.

θ

6 in.

1 in.

Probs. 10–34/35

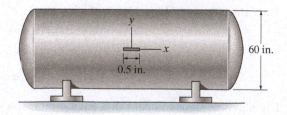

y

x

60 in.

0.5 in.

Prob. 10–38

10–39. The principal strains at a point on the aluminum fuselage of a jet aircraft are $\epsilon_1 = 780(10^{-6})$ and $\epsilon_2 = 400(10^{-6})$. Determine the associated principal stresses at the point in the same plane. $E_{al} = 10(10^3)$ ksi, $\nu_{al} = 0.33$. *Hint:* See Prob. 10–30.

*10–40.** The strain in the x direction at point A on the A-36 structural-steel beam is measured and found to be $\epsilon_x = 200(10^{-6})$. Determine the applied load P. What is the shear strain γ_{xy} at point A?

10–41. If a load of $P = 3$ kip is applied to the A-36 structural-steel beam, determine the strain ϵ_x and γ_{xy} at point A.

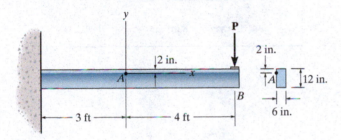

Probs. 10–40/41

10–42. The cube of aluminum is subjected to the three stresses shown. Determine the principal strains. Take $E_{al} = 10(10^3)$ ksi and $\nu_{al} = 0.33$.

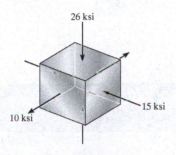

Prob. 10–42

10–43. The principal strains at a point on the aluminum surface of a tank are $\epsilon_1 = 630(10^{-6})$ and $\epsilon_2 = 350(10^{-6})$. If this is a case of plane stress, determine the associated principal stresses at the point in the same plane. $E_{al} = 10(10^3)$ ksi, $\nu_{al} = 0.33$. *Hint:* See Prob. 10–30.

*10–44.** A uniform edge load of 500 lb/in. and 350 lb/in. is applied to the polystyrene specimen. If the specimen is originally square and has dimensions of $a = 2$ in., $b = 2$ in., and a thickness of $t = 0.25$ in., determine its new dimensions a', b', and t' after the load is applied. $E_p = 597(10^3)$ psi and $\nu_p = 0.25$.

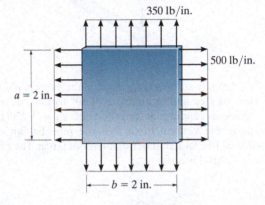

Prob. 10–44

10–45. A material is subjected to principal stresses σ_x and σ_y. Determine the orientation θ of the strain gage so that its reading of normal strain responds only to σ_y and not σ_x. The material constants are E and ν.

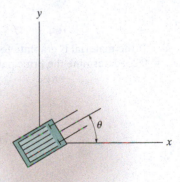

Prob. 10–45

10–46. A single strain gage, placed in the vertical plane on the outer surface and at an angle 60° to the axis of the pipe, gives a reading at point A of $\epsilon_A = -250(10^{-6})$. Determine the principal strains in the pipe at this point. The pipe has an outer diameter of 1 in. and an inner diameter of 0.6 in. and is made of C86100 bronze.

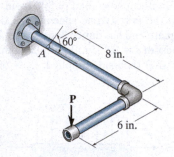

Prob. 10–46

10–47. A single strain gage, placed in the vertical plane on the outer surface and at an angle of 60° to the axis of the pipe, gives a reading at point A of $\epsilon_A = -250(10^{-6})$. Determine the vertical force P if the pipe has an outer diameter of 1 in and an inner diameter of 0.6 in. The pipe is made of C86100 bronze.

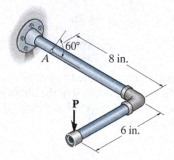

Prob. 10–47

*****10–48.** If the material is graphite for which $E_g = 800$ ksi and $v_g = 0.23$, determine the principal strains.

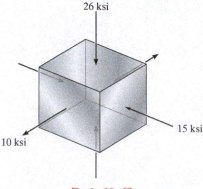

Prob. 10–48

10–49. Initially, gaps between the A-36 steel plate and the rigid constraint are as shown. Determine the normal stresses σ_x and σ_y in the plate if the temperature is increased by $\Delta T = 100°F$. *Hint:* To solve, add the thermal strain $\alpha \Delta T$ to the equations for Hooke's law.

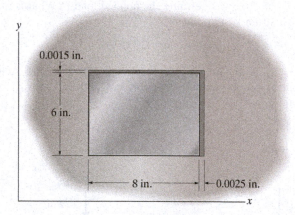

Prob. 10–49

10–50. The steel shaft has a radius of 15 mm. Determine the torque T in the shaft if the two strain gages, attached to the surface of the shaft, report strains of $\epsilon_{x'} = -80(10^{-6})$ and $\epsilon_{y'} = 80(10^{-6})$. Also, determine the strains acting in the x and y directions. $E_{st} = 200$ GPa, $v_{st} = 0.3$.

10–51. The shaft has a radius of 15 mm and is made of L2 tool steel. Determine the strains in the x' and y' direction if a torque $T = 2$ kN · m is applied to the shaft.

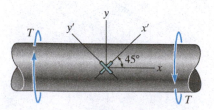

Probs. 10–50/51

*10–52.   The A-36 steel pipe is subjected to the axial loading of 60 kN. Determine the change in volume of the material after the load is applied.

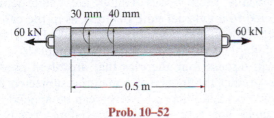

30 mm 40 mm

60 kN ← | | → 60 kN

0.5 m

Prob. 10–52

10–53.   Air is pumped into the steel thin-walled pressure vessel at C. If the ends of the vessel are closed using two pistons connected by a rod AB, determine the increase in the diameter of the pressure vessel when the internal gage pressure is 5 MPa. Also, what is the tensile stress in rod AB if it has a diameter of 100 mm? The inner radius of the vessel is 400 mm, and its thickness is 10 mm. $E_{st} = 200$ GPa and $\nu_{st} = 0.3$.

10–54.   Determine the increase in the diameter of the pressure vessel in Prob. 10–53 if the pistons are replaced by walls connected to the ends of the vessel.

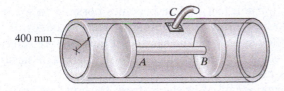

400 mm

C

A B

Probs. 10–53/54

10–55.   A thin-walled spherical pressure vessel having an inner radius r and thickness t is subjected to an internal pressure p. Show that the increase in the volume within the vessel is $\Delta V = (2p\pi r^4/Et)(1 - \nu)$. Use a small-strain analysis.

*10–56.   The thin-walled cylindrical pressure vessel of inner radius r and thickness t is subjected to an internal pressure p. If the material constants are E and ν, determine the strains in the circumferential and longitudinal directions. Using these results, calculate the increase in both the diameter and the length of a steel pressure vessel filled with air and having an internal gage pressure of 15 MPa. The vessel is 3 m long, and has an inner radius of 0.5 m and a thickness of 10 mm. $E_{st} = 200$ GPa, $\nu_{st} = 0.3$.

10–57.   Estimate the increase in volume of the pressure vessel in Prob. 10–56.

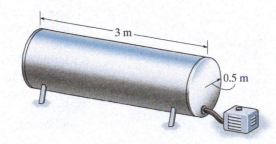

3 m

0.5 m

Probs. 10–56/57

10–58.   A soft material is placed within the confines of a rigid cylinder which rests on a rigid support. Assuming that $\epsilon_x = 0$ and $\epsilon_y = 0$, determine the factor by which the stiffness of the material, or the apparent modulus of elasticity, will be increased when a load is applied, if $\nu = 0.3$ for the material.

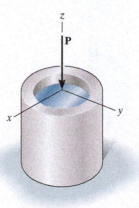

z

P

x y

Prob. 10–58

10

*10.7   THEORIES OF FAILURE

When an engineer is faced with the problem of design using a specific material, it becomes important to place an upper *limit* on the state of stress that defines the material's failure. If the material is *ductile*, failure is usually specified by the initiation of *yielding*, whereas if the material is *brittle*, it is specified by *fracture*. These modes of failure are readily defined if the member is subjected to a uniaxial state of stress, as in the case of simple tension; however, if the member is subjected to biaxial or triaxial stress, the criterion for failure becomes more difficult to establish.

In this section we will discuss four theories that are often used in engineering practice to predict the failure of a material subjected to a *multiaxial* state of stress. No single theory, however, can be applied to a specific material at *all times*. This is because the material may behave in either a ductile or brittle manner depending on the temperature, rate of loading, chemical environment, or the way the material is shaped or formed. When using a particular theory of failure, it is first necessary to calculate the normal and shear stress at points where they are the largest in the member. Once this state of stress is established, the *principal stresses* at these critical points must then be determined, since each of the following theories is based on knowing the principal stress.

Ductile Materials

Maximum Shear Stress Theory.   The most common type of *yielding of a ductile material*, such as steel, is caused by *slipping*, which occurs between the contact planes of randomly ordered crystals that make up the material. If a specimen is made into a highly polished thin strip and subjected to a simple tension test, it then becomes possible to see how this slipping causes the material to *yield*, Fig. 10–26. The edges of the planes of slipping as they appear on the surface of the strip are referred to as **Lüder's lines**. These lines clearly indicate the slip planes in the strip, which occur at approximately 45° as shown.

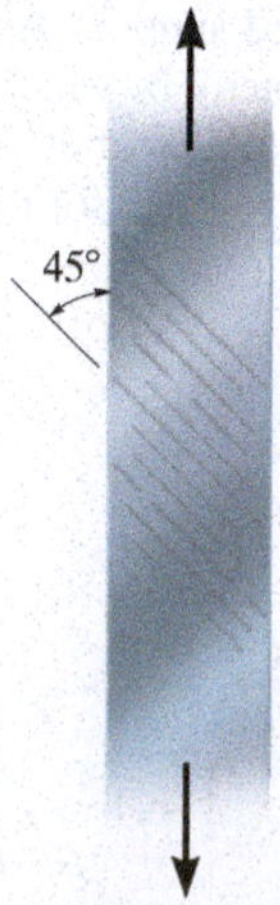

The slipping that occurs is caused by shear stress. To show this, consider an element of the material taken from a tension specimen, Fig. 10–27*a*, when the specimen is subjected to the yield stress σ_Y. The maximum shear stress can be determined from Mohr's circle, Fig. 10–27*b*. The results indicate that

45°

Lüder's lines on
mild steel strip

Fig. 10–26

$$\tau_{max} = \frac{\sigma_Y}{2} \qquad (10\text{–}26)$$

Furthermore, this shear stress acts on planes that are 45° from the planes of principal stress, Fig. 10–27c, and since these planes *coincide* with the direction of the Lüder lines shown on the specimen, this indeed indicates that failure occurs by shear.

Realizing that ductile materials fail by shear, in 1868 Henri Tresca proposed the **maximum shear stress theory** or **Tresca yield criterion**. This theory states that regardless of the loading, yielding of the material begins when the absolute maximum shear stress in the material reaches the shear stress that causes the same material to yield when it is subjected *only* to axial tension. Therefore, to avoid failure, it is required that $\tau_{\max}^{\text{abs}}$ in the material must be less than or equal to $\sigma_Y/2$, where σ_Y is determined from a simple tension test.

For plane stress we will use the ideas discussed in Sec. 9.5 and express the absolute maximum shear stress in terms of the *principal stresses* σ_1 and σ_2. If these two principal stresses have the *same sign*, i.e., they are both tensile or both compressive, then failure will occur *out of the plane*, and from Eq. 9–13,

$$\tau_{\max}^{\text{abs}} = \frac{\sigma_1}{2}$$

If instead the principal stresses are of *opposite signs*, then failure occurs in the plane, and from Eq. 9–14,

$$\tau_{\max}^{\text{abs}} = \frac{\sigma_1 - \sigma_2}{2}$$

With these two equations and Eq. 10–26, the maximum shear stress theory for *plane stress* can therefore be expressed by the following criteria:

$$\left.\begin{array}{l} |\sigma_1| = \sigma_Y \\ |\sigma_2| = \sigma_Y \end{array}\right\} \quad \sigma_1, \sigma_2 \text{ have same signs}$$

$$|\sigma_1 - \sigma_2| = \sigma_Y\} \quad \sigma_1, \sigma_2 \text{ have opposite signs}$$

(10–27)

A graph of these equations is shown in Fig. 10–28. Therefore, if any point of the material is subjected to plane stress represented by the coordinates (σ_1, σ_2) that fall *on the boundary* or *outside* the shaded hexagonal area, the material will yield at the point and failure is said to occur.

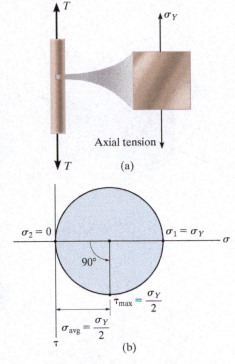

Axial tension

(a)

$\sigma_2 = 0$ $\sigma_1 = \sigma_Y$

$90°$

$\tau_{\max} = \frac{\sigma_Y}{2}$

$\sigma_{\text{avg}} = \frac{\sigma_Y}{2}$

(b)

$\tau_{\max} = \frac{\sigma_Y}{2}$ $\sigma_{\text{avg}} = \frac{\sigma_Y}{2}$

$45°$

(c)

Fig. 10–27

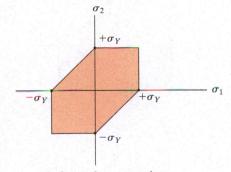

Maximum shear stress theory

Fig. 10–28

Maximum Distortion Energy Theory.   It was stated in Sec. 3.5 that an external load will deform a material, causing it to store energy *internally* throughout its volume. The energy per unit volume of material is called the **strain-energy density**, and if the material is subjected to a uniaxial stress the strain-energy density, defined by Eq. 3–6, becomes

$$u = \frac{1}{2}\sigma\epsilon \tag{10–28}$$

If the material is subjected to triaxial stress, Fig. 10–29a, then each principal stress contributes a portion of the total strain-energy density, so that

$$u = \frac{1}{2}\sigma_1\,\epsilon_1 + \frac{1}{2}\sigma_2\,\epsilon_2 + \frac{1}{2}\sigma_3\,\epsilon_3$$

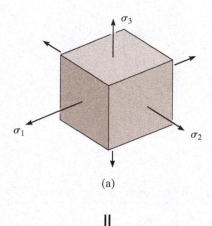

(a)

||

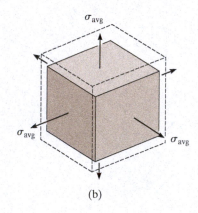

(b)

+

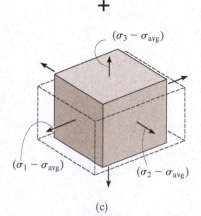

(c)

Fig. 10–29

Furthermore, if the material behaves in a linear elastic manner, then Hooke's law applies. Therefore, substituting Eq. 10–18 into the above equation and simplifying, we get

$$u = \frac{1}{2E}\left[\sigma_1{}^2 + \sigma_2{}^2 + \sigma_3{}^2 - 2\nu(\sigma_1\,\sigma_2 + \sigma_1\,\sigma_3 + \sigma_3\,\sigma_2)\right] \tag{10–29}$$

This strain-energy density can be considered as the sum of two parts. One part is the energy needed to cause a *volume change* of the element with no change in shape, and the other part is the energy needed to *distort* the element. Specifically, the energy stored in the element as a result of its volume being changed is caused by application of the average principal stress, $\sigma_{avg} = (\sigma_1 + \sigma_2 + \sigma_3)/3$, since this stress causes equal principal strains in the material, Fig. 10–29b. The remaining portion of the stress, $(\sigma_1 - \sigma_{avg})$, $(\sigma_2 - \sigma_{avg})$, $(\sigma_3 - \sigma_{avg})$, causes the energy of distortion, Fig. 10–29c.

Experimental evidence has shown that materials do not yield when they are subjected to a uniform (hydrostatic) stress, such as σ_{avg}. As a result, in 1904, M. Huber proposed that yielding in a ductile material occurs when the *distortion energy* per unit volume of the material equals or exceeds the distortion energy per unit volume of the same material when it is subjected to yielding in a simple tension test. This theory is called the **maximum distortion energy theory**, and since it was later redefined independently by R. von Mises and H. Hencky, it sometimes also bears their names.

To obtain the distortion energy per unit volume, we must substitute the stresses $(\sigma_1 - \sigma_{avg})$, $(\sigma_2 - \sigma_{avg})$, and $(\sigma_3 - \sigma_{avg})$ for σ_1, σ_2, and σ_3, respectively, into Eq. 10–29, realizing that $\sigma_{avg} = (\sigma_1 + \sigma_2 + \sigma_3)/3$. Expanding and simplifying, we obtain

$$u_d = \frac{1 + \nu}{6E}\left[(\sigma_1 - \sigma_2)^2 + (\sigma_2 - \sigma_3)^2 + (\sigma_3 - \sigma_1)^2\right]$$

In the case of *plane stress*, $\sigma_3 = 0$, and this equation reduces to

$$u_d = \frac{1 + \nu}{3E}\left(\sigma_1^2 - \sigma_1\sigma_2 + \sigma_2^2\right)$$

For a *uniaxial* tension test, $\sigma_1 = \sigma_Y, \sigma_2 = \sigma_3 = 0$, and so

$$(u_d)_Y = \frac{1 + \nu}{3E}\sigma_Y^2$$

Since the maximum distortion energy theory requires $u_d = (u_d)_Y$, then for the case of plane or biaxial stress, we have

$$\boxed{\sigma_1^2 - \sigma_1\sigma_2 + \sigma_2^2 = \sigma_Y^2} \tag{10–30}$$

This is the equation of an ellipse, Fig. 10–30. Thus, if a point in the material is stressed such that (σ_1, σ_2) is plotted on the boundary or outside the shaded area, the material is said to fail.

A comparison of the above two failure criteria is shown in Fig. 10–31. Note that both theories give the same results when the principal stresses are equal in magnitude, i.e., $\sigma_1 = \sigma_2 = \sigma_Y$, or when one of the principal stresses is zero and the other has a magnitude of σ_Y. If the material is subjected to pure shear, τ, then the theories have the largest discrepancy in predicting failure. The stress coordinates of these points on the curves can be determined by considering the element shown in Fig. 10–32a. From Mohr's circle representing this state of stress, Fig. 10–32b, we obtain principal stresses $\sigma_1 = \tau$ and $\sigma_2 = -\tau$. Thus, with $\sigma_1 = -\sigma_2$, then from Eq. 10–27, the maximum shear stress theory gives $(\sigma_Y/2, -\sigma_Y/2)$, and from Eq. 10–30, the maximum distortion energy theory gives $(\sigma_Y/\sqrt{3}, -\sigma_Y/\sqrt{3})$, Fig. 10–31.

By performing torsion tests, which develop pure shear in a ductile specimen, it has been shown that the maximum distortion energy theory gives more accurate results for pure-shear failure than the maximum shear stress theory. In fact, since $(\sigma_Y/\sqrt{3})/(\sigma_Y/2) = 1.15$, the shear stress for yielding of the material, as given by the maximum distortion energy theory, is 15% more accurate than that given by the maximum shear stress theory.

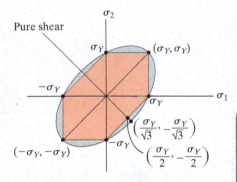

Maximum distortion energy theory

Fig. 10–30

Pure shear

(σ_Y, σ_Y)

$\left(\dfrac{\sigma_Y}{\sqrt{3}}, -\dfrac{\sigma_Y}{\sqrt{3}}\right)$

$(-\sigma_Y, -\sigma_Y)$ $\left(\dfrac{\sigma_Y}{2}, -\dfrac{\sigma_Y}{2}\right)$

Fig. 10–31

10

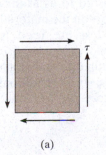

(a)

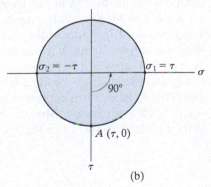

(b)

Fig. 10–32

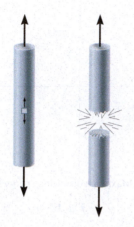

Failure of a brittle material
in tension

(a)

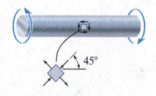

Failure of a brittle material
in torsion

(b)

Fig. 10–33

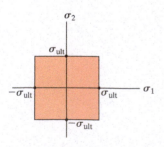

Maximum normal stress theory

Fig. 10–34

Brittle Materials

Maximum Normal Stress Theory. It was previously stated that brittle materials, such as gray cast iron, tend to fail suddenly by *fracture* with no apparent yielding. In a *tension test*, the fracture occurs when the normal stress reaches the ultimate stress σ_{ult}, Fig. 10–33a. Also, brittle fracture occurs in a torsion test due to tension, since the plane of fracture for an element is at 45° to the shear direction, Fig. 10–33b. The fracture surface is therefore helical as shown.* Experiments have further shown that during torsion the material's strength is only slightly affected by the presence of the associated principal compressive stress being at right angles to the principal tensile stress. Consequently, the tensile stress needed to fracture a specimen during a torsion test is approximately the same as that needed to fracture a specimen in simple tension. Because of this, the **maximum normal stress theory** states that when a brittle material is subjected to a multiaxial state of stress, the material will fail when a principal tensile stress in the material reaches a value that is equal to the ultimate normal stress the material can sustain when it is subjected to simple tension. Therefore, if the material is subjected to *plane stress*, we require that

$$|\sigma_1| = \sigma_{ult}$$
$$|\sigma_2| = \sigma_{ult}$$

(10–31)

These equations are shown graphically in Fig. 10–34. Here the stress coordinates (σ_1, σ_2) at a point in the material must not fall on the boundary or outside the shaded area, otherwise the material is said to fracture. This theory is generally credited to W. Rankine, who proposed it in the mid-1800s. Experimentally it has been found to be in close agreement with the behavior of brittle materials that have stress–strain diagrams that are *similar* in both tension and compression.

Mohr's Failure Criterion. In some brittle materials the tension and compression properties are *different*. When this occurs a criterion based on the use of Mohr's circle may be used to predict failure. This method was developed by Otto Mohr and is sometimes referred to as **Mohr's failure criterion.** To apply it, one first performs *three tests* on the material. A uniaxial tensile test and uniaxial compressive test are used to determine the ultimate tensile and compressive stresses $(\sigma_{ult})_t$ and $(\sigma_{ult})_c$, respectively. Also a torsion test is performed to determine the material's ultimate shear stress τ_{ult}. Mohr's circle for each of these stress conditions is then plotted as shown in Fig. 10–35. These three circles are contained

*A stick of blackboard chalk fails in this way when its ends are twisted with the fingers.

within a "failure envelope" indicated by the extrapolated colored curve that is drawn tangent to all three circles. If a plane-stress condition at a point is represented by a circle that has a point of tangency with the envelope, or if it extends beyond the envelope's boundary, then failure is said to occur.

We may also represent this criterion on a graph of principal stresses σ_1 and σ_2. This is shown in Fig. 10–36. Here failure occurs when the absolute value of either one of the principal stresses reaches a value equal to or greater than $(\sigma_{ult})_t$ or $(\sigma_{ult})_c$, or in general, if the state of stress at a point defined by the stress coordinates (σ_1, σ_2) is plotted on the boundary or outside the shaded area.

Either the maximum normal stress theory or Mohr's failure criterion can be used in practice to predict the failure of a brittle material. However, it should be realized that their usefulness is quite limited. A tensile fracture occurs very suddenly, and its initiation generally depends on stress concentrations developed at microscopic imperfections of the material, such as inclusions or voids, surface indentations, and small cracks. Unfortunately each of these irregularities varies from specimen to specimen, and so it becomes difficult to specify fracture on the basis of a single test.

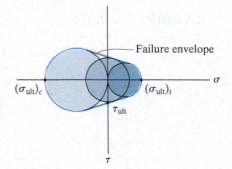

Fig. 10–35

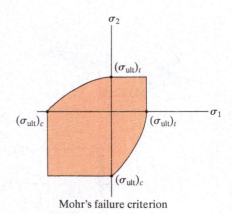

Mohr's failure criterion

Fig. 10–36

IMPORTANT POINTS

- If a material is *ductile*, failure is specified by the initiation of *yielding*, whereas if it is *brittle*, it is specified by *fracture*.

- *Ductile failure* can be defined when *slipping* occurs between the crystals that compose the material. This slipping is due to *shear stress* and the *maximum shear stress theory* is based on this idea.

- *Strain energy* is stored in a material when it is subjected to normal stress. The *maximum distortion energy theory* depends on the *strain energy* that *distorts* the material, and not the strain energy that increases its volume.

- The fracture of a *brittle material* is caused only by the *maximum tensile stress* in the material, and not the compressive stress. This is the basis of the *maximum normal stress theory*, and it is applicable if the stress–strain diagram is *similar* in tension and compression.

- If a *brittle material* has a stress–strain diagram that is *different* in tension and compression, then *Mohr's failure criterion* may be used to predict failure.

- Due to material imperfections, *tensile fracture* of a brittle material is *difficult to predict*, and so theories of failure for brittle materials should be used with caution.

EXAMPLE 10.12

The solid shaft shown in Fig. 10–37a has a radius of 0.5 in. and is made of steel having a yield stress of $\sigma_Y = 36$ ksi. Determine if the loadings cause the shaft to fail if a design code is specified using the maximum shear stress theory. What would be the result if the maximum distortion energy theory was specified?

SOLUTION

The state of stress in the shaft is caused by both the axial force and the torque. Since maximum shear stress caused by the torque occurs in the material at the outer surface, we have

$$\sigma_x = \frac{N}{A} = \frac{-15 \text{ kip}}{\pi (0.5 \text{ in.})^2} = -19.10 \text{ ksi}$$

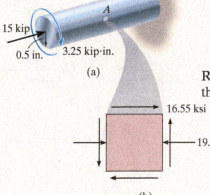

(a)

$$\tau_{xy} = \frac{Tc}{J} = \frac{3.25 \text{ kip} \cdot \text{in.} (0.5 \text{ in.})}{\frac{\pi}{2}(0.5 \text{ in.})^4} = 16.55 \text{ ksi}$$

The stress components are shown on an element at point A in Fig. 10–37b. Rather than using Mohr's circle, the principal stresses will be obtained using the stress transformation Eq. 9–5.

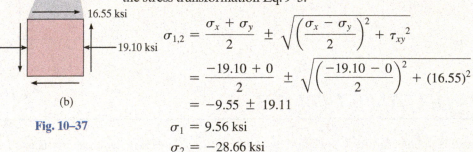

(b)

Fig. 10–37

$$\sigma_{1,2} = \frac{\sigma_x + \sigma_y}{2} \pm \sqrt{\left(\frac{\sigma_x - \sigma_y}{2}\right)^2 + \tau_{xy}^2}$$

$$= \frac{-19.10 + 0}{2} \pm \sqrt{\left(\frac{-19.10 - 0}{2}\right)^2 + (16.55)^2}$$

$$= -9.55 \pm 19.11$$

$$\sigma_1 = 9.56 \text{ ksi}$$

$$\sigma_2 = -28.66 \text{ ksi}$$

Maximum Shear Stress Theory. Since the principal stresses have *opposite signs*, then applying the second of Eqs. 10–27, we have

$$|\sigma_1 - \sigma_2| \leq \sigma_Y$$

$$|9.56 - (-28.66)| \overset{?}{\leq} 36$$

$$38.2 > 36$$

Thus, shear failure of the material will occur according to this theory.

Maximum Distortion Energy Theory.   Applying Eq. 10–30, we have

$$(\sigma_1^2 - \sigma_1 \sigma_2 + \sigma_2^2) \leq \sigma_Y^2$$

$$\left[(9.56)^2 - (9.56)(-28.66) + (-28.66)^2\right] \overset{?}{\leq} (36)^2$$

$$1187 \leq 1296$$

Using this theory, failure will not occur.

EXAMPLE 10.13

The solid cast-iron shaft shown in Fig. 10–38a is subjected to a torque of $T = 400$ lb·ft. Determine its smallest radius so that it does not fail according to the maximum normal stress theory. A specimen of cast iron, tested in tension, has an ultimate stress of $(\sigma_{ult})_t = 20$ ksi.

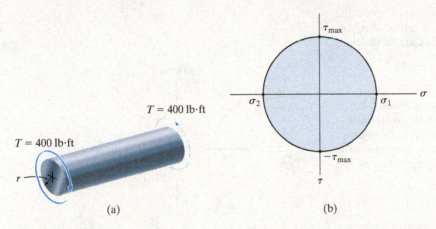

$T = 400$ lb·ft

$T = 400$ lb·ft

r

(a)

(b)

Fig. 10–38

SOLUTION

The maximum or critical stress occurs at a point located on the surface of the shaft. Assuming the shaft to have a radius r, the shear stress is

$$\tau_{max} = \frac{Tc}{J} = \frac{(400 \text{ lb} \cdot \text{ft})(12 \text{ in.}/\text{ft})r}{(\pi/2)r^4} = \frac{3055.8 \text{ lb} \cdot \text{in.}}{r^3}$$

Mohr's circle for this state of stress (pure shear) is shown in Fig. 10–38b. Since $R = \tau_{max}$, then

$$\sigma_1 = -\sigma_2 = \tau_{max} = \frac{3055.8 \text{ lb} \cdot \text{in.}}{r^3}$$

The maximum normal stress theory, Eq. 10–31, requires

$$|\sigma_1| \leq \sigma_{ult}$$

$$\frac{3055.8 \text{ lb} \cdot \text{in.}}{r^3} \leq 20\,000 \text{ lb/in}^2$$

Thus, the smallest radius of the shaft is

$$\frac{3055.8 \text{ lb} \cdot \text{in.}}{r^3} = 20\,000 \text{ lb/in}^2$$

$$r = 0.535 \text{ in.} \qquad \textit{Ans.}$$

PROBLEMS

10–59. A material is subjected to plane stress. Express the distortion energy theory of failure in terms of σ_x, σ_y, and τ_{xy}.

*10–60. A material is subjected to plane stress. Express the maximum shear stress theory of failure in terms of σ_x, σ_y, and τ_{xy}. Assume that the principal stresses are of different algebraic signs.

10–61. The yield stress for a zirconium-magnesium alloy is $\sigma_Y = 15.3$ ksi. If a machine part is made of this material and a critical point in the material is subjected to in-plane principal stresses σ_1 and $\sigma_2 = -0.5\sigma_1$, determine the magnitude of σ_1 that will cause yielding according to the maximum shear stress theory.

10–62. Solve Prob. 10–61 using the maximum distortion energy theory.

10–63. If a machine part is made of tool L2 steel and a critical point in the material is subjected to in-plane principal stresses σ_1 and $\sigma_2 = -0.5\sigma_1$, determine the magnitude of σ_1 in ksi that will cause yielding according to the maximum shear stress theory.

*10–64. Solve Prob. 10–63 using the maximum distortion energy theory.

10–65. Derive an expression for an equivalent torque T_e that, if applied alone to a solid bar with a circular cross section, would cause the same energy of distortion as the combination of an applied bending moment M and torque T.

10–66. If a shaft is made of a material for which $\sigma_Y = 75$ ksi, determine the maximum torsional shear stress required to cause yielding using the maximum distortion energy theory.

10–67. Solve Prob. 10–66 using the maximum shear stress theory.

*10–68. If the material is machine steel having a yield stress of $\sigma_Y = 700$ MPa, determine the factor of safety with respect to yielding if the maximum shear stress theory is considered.

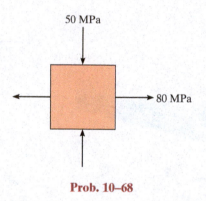

Prob. 10–68

10–69. The short concrete cylinder having a diameter of 50 mm is subjected to a torque of 500 N·m and an axial compressive force of 2 kN. Determine if it fails according to the maximum normal stress theory. The ultimate stress of the concrete is $\sigma_{\text{ult}} = 28$ MPa.

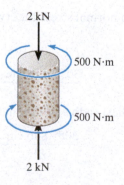

Prob. 10–69

10–70. Derive an expression for an equivalent bending moment M_e that, if applied alone to a solid bar with a circular cross section, would cause the same energy of distortion as the combination of an applied bending moment M and torque T.

10–71. The plate is made of Tobin bronze, which yields at $\sigma_Y = 25$ ksi. Using the maximum shear stress theory, determine the maximum tensile stress σ_x that can be applied to the plate if a tensile stress $\sigma_y = 1.5\sigma_x$ is also applied.

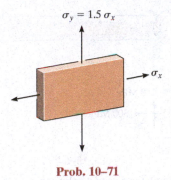

Prob. 10–71

*10–72. The plate is made of Tobin bronze, which yields at $\sigma_Y = 25$ ksi. Using the maximum distortion energy theory, determine the maximum tensile stress σ_x that can be applied to the plate if a tensile stress $\sigma_y = 1.5\sigma_x$ is also applied.

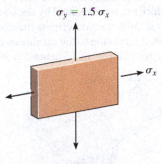

Prob. 10–72

10–73. An aluminum alloy is to be used for a solid drive shaft such that it transmits 30 hp at 1200 rev/min. Using a factor of safety of 2.5 with respect to yielding, determine the smallest-diameter shaft that can be selected based on the maximum shear stress theory. $\sigma_Y = 10$ ksi.

10–74. If a machine part is made of titanium (Ti-6Al-4V) and a critical point in the material is subjected to plane stress, such that the principal stresses are σ_1 and $\sigma_2 = 0.5\sigma_1$, determine the magnitude of σ_1 in MPa that will cause yielding according to (a) the maximum shear stress theory, and (b) the maximum distortion energy theory.

10–75. The components of plane stress at a critical point on a thin steel shell are shown. Determine if failure (yielding) has occurred on the basis of the maximum distortion energy theory. The yield stress for the steel is $\sigma_Y = 700$ MPa.

*10–76. Solve Prob. 10–75 using the maximum shear stress theory.

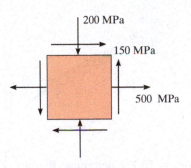

Probs. 10–75/76

10–77. The 304-stainless-steel cylinder has an inner diameter of 4 in. and a wall thickness of 0.1 in. If it is subjected to an internal pressure of $p = 80$ psi, axial load of 500 lb, and a torque of 70 lb · ft, determine if yielding occurs according to the maximum distortion energy theory.

10–78. The 304-stainless-steel cylinder has an inner diameter of 4 in. and a wall thickness of 0.1 in. If it is subjected to an internal pressure of $p = 80$ psi, axial load of 500 lb, and a torque of 70 lb · ft, determine if yielding occurs according to the maximum shear stress theory.

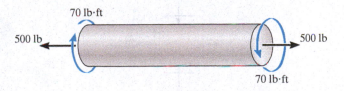

Probs. 10–77/78

10–79. If the 2-in.-diameter shaft is made from brittle material having an ultimate stress of $\sigma_{ult} = 50$ ksi, for both tension and compression, determine if the shaft fails according to the maximum normal stress theory. Use a factor of safety of 1.5 against rupture.

*10–80. If the 2-in.-diameter shaft is made from cast iron having tensile and compressive ultimate stress of $(\sigma_{ult})_t = 50$ ksi and $(\sigma_{ult})_c = 75$ ksi, respectively, determine if the shaft fails according to Mohr's failure criterion.

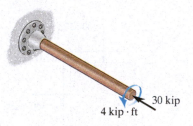

Probs. 10–79/80

10–81. If $\sigma_Y = 50$ ksi, determine the factor of safety for this loading against yielding based on (a) the maximum shear stress theory and (b) the maximum distortion energy theory.

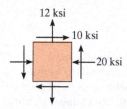

Prob. 10–81

10–82. The state of plane stress at a critical point in a steel machine bracket is shown. If the yield stress for steel is $\sigma_Y = 36$ ksi, determine if yielding occurs using the maximum distortion energy theory.

10–83. Solve Prob. 10–82 using the maximum shear stress theory.

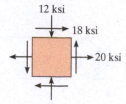

Probs. 10–82/83

*10–84. The state of stress acting at a critical point on a wrench is shown. Determine the smallest yield stress for steel that might be selected for the part, based on the maximum distortion energy theory.

10–85. The state of stress acting at a critical point on a wrench is shown in the figure. Determine the smallest yield stress for steel that might be selected for the part, based on the maximum shear stress theory.

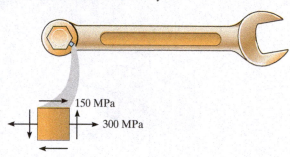

Probs. 10–84/85

10–86. The shaft consists of a solid segment AB and a hollow segment BC, which are rigidly joined by the coupling at B. If the shaft is made from A-36 steel, determine the maximum torque T that can be applied according to the maximum shear stress theory. Use a factor of safety of 1.5 against yielding.

10–87. The shaft consists of a solid segment AB and a hollow segment BC, which are rigidly joined by the coupling at B. If the shaft is made from A-36 steel, determine the maximum torque T that can be applied according to the maximum distortion energy theory. Use a factor of safety of 1.5 against yielding.

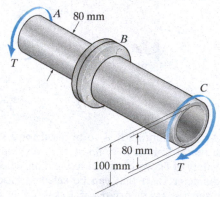

Probs. 10–86/87

*10–88. The principal stresses acting at a point on a thin-walled cylindrical pressure vessel are $\sigma_1 = pr/t$, $\sigma_2 = pr/2t$, and $\sigma_3 = 0$. If the yield stress is σ_Y, determine the maximum value of p based on (a) the maximum shear stress theory and (b) the maximum distortion energy theory.

10–89. If $\sigma_Y = 50$ ksi, determine the factor of safety for this loading based on (a) the maximum shear stress theory and (b) the maximum distortion energy theory.

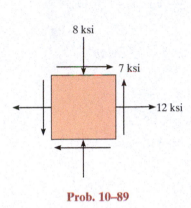

Prob. 10–89

10–90. The gas tank is made from A-36 steel and has an inner diameter of 1.50 m. If the tank is designed to withstand a pressure of 5 MPa, determine the required minimum wall thickness to the nearest millimeter using (a) the maximum shear stress theory, and (b) maximum distortion energy theory. Apply a factor of safety of 1.5 against yielding.

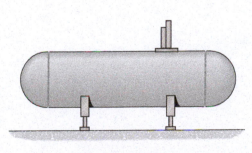

Prob. 10–90

10–91. The internal loadings at a critical section along the steel drive shaft of a ship are calculated to be a torque of 2300 lb · ft, a bending moment of 1500 lb · ft, and an axial thrust of 2500 lb. If the yield points for tension and shear are $\sigma_Y = 100$ ksi and $\tau_Y = 50$ ksi, respectively, determine the required diameter of the shaft using the maximum shear stress theory.

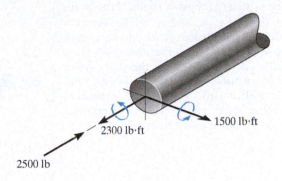

Prob. 10–91

*10–92. If the material is machine steel having a yield stress of $\sigma_Y = 750$ MPa, determine the factor of safety with respect to yielding using the maximum distortion energy theory.

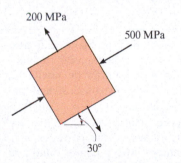

Prob. 10–92

10–93. If the material is machine steel having a yield stress of $\sigma_Y = 750$ MPa, determine the factor of safety with respect to yielding if the maximum shear stress theory is considered.

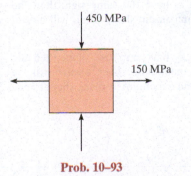

Prob. 10–93

CHAPTER REVIEW

When an element of material is subjected to deformations that only occur in a single plane, then it undergoes plane strain. If the strain components ϵ_x, ϵ_y, and γ_{xy} are known for a specified orientation of the element, then the strains acting for some other orientation can be determined using the plane-strain transformation equations. Likewise, the principal normal strains and maximum in-plane shear strain can be determined using transformation equations.

$$\epsilon_{x'} = \frac{\epsilon_x + \epsilon_y}{2} + \frac{\epsilon_x - \epsilon_y}{2}\cos 2\theta + \frac{\gamma_{xy}}{2}\sin 2\theta$$

$$\epsilon_{y'} = \frac{\epsilon_x + \epsilon_y}{2} - \frac{\epsilon_x - \epsilon_y}{2}\cos 2\theta - \frac{\gamma_{xy}}{2}\sin 2\theta$$

$$\frac{\gamma_{x'y'}}{2} = -\left(\frac{\epsilon_x - \epsilon_y}{2}\right)\sin 2\theta + \frac{\gamma_{xy}}{2}\cos 2\theta$$

$$\epsilon_{1,2} = \frac{\epsilon_x + \epsilon_y}{2} \pm \sqrt{\left(\frac{\epsilon_x - \epsilon_y}{2}\right)^2 + \left(\frac{\gamma_{xy}}{2}\right)^2}$$

Principal strains

$$\frac{\gamma_{\substack{\max \\ \text{in-plane}}}}{2} = \sqrt{\left(\frac{\epsilon_x - \epsilon_y}{2}\right)^2 + \left(\frac{\gamma_{xy}}{2}\right)^2}$$

Maximum in-plane shear stress

$$\epsilon_{\text{avg}} = \frac{\epsilon_x + \epsilon_y}{2}$$

Strain transformation problems can also be solved in a semigraphical manner using Mohr's circle. To draw the circle, the ϵ and $\gamma/2$ axes are established, and the center of the circle $C\left[(\epsilon_x + \epsilon_y)/2, 0\right]$ and the "reference point" A $(\epsilon_x, \gamma_{xy}/2)$ are plotted. The radius of the circle extends between these two points and is determined from trigonometry.

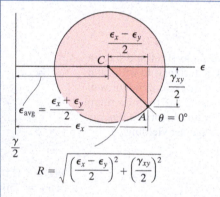

$$R = \sqrt{\left(\frac{\epsilon_x - \epsilon_y}{2}\right)^2 + \left(\frac{\gamma_{xy}}{2}\right)^2}$$

If ϵ_1 and ϵ_2 have the same sign, then the absolute maximum shear strain will be out of plane.

If ϵ_1 and ϵ_2 have opposite signs, then the absolute maximum shear strain will be equal to the maximum in-plane shear strain.

$$\gamma_{\substack{\text{abs} \\ \max}} = \epsilon_1$$

$$\gamma_{\substack{\text{abs} \\ \max}} = \gamma_{\substack{\text{in-plane} \\ \max}} = \epsilon_1 - \epsilon_2$$

If the material is subjected to triaxial stress, then the strain in each direction is influenced by the strain produced by all three stresses. Hooke's law then involves the material properties E and ν.	$$\epsilon_x = \frac{1}{E}[\sigma_x - \nu(\sigma_y + \sigma_z)]$$ $$\epsilon_y = \frac{1}{E}[\sigma_y - \nu(\sigma_x + \sigma_z)]$$ $$\epsilon_z = \frac{1}{E}[\sigma_z - \nu(\sigma_x + \sigma_y)]$$
If E and ν are known, then G can be determined.	$$G = \frac{E}{2(1 + \nu)}$$
The dilatation is a measure of volumetric strain.	$$e = \frac{1 - 2\nu}{E}(\sigma_x + \sigma_y + \sigma_z)$$
The bulk modulus is used to measure the stiffness of a volume of material.	$$k = \frac{E}{3(1 - 2\nu)}$$
If the principal stresses at a critical point in the material are known, then a theory of failure can be used as a basis for design. *Ductile materials fail in shear*, and here the maximum shear stress theory or the maximum distortion energy theory can be used to predict failure. Both of these theories make comparison to the yield stress of a specimen subjected to a uniaxial tensile stress. *Brittle materials fail in tension*, and so the maximum normal stress theory or Mohr's failure criterion can be used to predict failure. Here comparisons are made with the ultimate tensile stress developed in a specimen.	

10

REVIEW PROBLEMS

R10–1. In the case of plane stress, where the in-plane principal strains are given by ϵ_1 and ϵ_2, show that the third principal strain can be obtained from

$$\epsilon_3 = \frac{-\nu(\epsilon_1 + \epsilon_2)}{(1 - \nu)}$$

where ν is Poisson's ratio for the material.

R10–2. The plate is made of material having a modulus of elasticity $E = 200\,\text{GPa}$ and Poisson's ratio $\nu = \frac{1}{3}$. Determine the change in width a, height b, and thickness t when it is subjected to the uniform distributed loading shown.

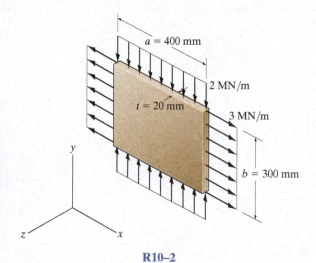

R10–2

R10–3. If the material is machine steel having a yield stress of $\sigma_Y = 500$ MPa, determine the factor of safety with respect to yielding if the maximum shear stress theory is considered.

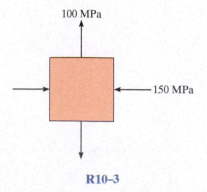

R10–3

*R10–4. The components of plane stress at a critical point on a thin steel shell are shown. Determine if yielding has occurred on the basis of the maximum distortion energy theory. The yield stress for the steel is $\sigma_Y = 650$ MPa.

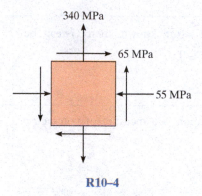

R10–4

R10–5. The 60° strain rosette is mounted on a beam. The following readings are obtained for each gage: $\epsilon_a = 600(10^{-6})$, $\epsilon_b = -700(10^{-6})$, and $\epsilon_c = 350(10^{-6})$. Determine (a) the in-plane principal strains and (b) the maximum in-plane shear strain and average normal strain. In each case show the deformed element due to these strains.

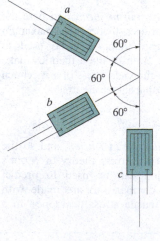

R10–5

R10–6. The state of strain at the point on the bracket has components $\epsilon_x = 350(10^{-6})$, $\epsilon_y = -860(10^{-6})$, $\gamma_{xy} = 250(10^{-6})$. Use the strain transformation equations to determine the equivalent in-plane strains on an element oriented at an angle of $\theta = 45°$ clockwise from the original position. Sketch the deformed element within the x–y plane due to these strains.

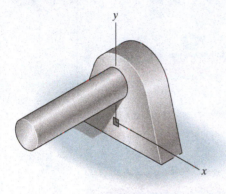

R10–6

R10–7. The A-36 steel post is subjected to the forces shown. If the strain gages a and b at point A give readings of $\epsilon_a = 300(10^{-6})$ and $\epsilon_b = 175(10^{-6})$, determine the magnitudes of $\mathbf{P}_1$ and $\mathbf{P}_2$.

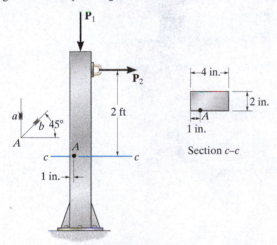

R10–7

***R10–8.** A differential element is subjected to plane strain that has the following components; $\epsilon_x = 950(10^{-6})$, $\epsilon_y = 420(10^{-6})$, $\gamma_{xy} = -325(10^{-6})$. Use the strain-transformation equations and determine (a) the principal strains and (b) the maximum in-plane shear strain and the associated average strain. In each case specify the orientation of the element and show how the strains deform the element.

R10–9. The state of strain at the point on the bracket has components $\epsilon_x = -130(10^{-6})$, $\epsilon_y = 280(10^{-6})$, $\gamma_{xy} = 75(10^{-6})$. Use the strain transformation equations to determine (a) the in-plane principal strains and (b) the maximum in-plane shear strain and average normal strain. In each case specify the orientation of the element and show how the strains deform the element within the x–y plane.

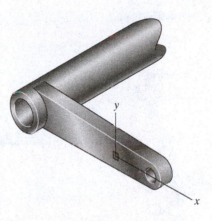

R10–9

R10–10. The state of plane strain on the element is $\epsilon_x = 400(10^{-6})$, $\epsilon_y = 200(10^{-6})$, and $\gamma_{xy} = -300(10^{-6})$. Determine the equivalent state of strain, which represents (a) the principal strains, and (b) the maximum in-plane shear strain and the associated average normal strain. Specify the orientation of the corresponding element at the point with respect to the original element. Sketch the results on the element.

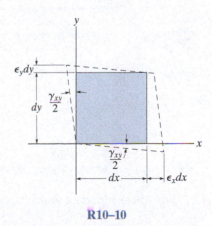

R10–10

CHAPTER 11

(© Olaf Speier/Alamy)

Beams are important structural members used to support roof and floor loadings.

DESIGN OF BEAMS AND SHAFTS

CHAPTER OBJECTIVES

■ In this chapter we will discuss how to design a prismatic beam so that it is able to resist both internal bending and shear. Also, we will present a method for determining the shape of a beam that is fully stressed along its length. At the end of the chapter, we will consider the design of shafts based on their resistance to both internal bending and torsion.

11.1 BASIS FOR BEAM DESIGN

Beams are said to be designed on the *basis of strength* when they can resist the internal shear and moment developed along their length. To design a beam in this way requires application of the shear and flexure formulas provided the material is homogeneous and has linear elastic behavior. Although some beams may also be subjected to an axial force, the effects of this force are often neglected in design since the axial stress is generally much smaller than the stress developed by shear and bending.

As shown in Fig. 11–1, the external loadings on a beam will create additional stresses in the beam *directly under the load*. Notably, a compressive stress σ_y will be developed, in addition to the bending stress σ_x and shear stress τ_{xy} discussed previously in Chapters 6 and 7. Using advanced methods of analysis, as treated in the theory of elasticity, it can be shown that σ_y diminishes rapidly throughout the beam's depth, and for *most* beam span-to-depth ratios used in engineering practice, the maximum value of σ_y remains small compared to the bending stress σ_x, that is, $\sigma_x \gg \sigma_y$. Furthermore, the direct application of concentrated loads is generally avoided in beam design. Instead, **bearing plates** are used to spread these loads more evenly onto the surface of the beam, thereby further reducing σ_y.

Beams must also be braced properly along their sides so that they do not sidesway or suddenly become unstable. In some cases they must also be designed to resist *deflection*, as when they support ceilings made of brittle materials such as plaster. Methods for finding beam deflections will be discussed in Chapter 12, and limitations placed on beam sidesway are often discussed in codes related to structural or mechanical design.

Knowing how the magnitude and direction of the principal stress change from point to point within a beam is important if the beam is made of a brittle material, because brittle materials, such as concrete, fail in tension. To give some idea as to how to determine this variation, let's consider the cantilever beam shown in Fig. 11–2a, which has a rectangular cross section and supports a load **P** at its end.

In general, at an arbitrary section *a–a* along the beam, Fig. 11–2b, the internal shear **V** and moment **M** create a *parabolic* shear-stress distribution and a *linear* normal-stress distribution, Fig. 11–2c. As a result, the stresses acting on elements located at points 1 through 5 along the section are shown in Fig. 11–2d. Note that elements 1 and 5 are subjected only to a maximum normal stress, whereas element 3, which is on the neutral axis, is subjected only to a maximum in-plane shear stress. The intermediate elements 2 and 4 must resist *both* normal and shear stress.

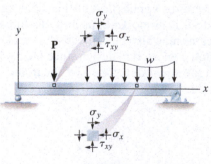

Fig. 11–1

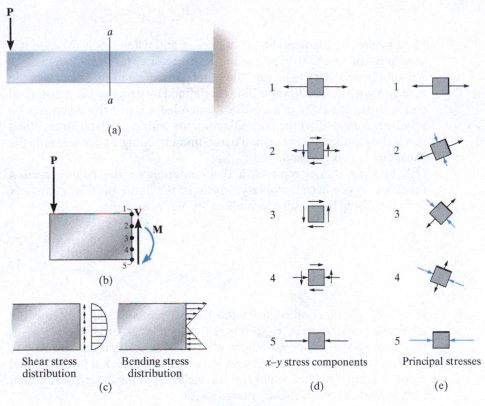

Shear stress
distribution

Bending stress
distribution

(c)

x–y stress components

(d)

Principal stresses

(e)

Fig. 11–2

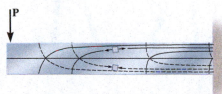

Stress trajectories for
cantilevered beam

Fig. 11–3

When these states of stress are transformed into *principal stresses*, using either the stress transformation equations or Mohr's circle, the results will look like those shown in Fig. 11–2e. If this analysis is extended to many vertical sections along the beam other than *a–a*, a profile of the results can be represented by curves called ***stress trajectories***. Each of these curves indicates the *direction* of a principal stress having a constant magnitude. Some of these trajectories are shown in Fig. 11–3. Here the solid lines represent the direction of the tensile principal stresses and the dashed lines represent the direction of the compressive principal stresses. As expected, the lines intersect the neutral axis at 45° angles (like element 3), and the solid and dashed lines will intersect at 90° because the principal stresses are always 90° apart. Once the directions of these lines are established, it can help engineers decide where and how to place reinforcement in a beam if it is made of brittle material, so that it does not fail.

11.2   PRISMATIC BEAM DESIGN

Most beams are made of ductile materials, and when this is the case it is generally not necessary to plot the stress trajectories for the beam. Instead, it is simply necessary to be sure the *actual* bending and shear stress in the beam do not exceed allowable limits as defined by structural or mechanical codes. In the majority of cases the suspended span of the beam will be relatively long, so that the internal moments within it will be large. When this is the case, the design is then based upon bending, and afterwards the shear strength is checked.

A bending design requires a determination of the beam's **section modulus**, a geometric property which is the ratio of I to c, that is, $S = I/c$. Using the flexure formula, $\sigma = Mc/I$, we have

$$S_{\text{req'd}} = \frac{M_{\text{max}}}{\sigma_{\text{allow}}} \tag{11–1}$$

Here M_{max} is determined from the beam's moment diagram, and the allowable bending stress, σ_{allow}, is specified in a design code. In many cases the beam's as yet unknown weight will be small, and can be neglected in comparison with the loads the beam must carry. However, if the additional moment caused by the weight is to be included in the design, a selection for S is made so that it slightly *exceeds* $S_{\text{req'd}}$.

Once $S_{\text{req'd}}$ is known, if the beam has a simple cross-sectional shape, such as a square, a circle, or a rectangle of known width-to-height proportions, its *dimensions* can be determined directly from $S_{\text{req'd}}$, since $S_{\text{req'd}} = I/c$. However, if the cross section is made from several elements, such as a wide-flange section, then an infinite number of web and flange dimensions can be determined that satisfy the value of $S_{\text{req'd}}$. In practice, however, engineers choose a particular beam meeting the requirement that $S > S_{\text{req'd}}$ from a table that lists the standard sizes available from manufacturers. Often several beams that have the same section modulus can be selected, and if deflections are not restricted, usually the beam having the smallest cross-sectional area is chosen, since it is made of less material, and is therefore both lighter and more economical than the others.

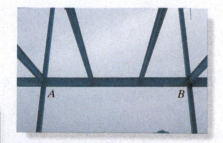

The two floor beams are connected to the beam *AB*, which transmits the load to the columns of this building frame. For design, all the connections can be considered to act as pins.

Once the beam has been selected, the shear formula can then be used to be sure the allowable shear stress is not exceeded, $\tau_{allow} \geq VQ/It$. Often this requirement will not present a problem; however, if the beam is "short" and supports large concentrated loads, the shear-stress limitation may dictate the size of the beam.

Steel Sections.
Most manufactured steel beams are produced by rolling a hot ingot of steel until the desired shape is formed. These so-called *rolled shapes* have properties that are tabulated in the American Institute of Steel Construction (AISC) manual. A representative listing of different cross sections taken from this manual is given in Appendix B. Here the wide-flange shapes are designated by their depth and weight per unit length; for example, W18 × 46 indicates a wide-flange cross section (W) having a depth of 18 in. and a weight of 46 lb/ft, Fig. 11–4. For any given selection, the weight per length, dimensions, cross-sectional area, moment of inertia, and section modulus are reported. Also included is the radius of gyration, r, which is a geometric property related to the section's buckling strength. This will be discussed in Chapter 13.

Typical profile view of a steel wide-flange beam

The large shear force that occurs at the support of this steel beam can cause localized buckling of the beam's flanges or web. To avoid this, a "stiffener" A is placed along the web to maintain stability.

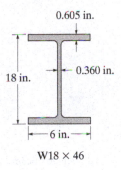

W18 × 46

Fig. 11–4

Wood Sections.
Most beams made of wood have rectangular cross sections because such beams are easy to manufacture and handle. Manuals, such as that of the National Forest Products Association, list the dimensions of lumber often used in the design of wood beams. Lumber is identified by its *nominal dimensions*, such as 2 × 4 (2 in. by 4 in.); however, its *actual* or "dressed" dimensions are smaller, being 1.5 in. by 3.5 in. The reduction in the dimensions occurs in order to obtain a smooth surface from lumber that is rough sawn. Obviously, the *actual dimensions* must be used whenever stress calculations are performed on wood beams.

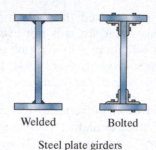

Welded   Bolted

Steel plate girders

Fig. 11–5

Wooden box beam

(a)

Glulam beam

(b)

Fig. 11–6

Built-up Sections.

A ***built-up section*** is constructed from two or more parts joined together to form a single unit. The capacity of this section to resist a moment will vary directly with its section modulus, since $S_{req'd} = M/\sigma_{allow}$. If $S_{req'd}$ is *increased*, then so is I because by definition $S_{req'd} = I/c$. For this reason, *most of the material* for a built-up section should be placed as far away from the neutral axis as practical. This, of course, is what makes a deep wide-flange beam so efficient in resisting a moment. For a very large load, however, an available rolled-steel section may not have a section modulus great enough to support the load. When this is the case, engineers will usually "build up" a beam made from plates and angles. A deep I-shaped section having this form is called a ***plate girder***. For example, the steel plate girder in Fig. 11–5 has two flange plates that are either welded or, using angles, bolted to the web plate.

Wood beams are also "built up," usually in the form of a box beam, Fig. 11–6*a*. They may also be made having plywood webs and larger boards for the flanges. For very large spans, ***glulam beams*** are used. These members are made from several boards glue-laminated together to form a single unit, Fig. 11–6*b*.

Just as in the case of rolled sections or beams made from a single piece, the design of built-up sections requires that the bending and shear stresses be checked. In addition, the shear stress in the fasteners, such as weld, glue, nails, etc., must be checked to be certain the beam performs as a single unit.

> ## IMPORTANT POINTS
>
> - Beams support loadings that are applied perpendicular to their axes. If they are designed on the basis of strength, they must resist their allowable shear and bending stresses.
>
> - The maximum bending stress in the beam is assumed to be much greater than the localized stresses caused by the application of loadings on the surface of the beam.

▶ PROCEDURE FOR ANALYSIS

Based on the previous discussion, the following procedure provides a rational method for the design of a beam on the basis of strength.

Shear and Moment Diagrams.

- Determine the maximum shear and moment in the beam. Often this is done by constructing the beam's shear and moment diagrams.

Bending Stress.

- If the beam is relatively long, it is designed by finding its section modulus using the flexure formula, $S_{req'd} = M_{max}/\sigma_{allow}$.
- Once $S_{req'd}$ is determined, the cross-sectional dimensions for simple shapes can then be calculated, since $S_{req'd} = I/c$.
- If rolled-steel sections are to be used, several possible beams can be selected from the tables in Appendix B that meet the requirement that $S \geq S_{req'd}$. Of these, choose the one having the smallest cross-sectional area, since this beam has the least weight and is therefore the most economical.
- Make sure that the selected section modulus, S, is *slightly greater* than $S_{req'd}$, so that the additional moment created by the beam's weight is considered.

Shear Stress.

- Normally beams that are short and carry large loads, especially those made of wood, are first designed to resist shear and then later checked against the allowable bending stress requirement.
- Using the shear formula, check to see that the allowable shear stress is not exceeded; that is, use $\tau_{allow} \geq V_{max}Q/It$.
- If the beam has a solid *rectangular* cross section, the shear formula becomes $\tau_{allow} \geq 1.5\,(V_{max}/A)$ (see Eq. 2 of Example 7.2.), and if the cross section is a *wide flange*, it is generally appropriate to assume that the shear stress is *constant* over the cross-sectional area of the beam's web so that $\tau_{allow} \geq V_{max}/A_{web}$, where A_{web} is determined from the product of the web's depth and its thickness. (See the hint at the end of Example 7.3.)

Adequacy of Fasteners.

- The adequacy of fasteners used on built-up beams depends upon the shear stress the fasteners can resist. Specifically, the required spacing of nails or bolts of a particular size is determined from the allowable shear flow, $q_{allow} = VQ/I$, calculated at points on the cross section where the fasteners are located. (See Sec. 7.3.)

11

EXAMPLE 11.1

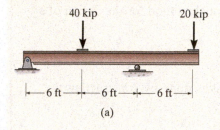

(a)

Fig. 11–7

A beam is to be made of steel that has an allowable bending stress of $\sigma_{\text{allow}} = 24$ ksi and an allowable shear stress of $\tau_{\text{allow}} = 14.5$ ksi. Select an appropriate W shape that will carry the loading shown in Fig. 11–7a.

SOLUTION

Shear and Moment Diagrams. The support reactions have been calculated, and the shear and moment diagrams are shown in Fig. 11–7b. From these diagrams, $V_{\text{max}} = 30$ kip and $M_{\text{max}} = 120$ kip · ft.

Bending Stress. The required section modulus for the beam is determined from the flexure formula,

$$S_{\text{req'd}} = \frac{M_{\text{max}}}{\sigma_{\text{allow}}} = \frac{120 \text{ kip} \cdot \text{ft (12 in./ft)}}{24 \text{ kip/in}^2} = 60 \text{ in}^3$$

Using the table in Appendix B, the following beams are adequate:

W18 × 40	$S = 68.4 \text{ in}^3$
W16 × 45	$S = 72.7 \text{ in}^3$
W14 × 43	$S = 62.7 \text{ in}^3$
W12 × 50	$S = 64.7 \text{ in}^3$
W10 × 54	$S = 60.0 \text{ in}^3$
W8 × 67	$S = 60.4 \text{ in}^3$

The beam having the least weight per foot is chosen,* i.e.,

$$\text{W18} \times 40$$

Shear Stress. Since the beam is a *wide-flange section*, the *average shear stress* within the web will be considered. (See Example 7.3.) Here the web is assumed to extend from the very top to the very bottom of the beam. From Appendix B, for a W18 × 40, $d = 17.90$ in., $t_w = 0.315$ in. Thus,

$$\tau_{\text{avg}} = \frac{V_{\text{max}}}{A_w} = \frac{30 \text{ kip}}{(17.90 \text{ in.})(0.315 \text{ in.})} = 5.32 \text{ ksi} < 14.5 \text{ ksi} \quad \text{OK}$$

Use a W18 × 40.   *Ans.*

---

*The additional moment caused by the weight of the beam, $(0.040 \text{ kip/ft}) (18 \text{ ft}) = 0.720$ kip, will only slightly increase $S_{\text{req'd}}$.

EXAMPLE 11.2

The laminated wooden beam shown in Fig. 11–8a supports a uniform distributed loading of 12 kN/m. If the beam is to have a height-to-width ratio of 1.5, determine its smallest width. Take $\sigma_{allow} = 9$ MPa, and $\tau_{allow} = 0.6$ MPa. Neglect the weight of the beam.

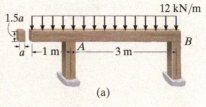

SOLUTION

Shear and Moment Diagrams. The support reactions at A and B have been calculated, and the shear and moment diagrams are shown in Fig. 11–8b. Here $V_{max} = 20$ kN, $M_{max} = 10.67$ kN·m.

Bending Stress. Applying the flexure formula,

$$S_{req'd} = \frac{M_{max}}{\sigma_{allow}} = \frac{10.67(10^3) \text{ N} \cdot \text{m}}{9(10^6) \text{ N/m}^2} = 0.00119 \text{ m}^3$$

Assuming that the width is a, then the height is $1.5a$, Fig. 11–8a. Thus,

$$S_{req'd} = \frac{I}{c} = 0.00119 \text{ m}^3 = \frac{\frac{1}{12}(a)(1.5a)^3}{(0.75a)}$$

$$a^3 = 0.003160 \text{ m}^3$$

$$a = 0.147 \text{ m}$$

Shear Stress. Applying the shear formula for rectangular sections (which is a special case of $\tau_{max} = VQ/It$, as shown in Example 7.2), we have

$$\tau_{max} = 1.5 \frac{V_{max}}{A} = (1.5) \frac{20(10^3) \text{ N}}{(0.147 \text{ m})(1.5)(0.147 \text{ m})}$$

$$= 0.929 \text{ MPa} > 0.6 \text{ MPa}$$

Since the design based on bending fails the shear criterion, the beam must be redesigned on the basis of shear.

$$\tau_{allow} = 1.5 \frac{V_{max}}{A}$$

$$600 \text{ kN/m}^2 = 1.5 \frac{20(10^3) \text{ N}}{(a)(1.5a)}$$

$$a = 0.183 \text{ m} = 183 \text{ mm} \qquad \textit{Ans.}$$

This larger section will also adequately resist the bending stress.

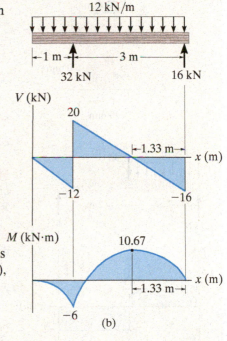

Fig. 11–8

11

EXAMPLE 11.3

The wooden T-beam shown in Fig. 11–9a is made from two 200 mm × 30 mm boards. If $\sigma_{allow} = 12$ MPa and $\tau_{allow} = 0.8$ MPa, determine if the beam can safely support the loading shown. Also, specify the maximum spacing of nails needed to hold the two boards together if each nail can safely resist 1.50 kN in shear.

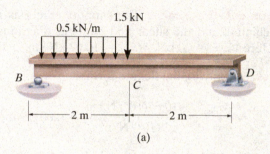

(a)

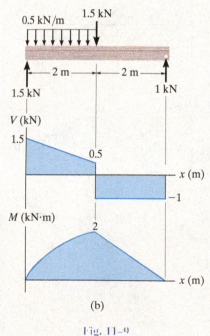

(b)

Fig. 11–9

SOLUTION

Shear and Moment Diagrams. The reactions on the beam are shown, and the shear and moment diagrams are drawn in Fig. 11–9b. Here $V_{max} = 1.5$ kN, $M_{max} = 2$ kN · m.

Bending Stress. The neutral axis (centroid) will be located from the bottom of the beam. Working in units of meters, we have

$$\bar{y} = \frac{\Sigma \bar{y}A}{\Sigma A}$$

$$= \frac{(0.1 \text{ m})(0.03 \text{ m})(0.2 \text{ m}) + 0.215 \text{ m}(0.03 \text{ m})(0.2 \text{ m})}{0.03 \text{ m}(0.2 \text{ m}) + 0.03 \text{ m}(0.2 \text{ m})} = 0.1575 \text{ m}$$

Thus,

$$I = \left[\frac{1}{12}(0.03 \text{ m})(0.2 \text{ m})^3 + (0.03 \text{ m})(0.2 \text{ m})(0.1575 \text{ m} - 0.1 \text{ m})^2 \right]$$

$$+ \left[\frac{1}{12}(0.2 \text{ m})(0.03 \text{ m})^3 + (0.03 \text{ m})(0.2 \text{ m})(0.215 \text{ m} - 0.1575 \text{ m})^2 \right]$$

$$= 60.125(10^{-6}) \text{ m}^4$$

Since $c = 0.1575$ m (not $0.230 \text{ m} - 0.1575 \text{ m} = 0.0725$ m), we require

$$\sigma_{allow} \geq \frac{M_{max}c}{I}$$

$$12(10^6) \text{ Pa} \geq \frac{2(10^3) \text{ N} \cdot \text{m}(0.1575 \text{ m})}{60.125(10^{-6}) \text{ m}^4} = 5.24(10^6) \text{ Pa} \qquad \text{OK}$$

Shear Stress. Maximum shear stress in the beam depends upon the magnitude of Q and t. It occurs at the neutral axis, since Q is a maximum there and at the neutral axis the thickness $t = 0.03$ m is the smallest for the cross section. For simplicity, we will use the rectangular area below the neutral axis to calculate Q, rather than a two-part composite area above this axis, Fig. 11–9c. We have

$$Q = \bar{y}'A' = \left(\frac{0.1575 \text{ m}}{2}\right)[(0.1575 \text{ m})(0.03 \text{ m})] = 0.372(10^{-3}) \text{ m}^3$$

so that

$$\tau_{allow} \geq \frac{V_{max}Q}{It}$$

$$800(10^3) \text{ Pa} \geq \frac{1.5(10^3) \text{ N}[0.372(10^{-3})] \text{ m}^3}{60.125(10^{-6}) \text{ m}^4(0.03 \text{ m})} = 309(10^3) \text{ Pa} \quad \text{OK}$$

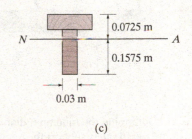

(c)

Nail Spacing. From the shear diagram it is seen that the shear varies over the entire span. Since the nail spacing depends on the magnitude of shear in the beam, for simplicity (and to be conservative), we will design the spacing on the basis of $V = 1.5$ kN for region BC, and $V = 1$ kN for region CD. Since the nails join the flange to the web, Fig. 11–9d, we have

$$Q = \bar{y}'A' = (0.0725 \text{ m} - 0.015 \text{ m})[(0.2 \text{ m})(0.03 \text{ m})] = 0.345(10^{-3}) \text{ m}^3$$

The shear flow for each region is therefore

$$q_{BC} = \frac{V_{BC}Q}{I} = \frac{1.5(10^3) \text{ N}[0.345(10^{-3}) \text{ m}^3]}{60.125(10^{-6}) \text{ m}^4} = 8.61 \text{ kN/m}$$

$$q_{CD} = \frac{V_{CD}Q}{I} = \frac{1(10^3) \text{ N}[0.345(10^{-3}) \text{ m}^3]}{60.125(10^{-6}) \text{ m}^4} = 5.74 \text{ kN/m}$$

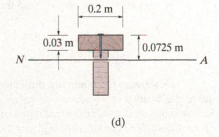

(d)

Fig. 11–9 (cont.)

One nail can resist 1.50 kN in shear, so the maximum spacing becomes

$$s_{BC} = \frac{1.50 \text{ kN}}{8.61 \text{ kN/m}} = 0.174 \text{ m}$$

$$s_{CD} = \frac{1.50 \text{ kN}}{5.74 \text{ kN/m}} = 0.261 \text{ m}$$

For ease of measuring, use

$$s_{BC} = 150 \text{ mm} \qquad \qquad \textit{Ans.}$$

$$s_{CD} = 250 \text{ mm} \qquad \qquad \textit{Ans.}$$

FUNDAMENTAL PROBLEMS

F11–1. Determine the minimum dimension a to the nearest mm of the beam's cross section to safely support the load. The wood has an allowable normal stress of $\sigma_{allow} = 10$ MPa and an allowable shear stress of $\tau_{allow} = 1$ MPa.

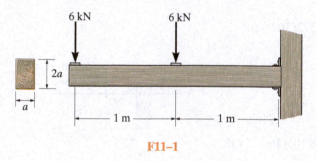

F11–1

F11–2. Determine the minimum diameter d to the nearest $\frac{1}{8}$ in. of the rod to safely support the load. The rod is made of a material having an allowable normal stress of $\sigma_{allow} = 20$ ksi and an allowable shear stress of $\tau_{allow} = 10$ ksi.

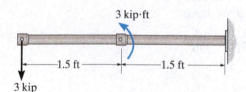

F11–2

F11–3. Determine the minimum dimension a to the nearest mm of the beam's cross section to safely support the load. The wood has an allowable normal stress of $\sigma_{allow} = 12$ MPa and an allowable shear stress of $\tau_{allow} = 1.5$ MPa.

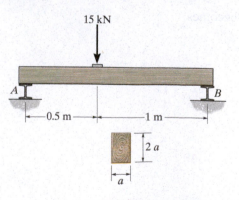

F11–3

F11–4. Determine the minimum dimension h to the nearest $\frac{1}{8}$ in. of the beam's cross section to safely support the load. The wood has an allowable normal stress of $\sigma_{allow} = 2$ ksi and an allowable shear stress of $\tau_{allow} = 200$ psi.

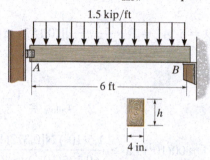

F11–4

F11–5. Determine the minimum dimension b to the nearest mm of the beam's cross section to safely support the load. The wood has an allowable normal stress of $\sigma_{allow} = 12$ MPa and an allowable shear stress of $\tau_{allow} = 1.5$ MPa.

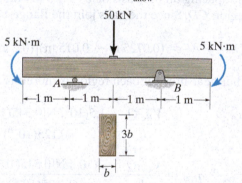

F11–5

F11–6. Select the lightest W410-shaped section that can safely support the load. The beam is made of steel having an allowable normal stress of $\sigma_{allow} = 150$ MPa and an allowable shear stress of $\tau_{allow} = 75$ MPa. Assume the beam is pinned at A and roller supported at B.

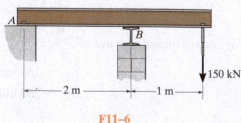

F11–6

PROBLEMS

11–1. The beam is made of timber that has an allowable bending stress of $\sigma_{allow} = 6.5$ MPa and an allowable shear stress of $\tau_{allow} = 500$ kPa. Determine its dimensions if it is to be rectangular and have a height-to-width ratio of 1.25. Assume the beam rests on smooth supports.

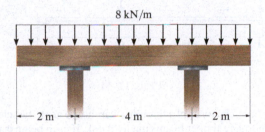

Prob. 11–1

11–2. Determine the minimum width of the beam to the nearest $\frac{1}{4}$ in. that will safely support the loading of $P = 8$ kip. The allowable bending stress is $\sigma_{allow} = 24$ ksi and the allowable shear stress is $\tau_{allow} = 15$ ksi.

11–3. Solve Prob. 11–2 if $P = 10$ kip.

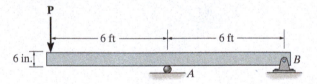

Probs. 11–2/3

*11–4. The brick wall exerts a uniform distributed load of 1.20 kip/ft on the beam. If the allowable bending stress is $\sigma_{allow} = 22$ ksi and the allowable shear stress is $\tau_{allow} = 12$ ksi, select the lightest wide-flange section with the shortest depth from Appendix B that will safely support the load. If there are several choices of equal weight, choose the one with the shortest height.

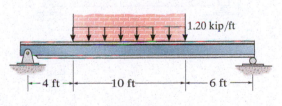

Prob. 11–4

11–5. Select the lightest-weight wide-flange beam from Appendix B that will safely support the machine loading shown. The allowable bending stress is $\sigma_{allow} = 24$ ksi and the allowable shear stress is $\tau_{allow} = 14$ ksi.

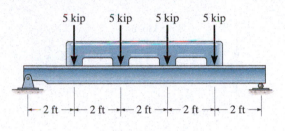

Prob. 11–5

11–6. The spreader beam AB is used to slowly lift the 3000-lb pipe that is centrally located on the straps at C and D. If the beam is a W12 × 45, determine if it can safely support the load. The allowable bending stress is $\sigma_{allow} = 22$ ksi and the allowable shear stress is $\tau_{allow} = 12$ ksi.

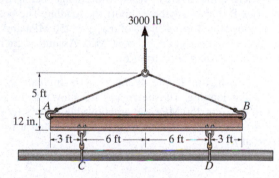

Prob. 11–6

11–7. Select the lightest-weight wide-flange beam with the shortest depth from Appendix B that will safely support the loading shown. The allowable bending stress is $\sigma_{allow} = 24$ ksi and the allowable shear stress of $\tau_{allow} = 14$ ksi.

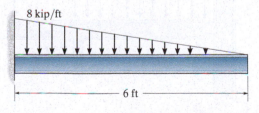

Prob. 11–7

11

*11–8. Select the lightest-weight wide-flange beam from Appendix B that will safely support the loading shown. The allowable bending stress $\sigma_{allow} = 24$ ksi and the allowable shear stress of $\tau_{allow} = 14$ ksi.

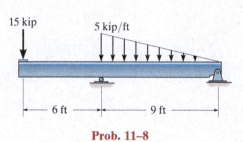

Prob. 11–8

11–9. Select the lightest W360 wide-flange beam from Appendix B that can safely support the loading. The beam has an allowable normal stress of $\sigma_{allow} = 150$ MPa and an allowable shear stress of $\tau_{allow} = 80$ MPa. Assume there is a pin at A and a roller support at B.

11–10. Investigate if the W250 × 58 beam can safely support the loading. The beam has an allowable normal stress of $\sigma_{allow} = 150$ MPa and an allowable shear stress of $\tau_{allow} = 80$ MPa. Assume there is a pin at A and a roller support at B.

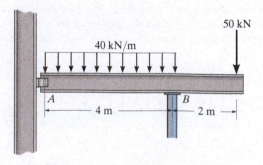

Probs. 11–9/10

11–11. The beam is constructed from two boards. If each nail can support a shear force of 200 lb, determine the maximum spacing of the nails, s, s', and s'', to the nearest $\frac{1}{8}$ inch for regions AB, BC, and CD, respectively.

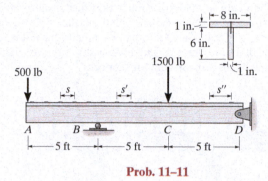

Prob. 11–11

*11–12. The joists of a floor in a warehouse are to be selected using square timber beams made of oak. If each beam is to be designed to carry 90 lb/ft over a simply supported span of 25 ft, determine the dimension a of its square cross section to the nearest $\frac{1}{4}$ in. The allowable bending stress is $\sigma_{allow} = 4.5$ ksi and the allowable shear stress is $\tau_{allow} = 125$ psi.

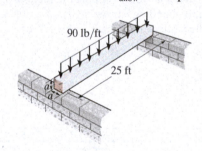

Prob. 11–12

11–13. The timber beam has a width of 6 in. Determine its height h so that it simultaneously reaches its allowable bending stress $\sigma_{allow} = 1.50$ ksi and an allowable shear stress of $\tau_{allow} = 50$ psi. Also, what is the maximum load P that the beam can then support?

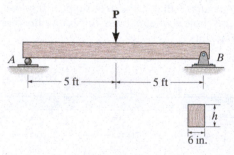

Prob. 11–13

11–14. The beam is constructed from four boards. If each nail can support a shear force of 300 lb, determine the maximum spacing of the nails, s, s' and s'', for regions AB, BC, and CD, respectively.

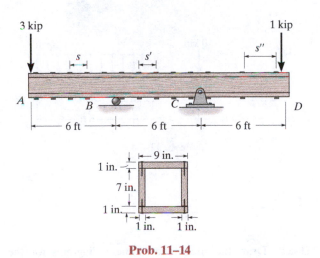

Prob. 11–14

11–15. The beam is constructed from two boards. If each nail can support a shear force of 200 lb, determine the maximum spacing of the nails, s, s', and s'', to the nearest $\frac{1}{8}$ in. for regions AB, BC, and CD, respectively.

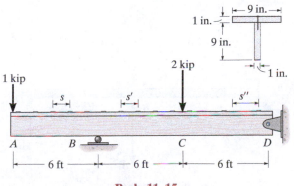

Prob. 11–15

*11–16. If the cable is subjected to a maximum force of $P = 50$ kN, select the lightest W310 wide-flange beam that can safely support the load. The beam has an allowable normal stress of $\sigma_{allow} = 150$ MPa and an allowable shear stress of $\tau_{allow} = 85$ MPa.

11–17. If the W360 × 45 wide-flange beam has an allowable normal stress of $\sigma_{allow} = 150$ MPa and an allowable shear stress of $\tau_{allow} = 85$ MPa, determine the maximum cable force P that can safely be supported by the beam.

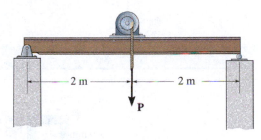

Probs. 11–16/17

11–18. If $P = 800$ lb, determine the minimum dimension a of the beam's cross section to the nearest $\frac{1}{8}$ in. to safely support the load. The wood has an allowable normal stress of $\sigma_{allow} = 1.5$ ksi and an allowable shear stress of $\tau_{allow} = 150$ psi.

11–19. If $a = 3$ in. and the wood has an allowable normal stress of $\sigma_{allow} = 1.5$ ksi, and an allowable shear stress of $\tau_{allow} = 150$ psi, determine the maximum allowable value of P that can act on the beam.

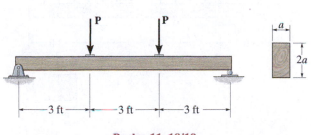

Probs. 11–18/19

*11–20. The beam is constructed from three plastic strips. If the glue can support a shear stress of $\tau_{allow} = 8$ kPa, determine the largest magnitude of the loads **P** that the beam can support.

11–21. If the allowable bending stress is $\sigma_{allow} = 6$ MPa, and the glue can support a shear stress of $\tau_{allow} = 8$ kPa, determine the largest magnitude of the loads **P** that can be applied to the beam.

11–23. Select the lightest-weight wide-flange beam from Appendix B that will safely support the loading. The allowable bending stress is $\sigma_{allow} = 24$ ksi and the allowable shear stress is $\tau_{allow} = 14$ ksi.

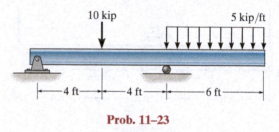

Prob. 11–23

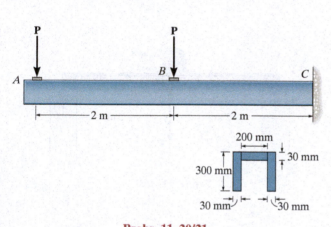

Probs. 11–20/21

*11–24. Draw the shear and moment diagrams for the shaft, and determine its required diameter to the nearest $\frac{1}{8}$ in. if $\sigma_{allow} = 30$ ksi and $\tau_{allow} = 15$ ksi. The journal bearings at A and C exert only vertical reactions on the shaft. Take $P = 6$ kip.

11–25. Draw the shear and moment diagrams for the shaft, and determine its required diameter to the nearest $\frac{1}{4}$ in. if $\sigma_{allow} = 30$ ksi and $\tau_{allow} = 15$ ksi. The journal bearings at A and C exert only vertical reactions on the shaft. Take $P = 12$ kip.

11–22. The beam is made of Douglas fir having an allowable bending stress of $\sigma_{allow} = 1.1$ ksi and an allowable shear stress of $\tau_{allow} = 0.70$ ksi. Determine the width b if the height $h = 2b$.

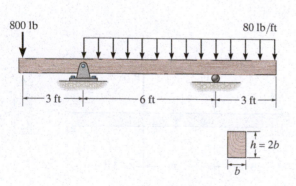

Prob. 11–22

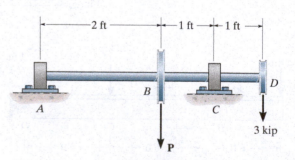

Probs. 11–24/25

11–26. Select the lightest-weight wide-flange beam from Appendix B that will safely support the loading. The allowable bending stress is $\sigma_{allow} = 22$ ksi and the allowable shear stress is $\tau_{allow} = 12$ ksi.

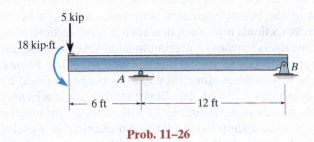

Prob. 11–26

11–27. Draw the shear and moment diagrams for the W14 $\times$ 22 wide-flange beam and check if the beam will safely support the loading. The allowable bending stress is $\sigma_{allow} = 30$ ksi and the allowable shear stress is $\tau_{allow} = 15$ ksi.

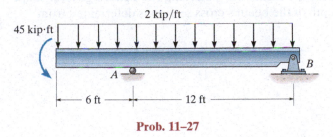

Prob. 11–27

***11–28.** Select the lightest-weight W16 wide-flange beam from Appendix B that will safely support the loading. The allowable bending stress is $\sigma_{allow} = 30$ ksi and the allowable shear stress is $\tau_{allow} = 15$ ksi.

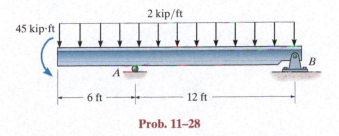

Prob. 11–28

11–29. The beam is to be used to support the machine, which exerts the forces of 6 kip and 8 kip. If the maximum bending stress is not to exceed $\sigma_{allow} = 22$ ksi, determine the required width b of the flanges.

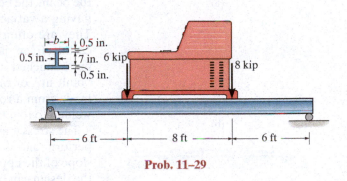

Prob. 11–29

11–30. The steel beam has an allowable bending stress $\sigma_{allow} = 140$ MPa and an allowable shear stress of $\tau_{allow} = 90$ MPa. Determine the maximum load that can safely be supported.

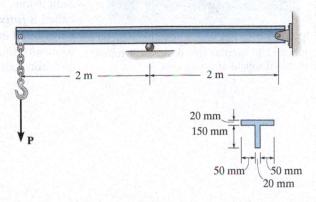

Prob. 11–30

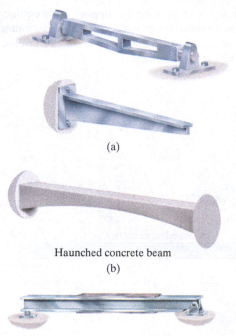

(a)

Haunched concrete beam
(b)

Wide-flange beam with cover plates
(c)

Fig. 11–10

The beam for this bridge pier has a variable moment of inertia. This design will reduce material weight and save cost.

*11.3 FULLY STRESSED BEAMS

Since the moment in a beam generally *varies* along its length, the choice of a prismatic beam is usually inefficient, because it is never fully stressed at points where the internal moment is less than the point of maximum moment. In order to fully use the strength of the material and thereby reduce the weight of the beam, engineers sometimes choose a beam having a *variable* cross-sectional area, such that at each cross section along the beam, the bending stress reaches its maximum allowable value. Beams having a variable cross-sectional area are called **nonprismatic beams**. They are often used in machines since they can be readily formed by casting. Examples are shown in Fig. 11–10a. In structures such beams may be "haunched" at their ends as shown in Fig. 11–10b. Also, beams may be "built up" or fabricated in a shop using plates. An example is a girder made from a rolled-shaped wide-flange beam, and having cover plates welded to it in the region where the moment is a maximum, Fig. 11–10c.

The stress analysis of a nonprismatic beam is generally very difficult to perform and is beyond the scope of this text. However, if the taper or slope of the upper or lower boundary of the beam is not too severe, then the design can be based on the flexure formula.

Although caution is advised when applying the flexure formula to nonprismatic beam design, we will show here how this formula can be used as an approximate means for obtaining the beam's general shape. To do this, the depth of the beam's cross section is determined from

$$S = \frac{M}{\sigma_{\text{allow}}}$$

If we express M in terms of its position x along the beam, then since σ_{allow} is a known constant, the section modulus S or the beam's dimensions become a function of x. A beam designed in this manner is called a **fully stressed beam**. Although *only* bending stresses have been considered in approximating its final shape, attention must also be given to ensure that the beam will resist shear, especially at the points where concentrated loads are applied.

EXAMPLE 11.4

Determine the shape of a fully stressed, simply supported beam that supports a concentrated force at its center, Fig. 11–11a. The beam has a rectangular cross section of constant width b, and the allowable stress is σ_{allow}.

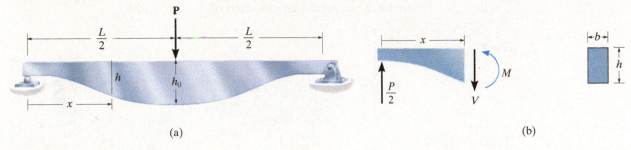

(a)   (b)

Fig. 11–11

SOLUTION

The internal moment in the beam, Fig. 11–11b, expressed as a function of position, $0 \le x < L/2$, is

$$M = \frac{P}{2}x$$

Therefore the required section modulus is

$$S = \frac{M}{\sigma_{allow}} = \frac{P}{2\sigma_{allow}}x$$

Since $S = I/c$, then for a cross-sectional area h by b we have

$$\frac{I}{c} = \frac{\frac{1}{12}bh^3}{h/2} = \frac{P}{2\sigma_{allow}}x$$

$$h^2 = \frac{3P}{\sigma_{allow}b}x$$

If $h = h_0$ at $x = L/2$, then

$$h_0{}^2 = \frac{3PL}{2\sigma_{allow}b}$$

so that

$$h^2 = \left(\frac{2h_0{}^2}{L}\right)x \qquad\qquad Ans.$$

By inspection, the depth h must therefore vary in a *parabolic* manner with the distance x.

NOTE: In practice, this shape is the basis for the design of leaf springs used to support the rear-end axles of most heavy trucks or train cars, as shown in the adjacent photo. Note that although this result indicates that $h = 0$ at $x = 0$, it is necessary that the beam resist shear stress at the supports, and so practically speaking, it must be required that $h > 0$ at the supports, Fig. 11–11a.

EXAMPLE 11.5

The cantilevered beam shown in Fig. 11–12a is formed into a trapezoidal shape having a depth h_0 at A, and a depth $3h_0$ at B. If it supports a load **P** at its end, determine the absolute maximum normal stress in the beam. The beam has a rectangular cross section of constant width b.

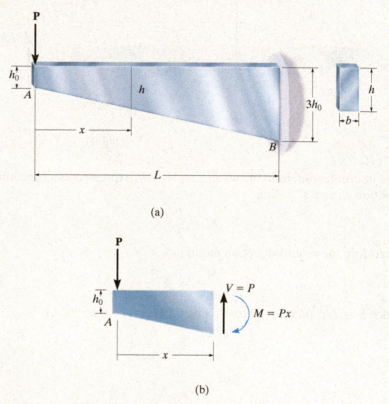

(a)

(b)

Fig. 11–12

SOLUTION

At any cross section, the maximum normal stress occurs at the top and bottom surface of the beam. However, since $\sigma_{max} = M/S$ and the section modulus S increases as x increases, then the absolute maximum normal stress will *not* necessarily occur at the wall B, where the moment is maximum. Using the flexure formula, we can express the maximum normal stress at an arbitrary section in terms of its position x, Fig. 11–12b. Here the internal moment is $M = Px$, and since the slope of the bottom of the beam is $2h_0/L$, Fig. 11–12a, the depth of the beam at position x is

$$h = \frac{2h_0}{L}x + h_0 = \frac{h_0}{L}(2x + L)$$

Applying the flexure formula, we have

$$\sigma = \frac{Mc}{I} = \frac{Px(h/2)}{\left(\frac{1}{12}bh^3\right)} = \frac{6PL^2x}{bh_0^2(2x+L)^2} \tag{1}$$

To determine the position x where the absolute maximum bending stress occurs, we must take the derivative of σ with respect to x and set it equal to zero. This gives

$$\frac{d\sigma}{dx} = \left(\frac{6PL^2}{bh_0^2}\right)\frac{1(2x+L)^2 - x(2)(2x+L)(2)}{(2x+L)^4} = 0$$

Thus,

$$4x^2 + 4xL + L^2 - 8x^2 - 4xL = 0$$

$$L^2 - 4x^2 = 0$$

$$x = \frac{1}{2}L$$

Substituting into Eq. 1 and simplifying, the absolute maximum normal stress is therefore

$$\sigma_{\substack{abs\\max}} = \frac{3}{4}\frac{PL}{bh_0^2} \qquad\qquad Ans.$$

By comparison, at the wall, B, the maximum normal stress is

$$(\sigma_{max})_B = \frac{Mc}{I} = \frac{PL(1.5h_0)}{\left[\frac{1}{12}b(3h_0)^3\right]} = \frac{2}{3}\frac{PL}{bh_0^2}$$

which is 11.1% smaller than $\sigma_{\substack{abs\\max}}$.

NOTE: Recall that the flexure formula was derived on the basis of assuming the beam to be *prismatic*. Since this is not the case here, some error is to be expected in this analysis and that of Example 11.4. A more exact mathematical analysis, using the theory of elasticity, reveals that application of the flexure formula as in the above example gives only small errors in the bending stress if the tapered angle of the beam is small. For example, if this angle is 15°, the stress calculated from the formula will be about 5% greater than that calculated by the more exact analysis. It may also be worth noting that the calculation of $(\sigma_{max})_B$ was done only for illustrative purposes, since, by Saint-Venant's principle, the actual stress distribution at the support (wall) is highly irregular.

*11.4 SHAFT DESIGN

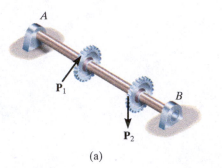

(a)

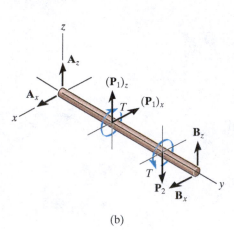

(b)

Shafts that have circular cross sections are often used in mechanical equipment and machinery. As a result, they can be subjected to cyclic or fatigue stress, which is caused by the combined bending and torsional loads they must transmit or resist. In addition to these loadings, stress concentrations may exist on a shaft due to keys, couplings, and sudden transitions in its cross section (Sec. 5.8). In order to design a shaft properly, it is therefore necessary to take all of these effects into account.

In this section we will consider the design of shafts required to transmit power. These shafts are often subjected to loads applied to attached pulleys and gears, such as the one shown in Fig. 11–13a. Since the loads can be applied to the shaft at various angles, the internal bending and torsion at any cross section can best be determined by replacing the loads by their statically equivalent loadings, and then resolving these loads into components in two perpendicular planes, Fig. 11–13b. The bending-moment diagrams for the loads *in each plane* can then be drawn, and the resultant internal moment at any section along the shaft is then determined by vector addition, $M = \sqrt{M_x^2 + M_z^2}$, Fig. 11–13c. In addition to this moment, segments of the shaft may also be subjected to different internal torques, Fig. 11–13b. To account for this general variation of torque along the shaft, a **torque diagram** may also be drawn, Fig. 11–13d.

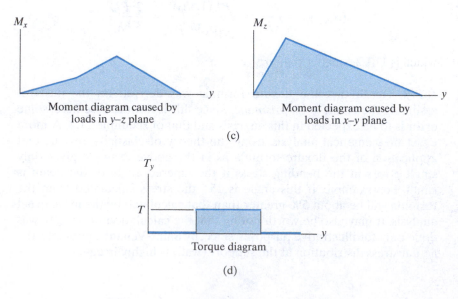

(c)

(d)

Fig. 11–13

Once the moment and torque diagrams have been established, it is then possible to investigate certain critical sections along the shaft where the *combination* of a resultant moment **M** and a torque **T** creates the critical stress situation. Since the moment of inertia of a circular shaft is the *same* about *any* diametrical axis, we can apply the flexure formula using the *resultant moment* to obtain the maximum bending stress. For example, as shown in Fig. 11–13e, this stress will occur on two elements, C and D, each located on the outer boundary of the shaft. If a torque **T** is also resisted at this section, then a maximum shear stress will also be developed on these elements, Fig. 11–13f. In addition, the external forces will also create shear stress in the shaft determined from $\tau = VQ/It$; however, this stress will generally contribute a much smaller stress distribution on the cross section compared with that developed by bending and torsion. In some cases, it must be investigated, but for simplicity, we will neglect its effect here. In general, then, the critical element C (or D) on the shaft is subjected to *plane stress* as shown in Fig. 11–13g, where

$$\sigma = \frac{Mc}{I} \quad \text{and} \quad \tau = \frac{Tc}{J}$$

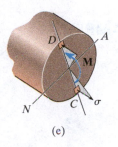

(e)

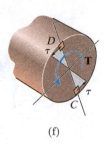

(f)

If the allowable normal or shear stress for the material is known, the size of the shaft is then based on the use of these equations and selection of an appropriate theory of failure. For example, if the material is ductile, then the maximum shear stress theory may be appropriate. As stated in Sec. 10.7, this theory requires the allowable shear stress, which is determined from the results of a simple tension test, to be equal to the maximum shear stress in the element. Using the stress transformation equation, Eq. 9–7, for the stress state in Fig. 11–13g, we get

$$\tau_{\text{allow}} = \sqrt{\left(\frac{\sigma}{2}\right)^2 + \tau^2}$$

$$= \sqrt{\left(\frac{Mc}{2I}\right)^2 + \left(\frac{Tc}{J}\right)^2}$$

Since $I = \pi c^4/4$ and $J = \pi c^4/2$, this equation becomes

$$\tau_{\text{allow}} = \frac{2}{\pi c^3} \sqrt{M^2 + T^2}$$

Solving for the radius of the shaft,

$$c = \left(\frac{2}{\pi \tau_{\text{allow}}} \sqrt{M^2 + T^2}\right)^{1/3} \tag{11–2}$$

Application of any other theory of failure will, of course, lead to a different formulation for c. However, in all cases it may be necessary to apply this result at various "critical sections" along the shaft, in order to determine the particular combination of M and T that gives the largest value for c.

The following example numerically illustrates the procedure.

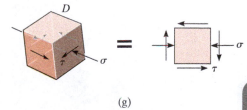

(g)

Fig. 11–13 (cont.)

EXAMPLE 11.6

The shaft in Fig. 11–14a is supported by journal bearings at A and B. Due to the transmission of power to and from the shaft, the belts on the pulleys are subjected to the tensions shown. Determine the smallest diameter of the shaft using the maximum shear stress theory, with $\tau_{allow} = 50$ MPa.

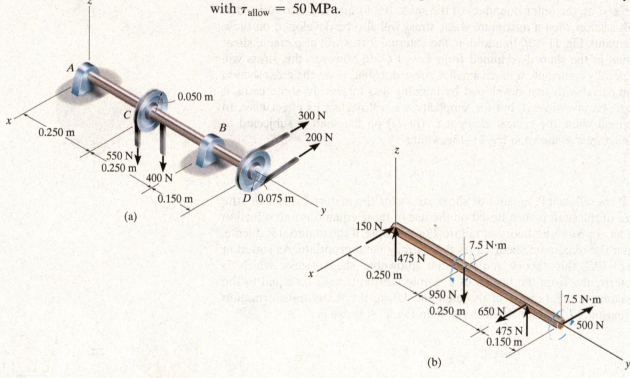

Fig. 11–14

SOLUTION

The support reactions have been calculated and are shown on the free-body diagram of the shaft, Fig. 11–14b. Bending-moment diagrams for M_x and M_z are shown in Figs. 11–14c and 11–14d. The torque diagram is shown in Fig. 11–14e. By inspection, critical points for bending moment occur either at C or B. Also, just to the right of C and at B the torque is 7.5 N·m. At C, the resultant moment is

$$M_C = \sqrt{(118.75 \text{ N} \cdot \text{m})^2 + (37.5 \text{ N} \cdot \text{m})^2} = 124.5 \text{ N} \cdot \text{m}$$

whereas at B it is smaller, namely

$$M_B = 75 \text{ N} \cdot \text{m}$$

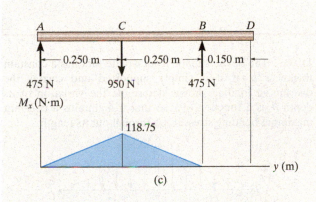

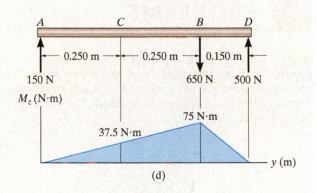

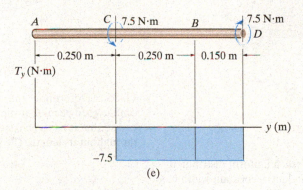

Fig. 11–14 (cont.)

Since the design is based on the maximum shear stress theory, Eq. 11–2 applies. The radical $\sqrt{M^2 + T^2}$ will be the largest at a section just to the right of C. We have

$$c = \left(\frac{2}{\pi \tau_{\text{allow}}} \sqrt{M^2 + T^2} \right)^{1/3}$$

$$= \left(\frac{2}{\pi (50)(10^6) \text{ N/m}^2} \sqrt{(124.5 \text{ N} \cdot \text{m})^2 + (7.5 \text{ N} \cdot \text{m})^2} \right)^{1/3}$$

$$= 0.0117 \text{ m}$$

Thus, the smallest allowable diameter is

$$d = 2(0.0117 \text{ m}) = 23.3 \text{ mm} \qquad \textit{Ans.}$$

PROBLEMS

11–31. Determine the variation in the width w as a function of x for the cantilevered beam that supports a concentrated force **P** at its end so that it has a maximum bending stress σ_{allow} throughout its length. The beam has a constant thickness t.

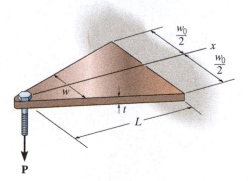

Prob. 11–31

***11–32.** The tapered beam supports a uniform distributed load w. If it is made from a plate and has a constant width b, determine the absolute maximum bending stress in the beam.

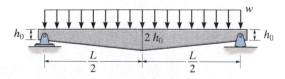

Prob. 11–32

11–33. The tapered beam supports the concentrated force **P** at its center. Determine the absolute maximum bending stress in the beam. The reactions at the supports are vertical.

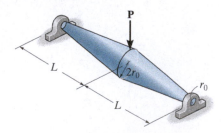

Prob. 11–33

11–34. The beam is made from a plate that has a constant thickness b. If it is simply supported and carries the distributed loading shown, determine the variation of its depth h as a function of x so that it maintains a constant maximum bending stress σ_{allow} throughout its length.

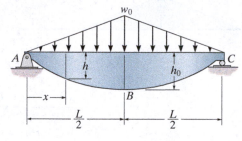

Prob. 11–34

11–35. Determine the variation in the depth d of a cantilevered beam that supports a concentrated force **P** at its end so that it has a constant maximum bending stress σ_{allow} throughout its length. The beam has a constant width b_0.

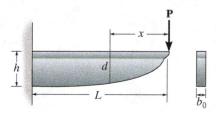

Prob. 11–35

***11–36.** Determine the variation of the radius r of the cantilevered beam that supports the uniform distributed load so that it has a constant maximum bending stress σ_{max} throughout its length.

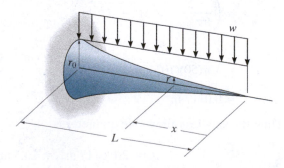

Prob. 11–36

11–37. The tapered beam supports a uniform distributed load w. If it is made from a plate that has a constant width b_0, determine the absolute maximum bending stress in the beam.

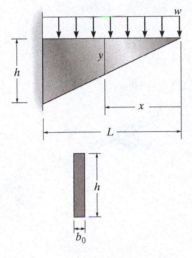

Prob. 11–37

11–38. Determine the variation in the width b as a function of x for the cantilevered beam that supports a uniform distributed load along its centerline so that it has the same maximum bending stress σ_{allow} throughout its length. The beam has a constant depth t.

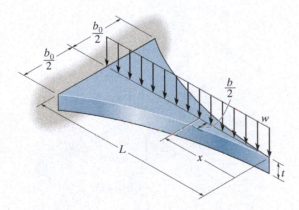

Prob. 11–38

11–39. The tubular shaft has an inner diameter of 15 mm. Determine to the nearest millimeter its minimum outer diameter if it is subjected to the gear loading. The bearings at A and B exert force components only in the y and z directions on the shaft. Use an allowable shear stress of $\tau_{allow} = 70$ MPa, and base the design on the maximum shear stress theory of failure.

*11–40. Determine to the nearest millimeter the minimum diameter of the solid shaft if it is subjected to the gear loading. The bearings at A and B exert force components only in the y and z directions on the shaft. Base the design on the maximum distortion energy theory of failure with $\sigma_{allow} = 150$ MPa.

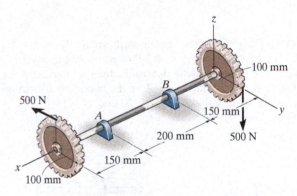

Probs. 11–39/40

11–41. The 50-mm-diameter shaft is supported by journal bearings at A and B. If the pulleys C and D are subjected to the loadings shown, determine the absolute maximum bending stress in the shaft.

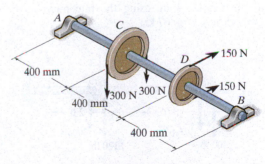

Prob. 11–41

11–42. The pulleys fixed to the shaft are loaded as shown. If the journal bearings at A and B exert only horizontal and vertical forces on the shaft, determine the required diameter of the shaft to the nearest $\frac{1}{8}$ in. using the maximum shear stress theory of failure, $\tau_{allow} = 12$ ksi.

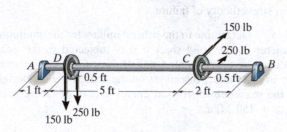

Prob. 11–42

11–43. The pulleys fixed to the shaft are loaded as shown. If the journal bearings at A and B exert only horizontal and vertical forces on the shaft, determine the required diameter of the shaft to the nearest $\frac{1}{8}$ in. Use the maximum distortion energy theory of failure, $\sigma_{allow} = 20$ ksi.

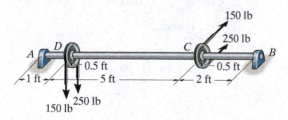

Prob. 11–43

*11–44. The two pulleys fixed to the shaft are loaded as shown. If the journal bearings at A and B exert only vertical forces on the shaft, determine the required diameter of the shaft to the nearest $\frac{1}{8}$ in. using the maximum distortion energy theory. $\sigma_{allow} = 67$ ksi.

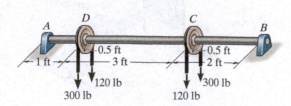

Prob. 11–44

11–45. The shaft is supported by journal bearings at A and B that exert force components only in the x and z directions. If the allowable normal stress for the shaft is $\sigma_{allow} = 15$ ksi, determine the smallest diameter of the shaft to the nearest $\frac{1}{8}$ in. Use the maximum distortion energy theory of failure.

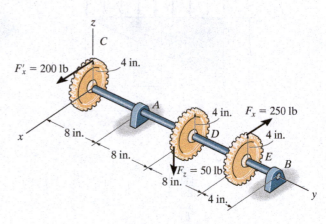

Prob. 11–45

11–46. The shaft is supported by journal bearings at A and B that exert force components only in the x and z directions. Determine the smallest diameter of the shaft to the nearest $\frac{1}{8}$ in. Use the maximum shear stress theory of failure with $\tau_{allow} = 6$ ksi.

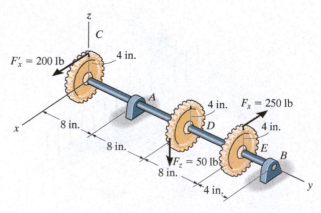

Prob. 11–46

CHAPTER REVIEW

Failure of a beam will occur where the internal shear or moment in the beam is a maximum. To resist these loadings, it is therefore important that the maximum shear and bending stress not exceed allowable values as stated in codes. Normally, the cross section of a beam is first designed to resist the allowable bending stress, $$\sigma_{allow} = \frac{M_{max}c}{I}$$	
Then the allowable shear stress is checked. For rectangular sections, $\tau_{allow} \geq 1.5(V_{max}/A)$, and for wide-flange sections it is appropriate to use $\tau_{allow} \geq V_{max}/A_{web}$. In general, use $$\tau_{allow} = \frac{V_{max}Q}{It}$$	
For built-up beams, the spacing of fasteners or the strength of glue or weld is determined using an allowable shear flow $$q_{allow} = \frac{VQ}{I}$$	
Fully stressed beams are nonprismatic, and designed such that the bending stress at each cross section along the beam will equal an allowable bending stress. This condition will define the shape of the beam.	
A shaft that transmits power is generally designed to resist both bending and torsion. Once the maximum bending and torsion stresses are determined, then depending upon the type of material, an appropriate theory of failure is used to compare the allowable stress to what is required.	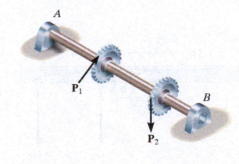

REVIEW PROBLEMS

R11–1. The cantilevered beam has a circular cross section. If it supports a force **P** at its end, determine its radius y as a function of x so that it is subjected to a constant maximum bending stress σ_{allow} throughout its length.

R11–3. The journal bearings at A and B exert only x and z components of force on the shaft. Determine the shaft's diameter to the nearest millimeter so that it can resist the loadings without exceeding an allowable shear stress of $\tau_{allow} = 80$ MPa. Use the maximum shear stress theory of failure.

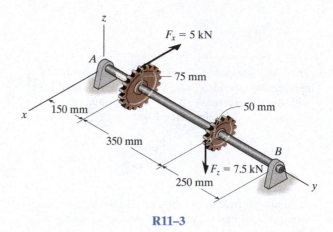

R11–3

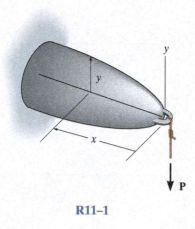

R11–1

R11–2. Select the lightest-weight wide-flange overhanging beam from Appendix B that will safely support the loading. Assume the support at A is a pin and the support at B is a roller. The allowable bending stress is $\sigma_{allow} = 24$ ksi and the allowable shear stress is $\tau_{allow} = 14$ ksi.

*R11–4. The journal bearings at A and B exert only x and z components of force on the shaft. Determine the shaft's diameter to the nearest millimeter so that it can resist the loadings. Use the maximum distortion energy theory of failure with $\sigma_{allow} = 200$ MPa.

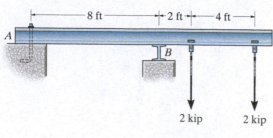

R11–2

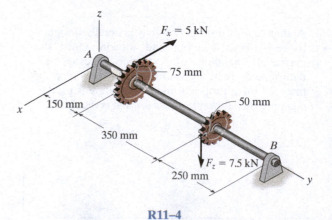

R11–4

R11–5. Select the lightest-weight wide-flange beam from Appendix B that will safely support the loading. The allowable bending stress is $\sigma_{allow} = 22$ ksi and the allowable shear stress is $\tau_{allow} = 22$ ksi.

R11–7. The simply supported joist is used in the construction of a floor for a building. In order to keep the floor low with respect to the sill beams C and D, the ends of the joist are notched as shown. If the allowable shear stress is $\tau_{allow} = 350$ psi and the allowable bending stress is $\sigma_{allow} = 1700$ psi, determine the smallest height h so that the beam will support a load of $P = 600$ lb. Also, will the entire joist safely support the load? Neglect the stress concentration at the notch.

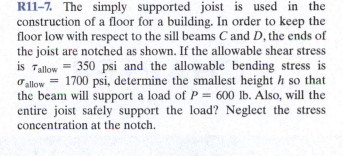

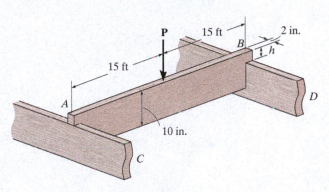

R11–7

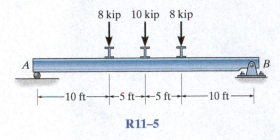

R11–5

R11–6. The simply supported joist is used in the construction of a floor for a building. In order to keep the floor low with respect to the sill beams C and D, the ends of the joists are notched as shown. If the allowable shear stress is $\tau_{allow} = 350$ psi and the allowable bending stress is $\sigma_{allow} = 1500$ psi, determine the height h that will cause the beam to reach both allowable stresses at the same time. Also, what load P causes this to happen? Neglect the stress concentration at the notch.

*R11–8. The overhang beam is constructed using two 2-in. by 4-in. pieces of wood braced as shown. If the allowable bending stress is $\sigma_{allow} = 600$ psi, determine the largest load P that can be applied. Also, determine the maximum spacing of nails, s, along the beam section AC if each nail can resist a shear force of 800 lb. Assume the beam is pin connected at A, B, and D. Neglect the axial force developed in the beam along DA.

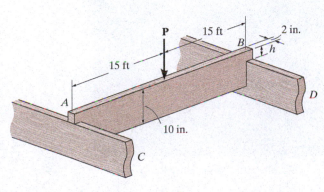

R11–6

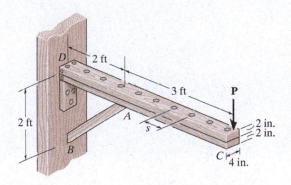

R11–8

11

CHAPTER 12

(© Michael Blann/Getty Images)

If the curvature of this pole is measured, it is then possible to determine the bending stress developed within it.

DEFLECTION OF BEAMS AND SHAFTS

CHAPTER OBJECTIVES

■ In this chapter we will discuss various methods for determining the deflection and slope of beams and shafts. The analytical methods include the integration method, the use of discontinuity functions, and the method of superposition. Also, a semigraphical technique, called the moment-area method, will be presented. At the end of the chapter, we will use these methods to solve for the support reactions on a beam or shaft that is statically indeterminate.

12.1 THE ELASTIC CURVE

The deflection of a beam or shaft must often be limited in order to provide stability, and for beams, to prevent the cracking of any attached brittle materials such as concrete or plaster. Most importantly, though, slopes and displacements must be determined in order to find the reactions if the beam is statically indeterminate. In this chapter we will find these slopes and displacements caused by the effects of bending. The additional rather small deflection caused by shear will be discussed in Chapter 14.

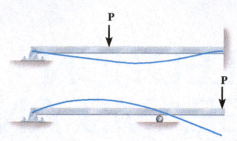

Fig. 12–1

Before finding the slope or displacement, it is often helpful to sketch the deflected shape of the beam, which is represented by its ***elastic curve***. This curve passes through the centroid of each cross section of the beam, and for most cases it can be sketched without much difficulty. When doing so, just remember that supports that resist a *force*, such as a pin, restrict *displacement*, and those that resist a *moment*, such as a fixed wall, restrict *rotation* or *slope* as well as displacement. Two examples of the elastic curves for loaded beams are shown in Fig. 12–1.

If the elastic curve for a beam seems difficult to establish, it is suggested that the moment diagram for the beam be drawn first. Using the beam sign convention established in Sec. 6.1, a positive internal moment tends to bend the beam concave upwards, Fig. 12–2*a*. Likewise, a negative moment tends to bend the beam concave downwards, Fig. 12–2*b*. Therefore, if the moment diagram is *known*, it will be easy to construct the elastic curve. For example, consider the beam in Fig. 12–3*a* with its associated moment diagram shown in Fig. 12–3*b*. Due to the roller and pin supports, the displacement at *B* and *D* must be zero. Within the region of negative moment, *AC*, Fig. 12–3*b*, the elastic curve must be concave downwards, and within the region of positive moment, *CD*, the elastic curve must be concave upwards. There is an *inflection point* at *C*, where the curve changes from concave up to concave down, since this is a point of zero moment. It should also be noted that the displacements Δ_A and Δ_E are especially critical. At point *E* the *slope* of the elastic curve is *zero*, and there the beam's *deflection* may be a *maximum*. Whether Δ_E is actually greater than Δ_A depends on the relative magnitudes of $\mathbf{P}_1$ and $\mathbf{P}_2$ and the location of the roller at *B*.

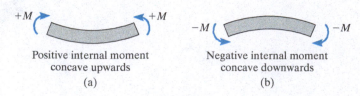

Positive internal moment concave upwards	Negative internal moment concave downwards
(a)	(b)

Fig. 12–2

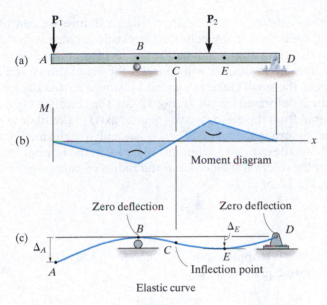

Fig. 12–3

Following these same principles, note how the elastic curve in Fig. 12–4 was constructed. Here the beam is cantilevered from a fixed support at A, and therefore the elastic curve must have both zero displacement and zero slope at this point. Also, the largest displacement will occur either at D, where the slope is zero, or at C.

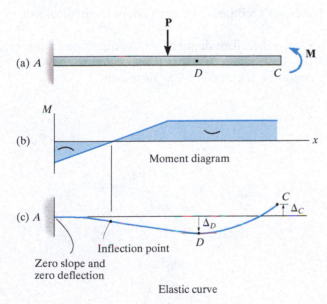

Fig. 12–4

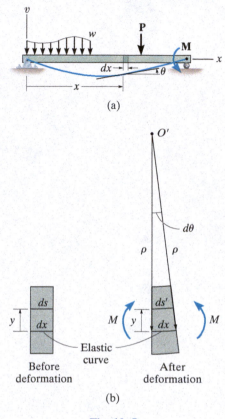

(a)

Elastic
curve

Before
deformation

After
deformation

(b)

Fig. 12–5

Moment–Curvature Relationship.

Before we can obtain the slope and deflection at any point on the elastic curve, it is first necessary to relate the internal moment to the radius of curvature ρ (rho) of the elastic curve. To do this, we will consider the beam shown in Fig. 12–5a, and remove the small element located a distance x from the left end and having an undeformed length dx, Fig. 12–5b. The "localized" y coordinate is measured from the elastic curve (neutral axis) to the fiber in the beam that has an original length of $ds = dx$ and a deformed length ds'. In Sec. 6.3 we developed a relationship between the normal strain in this fiber and the internal moment and the radius of curvature of the beam element, Fig. 12–5b. It is

$$\frac{1}{\rho} = -\frac{\epsilon}{y} \qquad (12\text{–}1)$$

Since Hooke's law applies, $\epsilon = \sigma/E$, and $\sigma = -My/I$, after substituting into the above equation, we get

$$\boxed{\frac{1}{\rho} = \frac{M}{EI}} \qquad (12\text{–}2)$$

Here

ρ = the radius of curvature at the point on the elastic curve
    ($1/\rho$ is referred to as the *curvature*)
M = the internal moment in the beam at the point
E = the material's modulus of elasticity
I = the beam's moment of inertia about the neutral axis

The sign for ρ therefore depends on the direction of the moment. As shown in Fig. 12–6, when M is *positive*, ρ extends *above* the beam, and when M is *negative*, ρ extends *below* the beam.

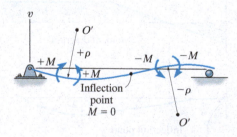

Fig. 12–6

12.2   SLOPE AND DISPLACEMENT BY INTEGRATION

The equation of the elastic curve in Fig. 12–5a will be defined by the coordinates v and x. And so to find the deflection $v = f(x)$ we must be able to represent the curvature $(1/\rho)$ in terms of v and x. In most calculus books it is shown that this relationship is

$$\frac{1}{\rho} = \frac{d^2v/dx^2}{[1 + (dv/dx)^2]^{3/2}} \qquad (12\text{--}3)$$

Substituting into Eq. 12–2, we have

$$\frac{d^2v/dx^2}{[1 + (dv/dx)^2]^{3/2}} = \frac{M}{EI} \qquad (12\text{--}4)$$

Apart from a few cases of simple beam geometry and loading, this equation is difficult to solve, because it represents a nonlinear second-order differential equation. Fortunately it can be modified, because most engineering design codes will restrict the maximum deflection of a beam or shaft. Consequently, the *slope* of the elastic curve, which is determined from dv/dx, will be *very small*, and its square will be negligible compared with unity.* Therefore the curvature, as defined in Eq. 12–3, can be *approximated* by $1/\rho = d^2v/dx^2$. With this simplification, Eq. 12–4 can now be written as

$$\frac{d^2v}{dx^2} = \frac{M}{EI} \qquad (12\text{--}5)$$

It is also possible to write this equation in two alternative forms. If we differentiate each side with respect to x and substitute $V = dM/dx$ (Eq. 6–2), we get

$$\frac{d}{dx}\left(EI\frac{d^2v}{dx^2} \right) = V(x) \qquad (12\text{--}6)$$

Differentiating again, using $w = dV/dx$ (Eq. 6–1), yields

$$\frac{d^2}{dx^2}\left(EI\frac{d^2v}{dx^2} \right) = w(x) \qquad (12\text{--}7)$$

*See Example 12.1.

TABLE 12–1

$v = 0$
Roller

$v = 0$
Pin

$v = 0$
Roller

$v = 0$
Pin

$\theta = 0$
$v = 0$
Fixed end

For most problems the *flexural rigidity* (EI) will be constant along the length of the beam. Assuming this to be the case, the above results may be reordered into the following set of three equations:

$$EI\frac{d^4v}{dx^4} = w(x) \tag{12–8}$$

$$EI\frac{d^3v}{dx^3} = V(x) \tag{12–9}$$

$$EI\frac{d^2v}{dx^2} = M(x) \tag{12–10}$$

Boundary Conditions.   Solution of any of these equations requires successive integrations to obtain v. For each integration, it is necessary to introduce a "constant of integration" and then solve for all the constants to obtain a unique solution for a particular problem. For example, if the distributed load w is expressed as a function of x and Eq. 12–8 is used, then four constants of integration must be evaluated; however, it is generally easier to determine the internal moment M as a function of x and use Eq. 12–10, so that only two constants of integration must be found.

Most often, the integration constants are determined from *boundary conditions* for the beam, Table 12–1. As noted, if the beam is supported by a roller or pin, then it is required that the displacement be *zero* at these points. At the fixed support, the slope and displacement are both zero.

Continuity Conditions.   Recall from Sec. 6.1 that if the loading on a beam is discontinuous, that is, it consists of a series of several distributed and concentrated loads, Fig. 12–7a, then several functions must be written for the internal moment, each valid within the region between two discontinuities. For example, the internal moment in regions AB, BC, and CD can be written in terms of the x_1, x_2, and x_3 coordinates selected as shown in Fig. 12–7b.

When each of these functions is integrated twice, it will produce two constants of integration, and since not all of these constants can be determined from the boundary conditions, some must be determined using *continuity conditions*. For example, consider the beam in Fig. 12–8. Here two x coordinates are chosen with origin at A. Once the functions for the slope and deflection are obtained, they must give the *same values* for the slope and deflection at point B so the elastic curve is physically *continuous*. Expressed mathematically, these continuity conditions are $\theta_1(a) = \theta_2(a)$ and $v_1(a) = v_2(a)$. They are used to evaluate the two constants of integration. Once these functions and the constants of integration are determined, they will then give the slope and deflection (elastic curve) for each region of the beam for which they are valid.

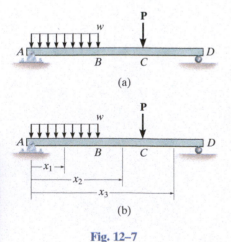

(a)

(b)

Fig. 12–7

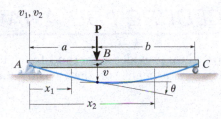

Fig. 12–8

Sign Convention and Coordinates.

When applying Eqs. 12–8 through 12–10, it is important to use the proper signs for w, V, or M as established for the derivation of these equations, Fig. 12–9a. Also, since *positive deflection*, v, is *upwards* then *positive slope* θ will be measured *counterclockwise* from the x axis when x is *positive to the right*, Fig. 12–9b. This is because a positive *increase dx* and dv creates an increased θ that is counterclockwise. By the same reason, if *positive x* is directed to the *left*, then θ will be *positive clockwise*, Fig. 12–9c.

Since we have considered $dv/dx \approx 0$, the original horizontal length of the beam's axis and the length of the arc of its elastic curve will almost be the same. In other words, ds in Figs. 12–9b and 12–9c is approximately equal to dx, since $ds = \sqrt{(dx)^2 + (dv)^2} = \sqrt{1 + (dv/dx)^2}\, dx \approx dx$. As a result, points on the elastic curve will only be *displaced vertically*, and not horizontally. Also, since the *slope* θ will be *very small*, its value in radians can be determined *directly* from $\theta \approx \tan \theta = dv/dx$.

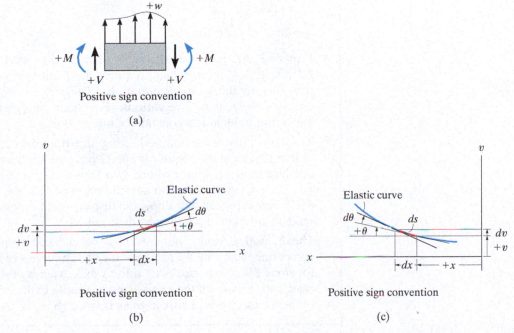

Positive sign convention

(a)

Positive sign convention

(b)

Positive sign convention

(c)

Fig. 12–9

PROCEDURE FOR ANALYSIS

The following procedure provides a method for determining the slope and deflection of a beam (or shaft) using the method of integration.

Elastic Curve.

- Draw an exaggerated view of the beam's elastic curve. Recall that zero slope and zero displacement occur at all fixed supports, and zero displacement occurs at all pin and roller supports.

- Establish the x and v coordinate axes. The x axis must be parallel to the undeflected beam and can have an origin at any point along the beam, with a positive direction either to the right or to the left. The positive v axis should be directed upwards.

- If several discontinuous loads are present, establish x coordinates that are valid for each region of the beam between the discontinuities. Choose these coordinates so that they will simplify subsequent algebraic work.

Load or Moment Function.

- For each region in which there is an x coordinate, express the loading w or the internal moment M as a function of x. In particular, *always* assume that M acts in the *positive direction* when applying the equation of moment equilibrium to determine $M = f(x)$.

Slope and Elastic Curve.

- Provided EI is constant, apply either the load equation $EI\, d^4v/dx^4 = w(x)$, which requires four integrations to get $v = v(x)$, or the moment equation $EI\, d^2v/dx^2 = M(x)$, which requires only two integrations. For each integration it is important to include a constant of integration.

- The constants are evaluated using the boundary conditions (Table 12–1) and the continuity conditions that apply to slope and displacement at points where two functions meet. Once the constants are evaluated and substituted back into the slope and deflection equations, the slope and displacement at *specific points* on the elastic curve can then be determined.

- The numerical values obtained can be checked graphically by comparing them with the sketch of the elastic curve. *Positive* values for *slope* are *counterclockwise* if the x axis extends *positive* to the *right*, and *clockwise* if the x axis extends *positive* to the *left*. In either of these cases, *positive displacement* is *upwards*.

12

EXAMPLE 12.1

The beam shown in Fig. 12–10a supports the triangular distributed loading. Determine its maximum deflection. EI is constant.

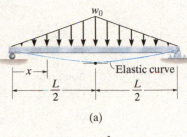

(a)

SOLUTION

Elastic Curve.   Due to symmetry, only one x coordinate is needed for the solution, in this case $0 \le x \le L/2$. The beam deflects as shown in Fig. 12–10a. The maximum deflection occurs at the center since the slope is zero at this point.

Moment Function.   A free-body diagram of the segment on the left is shown in Fig. 12–10b. The equation for the distributed loading is

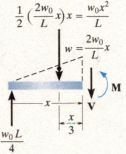

(b)

Fig. 12–10

$$w = \frac{2w_0}{L}x \qquad (1)$$

Hence,

$$\zeta + \Sigma M_{NA} = 0; \qquad M + \frac{w_0 x^2}{L}\left(\frac{x}{3}\right) - \frac{w_0 L}{4}(x) = 0$$

$$M = -\frac{w_0 x^3}{3L} + \frac{w_0 L}{4}x$$

Slope and Elastic Curve.   Using Eq. 12–10 and integrating twice, we have

$$EI\frac{d^2 v}{dx^2} = M = -\frac{w_0}{3L}x^3 + \frac{w_0 L}{4}x \qquad (2)$$

$$EI\frac{dv}{dx} = -\frac{w_0}{12L}x^4 + \frac{w_0 L}{8}x^2 + C_1$$

$$EIv = -\frac{w_0}{60L}x^5 + \frac{w_0 L}{24}x^3 + C_1 x + C_2$$

The constants of integration are obtained by applying the boundary condition $v = 0$ at $x = 0$ and the symmetry condition that $dv/dx = 0$ at $x = L/2$. This leads to

$$C_1 = -\frac{5w_0 L^3}{192} \qquad C_2 = 0$$

Hence,

$$EI\frac{dv}{dx} = -\frac{w_0}{12L}x^4 + \frac{w_0 L}{8}x^2 - \frac{5w_0 L^3}{192}$$

$$EIv = -\frac{w_0}{60L}x^5 + \frac{w_0 L}{24}x^3 - \frac{5w_0 L^3}{192}x$$

Determining the maximum deflection at $x = L/2$, we get

$$v_{max} = -\frac{w_0 L^4}{120EI} \qquad \qquad \textit{Ans.}$$

12

EXAMPLE 12.2

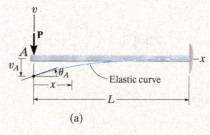

(a)

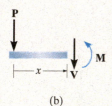

(b)

Fig. 12–11

The cantilevered beam shown in Fig. 12–11a is subjected to a vertical load **P** at its end. Determine the equation of the elastic curve. *EI* is constant.

SOLUTION I

Elastic Curve. The load tends to deflect the beam as shown in Fig. 12–11a. By inspection, the internal moment can be represented throughout the beam using a single *x* coordinate.

Moment Function. From the free-body diagram, with **M** acting in the *positive direction*, Fig. 12–11b, we have

$$M = -Px$$

Slope and Elastic Curve. Applying Eq. 12–10 and integrating twice yields

$$EI\frac{d^2v}{dx^2} = -Px \tag{1}$$

$$EI\frac{dv}{dx} = -\frac{Px^2}{2} + C_1 \tag{2}$$

$$EIv = -\frac{Px^3}{6} + C_1x + C_2 \tag{3}$$

Using the boundary conditions $dv/dx = 0$ at $x = L$ and $v = 0$ at $x = L$, Eqs. 2 and 3 become

$$0 = -\frac{PL^2}{2} + C_1$$

$$0 = -\frac{PL^3}{6} + C_1L + C_2$$

Thus, $C_1 = PL^2/2$ and $C_2 = -PL^3/3$. Substituting these results into Eqs. 2 and 3 with $\theta = dv/dx$, we get

$$\theta = \frac{P}{2EI}(L^2 - x^2)$$

$$v = \frac{P}{6EI}(-x^3 + 3L^2x - 2L^3) \qquad \textit{Ans.}$$

Maximum slope and displacement occur at $A(x = 0)$, for which

$$\theta_A = \frac{PL^2}{2EI} \tag{4}$$

$$v_A = -\frac{PL^3}{3EI} \tag{5}$$

The *positive* result for θ_A indicates *counterclockwise* rotation and the *negative* result for v_A indicates that v_A is *downward*. This agrees with the results sketched in Fig. 12–11a.

In order to obtain some idea as to the actual *magnitude* of the slope and displacement at the end A, consider the beam in Fig. 12–11a to have a length of 15 ft, support a load of $P = 6$ kip, and be made of A-36 steel having $E_{st} = 29(10^3)$ ksi. Using the methods of Sec. 11.2, if this beam was designed without a factor of safety by assuming the allowable normal stress is equal to the yield stress $\sigma_{allow} = 36$ ksi, then a W12 × 26 would be found to be adequate ($I = 204$ in^4). From Eqs. 4 and 5 we get

$$\theta_A = \frac{6 \text{ kip}(15 \text{ ft})^2(12 \text{ in.}/\text{ft})^2}{2[29(10^3) \text{ kip}/\text{in}^2](204 \text{ in}^4)} = 0.0164 \text{ rad}$$

$$v_A = -\frac{6 \text{ kip}(15 \text{ ft})^3(12 \text{ in.}/\text{ft})^3}{3[29(10^3) \text{ kip}/\text{in}^2](204 \text{ in}^4)} = -1.97 \text{ in.}$$

Since $\theta_A^2 = (dv/dx)^2 = 0.000270 \text{ rad}^2 \ll 1$, this justifies the use of Eq. 12–10, rather than applying the more exact Eq. 12–4. Also, since this numerical application is for a *cantilevered beam*, we have obtained *larger values* for θ and v than would have been obtained if the beam were supported using pins, rollers, or other fixed supports.

SOLUTION II

This problem can also be solved using Eq. 12–8, $EI\, d^4v/dx^4 = w(x)$. Here $w(x) = 0$ for $0 \le x \le L$, Fig. 12–11a, so that upon integrating once we get the form of Eq. 12–9, i.e.,

$$EI\frac{d^4v}{dx^4} = 0$$

$$EI\frac{d^3v}{dx^3} = C_1' = V$$

The shear constant C_1' can be evaluated at $x = 0$, since $V_A = -P$ (negative according to the beam sign convention, Fig. 12–9a). Thus, $C_1' = -P$. Integrating again yields the form of Eq. 12–10, i.e.,

$$EI\frac{d^3v}{dx^3} = -P$$

$$EI\frac{d^2v}{dx^2} = -Px + C_2' = M$$

Here $M = 0$ at $x = 0$, so $C_2' = 0$, and as a result one obtains Eq. 1, and the solution proceeds as before.

12

EXAMPLE 12.3

The simply supported beam shown in Fig. 12–12a is subjected to the concentrated force. Determine the maximum deflection of the beam. EI is constant.

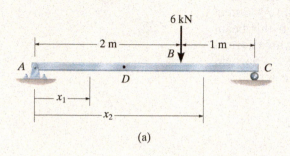

(a)

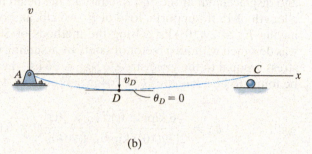

(b)

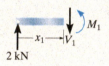

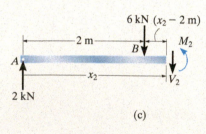

(c)

Fig. 12–12

SOLUTION

Elastic Curve. The beam deflects as shown in Fig. 12–12b. Two coordinates must be used, since the moment function will change at B. Here we will take x_1 and x_2, having the *same origin* at A.

Moment Function. From the free-body diagrams shown in Fig. 12–12c,

$$M_1 = 2x_1$$

$$M_2 = 2x_2 - 6(x_2 - 2) = 4(3 - x_2)$$

Slope and Elastic Curve. Applying Eq. 12–10 for M_1, for $0 \leq x_1 < 2$ m, and integrating twice yields

$$EI\frac{d^2v_1}{dx_1^2} = 2x_1$$

$$EI\frac{dv_1}{dx_1} = x_1^2 + C_1 \qquad (1)$$

$$EIv_1 = \frac{1}{3}x_1^3 + C_1 x_1 + C_2 \qquad (2)$$

Likewise for M_2, for 2 m $< x_2 \leq 3$ m,

$$EI\frac{d^2v_2}{dx_2^2} = 4(3 - x_2)$$

$$EI\frac{dv_2}{dx_2} = 4\left(3x_2 - \frac{x_2^2}{2}\right) + C_3 \qquad (3)$$

$$EIv_2 = 4\left(\frac{3}{2}x_2^2 - \frac{x_2^3}{6}\right) + C_3 x_2 + C_4 \qquad (4)$$

The four constants are evaluated using *two* boundary conditions, namely, $x_1 = 0$, $v_1 = 0$ and $x_2 = 3$ m, $v_2 = 0$. Also, *two* continuity conditions must be applied at B, that is, $dv_1/dx_1 = dv_2/dx_2$ at $x_1 = x_2 = 2$ m and $v_1 = v_2$ at $x_1 = x_2 = 2$ m. Therefore

$v_1 = 0$ at $x_1 = 0$; $0 = 0 + 0 + C_2$

$v_2 = 0$ at $x_2 = 3$ m; $0 = 4\left(\dfrac{3}{2}(3)^2 - \dfrac{(3)^3}{6}\right) + C_3(3) + C_4$

$\left.\dfrac{dv_1}{dx_1}\right|_{x=2m} = \left.\dfrac{dv_2}{dx_2}\right|_{x=2m}$; $(2)^2 + C_1 = 4\left(3(2) - \dfrac{(2)^2}{2}\right) + C_3$

$v_1(2\text{ m}) = v_2(2\text{ m})$;

$\dfrac{1}{3}(2)^3 + C_1(2) + C_2 = 4\left(\dfrac{3}{2}(2)^2 - \dfrac{(2)^3}{6}\right) + C_3(2) + C_4$

Solving, we get

$$C_1 = -\frac{8}{3} \qquad C_2 = 0$$

$$C_3 = -\frac{44}{3} \qquad C_4 = 8$$

Thus Eqs. 1–4 become

$$EI\frac{dv_1}{dx_1} = x_1^2 - \frac{8}{3} \qquad (5)$$

$$EIv_1 = \frac{1}{3}x_1^3 - \frac{8}{3}x_1 \qquad (6)$$

$$EI\frac{dv_2}{dx_2} = 12x_2 - 2x_2^2 - \frac{44}{3} \qquad (7)$$

$$EIv_2 = 6x_2^2 - \frac{2}{3}x_2^3 - \frac{44}{3}x_2 + 8 \qquad (8)$$

By inspection of the elastic curve, Fig. 12–12b, the maximum deflection occurs at D, somewhere within region AB. Here the slope must be zero. From Eq. 5,

$$x_1^2 - \frac{8}{3} = 0$$

$$x_1 = 1.633$$

Substituting into Eq. 6,

$$v_{max} = -\frac{2.90\text{ kN} \cdot \text{m}^3}{EI} \qquad \qquad Ans.$$

The negative sign indicates that the deflection is downwards.

EXAMPLE   12.4

The beam in Fig. 12–13a is subjected to a load at its end. Determine the displacement at C. EI is constant.

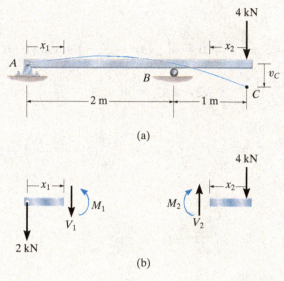

(a)

(b)

Fig. 12–13

SOLUTION

Elastic Curve.   The beam deflects into the shape shown in Fig. 12–13a. Due to the loading, two x coordinates will be considered, namely, $0 \leq x_1 < 2$ m and $0 \leq x_2 < 1$ m, where x_2 is directed to the left from C, since the internal moment is easy to formulate.

Moment Functions.   Using the free-body diagrams shown in Fig. 12–13b, we have

$$M_1 = -2x_1 \qquad M_2 = -4x_2$$

Slope and Elastic Curve.   Applying Eq. 12–10,

For $0 \leq x_1 \leq 2$: $EI\dfrac{d^2v_1}{dx_1^{\,2}} = -2x_1$

$$EI\frac{dv_1}{dx_1} = -x_1^2 + C_1 \qquad (1)$$

$$EIv_1 = -\frac{1}{3}x_1^3 + C_1x_1 + C_2 \qquad (2)$$

For $0 \le x_2 \le 1$ m:

$$EI\frac{d^2v_2}{dx_2^2} = -4x_2$$

$$EI\frac{dv_2}{dx_2} = -2x_2^2 + C_3 \qquad (3)$$

$$EIv_2 = -\frac{2}{3}x_2^3 + C_3x_2 + C_4 \qquad (4)$$

The *four* constants of integration are determined using *three* boundary conditions, namely, $v_1 = 0$ at $x_1 = 0$, $v_1 = 0$ at $x_1 = 2$ m, and $v_2 = 0$ at $x_2 = 1$ m, and *one* continuity equation. Here the continuity of slope at the roller requires $dv_1/dx_1 = -dv_2/dx_2$ at $x_1 = 2$ m and $x_2 = 1$ m. There is a negative sign in this equation because the slope is measured positive counterclockwise from the right, and positive clockwise from the left, Fig. 12–9. (Continuity of displacement at B has been indirectly considered in the boundary conditions, since $v_1 = v_2 = 0$ at $x_1 = 2$ m and $x_2 = 1$ m.) Applying these four conditions yields

$v_1 = 0$ at $x_1 = 0$;

$$0 = 0 + 0 + C_2$$

$v_1 = 0$ at $x_1 = 2$ m;

$$0 = -\frac{1}{3}(2)^3 + C_1(2) + C_2$$

$v_2 = 0$ at $x_2 = 1$ m;

$$0 = -\frac{2}{3}(1)^3 + C_3(1) + C_4$$

$\dfrac{dv_1}{dx_1}\Big|_{x=2\,\text{m}} = \dfrac{dv_2}{dx_2}\Big|_{x=1\,\text{m}}$;

$$-(2)^2 + C_1 = -(-2(1)^2 + C_3)$$

Solving, we obtain

$$C_1 = \frac{4}{3} \qquad C_2 = 0 \qquad C_3 = \frac{14}{3} \qquad C_4 = -4$$

Substituting C_3 and C_4 into Eq. 4 gives

$$EIv_2 = -\frac{2}{3}x_2^3 + \frac{14}{3}x_2 - 4$$

The displacement at C is determined by setting $x_2 = 0$. We get

$$v_C = -\frac{4\,\text{kN}\cdot\text{m}^3}{EI} \qquad\qquad Ans.$$

12

PRELIMINARY PROBLEMS

P12–1. In each case, determine the internal bending moment as a function of x, and state the necessary boundary and/or continuity conditions used to determine the elastic curve for the beam.

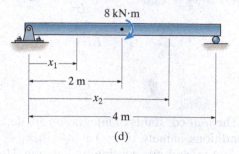

(d)

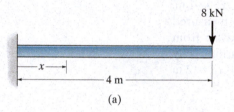

(a)

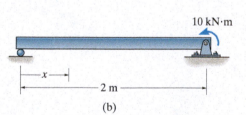

(b)

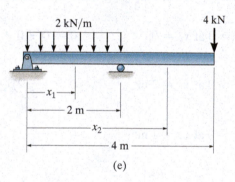

(e)

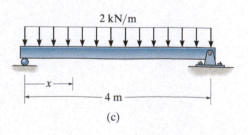

(c)

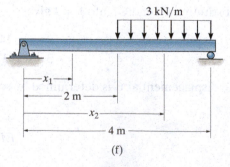

(f)

Prob. 12–1

FUNDAMENTAL PROBLEMS

F12–1. Determine the slope and deflection of end A of the cantilevered beam. $E = 200$ GPa and $I = 65.0(10^6)$ mm^4.

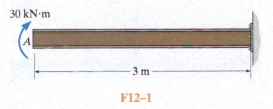

F12–1

F12–2. Determine the slope and deflection of end A of the cantilevered beam. $E = 200$ GPa and $I = 65.0(10^6)$ mm^4.

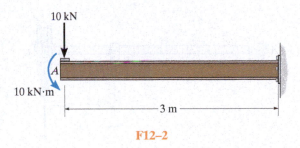

F12–2

F12–3. Determine the slope of end A of the cantilevered beam. $E = 200$ GPa and $I = 65.0(10^6)$ mm^4.

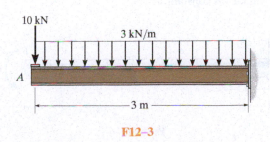

F12–3

F12–4. Determine the maximum deflection of the simply supported beam. The beam is made of wood having a modulus of elasticity of $E_w = 1.5(10^3)$ ksi and a rectangular cross section of width $b = 3$ in. and height $h = 6$ in.

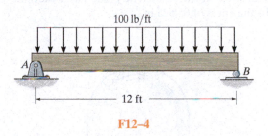

F12–4

F12–5. Determine the maximum deflection of the simply supported beam. $E = 200$ GPa and $I = 39.9(10^{-6})$ m^4.

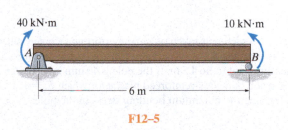

F12–5

F12–6. Determine the slope of the simply supported beam at A. $E = 200$ GPa and $I = 39.9(10^{-6})$ m^4.

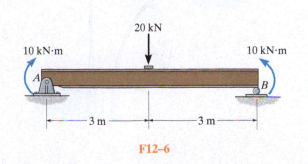

F12–6

12

PROBLEMS

12–1. An L2 steel strap having a thickness of 0.125 in. and a width of 2 in. is bent into a circular arc of radius 600 in. Determine the maximum bending stress in the strap.

12–2. The L2 steel blade of the band saw wraps around the pulley having a radius of 12 in. Determine the maximum normal stress in the blade. The blade has a width of 0.75 in. and a thickness of 0.0625 in.

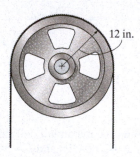

12 in.

Prob. 12–2

12–3. A picture is taken of a man performing a pole vault, and the minimum radius of curvature of the pole is estimated by measurement to be 4.5 m. If the pole is 40 mm in diameter and it is made of a glass-reinforced plastic for which $E_g = 131$ GPa, determine the maximum bending stress in the pole.

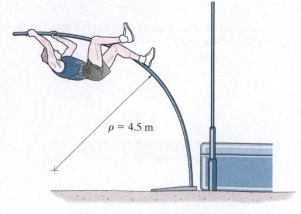

$\rho = 4.5$ m

Prob. 12–3

*12–4. Determine the equation of the elastic curve for the beam using the x coordinate that is valid for $0 \leq x < L/2$. Specify the slope at A and the beam's maximum deflection. EI is constant.

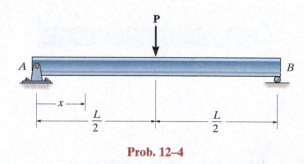

P

A   B

x

$\dfrac{L}{2}$ $\dfrac{L}{2}$

Prob. 12–4

12–5. Determine the deflection of end C of the 100-mm-diameter solid circular shaft. Take $E = 200$ GPa.

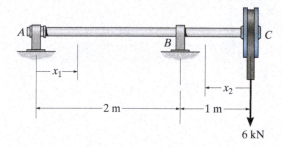

A   C

B

x_1

x_2

2 m   1 m

6 kN

Prob. 12–5

12–6. Determine the elastic curve for the cantilevered beam, which is subjected to the couple moment $\mathbf{M}_0$. Also calculate the maximum slope and maximum deflection of the beam. EI is constant.

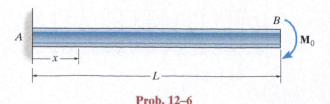

B

A $\mathbf{M}_0$

x

L

Prob. 12–6

12

12–7. The A-36 steel beam has a depth of 10 in. and is subjected to a constant moment M_0, which causes the stress at the outer fibers to become $\sigma_Y = 36$ ksi. Determine the radius of curvature of the beam and the beam's maximum slope and deflection.

12–10. Determine the equations of the elastic curve using the coordinates x_1 and x_2. What is the slope at C and displacement at B? EI is constant.

12–11. Determine the equations of the elastic curve using the coordinates x_1 and x_3. What is the slope at B and deflection at C? EI is constant.

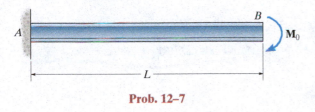

Prob. 12–7

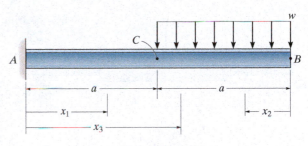

Probs. 12–10/11

*12–8. Determine the equations of the elastic curve using the coordinates x_1 and x_2. EI is constant.

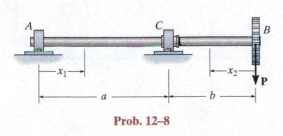

Prob. 12–8

*12–12. Draw the bending-moment diagram for the shaft and then, from this diagram, sketch the deflection or elastic curve for the shaft's centerline. Determine the equations of the elastic curve using the coordinates x_1 and x_2. EI is constant.

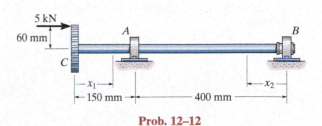

Prob. 12–12

12–9. Determine the equations of the elastic curve for the beam using the x_1 and x_2 coordinates. EI is constant.

12–13. Determine the maximum deflection of the beam and the slope at A. EI is constant.

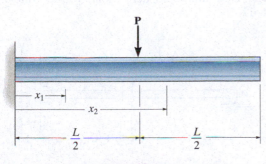

Prob. 12–9

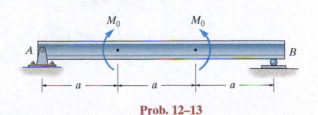

Prob. 12–13

12

12–14. The simply supported shaft has a moment of inertia of $2I$ for region BC and a moment of inertia I for regions AB and CD. Determine the maximum deflection of the shaft due to the load **P**.

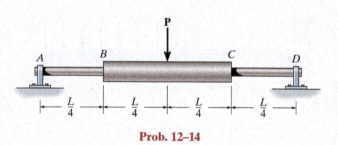

Prob. 12–14

12–15. A torque wrench is used to tighten the nut on a bolt. If the dial indicates that a torque of 60 lb · ft is applied when the bolt is fully tightened, determine the force P acting at the handle and the distance s the needle moves along the scale. Assume only the portion AB of the beam distorts. The cross section is square having dimensions of 0.5 in. by 0.5 in. $E = 29(10^3)$ ksi.

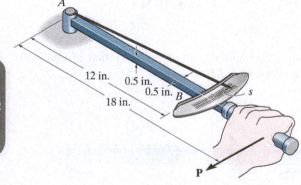

Prob. 12–15

*12–16. The pipe can be assumed roller supported at its ends and by a rigid saddle C at its center. The saddle rests on a cable that is connected to the supports. Determine the force that should be developed in the cable if the saddle keeps the pipe from sagging or deflecting at its center. The pipe and fluid within it have a combined weight of 12.5 lb/ft. EI is constant.

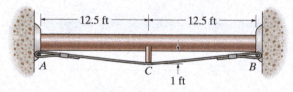

Prob. 12–16

12–17. Determine the equations of the elastic curve for the beam using the x_1 and x_2 coordinates. Specify the beam's maximum deflection. EI is constant.

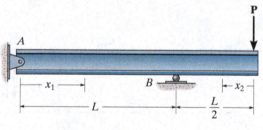

Prob. 12–17

12–18. The bar is supported by a roller constraint at B, which allows vertical displacement but resists axial load and moment. If the bar is subjected to the loading shown, determine the slope at A and the deflection at C. EI is constant.

12–19. Determine the deflection at B of the bar in Prob. 12–18.

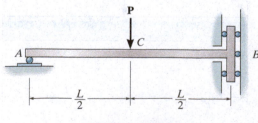

Probs. 12–18/19

*12–20. Determine the equations of the elastic curve using the x_1 and x_2 coordinates. What is the slope at A and the deflection at C? EI is constant.

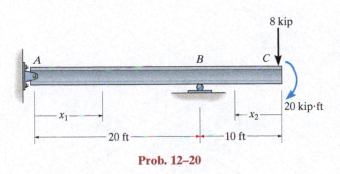

Prob. 12–20

12–21. Determine the maximum deflection of the solid circular shaft. The shaft is made of steel having $E = 200$ GPa. It has a diameter of 100 mm.

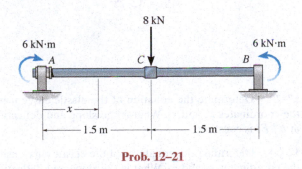

Prob. 12–21

12–22. Determine the elastic curve for the cantilevered W14 × 30 beam using the x coordinate. Specify the maximum slope and maximum deflection. $E = 29(10^3)$ ksi.

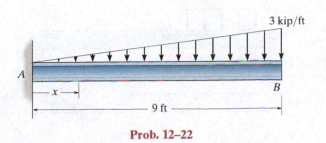

Prob. 12–22

12–23. Determine the equations of the elastic curve using the coordinates x_1 and x_2. What is the deflection and slope at C? EI is constant.

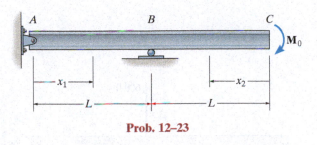

Prob. 12–23

*12–24. Determine the equations of the elastic curve using the coordinates x_1 and x_2. What is the slope at A? EI is constant.

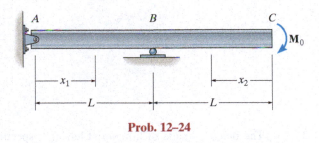

Prob. 12–24

12–25. The floor beam of the airplane is subjected to the loading shown. Assuming that the fuselage exerts only vertical reactions on the ends of the beam, determine the maximum deflection of the beam. EI is constant.

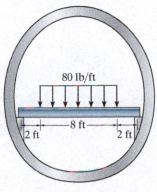

Prob. 12–25

12–26. Determine the maximum deflection of the simply supported beam. The beam is made of wood having a modulus of elasticity of $E = 1.5 \, (10^3)$ ksi.

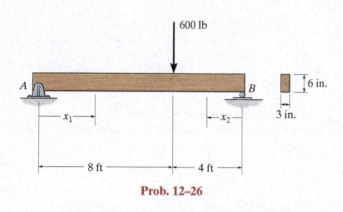

Prob. 12–26

12–27. The beam is made of a material having a specific weight γ. Determine the displacement and slope at its end A due to its weight. The modulus of elasticity for the material is E.

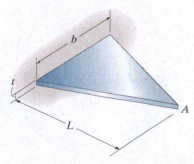

Prob. 12–27

*12–28. Determine the slope at end B and the maximum deflection of the cantilever triangular plate of constant thickness t. The plate is made of material having a modulus of elasticity of E.

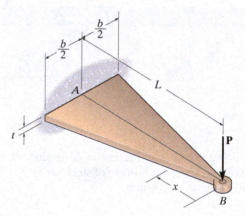

Prob. 12–28

12–29. Determine the equation of the elastic curve using the coordinates x_1 and x_2. What is the slope and deflection at B? EI is constant.

12–30. Determine the equations of the elastic curve using the coordinates x_1 and x_3. What is the slope and deflection at point B? EI is constant.

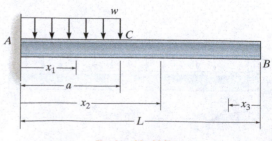

Probs. 12–29/30

*12.3 DISCONTINUITY FUNCTIONS

The method of integration, used to find the equation of the elastic curve for a beam or shaft, is convenient if the load or internal moment can be expressed as a continuous function throughout the beam's entire length. If several different loadings act on the beam, however, this method can become tedious to apply, because separate loading or moment functions must be written for each region of the beam. Furthermore, as noted in Examples 12.3 and 12.4, integration of these functions requires the evaluation of integration constants using both the boundary and continuity conditions.

In this section, we will discuss a method for finding the equation of the elastic curve using a *single expression*, either formulated directly from the loading on the beam, $w = w(x)$, or from the beam's internal moment, $M = M(x)$. Then when this expression for w is substituted into $EI\, d^4v/dx^4 = w(x)$ and integrated four times, or if the expression for M is substituted into $EI\, d^2v/dx^2 = M(x)$ and integrated twice, the constants of integration will only have to be determined from the boundary conditions.

Discontinuity Functions. In order to express the load on the beam or the internal moment within it using a single expression, we will use two types of mathematical operators known as *discontinuity functions*.

For safety, the beams supporting these bags of cement must be designed for strength and a restricted amount of deflection.

12

Macaulay Functions. For purposes of beam or shaft deflection, Macaulay functions, named after the mathematician W. H. Macaulay, can be used to describe *distributed loadings*. These functions can be written in general form as

$$\langle x - a \rangle^n = \begin{cases} 0 & \text{for } x < a \\ (x - a)^n & \text{for } x \geq a \end{cases} \tag{12–11}$$

$$n \geq 0$$

Here x represents the location of a point on the beam, and a is the location where the distributed loading *begins*. The Macaulay function $\langle x - a \rangle^n$ is written with angle or Macaulay brackets to distinguish it from an ordinary function $(x - a)^n$ written with parentheses. As stated by the equation, only when $x \geq a$ is $\langle x - a \rangle^n = (x - a)^n$; otherwise it is zero. Furthermore, this function is valid only for exponential values $n \geq 0$. Integration of the Macaulay function follows the same rules as for ordinary functions, i.e.,

$$\int \langle x - a \rangle^n dx = = \frac{\langle x - a \rangle^{n+1}}{n + 1} + C \tag{12–12}$$

Macaulay functions for a uniform and triangular load are shown in Table 12–2. Using integration, the Macaulay functions for shear, $V = \int w(x)\, dx$, and moment, $M = \int V\, dx$, are also shown in the table.

TABLE 12–2

Loading	Loading Function $w = w(x)$	Shear $V = \int w(x)dx$	Moment $M = \int V dx$
M_0	$w = M_0 \langle x-a \rangle^{-2}$	$V = M_0 \langle x-a \rangle^{-1}$	$M = M_0 \langle x-a \rangle^0$
P	$w = P \langle x-a \rangle^{-1}$	$V = P \langle x-a \rangle^0$	$M = P \langle x-a \rangle^1$
w_0	$w = w_0 \langle x-a \rangle^0$	$V = w_0 \langle x-a \rangle^1$	$M = \frac{w_0}{2} \langle x-a \rangle^2$
slope $= m$	$w = m \langle x-a \rangle^1$	$V = \frac{m}{2} \langle x-a \rangle^2$	$M = \frac{m}{6} \langle x-a \rangle^3$

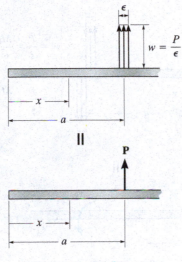

Fig. 12–14

Singularity Functions. These functions are used to describe concentrated forces or couple moments acting on a beam or shaft. Specifically, a concentrated force **P** can be considered a special case of a distributed loading having an intensity of $w = P/\epsilon$ when its length $\epsilon \rightarrow 0$, Fig. 12–14. The area under this loading diagram is equivalent to P, *positive upwards*, and has this value only when $x = a$. We will use a symbolic representation to express this result, namely

$$w = P\langle x - a\rangle^{-1} = \begin{cases} 0 & \text{for } x \neq a \\ P & \text{for } x = a \end{cases} \qquad (12\text{–}13)$$

This expression is referred to as a *singularity function*, since it takes on the value P only at the point $x = a$ where the load acts, otherwise it is zero.*

*It is also referred to as a unit impulse function or the Dirac delta.

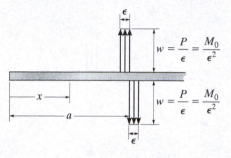

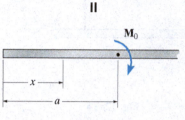

Fig. 12–15

In a similar manner, a couple moment M_0, considered *positive clockwise*, is a limit as $\epsilon \to 0$ of two distributed loadings, as shown in Fig. 12–15. Here the following function describes its value.

$$w = M_0 \langle x - a \rangle^{-2} = \begin{cases} 0 & \text{for } x \neq a \\ M_0 & \text{for } x = a \end{cases} \qquad (12\text{–}14)$$

The exponent $n = -2$ in order to ensure that the units of w, force per length, are maintained.

Integration of the above two functions follows the rules of calculus and yields results that are *different* from those of the Macaulay function. Specifically,

$$\int \langle x - a \rangle^n dx = \langle x - a \rangle^{n+1}, n = -1, -2 \qquad (12\text{–}15)$$

Using this formula, notice how M_0 and P, described in Table 12–2, are integrated once, then twice, to obtain the internal shear and moment in the beam.

Application of Eqs. 12–11 through 12–15 provides a direct means for expressing the loading or the internal moment in a beam as a function of x. Close attention, however, must be paid to the signs of the external loadings. As stated above, and as shown in Table 12–2, *concentrated forces and distributed loads are positive upwards, and couple moments are positive clockwise*. If this sign convention is followed, then the internal shear and moment will be in accordance with the beam sign convention established in Sec. 6.1.

Application. As an example of how to apply discontinuity functions to describe the loading or internal moment, we will consider the beam in Fig. 12–16a. Here the reactive 2.75-kN force created by the roller, Fig. 12–16b, is positive since it acts upwards, and the 1.5-kN·m couple moment is also positive since it acts clockwise. Finally, the trapezoidal loading is negative and by superposition has been separated into triangular and uniform loadings. From Table 12–2, the loading at any point x on the beam is therefore

$$w = 2.75 \text{ kN} \langle x - 0 \rangle^{-1} + 1.5 \text{ kN} \cdot \text{m} \langle x - 3 \text{ m} \rangle^{-2} - 3 \text{ kN/m} \langle x - 3 \text{ m} \rangle^{0} - 1 \text{ kN/m}^2 \langle x - 3 \text{ m} \rangle^{1}$$

The reactive force at B is not included here since x is never greater than 6 m. In the same manner, we can determine the moment expression directly from Table 12–2. It is

$$M = 2.75 \text{ kN} \langle x - 0 \rangle^{1} + 1.5 \text{ kN} \cdot \text{m} \langle x - 3 \text{ m} \rangle^{0} - \frac{3 \text{ kN/m}}{2} \langle x - 3 \text{ m} \rangle^{2} - \frac{1 \text{ kN/m}^2}{6} \langle x - 3 \text{ m} \rangle^{3}$$

$$= 2.75x + 1.5 \langle x - 3 \rangle^{0} - 1.5 \langle x - 3 \rangle^{2} - \frac{1}{6} \langle x - 3 \rangle^{3}$$

The deflection of the beam can now be determined after this equation is integrated two successive times, and the constants of integration are evaluated using the boundary conditions of zero displacement at A and B.

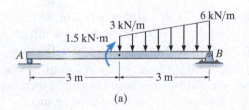

(a)

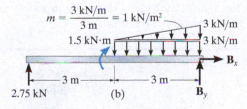

(b)

Fig. 12–16

PROCEDURE FOR ANALYSIS

The following procedure provides a method for using discontinuity functions to determine a beam's elastic curve. This method is particularly advantageous for solving problems involving beams or shafts subjected to *several loadings*, since the constants of integration can be evaluated by using *only* the boundary conditions, while the compatibility conditions are automatically satisfied.

Elastic Curve.

* Sketch the beam's elastic curve and identify the boundary conditions at the supports.
* Zero displacement occurs at all pin and roller supports, and zero slope and zero displacement occur at fixed supports.
* Establish the x axis so that it extends to the right and has its origin at the beam's left end.

Load or Moment Function.

* Calculate the support reactions and then use the discontinuity functions in Table 12–2 to express either the loading w or the internal moment M as a function of x. Make sure to follow the sign convention for each loading.
* Note that the distributed loadings must extend all the way to the beam's right end to be valid. If this does not occur, use the method of superposition, which is illustrated in Example 12.6.

Slope and Elastic Curve.

* Substitute w into $EI\, d^4v/dx^4 = w(x)$, or M into the moment curvature relation $EI\, d^2v/dx^2 = M$, and integrate to obtain the equations for the beam's slope and deflection.
* Evaluate the constants of integration using the boundary conditions, and substitute these constants into the slope and deflection equations to obtain the final results.
* When the slope and deflection equations are evaluated at any point on the beam, a *positive slope* is *counterclockwise*, and a *positive displacement* is *upwards*.

EXAMPLE 12.5

Determine the equation of the elastic curve for the cantilevered beam shown in Fig. 12–17a. EI is constant.

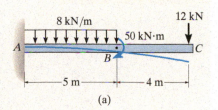

(a)

SOLUTION

Elastic Curve. The loads cause the beam to deflect as shown in Fig. 12–17a. The boundary conditions require zero slope and displacement at A.

Loading Function. The support reactions at A have been calculated and are shown on the free-body diagram in Fig. 12–17b. Since the distributed loading in Fig. 12–17a does not extend to C as required, we will use the superposition of loadings shown in Fig. 12–17b to represent the same effect. By our sign convention, the beam's loading is therefore

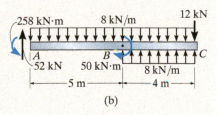

(b)

Fig. 12–17

$$w = 52 \text{ kN} \langle x - 0 \rangle^{-1} - 258 \text{ kN} \cdot \text{m} \langle x - 0 \rangle^{-2} - 8 \text{ kN/m} \langle x - 0 \rangle^0$$
$$+ 50 \text{ kN} \cdot \text{m} \langle x - 5 \text{ m} \rangle^{-2} + 8 \text{ kN/m} \langle x - 5 \text{ m} \rangle^0$$

The 12-kN load is *not included* here, since x cannot be greater than 9 m. Because $dV/dx = w(x)$, then by integrating, and neglecting the constant of integration since the reactions at A are included in the load function, we get

$$V = 52 \langle x - 0 \rangle^0 - 258 \langle x - 0 \rangle^{-1} - 8 \langle x - 0 \rangle^1 + 50 \langle x - 5 \rangle^{-1} + 8 \langle x - 5 \rangle^1$$

Furthermore, $dM/dx = V$, so that integrating again yields

$$M = -258 \langle x - 0 \rangle^0 + 52 \langle x - 0 \rangle^1 - \frac{1}{2}(8) \langle x - 0 \rangle^2 + 50 \langle x - 5 \rangle^0 + \frac{1}{2}(8) \langle x - 5 \rangle^2$$

$$= (-258 + 52x - 4x^2 + 50 \langle x - 5 \rangle^0 + 4 \langle x - 5 \rangle^2) \text{ kN} \cdot \text{m}$$

This same result can be obtained *directly* from Table 12–2.

Slope and Elastic Curve. Applying Eq. 12–10 and integrating twice, we have

$$EI \frac{d^2v}{dx^2} = -258 + 52x - 4x^2 + 50 \langle x - 5 \rangle^0 + 4 \langle x - 5 \rangle^2$$

$$EI \frac{dv}{dx} = -258x + 26x^2 - \frac{4}{3}x^3 + 50 \langle x - 5 \rangle^1 + \frac{4}{3} \langle x - 5 \rangle^3 + C_1$$

$$EIv = -129x^2 + \frac{26}{3}x^3 - \frac{1}{3}x^4 + 25 \langle x - 5 \rangle^2 + \frac{1}{3} \langle x - 5 \rangle^4 + C_1 x + C_2$$

Since $dv/dx = 0$ at $x = 0$, $C_1 = 0$; and $v = 0$ at $x = 0$, so $C_2 = 0$. Thus,

$$v = \frac{1}{EI} \left(-129x^2 + \frac{26}{3}x^3 - \frac{1}{3}x^4 + 25 \langle x - 5 \rangle^2 + \frac{1}{3} \langle x - 5 \rangle^4 \right) \text{ m} \quad \textit{Ans.}$$

12

EXAMPLE   12.6

Determine the maximum deflection of the beam shown in Fig. 12–18a. EI is constant.

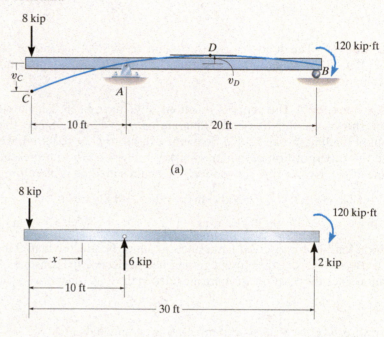

(a)

(b)

Fig. 12–18

SOLUTION

Elastic Curve.   The beam deflects as shown in Fig. 12–18a. The boundary conditions require zero displacement at A and B.

Loading Function.   The reactions have been calculated and are shown on the free-body diagram in Fig. 12–18b. The loading function for the beam is

$$w = -8 \text{ kip} \langle x - 0 \rangle^{-1} + 6 \text{ kip} \langle x - 10 \text{ ft} \rangle^{-1}$$

The couple moment and force at B are not included here, since they are located at the right end of the beam, and x cannot be greater than 30 ft. Integrating $dV/dx = w(x)$, we get

$$V = -8 \langle x - 0 \rangle^0 + 6 \langle x - 10 \rangle^0$$

In a similar manner, $dM/dx = V$ yields

$$M = -8 \langle x - 0 \rangle^1 + 6 \langle x - 10 \rangle^1$$
$$= (-8x + 6 \langle x - 10 \rangle^1) \text{ kip} \cdot \text{ft}$$

Notice how this equation can also be established *directly* using the results of Table 12–2 for moment.

Slope and Elastic Curve. Integrating twice yields

$$EI\frac{d^2v}{dx^2} = -8x + 6\langle x - 10\rangle^1$$

$$EI\frac{dv}{dx} = -4x^2 + 3\langle x - 10\rangle^2 + C_1$$

$$EIv = -\frac{4}{3}x^3 + \langle x - 10\rangle^3 + C_1 x + C_2 \qquad (1)$$

From Eq. 1, the boundary condition $v = 0$ at $x = 10$ ft and $v = 0$ at $x = 30$ ft gives

$$0 = -1333 + (10 - 10)^3 + C_1(10) + C_2$$

$$0 = -36\,000 + (30 - 10)^3 + C_1(30) + C_2$$

Solving these equations simultaneously for C_1 and C_2, we get $C_1 = 1333$ and $C_2 = -12\,000$. Thus,

$$EI\frac{dv}{dx} = -4x^2 + 3\langle x - 10\rangle^2 + 1333 \qquad (2)$$

$$EIv = -\frac{4}{3}x^3 + \langle x - 10\rangle^3 + 1333x - 12\,000 \qquad (3)$$

From Fig. 12–18a, maximum displacement can occur either at C, or at D where the slope $dv/dx = 0$. To obtain the displacement of C, set $x = 0$ in Eq. 3. We get

$$v_C = -\frac{12\,000 \text{ kip} \cdot \text{ft}^3}{EI} \qquad \qquad \textit{Ans.}$$

The *negative* sign indicates that the displacement is *downwards* as shown in Fig. 12–17a. To locate point D, use Eq. 2 with $x > 10$ ft and $dv/dx = 0$. This gives

$$0 = -4x_D{}^2 + 3(x_D - 10)^2 + 1333$$

$$x_D{}^2 + 60x_D - 1633 = 0$$

Solving for the positive root,

$$x_D = 20.3 \text{ ft}$$

Hence, from Eq. 3,

$$EIv_D = -\frac{4}{3}(20.3)^3 + (20.3 - 10)^3 + 1333(20.3) - 12\,000$$

$$v_D = \frac{5006 \text{ kip} \cdot \text{ft}^3}{EI}$$

Comparing this value with v_C, we see that $v_{\max} = v_C$.

PROBLEMS

12–31. The shaft is supported at A by a journal bearing and at C by a thrust bearing. Determine the equation of the elastic curve. EI is constant.

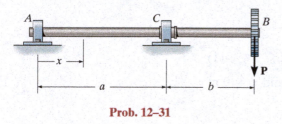

Prob. 12–31

*12–32. The shaft supports the two pulley loads shown. Determine the equation of the elastic curve. EI is constant.

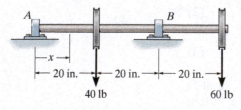

Prob. 12–32

12–33. The beam is made of a ceramic material. If it is subjected to the elastic loading shown, and the moment of inertia is I and the beam has a measured maximum deflection Δ at its center, determine the modulus of elasticity, E. The supports at A and D exert only vertical reactions on the beam.

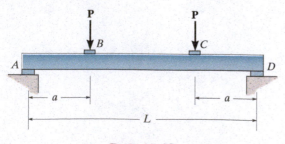

Prob. 12–33

12–34. Determine the equation of the elastic curve, the maximum deflection in region AB, and the deflection of end C. EI is constant.

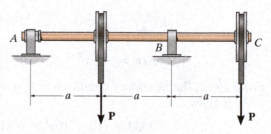

Prob. 12–34

12–35. The beam is subjected to the load shown. Determine the equation of the elastic curve. EI is constant.

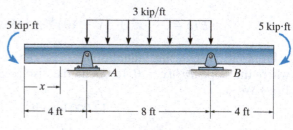

Prob. 12–35

*12–36. Determine the equation of the elastic curve, the slope at A, and the deflection at B. EI is constant.

12–37. Determine the equation of the elastic curve and the maximum deflection of the simply supported beam. EI is constant.

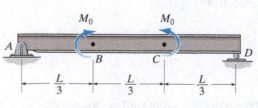

Probs. 12–36/37

12–38. The shaft supports the two pulley loads. Determine the equation of the elastic curve. *EI* is constant.

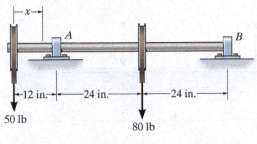

Prob. 12–38

12–39. Determine the maximum deflection of the cantilevered beam. Take $E = 200$ GPa and $I = 65.0(10^6)$ mm^4.

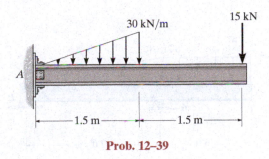

Prob. 12–39

***12–40.** Determine the slope at A and the deflection of end C of the overhang beam. Take $E = 29(10^3)$ ksi and $I = 204$ in^4.

12–41. Determine the maximum deflection in region AB of the overhang beam. Take $E = 29(10^3)$ ksi and $I = 204$ in^4.

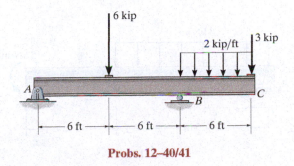

Probs. 12–40/41

12–42. The shaft supports the two pulley loads shown. Determine the slope of the shaft at A and B. The bearings exert only vertical reactions on the shaft. *EI* is constant.

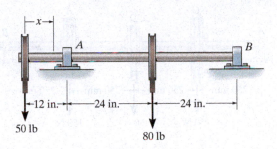

Prob. 12–42

12–43. Determine the equation of the elastic curve. *EI* is constant.

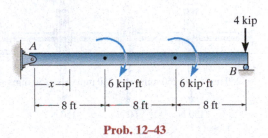

Prob. 12–43

***12–44.** Determine the equation of the elastic curve. *EI* is constant.

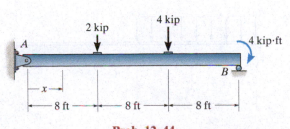

Prob. 12–44

12

12–45. Determine the deflection at each of the pulleys C, D, and E. The shaft is made of steel and has a diameter of 30 mm. $E_{st} = 200$ GPa.

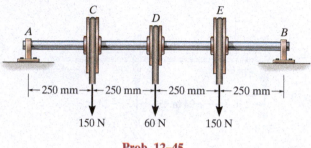

Prob. 12–45

12–46. Determine the slope of the shaft at A and B. The shaft is made of steel and has a diameter of 30 mm. The bearings only exert vertical reactions on the shaft. $E_{st} = 200$ GPa.

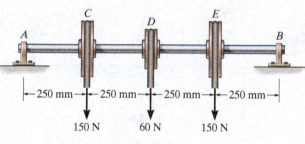

Prob. 12–46

12–47. Determine the equation of the elastic curve. Specify the slopes at A and B. EI is constant.

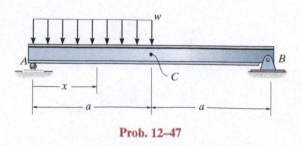

Prob. 12–47

*12–48.** Determine the value of a so that the displacement at C is equal to zero. EI is constant.

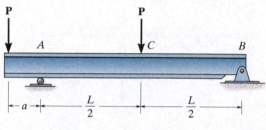

Prob. 12–48

12–49. Determine the displacement at C and the slope at A. EI is constant.

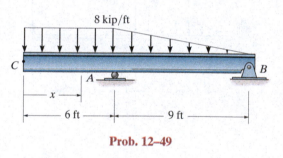

Prob. 12–49

12–50. Determine the equations of the slope and elastic curve. EI is constant.

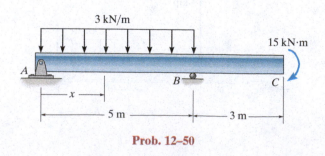

Prob. 12–50

*12.4 SLOPE AND DISPLACEMENT BY THE MOMENT-AREA METHOD

The moment-area method provides a semigraphical technique for finding the slope and displacement at specific points on the elastic curve of a beam or shaft. Application of the method requires calculating area segments of the beam's moment diagram; and so if this diagram consists of simple shapes, the method is very convenient to use.

To develop the moment-area method we will make the same assumptions we used for the method of integration: The beam is initially straight, it is elastically deformed by the loads, such that the slope and deflection of the elastic curve are very small, and the deformations are only caused by bending. The moment-area method is based on two theorems, one used to determine the slope and the other to determine the displacement.

Theorem 1. Consider the simply supported beam with its associated elastic curve, shown in Fig. 12–19a. A differential segment dx of the beam is shown in Fig. 12–19b. Here the beam's internal moment M deforms the element such that the *tangents* to the elastic curve at each side intersect at an angle $d\theta$. This angle can be determined from Eq. 12–10, written as

$$EI\frac{d^2v}{dx^2} = EI\frac{d}{dx}\left(\frac{dv}{dx}\right) = M$$

Since the *slope* is *small*, then $\theta = dv/dx$, and therefore

$$d\theta = \frac{M}{EI}dx \qquad (12\text{–}16)$$

If the moment diagram for the beam is constructed and divided by the flexural rigidity, EI, Fig. 12–19c, then this equation indicates that $d\theta$ is equal to the *area* under the "M/EI diagram" for the beam segment dx. Integrating from a selected point A on the elastic curve to another point B, we have

$$\theta_{B/A} = \int_A^B \frac{M}{EI}dx \qquad (12\text{–}17)$$

This result forms the basis for the first moment-area theorem.

The angle, measured in radians, between the tangents at any two points on the elastic curve equals the area under the M/EI diagram between these two points.

The notation $\theta_{B/A}$ refers to the angle of the tangent at B measured *with respect to* the tangent at A. From the proof it should be evident that this angle is measured *counterclockwise*, from tangent A to tangent B, if the area under the M/EI diagram is *positive*. Conversely, if the area is *negative*, or lies below the x axis, the angle $\theta_{B/A}$ is measured clockwise from tangent A to tangent B.

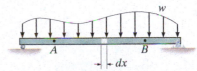

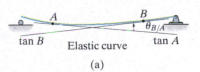

(a)

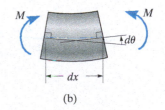

(b)

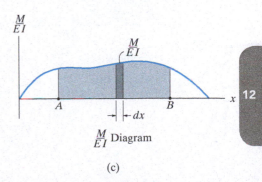

(c)

Fig. 12–19

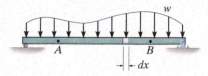

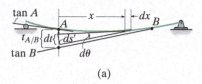

(a)

Theorem 2. The second moment-area theorem is based on the relative *deviation* of tangents to the elastic curve. Figure 12–20*a* shows a greatly exaggerated view of the vertical deviation *dt* of the tangents on each side of the differential element *dx*. This deviation is caused by the curvature of the element and has been measured along a vertical line passing through point *A* on the elastic curve. Since the slope of the elastic curve and its deflection are assumed to be very small, it is satisfactory to approximate the length of each tangent line by *x* and the arc *ds′* by *dt*. Using the circular-arc formula $s = \theta r$, where *r* is the length *x* and *s* is *dt*, we can write $dt = x\, d\theta$. Substituting Eq. 12–16 into this equation and integrating from *A* to *B*, the vertical deviation of the tangent at *A* *with respect to* the tangent at *B* becomes

$$t_{A/B} = \int_A^B x \frac{M}{EI} dx \qquad (12\text{–}18)$$

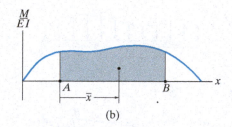

(b)

Since the centroid of an area is found from $\bar{x} \int dA = \int x\, dA$, and $\int (M/EI)\, dx$ represents the area under the *M/EI* diagram, we can also write

$$t_{A/B} = \bar{x} \int_A^B \frac{M}{EI} dx \qquad (12\text{–}19)$$

Here $\bar{x}$ is the distance from *A* to the *centroid* of the area under the *M/EI* diagram between *A* and *B*, Fig. 12–20*b*.

The second moment-area theorem can now be stated as follows:

The vertical distance between the tangent at a point (A) on the elastic curve and the tangent extended from another point (B) equals the moment of the area under the M/EI diagram between these two points (A and B). This moment is calculated about the point (A) where the vertical distance ($t_{A/B}$) is to be determined.

Note that $t_{A/B}$ is *not* equal to $t_{B/A}$, which is shown in Fig. 12–20*c*. This is because the moment of the area under the *M/EI* diagram between *A* and *B* is calculated about point *A* to determine $t_{A/B}$, Fig. 12–20*b*, and it is calculated about point *B* to determine $t_{B/A}$, Fig. 12–20*c*.

If $t_{A/B}$ is calculated from the moment of a *positive M/EI* area between *A* and *B*, it indicates that *A* is *above* the tangent extended from *B*, Fig. 12–20*a*. Similarly, *negative M/EI* areas indicate that *A* will be *below* the tangent extended from *B*. This same rule applies for $t_{B/A}$.

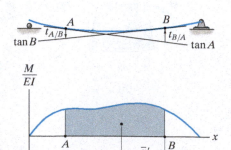

(c)

Fig. 12–20

PROCEDURE FOR ANALYSIS

The following procedure provides a method that may be used to apply the two moment-area theorems.

M/EI Diagram.

- Determine the support reactions and draw the beam's M/EI diagram. If the beam is loaded with concentrated forces and couple moments, the M/EI diagram will consist of a series of straight line segments, and the areas and their moments required for the moment-area theorems will be relatively easy to calculate. If the loading consists of a series of distributed loads, the M/EI diagram will consist of parabolic or perhaps higher-order curves, and it is suggested that the table on the inside front cover be used to locate the area and centroid under each curve.

Elastic Curve.

- Draw an exaggerated view of the beam's elastic curve. Recall that points of zero slope and zero displacement always occur at a fixed support, and zero displacement occurs at all pin and roller supports.
- The unknown displacement and slope to be determined should be indicated on the curve.
- Since the moment-area theorems apply *only between two tangents*, attention should be given as to which tangents should be constructed so that the angles or vertical distance between them will lead to the solution of the problem. In particular, *the tangents at the supports should be considered*, since the beam has zero displacement and/or zero slope at the supports.

Moment-Area Theorems.

- Apply Theorem 1 to determine the *angle* between any two tangents on the elastic curve and Theorem 2 to determine the vertical distance between the tangents.
- A *positive* $\theta_{B/A}$ represents a *counterclockwise* rotation of the tangent at B with respect to the tangent at A, and a *positive* $t_{B/A}$ indicates that B on the elastic curve lies *above* the extended tangent from A.

EXAMPLE 12.7

Determine the slope of the beam shown in Fig. 12–21a at point B. EI is constant.

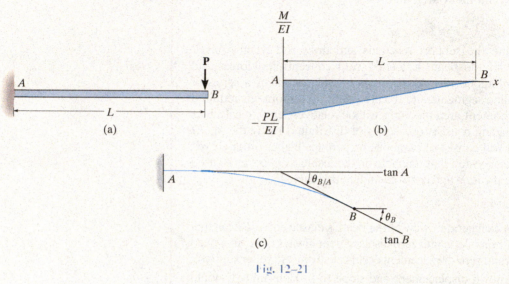

(a)

(b)

(c)

Fig. 12–21

SOLUTION

M/EI Diagram. See Fig. 12–21b.

Elastic Curve. The force **P** causes the beam to deflect as shown in Fig. 12–21c. The tangent at B is indicated since we are required to find θ_B. Also, the tangent at the support (A) is shown. This tangent has a *known* zero slope. By the construction, the angle between tan A and tan B is equivalent to θ_B. Thus,

$$\theta_B = \theta_{B/A}$$

Moment-Area Theorem. Applying Theorem 1, $\theta_{B/A}$ is equal to the area under the M/EI diagram between A and B; that is,

$$\theta_B = \theta_{B/A} = \frac{1}{2}\left(-\frac{PL}{EI}\right)L$$

$$= -\frac{PL^2}{2EI} \qquad\qquad Ans.$$

The *negative sign* indicates that the angle measured from the tangent at A to the tangent at B is *clockwise*. This checks, since the beam slopes downwards at B.

EXAMPLE 12.8

Determine the displacement of points B and C of the beam shown in Fig. 12–22a. EI is constant.

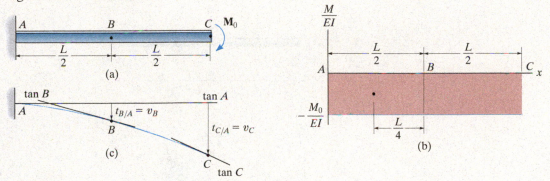

Fig. 12–22

SOLUTION

M/EI Diagram.   See Fig. 12–22b.

Elastic Curve.   The couple moment at C causes the beam to deflect as shown in Fig. 12–22c. The tangents at B and C are indicated since we are required to find v_B and v_C. Also, the tangent at the support (A) is shown since it is horizontal. The required displacements can now be related directly to the vertical distance between the tangents at B and A and C and A. Specifically,

$$v_B = t_{B/A}$$
$$v_C = t_{C/A}$$

Moment-Area Theorem.   Applying Theorem 2, $t_{B/A}$ is equal to the moment of the shaded area under the M/EI diagram between A and B calculated about B (the point on the elastic curve), since this is the point where the vertical distance is to be determined. Hence, from Fig. 12–22b,

$$v_B = t_{B/A} = \left(\frac{L}{4}\right)\left[\left(-\frac{M_0}{EI}\right)\left(\frac{L}{2}\right)\right] = -\frac{M_0 L^2}{8EI} \qquad \textit{Ans.}$$

Likewise, for $t_{C/A}$ we must determine the moment of the area under the *entire* M/EI diagram from A to C about C (the point on the elastic curve). We have

$$v_C = t_{C/A} = \left(\frac{L}{2}\right)\left[\left(-\frac{M_0}{EI}\right)(L)\right] = -\frac{M_0 L^2}{2EI} \qquad \textit{Ans.}$$

Since both answers are *negative*, they indicate that B and C lie *below* the tangent at A. This checks with Fig. 12–22c.

EXAMPLE 12.9

Determine the slope at point C on the shaft in Fig. 12–23a. EI is constant.

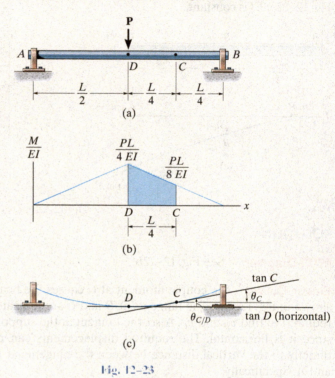

Fig. 12–23

SOLUTION

M/EI Diagram.   See Fig. 12–23b.

Elastic Curve.   Since the loading is applied symmetrically to the shaft, the elastic curve is symmetric, and the tangent at D is horizontal, Fig. 12–23c. Also the tangent at C is drawn, since we must find the slope θ_C. By the construction, the angle between tan D and tan C is equal to θ_C; that is,

$$\theta_C = \theta_{C/D}$$

Moment-Area Theorem.   Using Theorem 1, $\theta_{C/D}$ is equal to the shaded area under the M/EI diagram between D and C. We have

$$\theta_C = \theta_{C/D} = \left(\frac{PL}{8EI}\right)\left(\frac{L}{4}\right) + \frac{1}{2}\left(\frac{PL}{4EI} - \frac{PL}{8EI}\right)\left(\frac{L}{4}\right) = \frac{3PL^2}{64EI} \qquad Ans.$$

What does the positive result indicate?

EXAMPLE 12.10

Determine the slope at point C on the steel beam in Fig. 12–24a. Take $E_{st} = 200$ GPa, $I = 17(10^6)$ mm^4.

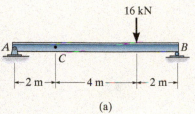

(a)

SOLUTION

M/EI Diagram. See Fig. 12–24b.

Elastic Curve. The elastic curve is shown in Fig. 12–24c. Here we are required to find θ_C. Tangents at the *supports* A and B are also shown in the figure. The slope at A, θ_A, in Fig. 12–24c can be found using $|\theta_A| = |t_{B/A}|/L_{AB}$. This equation is valid since $t_{B/A}$ is actually very small, so that $t_{B/A}$ can be approximated by the length of a circular arc defined by a radius of $L_{AB} = 8$ m and a sweep of θ_A in radians. (Recall that $s = \theta r$.) From the geometry of Fig. 12–24c, we have

$$|\theta_C| = |\theta_A| - |\theta_{C/A}| = \left|\frac{t_{B/A}}{8}\right| - |\theta_{C/A}| \qquad (1)$$

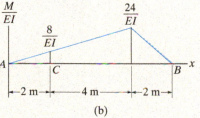

(b)

Moment-Area Theorems. Using Theorem 1, $\theta_{C/A}$ is equivalent to the area under the M/EI diagram between points A and C; that is,

$$\theta_{C/A} = \frac{1}{2}(2\text{ m})\left(\frac{8\text{ kN}\cdot\text{m}}{EI}\right) = \frac{8\text{ kN}\cdot\text{m}^2}{EI}$$

Applying Theorem 2, $t_{B/A}$ is equivalent to the moment of the area under the M/EI diagram between B and A about point B (the point on the elastic curve), since this is the point where the vertical distance is to be determined. We have

$$t_{B/A} = \left(2\text{ m} + \frac{1}{3}(6\text{ m})\right)\left[\frac{1}{2}(6\text{ m})\left(\frac{24\text{ kN}\cdot\text{m}}{EI}\right)\right]$$
$$+ \left(\frac{2}{3}(2\text{ m})\right)\left[\frac{1}{2}(2\text{ m})\left(\frac{24\text{ kN}\cdot\text{m}}{EI}\right)\right]$$
$$= \frac{320\text{ kN}\cdot\text{m}^3}{EI}$$

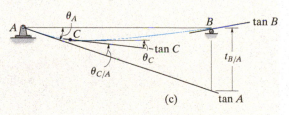

(c)

Fig. 12–24

Substituting these results into Eq. 1, we get

$$\theta_C = \frac{320\text{ kN}\cdot\text{m}^2}{(8\text{ m})EI} - \frac{8\text{ kN}\cdot\text{m}^2}{EI} = \frac{32\text{ kN}\cdot\text{m}^2}{EI} \,\Large\circlearrowright$$

Using a consistent set of units, we have

$$\theta_C = \frac{32\text{ kN}\cdot\text{m}^2}{[200(10^6)\text{ kN/m}^2[17(10^{-6})\text{ m}^4]} = 0.00941\text{ rad} \,\Large\circlearrowright \qquad \textit{Ans.}$$

12

EXAMPLE 12.11

Determine the displacement at C on the beam shown in Fig. 12–25a. EI is constant.

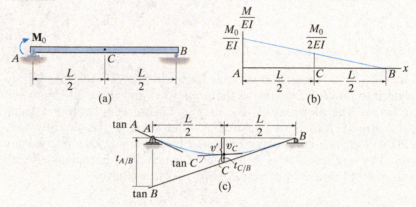

Fig. 12–25

SOLUTION

M/EI Diagram. See Fig. 12–25b.

Elastic Curve. The tangent at C is drawn on the elastic curve since we are required to find v_C, Fig. 12–25c. (Note that C is *not* the location of the maximum deflection of the beam, because the loading and hence the elastic curve are *not symmetric*.) Also indicated in Fig. 12–25c are the tangents at the supports A and B. It is seen that $v_C = v' - t_{C/B}$. If $t_{A/B}$ is determined, then v' can be found from proportional triangles, that is, $v'/(L/2) = t_{A/B}/L$ or $v' = t_{A/B}/2$. Hence,

$$v_C = \frac{t_{A/B}}{2} - t_{C/B} \qquad (1)$$

Moment-Area Theorem. Applying Theorem 2 to determine $t_{A/B}$ and $t_{C/B}$, we have

$$t_{A/B} = \left(\frac{1}{3}(L)\right)\left[\frac{1}{2}(L)\left(\frac{M_0}{EI}\right)\right] = \frac{M_0 L^2}{6EI}$$

$$t_{C/B} = \left(\frac{1}{3}\left(\frac{L}{2}\right)\right)\left[\frac{1}{2}\left(\frac{L}{2}\right)\left(\frac{M_0}{2EI}\right)\right] = \frac{M_0 L^2}{48EI}$$

Substituting these results into Eq. 1 gives

$$v_C = \frac{1}{2}\left(\frac{M_0 L^2}{6EI}\right) - \left(\frac{M_0 L^2}{48EI}\right)$$

$$= \frac{M_0 L^2}{16EI}\downarrow \qquad\qquad\qquad Ans.$$

12

EXAMPLE 12.12

Determine the displacement at point C on the steel overhanging beam shown in Fig. 12–26a. Take $E_{st} = 29(10^3)$ ksi, $I = 125$ in^4.

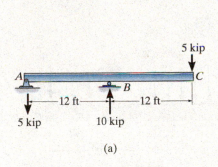

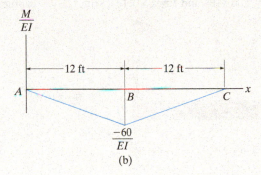

(a)   (b)

SOLUTION

M/EI Diagram.   See Fig. 12–26b.

Elastic Curve.   The loading causes the beam to deflect as shown in Fig. 12–26c. We are required to find v_C. By constructing tangents at C and at the supports A and B, it is seen that $v_C = |t_{C/A}| - v'$. However, v' can be related to $t_{B/A}$ by proportional triangles; that is, $v'/24 = |t_{B/A}|/12$ or $v' = 2|t_{B/A}|$. Hence

$$v_C = |t_{C/A}| - 2|t_{B/A}| \qquad (1)$$

Moment-Area Theorem.   Applying Theorem 2 to determine $t_{C/A}$ and $t_{B/A}$, we have

$$t_{C/A} = (12 \text{ ft})\left(\frac{1}{2}(24 \text{ ft})\left(-\frac{60 \text{ kip} \cdot \text{ft}}{EI}\right)\right)$$

$$= -\frac{8640 \text{ kip} \cdot \text{ft}^3}{EI}$$

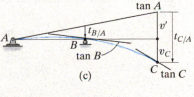

(c)

Fig. 12–26

$$t_{B/A} = \left(\frac{1}{3}(12 \text{ ft})\right)\left[\frac{1}{2}(12 \text{ ft})\left(-\frac{60 \text{ kip} \cdot \text{ft}}{EI}\right)\right] = -\frac{1440 \text{ kip} \cdot \text{ft}^3}{EI}$$

Why are these terms negative? Substituting the results into Eq. 1 yields

$$v_C = \frac{8640 \text{ kip} \cdot \text{ft}^3}{EI} - 2\left(\frac{1440 \text{ kip} \cdot \text{ft}^3}{EI}\right) = \frac{5760 \text{ kip} \cdot \text{ft}^3}{EI} \downarrow$$

Using a consistent set of units, we have

$$v_C = \frac{5760 \text{ kip} \cdot \text{ft}^3(1728 \text{ in}^3/\text{ft}^3)}{[29(10^3) \text{ kip/in}^2](125 \text{ in}^4)} = 2.75 \text{ in.} \downarrow \qquad \textit{Ans.}$$

12

FUNDAMENTAL PROBLEMS

F12–7. Determine the slope and deflection of end A of the cantilevered beam. $E = 200$ GPa and $I = 65.0(10^{-6})$ m^4.

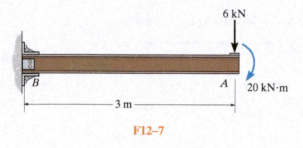

F12–7

F12–8. Determine the slope and deflection of end A of the cantilevered beam. $E = 200$ GPa and $I = 126(10^{-6})$ m^4.

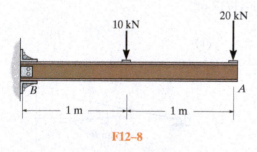

F12–8

F12–9. Determine the slope and deflection of end A of the cantilevered beam. $E = 200$ GPa and $I = 121(10^{-6})$ m^4.

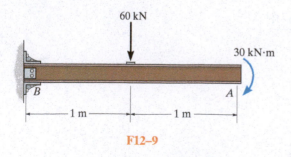

F12–9

F12–10. Determine the slope and deflection at A of the cantilevered beam. $E = 29(10^3)$ ksi, $I = 245$ in^4.

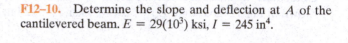

F12–10

F12–11. Determine the maximum deflection of the simply supported beam. $E = 200$ GPa and $I = 42.8(10^{-6})$ m^4.

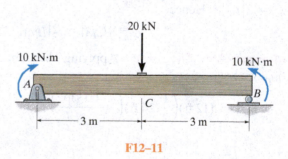

F12–11

F12–12. Determine the maximum deflection of the simply supported beam. $E = 200$ GPa and $I = 39.9(10^{-6})$ m^4.

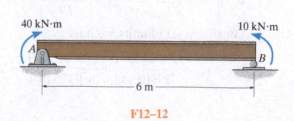

F12–12

PROBLEMS

12–51. Determine the slope and deflection at C. EI is constant.

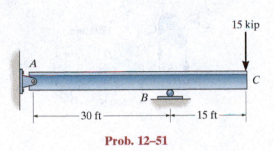

Prob. 12–51

15 kip
A
B
C
30 ft
15 ft

*12–52.** Determine the slope and deflection at C. EI is constant.

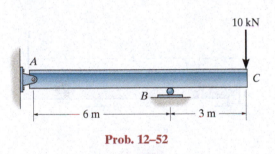

Prob. 12–52

10 kN
A
B
C
6 m
3 m

12–53. Determine the deflection of end B of the cantilever beam. EI is constant.

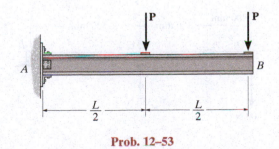

Prob. 12–53

P   P
A   B
$\frac{L}{2}$ $\frac{L}{2}$

12–54. Determine the slope at B and the deflection at C. EI is constant.

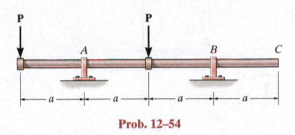

Prob. 12–54

P   P
A   B   C
a   a   a   a

12–55. The composite simply supported steel shaft is subjected to a force of 10 kN at its center. Determine its maximum deflection. $E_{st} = 200$ GPa.

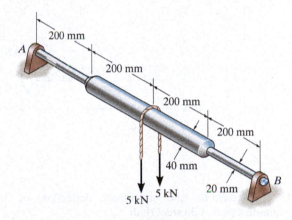

Prob. 12–55

200 mm
A
200 mm
200 mm
200 mm
40 mm
20 mm   B
5 kN   5 kN

*12–56.** Determine the slope of the shaft at A and the displacement at D. EI is constant.

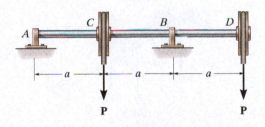

Prob. 12–56

C   B   D
A
a   a   a
P   P

12–57. The simply supported shaft has a moment of inertia of $2I$ for region BC and a moment of inertia I for regions AB and CD. Determine the maximum deflection of the shaft due to the load **P**. The modulus of elasticity is E.

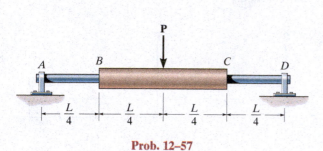

Prob. 12–57

12–58. Determine the deflection at C and the slope of the beam at A, B, and C. EI is constant.

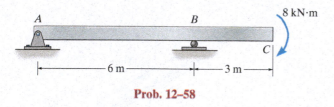

Prob. 12–58

12–59. Determine the maximum deflection of the 50-mm-diameter A-36 steel shaft.

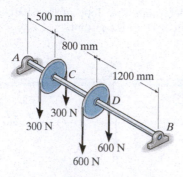

Prob. 12–59

*12–60. Determine the slope of the 50-mm-diameter A-36 steel shaft at the journal bearings at A and B. The bearings exert only vertical reactions on the shaft.

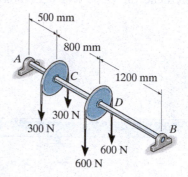

Prob. 12–60

12–61. Determine the position a of the roller support B in terms of L so that the deflection at end C is the same as the maximum deflection of region AB of the overhang beam. EI is constant.

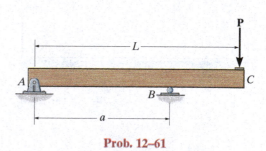

Prob. 12–61

12–62. Determine the slope of the 20-mm-diameter A-36 steel shaft at the journal bearings A and B.

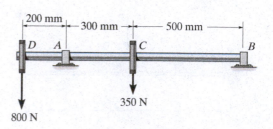

Prob. 12–62

12–63. Determine the slope and the deflection of end B of the cantilever beam. EI is constant.

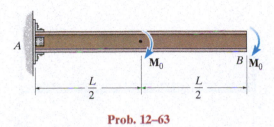

Prob. 12–63

*12–64. The two A-36 steel bars have a thickness of 1 in. and a width of 4 in. They are designed to act as a spring for the machine which exerts a force of 4 kip on them at A and B. If the supports exert only vertical forces on the bars, determine the maximum deflection of the bottom bar.

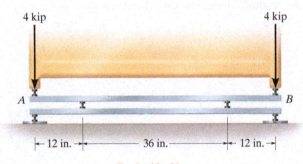

Prob. 12–64

12–65. Determine the slope at A and the displacement at C. Assume the support at A is a pin and B is a roller. EI is constant.

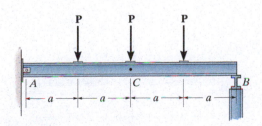

Prob. 12–65

12–66. Determine the deflection at C and the slopes at the bearings A and B. EI is constant.

12–67. Determine the maximum deflection within region AB. EI is constant.

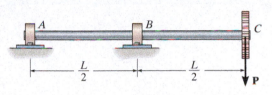

Probs. 12–66/67

*12–68. Determine the slope at A and the maximum deflection of the simply supported beam. EI is constant.

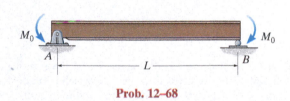

Prob. 12–68

12–69. Determine the slope at C and the deflection at B. EI is constant.

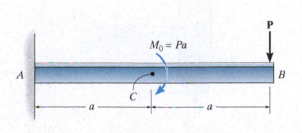

Prob. 12–69

12

12–70. Determine the slope at A and the maximum deflection of the beam. EI is constant.

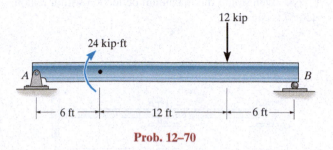

Prob. 12–70

12–71. Determine the displacement of the 20-mm-diameter A-36 steel shaft at D.

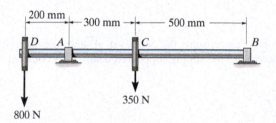

Prob. 12–71

12–72. The two force components act on the tire of the automobile. The tire is fixed to the axle, which is supported by bearings at A and B. Determine the maximum deflection of the axle. Assume that the bearings resist only vertical loads. The thrust on the axle is resisted at C. The axle has a diameter of 1.25 in. and is made of A-36 steel. Neglect the effect of axial load on deflection.

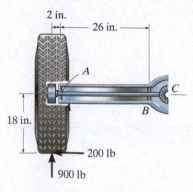

Prob. 12–72

12–73. At what distance a should the journal bearing supports at A and B be placed so that the deflection at the center of the shaft is equal to the deflection at its ends? EI is constant.

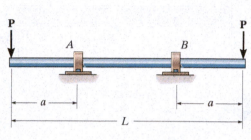

Prob. 12–73

12–74. The rod is constructed from two shafts for which the moment of inertia of AB is I and for BC it is $2I$. Determine the maximum slope and deflection of the rod due to the loading. The modulus of elasticity is E.

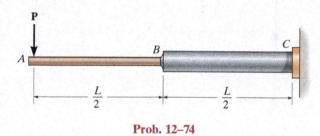

Prob. 12–74

12–75. Determine the slope at B and the deflection at C of the beam. $E = 200$ GPa and $I = 65.0(10^6)$ mm⁴.

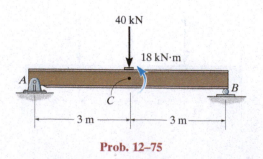

Prob. 12–75

*12–76. Determine the slope at point A and the maximum deflection of the simply supported beam. The beam is made of material having a modulus of elasticity E. The moment of inertia of segments AB and CD of the beam is I, and the moment of inertia of segment BC is $2I$.

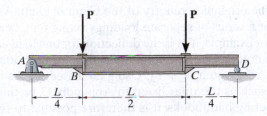

Prob. 12–76

12–77. Determine the position a of roller support B in terms of L so that deflection at end C is the same as the maximum deflection in region AB of the overhang beam. EI is constant.

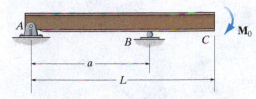

Prob. 12–77

12–78. Determine the slope at B and deflection at C. EI is constant.

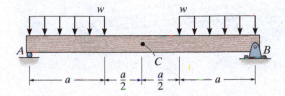

Prob. 12–78

12–79. Determine the slope and displacement at C. EI is constant.

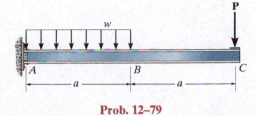

Prob. 12–79

*12–80. Determine the slope at C and deflection at B. EI is constant.

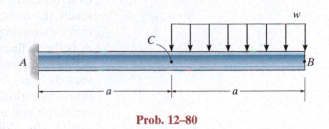

Prob. 12–80

12–81. The two bars are pin connected at D. Determine the slope at A and the deflection at D. EI is constant.

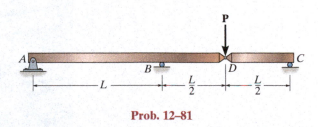

Prob. 12–81

12–82. Determine the maximum deflection of the beam. EI is constant.

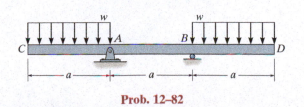

Prob. 12–82

12

12.5   METHOD OF SUPERPOSITION

The differential equation $EI\, d^4v/dx^4 = w(x)$ satisfies the two necessary requirements for applying the principle of superposition; i.e., the load $w(x)$ is linearly related to the deflection $v(x)$, and the load is assumed not to significantly change the original geometry of the beam or shaft. As a result, the deflections for a series of separate loadings acting on a beam may be superimposed. For example, if v_1 is the deflection for one load and v_2 is the deflection for another load, the total deflection for both loads acting together is the algebraic sum $v_1 + v_2$. Using tabulated results for various beam loadings, such as the ones listed in Appendix C, or those found in various engineering handbooks, it is therefore possible to find the slope and displacement at a point on a beam subjected to several loadings by adding the effects of each loading.

The following examples numerically illustrate how to do this.

The resultant deflection at any point on this beam can be determined from the superposition of the deflections caused by each of the separate loadings acting on the beam.

EXAMPLE 12.13

Determine the displacement at point C and the slope at the support A of the beam shown in Fig. 12–27a. EI is constant.

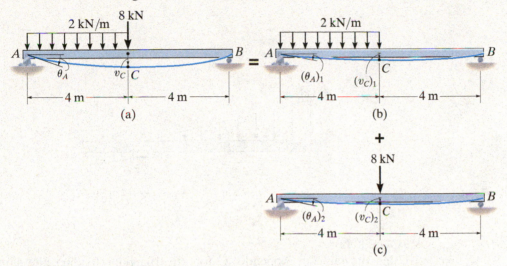

Fig. 12–27

SOLUTION

The loading can be separated into two component parts as shown in Figs. 12–27b and 12–27c. The displacement at C and slope at A are found using the table in Appendix C for each part.

For the distributed loading,

$$(\theta_A)_1 = \frac{3wL^3}{128EI} = \frac{3(2 \text{ kN/m})(8 \text{ m})^3}{128EI} = \frac{24 \text{ kN} \cdot \text{m}^2}{EI} \curvearrowright$$

$$(v_C)_1 = \frac{5wL^4}{768EI} = \frac{5(2 \text{ kN/m})(8 \text{ m})^4}{768EI} = \frac{53.33 \text{ kN} \cdot \text{m}^3}{EI} \downarrow$$

For the 8-kN concentrated force,

$$(\theta_A)_2 = \frac{PL^2}{16EI} = \frac{8 \text{ kN}(8 \text{ m})^2}{16EI} = \frac{32 \text{ kN} \cdot \text{m}^2}{EI} \curvearrowright$$

$$(v_C)_2 = \frac{PL^3}{48EI} = \frac{8 \text{ kN}(8 \text{ m})^3}{48EI} = \frac{85.33 \text{ kN} \cdot \text{m}^3}{EI} \downarrow$$

The displacement at C and the slope at A are the algebraic sums of these components. Hence,

$$(+\curvearrowright) \qquad \theta_A = (\theta_A)_1 + (\theta_A)_2 = \frac{56 \text{ kN} \cdot \text{m}^2}{EI} \curvearrowright \qquad \textit{Ans.}$$

$$(+\downarrow) \qquad v_C = (v_C)_1 + (v_C)_2 = \frac{139 \text{ kN} \cdot \text{m}^3}{EI} \downarrow \qquad \textit{Ans.}$$

EXAMPLE 12.14

Determine the displacement at the end *C* of the cantilever beam shown in Fig. 12–28. *EI* is constant.

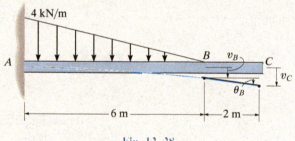

Fig. 12–28

SOLUTION
Using the table in Appendix C for the triangular loading, the slope and displacement at point *B* are

$$\theta_B = \frac{w_0 L^3}{24EI} = \frac{4 \text{ kN/m}(6 \text{ m})^3}{24EI} = \frac{36 \text{ kN} \cdot \text{m}^2}{EI}$$

$$v_B = \frac{w_0 L^4}{30EI} = \frac{4 \text{ kN/m}(6 \text{ m})^4}{30EI} = \frac{172.8 \text{ kN} \cdot \text{m}^3}{EI}$$

The unloaded region *BC* of the beam remains straight, as shown in Fig. 12–28. Since θ_B is small, the displacement at *C* becomes

$$(+\downarrow) \qquad v_C = v_B + \theta_B(L_{BC})$$

$$= \frac{172.8 \text{ kN} \cdot \text{m}^3}{EI} + \frac{36 \text{ kN} \cdot \text{m}^2}{EI}(2 \text{ m})$$

$$= \frac{244.8 \text{ kN} \cdot \text{m}^3}{EI} \downarrow \qquad\qquad \textit{Ans.}$$

EXAMPLE 12.15

Determine the displacement at the end C of the overhanging beam shown in Fig. 12–29a. EI is constant.

SOLUTION

Since the table in Appendix C *does not* include beams with overhangs, the beam will be separated into a simply supported and a cantilevered portion. First we will calculate the slope at B, as caused by the distributed load acting on the simply supported span, Fig. 12–29b.

$$(\theta_B)_1 = \frac{wL^3}{24EI} = \frac{5 \text{ kN/m}(4 \text{ m})^3}{24EI} = \frac{13.33 \text{ kN} \cdot \text{m}^2}{EI} \curvearrowright$$

Since this angle is *small*, the vertical displacement at point C is

$$(v_C)_1 = (2 \text{ m})\left(\frac{13.33 \text{ kN} \cdot \text{m}^2}{EI}\right) = \frac{26.67 \text{ kN} \cdot \text{m}^3}{EI}\uparrow$$

Next, the 10-kN load on the overhang causes a statically equivalent force of 10 kN and couple moment of 20 kN · m at the support B of the simply supported span, Fig. 12–29c. The 10-kN force does not cause a slope at B; however, the 20-kN · m couple moment does cause a slope. This slope is

$$(\theta_B)_2 = \frac{M_0 L}{3EI} = \frac{20 \text{ kN} \cdot \text{m}(4 \text{ m})}{3EI} = \frac{26.67 \text{ kN} \cdot \text{m}^2}{EI}\curvearrowleft$$

so that the displacement of point C is

$$(v_C)_2 = (2 \text{ m})\left(\frac{26.7 \text{ kN} \cdot \text{m}^2}{EI}\right) = \frac{53.33 \text{ kN} \cdot \text{m}^3}{EI}\downarrow$$

Finally, the cantilevered portion BC is displaced by the 10-kN force, Fig. 12–29d. We have

$$(v_C)_3 = \frac{PL^3}{3EI} = \frac{10 \text{ kN}(2 \text{ m})^3}{3EI} = \frac{26.67 \text{ kN} \cdot \text{m}^3}{EI}\downarrow$$

Summing these results algebraically, we get

$$(+\downarrow)\qquad v_C = -\frac{26.7}{EI} + \frac{53.3}{EI} + \frac{26.7}{EI} = \frac{53.3 \text{ kN} \cdot \text{m}^3}{EI}\downarrow \qquad \textit{Ans.}$$

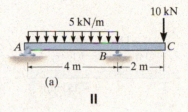

(a)

$\parallel$

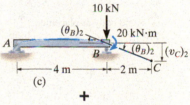

(b)

$+$

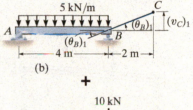

(c)

$+$

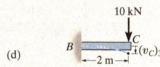

(d)

Fig. 12–29

EXAMPLE 12.16

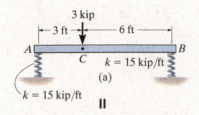

(a)

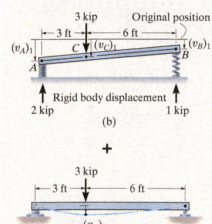

(b)

+

(c)

Fig. 12–30

The steel bar shown in Fig. 12–30a is supported by two springs at its ends A and B. Each spring has a stiffness of $k = 15$ kip/ft and is originally unstretched. If the bar is loaded with a force of 3 kip at point C, determine the vertical displacement of the force. Neglect the weight of the bar and take $E_{st} = 29(10^3)$ ksi, $I = 12$ in^4.

SOLUTION

The end reactions at A and B are calculated and shown in Fig. 12–30b. Each spring deflects by an amount

$$(v_A)_1 = \frac{2 \text{ kip}}{15 \text{ kip/ft}} = 0.1333 \text{ ft}$$

$$(v_B)_1 = \frac{1 \text{ kip}}{15 \text{ kip/ft}} = 0.0667 \text{ ft}$$

If the bar is considered to be *rigid*, then the vertical displacement at C is

$$(v_C)_1 = (v_B)_1 + \frac{6 \text{ ft}}{9 \text{ ft}}[(v_A)_1 - (v_B)_1]$$

$$= 0.0667 \text{ ft} + \frac{2}{3}[0.1333 \text{ ft} - 0.0667 \text{ ft}] = 0.1111 \text{ ft} \downarrow$$

We can find the displacement at C caused by the *deformation* of the bar, Fig. 12–30c, by using the table in Appendix C. We have

$$(v_C)_2 = \frac{Pab}{6EIL}(L^2 - b^2 - a^2)$$

$$= \frac{3 \text{ kip}(3 \text{ ft})(6 \text{ ft})[(9 \text{ ft})^2 - (6 \text{ ft})^2 - (3 \text{ ft})^2]}{6[29(10^3)\text{kip/in}^2](144 \text{ in}^2/1 \text{ ft}^2)(12 \text{ in}^4)(1 \text{ ft}^4/20 \text{ 736 in}^4)(9 \text{ ft})}$$

$$= 0.0149 \text{ ft} \downarrow$$

Adding the two displacement components, we get

$$(+\downarrow) \qquad v_C = 0.1111 \text{ ft} + 0.0149 \text{ ft} = 0.126 \text{ ft} = 1.51 \text{ in.} \downarrow \qquad Ans.$$

PROBLEMS

12–83. The W10 × 15 cantilevered beam is made of A-36 steel and is subjected to the loading shown. Determine the slope and displacement at its end B.

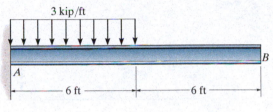

3 kip/ft

A B

|— 6 ft —|— 6 ft —|

Prob. 12–83

*__12–84.__ The W10 × 15 cantilevered beam is made of A-36 steel and is subjected to the loading shown. Determine the displacement at B and the slope at B.

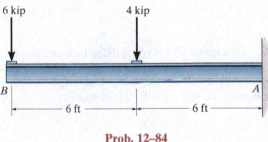

6 kip   4 kip

B A

|— 6 ft —|— 6 ft —|

Prob. 12–84

12–85. The W14 × 43 simply supported beam is made of A992 steel and is subjected to the loading shown. Determine the deflection at its center C.

12–86. The W14 × 43 simply supported beam is made of A992 steel and is subjected to the loading shown. Determine the slope at A and B.

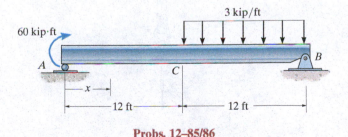

60 kip·ft

3 kip/ft

A C B

|— x —|

|— 12 ft —|— 12 ft —|

Probs. 12–85/86

12–87. The W14 × 43 simply supported beam is made of A-36 steel and is subjected to the loading shown. Determine the deflection at its center C.

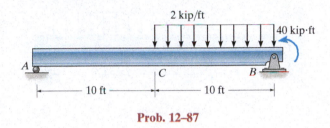

2 kip/ft

40 kip·ft

A C B

|— 10 ft —|— 10 ft —|

Prob. 12–87

*__12–88.__ The W14 × 43 simply supported beam is made of A-36 steel and is subjected to the loading shown. Determine the slope at A and B.

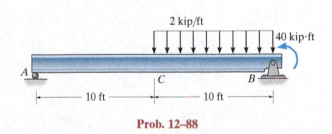

2 kip/ft

40 kip·ft

A C B

|— 10 ft —|— 10 ft —|

Prob. 12–88

12–89. The W8 × 48 cantilevered beam is made of A-36 steel and is subjected to the loading shown. Determine the displacement at C and the slope at A.

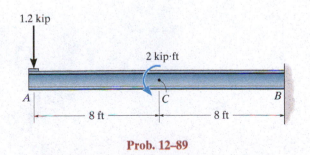

1.2 kip

2 kip·ft

A C B

|— 8 ft —|— 8 ft —|

Prob. 12–89

12

12–90. The beam supports the loading shown. Code restrictions, due to a plaster ceiling, require the maximum deflection not to exceed 1/360 of the span length. Select the lightest-weight A-36 steel wide-flange beam from Appendix B that will satisfy this requirement and safely support the load. The allowable bending stress is $\sigma_{allow} = 24$ ksi and the allowable shear stress is $\tau_{allow} = 14$ ksi. Assume A is a roller and B is a pin.

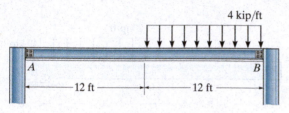

Prob. 12–90

12–91. The W24 × 104 A-36 steel beam is used to support the uniform distributed load and a concentrated force which is applied at its end. If the force acts at an angle with the vertical as shown, determine the horizontal and vertical displacement at A.

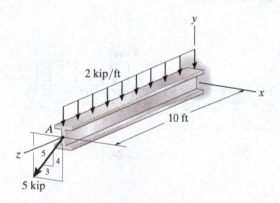

Prob. 12–91

*12–92. The W8 × 48 cantilevered beam is made of A-36 steel and is subjected to the loading shown. Determine the displacement at its end A.

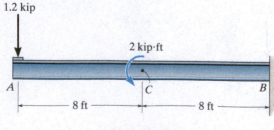

Prob. 12–92

12–93. The rod is pinned at its end A and attached to a torsional spring having a stiffness k, which measures the torque per radian of rotation of the spring. If a force $\mathbf{P}$ is always applied perpendicular to the end of the rod, determine the displacement of the force. EI is constant.

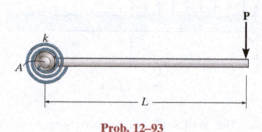

Prob. 12–93

12–94. Determine the vertical deflection and the change in angle at the end A of the bracket. Assume that the bracket is fixed supported at its base, and neglect the axial deformation of segment AB. EI is constant.

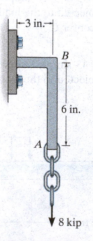

Prob. 12–94

12–95. The pipe assembly consists of three equal-sized pipes with flexibility stiffness EI and torsional stiffness GJ. Determine the vertical deflection at A.

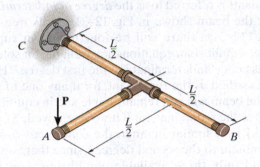

Prob. 12–95

12–97. The relay switch consists of a thin metal strip or armature AB that is made of red brass C83400 and is attracted to the solenoid S by a magnetic field. Determine the smallest force F required to attract the armature at C in order that contact is made at the free end B. Also, what should the distance a be for this to occur? The armature is fixed at A and has a moment of inertia of $I = 0.18(10^{-12})$ m⁴.

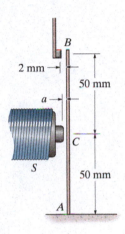

Prob. 12–97

*12–96. The assembly consists of a cantilevered beam CB and a simply supported beam AB. If each beam is made of A-36 steel and has a moment of inertia about its principal axis of $I_x = 118$ in⁴, determine the displacement at the center D of beam BA.

12–98. Determine the moment M_0 in terms of the load P and dimension a so that the deflection at the center of the shaft is zero. EI is constant.

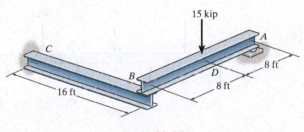

Prob. 12–96

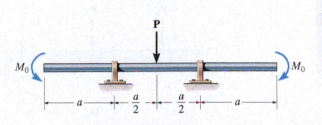

Prob. 12–98

12.6   STATICALLY INDETERMINATE BEAMS AND SHAFTS

In this section we will illustrate a general method for determining the reactions on a statically indeterminate beam or shaft. Specifically, a member is *statically indeterminate* if the number of unknown reactions *exceeds* the available number of equilibrium equations.

The additional support reactions on a beam (or shaft) that are *not needed* to keep it in stable equilibrium are called *redundants*, and the number of these redundants is referred to as the *degree of indeterminacy*. For example, consider the beam shown in Fig. 12–31a. If its free-body diagram is drawn, Fig. 12–31b, there will be four unknown support reactions, and since three equilibrium equations are available for solution, the beam is classified as being "indeterminate to the first degree." Either A_y, B_y, or M_A can be classified as the redundant, for if any one of these reactions is removed, the beam will still remain stable and in equilibrium (A_x cannot be classified as the redundant, for if it were removed, $\Sigma F_x = 0$ would not be satisfied.) In a similar manner, the *continuous beam* in Fig. 12–32a is "indeterminate to the second degree," since there are five unknown reactions and only three available equilibrium equations, Fig. 12–32b. Here any two redundant support reactions can be chosen among A_y, B_y, C_y, and D_y.

The reactions on a beam that is statically indeterminate must satisfy both the equations of equilibrium and the compatibility requirements at the supports. In the following sections we will illustrate how this is done using the method of integration, Sec. 12.7; the moment-area method, Sec. 12.8; and the method of superposition, Sec. 12.9.

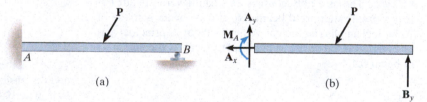

(a)   (b)

Fig. 12–31

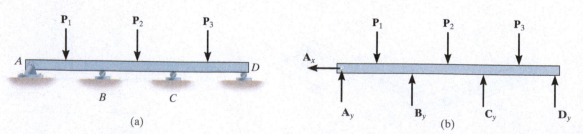

(a)   (b)

Fig. 12–32

12.7 STATICALLY INDETERMINATE BEAMS AND SHAFTS—METHOD OF INTEGRATION

The method of integration, discussed in Sec. 12.2, requires applying the load–displacement relationship, $d^2v/dx^2 = M/EI$, to obtain the elastic curve for the beam. If the beam is statically indeterminate, then M will be expressed in terms of both its position x *and* some of the unknown support reactions. Although this will occur, there will be additional boundary conditions available for solution.

The following example problems illustrate applications of the integration method using the procedure for analysis outlined in Sec. 12.2.

An example of a statically indeterminate beam used to support a bridge deck.

12

EXAMPLE 12.17

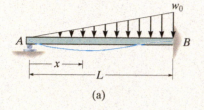

(a)

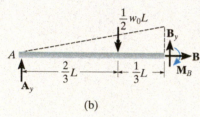

(b)

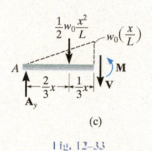

(c)

Fig. 12-33

The beam is subjected to the distributed loading shown in Fig. 12–33a. Determine the reaction at A. EI is constant.

SOLUTION

Elastic Curve.   The beam deflects as shown in Fig. 12–33a. Only one coordinate x is needed. For convenience we will take it directed to the right, since the internal moment is easy to formulate.

Moment Function.   The beam is indeterminate to the first degree as indicated from the free-body diagram, Fig. 12–33b. If we choose the segment shown in Fig. 12–33c, then the internal moment M will be a function of x, written in terms of the redundant force A_y.

$$M = A_y x - \frac{1}{6} w_0 \frac{x^3}{L}$$

Slope and Elastic Curve.   Applying Eq. 12–10, we have

$$EI \frac{d^2v}{dx^2} = A_y x - \frac{1}{6} w_0 \frac{x^3}{L}$$

$$EI \frac{dv}{dx} = \frac{1}{2} A_y x^2 - \frac{1}{24} w_0 \frac{x^4}{L} + C_1$$

$$EIv = \frac{1}{6} A_y x^3 - \frac{1}{120} w_0 \frac{x^5}{L} + C_1 x + C_2$$

The *three* unknowns A_y, C_1, and C_2 are determined from the *three* boundary conditions $x = 0$, $v = 0$; $x = L$, $dv/dx = 0$; and $x = L$, $v = 0$. Applying these conditions we get

$$x = 0, v = 0; \qquad 0 = 0 - 0 + 0 + C_2$$

$$x = L, \frac{dv}{dx} = 0; \qquad 0 = \frac{1}{2} A_y L^2 - \frac{1}{24} w_0 L^3 + C_1$$

$$x = L, v = 0; \qquad 0 = \frac{1}{6} A_y L^3 - \frac{1}{120} w_0 L^4 + C_1 L + C_2$$

Solving,

$$A_y = \frac{1}{10} w_0 L \qquad\qquad Ans.$$

$$C_1 = -\frac{1}{120} w_0 L^3 \quad C_2 = 0$$

NOTE: Using the result for A_y, the reactions at B can now be determined from the equations of equilibrium, Fig. 12–33b. Show that $B_x = 0$, $B_y = 2w_0 L/5$, and $M_B = w_0 L^2/15$.

EXAMPLE 12.18

The beam in Fig. 12–34a is fixed supported at both ends and is subjected to the uniform loading shown. Determine the reactions at the supports. Neglect the effect of axial load.

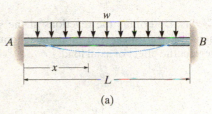

(a)

SOLUTION

Elastic Curve. The beam deflects as shown in Fig. 12–34a. As in the previous problem, only one x coordinate is necessary for the solution since the loading is continuous across the span.

Moment Function. From the free-body diagram, Fig. 12–34b, the respective shear and moment reactions at A and B must be equal, since there is symmetry of both loading and geometry. Because of this, the equation of equilibrium, $\Sigma F_y = 0$, requires

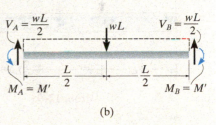

(b)

$$V_A = V_B = \frac{wL}{2} \qquad \text{Ans.}$$

The beam is indeterminate to the first degree, with M' the redundant at each end. Using the beam segment shown in Fig. 12–34c, the internal moment becomes

$$M = \frac{wL}{2}x - \frac{w}{2}x^2 - M'$$

Slope and Elastic Curve. Applying Eq. 12–10, we have

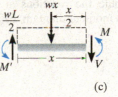

(c)

Fig. 12–34

$$EI\frac{d^2v}{dx^2} = \frac{wL}{2}x - \frac{w}{2}x^2 - M'$$

$$EI\frac{dv}{dx} = \frac{wL}{4}x^2 - \frac{w}{6}x^3 - M'x + C_1$$

$$EIv = \frac{wL}{12}x^3 - \frac{w}{24}x^4 - \frac{M'}{2}x^2 + C_1 x + C_2$$

The *three* unknowns, M', C_1, and C_2, can be determined from the *three* boundary conditions $v = 0$ at $x = 0$, which yields $C_2 = 0$; $dv/dx = 0$ at $x = 0$, which yields $C_1 = 0$; and $v = 0$ at $x = L$, which yields

$$M' = \frac{wL^2}{12} \qquad \text{Ans.}$$

Notice that because of symmetry the remaining boundary condition $dv/dx = 0$ at $x = L$ is automatically satisfied.

12

PROBLEMS

12–99. Determine the reactions at the supports A and B, then draw the shear and moment diagram. EI is constant. Neglect the effect of axial load.

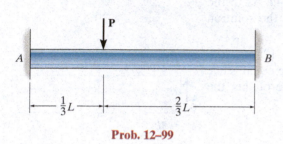

Prob. 12–99

*12–100. Determine the reactions at the supports, then draw the shear and moment diagram. EI is constant.

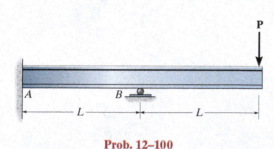

Prob. 12–100

12–101. Determine the reactions at the supports A, B, and C, then draw the shear and moment diagrams. EI is constant.

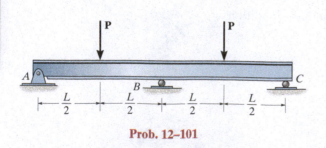

Prob. 12–101

12–102. Determine the reactions at the supports A and B, then draw the shear and moment diagrams. EI is constant.

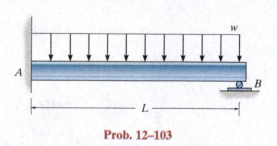

Prob. 12–102

12–103. Determine the reactions at the supports A and B, then draw the shear and moment diagrams. EI is constant.

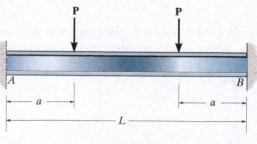

Prob. 12–103

*12–104. Determine the moment reactions at the supports A and B. EI is constant.

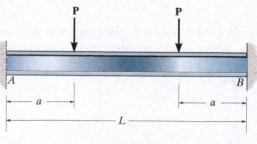

Prob. 12–104

12–105. Determine the reactions at the supports A and B, then draw the moment diagram. EI is constant.

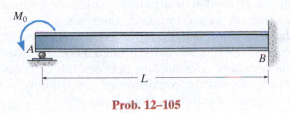

Prob. 12–105

12–106. Determine the reactions at the support A and B. EI is constant.

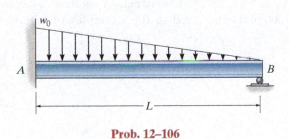

Prob. 12–106

12–107. Determine the reactions at roller support A and fixed support B.

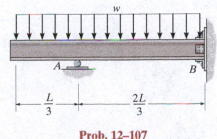

Prob. 12–107

***12–108.** Determine the moment reactions at the supports A and B, then draw the shear and moment diagrams. Solve by expressing the internal moment in the beam in terms of A_y and M_A. EI is constant.

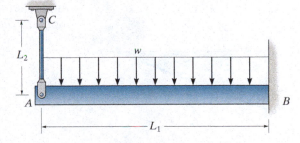

Prob. 12–108

12–109. The beam has a constant $E_1 I_1$ and is supported by the fixed wall at B and the rod AC. If the rod has a cross-sectional area A_2 and the material has a modulus of elasticity E_2, determine the force in the rod.

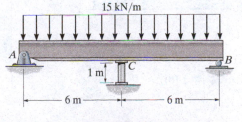

Prob. 12–109

12–110. The beam is supported by a pin at A, a roller at B, and a post having a diameter of 50 mm at C. Determine the support reactions at A, B, and C. The post and the beam are made of the same material having a modulus of elasticity $E = 200$ GPa, and the beam has a constant moment of inertia $I = 255(10^6)$ mm^4.

Prob. 12–110

*12.8   STATICALLY INDETERMINATE BEAMS AND SHAFTS— MOMENT-AREA METHOD

If the moment-area method is used to determine the unknown redundants of a statically indeterminate beam or shaft, then when the M/EI diagram is drawn, the redundants will be represented as unknowns on this diagram. However, by applying the moment-area theorems it will be possible to obtain the necessary relationships between the tangents on the elastic curve in order to meet the conditions of compatibility at the supports, and thereby obtain a solution for the redundants.

Moment Diagrams Constructed by the Method of Superposition.

Since application of the moment-area theorems requires calculation of both the area under the M/EI diagram and the centroidal location of this area, it is often convenient to use the method of superposition and use *separate M/EI diagrams* for *each* of the loads on the beam, rather than using the *resultant diagram* to calculate these geometric quantities.

Most loadings on cantilevered *beams* will be a combination of the four loadings shown in Fig. 12–35. Construction of their associated moment diagrams has been discussed in the examples of Chapter 6.

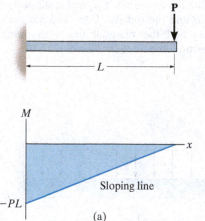

Sloping line

$-PL$

(a)

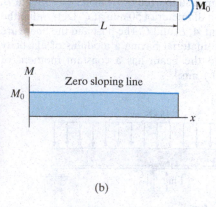

Zero sloping line

M_0

(b)

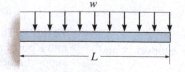

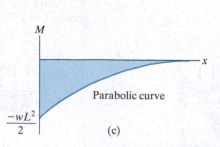

Parabolic curve

$-\dfrac{wL^2}{2}$

(c)

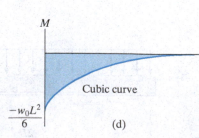

Cubic curve

$-\dfrac{w_0L^2}{6}$

(d)

Fig. 12–35

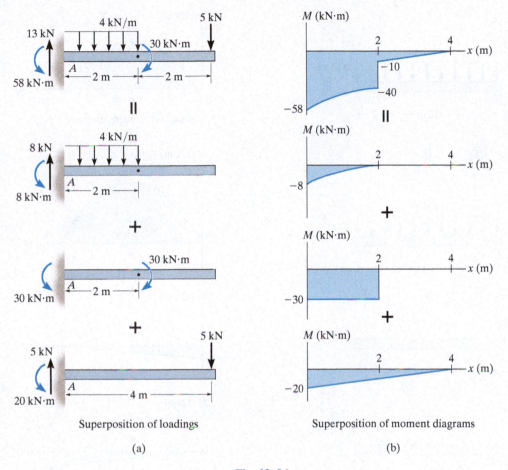

Superposition of loadings   Superposition of moment diagrams

(a)   (b)

Fig. 12–36

With these results, the method of superposition can then be used to represent a moment diagram for a beam by a series of separate moment diagrams. For example, the three loadings on the cantilevered beam shown in Fig. 12–36a are statically equivalent to the three separate loadings on the cantilevered beams below it. Thus, if the moment diagrams for each separate beam are drawn, Fig. 12–36b, the superposition of these diagrams will yield the moment diagram for the beam shown at the top. Obviously, it is easier to find the area and location of the centroid for each part rather than doing this for the resultant diagram.

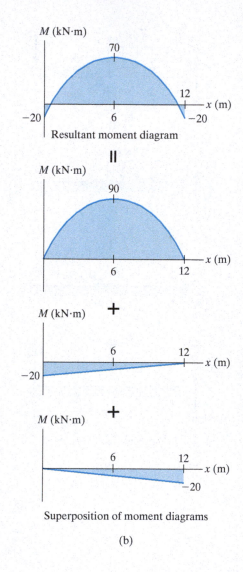

Fig. 12–37

In a similar manner, we can represent the resultant moment diagram for a *simply supported beam* by using the superposition of the moment diagrams for each of its loadings. For example, the beam loading shown at the top of Fig. 12–37a is equivalent to the sum of the beam loadings shown below it. Again, it is easier to sum the calculations of the areas and centroidal locations for the three moment diagrams rather than doing this for the moment diagram shown at the top of Fig. 12–37b.

The examples that follow should also clarify some of these points and illustrate how to use the moment-area theorems to directly obtain a specific redundant reaction on a statically indeterminate beam. The solutions follow the procedure for analysis outlined in Sec. 12.4.

EXAMPLE 12.19

The beam is subjected to the concentrated force shown in Fig. 12–38a. Determine the reactions at the supports. EI is constant.

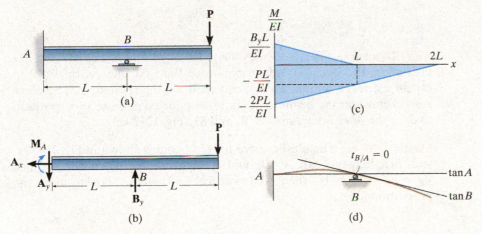

Fig. 12–38

SOLUTION

M/EI Diagram. The free-body diagram is shown in Fig. 12–38b. Assuming the beam to be cantilevered from A, and using the method of superposition, the separate M/EI diagrams for the redundant reaction $\mathbf{B}_y$ and the load $\mathbf{P}$ are shown in Fig. 12–38c.

Elastic Curve. The elastic curve for the beam is shown in Fig. 12–38d. The tangents at the supports A and B have been constructed. Since $v_B = 0$, then

$$t_{B/A} = 0$$

Moment-Area Theorem. Applying Theorem 2, we have

$$t_{B/A} = \left(\frac{2}{3}L\right)\left[\frac{1}{2}\left(\frac{B_yL}{EI}\right)L\right] + \left(\frac{L}{2}\right)\left[\frac{-PL}{EI}(L)\right]$$
$$+ \left(\frac{2}{3}L\right)\left[\frac{1}{2}\left(\frac{-PL}{EI}\right)(L)\right] = 0$$

$$B_y = 2.5P \qquad\qquad Ans.$$

Equations of Equilibrium. Using this result, the reactions at A on the free-body diagram, Fig. 12–38b, are

$$\xrightarrow{+} \Sigma F_x = 0; \qquad\qquad A_x = 0 \qquad\qquad Ans.$$
$$+\uparrow \Sigma F_y = 0; \qquad -A_y + 2.5P - P = 0$$
$$A_y = 1.5P \qquad\qquad Ans.$$
$$\zeta + \Sigma M_A = 0; \qquad -M_A + 2.5P(L) - P(2L) = 0$$
$$M_A = 0.5PL \qquad\qquad Ans.$$

12

EXAMPLE 12.20

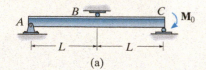

(a)

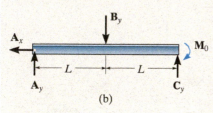

(b)

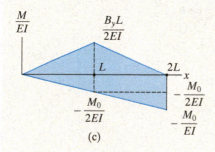

(c)

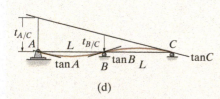

(d)

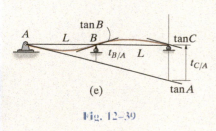

(e)

Fig. 12–39

The beam is subjected to the couple moment at its end C as shown in Fig. 12–39a. Determine the reaction at B. EI is constant.

SOLUTION

M/EI Diagram. The free-body diagram is shown in Fig. 12–39b. By inspection, the beam is indeterminate to the first degree. In order to obtain a direct solution, we will choose $\mathbf{B}_y$ as the redundant. Therefore we will consider the beam to be simply supported and use superposition to draw the M/EI diagrams for $\mathbf{B}_y$ and $\mathbf{M}_0$, Fig. 12–39c.

Elastic Curve. The elastic curve for the beam is shown in Fig. 12–39d, and the tangents at A, B, and C have been established. Since $v_A = v_B = v_C = 0$, then the vertical distances shown must be proportional; i.e.,

$$t_{B/C} = \frac{1}{2} t_{A/C} \qquad (1)$$

From Fig. 12–39c, we have

$$t_{B/C} = \left(\frac{1}{3}L\right)\left[\frac{1}{2}\left(\frac{B_yL}{2EI}\right)(L)\right] + \left(\frac{2}{3}L\right)\left[\frac{1}{2}\left(\frac{-M_0}{2EI}\right)(L)\right]$$
$$+ \left(\frac{L}{2}\right)\left[\left(\frac{-M_0}{2EI}\right)(L)\right]$$

$$t_{A/C} = (L)\left[\frac{1}{2}\left(\frac{B_yL}{2EI}\right)(2L)\right] + \left(\frac{2}{3}(2L)\right)\left[\frac{1}{2}\left(\frac{-M_0}{EI}\right)(2L)\right]$$

Substituting into Eq. 1 and simplifying yields

$$B_y = \frac{3M_0}{2L} \qquad Ans.$$

Equations of Equilibrium. The reactions at A and C can now be determined from the equations of equilibrium, Fig. 12–39b. Show that $A_x = 0$, $C_y = 5M_0/4L$, and $A_y = M_0/4L$.

Note from Fig. 12–39e that this problem can also be worked in terms of the vertical distances,

$$t_{B/A} = \frac{1}{2} t_{C/A}$$

PROBLEMS

12–111. Determine the moment reactions at the supports A and B. EI is constant.

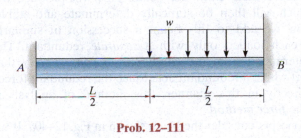

Prob. 12–111

*12–112. Determine the reaction at the supports, then draw the shear and moment diagrams. EI is constant.

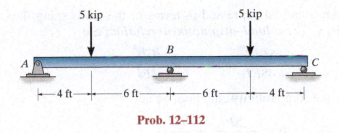

Prob. 12–112

12–113. Determine the vertical reaction at the journal bearing support A and at the fixed support B, then draw the shear and moment diagrams for the beam. EI is constant.

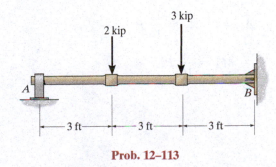

Prob. 12–113

12–114. Determine the reactions at the supports A and B, then draw the shear and moment diagrams. EI is constant.

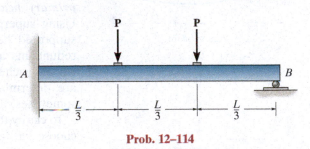

Prob. 12–114

12–115. Determine the reactions at the supports. EI is constant.

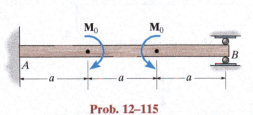

Prob. 12–115

*12–116. Determine the vertical reaction at the journal bearing support A and at the fixed support B, then draw the shear and moment diagrams for the shaft. EI is constant.

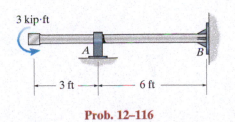

Prob. 12–116

12

12.9 STATICALLY INDETERMINATE BEAMS AND SHAFTS—METHOD OF SUPERPOSITION

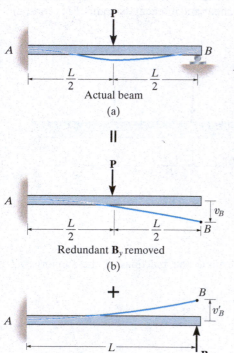

Actual beam
(a)

$\parallel$

Redundant $\mathbf{B}_y$ removed
(b)

$+$

Only redundant $\mathbf{B}_y$ applied
(c)

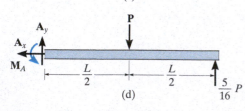

(d)

Fig. 12–40

In order to use the method of superposition to solve for the reactions on a statically indeterminate beam, it is first necessary to identify the redundants and remove them from the beam. This will produce the *primary beam*, which will then be statically determinate and stable. Using superposition, we add to this beam a succession of similarly supported beams, each loaded only with a *separate* redundant. The redundants are determined from the *conditions of compatibility* that exist at each support where a redundant acts. Since the redundant forces are determined directly in this manner, this method of analysis is sometimes called the *force method*.

To clarify these concepts, consider the beam shown in Fig. 12–40a. If we choose the reaction $\mathbf{B}_y$ at the roller as the redundant, then the primary beam is shown in Fig. 12–40b, and the beam with the redundant $\mathbf{B}_y$ acting on it is shown in Fig. 12–40c. The displacement at the roller is to be zero, and since the displacement of B on the primary beam is v_B, and $\mathbf{B}_y$ causes B to be displaced upward v'_B, we can write the compatibility equation at B as

$$(+\uparrow) \qquad\qquad 0 = -v_B + v'_B$$

These displacements can be expressed in terms of the loads using the table in Appendix C. These *load–displacement relations* are

$$v_B = \frac{5PL^3}{48EI} \quad \text{and} \quad v'_B = \frac{B_yL^3}{3EI}$$

Substituting into the compatibility equation, we get

$$0 = -\frac{5PL^3}{48EI} + \frac{B_y L^3}{3EI}$$

$$B_y = \frac{5}{16}P$$

Now that $\mathbf{B}_y$ is known, the reactions at the wall are determined from the three equations of equilibrium applied to the free-body diagram of the beam, Fig. 12–40d. The results are

$$A_x = 0 \quad A_y = \frac{11}{16}P$$

$$M_A = \frac{3}{16}PL$$

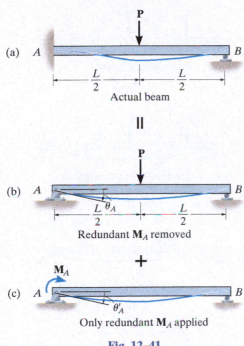

Fig. 12–41

As stated in Sec. 12.6, choice of a redundant is *arbitrary*, provided the primary beam remains stable. For example, the moment at A for the beam in Fig. 12–41a can also be chosen as the redundant. In this case the capacity of the beam to resist M_A is removed, and so the primary beam is then pin supported at A, Fig. 12–41b. To it we add the beam subjected only to the redundant, Fig. 12–41c. Referring to the slope at A caused by the load $\mathbf{P}$ as θ_A, and the slope at A caused by the redundant $\mathbf{M}_A$ as θ'_A, the compatibility equation for the slope at A requires

$$(\circlearrowleft +) \qquad\qquad 0 = \theta_A + \theta'_A$$

Again using the table in Appendix C to relate these rotations to the loads, we have

$$\theta_A = \frac{PL^2}{16EI} \quad \text{and} \quad \theta'_A = \frac{M_A L}{3EI}$$

Thus,

$$0 = \frac{PL^2}{16EI} + \frac{M_A L}{3EI}$$

$$M_A = -\frac{3}{16}PL$$

which is the same result determined previously. Here, however, the negative sign for M_A simply means that $\mathbf{M}_A$ acts in the opposite sense of direction to that shown in Fig. 12–41c.

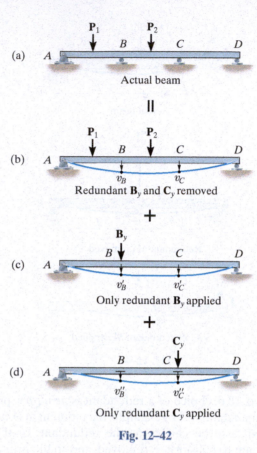

Fig. 12–42

A final example that illustrates this method is shown in Fig. 12–42a. In this case the beam is indeterminate to the second degree, and therefore two redundant reactions must be removed from the beam. We will choose the forces at the roller supports B and C as redundants. The primary (statically determinate) beam deforms as shown in Fig. 12–42b, and each redundant force deforms the beam as shown in Figs. 12–42c and 12–42d. By superposition, the compatibility equations for the displacements at B and C are therefore

$$(+\downarrow) \qquad\qquad 0 = v_B + v_B' + v_B''$$

$$(+\downarrow) \qquad\qquad 0 = v_C + v_C' + v_C'' \qquad\qquad (12\text{–}20)$$

Using the table in Appendix C, all these displacement components can be expressed in terms of the known and unknown loads. Once this is done, the equations can then be solved simultaneously for the two unknowns B_y and C_y.

PROCEDURE FOR ANALYSIS

The following procedure provides a means for applying the method of superposition (or the force method) to determine the reactions on statically indeterminate beams or shafts.

Elastic Curve.

- Specify the unknown redundant forces or moments that must be removed from the beam in order to make it statically determinate and stable.

- Using the principle of superposition, draw the statically indeterminate beam and show it equal to a sequence of corresponding *statically determinate beams*.

- The first of these beams, the primary beam, supports the same external loads as the statically indeterminate beam, and each of the other beams "added" to the primary beam shows the beam loaded with a separate redundant force or moment.

- Sketch the deflection curve for each beam and indicate the displacement (slope) at the point of each redundant force (moment).

Compatibility Equations.

- Write a compatibility equation for the displacement (slope) at each point where there is a redundant force (moment).

Load–Displacement Equations.

- Relate all the displacements or slopes to the forces or moments using the formulas in Appendix C.

- Substitute the results into the compatibility equations and solve for the unknown redundants.

- If a numerical value for a redundant is *positive*, it has the *same sense of direction* as originally assumed. A *negative* numerical value indicates the redundant acts *opposite* to its assumed *sense of direction*.

Equilibrium Equations.

- Once the redundant forces and/or moments have been determined, the remaining unknown reactions can be found from the equations of equilibrium applied to the loadings shown on the beam's free-body diagram.

12

EXAMPLE 12.21

The beam in Fig. 12–43a is fixed supported to the wall at A and pin connected to a $\frac{1}{2}$-in.-diameter rod BC. If $E = 29(10^3)$ ksi for both members, determine the force developed in the rod due to the loading. The moment of inertia of the beam about its neutral axis is $I = 475$ in^4.

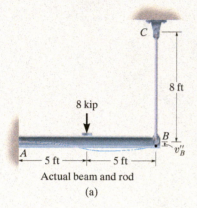

Actual beam and rod

(a)

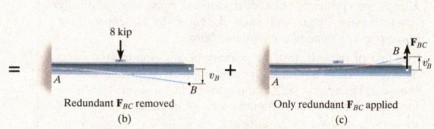

Redundant $\mathbf{F}_{BC}$ removed

(b)

Only redundant $\mathbf{F}_{BC}$ applied

(c)

Fig. 12–43

SOLUTION I

Principle of Superposition. By inspection, this problem is indeterminate to the first degree. Here B will undergo an unknown displacement v_B'', since the rod will stretch. The rod will be treated as the redundant and hence the force of the rod is removed from the beam at B, Fig. 12–43b, and then reapplied, Fig. 12–43c.

Compatibility Equation. At point B we require

$$(+\downarrow) \qquad\qquad\qquad v_B'' = v_B - v_B' \qquad\qquad\qquad (1)$$

Load–Displacement Equations. The displacements v_B and v_B' are related to the loads using Appendix C. The displacement v_B'' is calculated from Eq. 4–2. Working in kilopounds and inches, we have

$$v_B'' = \frac{PL}{AE} = \frac{F_{BC}\,(8\text{ ft})(12\text{ in./ft})}{(\pi/4)\left(\frac{1}{2}\text{ in.}\right)^2[29(10^3)\text{ kip/in}^2]} = 0.01686F_{BC} \downarrow$$

$$v_B = \frac{5PL^3}{48EI} = \frac{5(8\text{ kip})(10\text{ ft})^3\,(12\text{ in./ft})^3}{48[29(10^3)\text{ kip/in}^2](475\text{ in}^4)} = 0.1045\text{ in.} \downarrow$$

$$v_B' = \frac{PL^3}{3EI} = \frac{F_{BC}\,(10\text{ ft})^3\,(12\text{ in./ft})^3}{3[29(10)^3\text{ kip/in}^2](475\text{ in}^4)} = 0.04181F_{BC} \uparrow$$

Thus, Eq. 1 becomes

$$(+\downarrow) \qquad\qquad 0.01686F_{BC} = 0.1045 - 0.04181F_{BC}$$

$$F_{BC} = 1.78\text{ kip} \qquad\qquad\qquad\textit{Ans.}$$

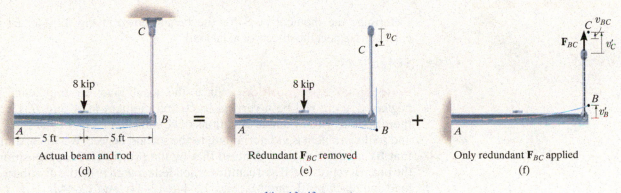

Actual beam and rod   Redundant $\mathbf{F}_{BC}$ removed   Only redundant $\mathbf{F}_{BC}$ applied

(d)   (e)   (f)

Fig. 12–43 (cont.)

SOLUTION II

Principle of Superposition. We can also solve this problem by removing the pin support at C and keeping the rod attached to the beam. In this case the 8-kip load will cause points B and C to be displaced downward the *same amount* v_C, Fig. 12–43e, since no force exists in rod BC. When the redundant force $\mathbf{F}_{BC}$ is applied at point C, it causes the end C of the rod to be displaced upward v_C' and the end B of the beam to be displaced upward v_B', Fig. 12–43f. The difference in these two displacements, v_{BC}, represents the stretch of the rod due to $\mathbf{F}_{BC}$, so that $v_C' = v_{BC} + v_B'$. Hence, from Figs. 12–43d, 12–43e, and 12–43f, the compatibility of displacement at point C is

$$(+\downarrow) \qquad\qquad 0 = v_C - (v_{BC} + v_B') \qquad\qquad (2)$$

From Solution I, we have

$$v_C = v_B = 0.1045 \text{ in. } \downarrow$$

$$v_{BC} = v_B'' = 0.01686 F_{BC} \uparrow$$

$$v_B' = 0.04181 F_{BC} \uparrow$$

Therefore, Eq. 2 becomes

$$(+\downarrow)\, 0 = 0.1045 - (0.01686 F_{BC} + 0.04181 F_{BC})$$

$$F_{BC} = 1.78 \text{ kip} \qquad\qquad\qquad \textit{Ans.}$$

12

EXAMPLE 12.22

Determine the moment at B for the beam shown in Fig. 12–44a. EI is constant. Neglect the effects of axial load.

SOLUTION

Principle of Superposition. Since the axial load on the beam is neglected, there will be a vertical force and moment at A and B. Here there are only two available equations of equilibrium ($\Sigma M = 0, \Sigma F_y = 0$), and so the problem is indeterminate to the second degree. We will assume that $\mathbf{B}_y$ and $\mathbf{M}_B$ are redundant, so that by the principle of superposition, the beam is represented as a cantilever, loaded *separately* by the distributed load and reactions $\mathbf{B}_y$ and $\mathbf{M}_B$, Figs. 12–44b, 12–44c, and 12–44d.

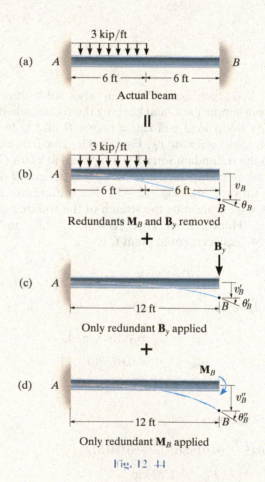

(a)

Actual beam

$\parallel$

(b)

Redundants $\mathbf{M}_B$ and $\mathbf{B}_y$ removed

$+$

(c)

Only redundant $\mathbf{B}_y$ applied

$+$

(d)

Only redundant $\mathbf{M}_B$ applied

Fig. 12–44

Compatibility Equations.   Referring to the displacement and slope at B, we require

$(\zeta+)$ $\qquad\qquad 0 = \theta_B + \theta'_B + \theta''_B$ $\qquad\qquad$ (1)

$(+\downarrow)$ $\qquad\qquad 0 = v_B + v'_B + v''_B$ $\qquad\qquad$ (2)

Load–Displacement Equations.   Using the table in Appendix C to relate the slopes and displacements to the loads, we have

$$\theta_B = \frac{wL^3}{48EI} = \frac{3 \text{ kip/ft }(12 \text{ ft})^3}{48EI} = \frac{108 \text{ kip}\cdot\text{ft}^2}{EI} \;\zeta$$

$$v_B = \frac{7wL^4}{384EI} = \frac{7(3 \text{ kip/ft})(12 \text{ ft})^4}{384EI} = \frac{1134 \text{ kip}\cdot\text{ft}^3}{EI} \;\downarrow$$

$$\theta'_B = \frac{PL^2}{2EI} = \frac{B_y(12 \text{ ft})^2}{2EI} = \frac{72B_y}{EI} \;\zeta$$

$$v'_B = \frac{PL^3}{3EI} = \frac{B_y(12 \text{ ft})^3}{3EI} = \frac{576B_y}{EI} \;\downarrow$$

$$\theta''_B = \frac{ML}{EI} = \frac{M_B(12 \text{ ft})}{EI} = \frac{12M_B}{EI} \;\zeta$$

$$v''_B = \frac{ML^2}{2EI} = \frac{M_B(12 \text{ ft})^2}{2EI} = \frac{72M_B}{EI} \;\downarrow$$

Substituting these values into Eqs. 1 and 2 and canceling out the common factor EI, we get

$(\zeta+)$ $\qquad\qquad 0 = 108 + 72B_y + 12M_B$

$(+\downarrow)$ $\qquad\qquad 0 = 1134 + 576B_y + 72M_B$

Solving these equations simultaneously gives

$$B_y = -3.375 \text{ kip}$$

$$M_B = 11.25 \text{ kip}\cdot\text{ft} \qquad\qquad \textit{Ans.}$$

NOTE: The reactions at A can now be determined from the equilibrium equations.

EXAMPLE 12.23

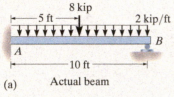

(a)   Actual beam

||

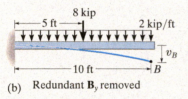

(b)   Redundant $\mathbf{B}_y$ removed

+

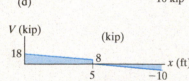

(c)   Only redundant $\mathbf{B}_y$ applied

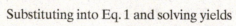

(d)

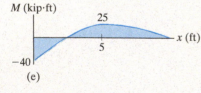

(e)

Fig. 12–45

Determine the reactions at the roller support B of the beam shown in Fig. 12–45a, then draw the shear and moment diagrams. EI is constant.

SOLUTION

Principle of Superposition. By inspection, the beam is statically indeterminate to the first degree. The roller support at B will be chosen as the redundant so that $\mathbf{B}_y$ will be determined *directly*. Figures 12–45b and 12–45c show application of the principle of superposition. Here we have assumed that $\mathbf{B}_y$ acts upward on the beam.

Compatibility Equation. Taking positive displacement as downward, the compatibility equation at B is

$$(+\downarrow) \qquad\qquad 0 = v_B - v'_B \qquad\qquad (1)$$

Load–Displacement Equations. These displacements are related to the loads using the table in Appendix C.

$$v_B = \frac{wL^4}{8EI} + \frac{5PL^3}{48EI}$$

$$= \frac{2\ \text{kip/ft}(10\ \text{ft})^4}{8EI} + \frac{5(8\ \text{kip})(10\ \text{ft})^3}{48EI} = \frac{3333\ \text{kip}\cdot\text{ft}^3}{EI} \downarrow$$

$$v'_B = \frac{PL^3}{3EI} = \frac{B_y\,(10\ \text{ft})^3}{3EI} = \frac{333.3\ \text{ft}^3\,B_y}{EI} \uparrow$$

Substituting into Eq. 1 and solving yields

$$0 = \frac{3333}{EI} - \frac{333.3B_y}{EI}$$

$$B_y = 10\ \text{kip} \qquad\qquad\qquad Ans.$$

Equilibrium Equations. Using this result and applying the three equations of equilibrium, we obtain the results shown on the beam's free-body diagram in Fig. 12–45d. The shear and moment diagrams are shown in Fig. 12–45e.

FUNDAMENTAL PROBLEMS

F12–13.   Determine the reactions at the fixed support A and the roller B. EI is constant.

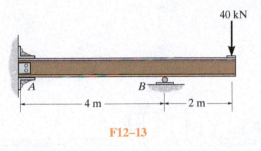

40 kN

A 4 m B 2 m

F12–13

F12–14.   Determine the reactions at the fixed support A and the roller B. EI is constant.

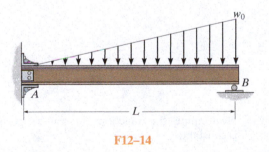

w_0

A L B

F12–14

F12–15.   Determine the reactions at the fixed support A and the roller B. Support B settles 2 mm. $E = 200$ GPa, $I = 65.0(10^{-6})$ m⁴.

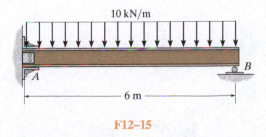

10 kN/m

A 6 m B

F12–15

F12–16.   Determine the reaction at the roller B. EI is constant.

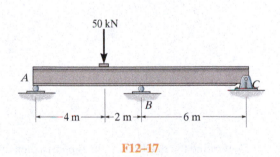

M_0

A B L L C

F12–16

F12–17.   Determine the reaction at the roller B. EI is constant.

50 kN

A 4 m 2 m B 6 m C

F12–17

F12–18.   Determine the reaction at the roller support B if it settles 5 mm. $E = 200$ GPa and $I = 65.0(10^{-6})$ m⁴.

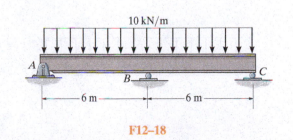

10 kN/m

A 6 m B 6 m C

F12–18

PROBLEMS

12–117. Determine the reactions at the journal bearing supports A, B, and C of the shaft, then draw the shear and moment diagrams. EI is constant.

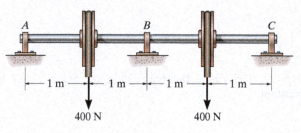

Prob. 12–117

12–119. Determine the reactions at the supports, then draw the shear and moment diagrams. EI is constant.

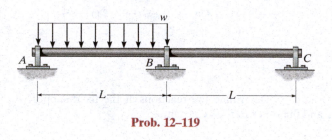

Prob. 12–119

12–118. Determine the reactions at the supports, then draw the shear and moment diagrams. EI is constant.

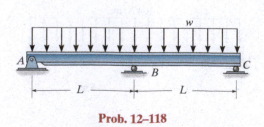

Prob. 12–118

*12–120. Determine the reactions at the supports A and B. EI is constant.

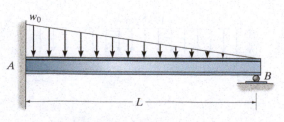

Prob. 12–120

12

12–121. The beam is used to support the 20-kip load. Determine the reactions at the supports. Assume A is fixed and B is a roller.

12–123. Determine the reactions at the supports A and B. EI is constant.

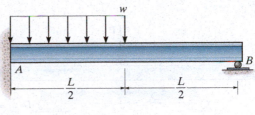

Prob. 12–123

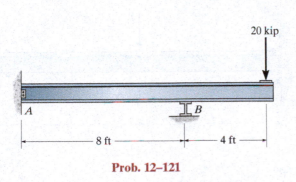

20 kip

8 ft   4 ft

Prob. 12–121

*12–124.** Before the uniform distributed load is applied to the beam, there is a small gap of 0.2 mm between the beam and the post at B. Determine the support reactions at A, B, and C. The post at B has a diameter of 40 mm, and the moment of inertia of the beam is $I = 875(10^6)$ mm^4. The post and the beam are made of material having a modulus of elasticity of $E = 200$ GPa.

12–122. Determine the reactions at the supports A and B. EI is constant.

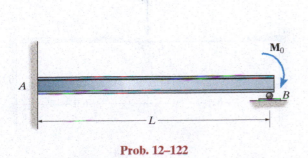

$\mathbf{M}_0$

L

Prob. 12–122

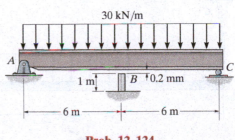

30 kN/m

1 m   0.2 mm

6 m   6 m

Prob. 12–124

12–125. The fixed supported beam AB is strengthened using the simply supported beam CD and the roller at F which is set in place just before application of the load **P**. Determine the reactions at the supports if EI is constant.

12–127. The beam is supported by the bolted supports at its ends. When loaded these supports initially do not provide an actual fixed connection, but instead allow a slight rotation α before becoming fixed after the load is fully applied. Determine the moment at the supports and the maximum deflection of the beam.

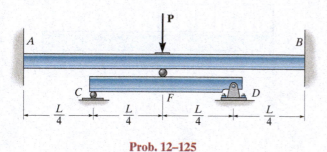

Prob. 12–125

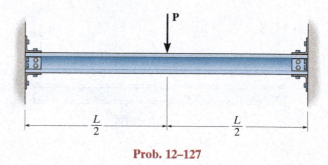

Prob. 12–127

12–126. The beam has a constant E_1I_1 and is supported by the fixed wall at B and the rod AC. If the rod has a cross-sectional area A_2 and the material has a modulus of elasticity E_2, determine the force in the rod.

*12–128. Each of the two members is made from 6061-T6 aluminum and has a square cross section 1 in. $\times$ 1 in. They are pin connected at their ends and a jack is placed between them and opened until the force it exerts on each member is 50 lb. Determine the greatest force P that can be applied to the center of the top member without causing either of the two members to yield. For the analysis neglect the axial force in each member. Assume the jack is rigid.

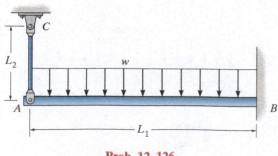

Prob. 12–126

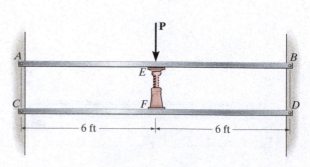

Prob. 12–128

12–129. The beam is made from a soft linear elastic material having a constant EI. If it is originally a distance Δ from the surface of its end support, determine the length a that rests on this support when it is subjected to the uniform load w_0, which is great enough to cause this to happen.

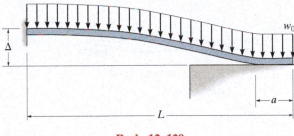

Prob. 12–129

12–130. The beam AB has a moment of inertia $I = 475$ in^4 and rests on the smooth supports at its ends. A 0.75-in.-diameter rod CD is welded to the center of the beam and to the fixed support at D. If the temperature of the rod is decreased by 150°F, determine the force developed in the rod. The beam and rod are both made of A-36 steel.

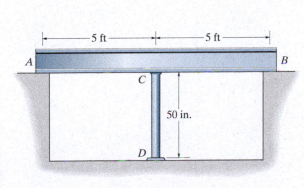

Prob. 12–130

12–131. The rim on the flywheel has a thickness t, width b, and specific weight γ. If the flywheel is rotating at a constant rate of ω, determine the maximum moment developed in the rim. Assume that the spokes do not deform. *Hint*: Due to symmetry of the loading, the slope of the rim at each spoke is zero. Consider the radius to be sufficiently large so that the segment AB can be considered as a straight beam fixed at both ends and loaded with a uniform centrifugal force per unit length. Show that this force is $w = bt\gamma\omega^2 r/g$.

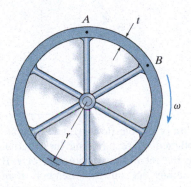

Prob. 12–131

*12–132. The box frame is subjected to a uniform distributed loading w along each of its sides. Determine the moment developed in each corner. Neglect the deflection due to axial load. EI is constant.

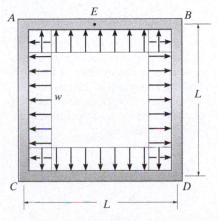

Prob. 12–132

12

CHAPTER REVIEW

The elastic curve represents the centerline deflection of a beam or shaft. Its shape can be determined using the moment diagram. Positive moments cause the elastic curve to be concave upwards and negative moments cause it to be concave downwards. The radius of curvature at any point is determined from

$$\frac{1}{\rho} = \frac{M}{EI}$$

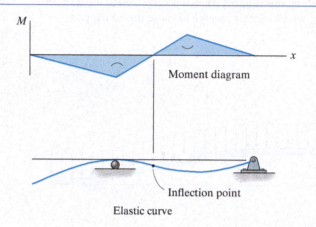

Moment diagram

Inflection point

Elastic curve

The equation of the elastic curve and its slope can be obtained by first finding the internal moment in the member as a function of x. If several loadings act on the member, then separate moment functions must be determined between each of the loadings. Integrating these functions once using $EI(d^2v/dx^2) = M(x)$ gives the equation for the slope of the elastic curve, and integrating again gives the equation for the deflection. The constants of integration are determined from the boundary conditions at the supports, or in cases where several moment functions are involved, continuity of slope and deflection at points where these functions join must be satisfied.

Boundary conditions

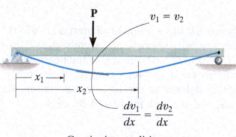

Continuity conditions

Discontinuity functions allow one to express the equation of the elastic curve as a continuous function, regardless of the number of loadings on the member. This method eliminates the need to use continuity conditions, since the two constants of integration can be determined from the two boundary conditions.

12

The moment-area method is a semigraphical technique for finding the slope of tangents or the vertical distance between tangents at specific points on the elastic curve. It requires finding area segments under the M/EI diagram, or the moment of these segments about points on the elastic curve. The method works well for M/EI diagrams composed of simple shapes, such as those produced by concentrated forces and couple moments.

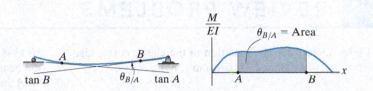

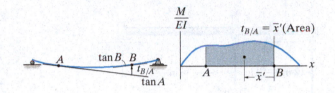

The deflection or slope at a point on a member subjected to combinations of loadings can be determined using the method of superposition. The table in Appendix C is available for this purpose.

Statically indeterminate beams and shafts have more unknown support reactions than available equations of equilibrium. To solve, one first identifies the redundant reactions. The method of integration or the moment-area theorems can then be used to solve for the unknown redundants. It is also possible to determine the redundants by using the method of superposition, where one considers the conditions of continuity at the redundant support.

12

REVIEW PROBLEMS

R12–1. Determine the equation of the elastic curve. Use discontinuity functions EI is constant.

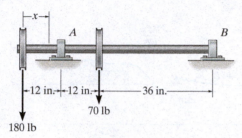

R12–1

R12–2. Draw the bending-moment diagram for the shaft and then, from this diagram, sketch the deflection or elastic curve for the shaft's centerline. Determine the equations of the elastic curve using the coordinates x_1 and x_2. Use the method of integration. EI is constant.

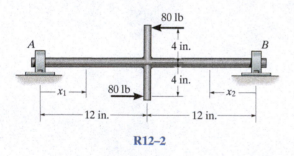

R12–2

R12–3. Determine the moment reactions at the supports A and B. Use the method of integration. EI is constant.

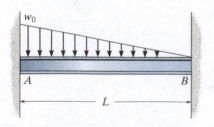

R12–3

*R12–4. Determine the equations of the elastic curve for the beam using the x_1 and x_2 coordinates. Specify the slope at A and the maximum deflection. Use the method of integration. EI is constant.

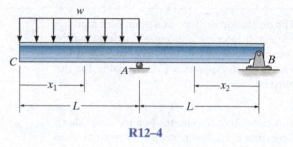

R12–4

R12–5. Determine the maximum deflection between the supports A and B. Use the method of integration. EI is constant.

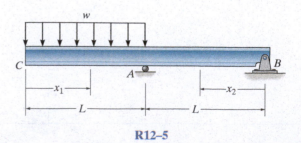

R12–5

R12–6. Determine the slope at B and the deflection at C. Use the moment-area theorems. EI is constant.

*R12–8. Using the method of superposition, determine the magnitude of $\mathbf{M_0}$ in terms of the distributed load w and dimension a so that the deflection at the center of the beam is zero. EI is constant.

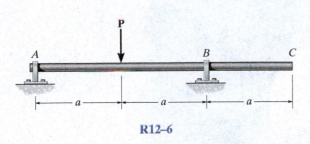

R12–6

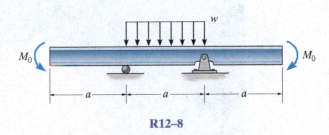

R12–8

R12–9. Using the method of superposition, determine the deflection at C of beam AB. The beams are made of wood having a modulus of elasticity of $E = 1.5(10^3)$ ksi.

R12–7. Determine the reactions, then draw the shear and moment diagrams. Use the moment-area theorems. EI is constant.

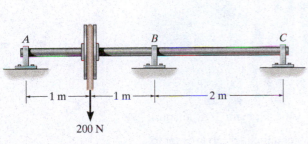

200 N

R12–7

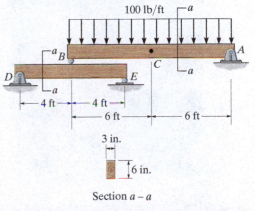

Section $a - a$

R12–9

12

CHAPTER 13

(© James Roman/Getty Images)

The columns of this water tank are braced at points along their length in order to reduce their chance of buckling.

BUCKLING OF COLUMNS

CHAPTER OBJECTIVES

■ In this chapter we will discuss the buckling behavior of a column subjected to axial and eccentric loads. Afterwards, some of the methods used to design columns made of common engineering materials will be presented.

13.1 CRITICAL LOAD

Not only must a member satisfy specific strength and deflection requirements but it must also be stable. Stability is particularly important if the member is long and slender, and it supports a compressive loading that becomes large enough to cause the member to suddenly deflect laterally or sidesway. These members are called *columns*, and the lateral deflection that occurs is called *buckling*. Quite often the buckling of a column can lead to a sudden and dramatic failure of a structure or mechanism, and as a result, special attention must be given to the design of columns so that they can safely support their intended loadings without buckling.

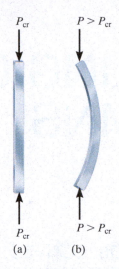

Fig. 13–1

The maximum axial load that a column can support when it is on the *verge of buckling* is called the **critical load**, P_{cr}, Fig. 13–1a. Any additional loading will cause the column to buckle and therefore deflect laterally as shown in Fig. 13–1b.

We can study the nature of this instability by considering the two-bar mechanism consisting of weightless rigid bars that are pin connected as shown in Fig. 13–2a. When the bars are in the vertical position, the spring, having a stiffness k, is unstretched, and a *small* vertical force **P** is applied at the top of one of the bars. To upset this equilibrium position the pin at A is displaced by a small amount Δ, Fig. 13–2b. As shown on the free-body diagram of the pin, Fig. 13–2c, the spring will produce a restoring force $F = k\Delta$ in order to resist the two horizontal components, $P_x = P \tan \theta$, which tend to push the pin (and the bars) further out of equilibrium. Since θ is small, $\Delta \approx \theta(L/2)$ and $\tan \theta \approx \theta$. Thus the *restoring* spring force becomes $F = k\theta(L/2)$, and the *disturbing* force is $2P_x = 2P\theta$.

If the restoring force is greater than the disturbing force, that is, $k\theta L/2 > 2P\theta$, then, noticing that θ cancels out, we can solve for P, which gives

$$P < \frac{kL}{4} \qquad \text{stable equilibrium}$$

This is a condition for **stable equilibrium**, since the force developed by the spring would be adequate to restore the bars back to their vertical position. However, if $k\theta(L/2) < 2P\theta$, or

$$P > \frac{kL}{4} \qquad \text{unstable equilibrium}$$

then the bars will be in **unstable equilibrium**. In other words, if this load is applied, and a slight displacement occurs at A, the bars will tend to move out of equilibrium and not be restored to their original position.

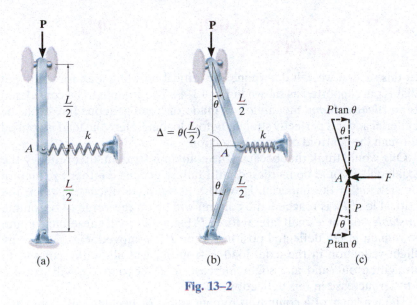

Fig. 13–2

The intermediate value of P, which requires $kL\theta/2 = 2P\theta$, is the *critical load*. Here

$$P_{cr} = \frac{kL}{4} \qquad \text{neutral equilibrium}$$

This loading represents a case of the bars being in **neutral equilibrium**. Since P_{cr} is *independent* of the (small) displacement θ of the bars, any slight disturbance given to the mechanism will not cause it to move further out of equilibrium, nor will it be restored to its original position. Instead, the bars will simply *remain* in the deflected position.

These three different states of equilibrium are represented graphically in Fig. 13–3. The transition point where the load is equal to its critical value $P = P_{cr}$ is called the **bifurcation point**. Here the bars will be in neutral equilibrium for any *small value* of θ. If a larger load P is placed on the bars, then they will undergo a larger deflection, so that the spring is compressed or elongated enough to hold them in equilibrium.

In a similar manner, if the load on an actual column exceeds its critical loading, then this loading will also require the column to undergo a *large* deflection; however, this is generally not tolerated in engineering structures or machines.

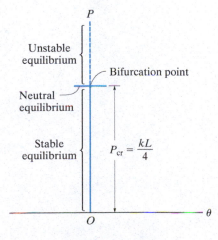

Fig. 13–3

13

13.2   IDEAL COLUMN WITH PIN SUPPORTS

In this section we will determine the critical buckling load for a column that is pin supported as shown in Fig. 13–4a. The column to be considered is an *ideal column*, meaning it is made of homogeneous linear elastic material and it is perfectly straight before loading. Here the load is applied through the centroid of the cross section.

One would think that because the column is straight, theoretically the axial load P could be increased until failure occurred either by fracture or yielding of the material. However, as we have discussed, when the critical load P_{cr} is reached, the column will be on the verge of becoming *unstable*, so that a small lateral force F, Fig. 13–4b, will cause the column to remain in the deflected position when F is removed, Fig. 13–4c. Any slight reduction in the axial load P from P_{cr} will allow the column to straighten out, and any slight increase in P, beyond P_{cr}, will cause a further increase in this deflection.

The tendency of a column to remain stable or become unstable when subjected to an axial load actually depends upon its ability to resist bending. Hence, in order to determine the critical load and the buckled shape of the column, we will apply Eq. 12–10, which relates the internal moment in the column to its deflected shape, i.e.,

$$EI\frac{d^2v}{dx^2} = M \tag{13-1}$$

The dramatic failure of this off-shore oil platform was caused by the horizontal forces of hurricane winds, which led to buckling of its supporting columns.

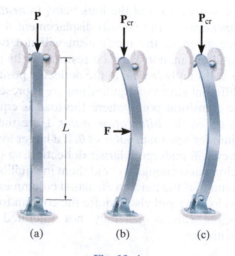

(a)   (b)   (c)

Fig. 13–4

A free-body diagram of a segment of the column in the deflected position is shown in Fig. 13–5a. Here both the displacement v and the internal moment M are shown in the *positive direction*. Since moment equilibrium requires $M = -Pv$, then Eq. 13–1 becomes

$$EI\frac{d^2v}{dx^2} = -Pv$$

$$\frac{d^2v}{dx^2} + \left(\frac{P}{EI}\right)v = 0 \tag{13–2}$$

This is a homogeneous, second-order, linear differential equation with constant coefficients. It can be shown by using the methods of differential equations, or by direct substitution into Eq. 13–2, that the general solution is

$$v = C_1 \sin\left(\sqrt{\frac{P}{EI}}\,x\right) + C_2 \cos\left(\sqrt{\frac{P}{EI}}\,x\right) \tag{13–3}$$

The two constants of integration can be determined from the boundary conditions at the ends of the column. Since $v = 0$ at $x = 0$, then $C_2 = 0$. And since $v = 0$ at $x = L$, then

$$C_1 \sin\left(\sqrt{\frac{P}{EI}}\,L\right) = 0$$

This equation is satisfied if $C_1 = 0$; however, then $v = 0$, which is a *trivial solution* that requires the column to always remain straight, even though the load may cause the column to become unstable. The other possibility requires

$$\sin\left(\sqrt{\frac{P}{EI}}\,L\right) = 0$$

which is satisfied if

$$\sqrt{\frac{P}{EI}}\,L = n\pi$$

or

$$P = \frac{n^2\pi^2 EI}{L^2} \quad n = 1, 2, 3, \ldots \tag{13–4}$$

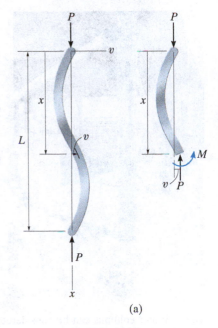

(a)

Fig. 13–5

These timber columns can be considered pinned at their bottom and fixed connected to the beams at their top.

The *smallest value* of P is obtained when $n = 1$, so the *critical load* for the column is therefore*

$$P_{cr} = \frac{\pi^2 EI}{L^2}$$

This load is sometimes referred to as the **Euler load**, named after the Swiss mathematician Leonhard Euler, who originally solved this problem in 1757. From Eq. 13–3, the corresponding buckled shape, shown in Fig. 13–5b, is therefore

$$v = C_1 \sin \frac{\pi x}{L}$$

The constant C_1 represents the maximum deflection, v_{max}, which occurs at the midpoint of the column, Fig. 13–5b. Unfortunately, a specific value for C_1 cannot be obtained once it has buckled. It is assumed, however, that this deflection is small.

As noted above, the critical load depends on the material's stiffness or modulus of elasticity E and not its yield stress. Therefore, a column made of high-strength steel offers no advantage over one made of lower-strength steel, since the modulus of elasticity for both materials is the same. Also note that P_{cr} will increase as the moment of inertia of the cross section increases. Thus, efficient columns are designed so that most of the column's cross-sectional area is located as far away as possible from the center of the section. This is why hollow sections such as tubes are more economical than solid sections. Furthermore, wide-flange sections, and columns that are "built up" from channels, angles, plates, etc., are better than sections that are solid and rectangular.

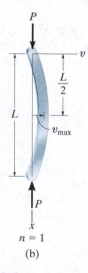

$n = 1$

(b)

Fig. 13–5 (cont.)

*n represents the number of curves in the deflected shape of the column. For example, if $n = 1$, then one curve appears as in Fig. 13–5b; if $n = 2$, then *two* curves appear as in Fig. 13–5a, etc.

Since P_{cr} is directly related to I, a column will buckle about the principal axis of the cross section having the ***least moment of inertia*** (the weakest axis), provided it is supported the same way about each axis. For example, a column having a rectangular cross section, like a meter stick, Fig. 13–6, will buckle about the a–a axis, not the b–b axis. Because of this, engineers usually try to achieve a balance, keeping the moments of inertia the same in all directions. Geometrically, then, circular tubes make excellent columns. Square tubes or those shapes having $I_x \approx I_y$ are also often selected for columns.

To summarize, the buckling equation for a pin-supported long slender column is

$$P_{cr} = \frac{\pi^2 EI}{L^2} \qquad (13\text{–}5)$$

where

P_{cr} = critical or maximum axial load on the column just before it begins to buckle. This load must *not* cause the stress in the column to exceed the proportional limit.

E = modulus of elasticity for the material

I = *least* moment of inertia for the column's cross-sectional area

L = unsupported length of the column, whose ends are pinned

For design purposes, the above equation can also be written in terms of stress, by using $I = Ar^2$, where A is the cross-sectional area and r is the ***radius of gyration*** of the cross-sectional area. We have,

$$P_{cr} = \frac{\pi^2 E(Ar^2)}{L^2}$$

$$\left(\frac{P}{A}\right)_{cr} = \frac{\pi^2 E}{(L/r)^2}$$

or

$$\sigma_{cr} = \frac{\pi^2 E}{(L/r)^2} \qquad (13\text{–}6)$$

Here

σ_{cr} = critical stress, which is an average normal stress in the column just before the column buckles. It is required that $\sigma_{cr} \le \sigma_Y$.

E = modulus of elasticity for the material

L = unsupported length of the column, whose ends are pinned

r = *smallest* radius of gyration of the column, determined from $r = \sqrt{I/A}$, where I is the *least* moment of inertia of the column's cross-sectional area A

The geometric ratio L/r in Eq. 13–6 is known as the ***slenderness ratio***. It is a measure of the column's flexibility, and as we will discuss later, it serves to classify columns as long, intermediate, or short.

Fig. 13–6

Failure of this crane boom was caused by the localized buckling of one of its tubular struts.

A graph of this equation for columns made of structural steel and an aluminum alloy is shown in Fig. 13–7. The curves are hyperbolic and are valid only for critical stresses that are below the material's yield point (proportional limit). Notice that for the steel the yield stress is $(\sigma_Y)_{st} = 36$ ksi $[E_{st} = 29(10^3)$ ksi], and for the aluminum it is $(\sigma_Y)_{al} = 27$ ksi $[E_{al} = 10(10^3)$ ksi]. If we substitute $\sigma_{cr} = \sigma_Y$ into Eq. 13–6, the *smallest* allowable slenderness ratios for the steel and aluminum columns then become $(L/r)_{st} = 89$ and $(L/r)_{al} = 60.5$, Fig. 13–7. Thus, for a steel column, if $(L/r)_{st} < 89$, the column's stress will exceed the yield point before buckling can occur, and so the Euler formula cannot be used.

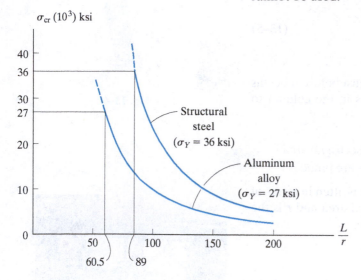

Fig. 13–7

> ## IMPORTANT POINTS
>
> - *Columns* are long slender members that are subjected to axial compressive loads.
>
> - The *critical load* is the maximum axial load that a column can support when it is on the verge of buckling. This loading represents a case of *neutral equilibrium*.
>
> - An *ideal column* is initially perfectly straight, made of homogeneous material, and the load is applied through the centroid of its cross section.
>
> - A pin-connected column will buckle about the principal axis of the cross section having the *least* moment of inertia.
>
> - The *slenderness ratio* is L/r, where r is the smallest radius of gyration of the cross section. Buckling will occur about the axis where this ratio gives the greatest value.

EXAMPLE 13.1

The A992 steel W8 × 31 member shown in Fig. 13–8 is to be used as a pin-connected column. Determine the largest axial load it can support before it either begins to buckle or the steel yields. Take $\sigma_Y = 50$ ksi.

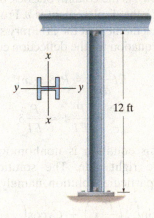

Fig. 13–8

SOLUTION

From the table in Appendix B, the column's cross-sectional area and moments of inertia are $A = 9.13$ in^2, $I_x = 110$ in^4, and $I_y = 37.1$ in^4. By inspection, buckling will occur about the y–y axis. Why? Applying Eq. 13–5, we have

$$P_{cr} = \frac{\pi^2 EI}{L^2} = \frac{\pi^2[29(10^3) \text{ kip/in}^2](37.1 \text{ in}^4)}{[12 \text{ ft}(12 \text{ in.}/\text{ft})]^2} = 512 \text{ kip}$$

When this load is applied, the average compressive stress in the column is

$$\sigma_{cr} = \frac{P_{cr}}{A} = \frac{512 \text{ kip}}{9.13 \text{ in}^2} = 56.1 \text{ ksi}$$

Since this stress exceeds the yield stress (50 ksi), the load P is determined from simple compression.

$$50 \text{ ksi} = \frac{P}{9.13 \text{ in}^2}$$

$$P = 456 \text{ kip} \qquad\qquad \textit{Ans.}$$

In actual practice, a factor of safety would be placed on this loading.

13.3 COLUMNS HAVING VARIOUS TYPES OF SUPPORTS

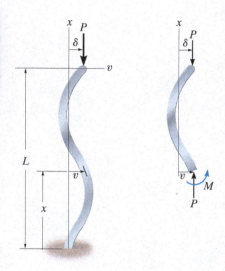

(a)   (b)

Fig. 13–9

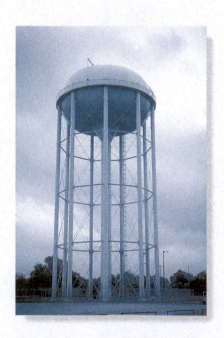

The tubular columns used to support this water tank have been braced at three locations along their length to prevent them from buckling.

The Euler load in Sec. 13.2 was derived for a column that is pin connected or free to rotate at its ends. Oftentimes, however, columns may be supported in other ways. For example, consider the case of a column fixed at its base and free at the top, Fig. 13–9a. As the column buckles, the load will sidesway and be displaced δ, while at x the displacement is v. From the free-body diagram in Fig. 13–9b, the internal moment at the arbitrary section is $M = P(\delta - v)$, and so the differential equation for the deflection curve is

$$EI\frac{d^2v}{dx^2} = P(\delta - v)$$

$$\frac{d^2v}{dx^2} + \frac{P}{EI}v = \frac{P}{EI}\delta \qquad (13\text{–}7)$$

Unlike Eq. 13–2, this equation is nonhomogeneous because of the nonzero term on the right side. The solution consists of both a complementary and a particular solution, namely,

$$v = C_1\sin\left(\sqrt{\frac{P}{EI}}x\right) + C_2\cos\left(\sqrt{\frac{P}{EI}}x\right) + \delta$$

The constants are determined from the boundary conditions. At $x = 0$, $v = 0$, and so $C_2 = -\delta$. Also,

$$\frac{dv}{dx} = C_1\sqrt{\frac{P}{EI}}\cos\left(\sqrt{\frac{P}{EI}}x\right) - C_2\sqrt{\frac{P}{EI}}\sin\left(\sqrt{\frac{P}{EI}}x\right)$$

and at $x = 0$, $dv/dx = 0$, then $C_1 = 0$. The deflection curve is therefore

$$v = \delta\left[1 - \cos\left(\sqrt{\frac{P}{EI}}x\right)\right] \qquad (13\text{–}8)$$

Finally, at the top of the column $x = L$, $v = \delta$, so that

$$\delta\cos\left(\sqrt{\frac{P}{EI}}L\right) = 0$$

The trivial solution $\delta = 0$ indicates that no buckling occurs, regardless of the load P. Instead,

$$\cos\left(\sqrt{\frac{P}{EI}}L\right) = 0 \qquad \text{or} \qquad \sqrt{\frac{P}{EI}}L = \frac{n\pi}{2}, \quad n = 1, 3, 5 \ldots$$

The smallest critical load occurs when $n = 1$, so that

$$P_{cr} = \frac{\pi^2EI}{4L^2} \qquad (13\text{–}9)$$

By comparison with Eq. 13–5, it is seen that a column fixed supported at its base and free at its top will support only one-fourth the critical load that can be applied to a column pin supported at both ends.

Other types of supported columns are analyzed in much the same way and will not be covered in detail here.* Instead, we will tabulate the results for the most common types of column support and show how to apply these results by writing Euler's formula in a general form.

Effective Length.   To use Euler's formula, Eq. 13–5, for columns having different types of supports, we will modify the column length L to represent the distance between the points of zero moment on the column. This distance is called the column's **effective length**, L_e. Obviously, for a pin-ended column $L_e = L$, Fig. 13–10a. For the fixed-end and free-ended column, the deflection curve is defined by Eq. 13–8. When plotted, its shape is equivalent to a pin-ended column having a length of $2L$, Fig. 13–10b, and so the effective length between the points of zero moment is $L_e = 2L$. Examples for two other columns with different end supports are also shown in Fig. 13–10. The column fixed at its ends, Fig. 13–10c, has inflection points or points of zero moment $L/4$ from each support. The effective length is therefore defined by the middle half of its length, that is, $L_e = 0.5L$. Finally, the pin- and fixed-ended column, Fig. 13–10d, has an inflection point at approximately $0.7L$ from its pinned end, so that $L_e = 0.7L$.

Rather than specifying the column's effective length, many design codes provide column formulas that employ a dimensionless coefficient K called the **effective-length factor**. This factor is defined from

$$L_e = KL \qquad (13\text{–}10)$$

Specific values of K are also given in Fig. 13–10. Based on this generality, we can therefore write Euler's formula as

$$P_{cr} = \frac{\pi^2 EI}{(KL)^2} \qquad (13\text{–}11)$$

or

$$\sigma_{cr} = \frac{\pi^2 E}{(KL/r)^2} \qquad (13\text{–}12)$$

Here (KL/r) is the column's **effective-slenderness ratio**.

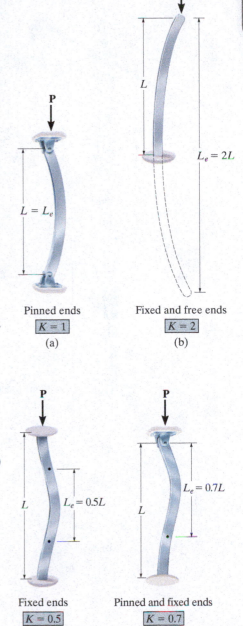

Pinned ends
$K = 1$
(a)

Fixed and free ends
$L_e = 2L$
$K = 2$
(b)

$L = L_e$

Fixed ends
$K = 0.5$
(c)

$L_e = 0.5L$

Pinned and fixed ends
$L_e = 0.7L$
$K = 0.7$
(d)

Fig. 13–10

*See Problems 13–43, 13–44, and 13–45.

EXAMPLE 13.2

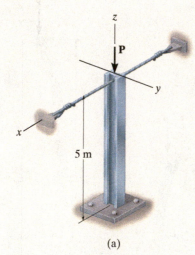

(a)

Fig. 13–11

The aluminum column in Fig. 13–11a is braced at its top by cables so as to prevent movement at the top along the x axis. If it is assumed to be fixed at its base, determine the largest allowable load **P** that can be applied. Use a factor of safety for buckling of F.S. = 3.0. Take E_{al} = 70 GPa, σ_Y = 215 MPa, A = 7.5(10^{-3}) m², I_x = 61.3(10^{-6}) m⁴, I_y = 23.2(10^{-6}) m⁴.

SOLUTION

Buckling about the x and y axes is shown in Figs. 13–11b and 13–11c. Using Fig. 13–10a, for x–x axis buckling, K = 2, so $(KL)_x$ = 2(5 m) = 10 m. For y–y axis buckling, K = 0.7, so $(KL)_y$ = 0.7(5 m) = 3.5 m.

Applying Eq. 13–11, the critical loads for each case are

$$(P_{cr})_x = \frac{\pi^2 E I_x}{(KL)_x^2} = \frac{\pi^2 [70(10^9)\ \text{N/m}^2](61.3(10^{-6})\ \text{m}^4)}{(10\ \text{m})^2}$$

$$= 424\ \text{kN}$$

$$(P_{cr})_y = \frac{\pi^2 E I_y}{(KL)_y^2} = \frac{\pi^2 [70(10^9)\ \text{N/m}^2](23.2(10^{-6})\ \text{m}^4)}{(3.5\ \text{m})^2}$$

$$= 1.31\ \text{MN}$$

By comparison, as P is increased the column will buckle about the x–x axis. The allowable load is therefore

$$P_{allow} = \frac{P_{cr}}{\text{F.S.}} = \frac{424\ \text{kN}}{3.0} = 141\ \text{kN} \qquad Ans.$$

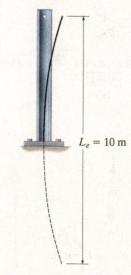

L_e = 10 m

Since

$$\sigma_{cr} = \frac{P_{cr}}{A} = \frac{424\ \text{kN}}{7.5(10^{-3})\ \text{m}^2} = 56.5\ \text{MPa} < 215\ \text{MPa}$$

Euler's equation is valid.

x–x axis buckling

(b)

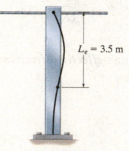

L_e = 3.5 m

y–y axis buckling

(c)

EXAMPLE   13.3

A W6 × 15 steel column is 24 ft long and is fixed at its ends as shown in Fig. 13–12a. Its load-carrying capacity is increased by bracing it about the y–y (weak) axis using struts that are assumed to be pin connected at its midheight. Determine the load the column can support so that it does not buckle nor the material exceed the yield stress. Take $E_{st} = 29(10^3)$ ksi and $\sigma_Y = 60$ ksi.

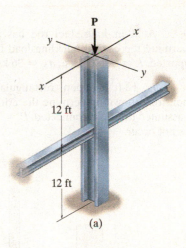

12 ft

12 ft

(a)

SOLUTION
The buckling behavior of the column will be *different* about the x–x and y–y axes due to the bracing. The buckled shape for each of these cases is shown in Figs. 13–12b and 13–12c. From Fig. 13–12b, the effective length for buckling about the x–x axis is $(KL)_x = 0.5(24 \text{ ft}) = 12 \text{ ft} = 144 \text{ in.}$, and from Fig. 13–12c, for buckling about the y–y axis $(KL)_y = 0.7(24 \text{ ft}/2) = 8.40 \text{ ft} = 100.8 \text{ in.}$ The moments of inertia for a W6 × 15 are found from the table in Appendix B. We have $I_x = 29.1 \text{ in}^4$, $I_y = 9.32 \text{ in}^4$.

Applying Eq. 13–11,

$$(P_{cr})_x = \frac{\pi^2 E I_x}{(KL)_x^2} = \frac{\pi^2 [29(10^3) \text{ ksi}]29.1 \text{ in}^4}{(144 \text{ in.})^2} = 401.7 \text{ kip} \quad (1)$$

$$(P_{cr})_y = \frac{\pi^2 E I_y}{(KL)_y^2} = \frac{\pi^2 [29(10^3) \text{ ksi}]9.32 \text{ in}^4}{(100.8 \text{ in.})^2} = 262.5 \text{ kip} \quad (2)$$

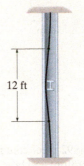

12 ft

x–x axis buckling

(b)

By comparison, buckling will occur about the y–y axis.

The area of the cross section is 4.43 in², so the average compressive stress in the column is

$$\sigma_{cr} = \frac{P_{cr}}{A} = \frac{262.5 \text{ kip}}{4.43 \text{ in}^2} = 59.3 \text{ ksi}$$

Since this stress is less than the yield stress, buckling will occur before the material yields. Thus,

$$P_{cr} = 263 \text{ kip} \qquad\qquad\qquad \textit{Ans.}$$

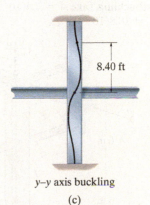

8.40 ft

y–y axis buckling

(c)

Fig. 13–12

FUNDAMENTAL PROBLEMS

F13–1. A 50-in.-long steel rod has a diameter of 1 in. Determine the critical buckling load if the ends are fixed supported. $E = 29(10^3)$ ksi, $\sigma_Y = 36$ ksi.

F13–2. A 12-ft wooden rectangular column has the dimensions shown. Determine the critical load if the ends are assumed to be pin connected. $E = 1.6(10^3)$ ksi. Yielding does not occur.

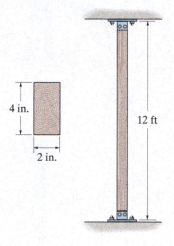

F13–2

F13–3. The A992 steel column can be considered pinned at its top and bottom and braced against its weak axis at the mid-height. Determine the maximum allowable force **P** that the column can support without buckling. Apply a F.S. = 2 against buckling. Take $A = 7.4(10^{-3})$ m^2, $I_x = 87.3(10^{-6})$ m^4, and $I_y = 18.8(10^{-6})$ m^4.

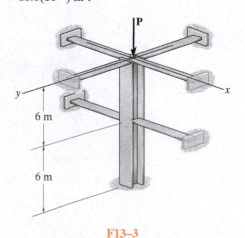

F13–3

F13–4. A steel pipe is fixed supported at its ends. If it is 5 m long and has an outer diameter of 50 mm and a thickness of 10 mm, determine the maximum axial load P that it can carry without buckling. $E_{st} = 200$ GPa, $\sigma_Y = 250$ MPa.

F13–5. Determine the maximum force **P** that can be supported by the assembly without causing member AC to buckle. The member is made of A992 steel and has a diameter of 2 in. Take F.S. = 2 against buckling.

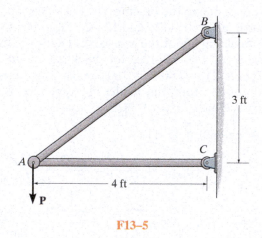

F13–5

F13–6. The A992 steel rod BC has a diameter of 50 mm and is used as a strut to support the beam. Determine the maximum intensity w of the uniform distributed load that can be applied to the beam without causing the strut to buckle. Take F.S. = 2 against buckling.

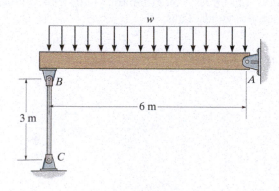

F13–6

PROBLEMS

13–1. Determine the critical buckling load for the column. The material can be assumed rigid.

13–3. The aircraft link is made from an A992 steel rod. Determine the smallest diameter of the rod, to the nearest $\frac{1}{16}$ in., that will support the load of 4 kip without buckling. The ends are pin connected.

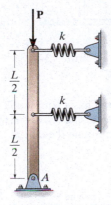

Prob. 13–1

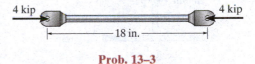

Prob. 13–3

13–2. The column consists of a rigid member that is pinned at its bottom and attached to a spring at its top. If the spring is unstretched when the column is in the vertical position, determine the critical load that can be placed on the column.

*13–4. Rigid bars AB and BC are pin connected at B. If the spring at D has a stiffness k, determine the critical load P_{cr} that can be applied to the bars.

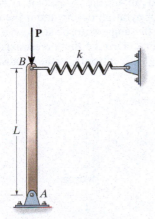

Prob. 13–2

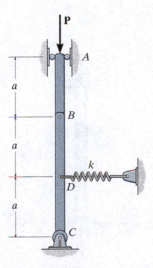

Prob. 13–4

13–5. A 2014-T6 aluminum alloy column has a length of 6 m and is fixed at one end and pinned at the other. If the cross-sectional area has the dimensions shown, determine the critical load. $\sigma_Y = 250$ MPa.

13–6. Solve Prob. 13–5 if the column is pinned at its top and bottom.

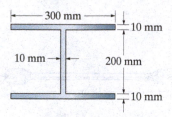

Probs. 13–5/6

13–7. The W12 × 50 is made of A992 steel and is used as a column that has a length of 20 ft. If its ends are assumed pin supported, and it is subjected to an axial load of 150 kip, determine the factor of safety with respect to buckling.

*13–8. The W12 × 50 is made of A992 steel and is used as a column that has a length of 20 ft. If the ends of the column are fixed supported, can the column support the critical load without yielding?

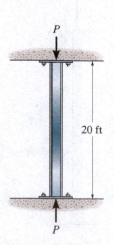

Probs. 13–7/8

13–9. A steel column has a length of 9 m and is fixed at both ends. If the cross-sectional area has the dimensions shown, determine the critical load. $E_{st} = 200$ GPa, $\sigma_Y = 250$ MPa.

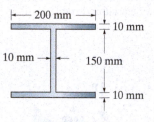

Prob. 13–9

13–10. A steel column has a length of 9 m and is pinned at its top and bottom. If the cross-sectional area has the dimensions shown, determine the critical load. $E_{st} = 200$ GPa, $\sigma_Y = 250$ MPa.

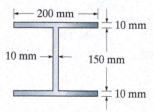

Prob. 13–10

13–11. The A992 steel angle has a cross-sectional area of $A = 2.48$ in^2 and a radius of gyration about the x axis of $r_x = 1.26$ in. and about the y axis of $r_y = 0.879$ in. The smallest radius of gyration occurs about the a–a axis and is $r_a = 0.644$ in. If the angle is to be used as a pin-connected 10-ft-long column, determine the largest axial load that can be applied through its centroid C without causing it to buckle.

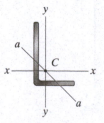

Prob. 13–11

*13–12. The 50-mm-diameter C86100 bronze rod is fixed supported at A and has a gap of 2 mm from the wall at B. Determine the increase in temperature ΔT that will cause the rod to buckle. Assume that the contact at B acts as a pin.

13–14. The W8 × 67 wide-flange A-36 steel column can be assumed fixed at its base and pinned at its top. Determine the largest axial force P that can be applied without causing it to buckle.

13–15. Solve Prob. 13–14 if the column is assumed fixed at its bottom and free at its top.

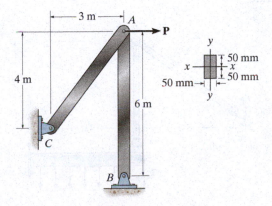

Prob. 13–12

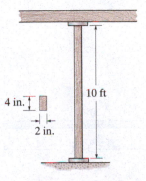

Probs. 13–14/15

13–13. Determine the maximum load P the frame can support without buckling member AB. Assume that AB is made of steel and is pinned at its ends for y–y axis buckling and fixed at its ends for x–x axis buckling. $E_{st} = 200$ GPa, $\sigma_Y = 360$ MPa.

*13–16. An A992 steel W200 × 46 column of length 9 m is fixed at one end and free at its other end. Determine the allowable axial load the column can support if F.S. = 2 against buckling.

13–17. The 10-ft wooden rectangular column has the dimensions shown. Determine the critical load if the ends are assumed to be pin connected. $E_w = 1.6(10^3)$ ksi, $\sigma_Y = 5$ ksi.

13–18. The 10-ft wooden column has the dimensions shown. Determine the critical load if the bottom is fixed and the top is pinned. $E_w = 1.6(10^3)$ ksi, $\sigma_Y = 5$ ksi.

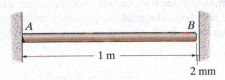

Prob. 13–13

Probs. 13–17/18

13

13–19. Determine the maximum force P that can be applied to the handle so that the A992 steel control rod AB does not buckle. The rod has a diameter of 1.25 in. It is pin connected at its ends.

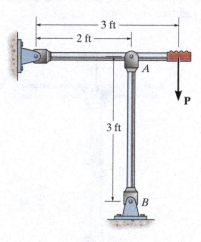

Prob. 13–19

*13–20. The A-36 steel pipe has an outer diameter of 2 in. and a thickness of 0.5 in. If it is held in place by a guywire, determine the largest vertical force P that can be applied without causing the pipe to buckle. Assume that the ends of the pipe are pin connected.

13–21. The A-36 steel pipe has an outer diameter of 2 in. If it is held in place by a guywire, determine its required inner diameter to the nearest $\frac{1}{8}$ in., so that it can support a maximum vertical load of $P = 4$ kip without causing the pipe to buckle. Assume the ends of the pipe are pin connected.

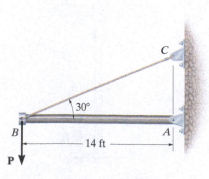

Probs. 13–20/21

13–22. The deck is supported by the two 40-mm-square columns. Column AB is pinned at A and fixed at B, whereas CD is pinned at C and D. If the deck is prevented from sidesway, determine the greatest weight of the load that can be applied without causing the deck to collapse. The center of gravity of the load is located at $d = 2$ m. Both columns are made from Douglas Fir.

13–23. The deck is supported by the two 40-mm-square columns. Column AB is pinned at A and fixed at B, whereas CD is pinned at C and D. If the deck is prevented from sidesway, determine the position d of the center of gravity of the load and the load's greatest magnitude without causing the deck to collapse. Both columns are made from Douglas Fir.

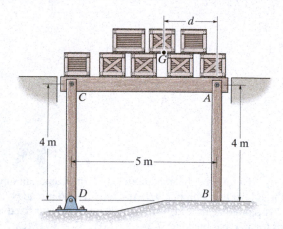

Probs. 13–22/23

*13–24. The beam is supported by the three pin-connected suspender bars, each having a diameter of 0.5 in. and made from A-36 steel. Determine the greatest uniform load w that can be applied to the beam without causing AB or CB to buckle.

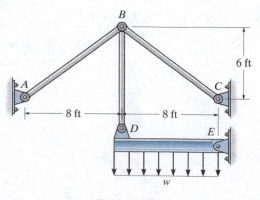

Prob. 13–24

13–25. The W14 × 30 A992 steel column is assumed pinned at both of its ends. Determine the largest axial force *P* that can be applied without causing it to buckle.

Prob. 13–25

13–26. The A992 steel bar *AB* has a square cross section. If it is pin connected at its ends, determine the maximum allowable load *P* that can be applied to the frame. Use a factor of safety with respect to buckling of 2.

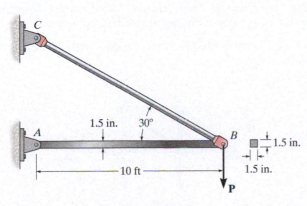

Prob. 13–26

13–27. The linkage is made using two A992 steel rods, each having a circular cross section. Determine the diameter of each rod to the nearest $\frac{1}{8}$ in. that will support a load of $P = 10$ kip. Assume that the rods are pin connected at their ends. Use a factor of safety with respect to buckling of 1.8.

*13–28. The linkage is made using two A992 steel rods, each having a circular cross section. If each rod has a diameter of 2 in., determine the largest load it can support without causing any rod to buckle if the factor of safety against buckling is 1.8. Assume that the rods are pin connected at their ends.

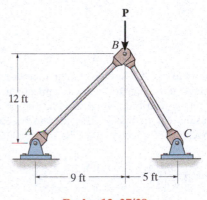

Probs. 13–27/28

13–29. The linkage is made using two A-36 steel rods, each having a circular cross section. Determine the diameter of each rod to the nearest $\frac{1}{8}$ in. that will support the 900-lb load. Assume that the rods are pin connected at their ends. Use a factor of safety with respect to buckling of F.S. = 1.8.

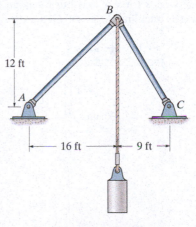

Prob. 13–29

13

13–30. The linkage is made using two A-36 steel rods, each having a circular cross section. If each rod has a diameter of $\frac{3}{4}$ in., determine the largest load it can support without causing any rod to buckle. Assume that the rods are pin connected at their ends.

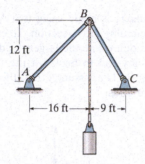

Prob. 13–30

13–31. The steel bar AB has a rectangular cross section. If it is pin connected at its ends, determine the maximum allowable intensity w of the distributed load that can be applied to BC without causing AB to buckle. Use a factor of safety with respect to buckling of 1.5. $E_{st} = 200$ GPa, $\sigma_Y = 360$ MPa.

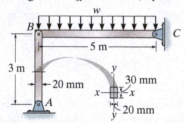

Prob. 13–31

*13–32. Determine if the frame can support a load of $P = 20$ kN if the factor of safety with respect to buckling of member AB is F.S. = 3. Assume that AB is made of steel and is pinned at its ends for x–x axis buckling and fixed at its ends for y–y axis buckling. $E_{st} = 200$ GPa, $\sigma_Y = 360$ MPa.

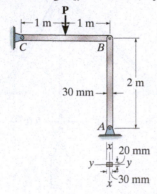

Prob. 13–32

13–33. Determine the maximum allowable load P that can be applied to member BC without causing member AB to buckle. Assume that AB is made of steel and is pinned at its ends for x–x axis buckling and fixed at its ends for y–y axis buckling. Use a factor of safety with respect to buckling of F.S. = 3. $E_{st} = 200$ GPa, $\sigma_Y = 360$ MPa.

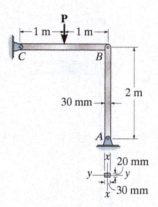

Prob. 13–33

13–34. A 6061-T6 aluminum alloy solid circular rod of length 4 m is pinned at both of its ends. If it is subjected to an axial load of 15 kN and F.S. = 2 against buckling, determine the minimum required diameter of the rod to the nearest mm.

13–35. A 6061-T6 aluminum alloy solid circular rod of length 4 m is pinned at one end while fixed at the other end. If it is subjected to an axial load of 15 kN and F.S. = 2 against buckling, determine the minimum required diameter of the rod to the nearest mm.

*13–36. The members of the truss are assumed to be pin connected. If member BD is an A992 steel rod of radius 2 in., determine the maximum load P that can be supported by the truss without causing the member to buckle.

13–37. Solve Prob. 13–36 for member AB, which has a radius of 2 in.

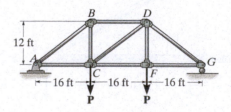

Probs. 13–36/37

13–38. The truss is made from A992 steel bars, each of which has a circular cross section with a diameter of 1.5 in. Determine the maximum force P that can be applied without causing any of the members to buckle. The members are pin connected at their ends.

13–39. The truss is made from A992 steel bars, each of which has a circular cross section. If the applied load $P = 10$ kip, determine the diameter of member AB to the nearest $\frac{1}{8}$ in. that will prevent this member from buckling. The members are pin connected at their ends.

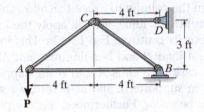

Probs. 13–38/39

*13–40. The steel bar AB of the frame is assumed to be pin connected at its ends for y–y axis buckling. If $P = 18$ kN, determine the factor of safety with respect to buckling about the y–y axis. $E_{st} = 200$ GPa, $\sigma_Y = 360$ MPa.

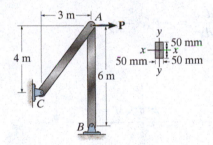

Prob. 13–40

13–41. The ideal column has a weight w (force/length) and is subjected to the axial load P. Determine the maximum moment in the column at midspan. EI is constant. *Hint:* Establish the differential equation for deflection, Eq. 13–1, with the origin at the midspan. The general solution is $v = C_1 \sin kx + C_2 \cos kx + (w/(2P))x^2 - (wL/(2P))x - (wEI/P^2)$ where $k^2 = P/EI$.

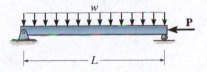

Prob. 13–41

13–42. The ideal column is subjected to the force $\mathbf{F}$ at its midpoint and the axial load $\mathbf{P}$. Determine the maximum moment in the column at midspan. EI is constant. *Hint:* Establish the differential equation for deflection, Eq. 13–1. The general solution is $v = C_1 \sin kx + C_2 \cos kx - c^2 x/k^2$, where $c^2 = F/2EI, k^2 = P/EI$.

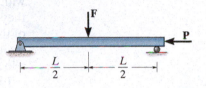

Prob. 13–42

13–43. The column with constant EI has the end constraints shown. Determine the critical load for the column.

Prob. 13–43

*13–44. Consider an ideal column as in Fig. 13–10c, having both ends fixed. Show that the critical load on the column is $P_{cr} = 4\pi^2 EI/L^2$. *Hint:* Due to the vertical deflection of the top of the column, a constant moment $\mathbf{M}'$ will be developed at the supports. Show that $d^2v/dx^2 + (P/EI)v = M'/EI$. The solution is of the form $v = C_1 \sin(\sqrt{P/EI}x) + C_2 \cos(\sqrt{P/EI}x) + M'/P$.

13–45. Consider an ideal column as in Fig. 13–10d, having one end fixed and the other pinned. Show that the critical load on the column is $P_{cr} = 20.19EI/L^2$. *Hint:* Due to the vertical deflection at the top of the column, a constant moment $\mathbf{M}'$ will be developed at the fixed support and horizontal reactive forces $\mathbf{R}'$ will be developed at both supports. Show that $d^2v/dx^2 + (P/EI)v = (R'/EI)(L - x)$. The solution is of the form $v = C_1\sin(\sqrt{P/EI}x) + C_2 \cos(\sqrt{P/EI}x) + (R'/P)(L - x)$. After application of the boundary conditions show that $\tan(\sqrt{P/EIL}) = \sqrt{P/EI}L$. Solve numerically for the smallest nonzero root.

*13.4   THE SECANT FORMULA

The Euler formula was derived assuming the load P is applied through the centroid of the column's cross-sectional area and that the column is perfectly straight. Actually this is quite unrealistic, since a manufactured column is never perfectly straight, nor is the application of the load known with great accuracy. In reality, then, columns never suddenly buckle; instead they begin to bend, although ever so slightly, immediately upon application of the load. As a result, the actual criterion for load application should be limited, either to a specified sidesway deflection of the column, or by not allowing the maximum stress in the column to exceed an allowable stress.

To study the effect of an eccentric loading, we will apply the load P to the column at a distance e from its centroid, Fig. 13–13a. This loading is statically equivalent to the axial load P and bending moment $M' = Pe$ shown in Fig. 13–13b. In both cases, the ends A and B are supported so that they are free to rotate (pin supported), and as before, we will only consider linear elastic material behavior. Furthermore, the x–v plane is a plane of symmetry for the cross-sectional area.

From the free-body diagram of the arbitrary section, Fig. 13–13c, the internal moment in the column is

$$M = -P(e + v) \tag{13–13}$$

And so the differential equation for the deflection curve becomes

$$EI\frac{d^2v}{dx^2} = -P(e + v)$$

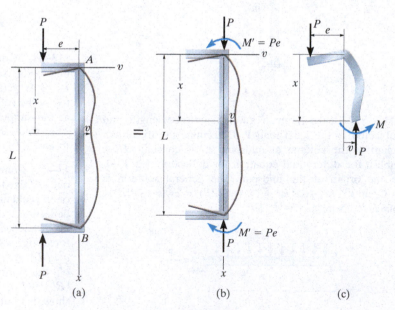

Fig. 13–13

or

$$\frac{d^2v}{dx^2} + \frac{P}{EI}v = -\frac{P}{EI}e$$

This equation is similar to Eq. 13–7, and its solution consists of both complementary and particular solutions, namely,

$$v = C_1 \sin\sqrt{\frac{P}{EI}}x + C_2 \cos\sqrt{\frac{P}{EI}}x - e \qquad (13\text{--}14)$$

To evaluate the constants we must apply the boundary conditions. At $x = 0, v = 0$, so $C_2 = e$. And at $x = L, v = 0$, which gives

$$C_1 = \frac{e[1 - \cos(\sqrt{P/EI}L)]}{\sin(\sqrt{P/EI}L)}$$

Since $1 - \cos(\sqrt{P/EI}L) = 2\sin^2(\sqrt{P/EI}L/2)$ and $\sin(\sqrt{P/EI}L) = 2\sin(\sqrt{P/EI}L/2)\cos(\sqrt{P/EI}L/2)$, we have

$$C_1 = e\tan\left(\sqrt{\frac{P}{EI}}\frac{L}{2}\right)$$

Hence, the deflection curve, Eq. 13–14, becomes

$$v = e\left[\tan\left(\sqrt{\frac{P}{EI}}\frac{L}{2}\right)\sin\left(\sqrt{\frac{P}{EI}}x\right) + \cos\left(\sqrt{\frac{P}{EI}}x\right) - 1\right] \qquad (13\text{--}15)$$

Maximum Deflection. Due to symmetry of loading, both the maximum deflection and maximum stress occur at the column's midpoint. Therefore, at $x = L/2$,

$$\boxed{v_{\max} = e\left[\sec\left(\sqrt{\frac{P}{EI}}\frac{L}{2}\right) - 1\right]} \qquad (13\text{--}16)$$

Notice that if e approaches zero, then $v_{\max}$ approaches zero. However, if the terms in the brackets approach infinity as e approaches zero, then $v_{\max}$ will have a nonzero value. Mathematically, this represents the behavior of an axially loaded column at failure when subjected to the critical load P_{cr}. Therefore, to find P_{cr} we require

$$\sec\left(\sqrt{\frac{P_{cr}}{EI}}\frac{L}{2}\right) = \infty$$

$$\sqrt{\frac{P_{cr}}{EI}}\frac{L}{2} = \frac{\pi}{2}$$

$$P_{cr} = \frac{\pi^2 EI}{L^2} \qquad (13\text{--}17)$$

which is the same result found from the Euler formula, Eq. 13–5.

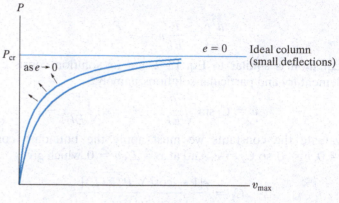

Fig. 13–14

If Eq. 13–16 is plotted for various values of eccentricity e, it results in a family of curves shown in Fig. 13–14. Here the critical load becomes an asymptote to the curves and represents the unrealistic case of an ideal column ($e = 0$). The results developed here apply only for small sidesway deflections, and so they certainly apply if the column is long and slender.

Notice that the curves in Fig. 13–14 show a *nonlinear* relationship between the load P and the deflection v. As a result, the principle of superposition *cannot be used* to determine the total deflection of a column. In other words, the deflection must be determined by applying the *total load* to the column, not a series of component loads. Furthermore, due to this nonlinear relationship, any factor of safety used for design purposes must be applied to the load and not to the stress.

The column supporting this crane is unusually long. It will be subjected not only to uniaxial load, but also a bending moment. To ensure it will not buckle, it should be braced at the roof as a pin connection.

The Secant Formula.

The maximum stress in an eccentrically loaded column is caused by both the axial load and the moment, Fig. 13–15a. Maximum moment occurs at the column's midpoint, and using Eqs. 13–13 and 13–16, it has a magnitude of

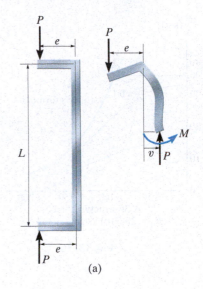

$$M = |P(e + v_{max})| \qquad M = Pe \sec\left(\sqrt{\frac{P}{EI}} \frac{L}{2}\right) \qquad (13\text{–}18)$$

As shown in Fig. 13–15b, the maximum stress in the column is therefore

$$\sigma_{max} = \frac{P}{A} + \frac{Mc}{I}; \qquad \sigma_{max} = \frac{P}{A} + \frac{Pec}{I}\sec\left(\sqrt{\frac{P}{EI}} \frac{L}{2}\right)$$

Since the radius of gyration is $r = \sqrt{I/A}$, the above equation can be written in a form called the *secant formula*:

$$\sigma_{max} = \frac{P}{A}\left[1 + \frac{ec}{r^2}\sec\left(\frac{L_e}{2r}\sqrt{\frac{P}{EA}}\right)\right] \qquad (13\text{–}19)$$

Axial stress

+

Bending stress

=

σ_{max}

Resultant stress

(b)

Fig. 13–15

Here

σ_{max} = maximum *elastic stress* in the column, which occurs at the inner concave side at the column's midpoint. This stress is compressive.

P = vertical load applied to the column. $P < P_{cr}$ unless $e = 0$; then $P = P_{cr}$ (Eq. 13–5).

e = eccentricity of the load P, measured from the centroidal axis of the column's cross-sectional area to the line of action of P

c = distance from the centroidal axis to the outer fiber of the column where the maximum compressive stress σ_{max} occurs

A = cross-sectional area of the column

L_e = unsupported length of the column *in the plane of bending*. Application is restricted to members that are pin connected, $L_e = L$, or have one end free and the other end fixed, $L_e = 2L$.

E = modulus of elasticity for the material

r = radius of gyration, $r = \sqrt{I/A}$, where I is calculated about the centroidal or bending axis

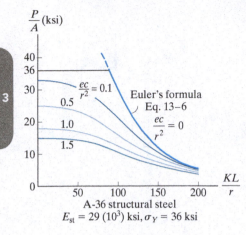

A-36 structural steel
$E_{st} = 29 (10^3)$ ksi, $\sigma_Y = 36$ ksi

Fig. 13–16

Graphs of Eq. 13–19 for various values of the *eccentricity ratio ec/r^2* are plotted in Fig. 13–16 for a structural-grade A-36 steel. Note that when $e \to 0$, or when $ec/r^2 \to 0$, Eq. 13–16 gives $\sigma_{max} = P/A$, where P is the critical load on the column, defined by Euler's formula. Since the results are valid only for elastic loadings, the stresses shown in the figure cannot exceed $\sigma_Y = 36$ ksi, represented here by the horizontal line.

By inspection, the curves indicate that changes in the eccentricity ratio have a marked effect on the load-carrying capacity of columns with *small* slenderness ratios. However, columns that have large slenderness ratios tend to fail at or near the Euler critical load regardless of the eccentricity ratio, since the curves bunch together. Therefore, when Eq. 13–19 is used for design purposes, it is important to have a somewhat accurate value for the eccentricity ratio for shorter-length columns.

Design.   Once the eccentricity ratio is specified, the column data can be substituted into Eq. 13–19. If a value of $\sigma_{max} = \sigma_Y$ is considered, then the corresponding load P_Y can be determined numerically, since the equation is transcendental and cannot be solved explicitly for P_Y. As a design aid, computer software, or graphs such as those in Fig. 13–16, can also be used to determine P_Y. Realize that due to the eccentric application of P_Y, this load will *always be smaller* than the critical load P_{cr}, which assumes (unrealistically) that the column is axially loaded.

IMPORTANT POINTS

- Due to imperfections in manufacturing or specific application of the load, a column will never suddenly buckle; instead, it begins to bend as it is loaded.

- The load applied to a column is related to its deflection in a nonlinear manner, and so the principle of superposition does not apply.

- As the slenderness ratio increases, eccentrically loaded columns tend to fail at or near the Euler buckling load.

EXAMPLE 13.4

The W8 × 40 A992 steel column shown in Fig. 13–17a is fixed at its base and braced at the top so that it is fixed from displacement, yet free to rotate about the y–y axis. Also, it can sway to the side in the y–z plane. Determine the maximum eccentric load the column can support before it either begins to buckle or the steel yields. Take $\sigma_Y = 50$ ksi.

SOLUTION

From the support conditions it is seen that about the y–y axis the column behaves as if it were pinned at its top and fixed at its bottom, and subjected to an axial load P, Fig. 13–17b. About the x–x axis the column is free at the top and fixed at the bottom, and it is subjected to both an axial load P and moment $M = P(9 \text{ in.})$, Fig. 13–17c.

y–y Axis Buckling. From Fig. 13–10d the effective length factor is $K_y = 0.7$, so $(KL)_y = 0.7(12)$ ft $= 8.40$ ft $= 100.8$ in. Using the table in Appendix B to determine I_y for the W8 × 40 section and applying Eq. 13–11, we have

$$(P_{cr})_y = \frac{\pi^2 E I_y}{(KL)_y^2} = \frac{\pi^2[29(10^3) \text{ ksi}](49.1 \text{ in}^4)}{(100.8 \text{ in.})^2} = 1383 \text{ kip}$$

x–x Axis Yielding. From Fig. 13–10b, $K_x = 2$, so $(KL)_x = 2(12)$ ft $= 24$ ft $= 288$ in. Again using the table in Appendix B, we have $A = 11.7 \text{ in}^2$, $c = 8.25 \text{ in.}/2 = 4.125$ in., and $r_x = 3.53$ in. Applying the secant formula,

$$\sigma_Y = \frac{P_x}{A}\left[1 + \frac{ec}{r_x^2}\sec\left(\frac{(KL)_x}{2r_x}\sqrt{\frac{P_x}{EA}}\right)\right]$$

$$50 \text{ ksi} = \frac{P_x}{11.7 \text{ in}^2}\left[1 + \frac{9 \text{ in. } (4.125 \text{ in.})}{(3.53 \text{ in.})^2}\sec\left(\frac{(288 \text{ in.})}{2\,(3.53 \text{ in.})}\sqrt{\frac{P_x}{29(10^3) \text{ ksi } (11.7 \text{ in}^2)}}\right)\right]$$

$$585 = P_x[1 + 2.979 \sec(0.0700\sqrt{P_x})]$$

Solving for P_x by trial and error, noting that the argument for the secant is in radians, we get

$$P_x = 115 \text{ kip} \qquad\qquad Ans.$$

Since this value is less than $(P_{cr})_y = 1383$ kip, failure will occur about the x–x axis.

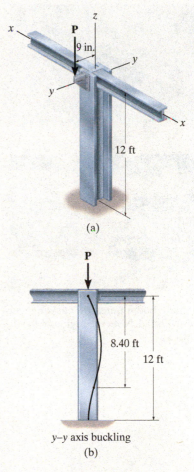

(a)

y–y axis buckling
(b)

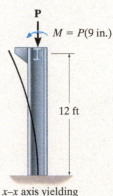

$M = P(9 \text{ in.})$

12 ft

x–x axis yielding
(c)

Fig. 13–17

13

This crane boom failed by buckling caused by an overload. Note the region of localized collapse.

*13.5   INELASTIC BUCKLING

In engineering practice, columns are generally classified according to the type of stresses developed within them at the time of failure. *Long slender columns* will become unstable when the compressive stress remains elastic. The failure that occurs is referred to as *elastic buckling*. *Intermediate columns* fail due to *inelastic buckling*, meaning that the compressive stress at failure is greater than the material's proportional limit. And *short columns*, sometimes called *posts*, do not become unstable; rather the material simply yields or fractures.

Application of the Euler equation requires that the stress in the column remain *below* the material's yield point (actually the proportional limit) when the column buckles, and so this equation applies only to long columns. In practice, however, many columns have intermediate lengths. One way to study the behavior of these columns is to modify the Euler equation so that it applies for inelastic buckling.

To show how this can be done, consider the material to have a stress–strain diagram as shown in Fig. 13–18a. Here the proportional limit is σ_{pl}, and the modulus of elasticity, or slope of the line AB, is E. If the column has a slenderness ratio that is *less* than its value at the proportional limit, $(KL/r)_{pl}$, then from the Euler equation, the critical stress in the column will be greater than σ_{pl} in order to buckle the column. For example, suppose a column has a slenderness ratio of $(KL/r)_1 < (KL/r)_{pl}$, with a corresponding critical stress $\sigma_D > \sigma_{pl}$, Fig. 13–18. When the column is *about to buckle*, the change in stress and strain that occurs in the column is within a *small range* $\Delta\sigma$ and $\Delta\epsilon$, so that the modulus of elasticity or stiffness for the material within this range can be taken as the *tangent modulus* $E_t = \Delta\sigma/\Delta\epsilon$. In other words, at the time of failure, the column behaves as if it were made of a material that has a *lower stiffness* than when it behaves elastically, $E_t < E$.

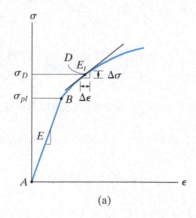

(a)

Fig. 13–18

In general, therefore, as the slenderness ratio (KL/r) continues to decrease, the *critical stress* for a column will continue to rise; and from the $\sigma{-}\epsilon$ diagram, the *tangent modulus* for the material will *decrease*. Using this idea, we can modify Euler's equation to include these cases of inelastic buckling by substituting the material's tangent modulus E_t for E, so that

$$\sigma_{cr} = \frac{\pi^2 E_t}{(KL/r)^2} \qquad (13\text{--}20)$$

This is the so-called **tangent modulus** or **Engesser equation**, proposed by F. Engesser in 1889. A plot of this equation for intermediate and short-length columns made of a material having the $\sigma{-}\epsilon$ diagram in Fig. 13–18a is shown in Fig. 13–18b.

As stated before, no *actual column* can be considered to either be perfectly straight or loaded along its centroidal axis, as assumed here, and therefore it is indeed very difficult to develop an expression that will provide a complete analysis of inelastic buckling. In spite of this, experimental testing of a large number of columns, each of which approximates the ideal column, has shown that Eq. 13–20 is *reasonably accurate* in predicting the column's critical stress. Furthermore, this tangent modulus approach to modeling inelastic column behavior is relatively easy to apply.*

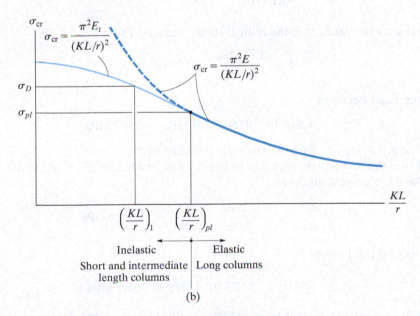

Fig. 13–18 (cont.)

*Other theories, such as Shanley's theory, have also been used to provide a description of inelastic buckling. Details can be found in books related to column stability.

EXAMPLE 13.5

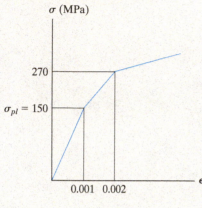

Fig. 13–19

A solid rod has a diameter of 30 mm and is 600 mm long. It is made of a material that can be modeled by the stress–strain diagram shown in Fig. 13–19. If it is used as a pin-supported column, determine the critical load.

SOLUTION

The radius of gyration is

$$r = \sqrt{\frac{I}{A}} = \sqrt{\frac{(\pi/4)(15 \text{ mm})^4}{\pi(15 \text{ mm})^2}} = 7.5 \text{ mm}$$

and therefore the slenderness ratio is

$$\frac{KL}{r} = \frac{1(600 \text{ mm})}{7.5 \text{ mm}} = 80$$

Applying Eq. 13–20 we have,

$$\sigma_{cr} = \frac{\pi^2 E_t}{(KL/r)^2} = \frac{\pi^2 E_t}{(80)^2} = 1.542(10^{-3})E_t \qquad (1)$$

First we will assume that the critical stress is elastic. From Fig. 13–19,

$$E = \frac{150 \text{ MPa}}{0.001} = 150 \text{ GPa}$$

Thus, Eq. 1 becomes

$$\sigma_{cr} = 1.542(10^{-3})[150(10^3)] \text{ MPa} = 231.3 \text{ MPa}$$

Since $\sigma_{cr} > \sigma_{pl} = 150$ MPa, inelastic buckling occurs.

From the second line segment of the $\sigma-\epsilon$ diagram, Fig. 13–19, we can obtain the tangent modulus.

$$E_t = \frac{\Delta\sigma}{\Delta\epsilon} = \frac{270 \text{ MPa} - 150 \text{ MPa}}{0.002 - 0.001} = 120 \text{ GPa}$$

Applying Eq. 1 yields

$$\sigma_{cr} = 1.542(10^{-3})[120(10^3)] \text{ MPa} = 185.1 \text{ MPa}$$

Since this value falls within the limits of 150 MPa and 270 MPa, it is indeed the critical stress.

The critical load on the rod is therefore

$$P_{cr} = \sigma_{cr}A = 185.1(10^6) \text{ Pa}[\pi(0.015 \text{ m})^2] = 131 \text{ kN} \qquad \textit{Ans.}$$

PROBLEMS

13–46. The wood column is fixed at its base and free at its top. Determine the load P that can be applied to the edge of the column without causing the column to fail either by buckling or by yielding. $E_w = 12$ GPa, $\sigma_Y = 55$ MPa.

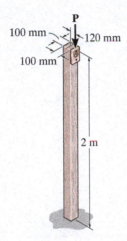

Prob. 13–46

13–47. The W10 × 12 structural A-36 steel column is used to support a load of 4 kip. If the column is fixed at the base and free at the top, determine the sidesway deflection at the top of the column due to the loading.

*13–48. The W10 × 12 structural A-36 steel column is used to support a load of 4 kip. If the column is fixed at its base and free at its top, determine the maximum stress in the column due to this loading.

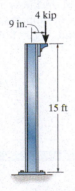

Probs. 13–47/48

13–49. The aluminum column is fixed at the bottom and free at the top. Determine the maximum force P that can be applied at A without causing it to buckle or yield. Use a factor of safety of 3 with respect to buckling and yielding. $E_{al} = 70$ GPa, $\sigma_Y = 95$ MPa.

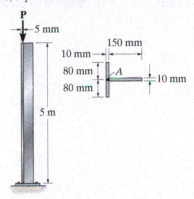

Prob. 13–49

13–50. A column of intermediate length buckles when the compressive stress is 40 ksi. If the slenderness ratio is 60, determine the tangent modulus.

13–51. The aluminum rod is fixed at its base and free at its top. If the eccentric load $P = 200$ kN is applied, determine the greatest allowable length L of the rod so that it does not buckle or yield. $E_{al} = 72$ GPa, $\sigma_Y = 410$ MPa.

*13–52. The aluminum rod is fixed at its base and free and at its top. If the length of the rod is $L = 2$ m, determine the greatest allowable load P that can be applied so that the rod does not buckle or yield. Also, determine the largest sidesway deflection of the rod due to the loading. $E_{al} = 72$ GPa, $\sigma_Y = 410$ MPa.

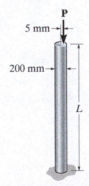

Probs. 13–50/51/52

13–53. Assume that the wood column is pin connected at its base and top. Determine the maximum eccentric load P that can be applied without causing the column to buckle or yield. $E_w = 1.8(10^3)$ ksi, $\sigma_Y = 8$ ksi.

13–54. Assume that the wood column is pinned top and bottom for movement about the x–y axis, and fixed at the bottom and free at the top for movement about the y–y axis. Determine the maximum eccentric load P that can be applied without causing the column to buckle or yield. $E_w = 1.8(10^3)$ ksi, $\sigma_Y = 8$ ksi.

13–57. The A992 steel rectangular hollow section column is pinned at both ends. If it has a length of $L = 14$ ft, determine the maximum allowable eccentric force P it can support without causing it to either buckle or yield.

13–58. The A992 steel rectangular hollow section column is pinned at both ends. If it is subjected to the eccentric force $P = 45$ kip, determine its maximum allowable length L without causing it to either buckle or yield.

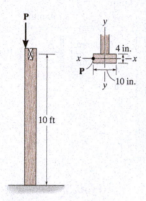

Probs. 13–53/54

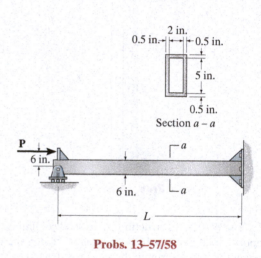

Probs. 13–57/58

13–55. The wood column is pinned at its base and top. If the eccentric force $P = 10$ kN is applied to the column, investigate whether the column is adequate to support this loading without buckling or yielding. Take $E = 10$ GPa and $\sigma_Y = 15$ MPa.

*13–56. The wood column is pinned at its base and top. Determine the maximum eccentric force P the column can support without causing it to either buckle or yield. Take $E = 10$ GPa and $\sigma_Y = 15$ MPa.

13–59. The tube is made of copper and has an outer diameter of 35 mm and a wall thickness of 7 mm. Determine the eccentric load P that it can support without failure. The tube is pin supported at its ends. $E_{cu} = 120$ GPa, $\sigma_Y = 750$ MPa.

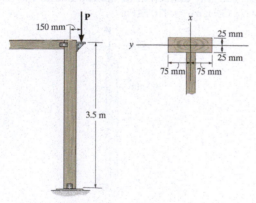

Probs. 13–55/56

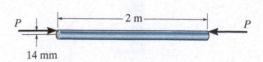

Prob. 13–59

*13–60. The wood column is pinned at its base and top. If $L = 5$ ft, determine the maximum eccentric load P that can be applied without causing the column to buckle or yield. $E_w = 1.8(10^3)$ ksi, $\sigma_Y = 8$ ksi.

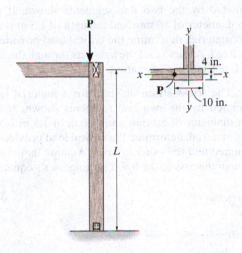

Prob. 13–60

13–61. The brass rod is fixed at one end and free at the other end. If the eccentric load $P = 200$ kN is applied, determine the greatest allowable length L of the rod so that it does not buckle or yield. $E_{br} = 101$ GPa, $\sigma_Y = 69$ MPa.

13–62. The brass rod is fixed at one end and free at the other end. If the length of the rod is $L = 2$ m, determine the greatest allowable load P that can be applied so that the rod does not buckle or yield. Also, determine the largest sideway deflection of the rod due to the loading. $E_{br} = 101$ GPa, $\sigma_Y = 69$ MPa.

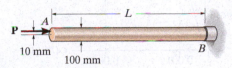

Probs. 13–61/62

13–63. A W12 × 26 structural A992 steel column is pin connected at its ends and has a length $L = 11.5$ ft. Determine the maximum eccentric load P that can be applied so the column does not buckle or yield. Compare this value with an axial critical load P' applied through the centroid of the column.

*13–64. A W14 × 30 structural A-36 steel column is pin connected at its ends and has a length $L = 10$ ft. Determine the maximum eccentric load P that can be applied so the column does not buckle or yield. Compare this value with an axial critical load P' applied through the centroid of the column.

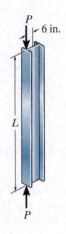

Probs. 13–63/64

13–65. Determine the maximum eccentric load P the 2014-T6-aluminum-alloy strut can support without causing it either to buckle or yield. The ends of the strut are pin connected.

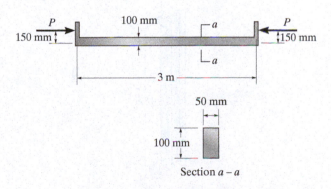

Prob. 13–65

13–66. The 6061-T6 aluminum alloy solid shaft is fixed at one end but free at the other end. If the shaft has a diameter of 100 mm, determine its maximum allowable length L if it is subjected to the eccentric force $P = 80$ kN.

13–67. The 6061-T6 aluminum alloy solid shaft is fixed at one end but free at the other end. If the length is $L = 3$ m, determine its minimum required diameter if it is subjected to the eccentric force $P = 60$ kN.

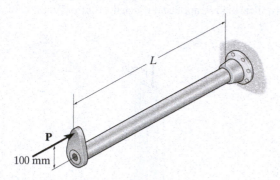

Probs. 13–66/67

*13–68.** The W14 × 53 structural A992 steel column is fixed at its base and free at its top. If $P = 75$ kip, determine the sidesway deflection at its top and the maximum stress in the column.

13–69. The W14 × 53 column is fixed at its base and free at its top. Determine the maximum eccentric load P that it can support without causing it to buckle or yield. $E_{st} = 29(10^3)$ ksi, $\sigma_Y = 50$ ksi.

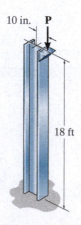

Probs. 13–68/69

13–70. The stress–strain diagram for a material can be approximated by the two line segments shown. If a bar having a diameter of 80 mm and a length of 1.5 m is made from this material, determine the critical load provided the ends are pinned. Assume that the load acts through the axis of the bar. Use Engesser's equation.

13–71. The stress–strain diagram for a material can be approximated by the two line segments shown. If a bar having a diameter of 80 mm and a length of 1.5 m is made from this material, determine the critical load provided the ends are fixed. Assume that the load acts through the axis of the bar. Use Engesser's equation.

*13–72.** The stress–strain diagram for a material can be approximated by the two line segments shown. If a bar having a diameter of 80 mm and length of 1.5 m is made from this material, determine the critical load provided one end is pinned and the other is fixed. Assume that the load acts through the axis of the bar. Use Engesser's equation.

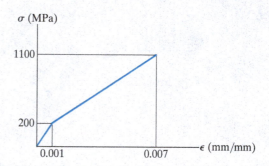

Probs. 13–70/71/72

13–73. The stress–strain diagram for the material of a column can be approximated as shown. Plot P/A vs. KL/r for the column.

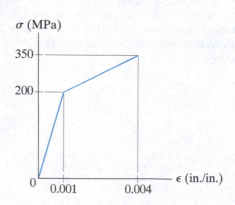

Prob. 13–73

13–74. Construct the buckling curve, P/A versus L/r, for a column that has a bilinear stress–strain curve in compression as shown.

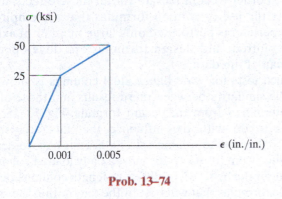

Prob. 13–74

13–75. The stress–strain diagram of the material can be approximated by the two line segments. If a bar having a diameter of 80 mm and a length of 1.5 m is made from this material, determine the critical load provided the ends are pinned. Assume that the load acts through the axis of the bar. Use Engesser's equation.

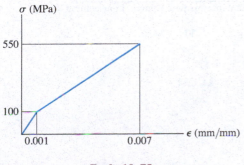

Prob. 13–75

*13–76. The stress–strain diagram of the material can be approximated by the two line segments. If a bar having a diameter of 80 mm and a length of 1.5 m is made from this material, determine the critical load provided the ends are fixed. Assume that the load acts through the axis of the bar. Use Engesser's equation.

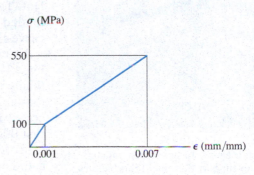

Prob. 13–76

13–77. The stress–strain diagram of the material can be approximated by the two line segments. If a bar having a diameter of 80 mm and a length of 1.5 m is made from this material, determine the critical load provided one end is pinned and the other is fixed. Assume that the load acts through the axis of the bar. Use Engesser's equation.

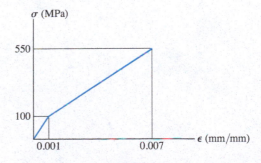

Prob. 13–77

13

These long unbraced timber columns are used to support the roof of this building.

*13.6   DESIGN OF COLUMNS FOR CONCENTRIC LOADING

Practically speaking, columns are not perfectly straight, and most have residual stresses in them, primarily due to nonuniform cooling during manufacture. Also, the supports for columns are less than exact, and the points of application and directions of loads are not known with absolute certainty. In order to compensate for all these effects, many design codes specify the use of column formulas that are empirical. Data found from experiments performed on a large number of axially loaded columns is plotted, and design formulas are developed by curve-fitting the mean of the data.

An example of such tests for wide-flange steel columns is shown in Fig. 13–20. Notice the similarity between these results and those of the family of curves determined from the secant formula, Fig. 13–16. The reason for this has to do with the influence that an "accidental" eccentricity ratio ec/r^2 has on the column's strength. Tests have indicated that this ratio will range from 0.1 to 0.6 for most axially loaded columns.

In order to account for the behavior of different-length columns, design codes usually provide formulas that will best fit the data within the short, intermediate, and long column range. The following examples of these formulas for steel, aluminum, and wood columns will now be discussed.

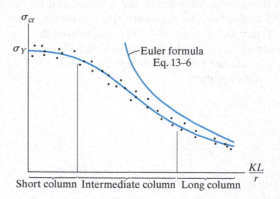

Fig. 13–20

Steel Columns.

Columns made of structural steel can be designed on the basis of formulas proposed by the Structural Stability Research Council (SSRC). Factors of safety have been applied to these formulas and adopted as specifications for building construction by the American Institute of Steel Construction (AISC). Basically these specifications provide two formulas for column design, each of which gives the maximum allowable stress in the column for a specific range of slenderness ratios.*

For long columns the Euler formula is proposed, i.e., $\sigma_{max} = \pi^2 E / (KL/r)^2$. Application of this formula requires that a factor of safety F.S. $= \frac{23}{12} \approx 1.92$ be applied. Thus, for design,

$$\sigma_{allow} = \frac{12\pi^2 E}{23(KL/r)^2} \qquad \left(\frac{KL}{r}\right)_c \leq \frac{KL}{r} \leq 200 \qquad (13\text{–}21)$$

As stated, this equation is applicable for a slenderness ratio bounded by 200 and a calculated value for $(KL/r)_c$. Through experiments it has been determined that compressive residual stresses can exist in rolled-formed steel sections that may be as much as one-half the yield stress. Since the Euler formula only can be used for elastic material behavior, then if the additional stress placed on the column is greater than $\frac{1}{2}\sigma_Y$, the equation will not apply. Therefore the lower bound value of $(KL/r)_c$ is determined as follows:

$$\frac{1}{2}\sigma_Y = \frac{\pi^2 E}{(KL/r)_c^2} \qquad \text{or} \qquad \left(\frac{KL}{r}\right)_c = \sqrt{\frac{2\pi^2 E}{\sigma_Y}} \qquad (13\text{–}22)$$

Columns having slenderness ratios less than $(KL/r)_c$ are designed using an empirical formula that is parabolic and has the form

$$\sigma_{max} = \left[1 - \frac{(KL/r)^2}{2(KL/r)_c^2}\right]\sigma_Y$$

Since there is more uncertainty in the use of this formula for longer columns, it is divided by a factor of safety defined as follows:

$$\text{F.S.} = \frac{5}{3} + \frac{3}{8}\frac{(KL/r)}{(KL/r)_c} - \frac{(KL/r)^3}{8(KL/r)_c^3}$$

Here it is seen that F.S. $= \frac{5}{3} \approx 1.67$ at $KL/r = 0$ and increases to F.S. $= \frac{23}{12} \approx 1.92$ at $(KL/r)_c$. Hence, for design purposes,

$$\sigma_{allow} = \frac{\left[1 - \dfrac{(KL/r)^2}{2(KL/r)_c^2}\right]\sigma_Y}{(5/3) + [(3/8)(KL/r)/(KL/r)_c] - \left[(KL/r)^3/8(KL/r)_c^3\right]}$$

$$(13\text{–}23)$$

For comparison, Eqs. 13–21 and 13–23 are plotted in Fig. 13–21.

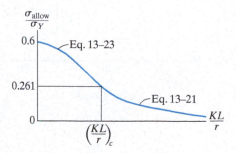

Fig. 13–21

*The current AISC code enables engineers to use one of two methods for design, namely, Load and Resistance Factor Design, and Allowable Stress Design. The latter is explained here.

Aluminum Columns.

Column design for structural aluminum is specified by the Aluminum Association using three equations, each applicable for a specific range of slenderness ratios. Since several types of aluminum alloy exist, there is a unique set of formulas for each type. For a common alloy (2014-T6) used in building construction, the formulas are

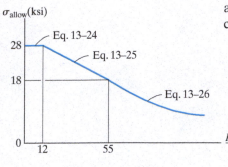

Fig. 13–22

$$\sigma_{allow} = 28 \text{ ksi} \quad 0 \le \frac{KL}{r} \le 12 \tag{13–24}$$

$$\sigma_{allow} = \left[30.7 - 0.23\left(\frac{KL}{r}\right)\right] \text{ksi} \quad 12 < \frac{KL}{r} < 55 \tag{13–25}$$

$$\sigma_{allow} = \frac{54\,000 \text{ ksi}}{(KL/r)^2} \quad 55 \le \frac{KL}{r} \tag{13–26}$$

These equations are plotted in Fig. 13–22. As shown, the first two represent straight lines and are used to model the effects of columns in the short and intermediate range. The third formula has the same form as the Euler formula and is used for long columns.

Timber Columns.

Columns used in timber construction often are designed using formulas published by the National Forest Products Association (NFPA) or the American Institute of Timber Construction (AITC). For example, the NFPA formulas for the allowable stress in short, intermediate, and long columns having a rectangular cross section of dimensions b and d, where $d < b$, are

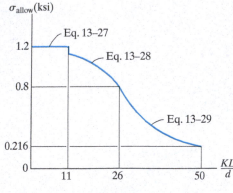

Fig. 13–23

$$\sigma_{allow} = 1.20 \text{ ksi} \quad 0 \le \frac{KL}{d} \le 11 \tag{13–27}$$

$$\sigma_{allow} = 1.20\left[1 - \frac{1}{3}\left(\frac{KL/d}{26.0}\right)^2\right] \text{ksi} \quad 11 < \frac{KL}{d} \le 26 \tag{13–28}$$

$$\sigma_{allow} = \frac{540 \text{ ksi}}{(KL/d)^2} \quad 26 < \frac{KL}{d} \le 50 \tag{13–29}$$

Here wood is assumed to have a modulus of elasticity of $E_w = 1.8(10^3)$ ksi and an allowable compressive stress of 1.2 ksi parallel to its grain. In particular, Eq. 13–29 has the same form as the Euler formula, having a factor of safety of 3. These three equations are plotted in Fig. 13–23.

EXAMPLE   13.6

An A992 steel W10 × 100 member is used as a pin-supported column, Fig. 13–24. Using the AISC column design formulas, determine the largest load that it can safely support.

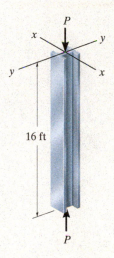

Fig. 13–24

SOLUTION

The following data for a W10 × 100 is taken from the table in Appendix B.

$$A = 29.4 \text{ in}^2 \quad r_x = 4.60 \text{ in.} \quad r_y = 2.65 \text{ in.}$$

Since $K = 1$ for both x and y axis buckling, the slenderness ratio is largest if r_y is used. Thus,

$$\frac{KL}{r} = \frac{1(16 \text{ ft})(12 \text{ in./ft})}{2.65 \text{ in.}} = 72.45$$

From Eq. 13–22, we have

$$\left(\frac{KL}{r} \right)_c = \sqrt{\frac{2\pi^2 E}{\sigma_Y}}$$

$$= \sqrt{\frac{2\pi^2[29(10^3) \text{ ksi}]}{50 \text{ ksi}}}$$

$$= 107$$

Here $0 < KL/r < (KL/r)_c$, so Eq. 13–23 applies.

$$\sigma_{\text{allow}} = \frac{\left[1 - \dfrac{(KL/r)^2}{2(KL/r)_c^2}\right]\sigma_Y}{(5/3) + [(3/8)(KL/r)/(KL/r)_c] - \left[(KL/r)^3/8(KL/r_c)^3\right]}$$

$$= \frac{[1 - (72.45)^2/2(107)^2]\, 50 \text{ ksi}}{(5/3) + [(3/8)(72.45/107)] - [(72.45)^3/8(107)^3]}$$

$$= 23.93 \text{ ksi}$$

The allowable load P on the column is therefore

$$\sigma_{\text{allow}} = \frac{P}{A}; \qquad 23.93 \text{ kip/in}^2 = \frac{P}{29.4 \text{ in}^2}$$

$$P = 704 \text{ kip} \qquad\qquad Ans.$$

EXAMPLE   13.7

The steel rod in Fig. 13–25 is to be used to support an axial load of 18 kip. If $E_{st} = 29(10^3)$ ksi and $\sigma_Y = 50$ ksi, determine the smallest diameter of the rod as allowed by the AISC specifications. The rod is fixed at both ends.

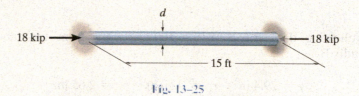

Fig. 13–25

SOLUTION

For a circular cross section the radius of gyration becomes

$$r = \sqrt{\frac{I}{A}} = \sqrt{\frac{(1/4)\pi(d/2)^4}{(1/4)\pi d^2}} = \frac{d}{4}$$

Applying Eq. 13–22, we have

$$\left(\frac{KL}{r}\right)_c = \sqrt{\frac{2\pi^2 E}{\sigma_Y}} = \sqrt{\frac{2\pi^2[29(10^3)\text{ ksi}]}{50\text{ ksi}}} = 107.0$$

Since the rod's radius of gyration is unknown, KL/r is unknown, and therefore a choice must be made as to whether Eq. 13–21 or Eq. 13–23 applies. We will consider Eq. 13–21. For a fixed-end column $K = 0.5$, and so

$$\sigma_{\text{allow}} = \frac{12\pi^2 E}{23(KL/r)^2}$$

$$\frac{18\text{ kip}}{(1/4)\pi d^2} = \frac{12\pi^2[29(10^3)\text{ kip/in}^2]}{23[0.5(15\text{ ft})(12\text{ in./ft})/(d/4)]^2}$$

$$\frac{22.92}{d^2} = 1.152d^2$$

$$d = 2.11\text{ in.}$$

Use

$$d = 2.25\text{ in.} = 2\tfrac{1}{4}\text{ in.} \qquad \textit{Ans.}$$

For this design, we must check the slenderness-ratio limits; i.e.,

$$\frac{KL}{r} = \frac{0.5(15\text{ ft})(12\text{ in./ft})}{(2.25\text{ in.}/4)} = 160$$

Since $107.0 < 160 < 200$, use of Eq. 13–21 is appropriate.

EXAMPLE 13.8

A bar having a length of 30 in. is used to support an axial compressive load of 12 kip, Fig. 13–26. It is pin supported at its ends and made of a 2014-T6 aluminum alloy. Determine the dimensions of its cross-sectional area if its width is to be twice its thickness.

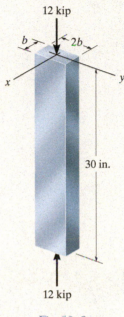

12 kip

30 in.

12 kip

Fig. 13–26

SOLUTION

Since $KL = 30$ in. is the same for both x and y axis buckling, the larger slenderness ratio is determined using the smaller radius of gyration, i.e., using $I_{min} = I_y$:

$$\frac{KL}{r_y} = \frac{KL}{\sqrt{I_y/A}} = \frac{1(30)}{\sqrt{(1/12)2b(b^3)/[2b(b)]}} = \frac{103.9}{b} \qquad (1)$$

Here we must apply Eq. 13–24, 13–25, or 13–26. Since we do not as yet know the slenderness ratio, we will begin by using Eq. 13–24.

$$\frac{P}{A} = 28 \text{ ksi}$$

$$\frac{12 \text{ kip}}{2b(b)} = 28 \text{ kip/in}^2$$

$$b = 0.463 \text{ in.}$$

Checking the slenderness ratio, we have

$$\frac{KL}{r} = \frac{103.9}{0.463} = 224.5 > 12$$

Try Eq. 13–26, which is valid for $KL/r \geq 55$,

$$\frac{P}{A} = \frac{54\,000 \text{ ksi}}{(KL/r)^2}$$

$$\frac{12}{2b(b)} = \frac{54\,000}{(103.9/b)^2}$$

$$b = 1.05 \text{ in.} \qquad \qquad \textit{Ans.}$$

From Eq. 1,

$$\frac{KL}{r} = \frac{103.9}{1.05} = 99.3 > 55 \quad \text{OK}$$

NOTE: Since $b \approx 1$ in., it would be satisfactory to choose a cross section with dimensions 1 in. by 2 in.

EXAMPLE 13.9

A board having cross-sectional dimensions of 5.5 in. by 1.5 in. is used to support an axial load of 5 kip, Fig. 13–27. If the board is assumed to be pin supported at its top and bottom, determine its *greatest* allowable length L as specified by the NFPA.

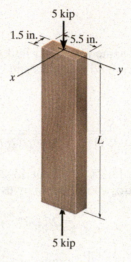

Fig. 13–27

SOLUTION

By inspection, the board will buckle about the y axis. In the NFPA equations, $d = 1.5$ in. Assuming that Eq. 13–29 applies, we have

$$\frac{P}{A} = \frac{540 \text{ ksi}}{(KL/d)^2}$$

$$\frac{5 \text{ kip}}{(5.5 \text{ in.})(1.5 \text{ in.})} = \frac{540 \text{ ksi}}{(1\,L/1.5 \text{ in.})^2}$$

$$L = 44.8 \text{ in.} \qquad \qquad \textit{Ans.}$$

Here

$$\frac{KL}{d} = \frac{1(44.8 \text{ in.})}{1.5 \text{ in.}} = 29.8$$

Since $26 < KL/d \le 50$, the solution is valid.

PROBLEMS

13–78. Determine the largest length of a W10 × 12 structural A992 steel section if it is pin supported and is subjected to an axial load of 28 kip. Use the AISC equations.

13–79. Using the AISC equations, select from Appendix B the lightest-weight wide-flange A992 steel column that is 14 ft long and supports an axial load of 40 kip. The ends are fixed.

*13–80. Using the AISC equations, select from Appendix B the lightest-weight wide-flange A992 steel column that is 14 ft long and supports an axial load of 40 kip. The ends are pinned. Take $\sigma_Y = 50$ ksi.

13–81. Determine the longest length of a W8 × 31 structural A992 steel section if it is pin supported and is subjected to an axial load of 130 kip. Use the AISC equations.

13–82. Using the AISC equations, select from Appendix B the lightest-weight wide-flange A992 steel column that is 12 ft long and supports an axial load of 20 kip. The ends are pinned.

13–83. Determine the longest length of a W10 × 12 structural A992 steel section if it is fixed supported and is subjected to an axial load of 28 kip. Use the AISC equations.

*13–84. Using the AISC equations, select from Appendix B the lightest-weight wide-flange A992 steel column that is 30 ft long and supports an axial load of 200 kip. The ends are fixed.

13–85. Determine the longest length of a W8 × 31 structural A992 steel section if it is pin supported and is subjected to an axial load of 18 kip. Use the AISC equations.

13–86. Using the AISC equations, select from Appendix B the lightest-weight wide-flange A992 steel column that is 12 ft long and supports an axial load of 40 kip. The ends are fixed.

13–87. A 5-ft-long rod is used in a machine to transmit an axial compressive load of 3 kip. Determine its smallest diameter if it is pin connected at its ends and is made of a 2014-T6 aluminum alloy.

*13–88. Determine the longest length of a W8 × 31 structural A992 steel column if it is to support an axial load of 10 kip. The ends are pinned.

13–89. Using the AISC equations, check if a column having the cross section shown can support an axial force of 1500 kN. The column has a length of 4 m, is made from A992 steel, and its ends are pinned.

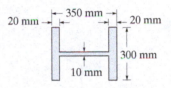

Prob. 13–89

13–90. The beam and column arrangement is used in a railroad yard for loading and unloading cars. If the maximum anticipated hoist load is 12 kip, determine if the W8 × 31 wide-flange A-36 steel column is adequate for supporting the load. The hoist travels along the bottom flange of the beam, 1 ft ≤ x ≤ 25 ft, and has negligible size. Assume the beam is pinned to the column at B and pin supported at A. The column is also pinned at C and it is braced so it will not buckle out of the plane of the loading.

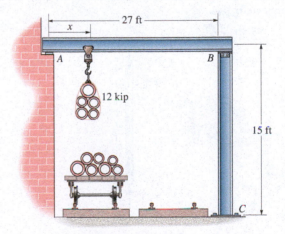

Prob. 13–90

13–91. The bar is made of a 2014-T6 aluminum alloy. Determine its smallest thickness b if its width is $5b$. Assume that it is pin connected at its ends.

*13–92. The bar is made of a 2014-T6 aluminum alloy. Determine its smallest thickness b if its width is $5b$. Assume that it is fixed connected at its ends.

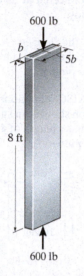

Probs. 13–91/92

13–93. The 1-in.-diameter rod is used to support an axial load of 5 kip. Determine its greatest allowable length L if it is made of 2014-T6 aluminum. Assume that the ends are pin connected.

13–94. The 1-in.-diameter rod is used to support an axial load of 5 kip. Determine its greatest allowable length L if it is made of 2014-T6 aluminum. Assume that the ends are fixed connected.

Probs. 13–93/94

13–95. The tube is 0.5 in. thick, is made of aluminum alloy 2014-T6, and is fixed connected at its ends. Determine the largest axial load that it can support.

*13–96. The tube is 0.5 in. thick, is made from aluminum alloy 2014-T6, and is fixed at its bottom and pinned at its top. Determine the largest axial load that it can support.

Probs. 13–95/96

13–97. The tube is 0.25 in. thick, is made of a 2014-T6 aluminum alloy, and is fixed at its bottom and pinned at its top. Determine the largest axial load that it can support.

13–98. The tube is 0.25 in. thick, is made of a 2014-T6 aluminum alloy, and is fixed connected at its ends. Determine the largest axial load that it can support.

13–99. The tube is 0.25 in. thick, is made of 2014-T6 aluminum alloy and is pin connected at its ends. Determine the largest axial load it can support.

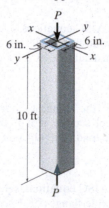

Probs. 13–97/98/99

*13–100. A rectangular wooden column has the cross section shown. If the column is 6 ft long and subjected to an axial force of $P = 15$ kip, determine the required minimum dimension a of its cross-sectional area to the nearest $\frac{1}{16}$ in. so that the column can safely support the loading. The column is pinned at both ends.

13–101. A rectangular wooden column has the cross section shown. If $a = 3$ in. and the column is 12 ft long, determine the allowable axial force P that can be safely supported by the column if it is pinned at its top and fixed at its base.

13–102. A rectangular wooden column has the cross section shown. If $a = 3$ in. and the column is subjected to an axial force of $P = 15$ kip, determine the maximum length the column can have to safely support the load. The column is pinned at its top and fixed at its base.

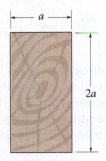

Probs. 13–100/101/102

13–103. The timber column has a square cross section and is assumed to be pin connected at its top and bottom. If it supports an axial load of 50 kip, determine its smallest side dimension a to the nearest $\frac{1}{2}$ in. Use the NFPA formulas.

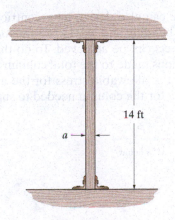

Prob. 13–103

*13–104. The bar is made of aluminum alloy 2014-T6. Determine its thickness b if its width is $1.5b$. Assume that it is fixed connected at its ends.

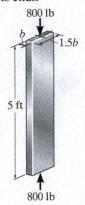

Prob. 13–104

13–105. The timber column has a length of 18 ft and is pin connected at its ends. Use the NFPA formulas to determine the largest axial force P that it can support.

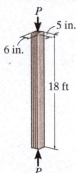

Prob. 13–105

13–106. The timber column has a length of 18 ft and is fixed connected at its ends. Use the NFPA formulas to determine the largest axial force P that it can support.

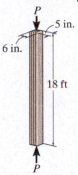

Prob. 13–106

13

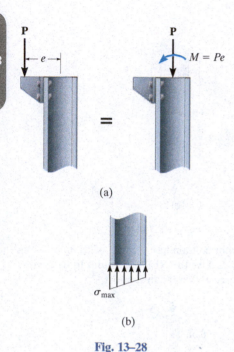

(a)

(b)

Fig. 13–28

*13.7   DESIGN OF COLUMNS FOR ECCENTRIC LOADING

When a column is required to support a load acting either at its edge or on an angle bracket or corbel attached to its side, Fig. 13–28a, then the bending moment $M = Pe$, which is caused by the eccentric loading, must be accounted for when the column is designed. There are several acceptable ways in which this is done in engineering practice. Here we will discuss two of the most common methods.

Use of Available Column Formulas. The stress distribution acting over the cross-sectional area of the column is shown in Fig. 13–28b. It is the result of a superposition of both the axial force P and the bending moment $M = Pe$. The maximum compressive stress is therefore

$$\sigma_{max} = \frac{P}{A} + \frac{Mc}{I} \tag{13–30}$$

If we conservatively *assume* that the entire cross section is subjected to the uniform stress σ_{max}, then we can compare σ_{max} with σ_{allow}, which is determined using one of the formulas given in Sec. 13.6. To be additionally conservative, calculation of σ_{allow} is done using the *largest* slenderness ratio for the column, regardless of the axis about which the column experiences bending. Then if

$$\sigma_{max} \le \sigma_{allow}$$

the column will support its intended loading. If this inequality does not hold, then the column's area A must be increased, and a new σ_{max} and σ_{allow} must be calculated. This method of design is rather simple to apply and works well for columns that are short or of intermediate length.

Interaction Formula. We can also design an eccentrically loaded column on the basis of how the bending and axial loads *interact*, so that a balance between these two effects can be achieved. To do this, we must consider the separate contributions made to the total column area by the axial force and the moment. If the allowable stress for the axial load is $(\sigma_a)_{allow}$, then the required area for the column needed to support P is

$$A_a = \frac{P}{(\sigma_a)_{allow}}$$

Similarly, if the allowable bending stress is $(\sigma_b)_{\text{allow}}$, then since $I = Ar^2$, the required area of the column needed to support the eccentric moment is determined from the flexure formula, that is,

$$A_b = \frac{Mc}{(\sigma_b)_{\text{allow}}\, r^2}$$

The total area A for the column needed to resist *both* the axial load and moment is therefore

$$A_a + A_b = \frac{P}{(\sigma_a)_{\text{allow}}} + \frac{Mc}{(\sigma_b)_{\text{allow}}\, r^2} \leq A$$

or

$$\frac{P/A}{(\sigma_a)_{\text{allow}}} + \frac{Mc/Ar^2}{(\sigma_b)_{\text{allow}}} \leq 1$$

$$\frac{\sigma_a}{(\sigma_a)_{\text{allow}}} + \frac{\sigma_b}{(\sigma_b)_{\text{allow}}} \leq 1 \qquad (13\text{--}31)$$

Here

σ_a = axial stress caused by the force P and determined from $\sigma_a = P/A$, where A is the cross-sectional area of the column

σ_b = bending stress caused by the eccentric load or applied moment. This stress is found from $\sigma_b = Mc/I$, where I is the moment of inertia of the cross-sectional area calculated about the bending or centroidal axis.

$(\sigma_a)_{\text{allow}}$ = allowable axial stress as defined by formulas given in Sec. 13.6 or by other design code specifications. For this purpose, always use the *largest* slenderness ratio for the column, regardless of the axis about which the column experiences bending.

$(\sigma_b)_{\text{allow}}$ = allowable bending stress as defined by code specifications

Each stress ratio in Eq. 13–31 indicates the contribution of axial load or bending moment. Since this equation shows how these loadings interact, this equation is sometimes referred to as the **interaction formula**. This design approach requires a trial-and-error procedure, where it is required that the designer *pick* an available column and then check to see if the inequality is satisfied. If it is not, a larger section is then chosen and the process repeated. An economical choice is made when the left side is close to but less than 1.

The interaction method is often specified in codes for the design of columns made of steel, aluminum, or timber. In particular, for allowable stress design, the American Institute of Steel Construction specifies the use of this equation only when the axial-stress ratio $\sigma_a/(\sigma_a)_{\text{allow}} \leq 0.15$. For other values of this ratio, a modified form of Eq. 13–31 is used.

Typical example of a column used to support an eccentric roof loading.

EXAMPLE 13.10

The column in Fig. 13–29 is made of aluminum alloy 2014-T6 and is used to support an eccentric load P. Determine the maximum allowable value of P that can be supported if the column is fixed at its base and free at its top. Use Eq. 13–30.

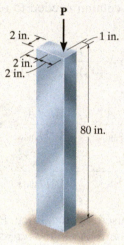

Fig. 13–29

SOLUTION

From Fig. 13–10b, $K = 2$. The largest slenderness ratio for the column is therefore

$$\frac{KL}{r} = \frac{2(80 \text{ in.})}{\sqrt{[(1/12)(4 \text{ in.})(2 \text{ in.})^3][(2 \text{ in.}) 4 \text{ in.}]}} = 277.1$$

By inspection, Eq. 13–26 must be used (277.1 > 55). Thus,

$$\sigma_{allow} = \frac{54\,000 \text{ ksi}}{(KL/r)^2} = \frac{54\,000 \text{ ksi}}{(277.1)^2} = 0.7031 \text{ ksi}$$

The maximum compressive stress in the column is determined from the combination of axial load and bending. We have

$$\sigma_{max} = \frac{P}{A} + \frac{(Pe)c}{I}$$

$$= \frac{P}{2 \text{ in.}(4 \text{ in.})} + \frac{P(1 \text{ in.})(2 \text{ in.})}{(1/12)(2 \text{ in.})(4 \text{ in.})^3}$$

$$= 0.3125P$$

Assuming that this stress is *uniform* over the cross section, we require

$$\sigma_{allow} = \sigma_{max}; \qquad 0.7031 = 0.3125P$$
$$P = 2.25 \text{ kip} \qquad\qquad \textit{Ans.}$$

EXAMPLE 13.11

The A-36 steel W6 × 20 column in Fig. 13–30 is pin connected at its ends and is subjected to the eccentric load P. Determine the maximum allowable value of P using the interaction method if the allowable bending stress is $(\sigma_b)_{allow} = 22$ ksi.

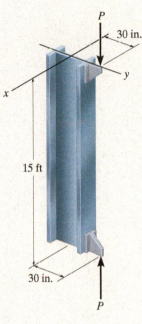

Fig. 13–30

SOLUTION

Here $K = 1$. The necessary geometric properties for the W6 × 20 are taken from the table in Appendix B.

$$A = 5.87 \text{ in}^2 \quad I_x = 41.4 \text{ in}^4 \quad r_y = 1.50 \text{ in.} \quad d = 6.20 \text{ in.}$$

We will consider r_y because this will lead to the *largest* value of the slenderness ratio. Also, I_x is needed since bending occurs about the x axis ($c = 6.20$ in./2 = 3.10 in.). To determine the allowable compressive stress, we have

$$\frac{KL}{r} = \frac{1[15 \text{ ft}(12 \text{ in./ft})]}{1.50 \text{ in.}} = 120$$

Since

$$\left(\frac{KL}{r}\right)_c = \sqrt{\frac{2\pi^2 E}{\sigma_Y}} = \sqrt{\frac{2\pi^2 [29(10^3) \text{ ksi}]}{36 \text{ ksi}}} = 126.1$$

then $KL/r < (KL/r)_c$ and so Eq. 13–23 must be used.

$$\sigma_{allow} = \frac{[1 - (KL/r)^2/2(KL/r)_c]^2 \sigma_Y}{(5/3) + [(3/8)(KL/r)/(KL/r)_c] - [(KL/r)^3/8(KL/r)_c^3]}$$

$$= \frac{[1 - (120)^2/2(126.1)^2] 36 \text{ ksi}}{(5/3) + [(3/8)(120)/(126.1)] - [(120)^3/8(126.1)^3]}$$

$$= 10.28 \text{ ksi}$$

Applying the interaction Eq. 13–31 yields

$$\frac{\sigma_a}{(\sigma_a)_{allow}} + \frac{\sigma_b}{(\sigma_b)_{allow}} \leq 1$$

$$\frac{P/5.87 \text{ in}^2}{10.28 \text{ ksi}} + \frac{P(30 \text{ in.})(3.10 \text{ in.})/(41.4 \text{ in}^4)}{22 \text{ ksi}} = 1$$

$$P = 8.43 \text{ kip} \qquad \qquad Ans.$$

Checking the application of the interaction method for the steel section, we require

$$\frac{\sigma_a}{(\sigma_a)_{allow}} = \frac{8.43 \text{ kip}/(5.87 \text{ in.})}{10.28 \text{ kip/in}^2} = 0.140 < 0.15 \qquad \qquad \text{OK}$$

EXAMPLE 13.12

The timber column in Fig. 13–31 is made from two boards nailed together so that the cross section has the dimensions shown. If the column is fixed at its base and free at its top, use Eq. 13–30 to determine the eccentric load P that can be supported.

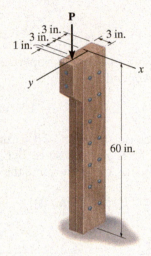

Fig. 13-31

SOLUTION

From Fig. 13–10b, $K = 2$. Here we must calculate KL/d to determine which of Eqs. 13–27 through 13–29 should be used. Since σ_{allow} is to be determined using the largest slenderness ratio, we choose $d = 3$ in. We have

$$\frac{KL}{d} = \frac{2(60 \text{ in.})}{3 \text{ in.}} = 40$$

Since $26 < KL/d < 50$ the allowable axial stress is determined using Eq. 13–29. Thus,

$$\sigma_{allow} = \frac{540 \text{ ksi}}{(KL/d)^2} = \frac{540 \text{ ksi}}{(40)^2} = 0.3375 \text{ ksi}$$

Applying Eq. 13–30 with $\sigma_{allow} = \sigma_{max}$, we have

$$\sigma_{allow} = \frac{P}{A} + \frac{Mc}{I}$$

$$0.3375 \text{ ksi} = \frac{P}{(3 \text{ in.})(6 \text{ in.})} + \frac{P(4 \text{ in.})(3 \text{ in.})}{(1/12)(3 \text{ in.})(6 \text{ in.})^3}$$

$$P = 1.22 \text{ kip} \qquad \textit{Ans.}$$

PROBLEMS

13–107. The W8 × 15 wide-flange A-36 steel column is assumed to be pinned at its top and bottom. Determine the largest eccentric load P that can be applied using Eq. 13–30 and the AISC equations of Sec. 13.6.

*13–108. Solve Prob. 13–107 if the column is fixed at its top and bottom.

13–109. Solve Prob. 13–107 if the column is fixed at its bottom and pinned at its top.

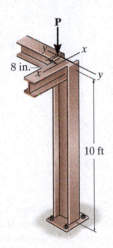

Probs. 13–107/108/109

13–110. The W12 × 50 wide-flange A-36 steel column is fixed at its bottom and free at its top. Determine the greatest eccentric load P that can be applied using Eq. 13–30 and the AISC equations of Sec. 13.6.

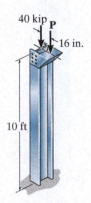

Prob. 13–110

13–111. The W8 × 15 wide-flange A992 steel column is fixed at its top and bottom. If it supports end moments of $M = 5$ kip · ft, determine the axial force P that can be applied. Bending is about the x–x axis. Use the AISC equations of Sec. 13.6 and Eq. 13–30.

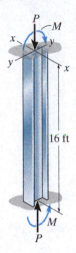

Prob. 13–111

*13–112. The W8 × 15 wide-flange A992 steel column is fixed at its top and bottom. If it supports end moments of $M = 23$ kip · ft, determine the axial force P that can be applied. Bending is about the x–x axis. Use the interaction formula with $(\sigma_b)_{allow} = 24$ ksi.

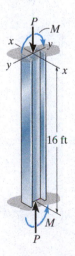

Prob. 13–112

13

13–113. The W12 × 22 wide-flange A-36 steel column is fixed at its bottom and free at its top. Determine the greatest eccentric load P that can be applied using Eq. 13–30 and the AISC equations of Sec. 13.6.

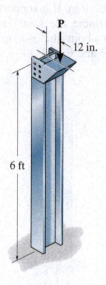

P

12 in.

6 ft

Prob. 13–113

13–114. The W10 × 15 wide-flange A-36 steel column is fixed at its bottom and free at its top. Determine the greatest eccentric load P that can be applied using Eq. 13–30 and the AISC equations of Sec. 13.6.

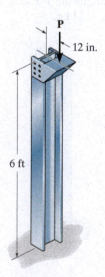

P

12 in.

6 ft

Prob. 13–114

13–115. The W10 × 15 wide-flange A-36 steel column is fixed at its bottom and free at its top. If it is subjected to a load $P = 2$ kip, determine if it is safe based on the AISC equations of Sec. 13.6 and Eq. 13–30.

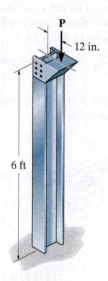

P

12 in.

6 ft

Prob. 13–115

*13–116. The W12 × 22 wide-flange A-36 steel column is fixed at its bottom and free at its top. If it is subjected to a load $P = 4$ kip, determine if it is safe based on the AISC equations of Sec. 13.6 and Eq. 13–30.

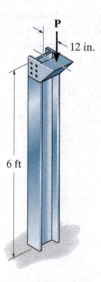

P

12 in.

6 ft

Prob. 13–116

13–117. A 20-ft-long column is made of aluminum alloy 2014-T6. If it is pinned at its top and bottom, and a compressive load **P** is applied at point A, determine the maximum allowable magnitude of **P** using the equations of Sec. 13.6 and Eq. 13–30.

13–119. The 2014-T6 aluminum hollow column is fixed at its base and free at its top. Determine the maximum eccentric force P that can be safely supported by the column. Use the allowable stress method. The thickness of the wall for the section is $t = 0.5$ in.

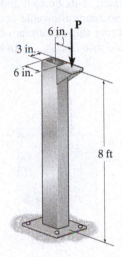

Prob. 13–119

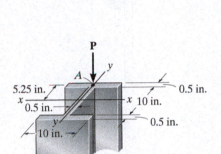

Prob. 13–117

13–118. A 20-ft-long column is made of aluminum alloy 2014-T6. If it is pinned at its top and bottom, and a compressive load **P** is applied at point A, determine the maximum allowable magnitude of **P** using the equations of Sec. 13.6 and the interaction formula with $(\sigma_b)_{allow} = 20$ ksi.

*13–120. The 2014-T6 aluminum hollow column is fixed at its base and free at its top. Determine the maximum eccentric force P that can be safely supported by the column. Use the interaction formula. The allowable bending stress is $(\sigma_b)_{allow} = 30$ ksi. The thickness of the wall for the section is $t = 0.5$ in.

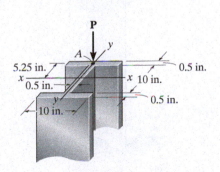

Prob. 13–118

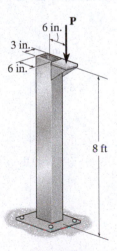

Prob. 13–120

13–121. The 10-ft-long bar is made of aluminum alloy 2014-T6. If it is fixed at its bottom and pinned at the top, determine the maximum allowable eccentric load **P** that can be applied using the formulas in Sec. 13.6 and Eq. 13–30.

13–122. The 10-ft-long bar is made of aluminum alloy 2014-T6. If it is fixed at its bottom and pinned at the top, determine the maximum allowable eccentric load **P** that can be applied using the equations of Sec. 13.6 and the interaction formula with $(\sigma_b)_{\text{allow}} = 18$ ksi.

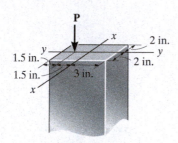

Probs. 13–121/122

13–123. Check if the wood column is adequate for supporting the eccentric load of $P = 800$ lb applied at its top. It is fixed at its base and free at its top. Use the NFPA equations of Sec. 13.6 and Eq. 13–30.

*13–124. Determine the maximum allowable eccentric load P that can be applied to the wood column. The column is fixed at its base and free at its top. Use the NFPA equations of Sec. 13.6 and Eq. 13–30.

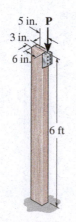

Probs. 13–123/124

13–125. The 10-in.-diameter utility pole supports the transformer that has a weight of 600 lb and center of gravity at G. If the pole is fixed to the ground and free at its top, determine if it is adequate according to the NFPA equations of Sec. 13.6 and Eq. 13–30.

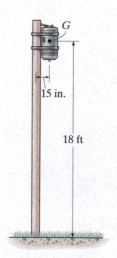

Prob. 13–125

13–126. Using the NFPA equations of Sec. 13.6 and Eq. 13–30, determine the maximum allowable eccentric load P that can be applied to the wood column. Assume that the column is pinned at both its top and bottom.

13–127. Using the NFPA equations of Sec. 13.6 and Eq. 13–30, determine the maximum allowable eccentric load P that can be applied to the wood column. Assume that the column is pinned at the top and fixed at the bottom.

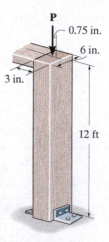

Probs. 13–126/127

CHAPTER REVIEW

Buckling is the sudden instability that occurs in columns or members that support an axial compressive load. The maximum axial load that a member can support just before buckling is called the critical load P_{cr}.		
The critical load for an ideal column is determined from Euler's formula, where $K = 1$ for pin supports, $K = 0.5$ for fixed supports, $K = 0.7$ for a pin and a fixed support, and $K = 2$ for a fixed support and a free end.	$$P_{cr} = \frac{\pi^2 EI}{(KL)^2}$$	
If the axial loading is applied eccentrically to the column, then the secant formula can be used to determine the maximum stress in the column.	$$\sigma_{max} = \frac{P}{A}\left[1 + \frac{ec}{r^2}\sec\left(\frac{L}{2r}\sqrt{\frac{P}{EA}}\right)\right]$$	
When the axial load causes yielding of the material, then the tangent modulus can be used with Euler's formula to determine the critical load for the column. This is referred to as Engesser's equation.	$$\sigma_{cr} = \frac{\pi^2 E_t}{(KL/r)^2}$$	
Empirical formulas based on experimental data have been developed for use in the design of steel, aluminum, and timber columns.		

REVIEW PROBLEMS

R13–1. The wood column has a thickness of 4 in. and a width of 6 in. Using the NFPA equations of Sec. 13.6 and Eq. 13–30, determine the maximum allowable eccentric load P that can be applied. Assume that the column is pinned at both its top and bottom.

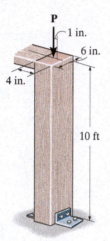

R13–1

R13–2. The wood column has a thickness of 4 in. and a width of 6 in. Using the NFPA equations of Sec. 13.6 and Eq. 13–30, determine the maximum allowable eccentric load P that can be applied. Assume that the column is pinned at the top and fixed at the bottom.

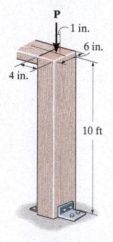

R13–2

R13–3. A steel column has a length of 5 m and is free at one end and fixed at the other end. If the cross-sectional area has the dimensions shown, determine the critical load. $E_{st} = 200$ GPa, $\sigma_Y = 360$ MPa.

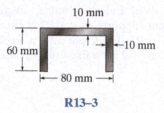

R13–3

*R13–4. The square structural A992 steel tubing has outer dimensions of 8 in. by 8 in. Its cross-sectional area is 14.40 in^2 and its moments of inertia are $I_x = I_y = 131$ in^4. Determine the maximum load P it can support. The column can be assumed fixed at its base and free at its top.

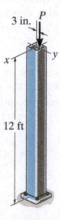

R13–4

R13–5. If the A-36 steel solid circular rod *BD* has a diameter of 2 in., determine the allowable maximum force **P** that can be supported by the frame without causing the rod to buckle. Use F.S. = 2 against buckling.

R13–6. If $P = 15$ kip, determine the required minimum diameter of the A992 steel solid circular rod *BD* to the nearest $\frac{1}{16}$ in. Use F.S. = 2 against buckling.

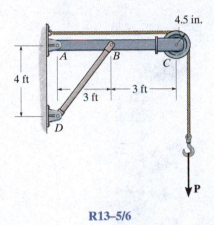

R13–5/6

R13–7. The steel pipe is fixed supported at its ends. If it is 4 m long and has an outer diameter of 50 mm, determine its required thickness so that it can support an axial load of $P = 100$ kN without buckling. $E_{st} = 200$ GPa, $\sigma_Y = 250$ MPa.

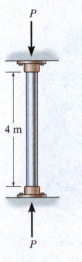

R13–7

*R13–8. The W200 × 46 wide-flange A992-steel column can be considered pinned at its top and fixed at its base. Also, the column is braced at its mid-height against weak axis buckling. Determine the maximum axial load the column can support without causing it to buckle.

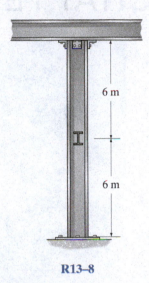

R13–8

R13–9. The wide-flange A992 steel column has the cross section shown. If it is fixed at the bottom and free at the top, determine the maximum force *P* that can be applied at *A* without causing it to buckle or yield. Use a factor of safety of 3 with respect to buckling and yielding.

R13–10. The wide-flange A992 steel column has the cross section shown. If it is fixed at the bottom and free at the top, determine if the column will buckle or yield when the load $P = 10$ kN is applied at *A*. Use a factor of safety of 3 with respect to buckling and yielding.

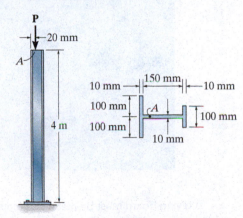

R13–9/10

CHAPTER 14

(© 68/Ocean/Corbis)

A diving board must be made of a material that can store a high value of elastic strain energy due to bending. This allows the board to have a large flexure, and thereby transfer this energy to the diver, as the board begins to straighten out.

ENERGY METHODS

CHAPTER OBJECTIVES

■ In this chapter we will show how to apply energy methods to solve problems involving deflection. We will begin with a discussion of work and strain energy; then, using the principle of conservation of energy, we will show how to determine the stress and deflection in a member subjected to impact. Finally, the method of virtual work and Castigliano's theorem will be used to determine the displacement and slope at points on a structural member or mechanical element.

14.1 EXTERNAL WORK AND STRAIN ENERGY

To use any of the energy methods developed in this chapter, we must first define the work caused by an external force and couple moment, and show how to express this work in terms of a body's strain energy.

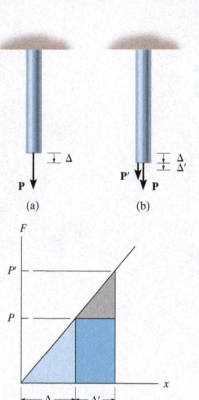

Fig. 14–1

Work of a Force.
A force does **work** when the force undergoes a displacement dx that is in the *same direction* as the force. Work is a scalar, defined as $dU_e = F\,dx$. If the total displacement is Δ, the work is

$$U_e = \int_0^\Delta F\,dx \tag{14–1}$$

Let's use this equation to calculate the work done by an axial force applied to the end of the bar shown in Fig. 14–1a. As the magnitude of the force is *gradually* increased from zero to some limiting value $F = P$, the final displacement of the end of the bar becomes Δ. If the material behaves in a linear elastic manner, then the force will be directly proportional to its displacement; and so, $F/x = P/\Delta$ or $F = (P/\Delta)x$. Substituting this into Eq. 14–1 and integrating from 0 to Δ, we get

$$U_e = \frac{1}{2}P\Delta \tag{14–2}$$

In other words, as the force is applied to the bar, its magnitude increases from zero to some value P, and consequently, the work done is equal to the *average force magnitude*, $P/2$, times the total displacement Δ. This work is represented graphically by the light-blue shaded area of the triangle in Fig. 14–1c.

Now suppose **P** is already applied to the bar and that *another force* **P'** is applied, so that the end of the bar is displaced *further* by an amount Δ', Fig. 14–1b. The work done by **P'** is equal to the gray shaded triangular area, and the additional work done by **P** is simply its magnitude P times the displacement Δ', i.e.,

$$U_e' = P\Delta' \tag{14–3}$$

This is represented by the dark-blue shaded *rectangular area* in Fig. 14–1c.

Work of a Couple Moment.

A couple moment **M** does work when it undergoes an angular displacement $d\theta$ along its line of action. The work is defined as $dU_e = M\,d\theta$, Fig. 14–2. If the total angular displacement is θ rad, the work becomes

$$U_e = \int_0^\theta M\,d\theta \qquad (14\text{–}4)$$

If a body has linear elastic behavior, and the magnitude of the couple moment is increased gradually from zero at $\theta = 0$ to M at θ, then, as in the case of a force, the work is

$$U_e = \frac{1}{2}M\theta \qquad (14\text{–}5)$$

However, if the couple moment is *already* applied to the body and other loadings further rotate the body by an amount θ', then the work is

$$U'_e = M\theta'$$

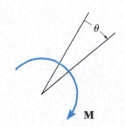

Fig. 14–2

Strain Energy.

When loads are applied to a body, they will deform the material, and provided no energy is lost in the form of heat, the external work done by the loads will be converted into internal work called **strain energy**. This energy is stored in the body and is caused by the action of either normal or shear stress.

Normal Stress.

To obtain the strain energy caused by the normal stress σ_z, consider the volume element shown in Fig. 14–3. The force created on the element's top and bottom surface will be $dF_z = \sigma_z\,dA = \sigma_z\,dx\,dy$. If this force (or stress) is applied gradually to the element, then the force magnitude will increase from zero to dF_z, while the element undergoes an elongation $d\Delta_z = \epsilon_z\,dz$. The work done by dF_z is therefore $dU_i = \frac{1}{2}dF_z\,d\Delta_z = \frac{1}{2}[\sigma_z\,dx\,dy]\epsilon_z\,dz$. Since the volume of the element is $dV = dx\,dy\,dz$, then

$$dU_i = \frac{1}{2}\sigma_z\epsilon_z\,dV \qquad (14\text{–}6)$$

This strain energy is *always positive*, even if σ_z is compressive, since σ_z and ϵ_z will always be in the same direction.

For a body of finite size, the strain energy in the body is therefore

$$U_i = \int_V \frac{\sigma\epsilon}{2}\,dV \qquad (14\text{–}7)$$

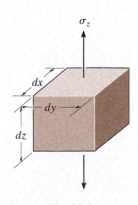

Fig. 14–3

When the material behaves in a linear elastic manner, then $\sigma = E\epsilon$, and we can then express the strain energy in terms of the normal stress as

$$U_i = \int_V \frac{\sigma^2}{2E} dV \tag{14–8}$$

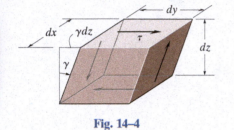

Fig. 14–4

Shear Stress.
A strain energy expression similar to that for normal stress can also be established for the material when it is subjected to shear stress. Here the element in Fig. 14–4 is subjected to the shear force $dF = \tau(dx\,dy)$, which acts on its top surface, causing this surface to be displaced $\gamma\,dz$ relative to its bottom surface. The vertical surfaces only rotate, and therefore the shear forces on these faces do no work. Hence, the strain energy stored in the element becomes

$$dU_i = \frac{1}{2}[\tau(dx\,dy)]\gamma\,dz$$

Or since $dV = dx\,dy\,dz$,

$$dU_i = \frac{1}{2}\tau\gamma\,dV \tag{14–9}$$

The strain energy stored in a body subjected to shear stress is therefore

$$U_i = \int_V \frac{\tau\gamma}{2} dV \tag{14–10}$$

Like the case for normal strain energy, shear strain energy is *always positive*, since τ and γ are always in the same direction. If the material is linear elastic, then $\gamma = \tau/G$, and we can express the strain energy in terms of the shear stress as

$$U_i = \int_V \frac{\tau^2}{2G} dV \tag{14–11}$$

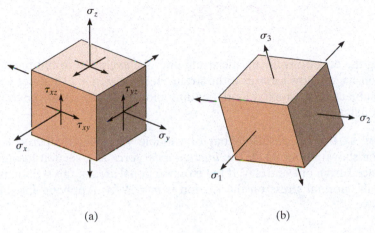

(a)

(b)

Fig. 14–5

Multiaxial Stress.

The previous development can be expanded to determine the strain energy in a body when it is subjected to a general state of stress, Fig. 14–5a. To do this, the strain energies associated with each of the six normal and shear stress components can be obtained from Eqs. 14–6 and 14–9, and since energy is a scalar, the total strain energy in the body becomes

$$U_i = \int_V \left[\frac{1}{2}\sigma_x \epsilon_x + \frac{1}{2}\sigma_y \epsilon_y + \frac{1}{2}\sigma_z \epsilon_z \right.$$

$$\left. + \frac{1}{2}\tau_{xy}\gamma_{xy} + \frac{1}{2}\tau_{yz}\gamma_{yz} + \frac{1}{2}\tau_{xz}\gamma_{xz} \right] dV \qquad (14\text{–}12)$$

The strains can be eliminated by using the generalized form of Hooke's law given by Eqs. 10–18 and 10–19. After substituting and combining terms, we have

$$U_i = \int_V \left[\frac{1}{2E}\left(\sigma_x^2 + \sigma_y^2 + \sigma_z^2\right) - \frac{\nu}{E}\left(\sigma_x \sigma_y + \sigma_y \sigma_z + \sigma_x \sigma_z\right) \right.$$

$$\left. + \frac{1}{2G}\left(\tau_{xy}^2 + \tau_{yz}^2 + \tau_{xz}^2\right) \right] dV \qquad (14\text{–}13)$$

If only the principal stresses $\sigma_1, \sigma_2, \sigma_3$ act on the element, Fig. 14–5b, this equation reduces to a simpler form, namely,

$$U_i = \int_V \left[\frac{1}{2E}\left(\sigma_1^2 + \sigma_2^2 + \sigma_3^2\right) - \frac{\nu}{E}\left(\sigma_1 \sigma_2 + \sigma_2 \sigma_3 + \sigma_1 \sigma_3\right) \right] dV \qquad (14\text{–}14)$$

14.2   ELASTIC STRAIN ENERGY FOR VARIOUS TYPES OF LOADING

Using the equations for elastic strain energy developed in the previous section, we will now formulate the strain energy stored in a member when it is subjected to an axial load, bending moment, transverse shear, and torsional moment.

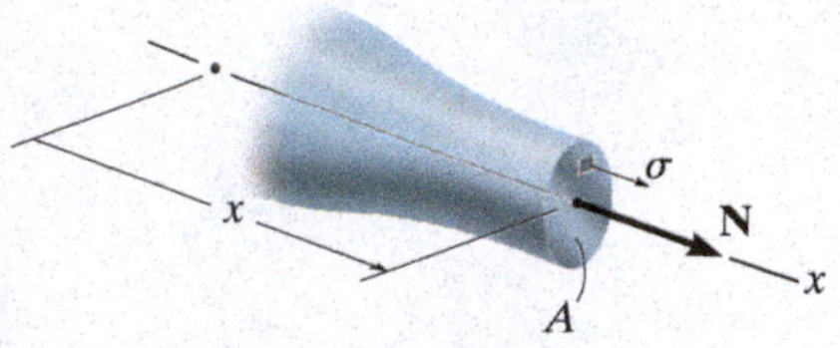

Axial Load.   Consider a bar of variable yet slightly tapered cross section shown in Fig. 14–6. The *internal axial force* at a section located a distance x from one end is N. If the cross-sectional area at this section is A, then the normal stress on the section is $\sigma = N/A$. Applying Eq. 14–8, we have

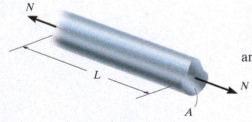

Fig. 14–6

$$U_i = \int_V \frac{\sigma_x^2}{2E} dV = \int_V \frac{N^2}{2EA^2} dV$$

If we choose a differential segment of the bar having a volume $dV = A\,dx$, the general formula for the strain energy in the bar is therefore

$$U_i = \int_0^L \frac{N^2}{2AE} dx \qquad (14\text{–}15)$$

For the most common case of a bar of constant cross-sectional area A and constant internal axial load N, Fig. 14–7, integration gives

$$U_i = \frac{N^2 L}{2AE} \qquad (14\text{–}16)$$

Fig. 14–7

Notice that the bar's elastic strain energy will *increase* if the length of the bar is increased, or if the modulus of elasticity or cross-sectional area is decreased. For example, an aluminum rod [$E_{al} = 10(10^3)$ ksi] will store approximately three times as much energy as a steel rod [$E_{st} = 29(10^3)$ ksi] having the same size and subjected to the same load. However, doubling the cross-sectional area of the rod will decrease its ability to store energy by one-half.

EXAMPLE   14.1

One of the two high-strength steel bolts A and B shown in Fig. 14–8 is to be chosen to support a sudden tensile loading. For the choice it is necessary to determine the greatest amount of elastic strain energy that each bolt can absorb. Bolt A has a diameter of 0.875 in. for 2 in. of its length and a root (or smallest) diameter of 0.731 in. within the 0.25-in. threaded region. Bolt B has "upset" threads, such that the diameter throughout its 2.25-in. length can be taken as 0.731 in. In both cases, neglect the extra material that makes up the threads. Take $E_{st} = 29(10^3)$ ksi, $\sigma_Y = 44$ ksi.

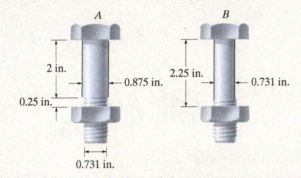

Fig. 14–8

SOLUTION
Bolt A.   If the bolt is subjected to its maximum tension, the maximum stress of $\sigma_Y = 44$ ksi will occur within the 0.25-in. region. This force is

$$P_{max} = \sigma_Y A = 44 \text{ ksi} \left[\pi \left(\frac{0.731 \text{ in.}}{2} \right)^2 \right] = 18.47 \text{ kip}$$

Applying Eq. 14–16 to each region of the bolt, we have

$$U_i = \sum \frac{N^2 L}{2AE}$$

$$= \frac{(18.47 \text{ kip})^2 (2 \text{ in.})}{2[\pi(0.875 \text{ in.}/2)^2][29(10^3) \text{ ksi}]} + \frac{(18.47 \text{ kip})^2 (0.25 \text{ in.})}{2[\pi(0.731 \text{ in.}/2)^2][29(10^3) \text{ ksi}]}$$

$$= 0.0231 \text{ in.} \cdot \text{kip} \qquad\qquad\qquad\qquad Ans.$$

Bolt B.   Here the bolt is assumed to have a uniform diameter of 0.731 in. throughout its 2.25-in. length. Also, from the calculation above, it can support a maximum force of $P_{max} = 18.47$ kip. Thus,

$$U_i = \frac{N^2 L}{2AE} = \frac{(18.47 \text{ kip})^2 (2.25 \text{ in.})}{2[\pi(0.731 \text{ in.}/2)^2][29(10^3) \text{ ksi}]} = 0.0315 \text{ in.} \cdot \text{kip} \quad Ans.$$

NOTE: By comparison, bolt B can absorb 37% more elastic energy than bolt A, because it has a smaller cross section along its length.

Bending Moment. Consider the axisymmetric beam shown in Fig. 14–9. Here the internal moment is M, which produces a normal stress of $\sigma = My/I$ on the arbitrary element located a distance y from the neutral axis. If the volume of this element is $dV = dA\ dx$, then the elastic strain energy in the beam is

$$U_i = \int_V \frac{\sigma^2}{2E}dV = \int_V \frac{1}{2E}\left(\frac{My}{I}\right)^2 dA\ dx$$

or

$$U_i = \int_0^L \frac{M^2}{2EI^2}\left(\int_A y^2\ dA\right) dx$$

Since the integral in parentheses represents the moment of inertia of the area about the neutral axis, the final result can be written as

$$U_i = \int_0^L \frac{M^2\ dx}{2EI} \tag{14–17}$$

To evaluate this strain energy, we must express the internal moment as a function of its position x along the beam, and then perform the integration over the beam's entire length.

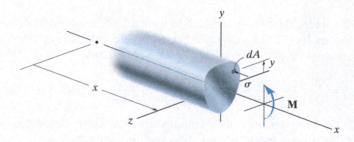

Fig. 14–9

EXAMPLE 14.2

Determine the elastic strain energy due to bending of the cantilevered beam in Fig. 14–10a. EI is constant.

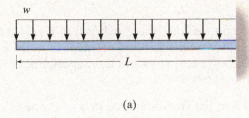

(a)

Fig. 14–10

SOLUTION

The internal moment in the beam is determined by establishing the x coordinate with origin at the left side. The left segment of the beam is shown in Fig. 14–10b. We have

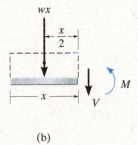

(b)

$$\zeta + \Sigma M_{NA} = 0; \qquad M + wx\left(\frac{x}{2}\right) = 0 \qquad M = -w\left(\frac{x^2}{2}\right)$$

Applying Eq. 14–17 yields

$$U_i = \int_0^L \frac{M^2\, dx}{2EI} = \int_0^L \frac{[-w(x^2/2)]^2\, dx}{2EI} = \frac{w^2}{8EI}\int_0^L x^4\, dx$$

or

$$U_i = \frac{w^2 L^5}{40EI} \qquad\qquad Ans.$$

We can also obtain the strain energy using an x coordinate having its origin at the right side of the beam and extending positive to the left, Fig. 14–10c. In this case,

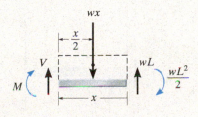

$$\zeta + \Sigma M_{NA} = 0; \qquad -M - wx\left(\frac{x}{2}\right) + wL(x) - \frac{wL^2}{2} = 0$$

$$M = -\frac{wL^2}{2} + wLx - w\left(\frac{x^2}{2}\right)$$

Applying Eq. 14–17, we will obtain the same result as before; however, more calculations are involved in this case.

(c)

EXAMPLE 14.3

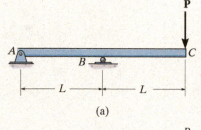

(a)

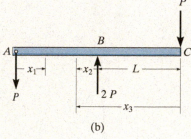

(b)

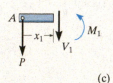

(c)

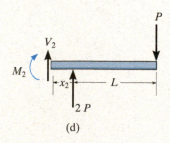

(d)

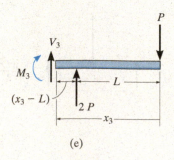

(e)

Fig. 14–11

Determine the bending strain energy in region AB of the beam shown in Fig. 14–11a. EI is constant.

SOLUTION

A free-body diagram of the beam is shown in Fig. 14–11b. To obtain the answer, we can express the internal moment in terms of any one of the indicated three "x" coordinates and then apply Eq. 14–17. Each of these solutions will now be considered.

$0 \leq x_1 \leq L$. From the free-body diagram of the section in Fig. 14–11c, we have

$$\zeta + \Sigma M_{NA} = 0; \qquad M_1 + Px_1 = 0$$

$$M_1 = -Px_1$$

$$U_i = \int \frac{M^2 \, dx}{2EI} = \int_0^L \frac{(-Px_1)^2 \, dx_1}{2EI} = \frac{P^2 L^3}{6EI} \qquad \textit{Ans.}$$

$0 \leq x_2 \leq L$. Using the free-body diagram of the section in Fig. 14–11d gives

$$\zeta + \Sigma M_{NA} = 0; \qquad -M_2 + 2P(x_2) - P(x_2 + L) = 0$$

$$M_2 = P(x_2 - L)$$

$$U_i = \int \frac{M^2 \, dx}{2EI} = \int_0^L \frac{[P(x_2 - L)]^2 \, dx_2}{2EI} = \frac{P^2 L^3}{6EI} \qquad \textit{Ans.}$$

$L \leq x_3 \leq 2L$. From the free-body diagram in Fig. 14–11e, we have

$$\zeta + \Sigma M_{NA} = 0; \qquad -M_3 + 2P(x_3 - L) - P(x_3) = 0$$

$$M_3 = P(x_3 - 2L)$$

$$U_i = \int \frac{M^2 \, dx}{2EI} = \int_L^{2L} \frac{[P(x_3 - 2L)]^2 \, dx_3}{2EI} = \frac{P^2 L^3}{6EI} \qquad \textit{Ans.}$$

NOTE: This and the previous example indicate that the strain energy for the beam can be found using *any* suitable x coordinate. It is only necessary to integrate over the range of the coordinate where the internal energy is to be determined. Here the choice of x_1 provides the simplest solution.

Transverse Shear.   Again we will consider an axisymmetric beam as shown in Fig. 14–12. If the internal shear at the section x is V, then the shear stress acting on the volume element of material having an area dA and length dx is $\tau = VQ/It$. Substituting this into Eq. 14–11, the strain energy for shear becomes

$$U_i = \int_V \frac{\tau^2}{2G} dV = \int_V \frac{1}{2G}\left(\frac{VQ}{It}\right)^2 dA\,dx$$

$$U_i = \int_0^L \frac{V^2}{2GI^2}\left(\int_A \frac{Q^2}{t^2} dA\right) dx$$

The integral in parentheses represents the **form factor** for shear, written as

$$f_s = \frac{A}{I^2}\int_A \frac{Q^2}{t^2} dA \qquad (14\text{–}18)$$

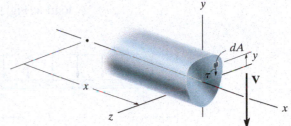

Fig. 14–12

Substituting this into the above equation, we get

$$U_i = \int_0^L \frac{f_s V^2\,dx}{2GA} \qquad (14\text{–}19)$$

From the way it is defined in Eq. 14–18, the form factor is a dimensionless number that is unique for each specific cross-sectional area. For example, if the beam has a rectangular cross section of width b and height h, Fig. 14–13, then

$$t = b$$
$$dA = b\,dy$$
$$I = \frac{1}{12}bh^3$$
$$Q = \bar{y}'A' = \left(y + \frac{(h/2) - y}{2}\right)b\left(\frac{h}{2} - y\right) = \frac{b}{2}\left(\frac{h^2}{4} - y^2\right)$$

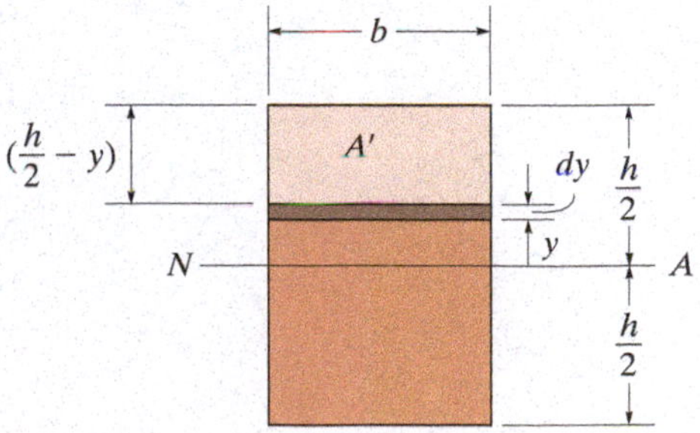

Substituting these terms into Eq. 14–18, we get

$$f_s = \frac{bh}{\left(\frac{1}{12}bh^3\right)^2}\int_{-h/2}^{h/2} \frac{b^2}{4b^2}\left(\frac{h^2}{4} - y^2\right)^2 b\,dy = \frac{6}{5} \qquad (14\text{–}20)$$

Fig. 14–13

EXAMPLE 14.4

Determine the strain energy in the cantilevered beam due to shear, if the beam has a square cross section and is subjected to a uniform distributed load w, Fig. 14–14a. EI and G are constant.

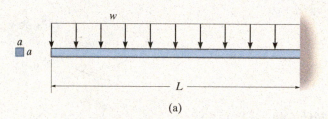

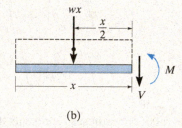

(a)

(b)

Fig. 14–14

SOLUTION

From the free-body diagram of an arbitrary section, Fig. 14–14b, we have

$$+\uparrow \Sigma F_y = 0; \qquad -V - wx = 0$$

$$V = -wx$$

Since the cross section is square, the form factor $f_s = \frac{6}{5}$ (Eq. 14–20), and therefore Eq. 14–19 becomes

$$(U_i)_s = \int_0^L \frac{\frac{6}{5}(-wx)^2\, dx}{2GA} = \frac{3w^2}{5GA}\int_0^L x^2\, dx$$

or

$$(U_i)_s = \frac{w^2 L^3}{5GA} \qquad\qquad Ans.$$

NOTE: Using the results of Example 14.2, with $A = a^2$, $I = \frac{1}{12}a^4$, the ratio of shear to bending strain energy is

$$\frac{(U_i)_s}{(U_i)_b} = \frac{w^2 L^3 / 5Ga^2}{w^2 L^5 / 40E\left(\frac{1}{12}a^4\right)} = \frac{2}{3}\left(\frac{a}{L}\right)^2 \frac{E}{G}$$

Since $G = E/2(1 + \nu)$ and $\nu \le \frac{1}{2}$ (Sec. 10.6), then as an *upper* limit, $E = 3G$, so that

$$\frac{(U_i)_s}{(U_i)_b} = 2\left(\frac{a}{L}\right)^2$$

Notice that this ratio will increase as L decreases. However, even for very short beams, where, say, $L = 5a$, the contribution due to shear strain energy is only 8% of the bending strain energy. It is for this reason the shear strain energy stored in beams is usually neglected in engineering analysis.

Torsional Moment.   For torsion, we will consider the slightly tapered shaft in Fig. 14–15, which is subjected to an internal torque T. On the arbitrary element of area dA and length dx, the shear stress is $\tau = T\rho/J$, and therefore the strain energy stored in the shaft is

$$U_i = \int_V \frac{\tau^2}{2G} dV = \int_V \frac{1}{2G}\left(\frac{T\rho}{J}\right)^2 dA\ dx = \int_0^L \frac{T^2}{2GJ^2}\left(\int_A \rho^2\ dA\right) dx$$

Since the integral in parentheses represents the polar moment of inertia J for the shaft at the section, the final result can be written as

$$U_i = \int_0^L \frac{T^2}{2GJ} dx \qquad (14\text{–}21)$$

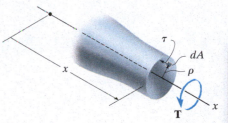

Fig. 14–15

The most common case occurs when the shaft (or tube) has a constant cross-sectional area and the applied torque is constant, Fig. 14–16. Integration of Eq. 14–21 then gives

$$U_i = \frac{T^2 L}{2GJ} \qquad (14\text{–}22)$$

Notice that the energy absorbing capacity of a torsionally loaded shaft is *decreased* by increasing the diameter of the shaft, since this increases J.

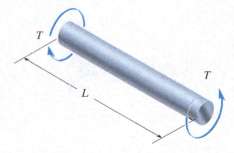

Fig. 14–16

> ## IMPORTANT POINTS

- A *force* does work when it moves through a *displacement*. When a force is applied to a body its magnitude is increased gradually from zero to F, and so the work is $U = (F/2)\Delta$. However, if a constant force acts on the body, and the body is given a displacement Δ, then the work becomes $U = F\Delta$.

- A *couple moment* does work when it is displaced through a *rotation*.

- *Strain energy* is caused by the internal work of the normal and shear stresses. It is always a *positive* quantity.

- The strain energy can be related to the internal loadings N, V, M, and T.

- As a beam becomes longer, the strain energy due to bending becomes much larger than the strain energy due to shear. For this reason, the *shear strain energy* in beams can generally be *neglected*.

EXAMPLE 14.5

The tubular shaft in Fig. 14–17a is fixed at the wall and subjected to two torques as shown. Determine the strain energy stored in the shaft due to this loading. $G = 75$ GPa.

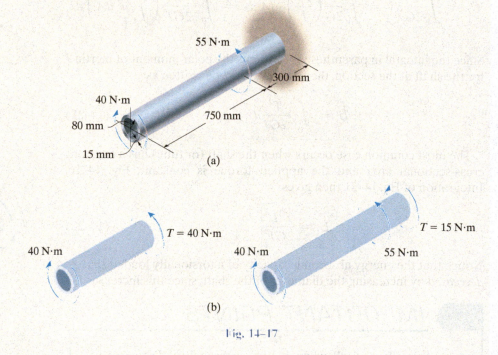

Fig. 14–17

SOLUTION

Using the method of sections, the internal torque is first determined within the two regions of the shaft where it is constant, Fig. 14–17b. Although these torques (40 N · m and 15 N · m) are in opposite directions, this will be of no consequence in determining the strain energy, since the torque is squared in Eq. 14–22. In other words, the strain energy is always positive. The polar moment of inertia for the shaft is

$$J = \frac{\pi}{2}[(0.08 \text{ m})^4 - (0.065 \text{ m})^4] = 36.30(10^{-6}) \text{ m}^4$$

Applying Eq. 14–22, we have

$$U_i = \sum \frac{T^2 L}{2GJ}$$

$$= \frac{(40 \text{ N} \cdot \text{m})^2(0.750 \text{ m})}{2[75(10^9) \text{ N/m}^2]36.30(10^{-6}) \text{ m}^4} + \frac{(15 \text{ N} \cdot \text{m})^2(0.300 \text{ m})}{2[75(10^9) \text{ N/m}^2]36.30(10^{-6}) \text{ m}^4}$$

$$= 233 \text{ } \mu\text{J} \hspace{4cm} \textit{Ans.}$$

PROBLEMS

14–1. A material is subjected to a general state of plane stress. Express the strain energy density in terms of the elastic constants E, G, and ν and the stress components σ_x, σ_y, and τ_{xy}.

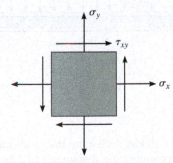

Prob. 14–1

14–2. The strain-energy density for plane stress must be the same whether the state of stress is represented by σ_x, σ_y, and τ_{xy}, or by the principal stresses σ_1 and σ_2. This being the case, equate the strain–energy expressions for each of these two cases and show that $G = E/[2(1 + \nu)]$.

14–3. The A-36 steel bar consists of two segments, one of circular cross section of radius r, and one of square cross section. If the bar is subjected to the axial loading of P, determine the dimensions a of the square segment so that the strain energy within the square segment is the same as in the circular segment.

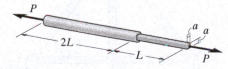

Prob. 14–3

*14–4. Determine the torsional strain energy in the A992 steel shaft. The shaft has a radius of 50 mm.

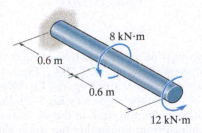

Prob. 14–4

14–5. Using bolts of the same material and cross-sectional area, two possible attachments for a cylinder head are shown. Compare the strain energy developed in each case, and then explain which design is better for resisting an axial shock or impact load.

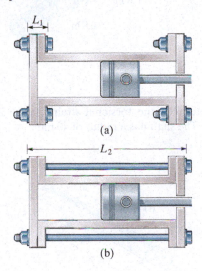

(a)

(b)

Prob. 14–5

14–6. If $P = 60$ kN, determine the total strain energy stored in the truss. Each member has a cross-sectional area of $2.5(10^3)$ mm^2 and is made of A-36 steel.

14–7. Determine the maximum force $\mathbf{P}$ and the corresponding maximum total strain energy stored in the truss without causing any of the members to have permanent deformation. Each member has the cross-sectional area of $2.5(10^3)$ mm^2 and is made of A-36 steel.

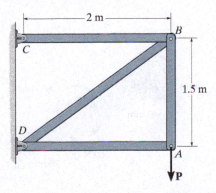

Probs. 14–6/7

*14–8. Determine the torsional strain energy in the A992 steel shaft. The shaft has a radius of 40 mm.

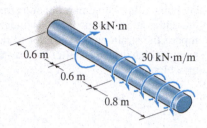

Prob. 14–8

14–9. Determine the torsional strain energy in the A-36 steel shaft. The shaft has a radius of 40 mm.

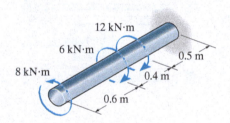

Prob. 14–9

14–10. The shaft assembly is fixed at C. The hollow segment BC has an inner radius of 20 mm and outer radius of 40 mm, while the solid segment AB has a radius of 20 mm. Determine the torsional strain energy stored in the shaft. The shaft is made of 2014-T6 aluminum alloy. The coupling at B is rigid.

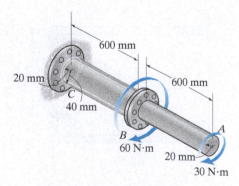

Prob. 14–10

14–11. Determine the total axial and bending strain energy in the A992 steel beam. $A = 2850$ mm^2, $I = 28.9(10^6)$ mm^4.

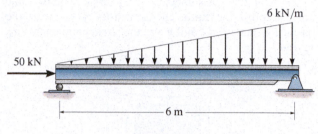

Prob. 14–11

*14–12. If $P = 10$ kip, determine the total strain energy in the truss. Each member has a diameter of 2 in. and is made of A992 steel.

14–13. Determine the maximum force $\mathbf{P}$ and the corresponding maximum total strain energy that can be stored in the truss without causing any of the members to have permanent deformation. Each member of the truss has a diameter of 2 in. and is made of A-36 steel.

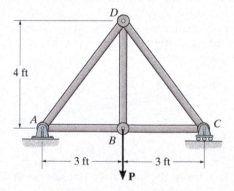

Probs. 14–12/13

14–14. Consider the thin-walled tube of Fig. 5–26. Use the formula for shear stress, $\tau_{avg} = T/2tA_m$, Eq. 5–18, and the general equation of shear strain energy, Eq. 14–11, to show that the twist of the tube is given by Eq. 5–20. *Hint*: Equate the work done by the torque T to the strain energy in the tube, determined from integrating the strain energy for a differential element, Fig. 14–4, over the volume of material.

14–15. Determine the bending strain energy in the A992 steel beam. $I = 156$ in^4.

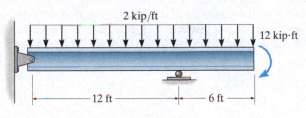

2 kip/ft

12 kip·ft

12 ft

6 ft

Prob. 14–15

*14–16.** Determine the bending strain energy in the beam. EI is constant.

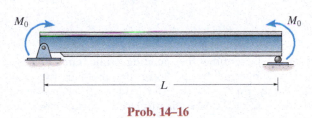

M_0

M_0

L

Prob. 14–16

14–17. Determine the bending strain energy in the simply supported beam. EI is constant.

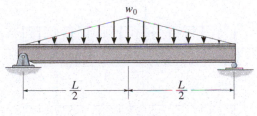

w_0

$\dfrac{L}{2}$

$\dfrac{L}{2}$

Prob. 14–17

14–18. The steel beam is supported on two springs, each having a stiffness of $k = 8$ MN/m. Determine the strain energy in each of the springs and the bending strain energy in the beam. $E_{st} = 200$ GPa, $I = 5(10^6)$ mm^4.

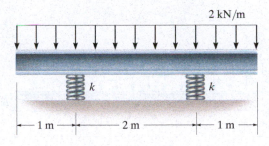

2 kN/m

k

k

1 m

2 m

1 m

Prob. 14–18

14–19. Determine the bending strain energy in the 2-in.-diameter A-36 steel rod.

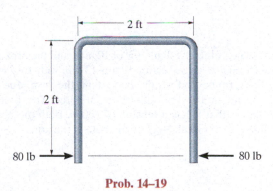

2 ft

2 ft

80 lb

80 lb

Prob. 14–19

*14–20.** Determine the total strain energy in the steel assembly. Consider the axial strain energy in the two 0.5-in.-diameter rods and the bending strain energy in the beam, which has a moment of inertia of $I = 43.4$ in^4. $E_{st} = 29(10^3)$ ksi.

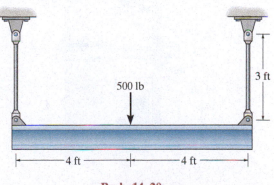

500 lb

3 ft

4 ft

4 ft

Prob. 14–20

14–21. Determine the bending strain energy in the beam. EI is constant.

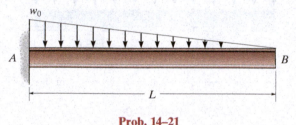

Prob. 14–21

14–22. The bolt has a diameter of 10 mm, and the arm AB has a rectangular cross section that is 12 mm wide by 7 mm thick. Determine the strain energy in the arm due to bending and in the bolt due to axial force. The bolt is tightened so that it has a tension of 500 N. Both members are made of A-36 steel. Neglect the hole in the arm.

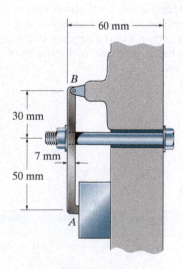

Prob. 14–22

14–23. Determine the bending strain energy in the cantilevered beam. Solve the problem two ways. (a) Apply Eq. 14–17. (b) The load $w\,dx$ acting on a segment dx of the beam is displaced a distance y, where $y = w(-x^4 + 4L^3 x - 3L^4)/(24EI)$, the equation of the elastic curve. Hence the internal strain energy in the differential segment dx of the beam is equal to the external work, i.e., $dU_i = \frac{1}{2}(w\,dx)(-y)$. Integrate this equation to obtain the total strain energy in the beam. EI is constant.

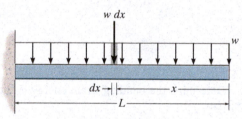

Prob. 14–23

*14–24. Determine the bending strain energy in the simply supported beam. Solve the problem two ways. (a) Apply Eq. 14–17. (b) The load $w\,dx$ acting on the segment dx of the beam is displaced a distance y, where $y = w(-x^4 + 2Lx^3 - L^3x)/(24EI)$, the equation of the elastic curve. Hence the internal strain energy in the differential segment dx of the beam is equal to the external work, i.e., $dU_i = \frac{1}{2}(w\,dx)(-y)$. Integrate this equation to obtain the total strain energy in the beam. EI is constant.

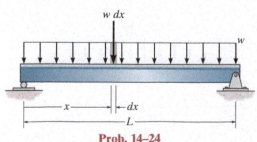

Prob. 14–24

14.3 CONSERVATION OF ENERGY

All energy methods are based on a balance of energy, often referred to as the conservation of energy. In this chapter, only mechanical energy will be considered for this energy balance; that is, the energy developed by heat, chemical reactions, and electromagnetic effects will be neglected. As a result, if a loading is applied *slowly* to a body, then these loads will do **external work** U_e as they are displaced. This external work is then transformed into **internal work** or strain energy U_i, which is stored in the body. When the loads are removed, the strain energy restores the body to its original undeformed position, provided the material's elastic limit is not exceeded. This conservation of energy for the body can be stated mathematically as

$$U_e = U_i \qquad\qquad (14\text{-}23)$$

Truss. To demonstrate how the conservation of energy applies, we will consider the truss in Fig. 14–18, which is subjected to the load **P**, causing the joint to be displaced Δ. Provided **P** is applied gradually, the external work done by **P** is determined from Eq. 14–2, that is, $U_e = \frac{1}{2}P\Delta$. Assuming that **P** develops an axial force **N** in a particular member, the strain energy stored in this member is determined from Eq. 14–16, that is, $U_i = N^2L/2AE$. Summing the strain energies for all the members of the truss, Eq. 14–23 then requires

$$\frac{1}{2}P\Delta = \sum \frac{N^2L}{2AE} \qquad\qquad (14\text{-}24)$$

Once the internal forces (N) in all the members of the truss are determined and the terms on the right calculated, it is then possible to determine the displacement Δ at the joint where **P** is applied.

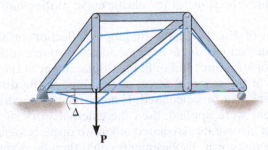

Fig. 14–18

Beam. Let us now consider finding the vertical displacement Δ under the load **P** acting on the beam in Fig. 14–19. Again, the external work is $U_e = \frac{1}{2}P\Delta$. In this case the strain energy is the result of internal shear and moment loadings caused by **P**. In particular, the contribution of strain energy due to shear is generally *neglected* in most beam deflection problems, unless the beam is short and supports a very large load. (See Example 14.4.) Consequently, the beam's strain energy will be determined only by the internal bending moment M, and therefore, using Eq. 14–17, the conservation of energy requires

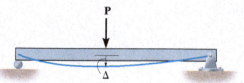

$$\frac{1}{2}P\Delta = \int_0^L \frac{M^2}{2EI}dx \qquad (14\text{–}25)$$

Fig. 14–19

Once M is expressed as a function of position x and the integral is evaluated, Δ can then be determined.

If the beam is subjected to a couple moment $\mathbf{M}_0$ as shown in Fig. 14–20, then this moment will cause the rotational displacement θ at its point of application. Since a couple moment only does work when it *rotates*, using Eq. 14–5, the external work is $U_e = \frac{1}{2}M_0\,\theta$, and so

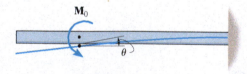

$$\frac{1}{2}M_0\,\theta = \int_0^L \frac{M^2}{2EI}dx \qquad (14\text{–}26)$$

Fig. 14–20

Here θ measures the *slope* of the elastic curve at the point where $\mathbf{M}_0$ is applied.

Application of Eq. 14–23 for finding a deflection or slope is quite limited, because only a *single* external force or couple moment acts on the member or structure, and only the displacement at the point and in the direction of the external force, or the slope in the direction of the couple moment, can be calculated. If more than one external force or couple moment were applied, then the external work of each loading would have to involve its associated unknown displacement. As a result, none of these unknown displacements could then be determined, since only the single equation $(U_e = U_i)$ is available for the solution.

EXAMPLE 14.6

The three-bar truss in Fig. 14–21a is subjected to a horizontal force of 5 kip. If the cross-sectional area of each member is 0.20 in², determine the horizontal displacement at point B. $E = 29(10^3)$ ksi.

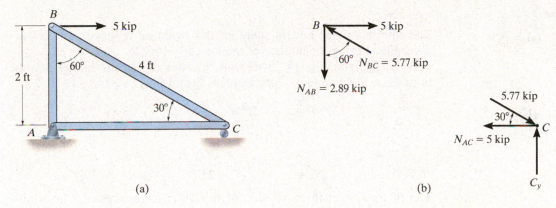

(a)   (b)

Fig. 14–21

SOLUTION

We can apply the conservation of energy to solve this problem, because only a *single* external force acts on the truss and the required displacement happens to be in the *same direction* as the force. Furthermore, the reactive forces on the truss do no work since they are not displaced.

Using the method of joints, the force in each member is determined as shown on the free-body diagrams of the pins at B and C, Fig. 14–21b.

Applying Eq. 14–24, we have

$$\frac{1}{2}P\Delta = \sum \frac{N^2 L}{2AE}$$

$$\frac{1}{2}(5 \text{ kip})(\Delta_B)_h = \frac{(2.89 \text{ kip})^2(2 \text{ ft})}{2AE} + \frac{(-5.77 \text{ kip})^2(4 \text{ ft})}{2AE}$$

$$+ \frac{(5 \text{ kip})^2(3.46 \text{ ft})}{2AE}$$

$$(\Delta_B)_h = \frac{47.32 \text{ kip} \cdot \text{ft}}{AE}$$

Substituting in the numerical data for A and E and solving, we get

$$(\Delta_B)_h = \frac{47.32 \text{ kip} \cdot \text{ft}(12 \text{ in./ft})}{(0.2 \text{ in}^2)[29(10^3) \text{ kip/in}^2]}$$

$$= 0.0979 \text{ in.} \rightarrow \qquad \qquad \textit{Ans.}$$

EXAMPLE 14.7

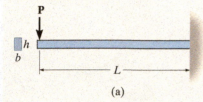

(a)

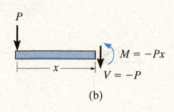

(b)

Fig. 14-22

The cantilevered beam in Fig. 14–22a has a rectangular cross section and is subjected to a load **P** at its end. Determine the displacement of the load. EI is constant.

SOLUTION

The internal shear and moment in the beam as a function of x are determined using the method of sections, Fig. 14–22b.

When applying Eq. 14–23 we will consider the strain energy due to both shear and bending. Using Eqs. 14–19 and 14–17, we have

$$\frac{1}{2}P\Delta = \int_0^L \frac{f_s V^2\, dx}{2GA} + \int_0^L \frac{M^2\, dx}{2EI}$$

$$= \int_0^L \frac{\left(\frac{6}{5}\right)(-P)^2 dx}{2GA} + \int_0^L \frac{(-Px)^2 dx}{2EI} = \frac{3P^2 L}{5GA} + \frac{P^2 L^3}{6EI} \qquad (1)$$

The first term on the right side of this equation represents the strain energy due to shear, while the second is the strain energy due to bending. As stated in Example 14.4, for most beams the shear strain energy is *much smaller* than the bending strain energy. To show when this is the case for the beam in Fig. 14–22a, we require

$$\frac{3}{5}\frac{P^2 L}{GA} \ll \frac{P^2 L^3}{6EI}$$

$$\frac{3}{5}\frac{P^2 L}{G(bh)} \ll \frac{P^2 L^3}{6E\left[\frac{1}{12}(bh^3)\right]}$$

$$\frac{3}{5G} \ll \frac{2L^2}{Eh^2}$$

Since $E \leq 3G$ (see Example 14.4), then

$$0.9 \ll \left(\frac{L}{h}\right)^2$$

Hence if L is relatively long compared with h, then the shear strain energy can be neglected. In other words, the *shear strain energy* becomes important only for *short, deep beams*. For example, if $L = 5h$, then approximately 28 times more bending strain energy will be absorbed in the beam than shear strain energy, so neglecting the shear strain energy represents an error of about 3.6%. With this in mind, Eq. 1 can be simplified to

$$\frac{1}{2}P\Delta = \frac{P^2 L^3}{6EI}$$

so that

$$\Delta = \frac{PL^3}{3EI} \qquad \qquad \textit{Ans.}$$

PROBLEMS

14–25. Determine the vertical displacement of joint D. AE is constant.

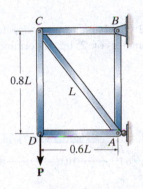

Prob. 14–25

14–26. Determine the horizontal displacement of joint C. AE is constant.

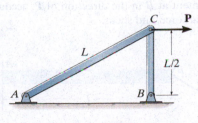

Prob. 14–26

14–27. Determine the horizontal displacement of joint A. Each bar is made of A992 steel and has a cross-sectional area of 1.5 in².

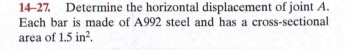
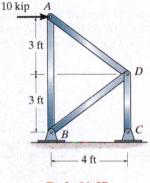

10 kip A

3 ft

D

3 ft

B C

4 ft

Prob. 14–27

*14–28. Determine the vertical displacement of joint C. AE is constant.

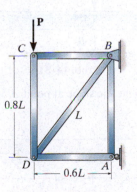

Prob. 14–28

14–29. Determine the vertical displacement of point C of the A992 steel beam. $I = 80(10^6)$ mm^4.

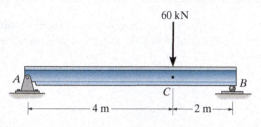

60 kN

A   C   B

|← 4 m →|← 2 m →|

Prob. 14–29

14–30. Determine the vertical displacement of end B of the cantilevered 6061-T6 aluminum alloy rectangular beam. Consider both shearing and bending strain energy.

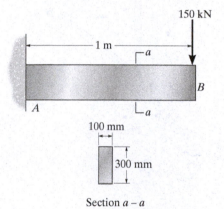

150 kN

1 m

a

A   a   B

100 mm

300 mm

Section a – a

Prob. 14–30

14–31. Determine the vertical displacement of point B on the A992 steel beam. $I = 80(10^6)$ mm^4.

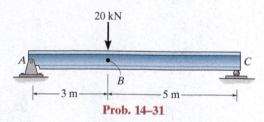

20 kN

A   B   C

|← 3 m →|← 5 m →|

Prob. 14–31

*14–32. Determine the slope at point A of the beam. EI is constant.

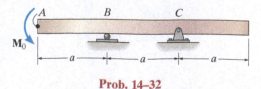

A   B   C

M_0

|← a →|← a →|← a →|

Prob. 14–32

14–33. The A992 steel bars are pin connected at C and D. If they each have the same rectangular cross section, with a height of 200 mm and a width of 100 mm, determine the vertical displacement at B. Neglect the axial load in the bars.

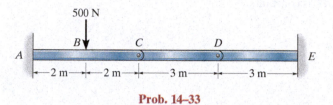

500 N

A   B   C   D   E

|← 2 m →|← 2 m →|← 3 m →|← 3 m →|

Prob. 14–33

14–34. The A992 steel bars are pin connected at C. If they each have a diameter of 2 in., determine the vertical displacement at E.

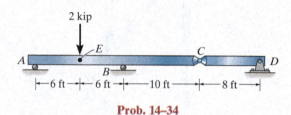

2 kip

A   E   B   C   D

|← 6 ft →|← 6 ft →|← 10 ft →|← 8 ft →|

Prob. 14–34

14–35. Determine the slope of the beam at the pin support A. Consider only bending strain energy. EI is constant.

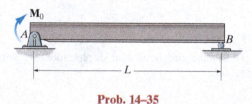

M_0

A   B

|←———— L ————→|

Prob. 14–35

*14–36. The cantilevered beam has a rectangular cross-sectional area A, a moment of inertia I, and a modulus of elasticity E. If a load P acts at point B as shown, determine the displacement at B in the direction of P, accounting for bending, axial force, and shear.

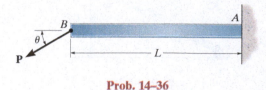

B   A

θ

P

|←———— L ————→|

Prob. 14–36

14–37. The rod has a circular cross section with a moment of inertia I. If a vertical force **P** is applied at A, determine the vertical displacement at this point. Only consider the strain energy due to bending. The modulus of elasticity is E.

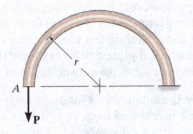

Prob. 14–37

14–38. The rod has a circular cross section with a moment of inertia I. If a vertical force **P** is applied at A, determine the vertical displacement at this point. Only consider the strain energy due to bending. The modulus of elastcity is E.

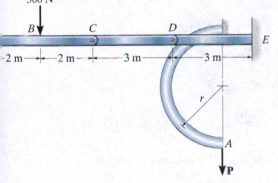

Prob. 14–38

14–39. Determine the vertical displacement of point B on the 2014-T6 aluminum beam.

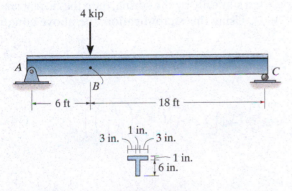

Prob. 14–39

*14–40. The rod has a circular cross section with a polar moment of inertia J and moment of inertia I. If a vertical force **P** is applied at A, determine the vertical displcement at this point. Consider the strain energy due to bending and torsion. The material constants are E and G.

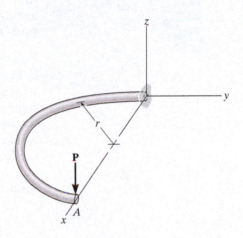

Prob. 14–40

14–41. Determine the vertical displacement of end B of the frame. Consider only bending strain energy. The frame is made using two A-36 steel W460 × 68 wide-flange sections.

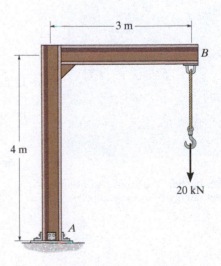

Prob. 14–41

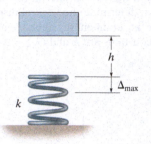

Fig. 14–23

14.4   IMPACT LOADING

Up to now we have considered all loadings to be applied to a body in a gradual manner, such that when they reach a maximum value the body remains static. Some loadings, however, are dynamic, such as when one object strikes another, producing large forces between them during a very short period of time. If we assume that during the collision no energy is lost due to heat, sound, or localized plastic deformations, then we can study the mechanics of this impact using the conservation of energy.

Falling Block.   Consider the simple block-and-spring system shown in Fig. 14–23. When the block is released from rest, it falls a distance h, striking the spring and compressing it a distance Δ_{max} before momentarily coming to rest. If we neglect the mass of the spring and assume that it responds *elastically*, then the conservation of energy requires that the work done by the block's weight, falling $h + \Delta_{max}$, be equal to the work needed to displace the end of the spring an amount Δ_{max}. Since the force in a spring is related to Δ_{max} by $F = k\Delta_{max}$, where k is the spring stiffness, then

$$U_e = U_i$$

$$W(h + \Delta_{max}) = \frac{1}{2}(k\Delta_{max})\,\Delta_{max}$$

$$W(h + \Delta_{max}) = \frac{1}{2}k\Delta_{max}^2 \qquad (14\text{--}27)$$

$$\Delta_{max}^2 - \frac{2W}{k}\Delta_{max} - 2\left(\frac{W}{k}\right)h = 0$$

Solving this quadratic equation for Δ_{max}, the maximum root is

$$\Delta_{max} = \frac{W}{k} + \sqrt{\left(\frac{W}{k}\right)^2 + 2\left(\frac{W}{k}\right)h}$$

If the weight W is supported statically by the spring, then the displacement of the block is $\Delta_{st} = W/k$. Using this simplification, the above equation becomes

$$\Delta_{max} = \Delta_{st} + \sqrt{(\Delta_{st})^2 + 2\Delta_{st}h}$$

or

$$\Delta_{max} = \Delta_{st}\left[1 + \sqrt{1 + 2\left(\frac{h}{\Delta_{st}}\right)}\right] \qquad (14\text{--}28)$$

This crash barrier is designed to absorb the impact energy of moving vehicles.

Once Δ_{max} is calculated, the maximum force applied to the spring can be determined from

$$F_{max} = k\Delta_{max} \qquad (14\text{–}29)$$

This force and its associated displacement occur only at an *instant*. Provided the block does not rebound off the spring, it will continue to vibrate until the motion dampens out and the block assumes the static position, Δ_{st}.

As a special case, if the block is held just above the spring and released, then from Eq. 14–28, with $h = 0$, the maximum displacement of the block will be

$$\Delta_{max} = 2\Delta_{st}$$

In other words, the displacement from the dynamic load is *twice* what it would be if the block were supported by the spring (a static load).

Sliding Block.   Using a similar analysis, it is also possible to determine the maximum displacement of the end of the spring if the block is sliding on a smooth horizontal surface, with a known velocity **v** just before it collides with the spring, Fig. 14–24. Here the block's kinetic energy,* $\frac{1}{2}(W/g)v^2$, will be transformed into stored energy in the spring. Hence,

$$U_e = U_i$$

$$\frac{1}{2}\left(\frac{W}{g}\right)v^2 = \frac{1}{2}k\Delta_{max}^2$$

$$\Delta_{max} = \sqrt{\frac{Wv^2}{gk}} \qquad (14\text{–}30)$$

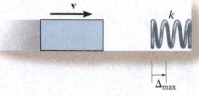

Fig. 14–24

Since the static displacement of a block resting on the spring is $\Delta_{st} = W/k$, then

$$\Delta_{max} = \sqrt{\frac{\Delta_{st}v^2}{g}} \qquad (14\text{–}31)$$

General Problem.   The results of this simplified analysis can be used to determine both the approximate deflection and the stress developed in an elastic member when it is subjected to impact. To do this we must make some necessary assumptions regarding the collision, so that the behavior of the colliding bodies is similar to the response of the block-and-spring models discussed above. Here we will consider the moving body to be *rigid* like the block, and the stationary body to be deformable like the spring. Also, like the spring, we will assume that the material behaves in a linear elastic manner and the mass of the elastic body can be neglected. Realize that all of these assumptions will lead to a *conservative* estimate of both the maximum stress and deflection of the elastic body. In other words, the calculated values will be *larger* than those that actually occur.

*Kinetic energy is "energy of motion." For the translation of a body it is $\frac{1}{2}mv^2$, where m is the body's mass, $m = W/g$.

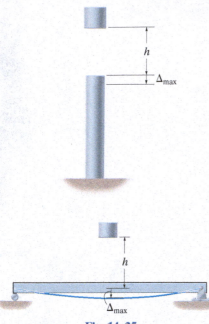

Fig. 14–25

The members of this crash guard must be designed to resist a prescribed impact loading in order to arrest the motion of a rail car.

A few examples of when this theory can be applied are shown in Fig. 14–25. Here a block of known weight is dropped onto a post and a beam, causing them to deform a maximum amount $\Delta_{\max}$. The energy of the falling block is transformed momentarily into axial strain energy in the post and bending strain energy in the beam.* In order to determine the deformation $\Delta_{\max}$, we could use the same approach as for the block–spring system, and that is to write the conservation of energy equation for the block and post or block and beam, and then solve for $\Delta_{\max}$. However, we can also solve these problems in a more direct manner by modeling the post and beam by an **equivalent spring**. For example, if a force **P** displaces the top of the post $\Delta = PL/AE$, then a spring having a stiffness $k = AE/L$ would be displaced the same amount by **P**, that is, $\Delta = P/k$. In a similar manner, from Appendix C, a force **P** applied to the center of a simply supported beam displaces the center $\Delta = PL^3/48EI$, and therefore an equivalent spring would have a stiffness of $k = 48EI/L^3$. However, to apply Eq. 14–28 or 14–30, it is not necessary to actually find this equivalent spring stiffness. All that is needed to determine $\Delta_{\max}$ is to calculate the *static displacement* Δ_{st} due to the weight $P_{st} = W$ of the block resting on the post or beam.

Once $\Delta_{\max}$ is determined, the maximum dynamic force can then be calculated from $P_{\max} = k\Delta_{\max}$. Then if we consider $P_{\max}$ to be an *equivalent static load*, the maximum stress in the member can be determined using statics and the theory of mechanics of materials. Of course this stress acts only for an *instant*, since the post or beam will begin to vibrate, thereby changing the stress within the material.

The ratio of the dynamic force $P_{\max}$ to the static force $P_{st} = W$ is called the **impact factor**, n. This factor represents the magnification of a statically applied load so that it can be treated dynamically. Since $P_{\max} = k\Delta_{\max}$ and $P_{st} = k\Delta_{st}$, then from Eq. 14–28, the impact factor becomes

$$n = 1 + \sqrt{1 + 2\left(\frac{h}{\Delta_{st}}\right)} \qquad (14\text{–}32)$$

For a complicated system of connected members, impact factors are determined from experience and experimental testing. Once n is determined, however, the dynamic stress and deflection $\Delta_{\max}$ at the point of impact are then found from the static stress σ_{st} and static deflection Δ_{st} caused by the load. They are $\sigma_{\max} = n\sigma_{st}$ and $\Delta_{\max} = n\Delta_{st}$.

---

*Strain energy due to shear is neglected for reasons discussed in Example 14.4.

IMPORTANT POINTS

- *Impact* occurs when a large force is developed between two objects which strike one another during a short period of time.

- We can analyze the effects of impact by assuming the moving body is rigid, the material of the stationary body is linear elastic, no energy is lost during collision, the bodies remain in contact during collision, and the mass of the elastic body is neglected.

- The dynamic load on a body can be determined by multiplying the static load by an *impact factor*.

EXAMPLE 14.8

The aluminum pipe shown in Fig. 14–26 is used to support a load of 150 kip. Determine the maximum displacement at the top of the pipe if the load is (a) applied gradually, and (b) applied by suddenly releasing it from the top of the pipe when $h = 0$. Take $E_{al} = 10(10^3)$ ksi and assume that the aluminum behaves elastically.

SOLUTION

Part (a). When the load is applied gradually, the work done by the weight is transformed into elastic strain energy in the pipe. Applying the conservation of energy, we have

$$U_e = U_i$$

$$\frac{1}{2}W\Delta_{st} = \frac{W^2 L}{2AE}$$

$$\Delta_{st} = \frac{WL}{AE} = \frac{150 \text{ kip}(12 \text{ in.})}{\pi[(3 \text{ in.})^2 - (2.5 \text{ in.})^2]\,10\,(10^3)\,\text{kip/in}^2}$$

$$= 0.02083 \text{ in.} = 0.0208 \text{ in.} \qquad \textit{Ans.}$$

Of course this result can also be obtained by direct application of $\Delta_{st} = NL/AE$.

Part (b). Here Eq. 14–28 can be applied, with $h = 0$. Hence,

$$\Delta_{max} = \Delta_{st}\left[1 + \sqrt{1 + 2\left(\frac{h}{\Delta_{st}}\right)}\right]$$

$$= 2\Delta_{st} = 2(0.02083 \text{ in.})$$

$$= 0.0417 \text{ in.} \qquad \textit{Ans.}$$

Hence, the displacement of the weight when applied dynamically is twice as great as when the load is applied statically. In other words, the impact factor is $n = 2$, Eq. 14–32.

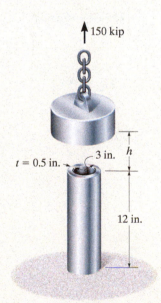

Fig. 14–26

EXAMPLE 14.9

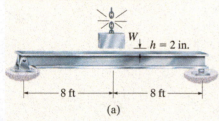

(a)

Fig. 14–27

The A992 steel beam shown in Fig. 14–27a is a W10 × 39. Determine the maximum bending stress in the beam and the beam's maximum deflection if the weight $W = 1.50$ kip is dropped from a height $h = 2$ in. onto the beam. $E_{st} = 29(10^3)$ ksi.

SOLUTION I

Here we will apply Eq. 14–28. First, however, we must calculate Δ_{st}. Using the table in Appendix C, and the data in Appendix B for the properties of a W10 × 39, we have

$$\Delta_{st} = \frac{WL^3}{48EI} = \frac{(1.50 \text{ kip})(16 \text{ ft})^3(12 \text{ in./ft})^3}{48[29(10^3) \text{ ksi}](209 \text{ in}^4)} = 0.03649 \text{ in.}$$

$$\Delta_{max} = \Delta_{st}\left[1 + \sqrt{1 + 2\left(\frac{h}{\Delta_{st}}\right)}\right]$$

$$= 0.03649 \text{ in.}\left[1 + \sqrt{1 + 2\left(\frac{2 \text{ in.}}{0.03649 \text{ in.}}\right)}\right] = 0.420 \text{ in.} \quad Ans.$$

With $k = 48EI/L^3$, the equivalent static load that causes this displacement is therefore

$$P_{max} = \frac{48EI}{L^3}\Delta_{max} = \frac{48(29(10^3) \text{ ksi})(209 \text{ in}^4)}{(16 \text{ ft})^3(12 \text{ in./ft})^3}(0.420 \text{ in.}) = 17.3 \text{ kip}$$

The internal moment caused by this load is maximum at the center of the beam, such that by the method of sections, Fig. 14–27b, $M_{max} = P_{max}L/4$. Applying the flexure formula to determine the bending stress, we have

$$\sigma_{max} = \frac{M_{max}c}{I} = \frac{P_{max}Lc}{4I} = \frac{12E\Delta_{max}c}{L^2}$$

$$= \frac{12[29(10^3) \text{ kip/in}^2](0.420 \text{ in.})(9.92 \text{ in.}/2)}{(16 \text{ ft})^2(12 \text{ in./ft})^2} = 19.7 \text{ ksi} \quad Ans.$$

SOLUTION II

It is also possible to obtain the dynamic or maximum deflection Δ_{max} from first principles. The external work of the falling weight W is $U_e = W(h + \Delta_{max})$. Since the beam deflects Δ_{max}, and $P_{max} = 48EI\Delta_{max}/L^3$, then

$$U_e = U_i$$

$$W(h + \Delta_{max}) = \frac{1}{2}\left(\frac{48EI\Delta_{max}}{L^3}\right)\Delta_{max}$$

$$(1.50 \text{ kip})(2 \text{ in.} + \Delta_{max}) = \frac{1}{2}\left[\frac{48[29(10^3)\text{kip/in}^2]209 \text{ in}^4}{(16 \text{ ft})^3(12 \text{ in./ft})^3}\right]\Delta_{max}^2$$

$$20.55\Delta_{max}^2 - 1.50\Delta_{max} - 3.00 = 0$$

Solving and choosing the positive root yields

$$\Delta_{max} = 0.420 \text{ in.} \quad Ans.$$

M_{max}

$V \longleftarrow \frac{L}{2} \longrightarrow \overset{\uparrow}{P_{max}} / 2$

(b)

EXAMPLE 14.10

A railroad car that is assumed to be rigid and has a mass of 80 Mg is moving forward at a speed of $v = 0.2$ m/s when it strikes a steel 200 mm by 200 mm post at A, Fig. 14–28a. If the post is fixed to the ground at C, determine the maximum horizontal displacement of its top B due to the impact. Take $E_{st} = 200$ GPa.

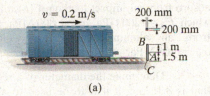

(a)

SOLUTION

Here the kinetic energy of the railroad car is transformed into internal bending strain energy only for region AC of the post. (Region BA is not subjected to an internal loading.)

We will solve for $(\Delta_A)_{max}$ using first principles rather than using Eq. 14–31. Assuming that point A is displaced $(\Delta_A)_{max}$, then the force P_{max} that causes this displacement can be determined from the table in Appendix C. We have

$$P_{max} = \frac{3EI(\Delta_A)_{max}}{L_{AC}^3} \qquad (1)$$

Thus,

$$U_e = U_i; \qquad \frac{1}{2}mv^2 = \frac{1}{2}P_{max}(\Delta_A)_{max}$$

$$\frac{1}{2}mv^2 = \frac{1}{2}\frac{3EI}{L_{AC}^3}(\Delta_A)_{max}^2; \qquad (\Delta_A)_{max} = \sqrt{\frac{mv^2L_{AC}^3}{3EI}}$$

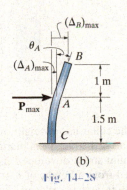

(b)

Fig. 14–28

Substituting in the numerical data yields

$$(\Delta_A)_{max} = \sqrt{\frac{80(10^3)\ \text{kg}(0.2\ \text{m/s})^2(1.5\ \text{m})^3}{3[200(10^9)\ \text{N/m}^2][\frac{1}{12}(0.2\ \text{m})^4]}} = 0.01162\ \text{m} = 11.62\ \text{mm}$$

Using Eq. 1, the force P_{max} is therefore

$$P_{max} = \frac{3[200(10^9)\ \text{N/m}^2][\frac{1}{12}(0.2\ \text{m})^4](0.01162\ \text{m})}{(1.5\ \text{m})^3} = 275.4\ \text{kN}$$

With reference to Fig. 14–28b, segment AB of the post remains straight. To determine the maximum displacement at B, we must first determine θ_A. Using the appropriate formula from the table in Appendix C, we have

$$\theta_A = \frac{P_{max}L_{AC}^2}{2EI} = \frac{275.4(10^3)\ \text{N}\ (1.5\ \text{m})^2}{2[200(10^9)\ \text{N/m}^2][\frac{1}{2}(0.2\ \text{m})]^4} = 0.01162\ \text{rad}$$

The maximum displacement at B is therefore

$$(\Delta_B)_{max} = (\Delta_A)_{max} + \theta_A L_{AB}$$

$$= 11.62\ \text{mm} + (0.01162\ \text{rad})1(10^3)\ \text{mm} = 23.2\ \text{mm} \quad Ans.$$

PROBLEMS

14–42. A bar is 4 m long and has a diameter of 30 mm. Determine the total amount of elastic energy that it can absorb from an impact loading if (a) it is made of steel for which $E_{st} = 200$ GPa, $\sigma_Y = 800$ MPa, and (b) it is made from an aluminum alloy for which $E_{al} = 70$ GPa, $\sigma_Y = 405$ MPa.

14–43. Determine the diameter of a red brass C83400 bar that is 8 ft long if it is to be used to absorb 800 ft·lb of energy in tension from an impact loading. No yielding occurs.

*14–44. Determine the speed v of the 50-Mg mass when it is just over the top of the steel post, if after impact, the maximun stress developed in the post is 550 MPa. The post has a length of $L = 1$ m and a cross-sectional area of 0.01 m². $E_{st} = 200$ GPa, $\sigma_Y = 600$ MPa.

Prob. 14–44

14–45. The collar has a weight of 50 lb and falls down the titanium bar. If the bar has a diameter of 0.5 in., determine the maximum stress developed in the bar if the weight is (a) dropped from a height of $h = 1$ ft, (b) released from a height $h \approx 0$, and (c) placed slowly on the flange at A. $E_{ti} = 16(10^3)$ ksi, $\sigma_Y = 60$ ksi.

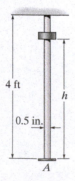

Prob. 14–45

14–46. The collar has a weight of 50 lb and falls down the titanium bar. If the bar has a diameter of 0.5 in., determine the largest height h at which the weight can be released and not permanently damage the bar after striking the flange at A. $E_{ti} = 16(10^3)$ ksi, $\sigma_Y = 60$ ksi.

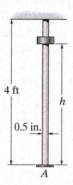

Prob. 14–46

14–47. A steel cable having a diameter of 0.4 in. wraps over a drum and is used to lower an elevator having a weight of 800 lb. The elevator is 150 ft below the drum and is descending at the constant rate of 2 ft/s when the drum suddenly stops. Determine the maximum stress developed in the cable when this occurs. $E_{st} = 29(10^3)$ ksi, $\sigma_Y = 50$ ksi.

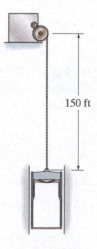

Prob. 14–47

*14–48. A steel cable having a diameter of 0.4 in. wraps over a drum and is used to lower an elevator having a weight of 800 lb. The elevator is 150 ft below the drum and is descending at the constant rate of 3 ft/s when the drum suddenly stops. Determine the maximum stress developed in the cable when this occurs. $E_{st} = 29(10^3)$ ksi, $\sigma_Y = 50$ ksi.

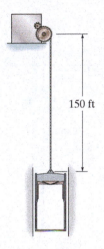

150 ft

Prob. 14–48

14–49. The A-36 steel bolt is required to absorb the energy of a 2-kg mass that falls $h = 30$ mm. If the bolt has a diameter of 4 mm, determine its required length L so the stress in the bolt does not exceed 150 MPa.

14–50. The A-36 steel bolt is required to absorb the energy of a 2-kg mass that falls $h = 30$ mm. If the bolt has a diameter of 4 mm and a length of $L = 200$ mm, determine if the stress in the bolt will exceed 175 MPa.

14–51. The A-36 steel bolt is required to absorb the energy of a 2-kg mass that falls along the 4-mm-diameter bolt shank that is 150 mm long. Determine the maximum height h of release so the stress in the bolt does not exceed 150 MPa.

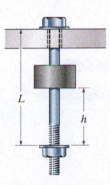

L

h

Probs. 14–49/50/51

*14–52. A cylinder having the dimensions shown is made from magnesium Am 1004-T61. If it is struck by a rigid block having a weight of 800 lb and traveling at 2 ft/s, determine the maximum stress in the cylinder. Neglect the mass of the cylinder.

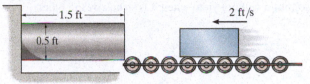

1.5 ft

0.5 ft

2 ft/s

Prob. 14–52

14–53. The composite aluminum 2014-T6 bar is made from two segments having diameters of 7.5 mm and 15 mm. Determine the maximum axial stress developed in the bar if the 10-kg collar is dropped from a height of $h = 100$ mm.

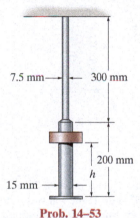

7.5 mm   300 mm

200 mm

h

15 mm

Prob. 14–53

14–54. The composite aluminum 2014-T6 bar is made from two segments having diameters of 7.5 mm and 15 mm. Determine the maximum height h from which the 10-kg collar should be dropped so that it produces a maximum axial stress in the bar of $\sigma_{max} = 300$ MPa.

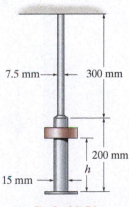

7.5 mm   300 mm

200 mm

h

15 mm

Prob. 14–54

14–55. When the 100-lb block is at $h = 3$ ft above the cylindrical post and spring assembly, it has a speed of $v = 20$ ft/s. If the post is made of 2014-T6 aluminum and the spring has the stiffness of $k = 250$ kip/in., determine the required minimum diameter d of the post to the nearest $\frac{1}{8}$ in. so that it will not yield when it is struck by the block.

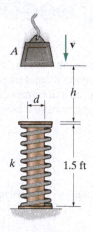

Prob. 14–55

14–57. The collar has a mass of 5 kg and falls down the titanium Ti-6A1-4V bar. If the bar has a diameter of 20 mm, determine if the weight can be released from rest at any point along the bar and not permanently damage the bar after striking the flange at A.

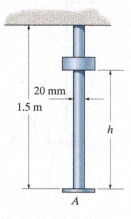

Prob. 14–57

*14–56.** The collar has a mass of 5 kg and falls down the titanium Ti-6A1-4V bar. If the bar has a diameter of 20 mm, determine the maximum stress developed in the bar if the weight is (a) dropped from a height of $h = 1$ m, (b) released from a height $h \approx 0$, and (c) placed slowly on the flange at A.

14–58. The tugboat has a weight of 120 000 lb and is traveling forward at 2 ft/s when it strikes the 12-in.-diameter fender post AB used to protect a bridge pier. If the post is made from treated white spruce and is assumed fixed at the river bed, determine the maximum horizontal distance the top of the post will move due to the impact. Assume the tugboat is rigid and neglect the effect of the water.

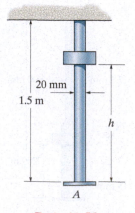

Prob. 14–56

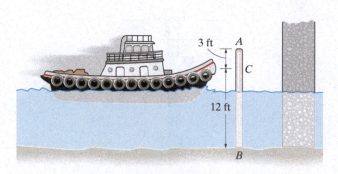

Prob. 14–58

14–59. The W10 × 12 beam is made from A-36 steel and is cantilevered from the wall at *B*. The spring mounted on the beam has a stiffness of *k* = 1000 lb/in. If a weight of 8 lb is dropped onto the spring from a height of 3 ft, determine the maximum bending stress developed in the beam.

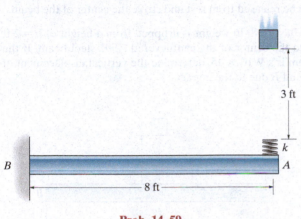

3 ft

k

B

A

8 ft

Prob. 14–59

14–62. Determine the maximum height *h* from which an 80-lb weight can be dropped onto the end of the A-36 steel W6 × 12 beam without exceeding the maximum elastic stress.

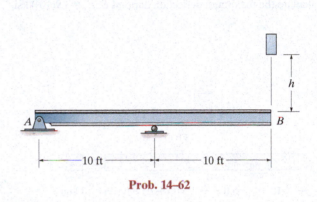

A

B

h

10 ft

10 ft

Prob. 14–62

*14–60. The weight of 175 lb is dropped from a height of 4 ft from the top of the A992 steel beam. Determine the maximum deflection and maximum stress in the beam if the supporting springs at *A* and *B* each have a stiffness of *k* = 500 lb/in. The beam is 3 in. thick and 4 in. wide.

14–61. The weight of 175 lb, is dropped from a height of 4 ft from the top of the A992 steel beam. Determine the load factor *n* if the supporting springs at *A* and *B* each have a stiffness of *k* = 500 lb/in. The beam is 3 in. thick and 4 in. wide.

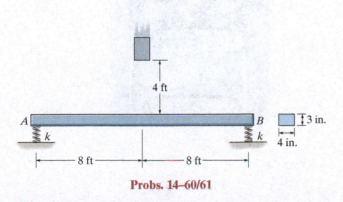

4 ft

A

B

k

k

8 ft

8 ft

3 in.

4 in.

Probs. 14–60/61

14–63. The 80-lb weight is dropped from rest at a height *h*=4 ft onto the end of the A-36 steel W6 × 12 beam. Determine the maximum bending stress developed in the beam.

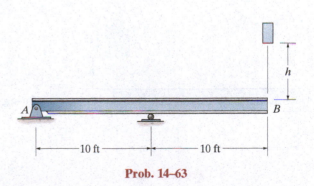

A

B

h

10 ft

10 ft

Prob. 14–63

14

*14–64. The 75-lb block has a downward velocity of 2 ft/s when it is 3 ft from the top of the beam. Determine the maximum bending stress in the beam due to the impact, and calculate the maximum deflection of its end D. $E_w = 1.9(10^3)$ ksi. Assume the material will not yield.

14–65. The 75-lb block has a downward velocity of 2 ft/s when it is 3 ft from the top of the beam. Determine the maximum bending stress in the beam due to the impact, and calculate the maximum deflection of point B. $E_w = 1.9(10^3)$ ksi.

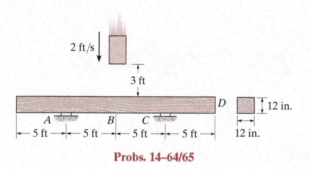

Probs. 14–64/65

14–66. The overhang beam is made of 2014-T6 aluminum. If the 75-kg block has a speed of $v = 3$ m/s at $h = 0.75$ m, determine the maximum bending stress in the beam.

14–67. The overhang beam is made of 2014-T6 aluminum. Determine the maximum height h from which the 100-kg block can be dropped from rest ($v = 0$), without causing the beam to yield.

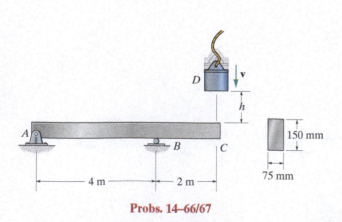

Probs. 14–66/67

*14–68. A 40-lb weight is dropped from a height of $h = 2$ ft onto the center of the cantilevered A992 steel beam. If the beam is a W10 × 15, determine the maximum bending stress in the beam.

14–69. If the maximum allowable bending stress for the W10 × 15 structural A992 steel beam is $\sigma_{allow} = 20$ ksi, determine the maximum height h from which a 50-lb weight can be released from rest and strike the center of the beam.

14–70. A 40-lb weight is dropped from a height of $h = 2$ ft onto the center of the cantilevered A992 steel beam. If the beam is a W10 × 15, determine the vertical displacement of its end B due to the impact.

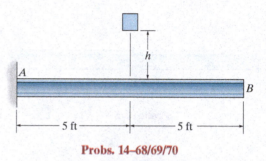

Probs. 14–68/69/70

14–71. The car bumper is made of polycarbonate-polybutylene terephthalate. If $E = 2.0$ GPa, determine the maximum deflection and maximum stress in the bumper if it strikes the rigid post when the car is coasting at $v = 0.75$ m/s. The car has a mass of 1.80 Mg, and the bumper can be considered simply supported on two spring supports connected to the rigid frame of the car. For the bumper take $I = 300(10^6)$ mm^4, $c = 75$ mm, $\sigma_Y = 30$ MPa and $k = 1.5$ MN/m.

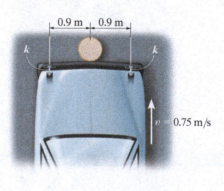

Prob. 14–71

*14.5   PRINCIPLE OF VIRTUAL WORK

The principle of virtual work was developed by John Bernoulli in 1717, and like other energy methods of analysis, it is based on the conservation of energy. Although this principle has many applications in mechanics, here we will use it to obtain the displacement and slope at a point on a deformable body.

To give a general idea as to how it is used, we will consider the body to be of arbitrary shape, as shown in Fig. 14–29b, and subjected to the "real loads" P_1, P_2, and P_3. Suppose that we want to find the displacement Δ of point A on the body. Since there is no force acting at A and in the direction of Δ, then no external work term at this point will be included when the conservation of energy principle is applied to the body. To get around this limitation, we will place an *imaginary* or "virtual" force P' on the body at A, such that P' acts in the *same direction* as Δ. Furthermore, this load will be applied *before* the real loads are applied, Fig. 14–29a. For convenience, we will choose P' to have a "unit" magnitude; that is, $P' = 1$. It should be emphasized that the term "*virtual*" is used here because it is an *imaginary load* and does not actually exist as part of the real loading. This external virtual load creates an internal virtual load u in a representative element or fiber of the body, as shown in Fig. 14–29a.

When we now apply the *real loads* P_1, P_2, and P_3, point A will be displaced Δ and the representative element will be elongated dL, Fig. 14–29b. The result of this creates *external virtual work** $1 \cdot \Delta$ on the body and *internal virtual work* $u \cdot dL$ on the element. If we consider *only* the conservation of *virtual energy*, then the external virtual work must be equal to the internal virtual work done on *all the elements* of the body. Therefore, the virtual-work equation becomes

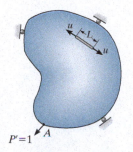

Application of virtual unit load
(a)

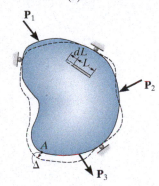

Application of real loads
(b)

Fig. 14–29

$$\underbrace{1 \cdot \Delta}_{\text{real displacements}} = \int \overbrace{u \cdot dL}^{\text{virtual loadings}} \qquad (14\text{–}33)$$

Here
$\quad P' = 1 =$ external virtual unit load acting in the direction of Δ
$\qquad u =$ internal virtual load acting on the element
$\qquad \Delta =$ displacement caused by the real loads
$\quad dL =$ displacement of the element in the direction of $\mathbf{u}$, caused by the real loads

Since we have chosen $P' = 1$, it can be seen that the solution for Δ follows directly, since $\Delta = \int u\, dL$.

---

*Prior to application of the real loads the body and the element will each undergo a virtual (imaginary) displacement, although we will *not* be interested in their magnitudes.

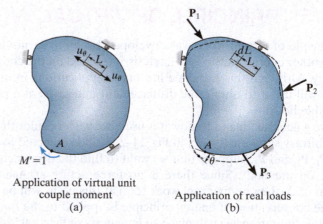

Application of virtual unit
couple moment
(a)

Application of real loads
(b)

Fig. 14–30

In a similar manner, if the angular displacement or slope of the tangent at a point on the body is to be determined at A, Fig. 14–30b, then a virtual *couple moment* $\mathbf{M}'$, having a "unit" magnitude, is applied at the point, Fig. 14–30a. As a result, this couple moment causes a virtual load u_θ in one of the elements of the body. Now applying the real loads $\mathbf{P}_1$, $\mathbf{P}_2$, $\mathbf{P}_3$, the element will be deformed an amount dL, and so the angular displacement θ can be found from the virtual-work equation

$$\overbrace{1 \cdot \theta}^{\text{virtual loadings}} = \underbrace{\int u_\theta \, dL}_{\text{real displacements}} \qquad (14\text{–}34)$$

Here

$M' = 1 =$ external virtual unit couple moment acting in the direction of θ

$u_\theta \ =$ internal virtual load acting on the element

$\theta \ =$ angular displacement in radians caused by the real loads

$dL =$ displacement of the element in the direction of $\mathbf{u}_\theta$ caused by the real loads

This method for applying the principle of virtual work is often referred to as the ***method of virtual forces***, since a *virtual force* is applied, resulting in a determination of an external *real displacement*. The equation of virtual work in this case represents a statement of *compatibility requirements* for the body.*

*We can also apply the principle of virtual work as a method of virtual displacements, that is, *virtual displacements* are imposed on the body when the body is subjected to *real loadings*. When it is used in this manner, the equation of virtual work is a statement of the *equilibrium requirements* for the body. See *Engineering Mechanics: Statics*, R. C. Hibbeler, Pearson Education, Inc.

TABLE 14–1

Deformation caused by	Strain energy	Internal virtual work
Axial load N	$\int_0^L \dfrac{N^2}{2EA}\,dx$	$\int_0^L \dfrac{nN}{EA}\,dx$
Shear V	$\int_0^L \dfrac{f_s V^2}{2GA}\,dx$	$\int_0^L \dfrac{f_s vV}{GA}\,dx$
Bending moment M	$\int_0^L \dfrac{M^2}{2EI}\,dx$	$\int_0^L \dfrac{mM}{EI}\,dx$
Torsional moment T	$\int_0^L \dfrac{T^2}{2GJ}\,dx$	$\int_0^L \dfrac{tT}{GJ}\,dx$

Internal Virtual Work.   If we assume that the material behaves in a linear elastic manner, and the stress does not exceed the proportional limit, we can then formulate the expressions for internal virtual work using the equations of elastic strain energy developed in Sec. 14.2. They are listed in the center column of Table 14–1. Recall that each of these expressions assumes that the internal loading N, V, M, or T was increased gradually from zero to its full value, and as a result, the work done by these resultants is shown in these expressions as *one-half* the product of the internal loading and its displacement. In the case of the virtual-force method, however, the virtual load is applied *before* the real loads cause displacements, and therefore the work of the virtual load is then the product of the virtual load and its real displacement (without the 1/2 factor). Referring to these internal virtual loadings (u) by the corresponding lowercase symbols n, v, m, and t, the virtual work due to each load is listed in the right-hand column of Table 14–1. Using these results, the virtual-work equation for a body subjected to a general loading can therefore be written as

$$1 \cdot \Delta = \int \frac{nN}{AE}\,dx + \int \frac{mM}{EI}\,dx + \int \frac{f_s vV}{GA}\,dx + \int \frac{tT}{GJ}\,dx \qquad (14\text{--}35)$$

In the following sections we will apply the above equation to problems involving the displacement of joints on trusses, and points on beams or shafts. We will also include a discussion of how to handle the effects of fabrication errors and differential temperature. For application it is important that a consistent set of units be used for all the terms. For example, if the real loads are expressed in kilo-newtons and the body's dimensions are in meters, a 1-kN virtual force or 1-kN · m virtual couple should be applied to the body. By doing so a calculated displacement Δ will be in meters, and a calculated slope will be in radians.

*14.6   METHOD OF VIRTUAL FORCES APPLIED TO TRUSSES

In this section we will show how to apply the method of virtual forces to determine the displacement of a truss joint. To illustrate, consider finding the vertical displacement of joint A of the truss shown in Fig. 14–31b. To do this, we must first place a virtual unit force at this joint, Fig. 14–31a, so that when the real loads $\mathbf{P}_1$ and $\mathbf{P}_2$ are applied to the truss, they cause the external virtual work $1 \cdot \Delta$. Since each member has a constant cross-sectional area A, and the virtual and real loads n and N are constant throughout the member's length, then from Table 14–1, the internal virtual work for each member is

$$\int_0^L \frac{nN}{AE}dx = \frac{nNL}{AE} \tag{14–36}$$

Therefore, the virtual-work equation for the entire truss is

$$\boxed{1 \cdot \Delta = \sum \frac{nNL}{AE}} \tag{14–37}$$

Here

$\quad 1 =$ external virtual unit load acting on the truss joint in the direction of Δ

$\quad \Delta =$ joint displacement caused by the real loads on the truss

$\quad n =$ internal virtual force in a truss member caused by the external virtual unit load

$\quad N =$ internal force in a truss member caused by the real loads

$\quad L =$ length of a member

$\quad A =$ cross-sectional area of a member

$\quad E =$ modulus of elasticity of a member

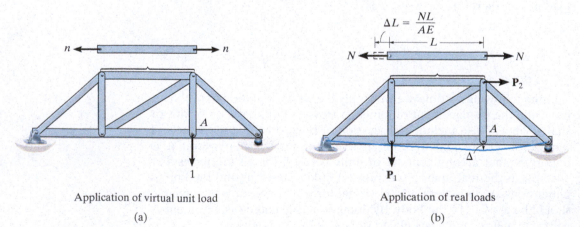

Application of virtual unit load

(a)

Application of real loads

(b)

Fig. 14–31

Temperature Change.

Truss members can change their length due to a change in temperature. If α is the coefficient of thermal expansion for a member and ΔT is the change in temperature, then the change in length of a member is $\Delta L = \alpha\, \Delta TL$ (Eq. 4–4). Hence, we can determine the displacement of a selected truss joint due to this temperature change using Eq. 14–33 written as

$$1 \cdot \Delta = \Sigma n\alpha\, \Delta TL \qquad (14\text{–}38)$$

Here

1 = external virtual unit load acting on the truss joint in the direction of Δ

Δ = joint displacement caused by the temperature change

n = internal virtual force in a truss member caused by the external virtual unit load

α = coefficient of thermal expansion of the material

ΔT = change in temperature of the member

L = length of the member

Fabrication Errors.

Occasionally errors in fabricating the lengths of the members of a truss may occur. If this happens, the displacement Δ in a particular direction of a truss joint from its expected position can be determined from the application of Eq. 14–33 written as

$$1 \cdot \Delta = \Sigma n\, \Delta L \qquad (14\text{–}39)$$

Here

1 = external virtual unit load acting on the truss joint in the direction of Δ

Δ = joint displacement caused by the fabrication errors

n = internal virtual force in a truss member caused by the external virtual unit load

ΔL = difference in length of the member from its intended length caused by a fabrication error

A combination of the right sides of Eqs. 14–37 through 14–39 will be necessary if external loads act on the truss and some of the members undergo a temperature change or have been fabricated with the wrong dimensions.

PROCEDURE FOR ANALYSIS

The following procedure provides a method that may be used to determine the displacement of any joint on a truss using the method of virtual forces.

Virtual Forces n.

- Place the virtual unit load on the truss at the joint where the displacement is to be determined. The load should be directed along the line of action of the displacement.

- With the unit load so placed and all the real loads *removed* from the truss, calculate the internal *n* force in each truss member. Assume that tensile forces are positive and compressive forces are negative.

Real Forces N.

- Determine the *N* forces in each member. These forces are caused only by the real loads acting on the truss. Again, assume that tensile forces are positive and compressive forces are negative.

Virtual-Work Equation.

- Apply the equation of virtual work to determine the desired displacement. It is important to retain the algebraic sign for each of the corresponding *n* and *N* forces when substituting these terms into the equation.

- If the resultant sum $\Sigma nNL/AE$ is positive, the displacement Δ is in the same direction as the virtual unit load. If a negative value results, Δ is opposite to the virtual unit load.

- When applying $1 \cdot \Delta = \Sigma n\alpha \, \Delta TL$, an *increase* in temperature, ΔT, will be *positive*, whereas a *decrease* in temperature will be *negative*.

- When applying $1 \cdot \Delta = \Sigma n \, \Delta L$, an increase in the length of a member, ΔL, is *positive*, whereas a *decrease* in length is *negative*.

EXAMPLE 14.11

Determine the vertical displacement of joint C of the steel truss shown in Fig. 14–32a. The cross-sectional area of each member is $A = 400$ mm^2 and $E_{st} = 200$ GPa.

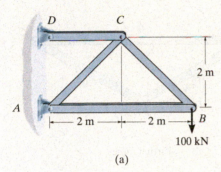

(a)

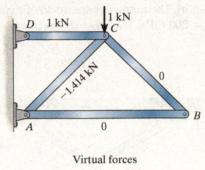

Virtual forces

(b)

Real forces

(c)

Fig. 14–32

SOLUTION

Virtual Forces n. Since the vertical displacement at joint C is to be determined, *only* a vertical 1-kN virtual load is placed at joint C, and the force in each member is calculated using the method of joints. The results are shown in Fig. 14–32b. Using our sign convention, positive numbers indicate tensile forces and negative numbers indicate compressive forces.

Real Forces N. The 100-kN load causes forces in the members that are also calculated using the method of joints. The results are shown in Fig. 14–32c.

Virtual-Work Equation. Arranging the data in tabular form, we have

Member	n	N	L	nNL
AB	0	-100	4	0
BC	0	141.4	2.828	0
AC	-1.414	-141.4	2.828	565.7
CD	1	200	2	400
				Σ 965.7 kN$^2 \cdot$ m

Thus,

$$1 \text{ kN} \cdot \Delta_{C_v} = \sum \frac{nNL}{AE} = \frac{965.7 \text{ kN}^2 \cdot \text{m}}{AE}$$

Substituting the numerical values for A and E, we have

$$1 \text{ kN} \cdot \Delta_{C_v} = \frac{965.7 \text{ kN}^2 \cdot \text{m}}{[400(10^{-6}) \text{ m}^2][200(10^6) \text{ kN/m}^2]}$$

$$\Delta_{C_v} = 0.01207 \text{ m} = 12.1 \text{ mm} \qquad Ans.$$

EXAMPLE 14.12

Determine the horizontal displacement of the roller at B of the truss shown in Fig. 14–33a. Due to radiant heating, member AB is subjected to an *increase* in temperature of $\Delta T = +60°C$, and this member has been fabricated 3 mm too short. The members are made of steel, for which $\alpha_{st} = 12(10^{-6})/°C$ and $E_{st} = 200$ GPa. The cross-sectional area of each member is 250 mm².

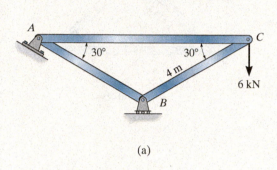

(a)

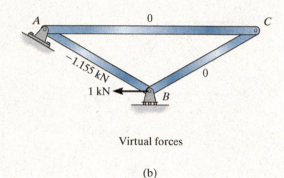

Virtual forces

(b)

SOLUTION

Virtual Forces n. A horizontal 1-kN virtual load is applied to the truss at joint B, and the forces in each member are calculated, Fig. 14–33b.

Real Forces N. Since the n forces in members AC and BC are *zero*, the N forces in these members do *not* have to be determined. Why? For completeness, though, the entire "real" force analysis is shown in Fig. 14–33c.

Virtual-Work Equation. The loads, temperature, and the fabrication error all affect the displacement of point B; therefore, Eqs. 14–37, 14–38, and 14–39 must be combined, which gives

$$1 \text{ kN} \cdot \Delta_{B_h} = \sum \frac{nNL}{AE} + \Sigma n\alpha\, \Delta TL + \Sigma n\Delta L$$

$$= 0 + 0 + \frac{(-1.155 \text{ kN})(-12 \text{ kN})(4 \text{ m})}{[250(10^{-6}) \text{ m}^2][200(10^6) \text{ kN/m}^2]}$$

$$+ (-1.155 \text{ kN})[12(10^{-6})/°C](60°C)(4 \text{ m})$$

$$+ (-1.155 \text{ kN})(-0.003 \text{ m})$$

$$\Delta_{B_h} = 0.00125 \text{ m}$$

$$= 1.25 \text{ mm} \leftarrow \qquad \text{Ans.}$$

Real forces

(c)

Fig. 14–33

PROBLEMS

*14–72. Determine the vertical displacement of joint A. Each A992 steel member has a cross-sectional area of 400 mm^2.

14–74. Determine the vertical displacement of joint B. Each A992 steel member has a cross-sectional area of 2 in^2.

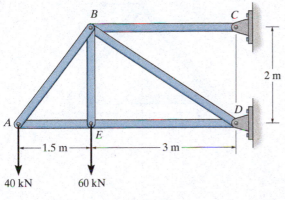

Prob. 14–72

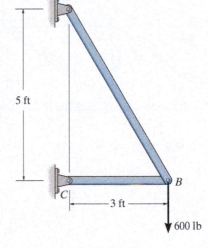

Prob. 14–74

14–73. Determine the horizontal displacement of joint B. Each A992 steel member has a cross-sectional area of 2 in^2.

14–75. Determine the vertical displacement of joint B. Each A992 steel member has a cross-sectional area of 1.5 in^2.

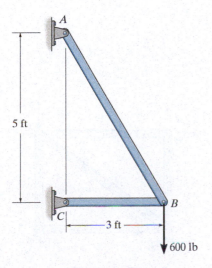

Prob. 14–73

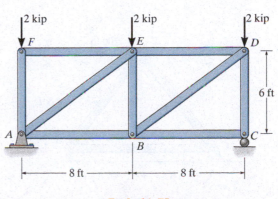

Prob. 14–75

*14–76. Determine the vertical displacement of joint *E*. Each A992 steel member has a cross-sectional area of 1.5 in².

14–78. Determine the vertical displacement of joint *B*. Each A-36 steel member has a cross-sectional area of 2 in².

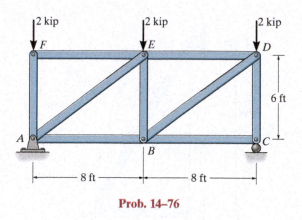

Prob. 14–76

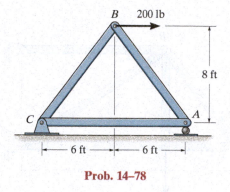

Prob. 14–78

14–77. Determine the horizontal displacement of joint *B*. Each A-36 steel member has a cross-sectional area of 2 in².

14–79. Determine the horizontal displacement of joint *B* of the truss. Each A992 steel member has a cross-sectional area of 400 mm².

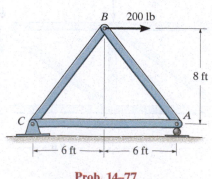

Prob. 14–77

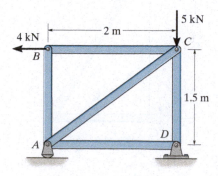

Prob. 14–79

*14–80.   Determine the vertical displacement of joint *C* of the truss. Each A992 steel member has a cross-sectional area of 400 mm².

14–83.   Determine the vertical displacement of joint *A*. The truss is made from A992 steel rods having a diameter of 30 mm.

*14–84.   Determine the vertical displacement of joint *D*. The truss is made from A992 steel rods having a diameter of 30 mm.

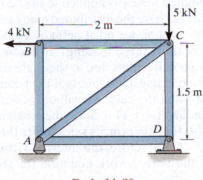

Prob. 14–80

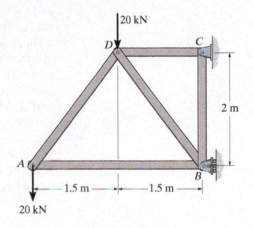

Probs. 14–83/84

14–81.   Determine the horizontal displacement of joint *C*. Each A-36 steel member has a cross-sectional area of 400 mm².

14–82.   Determine the vertical displacement of joint *D*. Each A-36 steel member has a cross-sectional area of 400 mm².

14–85.   Determine the horizontal displacement of joint *D*. Each A-36 steel member has a cross-sectional area of 300 mm².

14–86.   Determine the horizontal displacement of joint *E*. Each A-36 steel member has a cross-sectional area of 300 mm².

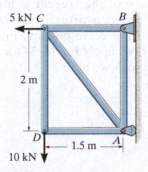

Probs. 14–81/82

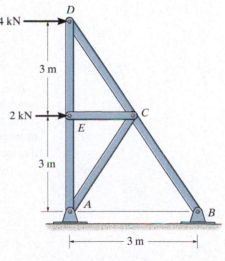

Probs. 14–85/86

*14.7   METHOD OF VIRTUAL FORCES APPLIED TO BEAMS

We can also apply the method of virtual forces to determine the displacement and slope at a point on a beam. For example, if we wish to determine the vertical displacement Δ of point A on the beam shown in Fig. 14–34b, we must first place a vertical unit load at this point, Fig. 14–34a, and then when the "real" distributed load w is applied to the beam it will cause external virtual work $1 \cdot \Delta$. Because the distributed load causes both a shear and moment within the beam, we must actually consider the internal virtual work due to both of these loadings. In Example 14.7, however, it was shown that beam deflections due to shear are negligible compared with those caused by bending, particularly if the beam is long and slender. Since this is most often the case, we will only consider the virtual strain energy due to bending, Table 14–1. Since the moment M will cause the element dx in Fig. 14–34b to deform, its sides rotate through an angle $d\theta = (M/EI)dx$, Eq. 12–16. Therefore, the internal virtual work is $m\,d\theta$. Applying Eq. 14–33, the virtual-work equation for the entire beam becomes

$$1 \cdot \Delta = \int_0^L \frac{mM}{EI}dx \qquad (14\text{–}40)$$

Here

$1 =$ external virtual unit load acting on the beam in the direction of Δ

$\Delta =$ displacement caused by the real loads acting on the beam

$m =$ internal virtual moment in the beam, expressed as a function of x and caused by the external virtual unit load

$M =$ internal moment in the beam, expressed as a function of x and caused by the real loads

$E =$ modulus of elasticity of the material

$I =$ moment of inertia of the cross-sectional area about the neutral axis

In a similar manner, if the slope θ of the tangent at a point on the beam's elastic curve is to be determined, a virtual unit couple moment must be applied at the point, and the corresponding internal virtual moment m_θ has to be determined. If we apply Eq. 14–34 for this case and neglect the effect of shear deformations, we have

$$1 \cdot \theta = \int_0^L \frac{m_\theta M}{EI}dx \qquad (14\text{–}41)$$

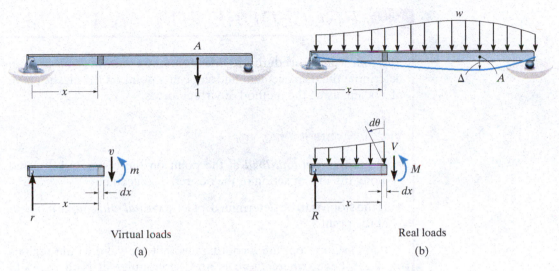

Virtual loads

(a)

Real loads

(b)

Fig. 14–34

When applying these equations, keep in mind that the integrals on the right side represent the amount of virtual bending strain energy that is *stored* in the beam. If a series of concentrated forces or couple moments act on the beam or the distributed load is discontinuous, a single integration *cannot* be performed across the beam's entire length. Instead, separate x coordinates must be chosen within regions that have no discontinuity of loading. Also, it is not necessary that each x have the same origin; however, the x selected for determining the real moment M in a particular region must be the *same* x selected for determining the virtual moment m or m_θ within this same region. For example, consider the beam in Fig. 14–35. In order to determine the displacement at D, we can use x_1 to determine the strain energy in region AB, x_2 for region BC, x_3 for region DE, and x_4 for region DC. For any problem, each x coordinate should be selected so that both M and m (or m_θ) can easily be formulated.

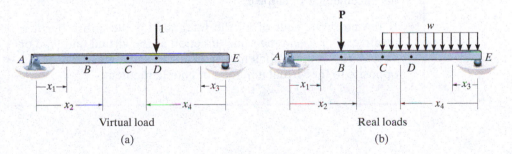

Virtual load

(a)

Real loads

(b)

Fig. 14–35

▶ PROCEDURE FOR ANALYSIS

The following procedure provides a method that may be used to determine the displacement and slope at a point on the elastic curve of a beam, using the method of virtual forces.

Virtual Moments m or m_θ.

- Place a *virtual unit load* at the point on the beam and directed along the line of action of the desired displacement.

- If the slope is to be determined, place a *virtual unit couple moment* at the point.

- Establish appropriate x coordinates that are valid within regions of the beam where there is no discontinuity of both real and virtual load.

- With the virtual load in place, and all the real loads *removed* from the beam, calculate the internal moment m or m_θ as a function of each x coordinate. When doing this, assume that m or m_θ acts in the positive direction according to the established beam sign convention for positive moment, Fig. 6–3.

Real Moments.

- Using the *same* x coordinates as those established for m or m_θ, determine the internal moments M caused by the real loads. Be sure M is also shown acting in the same positive direction as m or m_θ.

Virtual-Work Equation.

- Apply the equation of virtual work to determine the desired displacement Δ or slope θ.

- If the algebraic sum of all the integrals for the entire beam is positive, Δ or θ is in the same direction as the virtual unit load or virtual unit couple moment. If a negative value results, Δ or θ is opposite to the virtual unit load or couple moment.

EXAMPLE 14.13

Determine the displacement of point B on the beam shown in Fig. 14–36a. EI is constant.

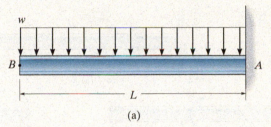

(a)

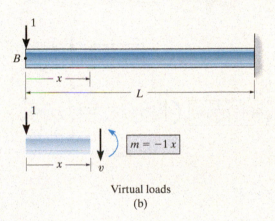

Virtual loads

(b)

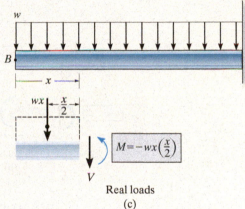

Real loads

(c)

Fig. 14–36

SOLUTION

Virtual Moment m.   The vertical displacement of point B is obtained by placing a virtual unit load at B, Fig. 14–36b. By inspection, there are no discontinuities of loading on the beam for *both* the real and virtual loads. Thus, a *single x* coordinate can be used to determine the virtual strain energy. This coordinate will be selected with its origin at B, so that the reactions at A do not have to be determined in order to find the internal moments m and M. Using the method of sections, the internal moment m is shown in Fig. 14–36b.

Real Moment M.   Using the *same x* coordinate, the internal moment M is shown in Fig. 14–36c.

Virtual-Work Equation.   The vertical displacement at B is thus

$$1 \cdot \Delta_B = \int \frac{mM}{EI}\,dx = \int_0^L \frac{(-1x)(-wx^2/2)\,dx}{EI}$$

$$\Delta_B = \frac{wL^4}{8EI} \qquad\qquad Ans.$$

EXAMPLE 14.14

Determine the slope at point B of the beam shown in Fig. 14–37a. EI is constant.

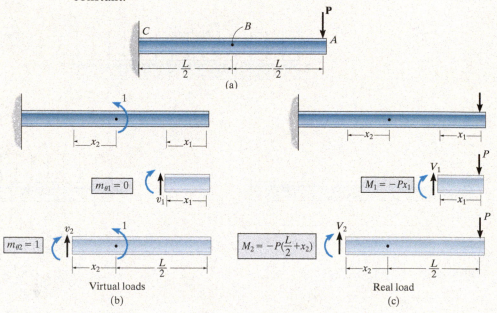

Virtual loads
(b)

Real load
(c)

Fig. 14–37

SOLUTION

Virtual Moments m_θ. The slope at B is determined by placing a virtual unit couple moment at B, Fig. 14–37b. Two x coordinates must be selected in order to determine the total virtual strain energy in the beam. Coordinate x_1 accounts for the strain energy within segment AB, and coordinate x_2 accounts for the strain energy in segment BC. Using the method of sections, the internal moments m_θ within each of these segments are shown in Fig. 14–37b.

Real Moments M. Using the *same coordinates* x_1 and x_2, the internal moments M are shown in Fig. 14–37c.

Virtual-Work Equation. The slope at B is thus

$$1 \cdot \theta_B = \int \frac{m_\theta M}{EI} dx$$

$$= \int_0^{L/2} \frac{0(-Px_1)\, dx_1}{EI} + \int_0^{L/2} \frac{1\{-P[(L/2) + x_2]\}\, dx_2}{EI}$$

$$\theta_B = -\frac{3PL^2}{8EI} \qquad\qquad\qquad Ans.$$

The *negative sign* indicates that θ_B is clockwise, that is, *opposite* to the direction of the virtual couple moment shown in Fig. 14–37b.

PROBLEMS

14–87. Determine the displacement at point C. EI is constant.

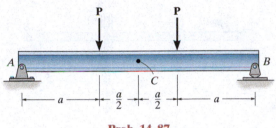

Prob. 14–87

*14–88.** The beam is made of southern pine for which $E_p = 13$ GPa. Determine the displacement at A.

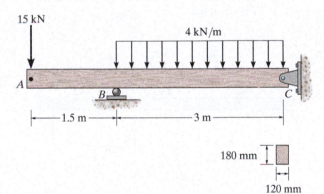

Prob. 14–88

14–89. Determine the displacement at point C. EI is constant.

14–90. Determine the slope at point C. EI is constant.

14–91. Determine the slope at point A. EI is constant.

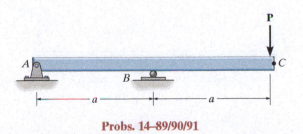

Probs. 14–89/90/91

*14–92.** Determine the displacement of point C of the beam made from A992 steel and having a moment of inertia of $I = 53.8$ in⁴.

14–93. Determine the slope at B of the beam made from A992 steel and having a moment of inertia of $I = 53.8$ in⁴.

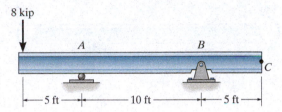

Probs. 14–92/93

14–94. The beam is made of Douglas fir. Determine the slope at C.

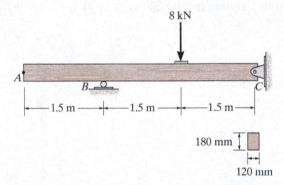

Prob. 14–94

14–95. Determine the displacement at pulley B. The A992 steel shaft has a diameter of 30 mm.

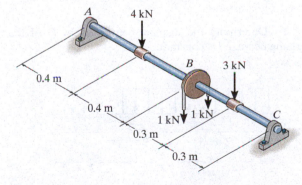

Prob. 14–95

*14–96. The A992 steel beam has a moment of inertia of $I = 125(10^6)$ mm^4. Determine the displacement at point D.

14–97. The A992 steel beam has a moment of inertia of $I = 125(10^6)$ mm^4. Determine the slope at A.

14–98. The A992 structural steel beam has a moment of inertia of $I = 125(10^6)$ mm^4. Determine the slope at B.

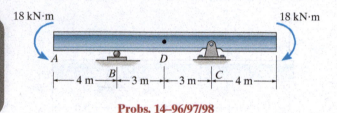

18 kN·m   18 kN·m

— 4 m — 3 m — 3 m — 4 m —

Probs. 14–96/97/98

14–99. Determine the displacement at point C of the shaft. EI is constant.

*14–100. Determine the slope at A of the shaft. EI is constant.

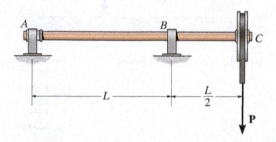

Probs. 14–99/100

14–101. Determine the slope of end C of the overhang beam. EI is constant.

14–102. Determine the displacement of point D of the overhang beam. EI is constant.

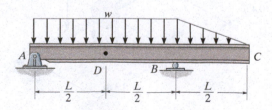

Probs. 14–101/102

14–103. Determine the slope at A of the 2014-T6 aluminum shaft having a diameter of 100 mm.

*14–104. Determine the displacement at point C of the 2014-T6 aluminum shaft having a diameter of 100 mm.

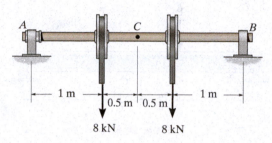

— 1 m — 0.5 m 0.5 m — 1 m —

8 kN   8 kN

Probs. 14–103/104

14–105. Determine the displacement at point C and the slope at B. EI is constant.

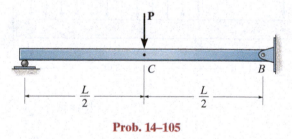

Prob. 14–105

14–106. Determine the displacement at point C of the W14 × 26 beam made from A992 steel.

14–107. Determine the slope at A of the W14 × 26 beam made from A992 steel.

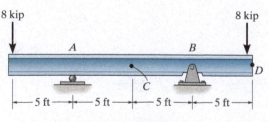

8 kip   8 kip

— 5 ft — 5 ft — 5 ft — 5 ft —

Probs. 14–106/107

*14–108.   Determine the slope at A. EI is constant.

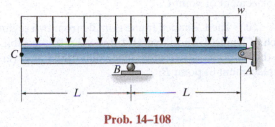

Prob. 14–108

14–109.   Determine the slope at C of the overhang white spruce beam.

14–110.   Determine the displacement at point D of the overhang white spruce beam.

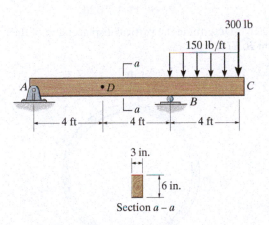

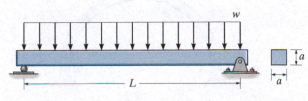

Probs. 14–109/110

14–111.   Determine the maximum deflection of the beam caused only by bending, and caused by bending and shear. Take E = 3G.

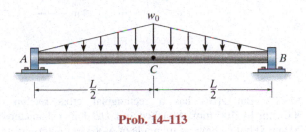

Prob. 14–111

*14–112.   The beam is made of oak, for which $E_o = 11$ GPa. Determine the slope and displacement at point A.

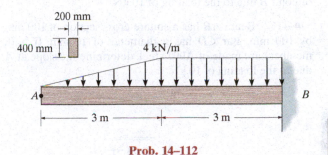

Prob. 14–112

14–113.   Determine the slope of the shaft at the bearing support A. EI is constant.

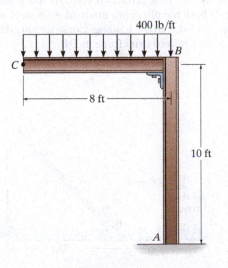

Prob. 14–113

14–114.   Determine the horizontal and vertical displacements of point C. There is a fixed support at A. EI is constant.

Prob. 14–114

14–115. Beam AB has a square cross section of 100 mm by 100 mm. Bar CD has a diameter of 10 mm. If both members are made of A992 steel, determine the vertical displacement of point B due to the loading of 10 kN.

*14–116. Beam AB has a square cross section of 100 mm by 100 mm. Bar CD has a diameter of 10 mm. If both members are made of A992 steel, determine the slope at A due to the loading of 10 kN.

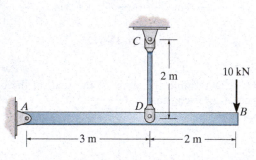

Probs. 14–115/116

14–117. Bar ABC has a rectangular cross section of 300 mm by 100 mm. Attached rod DB has a diameter of 20 mm. If both members are made of A-36 steel, determine the vertical displacement of point C due to the loading. Consider only the effect of bending in ABC and axial force in DB.

14–118. Bar ABC has a rectangular cross section of 300 mm by 100 mm. Attached rod DB has a diameter of 20 mm. If both members are made of A-36 steel, determine the slope at A due to the loading. Consider only the effect of bending in ABC and axial force in DB.

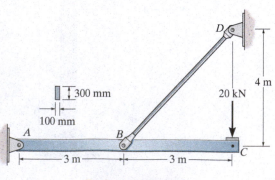

Probs. 14–117/118

14–119. The L-shaped frame is made from two segments, each of length L and flexural stiffness EI. If it is subjected to the uniform distributed load, determine the horizontal displacement of point C.

*14–120. The L-shaped frame is made from two segments, each of length L and flexural stiffness EI. If it is subjected to the uniform distributed load, determine the vertical displacement of point B.

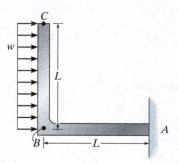

Probs. 14–119/120

14–121. Determine the vertical displacement of the ring at point B. EI is constant.

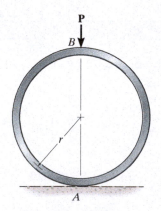

Prob. 14–121

14–122. Determine the horizontal displacement at the roller at A due to the loading. EI is constant.

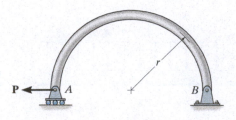

Prob. 14–122

*14.8   CASTIGLIANO'S THEOREM

In 1879, Alberto Castigliano, an Italian railroad engineer, published a book in which he outlined a method for determining the displacement and slope at a point in a body. This method, which is referred to as Castigliano's second theorem, applies only to bodies that have constant temperature and are made of linear elastic material. If the displacement at a point is to be determined, the theorem states that the displacement is equal to the first partial derivative of the strain energy in the body with respect to a force acting at the point and in the direction of displacement. In a similar manner, the slope of the tangent at a point in a body is equal to the first partial derivative of the strain energy in the body with respect to a couple moment acting at the point and in the direction of the slope angle.

This theorem considers a body of arbitrary shape, which is subjected to a series of n forces $\mathbf{P}_1, \mathbf{P}_2, \ldots, \mathbf{P}_n$, Fig. 14–38. According to the conservation of energy, the external work done by these forces must be equal to the internal strain energy stored in the body. However, the external work is a function of the external loads, $U_e = \Sigma \int P \, dx$, Eq. 14–1, so the internal work is also a function of the external loads. Thus,

$$U_i = U_e = f(P_1, P_2, \ldots, P_n) \qquad (14\text{–}42)$$

Now, if any one of the external forces, say P_j, is *increased* by a differential amount dP_j, the internal work will also be increased, such that the strain energy becomes

$$U_i + dU_i = U_i + \frac{\partial U_i}{\partial P_j} dP_j \qquad (14\text{–}43)$$

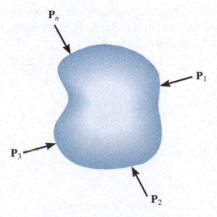

Fig. 14–38

This value, however, will not depend on the sequence in which the n forces are applied to the body. For example, we could also apply the increase $d\mathbf{P}_j$ to the body *first*, then apply the loads $\mathbf{P}_1, \mathbf{P}_2, \ldots, \mathbf{P}_n$. If we do this, $d\mathbf{P}_j$ would cause the body to displace a differential amount $d\Delta_j$ in the direction of $d\mathbf{P}_j$. By Eq. 14–2 ($U_e = \frac{1}{2}P_j\Delta_j$), the increment of strain energy would then be $\frac{1}{2}dP_j\,d\Delta_j$. This is a second-order differential and may be neglected. Application of the loads $\mathbf{P}_1, \mathbf{P}_2, \ldots, \mathbf{P}_n$ causes $d\mathbf{P}_j$ to move further, through the displacement Δ_j, so that now the strain energy becomes

$$U_i + dU_i = U_i + dP_j\,\Delta_j \tag{14–44}$$

Here U_i is the internal strain energy in the body, caused by the loads $\mathbf{P}_1, \mathbf{P}_2, \ldots, \mathbf{P}_n$, and $dP_j\,\Delta_j$ is the *additional* strain energy caused by $d\mathbf{P}_j$.

To summarize, Eq. 14–43 represents the strain energy in the body determined by first applying the loads $\mathbf{P}_1, \mathbf{P}_2, \ldots, \mathbf{P}_n$, then $d\mathbf{P}_j$; Eq. 14–44 represents the strain energy determined by first applying $d\mathbf{P}_j$ and then the loads $\mathbf{P}_1, \mathbf{P}_2, \ldots, \mathbf{P}_n$. Since these two equations must be equal, we require

$$\Delta_j = \frac{\partial U_i}{\partial P_j} \tag{14–45}$$

which proves the theorem; i.e., the displacement Δ_j in the direction of $\mathbf{P}_j$ is equal to the first partial derivative of the strain energy with respect to $\mathbf{P}_j$.

Castigliano's second theorem is a statement regarding the body's *compatibility requirements*, since it is a condition related to displacement.* The above derivation requires that *only conservative forces* be considered for the analysis. These forces can be applied in any order, and they do work that is independent of the path, and therefore create no energy loss. As long as the material has linear elastic behavior, the applied forces will be conservative and the theorem is valid.*

---

*Castigliano also stated a first theorem, which is similar; however, it relates the load P_j to the partial derivative of the strain energy with respect to the corresponding displacement, that is, $P_j = \partial U_i/\partial \Delta_j$. This theorem is another way of expressing the *equilibrium requirements* for the body.

*14.9 CASTIGLIANO'S THEOREM APPLIED TO TRUSSES

Since a truss member is only subjected to an axial load, the strain energy for the member is given by Eq. 14–16, $U_i = N^2 L/2AE$. Substituting this equation into Eq. 14–45 and omitting the subscript i, we have

$$\Delta = \frac{\partial}{\partial P} \sum \frac{N^2 L}{2AE}$$

It is generally easier to perform the differentiation prior to summation. Also, in general L, A, and E will be constant for a given member, and therefore we can write

$$\Delta = \sum N \left(\frac{\partial N}{\partial P} \right) \frac{L}{AE} \qquad (14\text{–}46)$$

Here

Δ = displacement of the truss joint

P = an external force of *variable magnitude* applied to the truss joint in the direction of Δ

N = internal axial force in a member caused by *both* force **P** and the actual loads on the truss

L = length of a member

A = cross-sectional area of a member

E = modulus of elasticity of the material

Note that the above equation is similar to that used for the method of virtual forces, Eq. 14–37 ($1 \cdot \Delta = \Sigma nNL/AE$), except n is replaced by $\partial N/\partial P$.

> # PROCEDURE FOR ANALYSIS

The following procedure provides a method that may be used to determine the displacement of any joint on a truss using Castigliano's second theorem.

External Force P.

- Place a force **P** on the truss at the joint where the displacement is to be determined. This force is assumed to have a *variable magnitude* and should be directed along the line of action of the displacement.

Internal Forces N.

- Determine the force N in each member in terms of both the actual (numerical) loads and the (variable) force P. Assume that tensile forces are positive and compressive forces are negative.
- Find the respective partial derivative $\partial N/\partial P$ for each member.
- *After N and $\partial N/\partial P$ have been determined, assign P its numerical value if it has actually replaced a real force on the truss. Otherwise, set P equal to zero.*

Castigliano's Second Theorem.

- Apply Castigliano's second theorem to determine the desired displacement Δ. It is important to retain the algebraic signs for corresponding values of N and $\partial N/\partial P$ when substituting these terms into the equation.
- If the resultant sum $\Sigma N(\partial N/\partial P)L/AE$ is positive, Δ is in the same direction as **P**. If a negative value results, Δ is opposite to **P**.

EXAMPLE 14.15

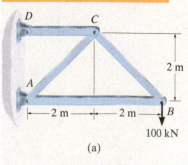

(a)

Fig. 14–39

Determine the vertical displacement of joint C of the steel truss shown in Fig. 14–39a. The cross-sectional area of each member is $A = 400 \text{ mm}^2$, and $E_{st} = 200 \text{ GPa}$.

SOLUTION

External Force P.   A vertical force **P** is applied to the truss at joint C, since this is where the vertical displacement is to be determined, Fig. 14–39b.

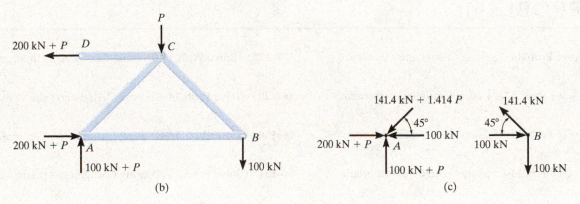

(b)   (c)

Fig. 14–39 (cont.)

Internal Forces N. The reactions at the truss supports A and D are calculated and the results are shown in Fig. 14–39b. Using the method of joints, the N forces in each member are determined, Fig. 14–39c.* For convenience, these results along with their partial derivatives $\partial N / \partial P$ are listed in tabular form. Note that since **P** does not actually exist as a real load on the truss, we require $P = 0$.

Member	N	$\dfrac{\partial N}{\partial P}$	$N(P = 0)$	L	$N\left(\dfrac{\partial N}{\partial P}\right)L$
AB	-100	0	-100	4	0
BC	141.4	0	141.4	2.828	0
AC	$-(141.4 + 1.414P)$	-1.414	-141.4	2.828	565.7
CD	$200 + P$	1	200	2	400

$$\Sigma \ 965.7 \ \text{kN} \cdot \text{m}$$

Castigliano's Second Theorem. Applying Eq. 14–46, we have

$$\Delta_{C_v} = \Sigma N\left(\frac{\partial N}{\partial P}\right)\frac{L}{AE} = \frac{965.7 \ \text{kN} \cdot \text{m}}{AE}$$

Substituting the numerical values for A and E, we get

$$\Delta_{C_v} = \frac{965.7 \ \text{kN} \cdot \text{m}}{[400(10^{-6}) \ \text{m}^2] \ 200(10^6) \ \text{kN}/\text{m}^2}$$

$$= 0.01207 \ \text{m} = 12.1 \ \text{mm} \qquad \qquad Ans.$$

This solution should be compared with that of Example 14.11 which uses the virtual-work method.

*It may be more convenient to analyze the truss with just the 100-kN load on it, then analyze the truss with the **P** load on it. The results can then be summed algebraically to give the N forces.

PROBLEMS

14–123. Solve Prob. 14–73 using Castigliano's theorem.

*14–124. Solve Prob. 14–74 using Castigliano's theorem.

14–125. Solve Prob. 14–75 using Castigliano's theorem.

14–126. Solve Prob. 14–76 using Castigliano's theorem.

14–127. Solve Prob. 14–77 using Castigliano's theorem.

*14–128. Solve Prob. 14–78 using Castigliano's theorem.

14–129. Solve Prob. 14–81 using Castigliano's theorem.

14–130. Solve Prob. 14–82 using Castigliano's theorem.

14–131. Solve Prob. 14–85 using Castigliano's theorem.

*14–132. Solve Prob. 14–86 using Castigliano's theorem.

*14.10   CASTIGLIANO'S THEOREM APPLIED TO BEAMS

The internal strain energy within a beam is caused by both bending and shear. However, as pointed out in Example 14.7, if the beam is long and slender, the strain energy due to shear can be neglected compared with that of bending. Assuming this to be the case, the strain energy is $U_i = \int M^2 \, dx / 2EI$, Eq. 14–17. Omitting the subscript i, Castigliano's second theorem, $\Delta_i = \partial U_i / \partial P_i$, becomes

$$\Delta = \frac{\partial}{\partial P} \int_0^L \frac{M^2 \, dx}{2EI}$$

Rather than squaring the expression for internal moment, integrating, and then taking the partial derivative, it is generally easier to differentiate prior to integration. Then we have

$$\Delta = \int_0^L M \left(\frac{\partial M}{\partial P} \right) \frac{dx}{EI} \tag{14–47}$$

Here

Δ = displacement of the point caused by the real loads acting on the beam

P = an external force of *variable magnitude* applied to the beam at the point and in the direction of Δ

M = internal moment in the beam, expressed as a function of x and caused by *both* the force P and the actual loads on the beam

E = modulus of elasticity of the material

I = moment of inertia of the cross-sectional area calculated about the neutral axis

If the slope of the tangent θ at a point on the elastic curve is to be determined, the partial derivative of the internal moment M with respect to an *external couple moment M'* acting at the point must be found. For this case,

$$\theta = \int_0^L M \left(\frac{\partial M}{\partial M'} \right) \frac{dx}{EI} \qquad (14\text{--}48)$$

The above equations are similar to those used for the method of virtual forces, Eqs. 14–40 and 14–41, except m and m_θ replace $\partial M / \partial P$ and $\partial M / \partial M'$, respectively.

In addition, if axial load, shear, and torsion cause significant strain energy within the member, then the effects of all these loadings should be included when applying Castigliano's theorem. To do this we must use the strain energy functions developed in Sec. 14.2, along with their associated partial derivatives. We have

$$\Delta = \Sigma N \left(\frac{\partial N}{\partial P} \right) \frac{L}{AE} + \int_0^L f_s V \left(\frac{\partial V}{\partial P} \right) \frac{dx}{GA} + \int_0^L M \left(\frac{\partial M}{\partial P} \right) \frac{dx}{EI} + \int_0^L T \left(\frac{\partial T}{\partial P} \right) \frac{dx}{GJ} \qquad (14\text{--}49)$$

The method of applying this general formulation is similar to that used to apply Eqs. 14–47 and 14–48.

14

PROCEDURE FOR ANALYSIS

The following procedure provides a method that may be used to apply Castigliano's second theorem.

External Force P or Couple Moment M'.

- Place a force **P** at the point on the beam and directed along the line of action of the desired displacement.

- If the slope of the tangent is to be determined at the point, place a couple moment **M'** at the point.

- Assume that both **P** and **M'** have a variable magnitude.

Internal Moments M.

- Establish appropriate x coordinates that are valid within regions of the beam where there is no discontinuity of force, distributed load, or couple moment.

- Determine the internal moments M as a function of P or M', and then find the partial derivatives $\partial M / \partial P$ or $\partial M / \partial M'$ for each x coordinate.

- *After M and $\partial M / \partial P$ or $\partial M / \partial M'$ have been determined*, assign P or M' its numerical value if it has actually replaced a real force or couple moment. Otherwise, set P or M' equal to zero.

Castigliano's Second Theorem.

- Apply Eq. 14–47 or 14–48 to determine the desired displacement Δ or slope θ. When doing so, it is important to retain the algebraic signs for corresponding values of M and $\partial M / \partial P$ or $\partial M / \partial M'$.

- If the resultant sum of all the definite integrals is positive, Δ or θ is in the same direction as **P** or **M'**. If a negative value results, Δ or θ is opposite to **P** or **M'**.

EXAMPLE 14.16

Determine the displacement of point B on the beam shown in Fig. 14–40a. EI is constant.

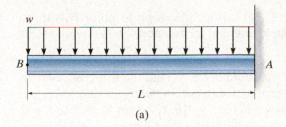

(a)

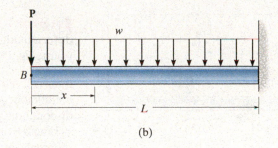

(b)

Fig. 14–40

SOLUTION

External Force P.   A vertical force **P** is placed on the beam at B as shown in Fig. 14–40b.

Internal Moments M.   A single x coordinate is needed for the solution, since there is no discontinuity of loading between A and B. Using the method of sections, Fig. 14–40c, the internal moment and its partial derivative are

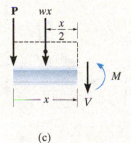

(c)

$$\zeta + \Sigma M_{NA} = 0; \quad M + wx\left(\frac{x}{2}\right) + P(x) = 0$$

$$M = -\frac{wx^2}{2} - Px$$

$$\frac{\partial M}{\partial P} = -x$$

Setting $P = 0$ gives

$$M = \frac{-wx^2}{2} \qquad \text{and} \qquad \frac{\partial M}{\partial P} = -x$$

Castigliano's Second Theorem.   Applying Eq. 14–47, we have

$$\Delta_B = \int_0^L M\left(\frac{\partial M}{\partial P}\right)\frac{dx}{EI} = \int_0^L \frac{(-wx^2/2)(-x)\,dx}{EI}$$

$$= \frac{wL^4}{8EI} \qquad\qquad\qquad\qquad\qquad Ans.$$

The similarity between this solution and that of the virtual-work method, Example 14.13, should be noted.

EXAMPLE 14.17

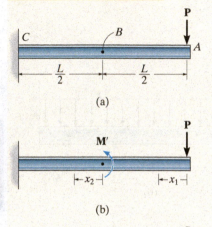

(a)

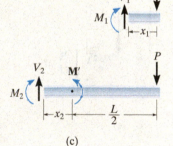

(b)

(c)

Fig. 14–41

Determine the slope at point B of the beam shown in Fig. 14–41a. EI is constant.

SOLUTION

External Couple Moment M'. Since the slope at point B is to be determined, an external couple moment M' is placed on the beam at this point, Fig. 14–41b.

Internal Moments M. Two coordinates, x_1 and x_2, must be used to completely describe the internal moments within the beam since there is a discontinuity, M', at B. As shown in Fig. 14–41b, x_1 ranges from A to B and x_2 ranges from B to C. Using the method of sections, Fig. 14–41c, the internal moments and the partial derivatives for x_1 and x_2 are

$$\zeta+\Sigma M_{NA} = 0; \qquad M_1 = -Px_1, \qquad \frac{\partial M_1}{\partial M'} = 0$$

$$\zeta+\Sigma M_{NA} = 0; \qquad M_2 = M' - P\left(\frac{L}{2} + x_2\right), \qquad \frac{\partial M_2}{\partial M'} = 1$$

Castigliano's Second Theorem. Setting $M' = 0$ and applying Eq. 14–48, we have

$$\theta_B = \int_0^L M\left(\frac{\partial M}{\partial M'}\right)\frac{dx}{EI} = \int_0^{L/2}\frac{(-Px_1)(0)\,dx_1}{EI} + \int_0^{L/2}\frac{-P[(L/2) + x_2](1)\,dx_2}{EI} = -\frac{3PL^2}{8EI} \quad \textit{Ans.}$$

Note the similarity between this solution and that of Example 14.14.

PROBLEMS

14–133. Solve Prob. 14–90 using Castigliano's theorem.

14–134. Solve Prob. 14–91 using Castigliano's theorem.

14–135. Solve Prob. 14–92 using Castigliano's theorem.

*14–136. Solve Prob. 14–93 using Castigliano's theorem.

14–137. Solve Prob. 14–95 using Castigliano's theorem.

14–138. Solve Prob. 14–96 using Castigliano's theorem.

14–139. Solve Prob. 14–97 using Castigliano's theorem.

*14–140. Solve Prob. 14–98 using Castigliano's theorem.

14–141. Solve Prob. 14–108 using Castigliano's theorem.

14–142. Solve Prob. 14–119 using Castigliano's theorem.

14–143. Solve Prob. 14–120 using Castigliano's theorem.

*14–144. Solve Prob. 14–105 using Castigliano's theorem.

CHAPTER REVIEW

When a force (couple moment) acts on a deformable body it will do external work when it is displaced (rotates). The internal stresses produced in the body also undergo displacement, thereby creating elastic strain energy that is stored in the material. The conservation of energy states that the external work done by the loading is equal to the internal elastic strain energy produced by the stresses in the body.	$$U_e = U_i$$
The conservation of energy can be used to solve problems involving elastic impact, which assumes the moving body is rigid and all the strain energy is stored in the stationary body. This concept allows us to find an impact factor n, which is a ratio of the dynamic load to the static load. It is used to determine the maximum stress and displacement of the body at the point of impact.	$$n = 1 + \sqrt{1 + 2\left(\frac{h}{\Delta_{st}}\right)}$$ $$\sigma_{max} = n\sigma_{st}$$ $$\Delta_{max} = n\Delta_{st}$$
The principle of virtual work can be used to determine the displacement of a joint on a truss, or the slope and the displacement of points on a beam. It requires placing an external virtual unit force (virtual unit couple moment) at the point where the displacement (rotation) is to be determined. The external virtual work that is produced by the external loading is then equated to the internal virtual strain energy in the structure.	$$1 \cdot \Delta = \sum \frac{nNL}{AE}$$ $$1 \cdot \Delta = \int_0^L \frac{mM}{EI} dx$$ $$1 \cdot \theta = \int_0^L \frac{m_\theta M}{EI} dx$$
Castigliano's second theorem can also be used to determine the displacement of a joint on a truss or the slope and the displacement at a point on a beam. Here a variable force P (couple moment M') is placed at the point where the displacement (slope) is to be determined. The internal loading is then determined as a function of P (M') and its partial derivative with respect to P (M') is determined. Castigliano's second theorem is then applied to obtain the desired displacement (rotation).	$$\Delta = \sum N\left(\frac{\partial N}{\partial P}\right)\frac{L}{AE}$$ $$\Delta = \int_0^L M\left(\frac{\partial M}{\partial P}\right)\frac{dx}{EI}$$ $$\theta = \int_0^L M\left(\frac{\partial M}{\partial M'}\right)\frac{dx}{EI}$$

REVIEW PROBLEMS

R14–1. Determine the total axial and bending strain energy in the A992 steel beam. $A = 2300 \text{ mm}^2$, $I = 9.5(10^6) \text{ mm}^4$.

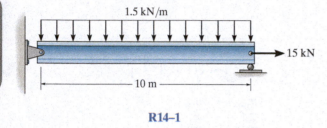

1.5 kN/m

15 kN

10 m

R14–1

R14–2. The 200-kg block D is dropped from a height $h = 1$ m onto end C of the A992 steel W200 × 36 overhang beam. If the spring at B has a stiffness $k = 200$ kN/m, determine the maximum bending stress developed in the beam.

R14–3. Determine the maximum height h from which the 200-kg block D can be dropped without causing the A992 steel W200 × 36 overhang beam to yield. The spring at B has a stiffness $k = 200$ kN/m.

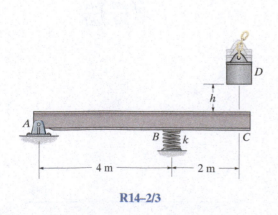

D

h

A

B   k

C

4 m

2 m

R14–2/3

*R14–4. The A992 steel bars are pin connected at B and C. If they each have a diameter of 30 mm, determine the slope at E.

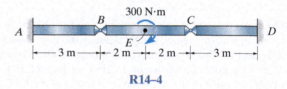

300 N·m

A   B   C   D

E

3 m   2 m   2 m   3 m

R14–4

R14–5. The steel chisel has a diameter of 0.5 in. and a length of 10 in. It is struck by a hammer that weighs 3 lb, and at the instant of impact it is moving at 12 ft/s. Determine the maximum compressive stress in the chisel, assuming that 80% of the impacting energy goes into the chisel. $E_{st} = 29(10^3)$ ksi, $\sigma_Y = 100$ ksi.

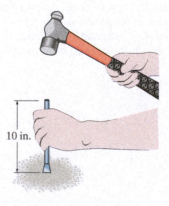

10 in.

R14–5

R14–6. Determine the total axial and bending strain energy in the A992 structural steel W8 × 58 beam.

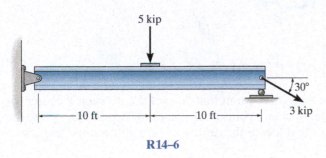

R14–6

R14–7. Determine the vertical displacement of joint C. The truss is made from A992 steel rods each having a diameter of 1 in.

***R14–8.** Determine the horizontal displacement of joint B. The truss is made from A992 steel rods each having a diameter of 1 in.

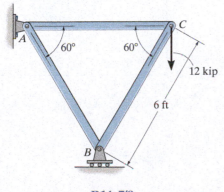

R14–7/8

R14–9. The cantilevered beam is subjected to a couple moment $\mathbf{M_0}$ applied at its end. Determine the slope of the beam at B. EI is constant. Use the method of virtual work.

R14–10. Solve Prob. R14–9 using Castigliano's theorem.

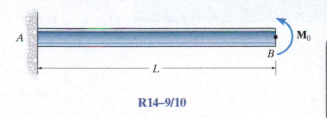

R14–9/10

R14–11. Determine the slope and displacement at point C. EI is constant.

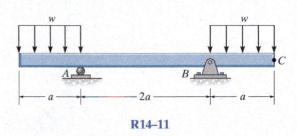

R14–11

***R14–12.** Determine the displacement at B. EI is constant.

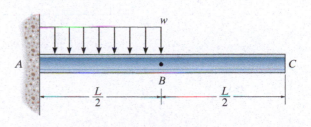

R14–12

GEOMETRIC PROPERTIES OF AN AREA

A.1 CENTROID OF AN AREA

The ***centroid*** of an area is the point that defines the geometric center for the area. If the area has an arbitrary shape, as shown in Fig. A–1a, the x and y coordinates that locate the centroid C are determined from

$$\bar{x} = \frac{\displaystyle\int_A x \, dA}{\displaystyle\int_A dA} \qquad \bar{y} = \frac{\displaystyle\int_A y \, dA}{\displaystyle\int_A dA} \tag{A–1}$$

The numerators in these equations represent the "moment" of the area element dA about the y and the x axis, respectively, Fig. A–1b, and the denominators represent the total area A of the shape.

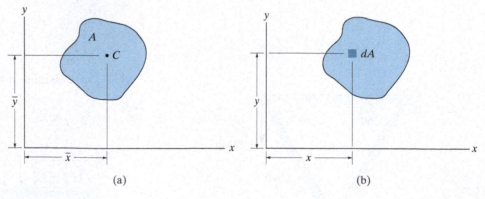

(a)   (b)

Fig. A–1

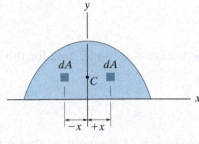

Fig. A–2

The location of the centroid for some areas may be partially or completely specified if the area is symmetric about an axis. Here, the centroid for the area will lie on this axis, Fig. A–2. If the area at the intersection of these axes, Fig. A–3. Based on this, or using Eq. A–1, the locations of the centroid for common area shapes are listed on the inside front cover.

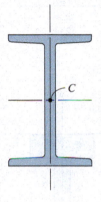

Fig. A–3

Composite Areas.   Often an area can be sectioned or divided into several parts having simpler shapes. Provided the area and location of the centroid of each of these "composite shapes" are known, one can eliminate the need for integration to determine the centroid for the entire area. In this case, equations analogous to Eq. A–1 must be used, except that finite summation signs replace the integrals; i.e.,

$$\bar{x} = \frac{\Sigma \tilde{x} A}{\Sigma A} \qquad \bar{y} = \frac{\Sigma \tilde{y} A}{\Sigma A} \qquad (A\text{–}2)$$

Here $\tilde{x}$ and $\tilde{y}$ represent the *algebraic distances* or *x, y* coordinates for the centroid of each composite part, and ΣA represents the sum of the areas of the composite parts or simply the *total area*. If a hole, or an empty region, is located within a composite part, the region is considered as an additional composite part having a *negative* area.

The following example illustrates application of Eq. A–2.

A

EXAMPLE   A.1

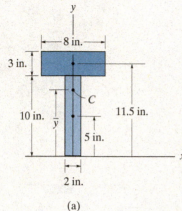

(a)

Locate the centroid C of the cross-sectional area for the T-beam shown in Fig. A–4a.

SOLUTION I

The y axis is placed along the axis of symmetry so that $\bar{x} = 0$, Fig. A–4a. To obtain $\bar{y}$ we will establish the x axis (reference axis) through the base of the area. The area is segmented into two rectangles as shown, and the centroidal location $\tilde{y}$ for each is established. Applying Eq. A–2, we have

$$\bar{y} = \frac{\Sigma \tilde{y} A}{\Sigma A} = \frac{[5 \text{ in.}](10 \text{ in.})(2 \text{ in.}) + [11.5 \text{ in.}](3 \text{ in.})(8 \text{ in.})}{(10 \text{ in.})(2 \text{ in.}) + (3 \text{ in.})(8 \text{ in.})}$$

$$= 8.55 \text{ in.} \qquad\qquad\qquad\qquad Ans.$$

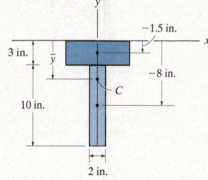

(b)

SOLUTION II

Using the same two segments, the x axis can be located at the top of the area, Fig. A–4b. Here

$$\bar{y} = \frac{\Sigma \tilde{y} A}{\Sigma A} = \frac{[-1.5 \text{ in.}](3 \text{ in.})(8 \text{ in.}) + [-8 \text{ in.}](10 \text{ in.})(2 \text{ in.})}{(3 \text{ in.})(8 \text{ in.}) + (10 \text{ in.})(2 \text{ in.})}$$

$$= -4.45 \text{ in.} \qquad\qquad\qquad\qquad Ans.$$

The negative sign indicates that C is located *below* the x axis, which is to be expected. Also note that from the two answers 8.55 in. + 4.45 in. = 13.0 in., which is the depth of the beam.

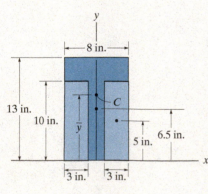

(c)

Fig. A–4

SOLUTION III

It is also possible to consider the cross-sectional area to be one large rectangle *less* two small rectangles shown shaded in Fig. A–4c. Here we have

$$\bar{y} = \frac{\Sigma \tilde{y} A}{\Sigma A} = \frac{[6.5 \text{ in.}](13 \text{ in.})(8 \text{ in.}) - 2[5 \text{ in.}](10 \text{ in.})(3 \text{ in.})}{(13 \text{ in.})(8 \text{ in.}) - 2(10 \text{ in.})(3 \text{ in.})}$$

$$= 8.55 \text{ in.} \qquad\qquad\qquad\qquad Ans.$$

A.2   MOMENT OF INERTIA FOR AN AREA

The **moment of inertia** of an area is a geometric property that is calculated about an axis, and for the x and y axes shown in Fig. A–5, it is defined as

$$I_x = \int_A y^2\, dA$$

$$I_y = \int_A x^2\, dA \qquad\qquad \text{(A–3)}$$

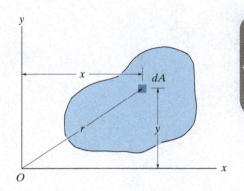

Fig. A–5

These integrals have no physical meaning, but they are so named because they are similar to the formulation of the moment of inertia of a mass, which is a dynamical property of matter.

We can also calculate the moment of inertia of an area about the pole O or z axis, Fig. A–5. This is referred to as the **polar moment of inertia**, which is defined as,

$$J_O = \int_A r^2\, dA = I_x + I_y \qquad\qquad \text{(A–4)}$$

Here r is the perpendicular distance from the pole (z axis) to the element dA. Since $r^2 = x^2 + y^2$, then $J_O = I_x + I_y$, Fig. A–5.

From the above formulations it is seen that I_x, I_y, and J_O will *always* be *positive*, since they involve the product of distance squared and area. Furthermore, the units for moment of inertia involve length raised to the fourth power, e.g., m^4, mm^4, or ft^4, in^4.

Using the above equations, the moments of inertia for some common area shapes are calculated about their *centroidal axes* and are listed on the inside front cover.

Parallel-Axis Theorem for an Area.

If the moment of inertia for an area is known about a centroidal axis, we can determine the moment of inertia of the area about a corresponding parallel axis using the **parallel-axis theorem**. To derive this theorem, consider finding the moment of inertia of the differential element dA in Fig. A–6, located at the arbitrary distance $y' + d_y$, from the x axis. It is $dI_x = (y' + d_y)^2\, dA$. Then for the entire area, we have

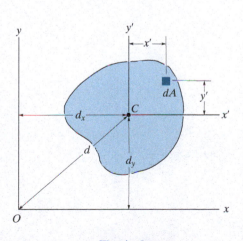

Fig. A–6

$$I_x = \int_A (y' + d_y)^2\, dA = \int_A y'^2\, dA + 2d_y \int_A y'\, dA + d_y^2 \int_A dA$$

The first term on the right represents the moment of inertia of the area about the x' axis, $\bar{I}_{x'}$. The second term is zero since the x' axis passes through the area's centroid C, that is, $\int y' \, dA = \bar{y}'A = 0$ since $\bar{y}' = 0$. The final result is therefore

$$I_x = \bar{I}_{x'} + A d_y^2 \qquad \text{(A–5)}$$

A similar expression can be written for I_y, that is,

$$I_y = \bar{I}_{y'} + A d_x^2 \qquad \text{(A–6)}$$

And finally, for the polar moment of inertia about an axis perpendicular to the x–y plane and passing through the pole O (z axis), Fig. A–6, we have

$$J_O = \bar{J}_C + A d^2 \qquad \text{(A–7)}$$

The form of each of the above equations states that *the moment of inertia of an area about an axis is equal to the area's moment of inertia about a parallel axis passing through the "centroid" of the area plus the product of the area and the square of the perpendicular distance between the axes.*

Composite Areas.   Many areas consist of a series of connected simpler shapes, such as rectangles, triangles, and semicircles. In order to properly determine the moment of inertia of this composite area about an axis, it is first necessary to divide the area into its parts and indicate the perpendicular distance from the axis to the parallel centroidal axis for each part. Using the table on the inside front cover of the book, the moment of inertia of each part is determined about the centroidal axis. If this axis does not coincide with the specified axis, then the moment of inertia of the part about the specified axis is determined using the parallel-axis theorem, $I = \bar{I} + A d^2$. The moment of inertia of the entire area about this axis is then found by summing the results of all its composite parts. In particular, if a composite part has an empty region (hole), the moment of inertia for the composite is found by "subtracting" the moment of inertia for the region from the moment of inertia of the entire area including the region.

EXAMPLE   A.2

Determine the moment of inertia of the cross-sectional area of the T-beam shown in Fig. A–7a about the centroidal x' axis.

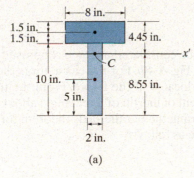

(a)

Fig. A–7

SOLUTION I

The area is segmented into two rectangles as shown in Fig. A–7a, and the distance from the x' axis and each centroidal axis is determined. Using the table on the inside front cover, the moment of inertia of a rectangle about its centroidal axis is $I = \frac{1}{12}bh^3$. Applying the parallel-axis theorem, Eq. A–5, to each rectangle and adding the results, we have

$$I = \Sigma(\bar{I}_{x'} + A\,d_y^{\,2})$$

$$= \left[\frac{1}{12}(2 \text{ in.})(10 \text{ in.})^3 + (2 \text{ in.})(10 \text{ in.})(8.55 \text{ in.} - 5 \text{ in.})^2 \right]$$

$$+ \left[\frac{1}{12}(8 \text{ in.})(3 \text{ in.})^3 + (8 \text{ in.})(3 \text{ in.})(4.45 \text{ in.} - 1.5 \text{ in.})^2 \right]$$

$$I = 646 \text{ in}^4 \qquad\qquad\qquad\qquad\qquad\qquad\qquad\qquad\qquad\quad \textit{Ans.}$$

SOLUTION II

The area can be considered as one large rectangle less two small rectangles, shown shaded in Fig. A–7b. We have

$$I = \Sigma(\bar{I}_{x'} + A\,d_y^{\,2})$$

$$= \left[\frac{1}{12}(8 \text{ in.})(13 \text{ in.})^3 + (8 \text{ in.})(13 \text{ in.})(8.55 \text{ in.} - 6.5 \text{ in.})^2 \right]$$

$$- 2\left[\frac{1}{12}(3 \text{ in.})(10 \text{ in.})^3 + (3 \text{ in.})(10 \text{ in.})(8.55 \text{ in.} - 5 \text{ in.})^2 \right]$$

$$I = 646 \text{ in}^4 \qquad\qquad\qquad\qquad\qquad\qquad \textit{Ans.}$$

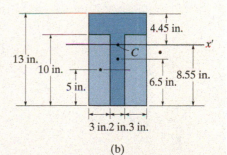

(b)

A

EXAMPLE A.3

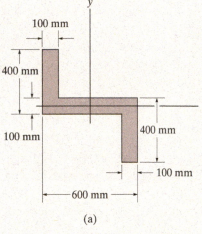

(a)

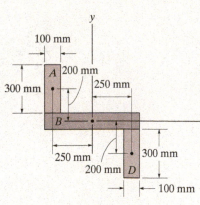

(b)

Fig. A-8

Determine the moments of inertia of the beam's cross-sectional area shown in Fig. A–8a about the x and y centroidal axes.

SOLUTION

The cross section can be considered as three composite rectangular areas A, B, and D shown in Fig. A–8b. For the calculation, the centroid of each of these rectangles is located in the figure. From the table on the inside front cover, the moment of inertia of a rectangle about its centroidal axis is $I = \frac{1}{12}bh^3$. Hence, using the parallel-axis theorem for rectangles A and D, the calculations are as follows:

Rectangle A:

$$I_x = \bar{I}_{x'} + Ad_y^2 = \frac{1}{12}(100 \text{ mm})(300 \text{ mm})^3 + (100 \text{ mm})(300 \text{ mm})(200 \text{ mm})^2$$

$$= 1.425(10^9) \text{ mm}^4$$

$$I_y = \bar{I}_{y'} + Ad_x^2 = \frac{1}{12}(300 \text{ mm})(100 \text{ mm})^3 + (100 \text{ mm})(300 \text{ mm})(250 \text{ mm})^2$$

$$= 1.90(10^9) \text{ mm}^4$$

Rectangle B:

$$I_x = \frac{1}{12}(600 \text{ mm})(100 \text{ mm})^3 = 0.05(10^9) \text{ mm}^4$$

$$I_y = \frac{1}{12}(100 \text{ mm})(600 \text{ mm})^3 = 1.80(10^9) \text{ mm}^4$$

Rectangle D:

$$I_x = \bar{I}_{x'} + Ad_y^2 = \frac{1}{12}(100 \text{ mm})(300 \text{ mm})^3 + (100 \text{ mm})(300 \text{ mm})(200 \text{ mm})^2$$

$$= 1.425(10^9) \text{ mm}^4$$

$$I_y = \bar{I}_{y'} + Ad_x^2 = \frac{1}{12}(300 \text{ mm})(100 \text{ mm})^3 + (100 \text{ mm})(300 \text{ mm})(250 \text{ mm})^2$$

$$= 1.90(10^9) \text{ mm}^4$$

The moments of inertia for the entire cross section are thus

$$I_x = 1.425(10^9) + 0.05(10^9) + 1.425(10^9)$$

$$= 2.90(10^9) \text{ mm}^4 \qquad \text{Ans.}$$

$$I_y = 1.90(10^9) + 1.80(10^9) + 1.90(10^9)$$

$$= 5.60(10^9) \text{ mm}^4 \qquad \text{Ans.}$$

A.3   PRODUCT OF INERTIA FOR AN AREA

In general, the moment of inertia for an area is different for every axis about which it is computed. In some applications it is necessary to know the orientation of those axes that give, respectively, the maximum and minimum moments of inertia for the area. The method for determining this is discussed in Sec. A.4. However, to use this method, it is necessary to first determine the product of inertia for the area as well as its moments of inertia referenced from the x, y axes.

The **product of inertia** for the area A shown in Fig. A–9 is defined as

$$I_{xy} = \int_A xy \, dA \qquad\qquad \text{(A–8)}$$

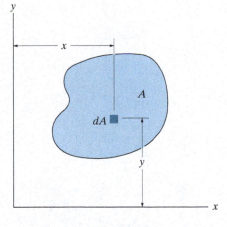

Fig. A–9

Like the moment of inertia, the product of inertia has units of length raised to the fourth power, e.g., m^4, mm^4 or ft^4, in^4. However, since x or y may be a negative quantity, while dA is always positive, the product of inertia may be positive, negative, or zero, depending on the location and orientation of the coordinate axes. For example, the product of inertia I_{xy} for an area will be *zero* if either the x or the y axis is an axis of *symmetry* for the area. To show this, consider the shaded area in Fig. A–10, where for every element dA located at point (x, y) there is a corresponding element dA located at $(x, -y)$. Since the products of inertia for these elements are, respectively, $xy \, dA$ and $-xy \, dA$, their algebraic sum or the integration of all such corresponding elements of area chosen in this way will cancel each other. Consequently, the product of inertia for the total area becomes zero.

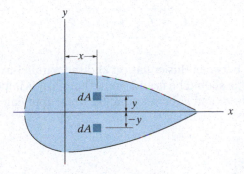

Fig. A–10

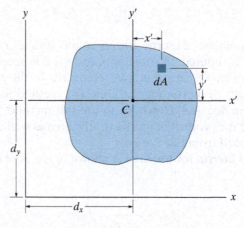

Fig. A–11

Parallel-Axis Theorem.

Consider the shaded area shown in Fig. A–11. Since the product of inertia of dA with respect to the x and y axes is $dI_{xy} = (x' + d_x)(y' + d_y)\,dA$, then for the entire area,

$$I_{xy} = \int_A (x' + d_x)(y' + d_y)\,dA$$

$$= \int_A x'y'\,dA + d_x \int_A y'\,dA + d_y \int_A x'\,dA + d_x d_y \int_A dA$$

The first term on the right represents the product of inertia of the area with respect to the centroidal axis, $\bar{I}_{x'y'}$. The second and third terms are zero since the moments of the area are taken about the centroidal x', y' axis. Realizing that the fourth integral represents the total area A, we therefore have

$$\boxed{I_{xy} = \bar{I}_{x'y'} + A\,d_x d_y} \qquad (A\text{–}9)$$

The similarity between this equation and the parallel-axis theorem for moments of inertia should be noted. Here though, it is important that the *algebraic signs* for d_x and d_y be maintained when applying Eq. A–9.

EXAMPLE   A.4

Determine the product of inertia of the beam's cross-sectional area, shown in Fig. A–12a, about the x and y centroidal axes.

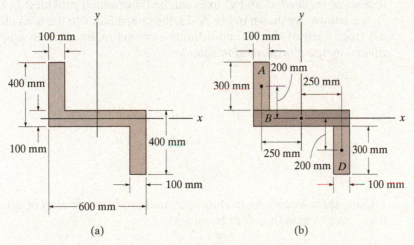

(a)   (b)

Fig. A–12

SOLUTION

As in Example A.3, the cross section can be considered as three composite rectangular areas A, B, and D, Fig. A–12b. The coordinates for the centroid of each of these rectangles are shown in the figure. Due to symmetry, the product of inertia of *each rectangle* is *zero* about a set of x', y' axes that pass through the rectangle's centroid. Hence, application of the parallel-axis theorem to each of the rectangles yields

Rectangle A:

$$I_{xy} = \bar{I}_{x'y'} + A d_x d_y$$
$$= 0 + (300 \text{ mm})(100 \text{ mm})(-250 \text{ mm})(200 \text{ mm})$$
$$= -1.50(10^9) \text{ mm}^4$$

Rectangle B:

$$I_{xy} = \bar{I}_{x'y'} + A d_x d_y$$
$$= 0 + 0$$
$$= 0$$

Rectangle D:

$$I_{xy} = \bar{I}_{x'y'} + A d_x d_y$$
$$= 0 + (300 \text{ mm})(100 \text{ mm})(250 \text{ mm})(-200 \text{ mm})$$
$$= -1.50(10^9) \text{ mm}^4$$

The product of inertia for the entire cross section is thus

$$I_{xy} = [-1.50(10^9) \text{ mm}^4] + 0 + [-1.50(10^9) \text{ mm}^4]$$
$$= -3.00(10^9) \text{ mm}^4 \qquad \qquad \textit{Ans.}$$

A.4 MOMENTS OF INERTIA FOR AN AREA ABOUT INCLINED AXES

The moments and product of inertia $I_{x'}, I_{y'}$ and $I_{x'y'}$ for an area with respect to a set of *inclined* x' and y' axes can be determined provided I_x, I_y, and I_{xy} are *known*. As shown in Fig. A–13, the coordinates to the area element dA from each of the two coordinate systems inclined at an angle are related by the *transformation equations*

$$x' = x \cos \theta + y \sin \theta$$

$$y' = y \cos \theta - x \sin \theta$$

Using these equations, the moments and product of inertia of dA about the x' and y' axes therefore become

$$dI_{x'} = y'^2\, dA = (y \cos \theta - x \sin \theta)^2\, dA$$

$$dI_{y'} = x'^2\, dA = (x \cos \theta + y \sin \theta)^2\, dA$$

$$dI_{x'y'} = x'y'\, dA = (x \cos \theta + y \sin \theta)(y \cos \theta - x \sin \theta)\, dA$$

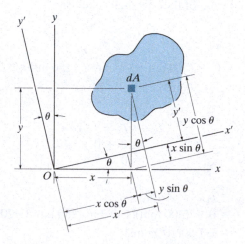

Fig. A–13

Expanding each expression and integrating, realizing that $I_x = \int y^2 \, dA$, $I_y = \int x^2 \, dA$, and $I_{xy} = \int xy \, dA$, we obtain

$$I_{x'} = I_x \cos^2 \theta + I_y \sin^2 \theta - 2I_{xy} \sin \theta \cos \theta$$

$$I_{y'} = I_x \sin^2 \theta + I_y \cos^2 \theta + 2I_{xy} \sin \theta \cos \theta$$

$$I_{x'y'} = I_x \sin \theta \cos \theta - I_y \sin \theta \cos \theta + I_{xy}(\cos^2 \theta - \sin^2 \theta)$$

These equations may be simplified by using the trigonometric identities $\sin 2\theta = 2 \sin \theta \cos \theta$ and $\cos 2\theta = \cos^2 \theta - \sin^2 \theta$, in which case

$$I_{x'} = \frac{I_x + I_y}{2} + \frac{I_x - I_y}{2} \cos 2\theta - I_{xy} \sin 2\theta$$

$$I_{y'} = \frac{I_x + I_y}{2} - \frac{I_x - I_y}{2} \cos 2\theta + I_{xy} \sin 2\theta \qquad \text{(A–10)}$$

$$I_{x'y'} = \frac{I_x - I_y}{2} \sin 2\theta + I_{xy} \cos 2\theta$$

Principal Moments of Inertia. Since $I_{x'}$, $I_{y'}$, and $I_{x'y'}$ depend on the angle of inclination, θ, of the x', y' axes, we can determine the orientation of these axes so that the moments of inertia for the area, $I_{x'}$ and $I_{y'}$, are maximum and minimum. This particular set of axes is called the *principal axes* of inertia for the area, and the corresponding moments of inertia with respect to these axes are called the *principal moments of inertia*. In general, there is a set of principal axes for every chosen origin O; however, in mechanics of materials the area's centroid is the most important location for O.

The angle $\theta = \theta_p$, which defines the orientation of the principal axes, can be found by differentiating the first of Eq. A–10 with respect to θ and setting the result equal to zero. We get

$$\frac{dI_{x'}}{d\theta} = -2\left(\frac{I_x - I_y}{2}\right) \sin 2\theta - 2I_{xy} \cos 2\theta = 0$$

Therefore, at $\theta = \theta_p$,

$$\tan 2\theta_p = \frac{-I_{xy}}{(I_x - I_y)/2} \qquad \text{(A–11)}$$

This equation has two roots, θ_{p_1} and θ_{p_2}, which are 90° apart and so specify the inclination of each principal axis.

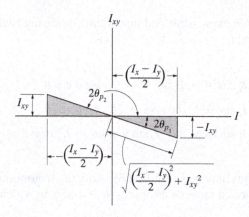

Fig. A–14

The sine and cosine of $2\theta_{p_1}$ and $2\theta_{p_2}$ can be obtained from the triangles shown in Fig. A–14, which are based on Eq. A–11. If these trigonometric relations are substituted into the first or second of Eq. A–10 and simplified, the result is

$$I^{max}_{min} = \frac{I_x + I_y}{2} \pm \sqrt{\left(\frac{I_x - I_y}{2}\right)^2 + I_{xy}^2} \qquad (A\text{–}12)$$

Depending on the sign chosen, this result gives the maximum or minimum moment of inertia for the area. Furthermore, if the above trigonometric relations for the sine and cosine of $2\theta_{p_1}$ and $2\theta_{p_2}$ are substituted into the third of Eq. A–10, it will be seen that $I_{x'y'} = 0$; that is, the *product of inertia with respect to the principal axes is zero.* Since it was indicated in Sec. A.3 that the product of inertia is zero with respect to any symmetrical axis, it therefore follows that *any symmetrical axis and the one perpendicular to it represent principal axes of inertia for the area.* The equations derived in this section are similar to those for stress and strain transformation developed in Chapters 9 and 10, respectively, and like stress and strain, we can also solve these equations using a semi-graphical technique called Mohr's circle of interia.*

*See *Engineering Mechanism: Statics*, 14th ed., R. C. Hibbeler, Pearson Education, Inc.

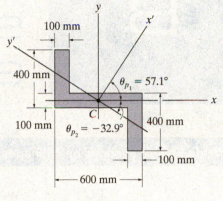

Fig. A-15

EXAMPLE A.5

Determine the principal moments of inertia for the beam's cross-sectional area shown in Fig. A–15 with respect to an axis passing through the centroid C.

SOLUTION

The moments and product of inertia of the cross section with respect to the x, y axes have been determined in Examples A.3 and A.4. The results are

$$I_x = 2.90(10^9) \text{ mm}^4 \qquad I_y = 5.60(10^9) \text{ mm}^4 \qquad I_{xy} = -3.00(10^9) \text{ mm}^4$$

Using Eq. A–11, the angles of inclination of the principal axes x' and y' are

$$\tan 2\theta_p = \frac{-I_{xy}}{(I_x - I_y)/2} = \frac{3.00(10^9)}{[2.90(10^9) - 5.60(10^9)]/2} = -2.22$$

$$2\theta_{p_1} = 114.2° \quad \text{and} \quad 2\theta_{p_2} = -65.8°$$

Thus, as shown in Fig. A–15,

$$\theta_{p1} = 57.1° \quad \text{and} \quad \theta_{p2} = -32.9°$$

The principal moments of inertia with respect to the x' and y' axes are determined by using Eq. A–12.

$$I_{\substack{max \\ min}} = \frac{I_x + I_y}{2} \pm \sqrt{\left(\frac{I_x - I_y}{2}\right)^2 + I_{xy}^2}$$

$$= \frac{2.90(10^9) + 5.60(10^9)}{2} \pm \sqrt{\left[\frac{2.90(10^9) - 5.60(10^9)}{2}\right]^2 + [-3.00(10^9)]^2}$$

$$= 4.25(10^9) \pm 3.29(10^9)$$

or

$$I_{max} = 7.54(10^9) \text{ mm}^4 \qquad I_{min} = 0.960(10^9) \text{ mm}^4 \qquad \textit{Ans.}$$

Specifically, the maximum moment of inertia, $I_{max} = 7.54(10^9) \text{ mm}^4$, occurs with respect to the x' axis (major axis), since *by inspection* most of the cross-sectional area is farthest away from this axis. To show this, substitute the data with $\theta = 57.1°$ into the first of Eq. A–10.

APPENDIX B

GEOMETRIC PROPERTIES OF STRUCTURAL SHAPES

Wide-Flange Sections or W Shapes FPS Units

Designation	Area A	Depth d	Web thickness t_w	Flange width b_f	Flange thickness t_f	x–x axis I	x–x axis S	x–x axis r	y–y axis I	y–y axis S	y–y axis r
in. × lb/ft	in^2	in.	in.	in.	in.	in^4	in^3	in.	in^4	in^3	in.
W24 × 104	30.6	24.06	0.500	12.750	0.750	3100	258	10.1	259	40.7	2.91
W24 × 94	27.7	24.31	0.515	9.065	0.875	2700	222	9.87	109	24.0	1.98
W24 × 84	24.7	24.10	0.470	9.020	0.770	2370	196	9.79	94.4	20.9	1.95
W24 × 76	22.4	23.92	0.440	8.990	0.680	2100	176	9.69	82.5	18.4	1.92
W24 × 68	20.1	23.73	0.415	8.965	0.585	1830	154	9.55	70.4	15.7	1.87
W24 × 62	18.2	23.74	0.430	7.040	0.590	1550	131	9.23	34.5	9.80	1.38
W24 × 55	16.2	23.57	0.395	7.005	0.505	1350	114	9.11	29.1	8.30	1.34
W18 × 65	19.1	18.35	0.450	7.590	0.750	1070	117	7.49	54.8	14.4	1.69
W18 × 60	17.6	18.24	0.415	7.555	0.695	984	108	7.47	50.1	13.3	1.69
W18 × 55	16.2	18.11	0.390	7.530	0.630	890	98.3	7.41	44.9	11.9	1.67
W18 × 50	14.7	17.99	0.355	7.495	0.570	800	88.9	7.38	40.1	10.7	1.65
W18 × 46	13.5	18.06	0.360	6.060	0.605	712	78.8	7.25	22.5	7.43	1.29
W18 × 40	11.8	17.90	0.315	6.015	0.525	612	68.4	7.21	19.1	6.35	1.27
W18 × 35	10.3	17.70	0.300	6.000	0.425	510	57.6	7.04	15.3	5.12	1.22
W16 × 57	16.8	16.43	0.430	7.120	0.715	758	92.2	6.72	43.1	12.1	1.60
W16 × 50	14.7	16.26	0.380	7.070	0.630	659	81.0	6.68	37.2	10.5	1.59
W16 × 45	13.3	16.13	0.345	7.035	0.565	586	72.7	6.65	32.8	9.34	1.57
W16 × 36	10.6	15.86	0.295	6.985	0.430	448	56.5	6.51	24.5	7.00	1.52
W16 × 31	9.12	15.88	0.275	5.525	0.440	375	47.2	6.41	12.4	4.49	1.17
W16 × 26	7.68	15.69	0.250	5.500	0.345	301	38.4	6.26	9.59	3.49	1.12
W14 × 53	15.6	13.92	0.370	8.060	0.660	541	77.8	5.89	57.7	14.3	1.92
W14 × 43	12.6	13.66	0.305	7.995	0.530	428	62.7	5.82	45.2	11.3	1.89
W14 × 38	11.2	14.10	0.310	6.770	0.515	385	54.6	5.87	26.7	7.88	1.55
W14 × 34	10.0	13.98	0.285	6.745	0.455	340	48.6	5.83	23.3	6.91	1.53
W14 × 30	8.85	13.84	0.270	6.730	0.385	291	42.0	5.73	19.6	5.82	1.49
W14 × 26	7.69	13.91	0.255	5.025	0.420	245	35.3	5.65	8.91	3.54	1.08
W14 × 22	6.49	13.74	0.230	5.000	0.335	199	29.0	5.54	7.00	2.80	1.04

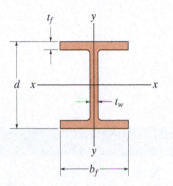

Wide-Flange Sections or W Shapes FPS Units

Designation	Area A	Depth d	Web thickness t_w	Flange width b_f	Flange thickness t_f	x–x axis I	x–x axis S	x–x axis r	y–y axis I	y–y axis S	y–y axis r
in. × lb/ft	in²	in.	in.	in.	in.	in⁴	in³	in.	in⁴	in³	in.
W12 × 87	25.6	12.53	0.515	12.125	0.810	740	118	5.38	241	39.7	3.07
W12 × 50	14.7	12.19	0.370	8.080	0.640	394	64.7	5.18	56.3	13.9	1.96
W12 × 45	13.2	12.06	0.335	8.045	0.575	350	58.1	5.15	50.0	12.4	1.94
W12 × 26	7.65	12.22	0.230	6.490	0.380	204	33.4	5.17	17.3	5.34	1.51
W12 × 22	6.48	12.31	0.260	4.030	0.425	156	25.4	4.91	4.66	2.31	0.847
W12 × 16	4.71	11.99	0.220	3.990	0.265	103	17.1	4.67	2.82	1.41	0.773
W12 × 14	4.16	11.91	0.200	3.970	0.225	88.6	14.9	4.62	2.36	1.19	0.753
W10 × 100	29.4	11.10	0.680	10.340	1.120	623	112	4.60	207	40.0	2.65
W10 × 54	15.8	10.09	0.370	10.030	0.615	303	60.0	4.37	103	20.6	2.56
W10 × 45	13.3	10.10	0.350	8.020	0.620	248	49.1	4.32	53.4	13.3	2.01
W10 × 39	11.5	9.92	0.315	7.985	0.530	209	42.1	4.27	45.0	11.3	1.98
W10 × 30	8.84	10.47	0.300	5.810	0.510	170	32.4	4.38	16.7	5.75	1.37
W10 × 19	5.62	10.24	0.250	4.020	0.395	96.3	18.8	4.14	4.29	2.14	0.874
W10 × 15	4.41	9.99	0.230	4.000	0.270	68.9	13.8	3.95	2.89	1.45	0.810
W10 × 12	3.54	9.87	0.190	3.960	0.210	53.8	10.9	3.90	2.18	1.10	0.785
W8 × 67	19.7	9.00	0.570	8.280	0.935	272	60.4	3.72	88.6	21.4	2.12
W8 × 58	17.1	8.75	0.510	8.220	0.810	228	52.0	3.65	75.1	18.3	2.10
W8 × 48	14.1	8.50	0.400	8.110	0.685	184	43.3	3.61	60.9	15.0	2.08
W8 × 40	11.7	8.25	0.360	8.070	0.560	146	35.5	3.53	49.1	12.2	2.04
W8 × 31	9.13	8.00	0.285	7.995	0.435	110	27.5	3.47	37.1	9.27	2.02
W8 × 24	7.08	7.93	0.245	6.495	0.400	82.8	20.9	3.42	18.3	5.63	1.61
W8 × 15	4.44	8.11	0.245	4.015	0.315	48.0	11.8	3.29	3.41	1.70	0.876
W6 × 25	7.34	6.38	0.320	6.080	0.455	53.4	16.7	2.70	17.1	5.61	1.52
W6 × 20	5.87	6.20	0.260	6.020	0.365	41.4	13.4	2.66	13.3	4.41	1.50
W6 × 16	4.74	6.28	0.260	4.030	0.405	32.1	10.2	2.60	4.43	2.20	0.966
W6 × 15	4.43	5.99	0.230	5.990	0.260	29.1	9.72	2.56	9.32	3.11	1.46
W6 × 12	3.55	6.03	0.230	4.000	0.280	22.1	7.31	2.49	2.99	1.50	0.918
W6 × 9	2.68	5.90	0.170	3.940	0.215	16.4	5.56	2.47	2.19	1.11	0.905

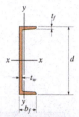

American Standard Channels or C Shapes FPS Units

Designation	Area A	Depth d	Web thickness t_w		Flange width b_f		Flange thickness t_f		x–x axis I	S	r	y–y axis I	S	r
in. × lb/ft	in^2	in.	in.		in.		in.		in^4	in^3	in.	in^4	in^3	in.
C15 × 50	14.7	15.00	0.716	11/16	3.716	$3\frac{3}{4}$	0.650	5/8	404	53.8	5.24	11.0	3.78	0.867
C15 × 40	11.8	15.00	0.520	1/2	3.520	$3\frac{1}{2}$	0.650	5/8	349	46.5	5.44	9.23	3.37	0.886
C15 × 33.9	9.96	15.00	0.400	3/8	3.400	$3\frac{3}{8}$	0.650	5/8	315	42.0	5.62	8.13	3.11	0.904
C12 × 30	8.82	12.00	0.510	1/2	3.170	$3\frac{1}{8}$	0.501	1/2	162	27.0	4.29	5.14	2.06	0.763
C12 × 25	7.35	12.00	0.387	3/8	3.047	3	0.501	1/2	144	24.1	4.43	4.47	1.88	0.780
C12 × 20.7	6.09	12.00	0.282	5/16	2.942	3	0.501	1/2	129	21.5	4.61	3.88	1.73	0.799
C10 × 30	8.82	10.00	0.673	11/16	3.033	3	0.436	7/16	103	20.7	3.42	3.94	1.65	0.669
C10 × 25	7.35	10.00	0.526	1/2	2.886	$2\frac{7}{8}$	0.436	7/16	91.2	18.2	3.52	3.36	1.48	0.676
C10 × 20	5.88	10.00	0.379	3/8	2.739	$2\frac{3}{4}$	0.436	7/16	78.9	15.8	3.66	2.81	1.32	0.692
C10 × 15.3	4.49	10.00	0.240	1/4	2.600	$2\frac{5}{8}$	0.436	7/16	67.4	13.5	3.87	2.28	1.16	0.713
C9 × 20	5.88	9.00	0.448	7/16	2.648	$2\frac{5}{8}$	0.413	7/16	60.9	13.5	3.22	2.42	1.17	0.642
C9 × 15	4.41	9.00	0.285	5/16	2.485	$2\frac{1}{2}$	0.413	7/16	51.0	11.3	3.40	1.93	1.01	0.661
C9 × 13.4	3.94	9.00	0.233	1/4	2.433	$2\frac{3}{8}$	0.413	7/16	47.9	10.6	3.48	1.76	0.962	0.669
C8 × 18.75	5.51	8.00	0.487	1/2	2.527	$2\frac{1}{2}$	0.390	3/8	44.0	11.0	2.82	1.98	1.01	0.599
C8 × 13.75	4.04	8.00	0.303	5/16	2.343	$2\frac{3}{8}$	0.390	3/8	36.1	9.03	2.99	1.53	0.854	0.615
C8 × 11.5	3.38	8.00	0.220	1/4	2.260	$2\frac{1}{4}$	0.390	3/8	32.6	8.14	3.11	1.32	0.781	0.625
C7 × 14.75	4.33	7.00	0.419	7/16	2.299	$2\frac{1}{4}$	0.366	3/8	27.2	7.78	2.51	1.38	0.779	0.564
C7 × 12.25	3.60	7.00	0.314	5/16	2.194	$2\frac{1}{4}$	0.366	3/8	24.2	6.93	2.60	1.17	0.703	0.571
C7 × 9.8	2.87	7.00	0.210	3/16	2.090	$2\frac{1}{8}$	0.366	3/8	21.3	6.08	2.72	0.968	0.625	0.581
C6 × 13	3.83	6.00	0.437	7/16	2.157	$2\frac{1}{8}$	0.343	5/16	17.4	5.80	2.13	1.05	0.642	0.525
C6 × 10.5	3.09	6.00	0.314	5/16	2.034	2	0.343	5/16	15.2	5.06	2.22	0.866	0.564	0.529
C6 × 8.2	2.40	6.00	0.200	3/16	1.920	$1\frac{7}{8}$	0.343	5/16	13.1	4.38	2.34	0.693	0.492	0.537
C5 × 9	2.64	5.00	0.325	5/16	1.885	$1\frac{7}{8}$	0.320	5/16	8.90	3.56	1.83	0.632	0.450	0.489
C5 × 6.7	1.97	5.00	0.190	3/16	1.750	$1\frac{3}{4}$	0.320	5/16	7.49	3.00	1.95	0.479	0.378	0.493
C4 × 7.25	2.13	4.00	0.321	5/16	1.721	$1\frac{3}{4}$	0.296	5/16	4.59	2.29	1.47	0.433	0.343	0.450
C4 × 5.4	1.59	4.00	0.184	3/16	1.584	$1\frac{5}{8}$	0.296	5/16	3.85	1.93	1.56	0.319	0.283	0.449
C3 × 6	1.76	3.00	0.356	3/8	1.596	$1\frac{5}{8}$	0.273	1/4	2.07	1.38	1.08	0.305	0.268	0.416
C3 × 5	1.47	3.00	0.258	1/4	1.498	$1\frac{1}{2}$	0.273	1/4	1.85	1.24	1.12	0.247	0.233	0.410
C3 × 4.1	1.21	3.00	0.170	3/16	1.410	$1\frac{3}{8}$	0.273	1/4	1.66	1.10	1.17	0.197	0.202	0.404

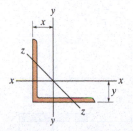

Angles Having Equal Legs FPS Units

Size and thickness	Weight per foot	Area A	x–x axis				y–y axis				z–z axis
			I	S	r	y	I	S	r	x	r
in.	lb	in²	in⁴	in³	in.	in.	in⁴	in³	in.	in.	in.
L8 × 8 × 1	51.0	15.0	89.0	15.8	2.44	2.37	89.0	15.8	2.44	2.37	1.56
L8 × 8 × $\frac{3}{4}$	38.9	11.4	69.7	12.2	2.47	2.28	69.7	12.2	2.47	2.28	1.58
L8 × 8 × $\frac{1}{2}$	26.4	7.75	48.6	8.36	2.50	2.19	48.6	8.36	2.50	2.19	1.59
L6 × 6 × 1	37.4	11.0	35.5	8.57	1.80	1.86	35.5	8.57	1.80	1.86	1.17
L6 × 6 × $\frac{3}{4}$	28.7	8.44	28.2	6.66	1.83	1.78	28.2	6.66	1.83	1.78	1.17
L6 × 6 × $\frac{1}{2}$	19.6	5.75	19.9	4.61	1.86	1.68	19.9	4.61	1.86	1.68	1.18
L6 × 6 × $\frac{3}{8}$	14.9	4.36	15.4	3.53	1.88	1.64	15.4	3.53	1.88	1.64	1.19
L5 × 5 × $\frac{3}{4}$	23.6	6.94	15.7	4.53	1.51	1.52	15.7	4.53	1.51	1.52	0.975
L5 × 5 × $\frac{1}{2}$	16.2	4.75	11.3	3.16	1.54	1.43	11.3	3.16	1.54	1.43	0.983
L5 × 5 × $\frac{3}{8}$	12.3	3.61	8.74	2.42	1.56	1.39	8.74	2.42	1.56	1.39	0.990
L4 × 4 × $\frac{3}{4}$	18.5	5.44	7.67	2.81	1.19	1.27	7.67	2.81	1.19	1.27	0.778
L4 × 4 × $\frac{1}{2}$	12.8	3.75	5.56	1.97	1.22	1.18	5.56	1.97	1.22	1.18	0.782
L4 × 4 × $\frac{3}{8}$	9.8	2.86	4.36	1.52	1.23	1.14	4.36	1.52	1.23	1.14	0.788
L4 × 4 × $\frac{1}{4}$	6.6	1.94	3.04	1.05	1.25	1.09	3.04	1.05	1.25	1.09	0.795
L3$\frac{1}{2}$ × 3$\frac{1}{2}$ × $\frac{1}{2}$	11.1	3.25	3.64	1.49	1.06	1.06	3.64	1.49	1.06	1.06	0.683
L3$\frac{1}{2}$ × 3$\frac{1}{2}$ × $\frac{1}{2}$	8.5	2.48	2.87	1.15	1.07	1.01	2.87	1.15	1.07	1.01	0.687
L3$\frac{1}{2}$ × 3$\frac{1}{2}$ × $\frac{1}{4}$	5.8	1.69	2.01	0.794	1.09	0.968	2.01	0.794	1.09	0.968	0.694
L3 × 3 × $\frac{1}{2}$	9.4	2.75	2.22	1.07	0.898	0.932	2.22	1.07	0.898	0.932	0.584
L3 × 3 × $\frac{3}{8}$	7.2	2.11	1.76	0.833	0.913	0.888	1.76	0.833	0.913	0.888	0.587
L3 × 3 × $\frac{1}{4}$	4.9	1.44	1.24	0.577	0.930	0.842	1.24	0.577	0.930	0.842	0.592
L2$\frac{1}{2}$ × 2$\frac{1}{2}$ × $\frac{1}{2}$	7.7	2.25	1.23	0.724	0.739	0.806	1.23	0.724	0.739	0.806	0.487
L2$\frac{1}{2}$ × 2$\frac{1}{2}$ × $\frac{3}{8}$	5.9	1.73	0.984	0.566	0.753	0.762	0.984	0.566	0.753	0.762	0.487
L2$\frac{1}{2}$ × 2$\frac{1}{2}$ × $\frac{1}{4}$	4.1	1.19	0.703	0.394	0.769	0.717	0.703	0.394	0.769	0.717	0.491
L2 × 2 × $\frac{3}{8}$	4.7	1.36	0.479	0.351	0.594	0.636	0.479	0.351	0.594	0.636	0.389
L2 × 2 × $\frac{1}{4}$	3.19	0.938	0.348	0.247	0.609	0.592	0.348	0.247	0.609	0.592	0.391
L2 × 2 × $\frac{1}{8}$	1.65	0.484	0.190	0.131	0.626	0.546	0.190	0.131	0.626	0.546	0.398

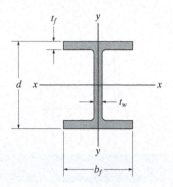

Wide-Flange Sections or W Shapes SI Units

Designation	Area A	Depth d	Web thickness t_w	Flange width b_f	Flange thickness t_f	x–x axis I	x–x axis S	x–x axis r	y–y axis I	y–y axis S	y–y axis r
mm × kg/m	mm²	mm	mm	mm	mm	10⁶ mm⁴	10³ mm³	mm	10⁶ mm⁴	10³ mm³	mm
W610 × 155	19 800	611	12.70	324.0	19.0	1 290	4 220	255	108	667	73.9
W610 × 140	17 900	617	13.10	230.0	22.2	1 120	3 630	250	45.1	392	50.2
W610 × 125	15 900	612	11.90	229.0	19.6	985	3 220	249	39.3	343	49.7
W610 × 113	14 400	608	11.20	228.0	17.3	875	2 880	247	34.3	301	48.8
W610 × 101	12 900	603	10.50	228.0	14.9	764	2 530	243	29.5	259	47.8
W610 × 92	11 800	603	10.90	179.0	15.0	646	2 140	234	14.4	161	34.9
W610 × 82	10 500	599	10.00	178.0	12.8	560	1 870	231	12.1	136	33.9
W460 × 97	12 300	466	11.40	193.0	19.0	445	1 910	190	22.8	236	43.1
W460 × 89	11 400	463	10.50	192.0	17.7	410	1 770	190	20.9	218	42.8
W460 × 82	10 400	460	9.91	191.0	16.0	370	1 610	189	18.6	195	42.3
W460 × 74	9 460	457	9.02	190.0	14.5	333	1 460	188	16.6	175	41.9
W460 × 68	8 730	459	9.14	154.0	15.4	297	1 290	184	9.41	122	32.8
W460 × 60	7 590	455	8.00	153.0	13.3	255	1 120	183	7.96	104	32.4
W460 × 52	6 640	450	7.62	152.0	10.8	212	942	179	6.34	83.4	30.9
W410 × 85	10 800	417	10.90	181.0	18.2	315	1 510	171	18.0	199	40.8
W410 × 74	9 510	413	9.65	180.0	16.0	275	1 330	170	15.6	173	40.5
W410 × 67	8 560	410	8.76	179.0	14.4	245	1 200	169	13.8	154	40.2
W410 × 53	6 820	403	7.49	177.0	10.9	186	923	165	10.1	114	38.5
W410 × 46	5 890	403	6.99	140.0	11.2	156	774	163	5.14	73.4	29.5
W410 × 39	4 960	399	6.35	140.0	8.8	126	632	159	4.02	57.4	28.5
W360 × 79	10 100	354	9.40	205.0	16.8	227	1 280	150	24.2	236	48.9
W360 × 64	8 150	347	7.75	203.0	13.5	179	1 030	148	18.8	185	48.0
W360 × 57	7 200	358	7.87	172.0	13.1	160	894	149	11.1	129	39.3
W360 × 51	6 450	355	7.24	171.0	11.6	141	794	148	9.68	113	38.7
W360 × 45	5 710	352	6.86	171.0	9.8	121	688	146	8.16	95.4	37.8
W360 × 39	4 960	353	6.48	128.0	10.7	102	578	143	3.75	58.6	27.5
W360 × 33	4 190	349	5.84	127.0	8.5	82.9	475	141	2.91	45.8	26.4

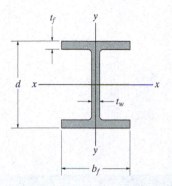

Wide-Flange Sections or W Shapes SI Units

Designation	Area A	Depth d	Web thickness t_w	Flange width b_f	Flange thickness t_f	x–x axis I	x–x axis S	x–x axis r	y–y axis I	y–y axis S	y–y axis r
mm × kg/m	mm²	mm	mm	mm	mm	10^6 mm⁴	10^3 mm³	mm	10^6 mm⁴	10^3 mm³	mm
W310 × 129	16 500	318	13.10	308.0	20.6	308	1940	137	100	649	77.8
W310 × 74	9 480	310	9.40	205.0	16.3	165	1060	132	23.4	228	49.7
W310 × 67	8 530	306	8.51	204.0	14.6	145	948	130	20.7	203	49.3
W310 × 39	4 930	310	5.84	165.0	9.7	84.8	547	131	7.23	87.6	38.3
W310 × 33	4 180	313	6.60	102.0	10.8	65.0	415	125	1.92	37.6	21.4
W310 × 24	3 040	305	5.59	101.0	6.7	42.8	281	119	1.16	23.0	19.5
W310 × 21	2 680	303	5.08	101.0	5.7	37.0	244	117	0.986	19.5	19.2
W250 × 149	19 000	282	17.30	263.0	28.4	259	1840	117	86.2	656	67.4
W250 × 80	10 200	256	9.40	255.0	15.6	126	984	111	43.1	338	65.0
W250 × 67	8 560	257	8.89	204.0	15.7	104	809	110	22.2	218	50.9
W250 × 58	7 400	252	8.00	203.0	13.5	87.3	693	109	18.8	185	50.4
W250 × 45	5 700	266	7.62	148.0	13.0	71.1	535	112	7.03	95	35.1
W250 × 28	3 620	260	6.35	102.0	10.0	39.9	307	105	1.78	34.9	22.2
W250 × 22	2 850	254	5.84	102.0	6.9	28.8	227	101	1.22	23.9	20.7
W250 × 18	2 280	251	4.83	101.0	5.3	22.5	179	99.3	0.919	18.2	20.1
W200 × 100	12 700	229	14.50	210.0	23.7	113	987	94.3	36.6	349	53.7
W200 × 86	11 000	222	13.00	209.0	20.6	94.7	853	92.8	31.4	300	53.4
W200 × 71	9 100	216	10.20	206.0	17.4	76.6	709	91.7	25.4	247	52.8
W200 × 59	7 580	210	9.14	205.0	14.2	61.2	583	89.9	20.4	199	51.9
W200 × 46	5 890	203	7.24	203.0	11.0	45.5	448	87.9	15.3	151	51.0
W200 × 36	4 570	201	6.22	165.0	10.2	34.4	342	86.8	7.64	92.6	40.9
W200 × 22	2 860	206	6.22	102.0	8.0	20.0	194	83.6	1.42	27.8	22.3
W150 × 37	4 730	162	8.13	154.0	11.6	22.2	274	68.5	7.07	91.8	38.7
W150 × 30	3 790	157	6.60	153.0	9.3	17.1	218	67.2	5.54	72.4	38.2
W150 × 22	2 860	152	5.84	152.0	6.6	12.1	159	65.0	3.87	50.9	36.8
W150 × 24	3 060	160	6.60	102.0	10.3	13.4	168	66.2	1.83	35.9	24.5
W150 × 18	2 290	153	5.84	102.0	7.1	9.19	120	63.3	1.26	24.7	23.5
W150 × 14	1 730	150	4.32	100.0	5.5	6.84	91.2	62.9	0.912	18.2	23.0

B

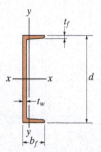

American Standard Channels or C Shapes SI Units

Designation	Area A	Depth d	Web thickness t_w	Flange width b_f	Flange thickness t_f	x–x axis I	x–x axis S	x–x axis r	y–y axis I	y–y axis S	y–y axis r
mm × kg/m	mm²	mm	mm	mm	mm	10^6 mm⁴	10^3 mm³	mm	10^6 mm⁴	10^3 mm³	mm
C380 × 74	9 480	381.0	18.20	94.4	16.50	168	882	133	4.58	61.8	22.0
C380 × 60	7 610	381.0	13.20	89.4	16.50	145	761	138	3.84	55.1	22.5
C380 × 50	6 430	381.0	10.20	86.4	16.50	131	688	143	3.38	50.9	22.9
C310 × 45	5 690	305.0	13.00	80.5	12.70	67.4	442	109	2.14	33.8	19.4
C310 × 37	4 740	305.0	9.83	77.4	12.70	59.9	393	112	1.86	30.9	19.8
C310 × 31	3 930	305.0	7.16	74.7	12.70	53.7	352	117	1.61	28.3	20.2
C250 × 45	5 690	254.0	17.10	77.0	11.10	42.9	338	86.8	1.61	27.1	17.0
C250 × 37	4 740	254.0	13.40	73.3	11.10	38.0	299	89.5	1.40	24.3	17.2
C250 × 30	3 790	254.0	9.63	69.6	11.10	32.8	258	93.0	1.17	21.6	17.6
C250 × 23	2 900	254.0	6.10	66.0	11.10	28.1	221	98.4	0.949	19.0	18.1
C230 × 30	3 790	229.0	11.40	67.3	10.50	25.3	221	81.7	1.01	19.2	16.3
C230 × 22	2 850	229.0	7.24	63.1	10.50	21.2	185	86.2	0.803	16.7	16.8
C230 × 20	2 540	229.0	5.92	61.8	10.50	19.9	174	88.5	0.733	15.8	17.0
C200 × 28	3 550	203.0	12.40	64.2	9.90	18.3	180	71.8	0.824	16.5	15.2
C200 × 20	2 610	203.0	7.70	59.5	9.90	15.0	148	75.8	0.637	14.0	15.6
C200 × 17	2 180	203.0	5.59	57.4	9.90	13.6	134	79.0	0.549	12.8	15.9
C180 × 22	2 790	178.0	10.60	58.4	9.30	11.3	127	63.6	0.574	12.8	14.3
C180 × 18	2 320	178.0	7.98	55.7	9.30	10.1	113	66.0	0.487	11.5	14.5
C180 × 15	1 850	178.0	5.33	53.1	9.30	8.87	99.7	69.2	0.403	10.2	14.8
C150 × 19	2 470	152.0	11.10	54.8	8.70	7.24	95.3	54.1	0.437	10.5	13.3
C150 × 16	1 990	152.0	7.98	51.7	8.70	6.33	83.3	56.4	0.360	9.22	13.5
C150 × 12	1 550	152.0	5.08	48.8	8.70	5.45	71.7	59.3	0.288	8.04	13.6
C130 × 13	1 700	127.0	8.25	47.9	8.10	3.70	58.3	46.7	0.263	7.35	12.4
C130 × 10	1 270	127.0	4.83	44.5	8.10	3.12	49.1	49.6	0.199	6.18	12.5
C100 × 11	1 370	102.0	8.15	43.7	7.50	1.91	37.5	37.3	0.180	5.62	11.5
C100 × 8	1 030	102.0	4.67	40.2	7.50	1.60	31.4	39.4	0.133	4.65	11.4
C75 × 9	1 140	76.2	9.04	40.5	6.90	0.862	22.6	27.5	0.127	4.39	10.6
C75 × 7	948	76.2	6.55	38.0	6.90	0.770	20.2	28.5	0.103	3.83	10.4
C75 × 6	781	76.2	4.32	35.8	6.90	0.691	18.1	29.8	0.082	3.32	10.2

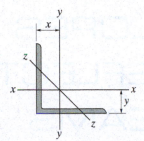

Angles Having Equal Legs SI Units

| Size and thickness | Mass per Meter | Area | x–x axis | | | | y–y axis | | | | z–z axis |
| | | | I | S | r | y | I | S | r | x | r |
mm	kg	mm²	10⁶ mm⁴	10⁶ mm³	mm	mm	10⁶ mm⁴	10⁶ mm³	mm	mm	mm
L203 × 203 × 25.4	75.9	9 680	36.9	258	61.7	60.1	36.9	258	61.7	60.1	39.6
L203 × 203 × 19.0	57.9	7 380	28.9	199	62.6	57.8	28.9	199	62.6	57.8	40.1
L203 × 203 × 12.7	39.3	5 000	20.2	137	63.6	55.5	20.2	137	63.6	55.5	40.4
L152 × 152 × 25.4	55.7	7 100	14.6	139	45.3	47.2	14.6	139	45.3	47.2	29.7
L152 × 152 × 19.0	42.7	5 440	11.6	108	46.2	45.0	11.6	108	46.2	45.0	29.7
L152 × 152 × 12.7	29.2	3 710	8.22	75.1	47.1	42.7	8.22	75.1	47.1	42.7	30.0
L152 × 152 × 9.5	22.2	2 810	6.35	57.4	47.5	41.5	6.35	57.4	47.5	41.5	30.2
L127 × 127 × 19.0	35.1	4 480	6.54	73.9	38.2	38.7	6.54	73.9	38.2	38.7	24.8
L127 × 127 × 12.7	24.1	3 060	4.68	51.7	39.1	36.4	4.68	51.7	39.1	36.4	25.0
L127 × 127 × 9.5	18.3	2 330	3.64	39.7	39.5	35.3	3.64	39.7	39.5	35.3	25.1
L102 × 102 × 19.0	27.5	3 510	3.23	46.4	30.3	32.4	3.23	46.4	30.3	32.4	19.8
L102 × 102 × 12.7	19.0	2 420	2.34	32.6	31.1	30.2	2.34	32.6	31.1	30.2	19.9
L102 × 102 × 9.5	14.6	1 840	1.84	25.3	31.6	29.0	1.84	25.3	31.6	29.0	20.0
L102 × 102 × 6.4	9.8	1 250	1.28	17.3	32.0	27.9	1.28	17.3	32.0	27.9	20.2
L89 × 89 × 12.7	16.5	2 100	1.52	24.5	26.9	26.9	1.52	24.5	26.9	26.9	17.3
L89 × 89 × 9.5	12.6	1 600	1.20	19.0	27.4	25.8	1.20	19.0	27.4	25.8	17.4
L89 × 89 × 6.4	8.6	1 090	0.840	13.0	27.8	24.6	0.840	13.0	27.8	24.6	17.6
L76 × 76 × 12.7	14.0	1 770	0.915	17.5	22.7	23.6	0.915	17.5	22.7	23.6	14.8
L76 × 76 × 9.5	10.7	1 360	0.726	13.6	23.1	22.5	0.726	13.6	23.1	22.5	14.9
L76 × 76 × 6.4	7.3	927	0.514	9.39	23.5	21.3	0.514	9.39	23.5	21.3	15.0
L64 × 64 × 12.7	11.5	1 450	0.524	12.1	19.0	20.6	0.524	12.1	19.0	20.6	12.4
L64 × 64 × 9.5	8.8	1 120	0.420	9.46	19.4	19.5	0.420	9.46	19.4	19.5	12.4
L64 × 64 × 6.4	6.1	766	0.300	6.59	19.8	18.2	0.300	6.59	19.8	18.2	12.5
L51 × 51 × 9.5	7.0	877	0.202	5.82	15.2	16.2	0.202	5.82	15.2	16.2	9.88
L51 × 51 × 6.4	4.7	605	0.146	4.09	15.6	15.1	0.146	4.09	15.6	15.1	9.93
L51 × 51 × 3.2	2.5	312	0.080	2.16	16.0	13.9	0.080	2.16	16.0	13.9	10.1

SLOPES AND DEFLECTIONS OF BEAMS

Simply Supported Beam Slopes and Deflections

Beam	Slope	Deflection	Elastic Curve
	$\theta_{max} = \dfrac{-PL^2}{16EI}$	$v_{max} = \dfrac{-PL^3}{48EI}$	$v = \dfrac{-Px}{48EI}(3L^2 - 4x^2)$ $0 \le x \le L/2$
	$\theta_1 = \dfrac{-Pab(L+b)}{6EIL}$ $\theta_2 = \dfrac{Pab(L+a)}{6EIL}$	$v\Big|_{x=a} = \dfrac{-Pba}{6EIL}(L^2 - b^2 - a^2)$	$v = \dfrac{-Pbx}{6EIL}(L^2 - b^2 - x^2)$ $0 \le x \le a$
	$\theta_1 = \dfrac{-M_0 L}{6EI}$ $\theta_2 = \dfrac{M_0 L}{3EI}$	$v_{max} = \dfrac{-M_0 L^2}{9\sqrt{3}\,EI}$ at $x = 0.5774L$	$v = \dfrac{-M_0 x}{6EIL}(L^2 - x^2)$
	$\theta_{max} = \dfrac{-wL^3}{24EI}$	$v_{max} = \dfrac{-5wL^4}{384EI}$	$v = \dfrac{-wx}{24EI}(x^3 - 2Lx^2 + L^3)$
	$\theta_1 = \dfrac{-3wL^3}{128EI}$ $\theta_2 = \dfrac{7wL^3}{384EI}$	$v\Big|_{x=L/2} = \dfrac{-5wL^4}{768EI}$ $v_{max} = -0.006563\dfrac{wL^4}{EI}$ at $x = 0.4598L$	$v = \dfrac{-wx}{384EI}(16x^3 - 24Lx^2 + 9L^3)$ $0 \le x \le L/2$ $v = \dfrac{-wL}{384EI}(8x^3 - 24Lx^2$ $\qquad + 17L^2 x - L^3)$ $L/2 \le x < L$
	$\theta_1 = \dfrac{-7w_0 L^3}{360EI}$ $\theta_2 = \dfrac{w_0 L^3}{45EI}$	$v_{max} = -0.00652\dfrac{w_0 L^4}{EI}$ at $x = 0.5193L$	$v = \dfrac{-w_0 x}{360EIL}(3x^4 - 10L^2 x^2 + 7L^4)$

Cantilevered Beam Slopes and Deflections

Beam	Slope	Deflection	Elastic Curve
	$\theta_{max} = \dfrac{-PL^2}{2EI}$	$v_{max} = \dfrac{-PL^3}{3EI}$	$v = \dfrac{-Px^2}{6EI}(3L - x)$
	$\theta_{max} = \dfrac{-PL^2}{8EI}$	$v_{max} = \dfrac{-5PL^3}{48EI}$	$v = \dfrac{-Px^2}{12EI}(3L - 2x) \quad 0 \le x \le L/2$ $v = \dfrac{-PL^2}{48EI}(6x - L) \quad L/2 \le x \le L$
	$\theta_{max} = \dfrac{-wL^3}{6EI}$	$v_{max} = \dfrac{-wL^4}{8EI}$	$v = \dfrac{-wx^2}{24EI}(x^2 - 4Lx + 6L^2)$
	$\theta_{max} = \dfrac{M_0 L}{EI}$	$v_{max} = \dfrac{M_0 L^2}{2EI}$	$v = \dfrac{M_0 x^2}{2EI}$
	$\theta_{max} = \dfrac{-wL^3}{48EI}$	$v_{max} = \dfrac{-7wL^4}{384EI}$	$v = \dfrac{-wx^2}{24EI}\left(x^2 - 2Lx + \tfrac{3}{2}L^2\right)$ $\qquad\qquad\qquad 0 \le x \le L/2$ $v = \dfrac{-wL^3}{384EI}(8x - L)$ $\qquad\qquad\qquad L/2 \le x \le L$
	$\theta_{max} = \dfrac{-w_0 L^3}{24EI}$	$v_{max} = \dfrac{-w_0 L^4}{30EI}$	$v = \dfrac{-w_0 x^2}{120EIL}(10L^3 - 10L^2 x + 5Lx^2 - x^3)$

Solutions and Answers for Preliminary Problems

P1–1a.

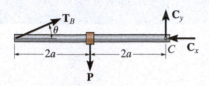

$$\zeta + \Sigma M_C = 0; \text{ get } T_B$$

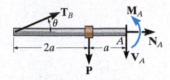

$$\xrightarrow{+} \Sigma F_x = 0; \text{ get } N_A$$
$$+\uparrow \Sigma F_y = 0; \text{ get } V_A$$
$$\zeta + \Sigma M_A = 0; \text{ get } M_A$$

P1–1b.

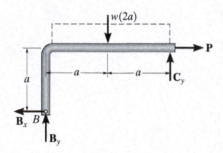

$$\zeta + \Sigma M_B = 0; \text{ get } C_y$$

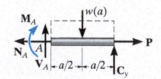

$$\xrightarrow{+} \Sigma F_x = 0; \text{ get } N_A$$
$$+\uparrow \Sigma F_y = 0; \text{ get } V_A$$
$$\zeta + \Sigma M_A = 0; \text{ get } M_A$$

P1–1c.

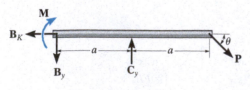

$$\zeta + \Sigma M_B = 0; \text{ get } C_y$$

P1–1d.

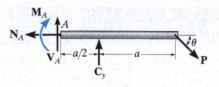

$$\xrightarrow{+} \Sigma F_x = 0; \text{ get } N_A$$
$$+\uparrow \Sigma F_y = 0; \text{ get } V_A$$
$$\zeta + \Sigma M_A = 0; \text{ get } M_A$$

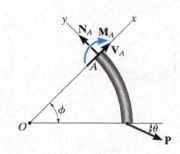

$$+\nwarrow \Sigma F_y = 0; \text{ get } N_A$$
$$+\nearrow \Sigma F_x = 0; \text{ get } V_A$$
$$\zeta + \Sigma M_O = 0 \text{ or } \Sigma M_A = 0; \text{ get } M_A$$

P1–1e.

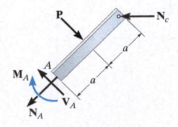

$$\zeta + \Sigma M_B = 0; \text{ get } N_C$$

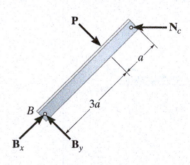

$$+\nearrow \Sigma F_x = 0; \text{ get } N_A$$
$$+\nwarrow \Sigma F_y = 0; \text{ get } V_A$$
$$\zeta + \Sigma M_A = 0; \text{ get } M_A$$

P1–1f.

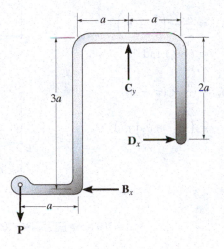

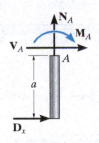

$+\uparrow \Sigma F_y = 0$; get $C_y \ (= P)$

$\zeta + \Sigma M_B = 0$; get D_x

$+\uparrow \Sigma F_y = 0$; get $N_A \ (= 0)$

$\xrightarrow{+} \Sigma F_x = 0$; get V_A

$\zeta + \Sigma M_A = 0$; get M_A

P1–2a.

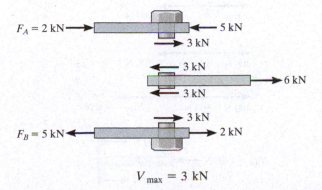

$V_{\text{max}} = 3 \text{ kN}$

P1–2b.

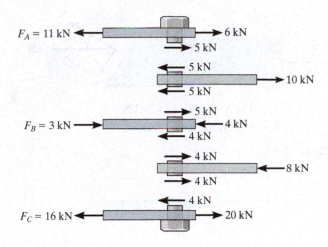

$V_{\text{max}} = 5 \text{ kN}$

P1–3.

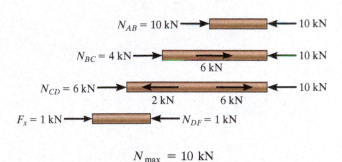

$N_{\text{max}} = 10 \text{ kN}$

P1–4.

$N_A = 24 \text{ kN}$

P1–5.

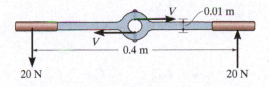

$\Sigma M = 0$; $20 \text{ N} \ (0.4 \text{ m}) - V(0.01 \text{ m}) = 0$

$V = 800 \text{ N}$

P1–6.

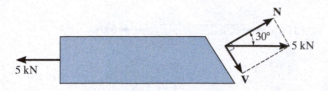

$$N = (5 \text{ kN}) \cos 30° = 4.33 \text{ kN}$$
$$V = (5 \text{ kN}) \sin 30° = 2.5 \text{ kN}$$

P2–1.

$$\frac{\Delta'}{L} = \frac{\Delta}{3L}, \quad \Delta' = \frac{\Delta}{3}$$

$$\epsilon_{AB} = \frac{\Delta/3}{L/2} = \frac{2\Delta}{3L}$$

$$\epsilon_{CD} = \frac{\Delta}{L}$$

P2–2.

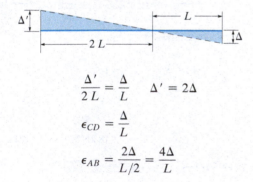

$$\frac{\Delta'}{2L} = \frac{\Delta}{L} \quad \Delta' = 2\Delta$$

$$\epsilon_{CD} = \frac{\Delta}{L}$$

$$\epsilon_{AB} = \frac{2\Delta}{L/2} = \frac{4\Delta}{L}$$

P2–3.

$$\epsilon_{AB} = \frac{L_{A'B} - L_{AB}}{L_{AB}}$$

P2–4.

$$\epsilon_{AB} = \frac{L_{AB'} - L_{AB}}{L_{AB}}, \quad \epsilon_{AC} = \frac{L_{AC'} - L_{AC}}{L_{AC}}$$

$$\epsilon_{BC} = \frac{L_{B'C'} - L_{BC}}{L_{BC}}, \quad (\gamma_A)_{xy} = \left(\frac{\pi}{2} - \theta\right) \text{ rad}$$

P2–5.

$$(\gamma_A)_{xy} = \frac{\pi}{2} - \left(\frac{\pi}{2} + \theta_1\right)$$
$$= (-\theta_1) \text{ rad}$$

$$(\gamma_B)_{xy} = \frac{\pi}{2} - (\pi - \theta_2)$$
$$= \left(-\frac{\pi}{2} + \theta_2\right) \text{ rad}$$

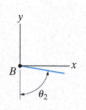

P4–1a.

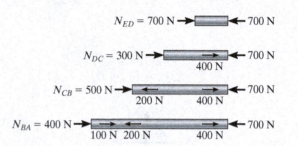

P4–1b.

P4–2.

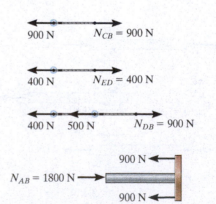

P4–3.

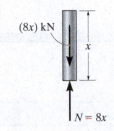

$(8x)$ kN

x

$N = 8x$

P4–4.

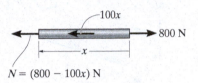

$100x$

800 N

x

$N = (800 - 100x)$ N

P4–5.

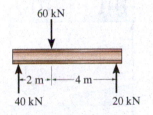

60 kN

-2 m$-$$-4$ m$-$

40 kN   20 kN

$$\Delta_B = \frac{PL}{AE} = \frac{20(10^3) \text{ N} (3 \text{ m})}{2(10^{-3}) \text{ m}^2 (60(10^9) \text{ N/m}^2)}$$

$$= 0.5(10^{-3}) \text{ m} = 0.5 \text{ mm}$$

P5–1.

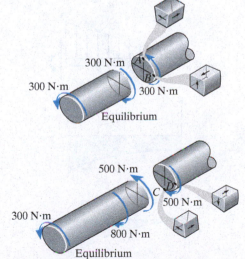

300 N·m A

300 N·m B

300 N·m

Equilibrium

500 N·m

C D

300 N·m   500 N·m

800 N·m

Equilibrium

P5–2.

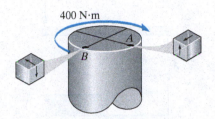

400 N·m

A

B

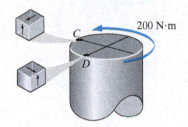

200 N·m

C

D

P5–3.

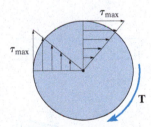

τ_{max}

τ_{max}

T

τ_{max}

T

τ_{max}

P5–4.

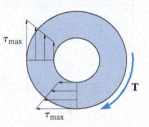

$$P = T\omega$$

$$(10 \text{ hp})\left(\frac{550 \text{ ft} \cdot \text{lb/s}}{1 \text{ hp}}\right) = T\left(1200 \frac{\text{rev}}{\text{min}}\right)\left(\frac{1 \text{ min}}{60 \text{ s}}\right)\frac{2\pi \text{ rad}}{1 \text{ rev}}$$

$$T = 43.8 \text{ lb} \cdot \text{ft}$$

P6–1a.

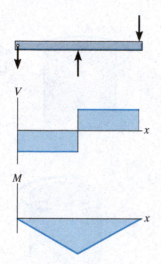

P6–1b.

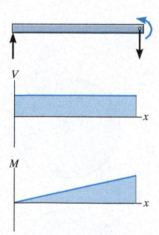

P6–1c.

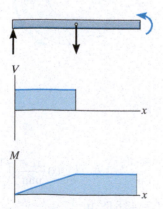

P6–1d.

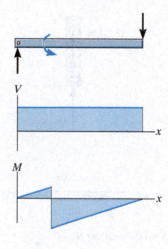

P6–1e.

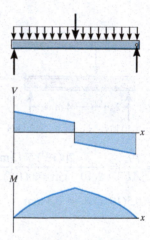

P6–1f.

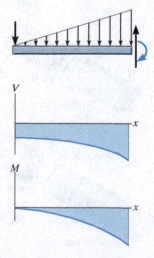

P6–1g.

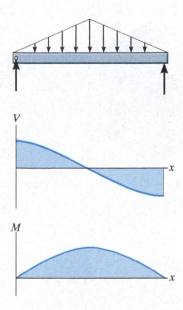

P6–1h.

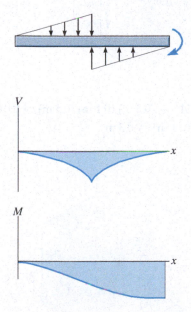

P6–2. $I = \left[\dfrac{1}{12}(0.2 \text{ m})(0.4 \text{ m})^3\right] - \left[\dfrac{1}{12}(0.1 \text{ m})(0.2 \text{ m})^3\right]$

$= 1.0 \, (10^{-3}) \text{ m}^4$

P6–3.

$$\bar{y} = \frac{\Sigma \tilde{y} A}{\Sigma A} = \frac{(0.05 \text{ m})(0.2 \text{ m})(0.1 \text{ m}) + (0.25 \text{ m})(0.1 \text{ m})(0.3 \text{ m})}{(0.2 \text{ m})(0.1 \text{ m}) + (0.1 \text{ m})(0.3 \text{ m})}$$

$$= 0.17 \text{ m}$$

$$I = \left[\frac{1}{12}(0.2 \text{ m})(0.1 \text{ m})^3 + (0.2 \text{ m})(0.1 \text{ m})(0.17 \text{ m} - 0.05 \text{ m})^2\right]$$

$$+ \left[\frac{1}{12}(0.1 \text{ m})(0.3 \text{ m})^3 + (0.1 \text{ m})(0.3 \text{ m})(0.25 \text{ m} - 0.17 \text{ m})^2\right]$$

$$= 0.722 \, (10^{-3}) \text{ m}^4$$

P6–4a.

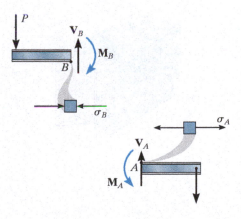

P6–4b.

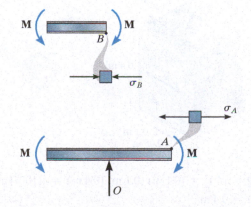

P6–5a.

P7–1b.

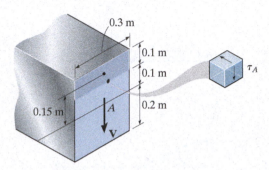

$$Q = \bar{y}'A' = (0.15 \text{ m}) (0.3 \text{ m}) (0.1 \text{ m}) = 4.5(10^{-3}) \text{ m}^3$$
$$t = 0.3 \text{ m}$$

P6–5b.

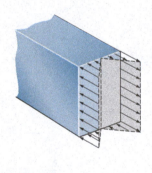

P7–1c.

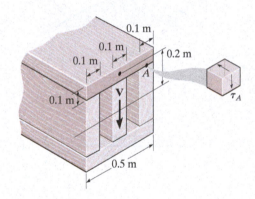

$$Q = \bar{y}'A' = (0.2 \text{ m}) (0.1 \text{ m}) (0.5 \text{ m}) = 0.01 \text{ m}^3$$
$$t = 3 (0.1 \text{ m}) = 0.3 \text{ m}$$

P7–1a.

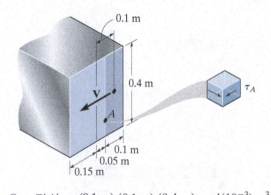

$$Q = \bar{y}'A' = (0.1 \text{ m}) (0.1 \text{ m}) (0.4 \text{ m}) = 4(10^{-3}) \text{ m}^3$$
$$t = 0.4 \text{ m}$$

P7–1d.

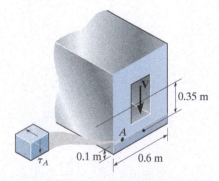

$$Q = \bar{y}'A' = (0.35 \text{ m}) (0.6 \text{ m}) (0.1 \text{ m}) = 0.021 \text{ m}^3$$
$$t = 0.6 \text{ m}$$

P7–1e.

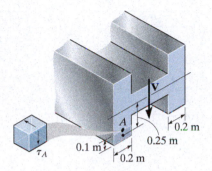

$$Q = \bar{y}'A' = (0.25 \text{ m})(0.2 \text{ m})(0.1 \text{ m}) = 5(10^{-3}) \text{ m}^3$$
$$t = 0.2 \text{ m}$$

P7–1f.

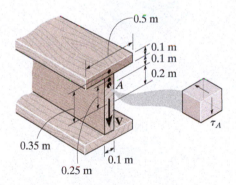

$$Q = \Sigma \bar{y}'A' = (0.25 \text{ m})(0.1 \text{ m})(0.1 \text{ m})$$
$$+ (0.35 \text{ m})(0.1 \text{ m})(0.5 \text{ m}) = 0.02 \text{ m}^3$$
$$t = 0.1 \text{ m}$$

P8–1a.

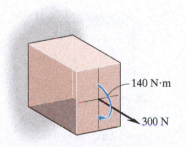

P8–1b.

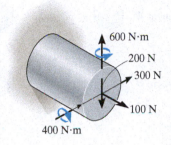

P8–1c.

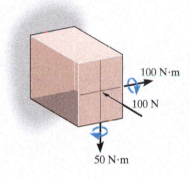

P8–1d.

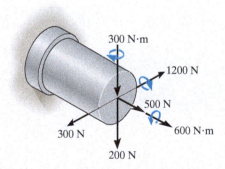

P8–2a.

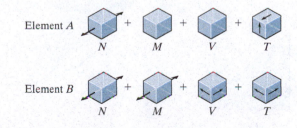

P8–2b.

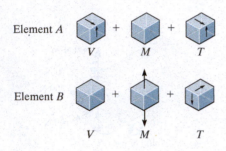

P9–2.

$$\tau_{max} = \sqrt{\left(\frac{\sigma_x - \sigma_y}{2}\right)^2 + \tau_{xy}^2} = \sqrt{\left(\frac{4 - (-4)}{2}\right)^2 + (0)^2}$$

$$= 4 \text{ MPa}$$

$$\sigma_{avg} = \frac{\sigma_x + \sigma_y}{2} = \frac{4 - 4}{2} = 0$$

$$\tan 2\theta_s = -\frac{(\sigma_x - \sigma_y)/2}{\tau_{xy}} = \frac{[4 - (-4)]/2}{0} = -\infty$$

$$\theta_s = -45°$$

$$\tau_{x'y'} = -\frac{\sigma_x - \sigma_y}{2} \sin 2\theta + \tau_{xy} \cos 2\theta$$

$$= -\frac{4 - (-4)}{2} \sin 2(-45°) + 0 = 4 \text{ MPa}$$

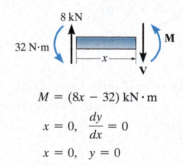

P9–1.

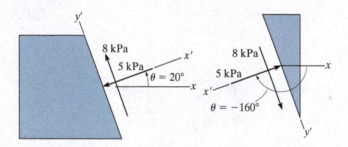

P9–1b.

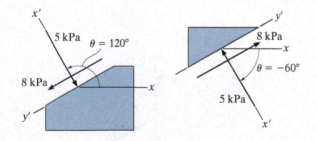

P12–1a.

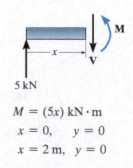

$$M = (8x - 32) \text{ kN} \cdot \text{m}$$

$$x = 0, \quad \frac{dy}{dx} = 0$$

$$x = 0, \quad y = 0$$

P9–1c.

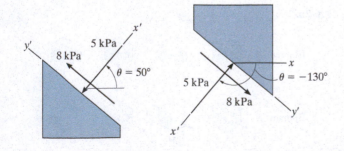

P12–1b.

$$M = (5x) \text{ kN} \cdot \text{m}$$

$$x = 0, \quad y = 0$$

$$x = 2 \text{ m}, \quad y = 0$$

P12–1c.

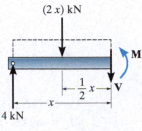

$$M = 4x - (2x)\left(\tfrac{1}{2}x\right)$$
$$M = (4x - x^2)\ \text{kN} \cdot \text{m}$$
$$x = 0, \quad y = 0$$
$$x = 4\ \text{m}, \ y = 0$$

P12–1d.

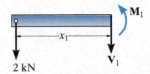

$$M_1 = (-2x_1)\ \text{kN} \cdot \text{m}$$

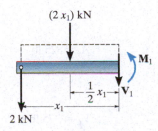

$$M_2 = (-2x + 8)\ \text{kN} \cdot \text{m}$$
$$x_1 = 0, \quad y_1 = 0$$
$$x_2 = 4\ \text{m}, \ y_2 = 0$$
$$x_1 = x_2 = 2\text{m}, \quad \frac{dy_1}{dx_1} = \frac{dy_2}{dx_2}$$
$$x_1 = x_2 = 2\text{m}, \quad y_1 = y_2$$

P12–1e.

$(2\,x_1)$ kN

M_1

V_1

2 kN

$$M_1 = -2x_1 - (2x_1)\left(\tfrac{1}{2}x_1\right)$$
$$M_1 = (-2x_1 - x_1^2)\ \text{kN} \cdot \text{m}$$

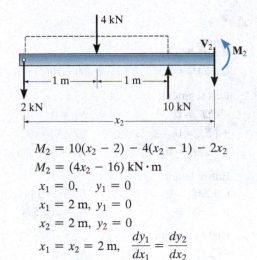

$$M_2 = 10(x_2 - 2) - 4(x_2 - 1) - 2x_2$$
$$M_2 = (4x_2 - 16)\ \text{kN} \cdot \text{m}$$
$$x_1 = 0, \quad y_1 = 0$$
$$x_1 = 2\ \text{m}, \ y_1 = 0$$
$$x_2 = 2\ \text{m}, \ y_2 = 0$$
$$x_1 = x_2 = 2\ \text{m}, \quad \frac{dy_1}{dx_1} = \frac{dy_2}{dx_2}$$

P12–1f.

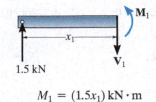

$$M_1 = (1.5x_1)\ \text{kN} \cdot \text{m}$$

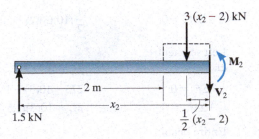

$$M_2 = 1.5x_2 - 3(x_2 - 2)\left(\frac{1}{2}\right)(x_2 - 2)$$
$$M_2 = -1.5x_2^2 + 7.5x_2 - 6$$
$$x_1 = 0, \ y_1 = 0$$
$$x_2 = 4\ \text{m}, \ y_2 = 0$$
$$x_1 = x_2 = 2\text{m}, \quad \frac{dy_1}{dx_1} = \frac{dy_2}{dx_2}$$
$$x_1 = x_2 = 2\text{m}, \quad y_1 = y_2$$

Fundamental Problems Partial Solutions and Answers

Chapter 1

F1–1 Entire beam:

$\zeta + \Sigma M_B = 0;$ $\qquad$ $60 - 10(2) - A_y(2) = 0$ $\qquad$ $A_y = 20$ kN

Left segment:

$\xrightarrow{+} \Sigma F_x = 0;$ $\qquad$ $N_C = 0$ $\qquad\qquad$ *Ans.*

$+\uparrow \Sigma F_y = 0;$ $\qquad$ $20 - V_C = 0$ $\qquad$ $V_C = 20$ kN $\qquad$ *Ans.*

$\zeta + \Sigma M_C = 0;$ $\qquad$ $M_C + 60 - 20(1) = 0$ $\qquad$ $M_C = -40$ kN $\cdot$ m $\qquad$ *Ans.*

F1–2 Entire beam:

$\zeta + \Sigma M_A = 0;$ $\qquad$ $B_y(3) - 100(1.5)(0.75) - 200(1.5)(2.25) = 0$

$\qquad\qquad\qquad\qquad\qquad$ $B_y = 262.5$ N

Right segment:

$\xrightarrow{+} \Sigma F_x = 0;$ $\qquad$ $N_C = 0$ $\qquad\qquad$ *Ans.*

$+\uparrow \Sigma F_y = 0;$ $\qquad$ $V_C + 262.5 - 200(1.5) = 0$ $\qquad$ $V_C = 37.5$ N $\qquad$ *Ans.*

$\zeta + \Sigma M_C = 0;$ $\qquad$ $262.5(1.5) - 200(1.5)(0.75) - M_C = 0$ $\qquad$ $M_C = 169$ N $\cdot$ m $\qquad$ *Ans.*

F1–3 Entire beam:

$\xrightarrow{+} \Sigma F_x = 0;$ $\qquad$ $B_x = 0$

$\zeta + \Sigma M_A = 0;$ $\qquad$ $20(2)(1) - B_y(4) = 0$ $\qquad$ $B_y = 10$ kN

Right segment:

$\xrightarrow{+} \Sigma F_x = 0;$ $\qquad$ $N_C = 0$ $\qquad\qquad$ *Ans.*

$+\uparrow \Sigma F_y = 0;$ $\qquad$ $V_C - 10 = 0$ $\qquad$ $V_C = 10$ kN $\qquad$ *Ans.*

$\zeta + \Sigma M_C = 0;$ $\qquad$ $-M_C - 10(2) = 0$ $\qquad$ $M_C = -20$ kN $\cdot$ m $\qquad$ *Ans.*

F1–4 Entire beam:

$\zeta + \Sigma M_B = 0;$ $\qquad$ $\dfrac{1}{2}(10)(3)(2) + 10(3)(4.5) - A_y(6) = 0$ $\qquad$ $A_y = 27.5$ kN

Left segment:

$\xrightarrow{+} \Sigma F_x = 0;$ $\qquad$ $N_C = 0$ $\qquad\qquad$ *Ans.*

$+\uparrow \Sigma F_y = 0;$ $\qquad$ $27.5 - 10(3) - V_C = 0$ $\qquad$ $V_C = -2.5$ kN $\qquad$ *Ans.*

$\zeta + \Sigma M_C = 0;$ $\qquad$ $M_C + 10(3)(1.5) - 27.5(3) = 0$ $\qquad$ $M_C = 37.5$ kN $\cdot$ m $\qquad$ *Ans.*

F1–5 Entire beam:

$\xrightarrow{+} \Sigma F_x = 0;$ $\qquad$ $A_x = 0$

$\zeta + \Sigma M_B = 0;$ $\qquad$ $300(6)(3) - \dfrac{1}{2}(300)(3)(1) - A_y(6) = 0$ $\qquad$ $A_y = 825$ lb

Left segment:

$\xrightarrow{+} \Sigma F_x = 0;$ $\qquad$ $N_C = 0$ $\qquad\qquad$ *Ans.*

$+\uparrow \Sigma F_y = 0;$ $\qquad$ $825 - 300(3) - V_C = 0$ $\qquad$ $V_C = -75$ lb $\qquad$ *Ans.*

$\zeta + \Sigma M_C = 0;$ $\qquad$ $M_C + 300(3)(1.5) - 825(3) = 0$ $\qquad$ $M_C = 1125$ lb $\cdot$ ft $\qquad$ *Ans.*

F1–6   Entire beam:

$\zeta + \Sigma M_A = 0;$ $\qquad F_{BD}\left(\dfrac{3}{5}\right)(4) - 5(6)(3) = 0$ $\qquad F_{BD} = 37.5$ kN

$\xrightarrow{+} \Sigma F_x = 0;$ $\qquad 37.5\left(\dfrac{4}{5}\right) - A_x = 0$ $\qquad A_x = 30$ kN

$+\uparrow \Sigma F_y = 0;$ $\qquad A_y + 37.5\left(\dfrac{3}{5}\right) - 5(6) = 0$ $\qquad A_y = 7.5$ kN

Left segment:

$\xrightarrow{+} \Sigma F_x = 0;$ $\qquad N_C - 30 = 0$ $\qquad N_C = 30$ kN $\qquad$ *Ans.*

$+\uparrow \Sigma F_y = 0;$ $\qquad 7.5 - 5(2) - V_C = 0$ $\qquad V_C = -2.5$ kN $\qquad$ *Ans.*

$\zeta + \Sigma M_C = 0;$ $\qquad M_C + 5(2)(1) - 7.5(2) = 0$ $\qquad M_C = 5$ kN · m $\qquad$ *Ans.*

F1–7   Beam:

$\Sigma M_A = 0;\ T_{CD} = 2w$

$\Sigma F_y = 0;\ T_{AB} = w$

Rod *AB*:

$\sigma = \dfrac{N}{A};\ 300(10^3) = \dfrac{w}{10};$

$w = 3$ N/m

Rod *CD*:

$\sigma = \dfrac{N}{A};\ 300(10^3) = \dfrac{2w}{15};$

$w = 2.25$ N/m $\qquad$ *Ans.*

F1–8 $A = \pi(0.1^2 - 0.08^2) = 3.6(10^{-3})\pi$ m^2

$\sigma_{avg} = \dfrac{N}{A} = \dfrac{300(10^3)}{3.6(10^{-3})\pi} = 26.5$ MPa $\qquad$ *Ans.*

F1–9 $A = 3[4(1)] = 12$ in^2

$\sigma_{avg} = \dfrac{N}{A} = \dfrac{15}{12} = 1.25$ ksi $\qquad$ *Ans.*

F1–10   Consider the cross section to be a rectangle and two triangles.

$\bar{y} = \dfrac{\Sigma \tilde{y}A}{\Sigma A} = \dfrac{0.15[(0.3)(0.12)] + (0.1)\left[\dfrac{1}{2}(0.16)(0.3)\right]}{0.3(0.12) + \dfrac{1}{2}(0.16)(0.3)}$

$= 0.13$ m $= 130$ mm $\qquad$ *Ans.*

$\sigma_{avg} = \dfrac{N}{A} = \dfrac{600(10^3)}{0.06} = 10$ MPa $\qquad$ *Ans.*

F1–11

$A_A = A_C = \dfrac{\pi}{4}(0.5^2) = 0.0625\pi$ in^2, $A_B = \dfrac{\pi}{4}(1^2) = 0.25\pi$ in^2

$\sigma_A = \dfrac{N_A}{A_A} = \dfrac{3}{0.0625\pi} = 15.3$ ksi (T) $\qquad$ *Ans.*

$\sigma_B = \dfrac{N_B}{A_B} = \dfrac{-6}{0.25\pi} = -7.64$ ksi $= 7.64$ ksi (C) $\qquad$ *Ans.*

$\sigma_C = \dfrac{N_C}{A_C} = \dfrac{2}{0.0625\pi} = 10.2$ ksi (T) $\qquad$ *Ans.*

F1–12   Pin at *A:*

$F_{AD} = 50(9.81)$ N $= 490.5$ N

$+\uparrow \Sigma F_y = 0;$ $\quad F_{AC}\left(\dfrac{3}{5}\right) - 490.5 = 0$ $\quad F_{AC} = 817.5$ N

$\xrightarrow{+} \Sigma F_x = 0;$ $\quad 817.5\left(\dfrac{4}{5}\right) - F_{AB} = 0$ $\quad F_{AB} = 654$ N

$A_{AB} = \dfrac{\pi}{4}(0.008^2) = 16(10^{-6})\pi$ m^2

$(\sigma_{AB})_{avg} = \dfrac{F_{AB}}{A_{AB}} = \dfrac{654}{16(10^{-6})\pi} = 13.0$ MPa $\qquad$ *Ans.*

F1–13   Ring *C:*

$+\uparrow \Sigma F_y = 0;$ $\quad 2F \cos 60° - 200(9.81) = 0$ $\quad F = 1962$ N

$(\sigma_{allow})_{avg} = \dfrac{F}{A};$ $\quad 150(10^6) = \dfrac{1962}{\dfrac{\pi}{4}d^2}$

$d = 0.00408$ m $= 4.08$ mm

Use $d = 5$ mm. $\qquad$ *Ans.*

F1–14   Entire frame:

$\Sigma F_y = 0; A_y = 600 \text{ lb}$

$\Sigma M_B = 0; A_x = 800 \text{ lb}$

$F_A = \sqrt{(600)^2 + (800)^2} = 1000 \text{ lb}$

$(\tau_A)_{avg} = \dfrac{F_A/2}{A} = \dfrac{1000/2}{\frac{\pi}{4}(0.25)^2} = 10.2 \text{ ksi}$ *Ans.*

F1–15   Center plate, bolts have double shear:

$\Sigma F_x = 0;\quad 4V - 10 = 0\quad V = 2.5 \text{ kip}$

$A = \dfrac{\pi}{4}\left(\dfrac{3}{4}\right)^2 = 0.140625\pi \text{ in}^2$

$\tau_{avg} = \dfrac{V}{A} = \dfrac{2.5}{0.140625\pi} = 5.66 \text{ ksi}$ *Ans.*

F1–16   Nails have single shear:

$\Sigma F_x = 0;\quad P - 3V = 0\quad V = \dfrac{P}{3}$

$A = \dfrac{\pi}{4}(0.004^2) = 4(10^{-6})\pi \text{ m}^2$

$(\tau_{avg})_{allow} = \dfrac{V}{A};\quad 60(10^6) = \dfrac{\frac{P}{3}}{4(10^{-6})\pi}$

$P = 2.262(10^3) \text{ N} = 2.26 \text{ kN}$ *Ans.*

F1–17   Strut:

$\xrightarrow{+}\Sigma F_x = 0;\quad V - P\cos 60° = 0\quad V = 0.5P$

$A = \left(\dfrac{0.05}{\sin 60°}\right)(0.025) = 1.4434(10^{-3}) \text{ m}^2$

$(\tau_{avg})_{allow} = \dfrac{V}{A};\quad 600(10^3) = \dfrac{0.5P}{1.4434(10^{-3})}$

$P = 1.732(10^3) \text{ N} = 1.73 \text{ kN}$ *Ans.*

F1–18   The resultant force on the pin is

$F = \sqrt{30^2 + 40^2} = 50 \text{ kN}.$

We have double shear:

$V = \dfrac{F}{2} = \dfrac{50}{2} = 25 \text{ kN}$

$A = \dfrac{\pi}{4}(0.03^2) = 0.225(10^{-3})\pi \text{ m}^2$

$\tau_{avg} = \dfrac{V}{A} = \dfrac{25(10^3)}{0.225(10^{-3})\pi} = 35.4 \text{ MPa}$ *Ans.*

F1–19   Eyebolt:

$\xrightarrow{+}\Sigma F_x = 0;\quad 30 - N = 0\quad N = 30 \text{ kN}$

$\sigma_{allow} = \dfrac{\sigma_Y}{\text{F.S.}} = \dfrac{250}{1.5} = 166.67 \text{ MPa}$

$\sigma_{allow} = \dfrac{N}{A};\quad 166.67(10^6) = \dfrac{30(10^3)}{\frac{\pi}{4}d^2}$

$d = 15.14 \text{ mm}$

Use $d = 16 \text{ mm}.$ *Ans.*

F1–20   Right segment through AB:

$\xrightarrow{+}\Sigma F_x = 0;\quad N_{AB} - 30 = 0\quad N_{AB} = 30 \text{ kip}$

Right segment through CB:

$\xrightarrow{+}\Sigma F_x = 0;\quad N_{BC} - 15 - 15 - 30 = 0\quad N_{BC} = 60 \text{ kip}$

$\sigma_{allow} = \dfrac{\sigma_Y}{\text{F.S.}} = \dfrac{50}{1.5} = 33.33 \text{ ksi}$

Segment AB:

$\sigma_{allow} = \dfrac{N_{AB}}{A_{AB}};\quad 33.33 = \dfrac{30}{h_1(0.5)}$

$h_1 = 1.8 \text{ in.}$

Segment BC:

$\sigma_{allow} = \dfrac{N_{BC}}{A_{BC}};\quad 33.33 = \dfrac{60}{h_2(0.5)}$

$h_2 = 3.6 \text{ in.}$

Use $h_1 = 1\dfrac{7}{8} \text{ in. and } h_2 = 3\dfrac{5}{8} \text{ in.}$ *Ans.*

F1–21 $N = P$

$\sigma_{allow} = \dfrac{\sigma_Y}{\text{F.S.}} = \dfrac{250}{2} = 125 \text{ MPa}$

$A_r = \dfrac{\pi}{4}(0.04^2) = 1.2566(10^{-3}) \text{ m}^2$

$A_{a-a} = 2(0.06 - 0.03)(0.05) = 3(10^{-3}) \text{ m}^2$

The rod will fail first.

$\sigma_{allow} = \dfrac{N}{A_r};\quad 125(10^6) = \dfrac{P}{1.2566(10^{-3})}$

$P = 157.08(10^3) \text{ N} = 157 \text{ kN}$ *Ans.*

F1–22   Pin has double shear:

$\xrightarrow{+}\Sigma F_x = 0;\quad 80 - 2V = 0\quad V = 40 \text{ kN}$

$\tau_{allow} = \dfrac{\tau_{fail}}{\text{F.S.}} = \dfrac{100}{2.5} = 40 \text{ MPa}$

$\tau_{allow} = \dfrac{V}{A};\quad 40(10^6) = \dfrac{40(10^3)}{\frac{\pi}{4}d^2}$

$d = 0.03568 \text{ m} = 35.68 \text{ mm}$

Use $d = 36 \text{ mm}.$ *Ans.*

F1–23

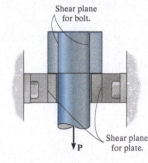

Shear plane for bolt.

Shear plane for plate.

P

$$V = P$$

$$\tau_{allow} = \frac{\tau_{fail}}{F.S.} = \frac{120}{2.5} = 48 \text{ MPa}$$

Area of shear plane for bolt head and plate:

$$A_b = \pi dt = \pi(0.04)(0.075) = 0.003\pi \text{ m}^2$$

$$A_p = \pi dt = \pi(0.08)(0.03) = 0.0024\pi \text{ m}^2$$

Since the area of shear plane for the plate is smaller,

$$\tau_{allow} = \frac{V}{A_p}; \qquad 48(10^6) = \frac{P}{0.0024\pi}$$

$$P = 361.91(10^3) \text{ N} = 362 \text{ kN} \qquad \text{Ans.}$$

F1–24   Support reaction at A:

$$\zeta + \Sigma M_B = 0; \quad \frac{1}{2}(300)(9)(6) - A_y(9) = 0 \quad A_y = 900 \text{ lb}$$

Each nail has single shear:

$$V = 900 \text{ lb}/6 = 150 \text{ lb}$$

$$\tau_{allow} = \frac{\tau_{fail}}{F.S.} = \frac{16}{2} = 8 \text{ ksi}$$

$$\tau_{allow} = \frac{V}{A}; \qquad 8(10^3) = \frac{150}{\frac{\pi}{4}d^2}$$

$$d = 0.1545 \text{ in.}$$

829 Use $d = \dfrac{3}{16}$ in.   *Ans.*

Chapter 2

F2–1 $\dfrac{\delta_C}{600} = \dfrac{0.2}{400}; \qquad \delta_C = 0.3 \text{ mm}$

$$\epsilon_{CD} = \frac{\delta_C}{L_{CD}} = \frac{0.3}{300} = 0.001 \text{ mm/mm} \qquad \text{Ans.}$$

F2–2

![F2-2 diagram]

$$\theta = \left(\frac{0.02°}{180°}\right)\pi \text{ rad} = 0.3491(10^{-3}) \text{ rad}$$

$$\delta_B = \theta L_{AB} = 0.3491(10^{-3})(600) = 0.2094 \text{ mm}$$

$$\delta_C = \theta L_{AC} = 0.3491(10^{-3})(1200) = 0.4189 \text{ mm}$$

$$\epsilon_{BD} = \frac{\delta_B}{L_{BD}} = \frac{0.2094}{400} = 0.524(10^{-3}) \text{ mm/mm} \qquad \text{Ans.}$$

$$\epsilon_{CE} = \frac{\delta_C}{L_{CE}} = \frac{0.4189}{600} = 0.698(10^{-3}) \text{ mm/mm} \qquad \text{Ans.}$$

F2–3

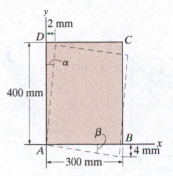

$$\alpha = \frac{2}{400} = 0.005 \text{ rad} \qquad \beta = \frac{4}{300} = 0.01333 \text{ rad}$$

$$(\gamma_A)_{xy} = \frac{\pi}{2} - \theta$$

$$= \frac{\pi}{2} - \left(\frac{\pi}{2} - \alpha + \beta\right)$$

$$= \alpha - \beta$$

$$= 0.005 - 0.01333$$

$$= -0.00833 \text{ rad} \qquad \text{Ans.}$$

F2–4

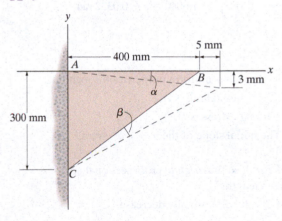

$$L_{BC} = \sqrt{300^2 + 400^2} = 500 \text{ mm}$$

$$L_{B'C} = \sqrt{(300 - 3)^2 + (400 + 5)^2} = 502.2290 \text{ mm}$$

$$\alpha = \frac{3}{405} = 0.007407 \text{ rad}$$

$$(\epsilon_{BC})_{avg} = \frac{L_{B'C} - L_{BC}}{L_{BC}} = \frac{502.2290 - 500}{500}$$

$$= 0.00446 \text{ mm/mm} \qquad \text{Ans.}$$

$$(\gamma_A)_{xy} = \frac{\pi}{2} - \theta = \frac{\pi}{2} - \left(\frac{\pi}{2} + \alpha\right) = -\alpha = -0.00741 \text{ rad} \quad \text{Ans.}$$

F2–5

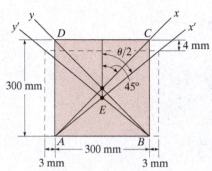

$$L_{AC} = \sqrt{L_{CD}^2 + L_{AD}^2} = \sqrt{300^2 + 300^2} = 424.2641 \text{ mm}$$

$$L_{A'C'} = \sqrt{L_{C'D'}^2 + L_{A'D'}^2} = \sqrt{306^2 + 296^2} = 425.7370 \text{ mm}$$

$$\frac{\theta}{2} = \tan^{-1}\left(\frac{L_{C'D'}}{L_{A'D'}}\right); \theta = 2\tan^{-1}\left(\frac{306}{296}\right) = 1.6040 \text{ rad}$$

$$(\epsilon_{AC})_{avg} = \frac{L_{A'C'} - L_{AC}}{L_{AC}} = \frac{425.7370 - 424.2641}{424.2641}$$

$$= 0.00347 \text{ mm/mm} \qquad\qquad Ans.$$

$$(\gamma_E)_{xy} = \frac{\pi}{2} - \theta = \frac{\pi}{2} - 1.6040 = -0.0332 \text{ rad} \qquad Ans.$$

Chapter 3

F3–1   Material has uniform properties throughout.   *Ans.*

F3–2   Proportional limit is A.   *Ans.*

        Ultimate stress is D.   *Ans.*

F3–3   The initial slope of the $\sigma - \epsilon$ diagram.   *Ans.*

F3–4   True.   *Ans.*

F3–5   False. Use the *original* cross-sectional area and length.   *Ans.*

F3–6   False. It will normally decrease.   *Ans.*

F3–7 $\epsilon = \dfrac{\sigma}{E} = \dfrac{N}{AE}$

$$\delta = \epsilon L = \frac{NL}{AE} = \frac{100(10^3)(0.100)}{\frac{\pi}{4}(0.015)^2 \, 200(10^9)}$$

$$= 0.283 \text{ mm} \qquad\qquad Ans.$$

F3–8 $\epsilon = \dfrac{\sigma}{E} = \dfrac{N}{AE}$

$$\delta = \epsilon L = \frac{NL}{AE}$$

$$0.003 = \frac{(10\,000)(8)}{12E}$$

$$E = 2.22(10^6) \text{ psi} \qquad\qquad Ans.$$

F3–9 $\epsilon = \dfrac{\sigma}{E} = \dfrac{N}{AE}$

$$\delta = \epsilon L = \frac{NL}{AE} = \frac{6(10^3)4}{\frac{\pi}{4}(0.01)^2 100(10^9)}$$

$$= 3.06 \text{ mm} \qquad\qquad Ans.$$

F3–10 $\sigma = \dfrac{N}{A} = \dfrac{100(10^3)}{\frac{\pi}{4}(0.02)^2} = 318.31 \text{ MPa}$

Since $\sigma < \sigma_Y = 450$ MPa, Hooke's Law is applicable.

$$E = \frac{\sigma_Y}{\epsilon_Y} = \frac{450(10^6)}{0.00225} = 200 \text{ GPa}$$

$$\epsilon = \frac{\sigma}{E} = \frac{318.31(10^6)}{200(10^9)} = 0.001592 \text{ mm/mm}$$

$$\delta = \epsilon L = 0.001592(50) = 0.0796 \text{ mm} \qquad Ans.$$

F3–11 $\sigma = \dfrac{N}{A} = \dfrac{150(10^3)}{\frac{\pi}{4}(0.02^2)} = 477.46 \text{ MPa}$

Since $\sigma > \sigma_Y = 450$ MPa, Hooke's Law is not applicable. From the geometry of the shaded triangle,

$$\frac{\epsilon - 0.00225}{0.03 - 0.00225} = \frac{477.46 - 450}{500 - 450}$$

$$\epsilon = 0.017493$$

When the load is removed, the strain recovers along a line AB which is parallel to the original elastic line.

Here $E = \dfrac{\sigma_Y}{\epsilon_Y} = \dfrac{450(10^6)}{0.00225} = 200$ GPa.

The elastic recovery is

$$\epsilon_r = \frac{\sigma}{E} = \frac{477.46(10^6)}{200(10^9)} = 0.002387 \text{ mm/mm}$$

$$\epsilon_p = \epsilon - \epsilon_r = 0.017493 - 0.002387$$

$$= 0.01511 \text{ mm/mm}$$

$$\delta_p = \epsilon_p L = 0.01511(50) = 0.755 \text{ mm} \qquad Ans.$$

F3–12 $\epsilon_{BC} = \dfrac{\delta_{BC}}{L_{BC}} = \dfrac{0.2}{300} = 0.6667(10^{-3})$ mm/mm

$\sigma_{BC} = E\epsilon_{BC} = 200(10^9)[0.6667(10^{-3})]$
$= 133.33$ MPa

Since $\sigma_{BC} < \sigma_Y = 250$ MPa, Hooke's Law is valid.

$\sigma_{BC} = \dfrac{F_{BC}}{A_{BC}};$ $133.33(10^6) = \dfrac{F_{BC}}{\dfrac{\pi}{4}(0.003^2)}$

$F_{BC} = 942.48$ N

$\zeta + \Sigma M_A = 0;$ $942.48(0.4) - P(0.6) = 0$
$P = 628.31$ N $= 628$ N   *Ans.*

F3–13 $\sigma = \dfrac{N}{A} = \dfrac{10(10^3)}{\dfrac{\pi}{4}(0.015)^2} = 56.59$ MPa

$\epsilon_{\text{long}} = \dfrac{\sigma}{E} = \dfrac{56.59(10^6)}{70(10^9)} = 0.808(10^{-3})$

$\epsilon_{\text{lat}} = -\nu\epsilon_{\text{long}} = -0.35(0.808(10^{-3}))$
$= -0.283(10^{-3})$

$\delta d = (-0.283(10^{-3}))(15 \text{ mm}) = -4.24(10^{-3})$ mm
     Ans.

F3–14 $\sigma = \dfrac{N}{A} = \dfrac{50(10^3)}{\dfrac{\pi}{4}(0.02^2)} = 159.15$ MPa

$\epsilon_{\text{long}} = \dfrac{\delta}{L} = \dfrac{1.40}{600} = 0.002333$ mm/mm

$E = \dfrac{\sigma}{\epsilon_{\text{long}}} = \dfrac{159.15(10^6)}{0.002333} = 68.2$ GPa   *Ans.*

$\epsilon_{\text{lat}} = \dfrac{d' - d}{d} = \dfrac{19.9837 - 20}{20} = -0.815(10^{-3})$ mm/mm

$\nu = -\dfrac{\epsilon_{\text{lat}}}{\epsilon_{\text{long}}} = -\dfrac{-0.815(10^{-3})}{0.002333} = 0.3493$

$G = \dfrac{E}{2(1+\nu)} = \dfrac{68.21}{2(1+0.3493)} = 25.3$ GPa   *Ans.*

F3–15

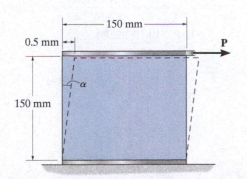

$\alpha = \dfrac{0.5}{150} = 0.003333$ rad

$\gamma = \dfrac{\pi}{2} - \theta = \dfrac{\pi}{2} - \left(\dfrac{\pi}{2} - \alpha\right)$
$= \alpha = 0.003333$ rad

$\tau = G\gamma = [26(10^9)](0.003333) = 86.67$ MPa

$\tau = \dfrac{V}{A};\ 86.67(10^6) = \dfrac{P}{0.15(0.02)}$

$P = 260$ kN   *Ans.*

F3–16

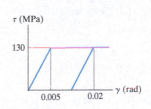

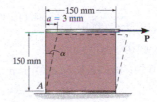

$\alpha = \dfrac{3}{150} = 0.02$ rad

$\gamma = \dfrac{\pi}{2} - \theta = \dfrac{\pi}{2} - \left(\dfrac{\pi}{2} - \alpha\right) = \alpha = 0.02$ rad

When P is removed, the shear strain recovers along a line parallel to the original elastic line.

$\gamma_r = \gamma_Y = 0.005$ rad

$\gamma_p = \gamma - \gamma_r = 0.02 - 0.005 = 0.015$ rad   *Ans.*

Chapter 4

F4–1 $A = \dfrac{\pi}{4}(0.02^2) = 0.1(10^{-3})\pi$ m^2

$N_{BC} = 40$ kN, $N_{AB} = -60$ kN

$\delta_C = \dfrac{1}{AE}\{40(10^3)(400) + [-60(10^3)(600)]\}$

$= \dfrac{-20(10^6) \text{ N} \cdot \text{mm}}{AE}$

$= -0.318$ mm   *Ans.*

F4–2

$$A_{AB} = A_{CD} = \frac{\pi}{4}(0.02^2) = 0.1(10^{-3})\pi \text{ m}^2$$

$$A_{BC} = \frac{\pi}{4}(0.04^2 - 0.03^2) = 0.175(10^{-3})\pi \text{ m}^2$$

$$N_{AB} = -10 \text{ kN}, N_{BC} = 10 \text{ kN}, N_{CD} = -20 \text{ kN}$$

$$\delta_{D/A} = \frac{[-10(10^3)](400)}{[0.1(10^{-3})\pi][68.9(10^9)]}$$

$$+ \frac{[10(10^3)](400)}{[0.175(10^{-3})\pi][68.9(10^9)]}$$

$$+ \frac{[-20(10^3)](400)}{[0.1(10^{-3})\pi][68.9(10^9)]}$$

$$= -0.449 \text{ mm} \qquad \qquad \textit{Ans.}$$

F4–3

$$A = \frac{\pi}{4}(0.03^2) = 0.225(10^{-3})\pi \text{ m}^2$$

$$N_{BC} = -90 \text{ kN}, N_{AB} = -90 + 2\left(\frac{4}{5}\right)(30) = -42 \text{ kN}$$

$$\delta_C = \frac{1}{0.225(10^{-3})\pi[200(10^9)]}\{[-42(10^3)(0.4)]$$

$$+ [-90(10^3)(0.6)]\}$$

$$= -0.501(10^{-3}) \text{ m} = -0.501 \text{ mm} \qquad \textit{Ans.}$$

F4–4

$$\delta_{A/B} = \frac{NL}{AE} = \frac{[60(10^3)](0.8)}{[0.1(10^{-3})\pi][200(10^9)]}$$

$$= 0.7639(10^{-3}) \text{ m} \downarrow$$

$$\delta_B = \frac{F_{sp}}{k} = \frac{60(10^3)}{50(10^6)} = 1.2(10^{-3}) \text{ m} \downarrow$$

$$+\downarrow \quad \delta_A = \delta_B + \delta_{A/B}$$

$$\delta_A = 1.2(10^{-3}) + 0.7639(10^{-3})$$

$$= 1.9639(10^{-3}) \text{ m} = 1.96 \text{ mm} \downarrow \qquad \textit{Ans.}$$

F4–5

30 kN/m

N(x)

x

$$A = \frac{\pi}{4}(0.02^2) = 0.1(10^{-3})\pi \text{ m}^2$$

Internal load $N(x) = 30(10^3)x$

$$\delta_A = \int \frac{N(x)dx}{AE}$$

$$= \frac{1}{[0.1(10^{-3})\pi][73.1(10^9)]} \int_0^{0.9 \text{ m}} 30(10^3)x \, dx$$

$$= 0.529(10^{-3}) \text{ m} = 0.529 \text{ mm} \qquad \textit{Ans.}$$

F4–6

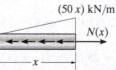

(50 x) kN/m

N(x)

x

Distributed load $N(x) = \frac{45(10^3)}{0.9}x = 50(10^3)x \text{ N/m}$

Internal load $N(x) = \frac{1}{2}(50(10^3))x(x) = 25(10^3)x^2$

$$\delta_A = \int_0^L \frac{N(x)dx}{AE}$$

$$= \frac{1}{[0.1(10^{-3})\pi][73.1(10^9)]} \int_0^{0.9 \text{ m}} [25(10^3)x^2]dx$$

$$= 0.265 \text{ mm} \qquad \qquad \textit{Ans.}$$

Chapter 5

F5–1

$$J = \frac{\pi}{2}(0.04^4) = 1.28(10^{-6})\pi \text{ m}^4$$

$$\tau_A = \tau_{\max} = \frac{Tc}{J} = \frac{5(10^3)(0.04)}{1.28(10^{-6})\pi} = 49.7 \text{ MPa} \qquad \textit{Ans.}$$

$$\tau_B = \frac{T\rho_B}{J} = \frac{5(10^3)(0.03)}{1.28(10^{-6})\pi} = 37.3 \text{ MPa} \qquad \textit{Ans.}$$

A 49.7 MPa

B 37.3 MPa

F5–2

$$J = \frac{\pi}{2}(0.06^4 - 0.04^4) = 5.2(10^{-6})\pi \text{ m}^4$$

$$\tau_B = \tau_{\max} = \frac{Tc}{J} = \frac{10(10^3)(0.06)}{5.2(10^{-6})\pi} = 36.7 \text{ MPa} \qquad \textit{Ans.}$$

$$\tau_A = \frac{T\rho_A}{J} = \frac{10(10^3)(0.04)}{5.2(10^{-6})\pi} = 24.5 \text{ MPa} \qquad \textit{Ans.}$$

A 24.5 MPa

B 36.7 MPa

F5–3 $J_{AB} = \dfrac{\pi}{2}(0.04^4 - 0.03^4) = 0.875(10^{-6})\pi \text{ m}^4$

$J_{BC} = \dfrac{\pi}{2}(0.04^4) = 1.28(10^{-6})\pi \text{ m}^4$

$(\tau_{AB})_{max} = \dfrac{T_{AB}\, c_{AB}}{J_{AB}} = \dfrac{[2(10^3)](0.04)}{0.875(10^{-6})\pi} = 29.1 \text{ MPa}$

$(\tau_{BC})_{max} = \dfrac{T_{BC}\, c_{BC}}{J_{BC}} = \dfrac{[6(10^3)](0.04)}{1.28(10^{-6})\pi}$

$= 59.7 \text{ MPa}$ *Ans.*

F5–4 $T_{AB} = 0,\ T_{BC} = 600 \text{ N} \cdot \text{m},\ T_{CD} = 0$

$J = \dfrac{\pi}{2}(0.02^4) = 80(10^{-9})\pi \text{ m}^4$

$\tau_{max} = \dfrac{Tc}{J} = \dfrac{600(0.02)}{80(10^{-9})\pi} = 47.7 \text{ MPa}$ *Ans.*

F5–5 $J_{BC} = \dfrac{\pi}{2}(0.04^4 - 0.03^4) = 0.875(10^{-6})\pi \text{ m}^4$

$(\tau_{BC})_{max} = \dfrac{T_{BC}\, c_{BC}}{J_{BC}} = \dfrac{2100(0.04)}{0.875(10^{-6})\pi}$

$= 30.6 \text{ MPa}$ *Ans.*

F5–6 $t = 5(10^3) \text{ N} \cdot \text{m/m}$

Internal torque is $T = 5(10^3)(0.8) = 4000 \text{ N} \cdot \text{m}$

$J = \dfrac{\pi}{2}(0.04^4) = 1.28(10^{-6})\pi \text{ m}^4$

$\tau_{AB} = \dfrac{T_A c}{J} = \dfrac{4000(0.04)}{1.28(10^{-6})\pi} = 39.8 \text{ MPa}$ *Ans.*

F5–7 $T_{AB} = 250 \text{ N} \cdot \text{m},\ T_{BC} = 175 \text{ N} \cdot \text{m},$
$T_{CD} = -150 \text{ N} \cdot \text{m}$

Maximum internal torque is in region AB.

$T_{AB} = 250 \text{ N} \cdot \text{m}$

$\tau^{abs}_{max} = \dfrac{T_{AB}\, c}{J} = \dfrac{250(0.025)}{\dfrac{\pi}{2}(0.025)^4} = 10.2 \text{ MPa}$ *Ans.*

F5–8 $P = T\omega;$ $3(550) \text{ ft} \cdot \text{lb/s} = T\left[150\left(\dfrac{2\pi}{60}\right) \text{rad/s}\right]$

$T = 105.04 \text{ ft} \cdot \text{lb}$

$\tau_{allow} = \dfrac{Tc}{J};$ $12(10^3) = \dfrac{105.04(12)(d/2)}{\dfrac{\pi}{2}(d/2)^4}$

$d = 0.812 \text{ in.}$

Use $d = \dfrac{7}{8} \text{ in.}$ *Ans.*

F5–9 $T_{AB} = -2 \text{ kN} \cdot \text{m},\ T_{BC} = 1 \text{ kN} \cdot \text{m}$

$J = \dfrac{\pi}{2}(0.03^4) = 0.405(10^{-6})\pi \text{ m}^4$

$\phi_{A/C} = \dfrac{-2(10^3)(0.6) + (10^3)(0.4)}{[0.405(10^{-6})\pi][75(10^9)]}$

$= -0.00838 \text{ rad} = -0.480°$ *Ans.*

F5–10 $T_{AB} = 600 \text{ N} \cdot \text{m}$

$J = \dfrac{\pi}{2}(0.02^4) = 80(10^{-9})\pi \text{ m}^4$

$\phi_{B/A} = \dfrac{600(0.45)}{[80(10^{-9})\pi][75(10^9)]}$

$= 0.01432 \text{ rad} = 0.821°$ *Ans.*

F5–11 $J = \dfrac{\pi}{2}(0.04^4 - 0.03^4) = 0.875(10^{-6})\pi \text{ m}^4$

$\phi_{A/B} = \dfrac{T_{AB}\, L_{AB}}{JG} = \dfrac{3(10^3)(0.9)}{[0.875(10^{-6})\pi][26(10^9)]}$

$= 0.03778 \text{ rad}$

$\phi_B = \dfrac{T_B}{k_B} = \dfrac{3(10^3)}{90(10^3)} = 0.03333 \text{ rad}$

$\phi_A = \phi_B + \phi_{A/B}$

$= 0.03333 + 0.03778$

$= 0.07111 \text{ rad} = 4.07°$ *Ans.*

F5–12 $T_{AB} = 600 \text{ N} \cdot \text{m},\ T_{BC} = -300 \text{ N} \cdot \text{m},$
$T_{CD} = 200 \text{ N} \cdot \text{m},\ T_{DE} = 500 \text{ N} \cdot \text{m}$

$J = \dfrac{\pi}{2}(0.02^4) = 80(10^{-9})\pi \text{ m}^4$

$\phi_{E/A} = \dfrac{[600 + (-300) + 200 + 500]0.2}{[80(10^{-9})\pi][75(10^9)]}$

$= 0.01061 \text{ rad} = 0.608°$ *Ans.*

F5–13

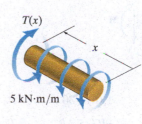

$J = \dfrac{\pi}{2}(0.04^4) = 1.28(10^{-6})\pi \text{ m}^4$

$t = 5(10^3) \text{ N} \cdot \text{m/m}$

Internal torque is $5(10^3)x$ N $\cdot$ m

$$\phi_{A/B} = \int_0^L \frac{T(x)dx}{JG}$$

$$= \frac{1}{[1.28(10^{-6})\pi][75(10^9)]}\int_0^{0.8\text{ m}} 5(10^3)x\,dx$$

$$= 0.00531 \text{ rad} = 0.304° \qquad \text{Ans.}$$

F5–14

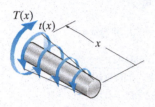

$T(x)$ $t(x)$ x

$$J = \frac{\pi}{2}(0.04^4) = 1.28(10^{-6})\pi \text{ m}^4$$

Distributed torque is $t = \dfrac{15(10^3)}{0.6}(x)$

$$= 25(10^3)x \text{ N} \cdot \text{m/m}$$

Internal torque in segment AB, $T(x) = \dfrac{1}{2}(25x)(10^3)(x)$

$$= 12.5(10^3)x^2 \text{ N} \cdot \text{m}$$

In segment BC,

$$T_{BC} = \frac{1}{2}[25(10^3)(0.6)](0.6) = 4500 \text{ N} \cdot \text{m}$$

$$\phi_{A/C} = \int_0^L \frac{T(x)dx}{JG} + \frac{T_{BC}L_{BC}}{JG}$$

$$= \frac{1}{[1.28(10^{-6})\pi][75(10^9)]}\left[\int_0^{0.6\text{ m}} 12.5(10^3)x^2\,dx + 4500(0.4)\right]$$

$$= 0.008952 \text{ rad} = 0.513° \qquad \text{Ans.}$$

Chapter 6

F6–1

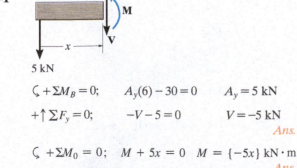

5 kN

$\zeta + \Sigma M_B = 0;$ $\qquad A_y(6) - 30 = 0$ $\qquad A_y = 5$ kN

$+\uparrow \Sigma F_y = 0;$ $\qquad -V - 5 = 0$ $\qquad V = -5$ kN $\qquad$ Ans.

$\zeta + \Sigma M_O = 0;$ $\quad M + 5x = 0$ $\quad M = \{-5x\} \text{ kN} \cdot \text{m}$ $\qquad$ Ans.

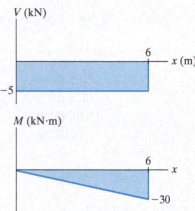

V (kN)

6 $\quad x$ (m)

-5

M (kN·m)

6 $\quad x$

-30

F6–2

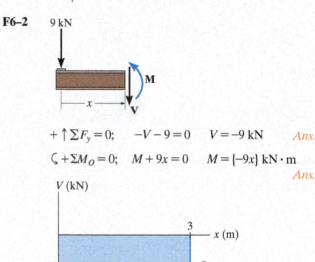

9 kN

M

x

V

$+\uparrow \Sigma F_y = 0;$ $\quad -V - 9 = 0$ $\quad V = -9$ kN $\qquad$ Ans.

$\zeta + \Sigma M_O = 0;$ $\quad M + 9x = 0$ $\quad M = \{-9x\} \text{ kN} \cdot \text{m}$ $\qquad$ Ans.

V (kN)

3 $\quad x$ (m)

-9

M (kN·m)

3 $\quad x$ (m)

-27

F6–3

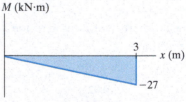

2 kip/ft

18 kip·ft $\qquad M$

x $\qquad V$

$+\uparrow \Sigma F_y = 0;$ $\quad -V - 2x = 0$ $\quad V = \{-2x\} \text{ kip}$ $\qquad$ Ans.

$\zeta + \Sigma M_O = 0;$ $\quad M + 2x\left(\dfrac{x}{2}\right) - 18 = 0$

$$M = \{18 - x^2\} \text{ kip} \cdot \text{ft} \qquad \text{Ans.}$$

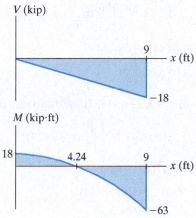

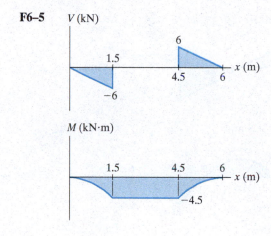

F6–5

F6–4

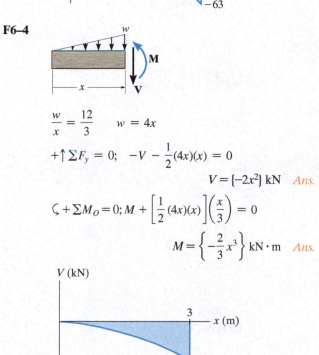

$$\frac{w}{x} = \frac{12}{3} \qquad w = 4x$$

$$+\uparrow \Sigma F_y = 0; \quad -V - \frac{1}{2}(4x)(x) = 0$$

$$V = \{-2x^2\} \text{ kN} \quad Ans.$$

$$\zeta + \Sigma M_O = 0; M + \left[\frac{1}{2}(4x)(x)\right]\left(\frac{x}{3}\right) = 0$$

$$M = \left\{-\frac{2}{3}x^3\right\} \text{ kN} \cdot \text{m} \quad Ans.$$

F6–6

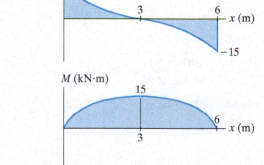

F6–7

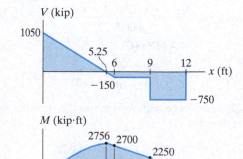

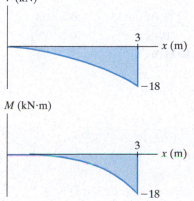

F6–8

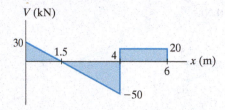

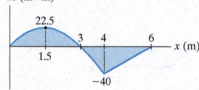

F6–9 Consider two vertical rectangles and a horizontal rectangle.

$$I = 2\left[\frac{1}{12}(0.02)(0.2^3)\right] + \frac{1}{12}(0.26)(0.02^3)$$

$$= 26.84(10^{-6}) \ \text{m}^4$$

$$\sigma_{max} = \frac{Mc}{I} = \frac{20(10^3)(0.1)}{26.84(10^{-6})} = 74.5 \ \text{MPa} \qquad Ans.$$

F6–10 See inside front cover.

$$\bar{y} = \frac{0.3}{3} = 0.1 \ \text{m}$$

$$I = \frac{1}{36}(0.3)(0.3^3) = 0.225(10^{-3}) \ \text{m}^4$$

$$(\sigma_{max})_c = \frac{Mc}{I} = \frac{50(10^3)(0.3 - 0.1)}{0.225(10^{-3})}$$

$$= 44.4 \ \text{MPa (C)} \qquad Ans.$$

$$(\sigma_{max})_t = \frac{My}{I} = \frac{50(10^3)(0.1)}{0.225(10^{-3})} = 22.2 \ \text{MPa (T)} \qquad Ans.$$

F6–11 Consider large rectangle minus the two side rectangles.

$$I = \frac{1}{12}(0.2)(0.3^3) - (2)\frac{1}{12}(0.09)(0.26^3)$$

$$= 0.18636(10^{-3}) \ \text{m}^4$$

$$\sigma_{max} = \frac{Mc}{I} = \frac{50(10^3)(0.15)}{0.18636(10^{-3})} = 40.2 \ \text{MPa} \qquad Ans.$$

F6–12 Consider two vertical rectangles and two horizontal rectangles.

$$I = 2\left[\frac{1}{12}(0.03)(0.4^3)\right] + 2\left[\frac{1}{12}(0.14)(0.03^3) + 0.14(0.03)(0.15^2)\right]$$

$$= 0.50963(10^{-3}) \ \text{m}^4$$

$$\sigma_{max} = \frac{Mc}{I} = \frac{10(10^3)(0.2)}{0.50963(10^{-3})} = 3.92 \ \text{MPa} \qquad Ans.$$

$$\sigma_A = 3.92 \ \text{MPa (C)}$$

$$\sigma_B = 3.92 \ \text{MPa (T)}$$

F6–13 Consider center rectangle and two side rectangles.

$$I = \frac{1}{12}(0.05)(0.4)^3 + 2\left[\frac{1}{12}(0.025)(0.3)^3\right]$$

$$= 0.37917(10^{-3}) \ \text{m}^4$$

$$\sigma_A = \frac{My_A}{I} = \frac{5(10^3)(-0.15)}{0.37917(10^{-3})} = 1.98 \ \text{MPa (T)} \qquad Ans.$$

F6–14 $\quad M_y = 50\left(\dfrac{4}{5}\right) = 40 \ \text{kN} \cdot \text{m}$

$$M_z = 50\left(\frac{3}{5}\right) = 30 \ \text{kN} \cdot \text{m}$$

$$I_y = \frac{1}{12}(0.3)(0.2^3) = 0.2(10^{-3}) \ \text{m}^4$$

$$I_z = \frac{1}{12}(0.2)(0.3^3) = 0.45(10^{-3}) \ \text{m}^4$$

$$\sigma = -\frac{M_z y}{I_z} + \frac{M_y z}{I_y}$$

$$\sigma_A = -\frac{[30(10^3)](-0.15)}{0.45(10^{-3})} + \frac{[40(10^3)](0.1)}{0.2(10^{-3})}$$

$$= 30 \ \text{MPa (T)} \qquad Ans.$$

$$\sigma_B = -\frac{[30(10^3)](0.15)}{0.45(10^{-3})} + \frac{[40(10^3)](0.1)}{0.2(10^{-3})}$$

$$= 10 \ \text{MPa (T)} \qquad Ans.$$

$$\tan \alpha = \frac{I_z}{I_y} \tan \theta$$

$$\tan \alpha = \left[\frac{0.45(10^{-3})}{0.2(10^{-3})}\right]\left(\frac{4}{3}\right)$$

$$\alpha = 71.6° \qquad Ans.$$

F6–15 Maximum stress occurs at D or A.

$$(\sigma_{max})_D = \frac{(50\cos 30°)12(3)}{\frac{1}{12}(4)(6)^3} + \frac{(50\sin 30°)12(2)}{\frac{1}{12}(6)(4)^3}$$

$$= 40.4 \ \text{psi} \qquad Ans.$$

Chapter 7

F7–1 Consider two vertical rectangles and a horizontal rectangle.

$$I = 2\left[\frac{1}{12}(0.02)(0.2^3)\right] + \frac{1}{12}(0.26)(0.02^3)$$

$$= 26.84(10^{-6})\ m^4$$

Take two rectangles above A.

$$Q_A = 2[0.055(0.09)(0.02)] = 198(10^{-6})\ m^3$$

$$\tau_A = \frac{VQ_A}{It} = \frac{100(10^3)[198(10^{-6})]}{[26.84(10^{-6})]2(0.02)}$$

$$= 18.4\ MPa \qquad\qquad Ans.$$

F7–2 Consider a vertical rectangle and two squares.

$$I = \frac{1}{12}(0.1)(0.3^3) + (2)\frac{1}{12}(0.1)(0.1^3)$$

$$= 0.24167(10^{-3})\ m^4$$

Take top half of area (above A).

$$Q_A = y'_1 A'_1 + y'_2 A'_2$$

$$= \left[\frac{1}{2}(0.05)\right](0.05)(0.3) + 0.1(0.1)(0.1)$$

$$= 1.375(10^{-3})\ m^3$$

$$\tau_A = \frac{VQ}{It} = \frac{600(10^3)[1.375(10^{-3})]}{[0.24167(10^{-3})](0.3)} = 11.4\ MPa\ \ Ans.$$

Take top square (above B).

$$Q_B = y'_2 A'_2 = 0.1(0.1)(0.1) = 1(10^{-3})\ m^3$$

$$\tau_B = \frac{VQ}{It} = \frac{600(10^3)[1(10^{-3})]}{[0.24167(10^{-3})](0.1)} = 24.8\ MPa\ \ Ans.$$

F7–3 $V_{max} = 4.5\ kip$

$$I = \frac{1}{12}(3)(6^3) = 54\ in^4$$

Take top half of area.

$$Q_{max} = y'A' = 1.5(3)(3) = 13.5\ in^3$$

$$(\tau_{max})_{abs} = \frac{V_{max}\ Q_{max}}{It} = \frac{4.5(10^3)(13.5)}{54(3)} = 375\ psi$$

$$Ans.$$

F7–4 Consider two vertical rectangles and two horizontal rectangles.

$$I = 2\left[\frac{1}{12}(0.03)(0.4^3)\right] + 2\left[\frac{1}{12}(0.14)(0.03^3)\right.$$

$$\left. + 0.14(0.03)(0.15^2)\right] = 0.50963(10^{-3})\ m^4$$

Take the top half of area.

$$Q_{max} = 2y'_1 A'_1 + y'_2 A'_2 = 2(0.1)(0.2)(0.03)$$

$$+ (0.15)(0.14)(0.03) = 1.83(10^{-3})\ m^3$$

$$\tau_{max} = \frac{VQ_{max}}{It} = \frac{20(10^3)[1.83(10^{-3})]}{0.50963(10^{-3})[2(0.03)]} = 1.20\ MPa$$

$$Ans.$$

F7–5 Consider one large vertical rectangle and two side rectangles.

$$I = \frac{1}{12}(0.05)(0.4)^3 + 2\left[\frac{1}{12}(0.025)(0.3)^3\right]$$

$$= 0.37917(10^{-3})\ m^4$$

Take the top half of area.

$$Q_{max} = 2y'_1 A'_1 + y'_2 A'_2 = 2(0.075)(0.025)(0.15)$$

$$+ (0.1)(0.05)(0.2) = 1.5625(10^{-3})\ m^3$$

$$\tau_{max} = \frac{VQ_{max}}{It} = \frac{20(10^3)[1.5625(10^{-3})]}{[0.37917(10^{-3})][2(0.025)]}$$

$$= 1.65\ MPa \qquad\qquad Ans.$$

F7–6 $I = \frac{1}{12}(0.3)(0.2^3) = 0.2(10^{-3})\ m^4$

Top (or bottom) board

$$Q = y'A' = 0.05(0.1)(0.3) = 1.5(10^{-3})\ m^3$$

Two rows of nails

$$q_{allow} = 2\left(\frac{F}{s}\right) = \frac{2[15(10^3)]}{s} = \frac{30(10^3)}{s}$$

$$q_{allow} = \frac{VQ}{I}; \qquad \frac{30(10^3)}{s} = \frac{50(10^3)[1.5(10^{-3})]}{0.2(10^{-3})}$$

$$s = 0.08\ m = 80\ mm \qquad Ans.$$

F7–7 Consider large rectangle minus two side rectangles.

$$I = \frac{1}{12}(0.2)(0.34^3) - (2)\frac{1}{12}(0.095)(0.28^3)$$

$$= 0.3075(10^{-3})\ m^4$$

Top plate

$$Q = y'A' = 0.16(0.02)(0.2) = 0.64(10^{-3})\ m^3$$

Two rows of bolts

$$q_{allow} = 2\left(\frac{F}{s}\right) = \frac{2[30(10^3)]}{s} = \frac{60(10^3)}{s}$$

$$q_{allow} = \frac{VQ}{I}; \qquad \frac{60(10^3)}{s} = \frac{300(10^3)[0.64(10^{-3})]}{0.3075(10^{-3})}$$

$$s = 0.09609\ m = 96.1\ mm$$

Use $s = 96\ mm$ \qquad\qquad Ans.

F7–8 Consider two large rectangles and two side rectangles.

$$I = 2\left[\frac{1}{12}(0.025)(0.3^3)\right] + 2\left[\frac{1}{12}(0.05)(0.2^3) + 0.05(0.2)(0.15^2)\right]$$

$$= 0.62917(10^{-3})\ \text{m}^4$$

Top center board is held onto beam by the top row of bolts.

$$Q = y'A' = 0.15(0.2)(0.05) = 1.5(10^{-3})\ \text{m}^3$$

Each bolt has two shearing surfaces.

$$q_{allow} = 2\left(\frac{F}{s}\right) = \frac{2[8(10^3)]}{s} = \frac{16(10^3)}{s}$$

$$q_{allow} = \frac{VQ}{I};\qquad \frac{16(10^3)}{s} = \frac{20(10^3)[1.5(10^{-3})]}{0.62917(10^{-3})}$$

$$s = 0.3356\ \text{m} = 335.56\ \text{mm}$$

Use $s = 335$ mm   *Ans.*

F7–9 Consider center board and four side boards.

$$I = \frac{1}{12}(1)(6^3) + 4\left[\frac{1}{12}(0.5)(4^3) + 0.5(4)(3^2)\right]$$

$$= 100.67\ \text{in}^4$$

Top-right board is held onto beam by a row of bolts.

$$Q = y'A' = 3(4)(0.5) = 6\ \text{in}^3$$

Bolts have one shear surface.

$$q_{allow} = \frac{F}{s} = \frac{6}{s}$$

$$q_{allow} = \frac{VQ}{I};\qquad \frac{6}{s} = \frac{15(6)}{100.67}$$

$$s = 6.711\ \text{in.}$$

Use $s = 6\frac{5}{8}$ in.   *Ans.*

Also, can consider the top *two* boards held onto beam by a row of bolts with two shearing surfaces.

Chapter 8

F8–1 $+\uparrow\Sigma F_z = (F_R)_z;\quad -500 - 300 = P$

$P = -800$ kN

$\Sigma M_x = 0;\qquad 300(0.05) - 500(0.1) = M_x$

$M_x = -35$ kN·m

$\Sigma M_y = 0;\qquad 300(0.1) - 500(0.1) = M_y$

$M_y = -20$ kN·m

$A = 0.3(0.3) = 0.09\ \text{m}^2$

$$I_x = I_y = \frac{1}{12}(0.3)(0.3^3) = 0.675(10^{-3})\ \text{m}^4$$

$$\sigma_A = \frac{-800(10^3)}{0.09} + \frac{[20(10^3)](0.15)}{0.675(10^{-3})} + \frac{[35(10^3)](0.15)}{0.675(10^{-3})}$$

$$= 3.3333\ \text{MPa} = 3.33\ \text{MPa (T)}\qquad \textit{Ans.}$$

$$\sigma_B = \frac{-800(10^3)}{0.09} + \frac{[20(10^3)](0.15)}{0.675(10^{-3})} - \frac{[35(10^3)](0.15)}{0.675(10^{-3})}$$

$$= -12.22\ \text{MPa} = 12.2\ \text{MPa (C)}\qquad \textit{Ans.}$$

F8–2 $+\uparrow\Sigma F_y = 0;\qquad V - 400 = 0\qquad V = 400$ kN

$\zeta + \Sigma M_A = 0; -M - 400(0.5) = 0\ \ M = -200$ kN·m

$$I = \frac{1}{12}(0.1)(0.3^3) = 0.225(10^{-3})\ \text{m}^4$$

Bottom segment

$$\sigma_A = \frac{My}{I} = \frac{[200(10^3)](-0.05)}{0.225(10^{-3})}$$

$$= -44.44\ \text{MPa} = 44.4\ \text{MPa (C)}\qquad \textit{Ans.}$$

$$Q_A = y'A' = 0.1(0.1)(0.1) = 1(10^{-3})\ \text{m}^3$$

$$\tau_A = \frac{VQ}{It} = \frac{400(10^3)[1(10^{-3})]}{0.225(10^{-3})(0.1)} = 17.8\ \text{MPa}\qquad \textit{Ans.}$$

F8–3 Left reaction is 20 kN.

Left segment:

$+\uparrow\Sigma F_y = 0;\qquad 20 - V = 0\qquad V = 20$ kN

$\zeta + \Sigma M_s = 0;\qquad M - 20(0.5) = 0\qquad M = 10$ kN·m

Consider large rectangle minus two side rectangles.

$$I = \frac{1}{12}(0.1)(0.2^3) - (2)\frac{1}{12}(0.045)(0.18^3)$$

$$= 22.9267(10^{-6})\ \text{m}^4$$

Top segment above A

$$Q_A = y'_1 A'_1 + y'_2 A'_2 = 0.07(0.04)(0.01)$$

$$+ 0.095(0.1)(0.01) = 0.123(10^{-3})\ \text{m}^3$$

$$\sigma_A = -\frac{My_A}{I} = -\frac{[10(10^3)](0.05)}{22.9267(10^{-6})}$$

$$= -21.81\ \text{MPa} = 21.8\ \text{MPa (C)}\qquad \textit{Ans.}$$

$$\tau_A = \frac{VQ_A}{It} = \frac{20(10^3)[0.123(10^{-3})]}{[22.9267(10^{-6})](0.01)}$$

$$= 10.7\ \text{MPa}\qquad \textit{Ans.}$$

F8–4   At the section through centroidal axis:

$N = P$

$V = 0$

$M = (2 + 1)P = 3P$

$\sigma = \dfrac{N}{A} + \dfrac{Mc}{I}$

$30 = \dfrac{P}{2(0.5)} + \dfrac{(3P)(1)}{\dfrac{1}{12}(0.5)(2)^3}$

$P = 3$ kip   *Ans.*

F8–5   At section through B:

$N = 500$ lb, $V = 400$ lb

$M = 400(10) = 4000$ lb $\cdot$ in.

Axial load:

$\sigma_x = \dfrac{N}{A} = \dfrac{500}{4(3)} = 41.667$ psi (T)

Shear load:

$\tau_{xy} = \dfrac{VQ}{It} = \dfrac{400[(1.5)(3)(1)]}{[\dfrac{1}{12}(3)(4)^3]3} = 37.5$ psi

Bending moment:

$\sigma_x = \dfrac{My}{I} = \dfrac{4000(1)}{\dfrac{1}{12}(3)(4)^3} = 250$ psi (C)

Thus

$\sigma_x = 41.667 - 250 = 208$ psi (C)   *Ans.*

$\sigma_y = 0$ *Ans.*

$\tau_{xy} = 37.5$ psi   *Ans.*

37.5 psi

208 psi

F8–6   Top segment:

$\Sigma F_y = 0; \quad V_y + 1000 = 0 \quad V_y = -1000$ N

$\Sigma F_x = 0; \quad V_x - 1500 = 0 \quad V_x = 1500$ N

$\Sigma M_z = 0; \quad T_z - 1500(0.4) = 0 \quad T_z = 600$ N $\cdot$ m

$\Sigma M_y = 0; \quad M_y - 1500(0.2) = 0 \quad M_y = 300$ N $\cdot$ m

$\Sigma M_x = 0; \quad M_x - 1000(0.2) = 0 \quad M_x = 200$ N $\cdot$ m

$I_y = I_x = \dfrac{\pi}{4}(0.02^4) = 40(10^{-9})\pi$ m^4

$J = \dfrac{\pi}{2}(0.02^4) = 80(10^{-9})\pi$ m^4

$(Q_y)_A = \dfrac{4(0.02)}{3\pi}\left[\dfrac{\pi}{2}(0.02^2)\right] = 5.3333(10^{-6})$ m^3

$\sigma_A = \dfrac{M_x y}{I_x} - \dfrac{M_y x}{I_y} = \dfrac{-200(0)}{40(10^{-9})\pi} - \dfrac{-300(0.02)}{40(10^{-9})\pi}$

$\quad = 47.7$ MPa (T)   *Ans.*

$[(\tau_{zy})_T]_A = \dfrac{T_z c}{J} = \dfrac{600(0.02)}{80(10^{-9})\pi} = 47.746$ MPa

$[(\tau_{zy})_V]_A = \dfrac{V_y(Q_y)_A}{I_x t} = \dfrac{1000[5.3333(10^{-6})]}{[40(10^{-9})\pi](0.04)}$

$\quad = 1.061$ MPa

Combining these two shear stress components,

$(\tau_{zy})_A = 47.746 + 1.061 = 48.8$ MPa   *Ans.*

47.7 MPa

48.8 MPa

F8–7   Right Segment:

$\Sigma F_z = 0; \quad V_z - 6 = 0 \quad\quad V_z = 6$ kN

$\Sigma M_y = 0; \quad T_y - 6(0.3) = 0 \quad T_y = 1.8$ kN $\cdot$ m

$\Sigma M_x = 0; \quad M_x - 6(0.3) = 0 \quad M_x = 1.8$ kN $\cdot$ m

$I_x = \dfrac{\pi}{4}(0.05^4 - 0.04^4) = 0.9225(10^{-6})\pi$ m^4

$J = \dfrac{\pi}{2}(0.05^4 - 0.04^4) = 1.845(10^{-6})\pi$ m^4

$(Q_z)_A = y_2' A_2' - y_1' A_1'$

$\quad = \dfrac{4(0.05)}{3\pi}\left[\dfrac{\pi}{2}(0.05^2)\right] - \dfrac{4(0.04)}{3\pi}\left[\dfrac{\pi}{2}(0.04^2)\right]$

$\quad = 40.6667(10^{-6})$ m^3

$\sigma_A = \dfrac{M_x z}{I_x} = \dfrac{1.8(10^3)(0)}{0.9225(10^{-6})\pi} = 0$ *Ans.*

$[(\tau_{yz})_T]_A = \dfrac{T_y c}{J} = \dfrac{[1.8(10^3)](0.05)}{1.845(10^{-6})\pi} = 15.53$ MPa

$[(\tau_{yz})_V]_A = \dfrac{V_z(Q_z)_A}{I_x t} = \dfrac{6(10^3)[40.6667(10^{-6})]}{[0.9225(10^{-6})\pi](0.02)}$

$\quad = 4.210$ MPa

Combining these two shear stress components,

$(\tau_{yz})_A = 15.53 - 4.210 = 11.3$ MPa   *Ans.*

11.3 MPa

F8–8   Left Segment:

$\Sigma F_z = 0;$ $V_z - 900 - 300 = 0$ $V_z = 1200$ N

$\Sigma M_y = 0;$ $T_y + 300(0.1) - 900(0.1) = 0$ $T_y = 60$ N·m

$\Sigma M_x = 0;$ $M_x + (900 + 300)0.3 = 0$ $M_x = -360$ N·m

$I_x = \dfrac{\pi}{4}(0.025^4 - 0.02^4) = 57.65625(10^{-9})\pi$ m⁴

$J = \dfrac{\pi}{2}(0.025^4 - 0.02^4) = 0.1153125(10^{-6})\pi$ m⁴

$(Q_y)_A = 0$

$\sigma_A = \dfrac{M_x y}{I_x} = \dfrac{(360)(0.025)}{57.65625(10^{-9})\pi} = 49.7$ MPa   *Ans.*

$[(\tau_{xy})_T]_A = \dfrac{T_y \rho_A}{J} = \dfrac{60(0.025)}{0.1153125(10^{-6})\pi} = 4.14$ MPa   *Ans.*

$[(\tau_{yz})_V]_A = \dfrac{V_z(Q_z)_A}{I_x t} = 0$ *Ans.*

4.14 MPa

49.7 MPa

Chapter 9

F9–1 $\theta = 120°$ $\sigma_x = 500$ kPa $\sigma_y = 0$ $\tau_{xy} = 0$

Apply Eqs. 9–1, 9–2.

$\sigma_{x'} = 125$ kPa   *Ans.*

$\tau_{x'y'} = 217$ kPa   *Ans.*

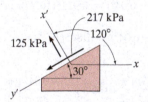

F9–2

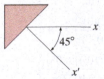

$\theta = -45°$ $\sigma_x = 0$ $\sigma_y = -400$ kPa

$\tau_{xy} = -300$ kPa

Apply Eqs. 9–1, 9–3, 9–2.

$\sigma_{x'} = 100$ kPa   *Ans.*

$\sigma_{y'} = -500$ kPa   *Ans.*

$\tau_{x'y'} = 200$ kPa   *Ans.*

F9–3 $\theta_x = 80$ kPa $\sigma_y = 0$ $\tau_{xy} = 30$ kPa

Apply Eqs. 9–5, 9–4.

$\sigma_1 = 90$ kPa $\sigma_2 = -10$ kPa   *Ans.*

$\theta_p = 18.43°$ and $108.43°$

From Eq. 9–1,

$\sigma_{x'} = \dfrac{80 + 0}{2} + \dfrac{80 - 0}{2}\cos 2(18.43°)$

$+ 30 \sin 2(18.43°)$

$= 90$ kPa $= \sigma_1$

Thus,

$(\theta_p)_1 = 18.4°$ for σ_1 *Ans.*

10 kPa

[diagram: 90 kPa, 18.4°]

F9–4 $\sigma_x = 100$ kPa $\sigma_y = 700$ kPa

$\tau_{xy} = -400$ kPa

Apply Eqs. 9–7, 9–8.

$\tau_{\substack{\max \\ \text{in-plane}}} = 500$ kPa   *Ans.*

$\sigma_{avg} = 400$ kPa   *Ans.*

F9–5   At the cross section through B:

$N = 4$ kN $V = 2$ kN

$M = 2(2) = 4$ kN·m

$\sigma_B = \dfrac{P}{A} + \dfrac{Mc}{I} = \dfrac{4(10^3)}{0.03(0.06)} + \dfrac{4(10^3)(0.03)}{\frac{1}{12}(0.03)(0.06)^3}$

$= 224$ MPa (T)

Note $\tau_B = 0$ since $Q = 0$.

Thus

$\sigma_1 = 224$ MPa   *Ans.*

$\sigma_2 = 0$

F9–6 $A_y = B_y = 12$ kN

Segment AC:

$V_C = 0$ $M_C = 24$ kN·m

$\tau_C = 0$ (since $V_C = 0$)

$\sigma_C = 0$ (since C is on neutral axis)

$\sigma_1 = \sigma_2 = 0$ *Ans.*

F9–7 $\sigma_{\text{avg}} = \dfrac{\sigma_x + \sigma_y}{2} = \dfrac{500 + 0}{2} = 250$ kPa

The coordinates of the center C of the circle and the reference point A are

$$A(500, 0) \qquad C(250, 0)$$

$R = CA = 500 - 250 = 250$ kPa

$\theta = 120°$ (counterclockwise). Rotate the radial line CA counterclockwise $2\theta = 240°$ to the coordinates of point $P(\sigma_{x'}, \tau_{x'y'})$.

$\alpha = 240° - 180° = 60°$

$\sigma_{x'} = 250 - 250\cos 60° = 125$ kPa   *Ans.*

$\tau_{x'y'} = 250\sin 60° = 217$ kPa   *Ans.*

F9–8 $\sigma_{\text{avg}} = \dfrac{\sigma_x + \sigma_y}{2} = \dfrac{80 + 0}{2} = 40$ kPa

The coordinates of the center C of the circle and the reference point A are

$$A(80, 30) \qquad C(40, 0)$$

$R = CA = \sqrt{(80 - 40)^2 + 30^2} = 50$ kPa

$\sigma_1 = 40 + 50 = 90$ kPa   *Ans.*

$\sigma_2 = 40 - 50 = -10$ kPa   *Ans.*

$\tan 2(\theta_p)_1 = \dfrac{30}{80 - 40} = 0.75$

$(\theta_p)_1 = 18.4°$ (counterclockwise)   *Ans.*

F9–9   The coordinates of the reference point A and the center C of the circle are

$$A(30, 40) \qquad C(0, 0)$$

$R = CA = 50$ MPa

$\sigma_1 = 50$ MPa

$\sigma_2 = -50$ MPa

F9–10 $J = \dfrac{\pi}{2}(0.04^4 - 0.03^4) = 0.875(10^{-6})\pi$ m^4

$\tau = \dfrac{Tc}{J} = \dfrac{4(10^3)(0.04)}{0.875(10^{-6})\pi} = 58.21$ MPa

$\sigma_x = \sigma_y = 0$ and $\tau_{xy} = -58.21$ MPa

$\sigma_{\text{avg}} = \dfrac{\sigma_x + \sigma_y}{2} = 0$

The coordinates of the reference point A and the center C of the circle are

$$A(0, -58.21) \qquad C(0, 0)$$

$R = CA = 58.21$ MPa

$\sigma_1 = 0 + 58.21 = 58.2$ MPa   *Ans.*

$\sigma_2 = 0 - 58.21 = -58.2$ MPa   *Ans.*

F9–11

$+\uparrow \Sigma F_y = 0; \qquad V - 30 = 0 \qquad V = 30$ kN

$\zeta + \Sigma M_O = 0; \qquad -M - 30(0.3) = 0 \qquad M = -9$ kN $\cdot$ m

$I = \dfrac{1}{12}(0.05)(0.15^3) = 14.0625(10^{-6})$ m^4

Segment above A,

$Q_A = y'A' = 0.05(0.05)(0.05) = 0.125(10^{-3})$ m^3

$\sigma_A = -\dfrac{My_A}{I} = \dfrac{[-9(10^3)](0.025)}{14.0625(10^{-6})} = 16$ MPa (T)

$\tau_A = \dfrac{VQ_A}{It} = \dfrac{30(10^3)[0.125(10^{-3})]}{14.0625(10^{-6})(0.05)} = 5.333$ MPa

$\sigma_x = 16$ MPa, $\sigma_y = 0$, and $\tau_{xy} = -5.333$ MPa

$\sigma_{\text{avg}} = \dfrac{\sigma_x + \sigma_y}{2} = \dfrac{16 + 0}{2} = 8$ MPa

The coordinates of the reference point A and the center C of the circle are

$$A(16, -5.333) \qquad C(8, 0)$$

$R = CA = \sqrt{(16 - 8)^2 + (-5.333)^2} = 9.615$ MPa

$\sigma_1 = 8 + 9.615 = 17.6$ MPa   *Ans.*

$\sigma_2 = 8 - 9.615 = -1.61$ MPa   *Ans.*

F9–12

$\zeta + \Sigma M_B = 0; \qquad 60(1) - A_y(1.5) = 0 \qquad A_y = 40$ kN

$+\uparrow \Sigma F_y = 0; \qquad 40 - V = 0 \qquad V = 40$ kN

$\zeta + \Sigma M_O = 0; \qquad M - 40(0.5) = 0 \qquad M = 20$ kN $\cdot$ m

Consider large rectangle minus two side rectangles.

$I = \dfrac{1}{12}(0.1)(0.2^3) - (2)\dfrac{1}{12}(0.045)(0.18^3) = 22.9267(10^{-6})$ m^4

Top rectangle,

$Q_A = y'A' = 0.095(0.01)(0.1) = 95(10^{-6})$ m^3

$\sigma_A = -\dfrac{My_A}{I} = -\dfrac{[20(10^3)](0.09)}{22.9267(10^{-6})} = -78.51$ MPa

$\qquad = 78.51$ MPa (C)

$\tau_A = \dfrac{VQ_A}{It} = \dfrac{40(10^3)[95(10^{-6})]}{[22.9267(10^{-6})](0.01)} = 16.57$ MPa

$\sigma_x = -78.51$ MPa, $\sigma_y = 0$, and $\tau_{xy} = -16.57$ MPa

$\sigma_{\text{avg}} = \dfrac{\sigma_x + \sigma_y}{2} = \dfrac{-78.51 + 0}{2} = -39.26$ MPa

The coordinates of the reference point A and the center C of the circle are

$$A(-78.51, -16.57) \qquad C(-39.26, 0)$$

$$R = CA = \sqrt{[-78.51 - (-39.26)]^2 + (-16.57)^2}$$
$$= 42.61 \text{ MPa}$$

$$\tau_{\substack{\max \\ \text{in-plane}}} = |R| = 42.6 \text{ MPa} \qquad \qquad \qquad Ans.$$

Chapter 11

F11–1

At support,

$$V_{\max} = 12 \text{ kN} \qquad M_{\max} = 18 \text{ kN} \cdot \text{m}$$

$$I = \frac{1}{12}(a)(2a)^3 = \frac{2}{3}a^4$$

$$\sigma_{\text{allow}} = \frac{M_{\max}c}{I}; \qquad 10(10^6) = \frac{18(10^3)(a)}{\frac{2}{3}a^4}$$

$$a = 0.1392 \text{ m} = 139.2 \text{ mm}$$

Use $a = 140$ mm   *Ans.*

$$I = \frac{2}{3}(0.14^4) = 0.2561(10^{-3}) \text{ m}^4$$

$$Q_{\max} = \frac{0.14}{2}(0.14)(0.14) = 1.372(10^{-3}) \text{ m}^3$$

$$\tau_{\max} = \frac{V_{\max}Q_{\max}}{It} = \frac{12(10^3)[1.372(10^{-3})]}{[0.2561(10^{-3})](0.14)}$$

$$= 0.459 \text{ MPa} < \tau_{\text{allow}} = 1 \text{ MPa (OK)}$$

F11–2

At support,

$$V_{\max} = 3 \text{ kip} \qquad M_{\max} = 12 \text{ kip} \cdot \text{ft}$$

$$I = \frac{\pi}{4}\left(\frac{d}{2}\right)^4 = \frac{\pi d^4}{64}$$

$$\sigma_{\text{allow}} = \frac{M_{\max}c}{I}; \qquad 20 = \frac{12(12)\left(\frac{d}{2}\right)}{\frac{\pi d^4}{64}}$$

$$d = 4.19 \text{ in.}$$

Use $d = 4\frac{1}{4}$ in.   *Ans.*

$$I = \frac{\pi}{64}(4.25^4) = 16.015 \text{ in}^4$$

Semicircle,

$$Q_{\max} = \frac{4(4.25/2)}{3\pi}\left[\frac{1}{2}\left(\frac{\pi}{4}\right)(4.25^2)\right] = 6.397 \text{ in}^3$$

$$\tau_{\max} = \frac{V_{\max}Q_{\max}}{It} = \frac{3(6.397)}{16.015(4.25)}$$

$$= 0.282 \text{ ksi} < \tau_{\text{allow}} = 10 \text{ ksi (OK)}$$

F11–3

At the supports,

$$V_{\max} = 10 \text{ kN}$$

Under 15-kN load,

$$M_{\max} = 5 \text{ kN} \cdot \text{m}$$

$$I = \frac{1}{12}(a)(2a)^3 = \frac{2}{3}a^4$$

$$\sigma_{\text{allow}} = \frac{M_{\max}c}{I}; \qquad 12(10^6) = \frac{5(10^3)(a)}{\frac{2}{3}a^4}$$

$$a = 0.0855 \text{ m} = 85.5 \text{ mm}$$

Use $a = 86$ mm   *Ans.*

$$I = \frac{2}{3}(0.086^4) = 36.4672(10^{-6}) \text{ m}^4$$

Top half of rectangle,

$$Q_{\max} = \frac{0.086}{2}(0.086)(0.086)$$

$$= 0.318028(10^{-3}) \text{ m}^3$$

$$\tau_{\max} = \frac{V_{\max}Q_{\max}}{It} = \frac{10(10^3)[0.318028(10^{-3})]}{[36.4672(10^{-6})](0.086)}$$

$$= 1.01 \text{ MPa} < \tau_{\text{allow}} = 1.5 \text{ MPa (OK)}$$

F11–4

At the supports,

$$V_{\max} = 4.5 \text{ kip}$$

At the center,

$$M_{\max} = 6.75 \text{ kip} \cdot \text{ft}$$

$$I = \frac{1}{12}(4)(h^3) = \frac{h^3}{3}$$

$$\sigma_{\text{allow}} = \frac{M_{\max}c}{I}; \qquad 2 = \frac{6.75(12)\left(\frac{h}{2}\right)}{\frac{h^3}{3}}$$

$$h = 7.794 \text{ in.}$$

Top half of rectangle,

$$Q_{\max} = y'A' = \frac{h}{4}\left(\frac{h}{2}\right)(4) = \frac{h^2}{2}$$

$$\tau_{\max} = \frac{V_{\max}Q_{\max}}{It}; \qquad 0.2 = \frac{4.5\left(\frac{h^2}{2}\right)}{\frac{h^3}{3}(4)}$$

$$h = 8.4375 \text{ in. (controls)}$$

Use $h = 8\frac{1}{2}$ in.   *Ans.*

F11–5

At the supports,

$V_{max} = 25$ kN

At the center,

$M_{max} = 20$ kN $\cdot$ m

$I = \dfrac{1}{12}(b)(3b)^3 = 2.25b^4$

$\sigma_{allow} = \dfrac{M_{max}c}{I}$; $12(10^6) = \dfrac{20(10^3)(1.5b)}{2.25b^4}$

$\qquad b = 0.1036$ m $= 103.6$ mm

Use $b = 104$ mm   *Ans.*

$I = 2.25(0.104^4) = 0.2632(10^{-3})$ m^4

Top half of rectangle,

$Q_{max} = 0.75(0.104)[1.5(0.104)(0.104)] = 1.2655(10^{-3})$ m^3

$\tau_{max} = \dfrac{V_{max}\,Q_{max}}{It} = \dfrac{25(10^3)[1.2655(10^{-3})]}{[0.2632(10^{-3})](0.104)}$

$\qquad = 1.156$ MPa $< \tau_{allow} = 1.5$ MPa (OK).

F11–6

Within the overhang,

$V_{max} = 150$ kN

At B,

$M_{max} = 150$ kN $\cdot$ m

$S_{reqd} = \dfrac{M_{max}}{\sigma_{allow}} = \dfrac{150(10^3)}{150(10^6)} = 0.001$ m$^3 = 1000(10^3)$ mm^3

Select W410 $\times$ 67 [$S_x = 1200(10^3)$ mm^3, $d = 410$ mm, and

$t_w = 8.76$ mm].   *Ans.*

$\tau_{max} = \dfrac{V}{t_w d} = \dfrac{150(10^3)}{0.00876(0.41)}$

$\qquad = 41.76$ MPa $< \tau_{allow} = 75$ MPa (OK)

Chapter 12

F12–1

Use left segment,

$M(x) = 30$ kN $\cdot$ m

$EI\dfrac{d^2v}{dx^2} = 30$

$EI\dfrac{dv}{dx} = 30x + C_1$

$EIv = 15x^2 + C_1 x + C_2$

At $x = 3$ m, $\dfrac{dv}{dx} = 0$.

$C_1 = -90$ kN $\cdot$ m^2

At $x = 3$ m, $v = 0$.

$C_2 = 135$ kN $\cdot$ m^3

$\dfrac{dv}{dx} = \dfrac{1}{EI}(30x - 90)$

$v = \dfrac{1}{EI}(15x^2 - 90x + 135)$

For end A, $x = 0$

$\theta_A = \dfrac{dv}{dx}\Big|_{x=0} = -\dfrac{90(10^3)}{200(10^9)[65.0(10^{-6})]} = -0.00692$ rad

Ans.

$v_A = v|_{x=0} = \dfrac{135(10^3)}{200(10^9)[65.0(10^{-6})]} = 0.01038$ m $= 10.4$ mm

Ans.

F12–2

Use left segment,

$M(x) = (-10x - 10)$ kN $\cdot$ m

$EI\dfrac{d^2v}{dx^2} = -10x - 10$

$EI\dfrac{dv}{dx} = -5x^2 - 10x + C_1$

$EIv = -\dfrac{5}{3}x^3 - 5x^2 + C_1 x + C_2$

At $x = 3$ m, $\dfrac{dv}{dx} = 0$.

$EI(0) = -5(3^2) - 10(3) + C_1$ $C_1 = 75$ kN $\cdot$ m^2

At $x = 3$ m, $v = 0$.

$EI(0) = -\dfrac{5}{3}(3^3) - 5(3^2) + 75(3) + C_2$ $C_2 = -135$ kN $\cdot$ m^3

$\dfrac{dv}{dx} = \dfrac{1}{EI}(-5x^2 - 10x + 75)$

$v = \dfrac{1}{EI}\left(-\dfrac{5}{3}x^3 - 5x^2 + 75x - 135\right)$

For end A, $x = 0$

$\theta_A = \dfrac{dv}{dx}\Big|_{x=0} = \dfrac{1}{EI}[-5(0) - 10(0) + 75]$

$\qquad = \dfrac{75(10^3)}{200(10^9)[65.0(10^{-6})]} = 0.00577$ rad   *Ans.*

$v_A = v|_{x=0} = \dfrac{1}{EI}\left[-\dfrac{5}{3}(0^3) - 5(0^2) + 75(0) - 135\right]$

$\qquad = -\dfrac{135(10^3)}{200(10^9)[65.0(10^{-6})]} = -0.01038$ m $= -10.4$ mm *Ans.*

F12–3

Use left segment,

$M(x) = \left(-\dfrac{3}{2}x^2 - 10x\right)$ kN $\cdot$ m

$$EI\frac{d^2v}{dx^2} = -\frac{3}{2}x^2 - 10x$$

$$EI\frac{dv}{dx} = -\frac{1}{2}x^3 - 5x^2 + C_1$$

At $x = 3$ m, $\dfrac{dv}{dx} = 0$.

$$EI(0) = -\frac{1}{2}(3^3) - 5(3^2) + C_1 \qquad C_1 = 58.5 \text{ kN} \cdot \text{m}^2$$

$$\frac{dv}{dx} = \frac{1}{EI}\left(-\frac{1}{2}x^3 - 5x^2 + 58.5\right)$$

For end A, $x = 0$

$$\theta_A = \frac{dv}{dx}\,\Big|_{x=0} = \frac{58.5(10^3)}{200(10^9)[65.0(10^{-6})]} = 0.0045 \text{ rad} \qquad Ans.$$

F12–4

$A_y = 600$ lb

Use left segment,

$$M(x) = (600x - 50x^2) \text{ lb} \cdot \text{ft}$$

$$EI\frac{d^2v}{dx^2} = 600x - 50x^2$$

$$EI\frac{dv}{dx} = 300x^2 - 16.667x^3 + C_1$$

$$EIv = 100x^3 - 4.1667x^4 + C_1x + C_2$$

At $x = 0$, $v = 0$.

$$EI(0) = 100(0^3) - 4.1667(0^4) + C_1(0) + C_2 \qquad C_2 = 0$$

At $x = 12$ ft, $v = 0$.

$$EI(0) = 100(12^3) - 4.1667(12^4) + C_1(12)$$

$$C_1 = -7200 \text{ lb} \cdot \text{ft}^2$$

$$\frac{dv}{dx} = \frac{1}{EI}(300x^2 - 16.667x^3 - 7200)$$

$$v = \frac{1}{EI}(100x^3 - 4.1667x^4 - 7200x)$$

v_{max} occurs where $\dfrac{dv}{dx} = 0$.

$$300x^2 - 16.667x^3 - 7200 = 0$$

$$x = 6 \text{ ft} \qquad\qquad\qquad Ans.$$

$$v = \frac{1}{EI}[100(6^3) - 4.1667(6^4) - 7200(6)]$$

$$= \frac{-27\,000(12 \text{ in.}/\text{ft})^3}{1.5(10^6)\left[\dfrac{1}{12}(3)(6^3)\right]}$$

$$= -0.576 \text{ in.} \qquad\qquad\qquad Ans.$$

F12–5

$A_y = -5$ kN

Use left segment,

$$M(x) = (40 - 5x) \text{ kN} \cdot \text{m}$$

$$EI\frac{d^2v}{dx^2} = 40 - 5x$$

$$EI\frac{dv}{dx} = 40x - 2.5x^2 + C_1$$

$$EIv = 20x^2 - 0.8333x^3 + C_1x + C_2$$

At $x = 0$, $v = 0$.

$$EI(0) = 20(0^2) - 0.8333(0^3) + C_1(0) + C_2 \qquad C_2 = 0$$

At $x = 6$ m, $v = 0$.

$$EI(0) = 20(6^2) - 0.8333(6^3) + C_1(6) + 0$$

$$C_1 = -90 \text{ kN} \cdot \text{m}^2$$

$$\frac{dv}{dx} = \frac{1}{EI}(40x - 2.5x^2 - 90)$$

$$v = \frac{1}{EI}(20x^2 - 0.8333x^3 - 90x)$$

v_{max} occurs where $\dfrac{dv}{dx} = 0$.

$$40x - 2.5x^2 - 90 = 0$$

$$x = 2.7085 \text{ m}$$

$$v = \frac{1}{EI}[20(2.7085^2) - 0.83333(2.7085^3) - 90(2.7085)]$$

$$= -\frac{113.60(10^3)}{200(10^9)[39.9(10^{-6})]} = -0.01424 \text{ m} = -14.2 \text{ mm} \quad Ans.$$

F12–6

$A_y = 10$ kN

Use left segment,

$$M(x) = (10x + 10) \text{ kN} \cdot \text{m}$$

$$EI\frac{d^2v}{dx^2} = 10x + 10$$

$$EI\frac{dv}{dx} = 5x^2 + 10x + C_1$$

Due to symmetry, $\dfrac{dv}{dx} = 0$ at $x = 3$ m.

$$EI(0) = 5(3^2) + 10(3) + C_1 \quad C_1 = -75 \text{ kN} \cdot \text{m}^2$$

$$\frac{dv}{dx} = \frac{1}{EI}[5x^2 + 10x - 75]$$

At $x = 0$,

$$\frac{dv}{dx} = \frac{-75(10^3)}{200(10^9)(39.9(10^{-6}))} = -9.40(10^{-3}) \text{ rad} \qquad Ans.$$

F12–7

Since B is a fixed support, $\theta_B = 0$.

$$\theta_A = |\theta_{A/B}| = \frac{1}{2}\left(\frac{38}{EI} + \frac{20}{EI}\right)(3) = \frac{87 \text{ kN} \cdot \text{m}^2}{EI}$$

$$= \frac{87(10^3)}{200(10^9)[65(10^{-6})]} = 0.00669 \text{ rad} \,\, \circlearrowleft \qquad Ans.$$

$$v_A = |t_{A/B}| = (1.5)\left[\frac{20}{EI}(3)\right] + 2\left[\frac{1}{2}\left(\frac{18}{EI}\right)(3)\right]$$

$$= \frac{144(10^3)}{200(10^9)[65(10^{-6})]} = 0.01108 \text{ m} = 11.1 \text{ mm} \downarrow \qquad Ans.$$

F12–8

Since B is a fixed support, $\theta_B = 0$.

$$\theta_A = |\theta_{A/B}| = \frac{1}{2}\left(\frac{50}{EI} + \frac{20}{EI}\right)(1) + \frac{1}{2}\left(\frac{20}{EI}\right)(1) = \frac{45 \text{ kN} \cdot \text{m}^2}{EI}$$

$$= \frac{45(10^3)}{200(10^9)[126(10^{-6})]} = 0.00179 \text{ rad} \,\, \circlearrowleft \qquad Ans.$$

$$v_A = |t_{A/B}| =$$

$$(1.6667)\left[\frac{1}{2}\left(\frac{30}{EI}\right)(1)\right] + 1.5\left[\frac{20}{EI}(1)\right] + 0.6667\left[\frac{1}{2}\left(\frac{20}{EI}\right)(1)\right]$$

$$= \frac{61.667 \text{ kN} \cdot \text{m}^3}{EI} = \frac{61.667(10^3)}{200(10^9)[126(10^{-6})]}$$

$$= 0.002447 \text{ m} = 2.48 \text{ mm} \downarrow \qquad Ans.$$

F12–9

Since B is a fixed support, $\theta_B = 0$.

$$\theta_A = |\theta_{A/B}| = \frac{1}{2}\left[\frac{60}{EI}(1)\right] + \frac{30}{EI}(2) = \frac{90 \text{ kN} \cdot \text{m}^2}{EI}$$

$$= \frac{90(10^3)}{200(10^9)[121(10^{-6})]} = 0.00372 \text{ rad} \,\, \circlearrowleft \qquad Ans.$$

$$v_A = |t_{A/B}| = 1.6667\left[\frac{1}{2}\left(\frac{60}{EI}\right)(1)\right] + (1)\left[\frac{30}{EI}(2)\right]$$

$$= \frac{110 \text{ kN} \cdot \text{m}^3}{EI}$$

$$= \frac{110(10^3)}{200(10^9)[121(10^{-6})]} = 0.004545 \text{ m} = 4.55 \text{ mm} \downarrow \qquad Ans.$$

F12–10

Since B is a fixed support, $\theta_B = 0$.

$$\theta_A = |\theta_{A/B}| = \frac{1}{2}\left(\frac{18}{EI}\right)(6) + \frac{1}{3}\left(\frac{9}{EI}\right)(3) = \frac{63 \text{ kip} \cdot \text{ft}^2}{EI}$$

$$= \frac{63(12^2)}{29(10^3)(245)} = 0.00128 \text{ rad} \,\, \circlearrowleft \qquad Ans.$$

$$v_A = |t_{A/B}| = 4\left[\frac{1}{2}\left(\frac{18}{EI}\right)(6)\right] + (3 + 2.25)\left[\frac{1}{3}\left(\frac{9}{EI}\right)(3)\right]$$

$$= \frac{263.25 \text{ kip} \cdot \text{ft}^3}{EI} = \frac{263.25(12^3)}{29(10^3)(245)} = 0.0640 \text{ in.} \downarrow \qquad Ans.$$

F12–11

Due to symmetry, the slope at the midspan of the beam (point C) is zero, i.e., $\theta_C = 0$.

$$v_{max} = v_C = |t_{A/C}| = (2)\left[\frac{1}{2}\left(\frac{30}{EI}\right)(3)\right] + 1.5\left[\frac{10}{EI}(3)\right]$$

$$= \frac{135 \text{ kN} \cdot \text{m}^3}{EI}$$

$$= \frac{135(10^3)}{200(10^9)[42.8(10^{-6})]} = 0.0158 \text{ m} = 15.8 \text{ mm} \downarrow \qquad Ans.$$

F12–12

$$t_{A/B} = 2\left[\frac{1}{2}\left(\frac{30}{EI}\right)(6)\right] + 3\left[\frac{10}{EI}(6)\right] = \frac{360}{EI}$$

$$\theta_B = \frac{|t_{A/B}|}{L} = \frac{\dfrac{360}{EI}}{6} = \frac{60}{EI}$$

The maximum deflection occurs at point C where the slope of the elastic curve is zero.

$$\theta_B = \theta_{B/C}$$

$$\frac{60}{EI} = \left(\frac{10}{EI}\right)x + \frac{1}{2}\left(\frac{5x}{EI}\right)x$$

$$2.5x^2 + 10x - 60 = 0$$

$$x = 3.2915 \text{ m}$$

$$v_{max} = |t_{B/C}| =$$

$$\frac{2}{3}(3.2915)\left\{\frac{1}{2}\left[\frac{5(3.2915)}{EI}\right](3.2915)\right\} + \frac{1}{2}(3.2915)\left[\frac{10}{EI}(3.2915)\right]$$

$$= \frac{113.60 \text{ kN} \cdot \text{m}^3}{EI}$$

$$= \frac{113.60(10^3)}{200(10^9)[39.9(10^{-6})]} = 0.01424 \text{ m} = 14.2 \text{ mm} \downarrow \qquad Ans.$$

F12–13

Remove B_y,

$$(v_B)_1 = \frac{Px^2}{6EI}(3L - x) = \frac{40(4^2)}{6EI}[3(6) - 4] = \frac{1493.33}{EI} \downarrow$$

Apply B_y,

$$(v_B)_2 = \frac{PL^3}{3EI} = \frac{B_y(4^3)}{3EI} = \frac{21.33B_y}{EI} \uparrow$$

$$(+\uparrow) \; v_B = 0 = (v_B)_1 + (v_B)_2$$

$$0 = -\frac{1493.33}{EI} + \frac{21.33B_y}{EI}$$

$B_y = 70$ kN   *Ans.*

For the beam,

$\xrightarrow{+} \Sigma F_x = 0;$ $A_x = 0$ *Ans.*

$+\uparrow \Sigma F_y = 0;$ $70 - 40 - A_y = 0$ $A_y = 30$ kN   *Ans.*

$\zeta + \Sigma M_A = 0;$ $70(4) - 40(6) - M_A = 0$

$M_A = 40$ kN $\cdot$ m   *Ans.*

F12–14

Remove B_y,

To use the deflection tables, consider loading as a superposition of uniform distributed load minus a triangular load.

$$(v_B)_1 = \frac{w_0 L^4}{8EI}\downarrow \qquad (v_B)_2 = \frac{w_0 L^4}{30EI}\uparrow$$

Apply B_y,

$$(+\uparrow) \quad (v_B)_3 = \frac{B_y L^3}{3EI}\uparrow \quad v_B = 0 = (v_B)_1 + (v_B)_2 + (v_B)_3$$

$$0 = -\frac{w_0 L^4}{8EI} + \frac{w_0 L^4}{30EI} + \frac{B_y L^3}{3EI}$$

$$B_y = \frac{11w_0 L}{40} \qquad\qquad\qquad Ans.$$

For the beam,

$\xrightarrow{+} \Sigma F_x = 0;$ $A_x = 0$ *Ans.*

$+\uparrow \Sigma F_y = 0;$ $A_y + \dfrac{11w_0 L}{40} - \dfrac{1}{2}w_0 L = 0$

$$A_y = \frac{9w_0 L}{40} \qquad\qquad\qquad Ans.$$

$\zeta + \Sigma M_A = 0;$ $M_A + \dfrac{11w_0 L}{40}(L) - \dfrac{1}{2}w_0 L\left(\dfrac{2}{3}L\right) = 0$

$$M_A = \frac{7w_0 L^2}{120} \qquad\qquad\qquad Ans.$$

F12–15

Remove B_y,

$$(v_B)_1 = \frac{wL^4}{8EI} = \frac{[10(10^3)](6^4)}{8[200(10^9)][65.0(10^{-6})]} = 0.12461 \text{ m} \downarrow$$

Apply B_y,

$$(v_B)_2 = \frac{B_y L^3}{3EI} = \frac{B_y (6^3)}{3[200(10^9)][65.0(10^{-6})]} = 5.5385(10^{-6})B_y \uparrow$$

$(+\downarrow) \quad v_B = (v_B)_1 + (v_B)_2$

$0.002 = 0.12461 - 5.5385(10^{-6})B_y$

$B_y = 22.314(10^3)$ N $= 22.1$ kN   *Ans.*

For the beam,

$\xrightarrow{+} \Sigma F_x = 0;$ $A_x = 0$ *Ans.*

$+\uparrow \Sigma F_y = 0;$ $A_y + 22.14 - 10(6) = 0$ $A_y = 37.9$ kN

                                                        Ans.

$\zeta + \Sigma M_A = 0;$ $M_A + 22.14(6) - 10(6)(3) = 0$

$M_A = 47.2$ kN $\cdot$ m   *Ans.*

F12–16

Remove B_y,

$$(v_B)_1 = \frac{M_O L}{6EI(2L)}[(2L)^2 - L^2] = \frac{M_O L^2}{4EI}\downarrow$$

Apply B_y,

$$(v_B)_2 = \frac{B_y (2L)^3}{48EI} = \frac{B_y L^3}{6EI}\uparrow$$

$(+\uparrow) \quad v_B = 0 = (v_B)_1 + (v_B)_2$

$$0 = -\frac{M_O L^2}{4EI} + \frac{B_y L^3}{6EI}$$

$$B_y = \frac{3M_O}{2L} \qquad\qquad\qquad Ans.$$

F12–17

Remove B_y,

$$(v_B)_1 = \frac{Pbx}{6EIL}(L^2 - b^2 - x^2) = \frac{50(4)(6)}{6EI(12)}(12^2 - 4^2 - 6^2)$$

$$= \frac{1533.3 \text{ kN} \cdot \text{m}^3}{EI}\downarrow$$

Apply B_y,

$$(v_B)_2 = \frac{B_y L^3}{48EI} = \frac{B_y (12^3)}{48EI} = \frac{36B_y}{EI}\uparrow$$

$(+\uparrow) \quad v_B = 0 = (v_B)_1 + (v_B)_2$

$$0 = -\frac{1533.3 \text{ kN} \cdot \text{m}^3}{EI} + \frac{36B_y}{EI}$$

$B_y = 42.6$ kN   *Ans.*

F12–18

Remove B_y,

$$(v_B)_1 = \frac{5wL^4}{384EI} = \frac{5[10(10^3)](12^4)}{384[200(10^9)][65.0(10^{-6})]} = 0.20769 \downarrow$$

Apply B_y,

$$(v_B)_2 = \frac{B_y L^3}{48EI} = \frac{B_y (12^3)}{48[200(10^9)][65.0(10^{-6})]}$$

$$= 2.7692(10^{-6})B_y \uparrow$$

$(+\uparrow) \quad v_B = (v_B)_1 + (v_B)_2$

$-0.005 = -0.20769 + 2.7692(10^{-6})B_y$

$B_y = 73.19(10^3)$ N $= 73.2$ kN   *Ans.*

Chapter 13

F13–1

$$P = \frac{\pi^2 EI}{(KL)^2} = \frac{\pi^2 \left[29(10^3)\right]\left[\frac{\pi}{4}(0.5)^4\right]}{[0.5(50)]^2} = 22.5 \text{ kip} \qquad Ans.$$

$$\sigma = \frac{P}{A} = \frac{22.5}{\pi(0.5)^2} = 28.6 \text{ ksi} < \sigma_Y \quad \text{OK}$$

F13–2

$$P = \frac{\pi^2 EI}{(KL)^2} = \frac{\pi^2 \left[1.6(10^3)\right]\left[\frac{1}{12}(4)(2)^3\right]}{[1(12)(12)]^2}$$

$$= 2.03 \text{ kip} \qquad Ans.$$

F13–3

For buckling about the x axis, $K_x = 1$ and $L_x = 12$ m.

$$P_{cr} = \frac{\pi^2 EI_x}{(K_x L_x)^2} = \frac{\pi^2 [200(10^9)][87.3(10^{-6})]}{[1(12)]^2} = 1.197(10^6) \text{ N}$$

For buckling about the y axis, $K_y = 1$ and $L_y = 6$ m.

$$P_{cr} = \frac{\pi^2 EI_y}{(K_y L_y)^2} = \frac{\pi^2 [200(10^9)][18.8(10^{-6})]}{[1(6)]^2}$$

$$= 1.031(10^6) \text{ N (controls)}$$

$$P_{allow} = \frac{P_{cr}}{\text{F.S.}} = \frac{1.031(10^6)}{2} = 515 \text{ kN} \qquad Ans.$$

$$\sigma_{cr} = \frac{P_{cr}}{A} = \frac{1.031(10^6)}{7.4(10^{-3})} = 139.30 \text{ MPa} < \sigma_Y = 345 \text{ MPa (OK)}$$

F13–4

$$A = \pi[(0.025)^2 - (0.015)^2] = 1.257(10^{-3}) \text{ m}^2$$

$$I = \frac{1}{4}\pi\left[(0.025)^4 - (0.015)^4\right] = 267.04(10^{-9}) \text{ m}^4$$

$$P = \frac{\pi^2 EI}{(KL)^2} = \frac{\pi^2 \left[200(10^9)\right]\left[267.04(10^{-9})\right]}{[0.5(5)]^2} = 84.3 \text{ kN} \qquad Ans.$$

$$\sigma = \frac{P}{A} = \frac{84.3(10^3)}{1.257(10^{-3})} = 67.1 \text{ MPa} < 250 \text{ MPa} \quad \text{(OK)}$$

F13–5

Joint A,

$$+\uparrow \Sigma F_y = 0; \qquad F_{AB}\left(\frac{3}{5}\right) - P = 0 \qquad F_{AB} = 1.6667P \text{ (T)}$$

$$\overset{+}{\rightarrow} \Sigma F_x = 0; \qquad 1.6667P\left(\frac{4}{5}\right) - F_{AC} = 0$$

$$F_{AC} = 1.3333P \text{ (C)}$$

$$A = \frac{\pi}{4}(2^2) = \pi \text{ in}^2 \qquad\qquad I = \frac{\pi}{4}(1^4) = \frac{\pi}{4} \text{ in}^4$$

$$P_{cr} = F(\text{F.S.}) = 1.3333P(2) = 2.6667P$$

$$P_{cr} = \frac{\pi^2 EI}{(KL)^2}$$

$$2.6667P = \frac{\pi^2 [29(10^3)]\left(\frac{\pi}{4}\right)}{[1(4)(12)]^2}$$

$$P = 36.59 \text{ kip} = 36.6 \text{ kip} \qquad Ans.$$

$$\sigma_{cr} = \frac{P_{cr}}{A} = \frac{2.6667(36.59)}{\pi} = 31.06 \text{ ksi} < \sigma_Y = 50 \text{ ksi}$$
$$\text{(OK)}$$

F13–6

Beam AB,

$$\zeta + \Sigma M_A = 0; \qquad w(6)(3) - F_{BC}(6) = 0 \qquad F_{BC} = 3w$$

Strut BC,

$$A_{BC} = \frac{\pi}{4}(0.05^2) = 0.625(10^{-3})\pi \text{ m}^2 \qquad I = \frac{\pi}{4}(0.025^4)$$

$$= 97.65625(10^{-9})\pi \text{ m}^4$$

$$P_{cr} = F_{BC}(\text{F.S.}) = 3w(2) = 6w$$

$$P_{cr} = \frac{\pi^2 EI}{(KL)^2}$$

$$6w = \frac{\pi^2 [200(10^9)][97.65625(10^{-9})\pi]}{[1(3)]^2}$$

$$w = 11.215(10^3) \text{ N/m} = 11.2 \text{ kN/m} \qquad Ans.$$

$$\sigma_{cr} = \frac{P_{cr}}{A} = \frac{6[11.215(10^3)]}{0.625(10^{-3})\pi} = 34.27 \text{ MPa} < \sigma_Y = 345 \text{ MPa}$$
$$\text{(OK)}$$

Selected Answers

Chapter 1

1–1. $N_E = 0, V_E = -200 \text{ lb}, M_E = -2.40 \text{ kip} \cdot \text{ft}$

1–2. (a) $N_a = 500 \text{ lb}, V_a = 0,$
(b) $N_b = 433 \text{ lb}, V_b = 250 \text{ lb}$

1–3. $V_{b-b} = 2.475 \text{ kip}, N_{b-b} = 0.390 \text{ kip},$
$M_{b-b} = 3.60 \text{ kip} \cdot \text{ft}$

1–5. $N_B = 0, V_B = 288 \text{ lb},$
$M_B = -1.15 \text{ kip} \cdot \text{ft}$

1–6. $N_D = 0.703 \text{ kN}, V_D = 0.3125 \text{ kN},$
$M_D = 0.3125 \text{ kN} \cdot \text{m}$

1–7. $N_F = 1.17 \text{ kN}, V_F = 0, M_F = 0, N_E = 0.703 \text{ kN},$
$V_E = -0.3125 \text{ kN}, M_E = 0.3125 \text{ kN} \cdot \text{m}$

1–9. $N_D = 0, V_D = -3.25 \text{ kN}, M_D = 5.625 \text{ kN} \cdot \text{m}$

1–10. $N_A = 0, V_A = 450 \text{ lb}, M_A = -1.125 \text{ kip} \cdot \text{ft},$
$N_B = 0, V_B = 850 \text{ lb}, M_B = -6.325 \text{ kip} \cdot \text{ft},$
$V_C = 0, N_C = -1.20 \text{ kip}, M_C = -8.125 \text{ kip} \cdot \text{ft}$

1–11. $N_D = -527 \text{ lb}, V_D = -373 \text{ lb}, M_D = -373 \text{ lb} \cdot \text{ft},$
$N_E = 75.0 \text{ lb}, V_E = 355 \text{ lb}, M_E = -727 \text{ lb} \cdot \text{ft}$

1–13. $N_{a-a} = -100 \text{ N}, V_{a-a} = 0, M_{a-a} = -15 \text{ N} \cdot \text{m}$

1–14. $N_{b-b} = -86.6 \text{ N}, V_{b-b} = 50 \text{ N}, M_{b-b} = -15 \text{ N} \cdot \text{m}$

1–15. $N_C = 0, V_C = -1.40 \text{ kip}, M_C = 8.80 \text{ kip} \cdot \text{ft}$

1–17. $(N_D)_x = 0, (V_D)_y = 154 \text{ N}, (V_D)_z = -171 \text{ N},$
$(T_D)_x = 0, (M_D)_y = -94.3 \text{ N} \cdot \text{m},$
$(M_D)_z = -149 \text{ N} \cdot \text{m}$

1–18. $(N_C)_x = 0, (V_C)_y = -246 \text{ N}, (V_C)_z = -171 \text{ N},$
$(T_C)_x = 0, (M_C)_y = -154 \text{ N} \cdot \text{m},$
$(M_C)_z = -123 \text{ N} \cdot \text{m}$

1–19. $(V_A)_x = 0, (N_A)_y = -25 \text{ lb}, (V_A)_z = 43.3 \text{ lb},$
$(M_A)_x = 303 \text{ lb} \cdot \text{in.}, (T_A)_y = -130 \text{ lb} \cdot \text{in.},$
$(M_A)_z = -75 \text{ lb} \cdot \text{in.}$

1–21. $N_E = -2.94 \text{ kN}, V_E = -2.94 \text{ kN},$
$M_E = -2.94 \text{ kN} \cdot \text{m}$

1–22. $F_{BC} = 1.39 \text{ kN}, F_A = 1.49 \text{ kN}, N_D = 120 \text{ N},$
$V_D = 0, M_D = 36.0 \text{ N} \cdot \text{m}$

1–23. $N_E = 0, V_E = 120 \text{ N}, M_E = 48.0 \text{ N} \cdot \text{m},$
Short link: $V = 0, N = 1.39 \text{ kN}, M = 0$

1–25. $V_B = 496 \text{ lb}, N_B = 59.8 \text{ lb}, M_B = 480 \text{ lb} \cdot \text{ft},$
$N_C = 495 \text{ lb}, V_C = 70.7 \text{ lb}, M_C = 1.59 \text{ kip} \cdot \text{ft}$

1–26. $N_F = 0, V_F = 80 \text{ lb}, M_F = 160 \text{ lb} \cdot \text{ft},$
$N_G = 16.7 \text{ lb}, V_G = 72.0 \text{ lb}, M_G = 108 \text{ lb} \cdot \text{ft}$

1–27. $(V_B)_x = -300 \text{ N}, (N_B)_y = -800 \text{ N}, (V_B)_z = 771 \text{ N},$
$(M_B)_x = 2.11 \text{ kN} \cdot \text{m}, (T_B)_y = -600 \text{ N} \cdot \text{m},$
$(M_B)_z = 600 \text{ N} \cdot \text{m}$

1–29. $V_B = 0.785 \, wr, N_B = 0, T_B = 0.0783 \, wr^2,$
$M_B = -0.293 \, wr^2$

1–31. $\tau_{\text{avg}} = 119 \text{ MPa}$

1–33. $\sigma_{\text{avg}} = \dfrac{P}{A} \sin^2 \theta, \ \tau_{\text{avg}} = \dfrac{P}{2A} \sin 2\theta$

1–34. $F = 22.5 \text{ kip}, d = 0.833 \text{ in.}$

1–35. $P_{\text{allow}} = 9.12 \text{ kip}$

1–37. $P = 40 \text{ MN}, d = 2.40 \text{ m}$

1–38. $\sigma = 2.92 \text{ psi}, \tau = 8.03 \text{ psi}$

1–39. $\sigma_{BC} = \left\{ \dfrac{1.528 \cos \theta}{\sin (45° + \theta/2)} \right\} \text{ksi}$

1–41. $P = 37.7 \text{ kN}$

1–42. $(\tau_{\text{avg}})_A = 50.9 \text{ MPa}$

1–43. $\tau_B = \tau_C = 81.9 \text{ MPa}, \tau_A = 88.1 \text{ MPa}$

1–45. $\sigma = (238 - 22.6z) \text{ kPa}$

1–46. $\sigma_{AB} = 333 \text{ MPa}, \sigma_{CD} = 250 \text{ MPa}$

1–47. $d = 1.20 \text{ m}$

1–49. $\tau_{\text{avg}} = 11.1 \text{ ksi}$

1–50. $\sigma_{a-a} = 90.0 \text{ kPa}, \tau_{a-a} = 52.0 \text{ kPa}$

1–51. $\sigma = 4.69 \text{ MPa}, \tau = 8.12 \text{ MPa}$

1–53. $\sigma = \{43.75 - 22.5x\} \text{ MPa}$

1–54. $\sigma = 66.7 \text{ psi}, \tau = 115 \text{ psi}$

1–55. $\sigma_{AB} = 127 \text{ MPa}, \sigma_{AC} = 129 \text{ MPa}$

1–57. $w = w_1 e^{(w_1^2 \gamma)z / (2P)}$

1–58. $\theta = 30.7°, \sigma = 152 \text{ MPa}$

1–59. $\sigma = \dfrac{m\omega^2}{8A} (L^2 - 4x^2)$

1–61. $\sigma = (32.5 - 20.0x) \text{ MPa}$

1–62. $\sigma = \dfrac{w_0}{2aA} (2a^2 - x^2)$

1–63. $\sigma = \dfrac{w_0}{2aA} (2a - x)^2$

1–65. $P = 9.375 \text{ kip}$

1–66. $P = 62.5 \text{ kN}$

1–67. $\omega = 6.85 \text{ rad/s}$

1–69. Use $h = 2\dfrac{3}{4} \text{ in.}$

1–70. $d = 5.71 \text{ mm}$

1–71. $P = 0.491 \text{ kip}$

1–73. $F = 3.09 \text{ kip}$

1–74. $F_H = 20.0 \text{ kN}, F_{BF} = F_{AG} = 15.0 \text{ kN},$
$d_{EF} = d_{CG} = 11.3 \text{ mm}$

1–75. For A': Use a 3 in. $\times$ 3 in. plate,
For B': Use a $4\frac{1}{2}$ in. $\times$ $4\frac{1}{2}$ in. plate

1–77. Use $d_A = \dfrac{5}{8} \text{ in.}, d_B = 1\dfrac{1}{16} \text{ in.}$

1–78. $d_{AB} = 4.81 \text{ mm}, d_{AC} = 5.22 \text{ mm}$

1–79. $P = 5.83 \text{ kN}$

1–81. $d_A = d_B = 5.20 \text{ mm}$

1–82. $(\text{F.S.})_{\text{st}} = 2.14, (\text{F.S.})_{\text{con}} = 3.53$

1–83. $W = 680 \text{ lb}$

1–85. $d_2 = 35.7 \text{ mm}, d_3 = 27.6 \text{ mm}, d_1 = 22.6 \text{ mm}$

1–86. $d_{AB} = 15.5 \text{ mm}, d_{AC} = 13.0 \text{ mm}$

1–87. $P = 7.54 \text{ kN}$

1–89. $P = 9.09 \text{ kip}$

1–90. $d_B = 6.11 \text{ mm}, d_w = 15.4 \text{ mm}$

1–91. $P = 55.0$ kN

1–93. $d_{AB} = 6.90$ mm, $d_{CD} = 6.20$ mm

1–94. $h = 1.74$ in.

1–95. $d = 0.620$ in., $P_{max} = 7.25$ kip

1–96. $t = 0.800$ in., $P_{max} = 24.0$ kip, $w = 2.50$ in.

R1–1. $N_D = -2.16$ kip, $V_D = 0$, $M_D = 2.16$ kip $\cdot$ ft,
$V_E = 0.540$ kip, $N_E = 4.32$ kip, $M_E = 2.16$ kip $\cdot$ ft

R1–2. $\sigma_s = 208$ MPa, $(\tau_{avg})_a = 4.72$ MPa,
$(\tau_{avg})_b = 45.5$ MPa

R1–3. Use $t = \dfrac{1}{4}$ in., $d_A = 1\dfrac{1}{8}$ in., $d_B = \dfrac{13}{16}$ in.

R1–5. $\tau_{avg} = 25.5$ MPa, $\sigma_b = 4.72$ MPa

R1–6. $\sigma_{a-a} = 200$ kPa, $\tau_{a-a} = 115$ kPa

R1–7. $\sigma_{40} = 3.98$ MPa, $\sigma_{30} = 7.07$ MPa,
$\tau_{avg} = 5.09$ MPa

Chapter 2

2–1. $\epsilon = 0.167$ in./in.

2–2. $\epsilon = 0.0472$ in./in.

2–3. $\epsilon_{CE} = 0.00250$ mm/mm, $\epsilon_{BD} = 0.00107$ mm/mm

2–5. $\gamma_{xy} = -0.0200$ rad

2–6. $(\gamma_A)_{xy} = -0.0262$ rad, $(\gamma_B)_{xy} = -0.205$ rad
$(\gamma_C)_{xy} = -0.205$ rad, $(\gamma_D)_{xy} = -0.0262$ rad

2–7. $(\epsilon_{avg})_{AC} = 6.04(10^{-3})$ mm/mm

2–9. $\epsilon_{AB} = \dfrac{0.5\Delta L}{L}$

2–10. $(\gamma_A)_{xy} = 27.8(10^{-3})$ rad, $(\gamma_B)_{xy} = 35.1(10^{-3})$ rad

2–11. $(\gamma_{xy})_C = 25.5(10^{-3})$ rad, $(\gamma_{xy})_D = 18.1(10^{-3})$ rad

2–13. $\epsilon_{AD} = 0.0566$ mm/mm, $\epsilon_{CF} = -0.0255$ mm/mm

2–14. $\epsilon_{AB} = 0.00418$ mm/mm

2–15. $\Delta_B = 6.68$ mm

2–17. $\epsilon = 2kx$

2–18. $(\gamma_B)_{xy} = 11.6(10^{-3})$ rad, $(\gamma_A)_{xy} = 11.6(10^{-3})$ rad

2–19. $(\gamma_C)_{xy} = 11.6(10^{-3})$ rad, $(\gamma_D)_{xy} = 11.6(10^{-3})$ rad

2–21. $\epsilon_x = -0.03$ in./in., $\epsilon_y = 0.02$ in./in.

2–22. $\gamma_{xy} = 0.00880$ rad

2–23. $\epsilon_x = 0.00443$ mm/mm

2–25. $\gamma_A = 0$, $\gamma_B = 0.199$ rad

2–26. $(\gamma_{x'y'})_A = -0.0672$ rad, $(\gamma_{x''y''})_B = 0.0672$ rad

2–27. $\epsilon_{AB} = -7.77(10^{-3})$ in./in., $\epsilon_{BD} = 0.025$ in./in.
$\epsilon_{AC} = -0.0417$ in./in.

2–29. $(\epsilon_{avg})_{AC} = 0.0168$ mm/mm, $(\gamma_A)_{xy} = 0.0116$ rad

2–30. $(\epsilon_{avg})_{BD} = 1.60(10^{-3})$ mm/mm,
$(\gamma_B)_{xy} = 0.0148$ rad

2–31. $(\Delta x)_C = \dfrac{kL}{\pi}$, $\epsilon_{avg} = \dfrac{2k}{\pi}$

2–33. $\epsilon_{AB} = \dfrac{v_B \sin\theta}{L} - \dfrac{u_A \cos\theta}{L}$

Chapter 3

3–1. $(\sigma_u)_{approx} = 110$ ksi, $(\sigma_f)_{approx} = 93.1$ ksi,
$(\sigma_Y)_{approx} = 55$ ksi, $E_{approx} = 32.0(10^3)$ ksi

3–2. $E = 55.3(10^3)$ ksi, $u_r = 9.96 \dfrac{\text{in} \cdot \text{lb}}{\text{in}^3}$

3–3. $(u_t)_{approx} = 85.0 \dfrac{\text{in} \cdot \text{lb}}{\text{in}^3}$

3–5. Elastic recovery $= 0.00350$ in.,
Permanent elongation $= 0.1565$ in.

3–6. $(u_r)_{approx} = 20.0 \dfrac{\text{in} \cdot \text{lb}}{\text{in}^3}$, $(u_t)_{approx} = 18.0 \dfrac{\text{in} \cdot \text{kip}}{\text{in}^3}$

3–7. $\delta_{AB} = 0.152$ in.

3–9. $\delta = 0.979$ in.

3–10. $E_{approx} = 10.0(10^3)$ ksi, $P_Y = 9.82$ kip,
$P_u = 13.4$ kip

3–11. Elastic recovery $= 0.012$ in.,
Permanent elongation $= 0.0680$ in.

3–13. $E = 28.6(10^3)$ ksi

3–14. $\delta_{BD} = 0.0632$ in.

3–15. $P = 570$ lb

3–17. $\delta_{AB} = 0.0913$ in.

3–18. $w = 228$ lb/ft

3–19. $\sigma_{YS} = 2.03$ MPa

3–21. $P = 15.0$ kip

3–22. $A_{BC} = 0.8$ in^2, $A_{BA} = 0.2$ in^2

3–23. $\delta_{AB} = 0.304$ in.

3–25. $\delta = 0.126$ mm, $\Delta d = -0.00377$ mm

3–26. $p = 741$ kPa, $\delta = 7.41$ mm

3–27. $\nu = 0.350$

3–29. $\gamma = 0.250$ rad

3–30. $\gamma = 3.06 (10^{-3})$ rad

3–31. $\gamma_P = 0.0189$ rad

3–33. $E = 32.5(10^3)$ ksi, $P = 2.45$ kip

3–34. $\delta = \dfrac{Pa}{2bhG}$

R3–1. $G_{al} = 4.31(10^3)$ ksi

R3–2. $d' = 0.4989$ in.

R3–3. $x = 1.53$ m, $d'_A = 30.008$ mm

R3–5. $P = 6.48$ kip

R3–6. $\epsilon = 0.000999$ in./in., $\epsilon_{unscr} = 0$

R3–7. $L = 10.17$ in.

R3–9. $\epsilon_b = 0.00227$ mm/mm, $\epsilon_s = 0.000884$ mm/mm

R3–10. $G = 5$ MPa

Chapter 4

4–1. $\delta_B = 2.93$ mm $\downarrow$, $\delta_A = 3.55$ mm $\downarrow$

4–2. $\delta_{A/D} = 0.111$ in. away from end D

4–3. $\sigma_{AB} = 22.2$ ksi (T), $\sigma_{BC} = 41.7$ ksi (C),
$\sigma_{CD} = 25.0$ ksi (C),
$\delta_{A/D} = 0.00157$ in. towards end D

4–5. $\delta_{A/E} = 0.697$ mm

4–6. $\sigma_A = 13.6$ ksi, $\sigma_B = 10.3$ ksi, $\sigma_C = 3.2$ ksi, $\delta_D = 2.99$ ft

4–7. $\delta_C = 0.0975$ mm $\rightarrow$

4–9. $\delta_F = 0.453$ mm

4–10. $P = 4.97$ kN

4–11. $\delta_l = 0.0260$ in.

4–13. $\delta_D = 17.3$ mm

4–14. $F = 8.00$ kN, $\delta_{A/B} = -0.311$ mm

4–15. $F = 4.00$ kN, $\delta_{A/B} = -0.259$ mm

4–17. $P = \dfrac{F_{max} L}{2}$, $\delta = \dfrac{F_{max} L^2}{3AE}$

4–18. $\delta_B = 0.262$ in.

4–19. $P = 57.3$ kip

4–21. $d_{AB} = 0.841$ in., $d_{CD} = 0.486$ in.

4–22. $x = 4.24$ ft, $w = 1.02$ kip/ft

4–23. $\delta_{A/D} = 0.129$ mm, $h' = 49.9988$ mm, $w' = 59.9986$ mm

4–25. $\delta_C = 0.00843$ in., $\delta_E = 0.00169$ in., $\delta_B = 0.0333$ in.

4–26. $P = 6.80$ kip

4–27. $P = 11.8$ kip

4–29. $\delta = 2.37$ mm

4–30. $\delta = \dfrac{2.63\, P}{\pi r E}$

4–31. $\sigma_{con} = 2.29$ ksi, $\sigma_{st} = 15.8$ ksi

4–33. $\sigma_{al} = 27.5$ MPa, $\sigma_{st} = 79.9$ MPa

4–34. $\sigma_{con} = 1.64$ ksi, $\sigma_{st} = 11.3$ ksi

4–35. $P = 114$ kip

4–37. $F_C = \left[\dfrac{9(8ka + \pi d^2 E)}{136ka + 18\pi d^2 E} \right] P$, $F_A = \left(\dfrac{64ka + 9\pi d^2 E}{136ka + 18\pi d^2 E} \right) P$

4–38. $T_{AC} = 0.806$ kip, $T_{AB} = 1.19$ kip

4–39. $A_{AB} = 0.0144$ in^2

4–41. $P = 126$ kN

4–42. $\sigma_{st} = 102$ MPa, $\sigma_{br} = 50.9$ MPa

4–43. $\sigma_{AB} = \sigma_{CD} = 26.5$ MPa, $\sigma_{EF} = 33.8$ MPa

4–45. $P_b = 14.4$ kN

4–46. $F_D = 20.4$ kN, $F_A = 180$ kN

4–47. $P = 198$ kN

4–49. $\delta_B = 0.0733$ in.

4–50. $\theta = 0.0875°$

4–51. $\delta_B = 0.00257$ in.

4–53. $F_D = 71.4$ kN, $F_C = 329$ kN

4–54. $F_D = 219$ kN, $F_C = 181$ kN

4–55. $\sigma_{BE} = 96.3$ MPa, $\sigma_{AD} = 79.6$ MPa, $\sigma_{CF} = 113$ MPa

4–57. $d_{AC} = 1.79$ mm

4–58. $F_B = 16.9$ kN, $F_A = 16.9$ kN

4–59. $\delta_{sp} = 0.0390$ mm

4–61. $s = 0.133(10^{-3})$ in.

4–62. $F_A = F_B = 25.6$ kN

4–63. $\delta_A = \delta_B = 4.42$ mm

4–65. $A'_1 = \left(\dfrac{E_1}{E_2} \right) A_1$

4–66. $A'_2 = \left(\dfrac{E_2}{E_1} \right) A_2$

4–67. $F_{AB} = 12.0$ kN (T), $F_{AC} = F_{AD} = 6.00$ kN (C)

4–69. $\sigma_{al} = 2.46$ ksi, $\sigma_{br} = 5.52$ ksi, $\sigma_{st} = 22.1$ ksi

4–70. $F = 0.510$ kip

4–71. $T_2 = 112°F$, $\sigma_{al} = \sigma_{cu} = 25.6$ ksi

4–73. $\sigma = 19.1$ ksi

4–74. $F = 7.60$ kip

4–75. $\delta = 0.348$ in., $F = 19.5$ kip

4–77. $F = \dfrac{\alpha A E}{2} (T_B - T_A)$

4–78. $\sigma = 180$ MPa

4–79. $\sigma = 105$ MPa

4–81. $F = 904$ N

4–82. $T_2 = 244°C$

4–83. $F_{AC} = F_{AB} = 10.0$ lb, $F_{AD} = 136$ lb

4–85. $F_{AB} = F_{EF} = 1.85$ kN

4–86. $d = \left(\dfrac{2E_2 + E_1}{3(E_2 + E_1)} \right) w$

4–87. $\sigma_{max} = 168$ MPa

4–89. $P = 49.1$ kN

4–90. $P = 77.1$ kN, $\delta = 0.429$ mm

4–91. $P = 1.34$ kip

4–93. $w = 2.32$ in.

4–94. $P = 16.8$ kip, $K = 1.29$

4–95. $P = 19$ kN, $K = 1.26$

4–97. $F_{st} = 444$ N, $F_{al} = 156$ N
$F_{st} = 480$ N, $F_{al} = 240$ N

4–98. $\delta_{tot} = 0.432$ in.

4–99. $\sigma_A = \sigma_B = \sigma_C = 53.3$ ksi, $\delta = 8.69$ in.

4–101. $F_{AB} = 3.14$ kN, $F_{CD} = 2.72$ kN, $\delta_{CD} = 0.324$ mm, $\delta_{AB} = 0.649$ mm

4–102. (a) $P = 2.62$ kN, (b) $P = 3.14$ kN

4–103. $\sigma_{st} = 36.0$ ksi, $\sigma_{al} = 19.8$ ksi

4–105. $(\sigma_{CF})_r = 17.7$ MPa (C), $(\sigma_{BE})_r = 53.2$ MPa (T) $(\sigma_{AD})_r = 35.5$ MPa (C)

4–106. $P = 92.8$ kN, $P = 181$ kN

4–107. $d_B = 17.8$ mm

4–109. $\delta = 0.0120$ in.

4–110. $w = 0.130$ kip/ft, $\delta = 0.0596$ in.

R4–1. $\sigma_b = 33.5$ MPa, $\sigma_r = 16.8$ MPa

R4–2. $T = 507°C$

R4–3. $F_{AB} = F_{AC} = F_{AD} = 58.9$ kN (C)

R4–5. At yielding of section AB:
$(F_{AB})_Y = 30$ kip, $(F_{BC})_Y = 120$ kip, $P = 150$ kip
R4–6. $F_B = 2.13$ kip, $F_A = 2.14$ kip
R4–7. $P = 4.85$ kip
R4–9. $\delta_{A/B} = 0.491$ mm

Chapter 5

5–1. $r' = 0.841r$
5–2. $r' = 0.707\, r$
5–3. $T = 19.6$ kN·m, $T' = 13.4$ kN·m
5–5. $\tau_A = 3.45$ ksi, $\tau_B = 2.76$ ksi
5–6. $(T_1)_{max} = 2.37$ kN·m, $(\tau_{max})_{CD} = 35.6$ MPa, $(\tau_{max})_{DE} = 23.3$ MPa
5–7. $\tau_{max}^{abs} = 44.8$ MPa
5–9. $\tau_B = 6.79$ MPa, $\tau_A = 7.42$ MPa
5–10. $\tau_{max} = 14.5$ MPa
5–11. $\tau_{AB} = 7.82$ ksi, $\tau_{BC} = 2.36$ ksi
5–13. $\tau_i = 34.5$ MPa, $\tau_o = 43.1$ MPa
5–14. Use $d = 1\frac{3}{4}$ in.
5–15. $(\tau_{AB})_{max} = 23.9$ MPa, $(\tau_{BC})_{max} = 15.9$ MPa
5–17. $\tau_{max} = 4.89$ ksi
5–18. $\tau_{max} = 7.33$ ksi
5–19. $\tau_A = 1.31$ ksi, $\tau_B = 2.62$ ksi
5–21. $\tau_A = 9.43$ MPa, $\tau_B = 14.1$ MPa
5–22. $\tau_{max}^{abs} = 0$ occurs at $x = 0.700$ m, $\tau_{max}^{abs} = 33.0$ MPa occurs at $x = 0$.
5–23. $d = 34.4$ mm
5–25. $d = 46.7$ mm
5–26. $T_B = \frac{2T_A + t_A L}{2}$, $\tau_{max} = \frac{(2T_A + t_A L)r_o}{\pi(r_o^4 - r_i^4)}$
5–27. $t = 0.174$ in.
5–29. $c = (2.98\, x)$ mm
5–30. Use $d = 1\frac{3}{8}$ in.
5–31. $(\tau_{AB})_{max} = 1.04$ MPa, $(\tau_{BC})_{max} = 3.11$ MPa
5–33. $\tau_{max} = 856$ psi
5–34. Use $d = \frac{1}{2}$ in.
5–35. $\tau_{max} = 1.07$ ksi
5–37. Use $d = \frac{11}{16}$ in.
5–38. $\tau_{max} = 6.57$ ksi
5–39. $(\tau_{max})_{CF} = 12.5$ MPa, $(\tau_{max})_{BC} = 7.26$ MPa
5–41. $t = 2.28$ mm
5–42. $\omega = 17.7$ rad/s
5–43. $\tau_{max} = \frac{2TL^3}{\pi[r_A(L-x) + r_B x]^3}$
5–45. $(\tau_{max})_B = 12.2$ ksi

5–46. Use $d_i = 1\frac{5}{8}$ in.
5–47. $\tau_{max} = 44.3$ MPa, $\phi = 11.9°$
5–49. $\phi_{B/A} = 0.730°$
5–50. $\tau_{max}^{abs} = 10.2$ MPa
5–51. $T = 5.09$ kN·m, $\phi_{A/C} = 3.53°$
5–53. $\phi_A = 1.57°$
5–54. Use $d = 22$ mm, $\phi_{A/D} = 2.54°$
5–55. Use $d = 25$ mm
5–57. $\tau_{max} = 9.12$ MPa, $\phi_{E/B} = 0.585°$
5–58. $\tau_{max} = 14.6$ MPa, $\phi_{B/E} = 1.11°$
5–59. $\phi_{B/D} = 1.15°$
5–61. $\tau_{max}^{abs} = 20.4$ MPa,
For $0 \le x < 0.5$ m,
$\phi(x) = \{0.005432\,(x^2 + x)\}$ rad
For 0.5 m $< x \le 1$ m,
$\phi(x) = \{-0.01086x^2 + 0.02173\,x - 0.004074\}$ rad
5–62. $\tau_{max}^{abs} = 24.3$ MPa, $\phi_{D/A} = 0.929°$
5–63. $\phi_{A/C} = 5.45°$
5–65. $\phi_{D/C} = 0.0823$ rad, $\tau_{max}^{abs} = 34.0$ ksi
5–66. $t = 7.53$ mm
5–67. $\omega = 131$ rad/s
5–69. $k = 1.20(10^6)$ N/m², $\phi = 3.56°$
5–70. $k = 12.3(10^3)$ N/m$^{2/3}$, $\phi = 2.97°$
5–71. $d_t = 201$ mm, $\phi = 3.30°$
5–73. $\phi_{F/E} = 0.999(10)^{-3}$ rad, $\phi_{F/D} = 0.999(10)^{-3}$ rad, $\tau_{max} = 3.12$ MPa
5–74. $\phi = \frac{t_0 L^2}{\pi c^4 G}$
5–75. $\phi_A = 0.432°$
5–77. $(\tau_{AC})_{max} = 14.3$ MPa, $(\tau_{CB})_{max} = 9.55$ MPa
5–78. $\tau_{max}^{abs} = 9.77$ MPa
5–79. $\tau_{max} = 29.3$ ksi
5–81. $\tau_{max} = 389$ psi
5–82. $(\tau_{max})_{AC} = 68.2$ MPa, $(\tau_{max})_{BC} = 90.9$ MPa
5–83. $(\tau_{AC})_{max} = 9.55$ MPa, $(\tau_{CB})_{max} = 6.37$ MPa
5–85. $(\tau_{st})_{max} = 86.5$ MPa, $(\tau_{mg})_{max} = 41.5$ MPa, $(\tau_{mg})|_{\rho=0.02m} = 20.8$ MPa
5–86. $T_B = 22.2$ N·m, $T_A = 55.6$ N·m
5–87. $\phi_E = 1.66°$
5–89. $\tau_{max}^{abs} = 64.1$ MPa
5–90. $(\tau_{BD})_{max} = 4.35$ ksi, $(\tau_{AC})_{max} = 2.17$ ksi
5–91. $\phi_{C/D} = 6.22°$
5–93. $T_B = \frac{37}{189}T, T_A = \frac{152}{189}T$
5–94. $T_B = \frac{7t_0 L}{12}, T_A = \frac{3t_0 L}{4}$
5–95. $T = 0.0820$ N·m, $\phi = 25.5$ rad
5–97. Factor of increase $= \frac{1}{k^2}$

5–98. $(\tau_{BC})_{max} = 0.955$ MPa, $(\tau_{AC})_{max} = 1.59$ MPa, $\phi_{B/A} = 0.207°$

5–99. $(\tau_{BC})_{max} = 0.955$ MPa, $(\tau_{AC})_{max} = 1.59$ MPa, $\phi_{B/C} = 0.0643°$

5–101. $T = 1.35$ kip·ft, $\phi = 3.42°$

5–102. $T_B = 26.7$ lb·ft, $T_A = 33.3$ lb·ft, $\phi_C = 0.0626°$

5–103. $\tau_{avg} = 3.35$ ksi

5–105. $\tau_{max} = 61.1$ MPa, $\phi_B = 0.700°$

5–106. Use $a = 47$ mm, $\phi_B = 0.897°$

5–107. $T = 6.65$ kip·ft

5–109. Factor of increase $= 1.66$

5–110. $a = 28.9$ mm

5–111. $\tau_{avg} = 1.25$ MPa

5–113. $t = 0.231$ in.

5–114. $T = 9.00$ kip·ft

5–115. $b = 0.773$ in.

5–117. $T = 3.36$ kN·m, $\phi = 11.6°$

5–118. $\tau_{avg} = 1.19$ MPa

5–119. $(\tau_{avg})_A = (\tau_{avg})_B = 357$ kPa

5–121. It is possible.

5–122. $P = 250$ kW

5–123. $\tau_{max} = 2.76$ ksi

5–125. $T_Y = 1.26$ kN·m, $\phi = 3.58°$, $\phi' = 4.86°$

5–126. $\rho_Y = 13.0$ mm

5–127. $T_P = 0.105$ N·m

5–129. $T = 20.8$ kN·m, $\phi = 34.4°$, $(\tau_r)_{max} = 56.7$ MPa, $\phi_r = 12.2°$

5–130. $T = 14.4$ kip·ft

5–131. $\rho_Y = 1.29$ in.

5–133. $T_P = 34.3$ kN·m, $\phi_r = 5.24°$, $(\tau_r)_o = 15.3$ MPa, $(\tau_r)_i = -17.3$ MPa

5–134. $T_P = 80.6$ kN·m, $\phi_r = 1.47°$

5–135. $\rho_Y = 1.00$ in., $\phi = 12.4°$

5–137. $T = 148$ kN·m

5–138. $T_P = 11.6$ kN·m, $\phi = 3.82°$

5–139. $\rho_Y = 26.7$ mm

5–141. $T = 193$ lb·ft, $\phi = 17.2°$

5–142. $T_P = 218$ lb·ft

5–143. $T_t = 7.39$ kN·m, $T_c = 7.61$ kN·m

R5–1. Use $d = 26$ mm, $\phi_{A/C} = 2.11°$

R5–2. Use $d = 28$ mm.

R5–3. $\tau = 88.3$ MPa, $\phi = 4.50°$

R5–5. The circular shaft will resist the largest torque.
For the square shaft: 73.7%,
For the triangular shaft: 62.2%

R5–6. $(\tau_{max})_{AB} = 3.60$ ksi, $(\tau_{max})_{BC} = 10.7$ ksi

R5–7. $P = 2.80$ kip

R5–9. $P = 1.10$ kW, $\tau_{max} = 825$ kPa

Chapter 6

6–1. For $0 \le x < 3$ ft:
$V = 170$ lb,
$M = \{170x\}$ lb·ft,
For 3 ft $< x < 5$ ft:
$V = -630$ lb,
$M = \{-630x + 2400\}$ lb·ft,
For 5 ft $< x \le 6$ ft:
$V = 500$ lb,
$M = \{500x - 3250\}$ lb·ft

6–2. For $0 < x \le 4$ ft:
$V = -250$ lb,
$M = \{-250x\}$ lb·ft,
For 4 ft $\le x \le 10$ ft: $V = \{1050 - 150x\}$ lb,
$M = \{-75x^2 + 1050x - 4000\}$ lb·ft,
For 10 ft $< x \le 14$ ft: $V = 250$ lb,
$M = \{250x - 3500\}$ lb·ft

6–3. For $0 \le x < 6$ ft:
$V = \{30.0 - 2x\}$ kip,
$M = \{-x^2 + 30.0x - 216\}$ kip·ft,
For 6 ft $< x \le 10$ ft:
$V = 8.00$ kip,
$M = \{8.00x - 120\}$ kip·ft

6–5. $V = \{-300 - 16.67x^2\}$ lb,
$M = \{-300x - 5.556x^3\}$ lb·ft

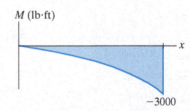

6–6. $V = 15.6$ N, $M = \{15.6x + 100\}$ N·m

6–7. For $0 \leq x < \dfrac{L}{2}$: $V = \dfrac{w_0 L}{24}, M = \dfrac{w_0 L}{24} x,$

For $\dfrac{L}{2} < x \leq L$: $V = \dfrac{w_0}{24L}\left[L^2 - 6(2x - L)^2\right],$

$M = \dfrac{w_0}{24L}\left[L^2 x - (2x - L)^3\right]$

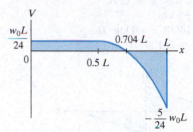

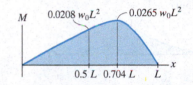

6–9. $T_1 = 250$ lb, $T_2 = 200$ lb

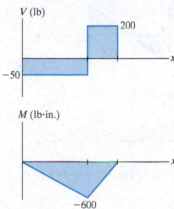

6–10.

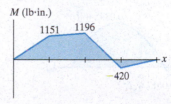

6–11.

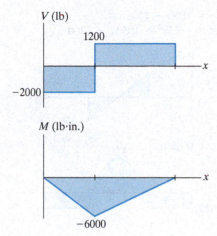

6–13. $V = -\dfrac{M_0}{L},$

For $0 \leq x < \dfrac{L}{2}$, $M = M_0 - \left(\dfrac{M_0}{L}\right)x,$

For $\dfrac{L}{2} < x \leq L$, $M = -\left(\dfrac{M_0}{L}\right)x$

6–14. For $0 \leq x < 5$ ft:
$V = \{-2x\}$ kip,
$M = \{-x^2\}$ kip $\cdot$ ft,
For 5 ft $< x < 10$ ft:
$V = -0.5$ kip,
$M = \{-22.5 - 0.5x\}$ kip $\cdot$ ft,
For 10 ft $< x \leq 15$ ft:
$V = -0.5$ kip,
$M = \{7.5 - 0.5x\}$ kip $\cdot$ ft

6–15.

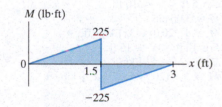

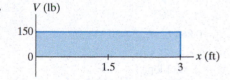

6–17. $V = 1050 - 150x$
$M = -75x^2 + 1050x - 3200$

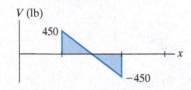

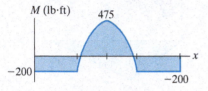

6–18.

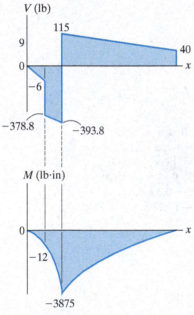

6–19. $a = 0.866 L, M_{\max} = 0.134 \, PL$

6–21.   For $0 \le x < 5$ ft:
$V = \{-2x\}$ kip,
$M = \{-20.0 - x^2\}$ kip·ft,
For 5 ft $< x < 10$ ft:
$V = 3.00$ kip,
$M = \{-20.0 + 3x\}$ kip·ft,
For 10 ft $< x \le 15$ ft:
$V = \{23 - 2x\}$ kip,
$M = \{-120 + 23x - x^2\}$ kip·ft

6–22.   For $0 \le x < 12$ ft:
$$V = \left\{10 - \frac{1}{8}x^2\right\} \text{ kip,}$$
$$M = \left\{10x - \frac{1}{24}x^3\right\} \text{ kip·ft,}$$
For 12 ft $< x \le 18$ ft:
$V = -8$ kip,
$M = [8(18 - x)]$ kip·ft

6–23. $M_{\max} = 281$ lb·ft

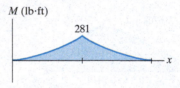

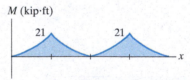

6–25.

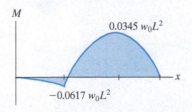

6–26. $V_{AB} = -1.625$ kip, $M_B = -18$ kip·ft

6–27. V

6–29. $V_B = -45$ kN, $M_B = -63$ kN $\cdot$ m

6–30. $V|_{x=15\,ft} = 1.12$ kip, $M|_{x=15} = -1.95$ kip $\cdot$ ft

6–31.

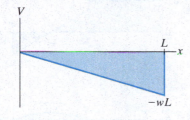

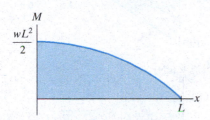

6–33.

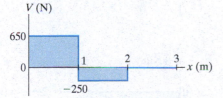

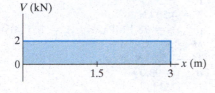

6–34.

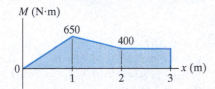

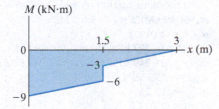

6–35. For $0 \le x < 3$ m: $V = 200$ N, $M = \{200x\}$ N $\cdot$ m,

For 3 m $< x \le 6$ m: $V = \left\{ -\dfrac{100}{3}x^2 + 500 \right\}$ N,

$$M = \left\{ -\dfrac{100}{9}x^3 + 500x - 600 \right\} \text{N} \cdot \text{m}$$

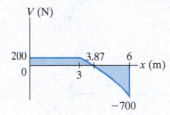

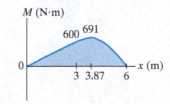

6–37.

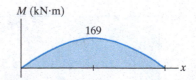

6–38.

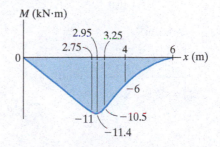

6–39.

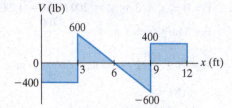

6–41.

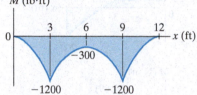

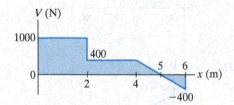

6–42.

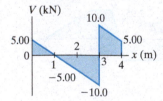

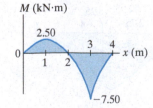

6–43.

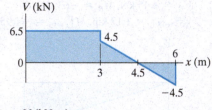

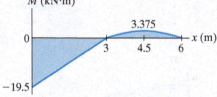

6–45.

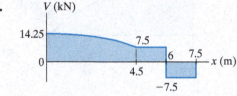

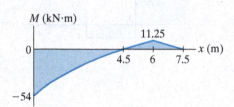

6–46.	$a = 0.207L$
6–47.	$r = 909$ mm, $M = 61.9$ N · m
6–49.	$(\sigma_t)_{max} = 3.72$ ksi, $(\sigma_c)_{max} = 1.78$ ksi
6–50.	$\sigma_{max} = 1.46$ ksi
6–51.	$F = 10.5$ kip
6–53.	$F = 4.56$ kN
6–54.	$(\sigma_{max})_c = 78.1$ MPa, $(\sigma_{max})_t = 165$ MPa
6–55.	$M = 50.3$ kN · m
6–57.	$M = 15.6$ kN · m, $\sigma_{max} = 12.0$ MPa
6–58.	$\sigma_{max} = 93.0$ psi
6–59.	$F = 753$ lb
6–61.	% of moment carried by web = 22.6%
6–62.	$\sigma_A = 199$ MPa, $\sigma_B = 66.2$ MPa
6–63.	$\sigma_{max} = 20.4$ ksi
6–65.	(a) $\sigma_{max} = 497$ kPa, (b) $\sigma_{max} = 497$ kPa
6–66.	(a) $\sigma_{max} = 249$ kPa, (b) $\sigma_{max} = 249$ kPa
6–67.	$\sigma_{max} = 158$ MPa
6–69.	$\sigma_{max} = 12.2$ ksi
6–70.	$\sigma_{max} = 2.70$ ksi
6–71.	$\sigma_{max} = 21.1$ ksi
6–73.	$d = 1.28$ in.

6–74. $\sigma_{max} = 45.1$ ksi

6–75. $\sigma_{max} = 52.8$ MPa

6–77. $(\sigma_{max})_c = 1.87$ ksi, $(\sigma_{max})_t = 1.37$ ksi

6–78. $M = 2.92$ kip · ft

6–79. $F_R = 4.23$ kip

6–81. $M = 123$ kN · m

6–82. $\sigma_{max} = 24.4$ ksi

6–83. Use $d = 3\frac{1}{16}$ in.

6–85. $w = 18.75$ kN/m

6–86. Use $b = 3\frac{5}{8}$ in.

6–87. $w_0 = 415$ lb/ft

6–89. $a = 66.9$ mm

6–90. $\sigma_{max} = \dfrac{23w_0 L^2}{36\, bh^2}$

6–91. $\sigma_{max} = 119$ MPa

6–93. $\sigma_{max}^{abs} = 24.0$ ksi

6–94. $\sigma_{max}^{abs} = 6.88$ ksi

6–95. Use $d = 1\frac{3}{8}$ in.

6–97. $P = 114$ kip

6–98. $\sigma_{max} = 7.59$ ksi

6–99. $\sigma_{max} = 22.1$ ksi

6–101. $d = 410$ mm

6–102. $w = 937.5$ N/m

6–103. $\sigma_{max}^{abs} = 10.7$ MPa

6–105. $\sigma_A = -119$ kPa, $\sigma_B = 446$ kPa, $\sigma_D = -446$ kPa, $\sigma_E = 119$ kPa

6–106. $a = 0,\ b = -\left(\dfrac{M_z I_y + M_y I_{yz}}{I_y I_z - I_{yz}^2}\right),\ c = \dfrac{M_y I_z + M_z I_{yz}}{I_y I_z - I_{yz}^2}$

6–107. $\sigma_A = 21.0$ ksi (C)

6–109. $d = 62.9$ mm

6–110. $\sigma_{max} = 163$ MPa

6–111. $\sigma_A = 20.6$ MPa (C)

6–113. $\sigma_D = 8.72$ ksi (T), $\sigma_B = 8.72$ ksi (C), $\alpha = -36.9°$

6–114. $M = 25.8$ kip · ft

6–115. $d = 28.9$ mm

6–117. $\sigma_A = 2.59$ MPa (T)

6–118. $\sigma_{max} = 151$ MPa, $\alpha = 72.5°$

6–119. $w = 4.37$ kN/m

6–121. $M = 128$ kN · m

6–122. $(\sigma_{max})_{st} = 22.6$ ksi, $(\sigma_{max})_{al} = 13.3$ ksi

6–123. $w = 0.875$ kip/ft

6–125. $M = 35.0$ kN · m

6–126. $(\sigma_{max})_{st} = 123$ MPa, $(\sigma_{max})_{w} = 5.14$ MPa

6–127. $(\sigma_{st})_{max} = 56.5$ MPa, $(\sigma_{w})_{max} = 3.70$ MPa

6–129. $(\sigma_{st})_{max} = 43.4$ MPa, $(\sigma_{w})_{max} = 1.81$ MPa

6–130. $M = 154$ lb · ft

6–131. $w_0 = 1.03$ kip/ft

6–133. $M = 58.8$ kN · m

6–134. $(\sigma_{max})_{st} = 154$ MPa, $(\sigma_{max})_{al} = 171$ MPa

6–135. $(\sigma_{max})_{st} = 4.55$ MPa, $(\sigma_{max})_{w} = 0.298$ MPa

6–137. % of error $= 22.3\%$

6–138. $M = 51.8$ kN · m

6–139. $\sigma_{max} = 842$ psi (T)

6–141. $\sigma_A = 43.7$ MPa (T), $\sigma_B = 7.77$ MPa (T), $\sigma_C = -65.1$ MPa (C)

6–142. $P = 6.91$ kN

6–143. $(\sigma_{max})_t = 204$ psi, $(\sigma_{max})_c = 120$ psi

6–145. $(\sigma_{max})_t = 0.978$ MPa (T), $(\sigma_{max})_c = 0.673$ MPa (C)

6–146. $\sigma_A = 144$ psi (T), $\sigma_B = 106$ psi (C), $\sigma_C = 6.11$ psi (C)

6–147. $\sigma_A = 8.48$ MPa (C), $\sigma_B = 5.04$ MPa (T). No, it is not the same.

6–149. $(\sigma_T)_{max} = 11.1$ MPa (T), $(\sigma_C)_{max} = 8.45$ MPa (C)

6–150. $\sigma_{max} = 187.5$ MPa

6–151. $M = 107$ N · m

6–153. $\sigma_{max} = 22.3$ ksi

6–154. $M = 347$ lb · ft

6–155. $M = 4.46$ kN · m

6–157. $\sigma_{max} = 34.6$ ksi

6–158. $k = 1.17$

6–159. $K = \dfrac{3h}{2}\left[\dfrac{4bt(h - t) + t(h - 2t)^2}{bh^3 - (b - t)(h - 2t)^3}\right]$

6–161. $M_Y = 271$ kN · m, $M_P = 460$ kN · m

6–162. $M_P = 392$ kip · ft

6–163. $M_P = 172$ kip · ft

6–165. $M_Y = 143$ kip · ft, $M_P = 243$ kip · ft

6–166. $k = 1.16$

6–167. Elastic: $P = 66.7$ kN, Plastic: $P = 100$ kN

6–169. $k = 1.71$

6–170. $k = 1.70$

6–171. $M_Y = 63.6$ kip · ft, $M_P = 180$ kip · ft

6–173. $M_Y = 50.7$ kN · m, $M_P = 86.25$ kN · m

6–174. $k = 1.58$

6–175. $k = 1.71$

6–177. $\sigma_{top} = \sigma_{bottom} = 43.5$ MPa

6–178. $M = 94.7$ N · m

6–179. Maximum elastic moment: $M = 35.0$ kip · ft, Ultimate moment: $M = 59.8$ kip · ft

6–181. $M = 81.7$ kip · ft

6–182. $w = 53.4$ kN/m

R6–1. $k = 1.22$

R6–2. $V = \dfrac{2wL}{27} - \dfrac{w}{2L}x^2,\ M = \dfrac{2wL}{27}x - \dfrac{w}{6L}x^3$

R6–3. $M = 14.9$ kN · m

R6–5. $\sigma_{max} = 8.41$ ksi

R6–6. $\sigma_A = 225$ kPa (C), $\sigma_B = 265$ kPa (T)

R6–7. $V = 20 - 2x$, $M = -x^2 + 20x - 166$

R6–9. $V|_{x=600 \text{ mm}} = -233$ N, $M|_{x=600 \text{ mm}} = -50$ N·m

R6–10. $\sigma_{max} = \dfrac{6M}{a^3}(\cos\theta + \sin\theta)$, $\theta = 45°$, $\alpha = 45°$

Chapter 7

7–1. $\tau_A = 2.56$ MPa
7–2. $\tau_{max} = 3.46$ MPa
7–3. $V_w = 19.0$ kN
7–5. $\tau_{max} = 3.91$ MPa
7–6. $V_{max} = 100$ kN
7–7. $\tau_{max} = 17.9$ MPa
7–9. $V_{max} = 32.1$ kip
7–10. $\tau_{max} = 4.48$ ksi
7–11. $\tau_{max} = 45.0$ MPa
7–13. $\tau_{max} = 4.22$ MPa
7–14. $V_{max} = 190$ kN
7–15. $V = 9.96$ kip
7–17. $\tau_A = 1.99$ MPa, $\tau_B = 1.65$ MPa
7–18. $\tau_{max} = 4.62$ MPa
7–19. $V_w = 27.1$ kN
7–21. $P_{max} = 1.28$ kip
7–22. $w_{max} = 2.15$ kip/ft
7–23. $\tau_{max} = 298$ psi
7–25. $\tau_{max} = 4.85$ MPa
7–26. $w = 11.3$ kip/ft, $\tau_{max} = 531$ psi
7–27. $\tau_{max} = 22.0$ MPa, $(\tau_{max})_s = 66.0$ MPa
7–29. $\tau_{max} = 1.05$ MPa
7–33. $F = 300$ lb
7–34. $s_t = 1.42$ in., $s_b = 1.69$ in.
7–35. $V = 4.97$ kip, $s_t = 1.14$ in.,
$s_b = 1.36$ in.
7–37. $V = 499$ kN
7–38. $P = 4.97$ kip.
For regions AC and BD, $s_{top} = 1.14$ in.,
$s_{bottom} = 1.36$ in.
For region CD, theoretically no nails are required.
7–39. $F = 12.5$ kN
7–41. $s = 71.3$ mm
7–42. $V_{max} = 8.82$ kip, use $s = 1\frac{1}{8}$ in.
7–43. $P_{max} = 317$ lb
7–45. $\tau_{max} = 1.83$ ksi
7–46. $w_0 = 983$ lb/ft
7–47. $s = 8.66$ in., $s' = 1.21$ in.
7–50. $q_A = 228$ kN/m, $q_B = 462$ kN/m
7–51. $q_C = 0$, $q_D = 601$ kN/m
7–53. $q_C = 38.6$ kN/m
7–54. $q_A = 1.39$ kN/m, $q_B = 1.25$ kN/m
7–55. $q_{max} = 1.63$ kN/m

7–57. $q_{max} = 2.79$ kip/in.
7–58. $q_A = 200$ kN/m
7–59. $\tau_{max} = 9.36$ MPa
7–61. $q_A = 196$ lb/in., $q_B = 452$ lb/in.,
$q_{max} = 641$ lb/in.
7–62. $q_B = 12.6$ kN/m, $q_{max} = 22.5$ kN/m
7–63. $e = 70$ mm
7–65. $V_{AB} = 1.47$ kip
7–66. $q_{AB} = 207 - 51.6x$, $q_{CD} = 44.3 - 22.1x$,
$F_{AB} = 413$ lb, $F_{CD} = 44.3$ lb
7–67. $q_A = 0$, $q_B = 417$ lb/in.
7–69. $e = \left[\dfrac{3(\pi + 4)}{4 + 3\pi}\right]r$
7–70. $e = 2r$
R7–1. $F_C = 197$ lb, $F_D = 1.38$ kip
R7–2. $V = 131$ kN
R7–3. $q_A = 0$, $q_B = 1.21$ kN/m, $q_C = 3.78$ kN/m
R7–5. $\tau_B = 795$ psi, $\tau_C = 596$ psi

Chapter 8

8–1. $t = 18.8$ mm
8–2. $r_o = 75.5$ in.
8–3. (a) $\sigma_1 = 1.04$ ksi, $\sigma_2 = 0$,
(b) $\sigma_1 = 1.04$ ksi, $\sigma_2 = 520$ psi
8–5. $\sigma_1 = 7.07$ MPa, $\sigma_2 = 0$
8–6. $P = 848$ N
8–7. (a) $\sigma_1 = 127$ MPa,
(b) $\sigma_1' = 79.1$ MPa,
(c) $(\tau_{avg})_b = 322$ MPa
8–9. $\sigma_{hoop} = 7.20$ ksi, $\sigma_{long} = 3.60$ ksi
8–10. $\sigma_1 = 1.60$ ksi, $p = 25$ psi, $\delta = 0.00140$ in.
8–11. $\sigma_2 = 11.5$ ksi, $\sigma_1 = 24$ ksi
8–13. $T_1 = 128°$F, $\sigma_1 = 12.1$ ksi, $p = 252$ psi
8–14. $\delta r_i = \dfrac{pr_i^2}{E(r_o - r_i)}$
8–15. $p = \dfrac{E(r_2 - r_3)}{\dfrac{r_2^2}{r_2 - r_1} + \dfrac{r_3^2}{r_4 - r_3}}$
8–17. $\sigma_{fil} = \dfrac{pr}{t + t'w/L} + \dfrac{T}{wt'}$, $\sigma_w = \dfrac{pr}{t + t'w/L} - \dfrac{T}{Lt}$
8–18. $d = 66.7$ mm
8–19. $d = 133$ mm
8–21. $\sigma_{max} = \sigma_B = 13.9$ ksi (T), $\sigma_A = 13.6$ ksi (C)
8–22. $P_{max} = 2.01$ kip
8–23. $\sigma_{max} = 22.4$ ksi (T)
8–25. $P_{max} = 128$ kN
8–26. $\sigma_{max} = 44.0$ ksi (T)
8–27. $\sigma_{max} = 44.0$ ksi (T), $\sigma_{min} = 41.3$ ksi (C)
8–29. $\sigma_A = 0.800$ ksi (T), $\sigma_B = 5.20$ ksi (C),
$\tau_A = 1.65$ ksi, $\tau_B = 0$

8–30. $\sigma_A = 25$ MPa (C), $\sigma_B = 0$, $\tau_A = 0$, $\tau_B = 5$ MPa

8–31. $\sigma_A = 28.8$ ksi, $\tau_A = 0$

8–33. $\sigma_A = 70.0$ MPa (C), $\sigma_B = 10.0$ MPa (C)

8–34. $\sigma_A = 70.0$ MPa (C), $\sigma_B = 10.0$ MPa (C), $\sigma_C = 50.0$ MPa (T), $\sigma_D = 10.0$ MPa (C)

8–35. $\sigma_A = 27.3$ ksi (T), $\sigma_B = 0.289$ ksi (T), $\tau_A = 0$, $\tau_B = 0.750$ ksi

8–37. $\sigma_B = 1.53$ MPa (C), $(\tau_{xz})_B = 0$, $(\tau_{xy})_B = 100$ MPa

8–38. $\sigma_D = -88.0$ MPa, $\tau_D = 0$

8–39. $\sigma_E = 57.8$ MPa, $\tau_E = 864$ kPa

8–41. $\sigma_B = 27.5$ MPa (C), $(\tau_{xz})_B = -8.81$ MPa, $(\tau_{xy})_B = 0$

8–42. $(\sigma_A)_y = 16.2$ ksi (T), $(\tau_A)_{yx} = -2.84$ ksi, $(\tau_A)_{yz} = 0$

8–43. $(\sigma_B)_y = 7.80$ ksi (T), $(\tau_B)_{yz} = 3.40$ ksi, $(\tau_B)_{yx} = 0$

8–45. $\sigma_A = 1.00$ ksi (C), $\sigma_B = 3.00$ ksi (C), $\sigma_C = 1.00$ ksi (C), $\sigma_D = 1.00$ ksi (T)

8–46. $(\sigma_t)_{max} = 103$ MPa (T), $(\sigma_c)_{max} = 117$ MPa (C)

8–47. $\sigma_A = 224$ MPa (T), $(\tau_{xz})_A = -30.7$ MPa, $(\tau_{xy})_A = 0$

8–49. $\sigma_C = 295$ MPa (C), $(\tau_{xy})_C = 25.9$ MPa, $(\tau_{xz})_C = 0$

8–50. $e = \dfrac{c}{4}$

8–51. $6e_y + 18e_z < 5a$

8–54. $\sigma_A = 15.9$ ksi (C), $\sigma_B = 44.6$ ksi (T), $\tau_A = \tau_B = 0$

8–55. $\sigma_A = 16.8$ ksi (C), $(\tau_{xz})_A = 4.07$ ksi, $(\tau_{xy})_A = 0$

8–57. $\sigma_A = 107$ MPa (T), $\tau_A = 15.3$ MPa, $\sigma_B = 0$, $\tau_B = 14.8$ MPa

8–58. $\sigma_C = 107$ MPa (C), $\tau_C = 15.3$ MPa, $\sigma_D = 0$, $\tau_D = 15.8$ MPa

8–59. $\sigma_A = 24.7$ ksi (T), $\tau_A = 0$

8–61. $(\sigma_{max})_t = 106$ MPa, $(\sigma_{max})_c = -159$ MPa

8–62. $P_{max} = 9.08$ kN

8–63. $\sigma_C = 15.6$ ksi (T), $\sigma_D = 124$ ksi (T), $(\tau_{xz})_D = 62.4$ ksi, $(\tau_{xy})_D = 0$, $(\tau_{xy})_C = -52.4$ ksi, $(\tau_{xz})_C = 0$

8–65. $\sigma_A = 15.3$ MPa, $\tau_A = 0$, $\sigma_B = 0$, $\tau_B = 0.637$ MPa

8–66. $\sigma_C = 15.3$ MPa, $\tau_C = 0$, $\sigma_D = 0$, $\tau_D = 0.637$ MPa

8–67. $\sigma_A = 94.4$ psi (T), $\sigma_B = 59.0$ psi (C)

8–69. $\sigma_B = 19.4$ MPa (C), $(\tau_{xy})_B = 0.509$ MPa, $(\tau_{xz})_B = 0$

8–70. $\tau_A = 0$, $\sigma_A = 30.2$ ksi (C)

8–71. $\sigma_B = 0$, $\tau_B = 0.377$ ksi

8–73. $\sigma = 1.48$ psi (T), $\tau = 384$ psi

R8–1. $(\sigma_t)_{max} = 15.8$ ksi, $(\sigma_c)_{max} = -10.5$ ksi

R8–2. $\sigma_E = 802$ kPa, $\tau_E = 69.8$ kPa

R8–3. $\sigma_F = 695$ kPa (C), $\tau_F = 31.0$ kPa

R8–5. $\sigma_{max} = 236$ psi (C)

R8–6. $\theta = 0.286°$

R8–7. $\sigma_C = 11.6$ ksi, $\tau_C = 0$, $\sigma_D = -23.2$ ksi, $\tau_D = 0$

Chapter 9

9–2. $\sigma_{x'} = 31.4$ MPa, $\tau_{x'y'} = 38.1$ MPa

9–3. $\sigma_{x'} = -388$ psi, $\tau_{x'y'} = 455$ psi

9–5. $\sigma_{x'} = 1.45$ ksi, $\tau_{x'y'} = 3.50$ ksi

9–6. $\sigma_{x'} = -4.05$ ksi, $\tau_{x'y'} = -0.404$ ksi

9–7. $\sigma_{x'} = -61.5$ MPa, $\tau_{x'y'} = 62.0$ MPa

9–9. $\sigma_{x'} = 36.0$ MPa, $\tau_{x'y'} = -37.0$ MPa

9–10. $\sigma_{x'} = 36.0$ MPa, $\tau_{x'y'} = -37.0$ MPa

9–11. $\sigma_{x'} = 47.5$ MPa, $\sigma_{y'} = 202$ MPa, $\tau_{x'y'} = -15.8$ MPa

9–13. $\sigma_{x'} = -62.5$ MPa, $\tau_{x'y'} = -65.0$ MPa

9–14. $\sigma_1 = 319$ MPa, $\sigma_2 = -219$ MPa, $\theta_{p1} = 10.9°$, $\theta_{p2} = -79.1°$, $\tau_{\substack{max \\ \text{in-plane}}} = 269$ MPa, $\theta_s = -34.1°$ and $55.9°$, $\sigma_{avg} = 50.0$ MPa

9–15. $\sigma_1 = 53.0$ MPa, $\sigma_2 = -68.0$ MPa, $\theta_{p1} = 14.9°$, $\theta_{p2} = -75.1°$, $\tau_{\substack{max \\ \text{in-plane}}} = 60.5$ MPa, $\sigma_{avg} = -7.50$ MPa, $\theta_s = -30.1°$ and $59.9°$

9–17. $\sigma_1 = 137$ MPa, $\sigma_2 = -86.8$ MPa, $\theta_{p1} = -13.3°$, $\theta_{p2} = 76.7°$, $\tau_{\substack{max \\ \text{in-plane}}} = 112$ MPa, $\theta_s = 31.7°$ and $122°$, $\sigma_{avg} = 25$ MPa

9–18. $\sigma_x = 33.0$ MPa, $\sigma_y = 137$ MPa, $\tau_{xy} = -30$ MPa

9–19. $\sigma_1 = 5.90$ MPa, $\sigma_2 = -106$ MPa, $\theta_{p1} = 76.7°$ and $\theta_{p2} = -13.3°$, $\tau_{\substack{max \\ \text{in-plane}}} = 55.9$ MPa, $\sigma_{avg} = -50$ MPa, $\theta_s = 31.7°$ and $122°$

9–21. $\tau_a = -1.96$ ksi, $\sigma_1 = 80.1$ ksi, $\sigma_2 = 19.9$ ksi

9–22. $\sigma_{x'} = -63.3$ MPa, $\tau_{x'y'} = 35.7$ MPa

9–23. $\sigma_{x'} = 19.5$ kPa, $\tau_{x'y'} = -53.6$ kPa

9–25. $\sigma_1 = 0$, $\sigma_2 = -36.6$ MPa, $\tau_{\substack{max \\ \text{in-plane}}} = 18.3$ MPa

9–26. $\sigma_1 = 16.6$ MPa, $\sigma_2 = 0$, $\tau_{\substack{max \\ \text{in-plane}}} = 8.30$ MPa

9–27. $\sigma_1 = 14.2$ MPa, $\sigma_2 = -8.02$ MPa, $\tau_{\substack{max \\ \text{in-plane}}} = 11.1$ MPa

9–29. Point D: $\sigma_1 = 7.56$ kPa, $\sigma_2 = -603$ kPa, Point E: $\sigma_1 = 395$ kPa, $\sigma_2 = -17.8$ kPa

9–30. Point A: $\sigma_1 = \sigma_y = 0$, $\sigma_2 = \sigma_x = -30.5$ MPa, Point B: $\sigma_1 = 0.541$ MPa, $\sigma_2 = -1.04$ MPa, $\theta_{p1} = -54.2°$, $\theta_{p2} = 35.8°$

9–31. $\tau_{x'y'} = 160$ kPa

9–33. $\sigma_1 = 4.38$ ksi, $\sigma_2 = -1.20$ ksi, $\tau_{\substack{max \\ \text{in-plane}}} = 2.79$ ksi

9–34. Point A: $\sigma_1 = 37.8$ kPa, $\sigma_2 = -10.8$ MPa
Point B: $\sigma_1 = 42.0$ MPa, $\sigma_2 = -10.6$ kPa

9–35. $\sigma_1 = 233$ psi, $\sigma_2 = -774$ psi, $\tau_{\max_{\text{in-plane}}} = 503$ psi

9–37. $\sigma_1 = \dfrac{4}{\pi d^2}\left(\dfrac{2PL}{d} - F\right),\ \sigma_2 = 0,$

$\tau_{\max_{\text{in-plane}}} = \dfrac{2}{\pi d^2}\left(\dfrac{2PL}{d} - F\right)$

9–38. $\sigma_1 = 838$ psi, $\sigma_2 = -37.8$ psi
9–39. $\sigma_1 = 628$ psi, $\sigma_2 = -166$ psi
9–41. $\sigma_1 = 1.37$ MPa, $\sigma_2 = -198$ MPa
9–42. $\sigma_1 = 111$ MPa, $\sigma_2 = 0$
9–43. $\sigma_1 = 2.40$ MPa, $\sigma_2 = -6.68$ MPa
9–45. $\sigma_{x'} = -388$ psi, $\tau_{x'y'} = 455$ psi
9–46. $\sigma_{x'} = -4.05$ ksi, $\tau_{x'y'} = -0.404$ ksi
9–47. $\sigma_{x'} = 47.5$ MPa, $\tau_{x'y'} = -15.8$ MPa,
$\sigma_{y'} = 202$ MPa

9–49. $\sigma_{\text{avg}} = 25$ MPa, $\sigma_1 = 54.2$ MPa, $\sigma_2 = -4.15$ MPa,
$\theta_p = -15.5°$, $\tau_{\max_{\text{in-plane}}} = 29.2$ MPa

9–51. $\sigma_{\text{avg}} = -40.0$ MPa, $\sigma_1 = 32.1$ MPa,
$\sigma_2 = -112$ MPa, $\theta_{p1} = 28.2°$, $\tau_{\max_{\text{in-plane}}} = 72.1$ MPa,
$\theta_s = -16.8°$

9–53. $\sigma_{x'} = -56.3$ ksi, $\sigma_{y'} = 56.3$ ksi, $\tau_{x'y'} = -32.5$ ksi
9–57. $\sigma_{\text{avg}} = 15.0$ ksi, $\sigma_1 = 32.5$ ksi, $\sigma_2 = -2.49$ ksi,

$\theta_{p1} = -15.5°$, $\tau_{\max_{\text{in-plane}}} = -17.5$ ksi, $\theta_s = 29.5°$

9–58. $\sigma_1 = 64.1$ MPa, $\sigma_2 = -14.1$ MPa, $\theta_P = 25.1°$,
$\sigma_{\text{avg}} = 25.0$ MPa, $\tau_{\max_{\text{in-plane}}} = 39.1$ MPa, $\theta_s = -19.9°$

9–59. $\theta_P = -14.9°$, $\sigma_1 = 227$ MPa, $\sigma_2 = -177$ MPa,
$\tau_{\max_{\text{in-plane}}} = 202$ MPa, $\sigma_{\text{avg}} = 25$ MPa, $\theta_s = 30.1°$

9–62. $\sigma_{x'} = 19.5$ kPa, $\tau_{x'y'} = -53.6$ kPa
9–63. $\tau_{\max_{\text{in-plane}}} = 41.0$ psi, $\sigma_1 = 0.976$ psi, $\sigma_2 = -81.0$ psi
9–65. $\sigma_1 = 29.4$ ksi, $\sigma_2 = -17.0$ ksi
9–66. $\sigma_{x'} = 75.3$ kPa, $\tau_{x'y'} = -78.5$ kPa
9–67. $\sigma_{x'} = -45.0$ kPa, $\tau_{x'y'} = 45.0$ kPa
9–69. $\sigma_1 = 4.71$ ksi, $\sigma_2 = -0.0262$ ksi
9–70. Mohr's circle is a point located at $(4.80, 0)$.
9–71. $\sigma_{x'} = 500$ MPa, $\tau_{x'y'} = -167$ MPa
9–73. $\sigma_1 = 1.15$ MPa, $\sigma_2 = -0.0428$ MPa
9–74. $\sigma_1 = 2.97$ ksi, $\sigma_2 = -2.97$ ksi, $\theta_{p1} = 45.0°$,
$\theta_{p2} = -45.0°$, $\tau_{\max_{\text{in-plane}}} = 2.97$ ksi, $\theta_s = 0$

9–75. $\sigma_1 = 2.59$ ksi, $\sigma_2 = -3.61$ ksi, $\theta_{p1} = -40.3°$,
$\theta_{p2} = -49.7°$, $\tau_{\max_{\text{in-plane}}} = 3.10$ ksi, $\theta_s = 4.73°$

9–81. $\sigma_1 = 24.4$ ksi, $\sigma_2 = 5.57$ ksi, $\tau_{\text{abs}_{\max}} = 12.2$ ksi
9–82. $\sigma_1 = 222$ MPa, $\sigma_2 = -102$ MPa, $\tau_{\text{abs}_{\max}} = 162$ MPa
9–83. $\sigma_1 = 6.73$ ksi, $\sigma_2 = -4.23$ ksi, $\tau_{\text{abs}_{\max}} = 5.48$ ksi

9–85. $\sigma_1 = 5.50$ MPa, $\sigma_2 = -0.611$ MPa,
$\sigma_1 = 1.29$ MPa, $\sigma_2 = -1.29$ MPa,
$\tau_{\text{abs}_{\max}} = 3.06$ MPa, $\tau_{\text{abs}_{\max}} = 1.29$ MPa

9–86. $\sigma_1 = 6.27$ kPa, $\sigma_2 = -806$ kPa,
$\tau_{\text{abs}_{\max}} = 406$ kPa

9–87. $\sigma_1 = 48.8$ ksi, $\sigma_2 = -25.4$ ksi,
$\tau_{\text{abs}_{\max}} = 31.1$ ksi

R9–1. $\sigma_1 = 119$ psi, $\sigma_2 = -119$ psi
R9–2. $\sigma_1 = 329$ psi, $\sigma_2 = -72.1$ psi
R9–3. $\sigma_{x'} = -0.611$ ksi, $\tau_{x'y'} = 7.88$ ksi, $\sigma_{y'} = -3.39$ ksi
R9–5. $\sigma_1 = 3.03$ ksi, $\sigma_2 = -33.0$ ksi,
$\theta_{p1} = -16.8°$ and $\theta_{p2} = 73.2°$,
$\tau_{\max_{\text{in-plane}}} = 18.0$ ksi, $\sigma_{\text{avg}} = -15$ ksi,
$\theta_s = 28.2°$ and $118°$

R9–6. $\sigma_1 = 3.29$ MPa, $\sigma_2 = -4.30$ MPa
R9–7. Point A: $\sigma_1 = 61.7$ psi, $\sigma_2 = 0$,
Point B: $\sigma_1 = 0$, $\sigma_2 = -46.3$ psi

R9–9. $\sigma_{x'} = -16.5$ ksi, $\tau_{x'y'} = 2.95$ ksi

Chapter 10

10–2. $\epsilon_{x'} = 248(10^{-6})$, $\gamma_{x'y'} = -233(10^{-6})$,
$\epsilon_{y'} = -348(10^{-6})$

10–3. $\epsilon_{x'} = 55.1(10^{-6})$, $\gamma_{x'y'} = 133(10^{-6})$,
$\epsilon_{y'} = 325(10^{-6})$

10–5. $\epsilon_{x'} = 77.4(10^{-6})$, $\gamma_{x'y'} = 1279(10^{-6})$,
$\epsilon_{y'} = 383(10^{-6})$

10–6. $\epsilon_{x'} = -116(10^{-6})$, $\epsilon_{y'} = 466(10^{-6})$,
$\gamma_{x'y'} = 393(10^{-6})$

10–7. $\epsilon_{x'} = 466(10^{-6})$, $\epsilon_{y'} = -116(10^{-6})$,
$\gamma_{x'y'} = -393(10^{-6})$

10–9. $\epsilon_1 = 188(10^{-6})$, $\epsilon_2 = -128(10^{-6})$,
$(\theta_P)_1 = -9.22°$, $(\theta_P)_2 = 80.8°$,
$\gamma_{\max_{\text{in-plane}}} = 316\left(10^{-6}\right)$,
$\epsilon_{\text{avg}} = 30(10^{-6})$,
$\theta_s = 35.8°$ and $-54.2°$

10–10. (a) $\epsilon_1 = 713(10^{-6})$, $\epsilon_2 = 36.6(10^{-6})$, $\theta_{p1} = 133°$,
(b) $\gamma_{\max_{\text{in-plane}}} = 677(10^{-6})$, $\epsilon_{\text{avg}} = 375(10^{-6})$,
$\theta_s = -2.12°$

10–11. $\epsilon_{x'} = 649(10^{-6})$, $\gamma_{x'y'} = -85.1(10^{-6})$,
$\epsilon_{y'} = 201(10^{-6})$

10–13. $\epsilon_1 = 17.7(10^{-6})$, $\epsilon_2 = -318(10^{-6})$,
$\theta_{p1} = 76.7°$ and $\theta_{p2} = -13.3°$,
$\gamma_{\max_{\text{in-plane}}} = 335(10^{-6})$, $\theta_s = 31.7°$ and $122°$,
$\epsilon_{\text{avg}} = -150(10^{-6})$

10–14. $\epsilon_1 = 368(10^{-6})$, $\epsilon_2 = 182(10^{-6})$,
$\theta_{p1} = -52.8°$ and $\theta_{p2} = 37.2°$,
$\gamma_{\substack{max \\ in\text{-}plane}} = 187(10^{-6})$, $\theta_s = -7.76°$ and $82.2°$,
$\epsilon_{avg} = 275(10^{-6})$

10–17. $\epsilon_{x'} = 55.1(10^{-6})$, $\gamma_{x'y'} = 133(10^{-6})$,
$\epsilon_{y'} = 325(10^{-6})$

10–18. $\epsilon_{x'} = 325(10^{-6})$, $\gamma_{x'y'} = -133(10^{-6})$,
$\epsilon_{y'} = 55.1(10^{-6})$

10–19. $\epsilon_{x'} = 77.4(10^{-6})$, $\gamma_{x'y'} = 1279(10^{-6})$,
$\epsilon_{y'} = 383(10^{-6})$

10–21. $\epsilon_{x'} = 466(10^{-6})$, $\gamma_{x'y'} = -393(10^{-6})$,
$\epsilon_{y'} = -116(10^{-6})$

10–22. (a) $\epsilon_1 = 773(10^{-6})$, $\epsilon_2 = 76.8(10^{-6})$,
(b) $\gamma_{\substack{max \\ in\text{-}plane}} = 696(10^{-6})$, (c) $\gamma_{\substack{abs \\ max}} = 773(10^{-6})$

10–23. $\epsilon_1 = 870(10^{-6})$, $\epsilon_2 = 405(10^{-6})$,
$\gamma_{\substack{max \\ in\text{-}plane}} = 465(10^{-6})$, $\gamma_{\substack{abs \\ max}} = 870(10^{-6})$

10–25. $\epsilon_1 = 380(10^{-6})$, $\epsilon_2 = -330(10^{-6})$

10–26. $\epsilon_1 = 517(10^{-6})$, $\epsilon_2 = -402(10^{-6})$

10–27. $\epsilon_1 = 862(10^{-6})$, $\epsilon_2 = -782(10^{-6})$,
$\theta_{p1} = 88.0°$ (clockwise),
$\epsilon_{avg} = 40.0(10^{-6})$, $\gamma_{\substack{max \\ in\text{-}plane}} = -1644(10^{-6})$,
$\theta_s = 43.0°$ (clockwise)

10–33. $E = 17.4$ GPa, $\Delta d = -12.6(10^{-6})$ mm

10–34. $\Delta \theta = -0.0103°$

10–35. $\nu_{pvc} = 0.164$

10–37. (a) $k_r = 3.33$ ksi, (b) $k_g = 5.13(10^3)$ ksi

10–38. $p = 0.967$ ksi, $\gamma_{\substack{max \\ in\text{-}plane}} = 1.30(10^{-3})$

10–39. $\sigma_1 = 10.2$ ksi, $\sigma_2 = 7.38$ ksi

10–41. $\epsilon_x = 23.0(10^{-6})$, $\gamma_{xy} = -3.16(10^{-6})$

10–42. $\epsilon_x = 2.35(10^{-3})$, $\epsilon_y = -0.972(10^{-3})$,
$\epsilon_z = -2.44(10^{-3})$

10–43. $\sigma_1 = 8.37$ ksi, $\sigma_2 = 6.26$ ksi

10–45. $\theta = \tan^{-1}\left(\dfrac{1}{\sqrt{\nu}}\right)$

10–46. $\epsilon_1 = 289(10^{-6})$, $\epsilon_2 = -289(10^{-6})$

10–47. $P = 83.0$ lb

10–49. $\sigma_x = -15.5$ ksi, $\sigma_y = -16.8$ ksi

10–50. $\epsilon_x = \epsilon_y = 0$, $\gamma_{xy} = -160(10^{-6})$, $T = 65.2$ N·m

10–51. $\epsilon_{x'} = -2.52(10^{-3})$, $\epsilon_{y'} = 2.52(10^{-3})$

10–53. $\Delta d = 0.800$ mm, $\sigma_{AB} = 315$ MPa

10–54. $\Delta d = 0.680$ mm

10–57. $\Delta V = 0.0168$ m^3

10–58. $k = 1.35$

10–59. $\sigma_x^2 + \sigma_y^2 - \sigma_x\sigma_y + 3\tau_{xy}^2 = \sigma_y^2$

10–61. $\sigma_1 = 10.2$ ksi

10–62. $\sigma_1 = 11.6$ ksi

10–63. $\sigma_1 = 68.0$ ksi

10–65. $T_e = \sqrt{\dfrac{4}{3}M^2 + T^2}$

10–66. $\tau = 43.3$ MPa

10–67. $\tau = 37.5$ ksi

10–69. No

10–70. $M_e = \sqrt{M^2 + \dfrac{3}{4}T^2}$

10–71. $\sigma_x = 16.7$ ksi

10–73. $d = 1.59$ in.

10–74. $\sigma_1 = 924$ MPa, $\sigma_1 = 1.07$ GPa

10–75. No

10–77. No

10–78. No

10–79. Yes

10–81. (a) F.S. = 1.32, (b) F.S. = 1.52

10–82. Yes

10–83. Yes

10–85. $\sigma_Y = 424$ MPa

10–86. $T_{max} = 8.38$ kN·m

10–87. $T_{max} = 9.67$ kN·m

10–89. (a) F.S. = 2.05, (b) F.S. = 2.35

10–90. (a) $t = 22.5$ mm, (b) $t = 19.5$ mm

10–91. $d = 1.50$ in.

10–93. F.S. = 1.25

R10–2. $\delta_a = 0.367$ mm, $\delta_b = -0.255$ mm,
$\delta_t = -0.00167$ mm

R10–3. F.S. = 2

R10–5. $\epsilon_{avg} = 83.3(10^{-6})$, $\epsilon_1 = 880(10^{-6})$,
$\epsilon_2 = -713(10^{-6})$, $\theta_p = 54.8°$ (clockwise),
$\gamma_{\substack{max \\ in\text{-}plane}} = -1593(10^{-6})$,
$\theta_s = 9.78°$ (clockwise)

R10–6. $\epsilon_{x'} = -380(10^{-6})$, $\epsilon_{y'} = -130(10^{-6})$,
$\gamma_{x'y'} = 1.21(10^{-3})$

R10–7. $P_2 = 11.4$ kip, $P_1 = 136$ kip

R10–9. $\epsilon_1 = 283(10^{-6})$, $\epsilon_2 = -133(10^{-6})$,
$\theta_{p1} = 84.8°$, $\theta_{p2} = -5.18°$, $\gamma_{\substack{max \\ in\text{-}plane}} = 417(10^{-6})$,
$\epsilon_{avg} = 75.0(10^{-6})$, $\theta_s = 39.8°$ and $130°$

R10–10. $\epsilon_1 = 480(10^{-6})$, $\epsilon_2 = 120(10^{-6})$,
$\theta_{p1} = 28.2°$ (clockwise),
$\gamma_{\substack{max \\ in\text{-}plane}} = -361(10^{-6})$,
$\theta_s = 16.8°$ (counterclockwise), $\epsilon_{avg} = 300(10^{-6})$

Chapter 11

11–1. $b = 211$ mm, $h = 264$ mm
11–2. Use $b = 4$ in.
11–3. Use $b = 5$ in.
11–5. Use W12 × 16.
11–6. Yes
11–7. Use W12 × 22.
11–9. Use W360 × 45.
11–10. Yes, it can.
11–11. Use $s = s'' = 2$ in.,
Use $s' = 1$ in.
11–13. $h = 8.0$ in., $P = 3.20$ kip
11–14. $s = 1.93$ in., $s' = 2.89$ in.,
$s'' = 5.78$ in.
11–15. Use $s = 1\frac{5}{8}$ in., $s' = 1\frac{1}{8}$ in.,
$s'' = 3\frac{1}{8}$ in.
11–17. $P = 103$ kN
11–18. Use $a = 3\frac{1}{8}$ in.
11–19. $P = 750$ lb
11–21. $P = 85.9$ N
11–22. $b = 3.40$ in.
11–23. Use W16 × 31.
11–25. Use $d = 3$ in.
11–26. Use W14 × 22.
11–27. The beam fails.
11–29. $b = 5.86$ in.
11–30. $P = 9.52$ kN
11–31. $w = \dfrac{w_0}{L} x$
11–33. $\sigma_{max} = \dfrac{8PL}{27\pi r_0^3}$
11–34. $h = \dfrac{h_0}{L^{3/2}} (3L^2 x - 4x^3)^{1/2}$
11–35. $d = h\sqrt{\dfrac{x}{L}}$
11–37. $\sigma_{max} = \dfrac{3wL^2}{b_0 h^2}$
11–38. $b = \dfrac{b_0}{L^2} x^2$
11–39. Use $d = 21$ mm.
11–41. $\sigma_{max} = 13.4$ MPa
11–42. Use $d = 1\frac{1}{2}$ in.
11–43. Use $d = 1\frac{5}{8}$ in.
11–45. Use $d = 1\frac{1}{8}$ in.

11–46. Use $d = 1\frac{1}{4}$ in.
R11–1. $y = \left[\dfrac{4P}{\pi \sigma_{allow}} x \right]^{1/3}$
R11–2. Use W10 × 12
R11–3. Use $d = 44$ mm.
R11–5. Use W18 × 50.
R11–6. $P = 556$ lb, $h = 0.595$ in.
R11–7. $h = 0.643$ in. Yes.

Chapter 12

12–1. $\sigma = 3.02$ ksi
12–2. $\sigma = 75.5$ ksi
12–3. $\sigma = 582$ MPa
12–5. $v_C = -6.11$ mm
12–6. $\theta_{max} = -\dfrac{M_0 L}{EI}$,
$v = -\dfrac{M_0 x^2}{2EI}$,
$v_{max} = -\dfrac{M_0 L^2}{2EI}$
12–7. $\rho = 336$ ft,
$\theta_{max} = \dfrac{M_0 L}{EI} \gtrdot$,
$v_{max} = -\dfrac{M_0 L^2}{2EI}$
12–9. $v_1 = \dfrac{P}{12EI}(2x_1^3 - 3Lx_1^2)$,
$v_2 = \dfrac{PL^2}{48EI}(-6x_2 + L)$
12–10. $v_1 = \dfrac{wax_1}{12EI}(2x^2 - 9ax_1)$,
$v_2 = \dfrac{w}{24EI}(-x_2^4 + 28a^3 x_2 - 41a^4)$,
$\theta_C = -\dfrac{wa^3}{EI}$, $v_B = -\dfrac{41wa^4}{24EI}$
12–11. $v_1 = \dfrac{wax_1}{12EI}(2x^2 - 9ax_1)$,
$v_3 = \dfrac{w}{24EI}(-x_3^4 + 8ax_3^3 - 24a^2 x_3^2 + 4a^3 x_3 - a^4)$,
$\theta_B = -\dfrac{7wa^3}{6EI}$, $v_C = -\dfrac{7wa^4}{12EI}$
12–13. $\theta_A = -\dfrac{M_0 a}{2EI}$, $v_{max} = -\dfrac{5M_0 a^2}{8EI}$
12–14. $v_{max} = -\dfrac{3PL^3}{256EI}$
12–15. $P = 40.0$ lb, $s = 0.267$ in.

12–17. $v_1 = \dfrac{Px_1}{12EI}\left(-x_1^2 + L^2\right),$

$v_2 = \dfrac{P}{24EI}\left(-4x_2^3 + 7L^2x_2 - 3L^3\right),$

$v_{\max} = \dfrac{PL^3}{8EI}$

12–18. $\theta_A = -\dfrac{3PL^2}{8EI}, v_C = -\dfrac{PL^3}{6EI}$

12–19. $v_B = -\dfrac{11PL^3}{48EI}$

12–21. $v_{\max} = -11.5$ mm

12–22. $v = \dfrac{1}{EI}(2.25x^3 - 0.002778x^5 - 40.5x^2)$ kip $\cdot$ ft³,

$\theta_{\max} = -0.00466$ rad, $v_{\max} = -0.369$ in.

12–23. $\theta_C = \dfrac{4M_0L}{3EI}\circlearrowleft, v_1 = \dfrac{M_0}{6EIL}\left(-x_1^3 + L^2x_1\right),$

$v_2 = \dfrac{M_0}{6EIL}\left(-3Lx_2^2 + 8L^2x_2 - 5L^3\right),$

$v_C = -\dfrac{5M_0L^2}{6EI}$

12–25. $v_{\max} = -\dfrac{18.8 \text{ kip} \cdot \text{ft}^3}{EI}$

12–26. $v_{\max} = -0.396$ in.

12–27. $\theta_A = \dfrac{2\gamma L^3}{3t^2E},$

$v_A = -\dfrac{\gamma L^4}{2t^2E}$

12–29. $\theta_B = -\dfrac{wa^3}{6EI},$

$v_1 = \dfrac{w}{24EI}\left(-x_1^4 + 4ax_1^3 - 6a^2x_1^2\right),$

$v_2 = \dfrac{wa^3}{24EI}(-4x_2 + a),$

$v_B = \dfrac{wa^3}{24EI}(-4L + a)$

12–30. $\theta_B = -\dfrac{wa^3}{6EI}, v_1 = \dfrac{wx_1^2}{24EI}\left(-x_1^2 + 4ax_1 - 6a^2\right),$

$v_2 = \dfrac{wa^3}{24EI}(4x_3 + a - 4L), v_B = \dfrac{wa^3}{24EI}(a - 4L)$

12–31. $v = \dfrac{1}{EI}\left[-\dfrac{Pb}{6a}x^3 + \dfrac{P(a+b)}{6a}\langle x - a\rangle^3 + \dfrac{Pab}{6}x\right]$

12–33. $E = \dfrac{Pa}{24\Delta I}(3L^2 - 4a^2)$

12–34. $v = \dfrac{P}{12EI}\left[-2\langle x - a\rangle^3 + 4\langle x - 2a\rangle^3 + a^2x\right],$

$(v_{\max})_{AB} = \dfrac{0.106Pa^3}{EI}, v_C = -\dfrac{3Pa^3}{4EI}$

12–35. $v = \dfrac{1}{EI}\Big[-2.5x^2 + 2\langle x - 4\rangle^3 - \dfrac{1}{8}\langle x - 4\rangle^4$

$+ 2\langle x - 12\rangle^3 + \dfrac{1}{8}\langle x - 12\rangle^4$

$- 24x + 136\Big]$ kip $\cdot$ ft³

12–37. $v = \dfrac{M_0}{6EI}\left[3\left\langle x - \dfrac{L}{3}\right\rangle^2 - 3\left\langle x - \dfrac{2}{3}L\right\rangle^2 - Lx\right],$

$v_{\max} = -\dfrac{5M_0L^2}{72EI}$

12–38. $v = \dfrac{1}{EI}\big[-8.33x^3 + 17.1\langle x - 12\rangle^3$

$- 13.3\langle x - 36\rangle^3 + 1680x - 5760\big]$ lb $\cdot$ in³

12–39. $v_{\max} = -12.9$ mm

12–41. $(v_{\max})_{AB} = 0.0867$ in.

12–42. $\theta_A = -\dfrac{1920}{EI}, \theta_B = \dfrac{6720 \text{ lb} \cdot \text{in}^2}{EI}$

12–43. $v = \dfrac{1}{EI}\Big[-0.0833x^3 + 3\langle x - 8\rangle^2$

$+ 3\langle x - 16\rangle^2 + 8.00x\Big]$ kip $\cdot$ ft³

12–45. $v_C = -0.501$ mm, $v_D = -0.698$ mm,

$v_E = -0.501$ mm

12–46. $\theta_A = -0.128°, \theta_B = 0.128°$

12–47. $\theta_A = -\dfrac{3wa^3}{16EI},$

$\theta_B = \dfrac{7wa^3}{48EI},$

$v = \dfrac{w}{48EI}\left[6ax^3 - 2x^4 + 2\langle x - a\rangle^4 - 9a^3x\right]$

12–49. $\theta_A = \dfrac{302}{EI}$ kip $\cdot$ ft², $v_C = -\dfrac{3110}{EI}$ kip $\cdot$ ft³

12–50. $\dfrac{dv}{dx} = \dfrac{1}{EI}\big[2.25x^2 - 0.5x^3 + 5.25\langle x - 5\rangle^2$

$+ 0.5\langle x - 5\rangle^3 - 3.125\big]$ kN $\cdot$ m²,

$v = \dfrac{1}{EI}\big[0.75x^3 - 0.125x^4 + 1.75\langle x - 5\rangle^3$

$+ 0.125\langle x - 5\rangle^4 - 3.125x\big]$ kN $\cdot$ m³

12–51. $\theta_C = -\dfrac{3937.5}{EI}, v_C = \dfrac{50\,625}{EI}\downarrow$

12–53. $v_B = \dfrac{7PL^3}{16EI}\downarrow$

12–54. $\theta_B = -\dfrac{Pa^2}{12EI}, v_C = \dfrac{Pa^3}{12EI}$

12–55. $v_{\max} = 12.2$ mm

12–57. $v_{\max} = \dfrac{3PL^3}{256EI}\downarrow$

12–58. $v_C = -\dfrac{84}{EI}, \theta_A = \dfrac{8}{EI}, \theta_B = -\dfrac{16}{EI}, \theta_C = -\dfrac{40}{EI}$

12–59. $v_{max} = 8.16 \text{ mm} \downarrow$

12–61. $a = 0.858L$

12–62. $\theta_A = 0.0181 \text{ rad}, \theta_B = 0.00592 \text{ rad}$

12–63. $\theta_B = -\dfrac{3M_0L}{2EI}, v_B = \dfrac{7M_0L^2}{8EI} \downarrow$

12–65. $\theta_A = -\dfrac{5Pa^2}{2EI}, v_C = \dfrac{19Pa^3}{6EI} \downarrow$

12–66. $v_C = \dfrac{PL^3}{12EI}, \theta_A = \dfrac{PL^2}{24EI}, \theta_B = -\dfrac{PL^2}{12EI}$

12–67. $v_{max} = \dfrac{0.00802PL^3}{EI}$

12–69. $\theta_C = -\dfrac{5Pa^2}{2EI}, v_B = \dfrac{25Pa^3}{6EI} \downarrow$

12–70. $\theta_A = -\dfrac{336 \text{ kip} \cdot \text{ft}^2}{EI}, v_{max} = \dfrac{3048 \text{ kip} \cdot \text{ft}^3}{EI} \downarrow$

12–71. $v_D = 4.98 \text{ mm} \downarrow$

12–73. $a = 0.152L$

12–74. $\theta_{max} = \dfrac{5PL^2}{16EI}, v_{max} = \dfrac{3PL^3}{16EI} \downarrow$

12–75. $\theta_B = 0.00658 \text{ rad}, v_C = 13.8 \text{ mm} \downarrow$

12–77. $a = 0.865 L$

12–78. $\theta_B = \dfrac{7wa^3}{12EI}, v_C = \dfrac{25wa^4}{48EI} \downarrow$

12–79. $\theta_C = -\dfrac{a^2}{6EI}(12P + wa),$

$v_C = \dfrac{a^3}{24EI}(64P + 7wa) \downarrow$

12–81. $\theta_A = \dfrac{PL^2}{12EI}, v_D = \dfrac{PL^3}{8EI} \downarrow$

12–82. $v_{max} = \dfrac{3wa^4}{8EI}$

12–83. $\theta_B = -0.00778 \text{ rad}, v_B = 0.981 \text{ in.} \downarrow$

12–85. $v_C = 1.20 \text{ in.} \downarrow$

12–86. $\theta_A = -0.822°, \theta_B = 0.806°$

12–87. $v_C = 0.429 \text{ in.} \downarrow$

12–89. $\theta_A = 0.00458 \text{ rad}, v_C = 0.187 \text{ in.} \downarrow$

12–90. Use W16 × 50.

12–91. $(v_A)_v = 0.0737 \text{ in.}, (v_A)_k = 0.230 \text{ in.}$

12–93. $v = PL^2 \left(\dfrac{1}{k} + \dfrac{L}{3EI} \right)$

12–94. $v_A = \dfrac{72}{EI} \downarrow, \theta_A = \dfrac{36}{EI} \curvearrowright$

12–95. $v_A = PL^3 \left(\dfrac{1}{12EI} + \dfrac{1}{8GJ} \right) \downarrow$

12–97. $F = 0.349 \text{ N}, a = 0.800 \text{ mm}$

12–98. $M_0 = \dfrac{Pa}{6}$

12–99. $A_x = B_x = 0, A_y = \dfrac{20}{27} P,$

$M_A = \dfrac{4}{27} PL, B_y = \dfrac{7}{27} P, M_B = \dfrac{2}{27} PL$

12–101. $A_x = 0, C_y = \dfrac{5}{16} P, B_y = \dfrac{11}{8} P, A_y = \dfrac{5}{16} P$

12–102. $A_x = 0, B_y = \dfrac{5}{16} P, A_y = \dfrac{11}{16} P, M_A = \dfrac{3PL}{16}$

12–103. $A_x = 0, B_y = \dfrac{3wL}{8}, A_y = \dfrac{5wL}{8}, M_A = \dfrac{wL^2}{8}$

12–105. $A_x = 0, A_y = \dfrac{3M_0}{2L}, B_y = \dfrac{3M_0}{2L}, M_B = \dfrac{M_0}{2}$

12–106. $A_x = 0, B_y = \dfrac{w_0L}{10}, A_y = \dfrac{2w_0L}{5}, M_A = \dfrac{w_0L^2}{15}$

12–107. $B_x = 0, A_y = \dfrac{17wL}{24}, B_y = \dfrac{7wL}{24}, M_B = \dfrac{wL^2}{36}$

12–109. $T_{AC} = \dfrac{3A_2E_2wL_1^4}{8(A_2E_2L_1^3 + 3E_1I_1L_2)}$

12–110. $A_x = 0, F_C = 112 \text{ kN}, A_y = 34.0 \text{ kN},$
$B_y = 34.0 \text{ kN}$

12–111. $M_A = \dfrac{5wL^2}{192}, M_B = \dfrac{11wL^2}{192}$

12–113. $A_y = 1.48 \text{ kip}, B_x = 0, B_y = 3.52 \text{ kip},$
$M_B = 7.67 \text{ kip} \cdot \text{ft}$

12–114. $B_y = \dfrac{2}{3}P, M_A = \dfrac{PL}{3}, A_y = \dfrac{4}{3}P, A_x = 0$

12–115. $A_x = 0, B_y = \dfrac{M_0}{6a}, A_y = \dfrac{M_0}{6a}, M_A = \dfrac{M_0}{2}$

12–117. $B_y = 550 \text{ N}, A_y = 125 \text{ N}, C_y = 125 \text{ N}$

12–118. $A_x = 0, B_y = \dfrac{5wL}{4}, C_y = \dfrac{3wL}{8}$

12–119. $B_y = \dfrac{5}{8}wL \uparrow, C_y = \dfrac{wL}{16} \downarrow, A_y = \dfrac{7}{16}wL \uparrow$

12–121. $A_x = 0, B_y = 35.0 \text{ kip}, A_y = 15.0 \text{ kip},$
$M_A = 40.0 \text{ kip} \cdot \text{ft}$

12–122. $A_x = 0, B_y = \dfrac{7P}{4}, A_y = \dfrac{3P}{4}, M_A = \dfrac{PL}{4}$

12–123. $A_x = 0, B_y = \dfrac{7wL}{128}, A_y = \dfrac{57wL}{128}, M_A = \dfrac{9wL^2}{128}$

12–125. $M_A = M_B = \dfrac{1}{24}PL, A_y = B_y = \dfrac{1}{6}P,$

$C_y = D_y = \dfrac{1}{3}P, D_x = 0$

12–126. $T_{AC} = \dfrac{3wA_2E_2L_1^4}{8(3E_1I_1L_2 + A_2E_2L_1^3)}$

12–127. $M = \dfrac{PL}{8} - \dfrac{2EI}{L}\alpha, \Delta_{max} = \dfrac{PL^3}{192EI} + \dfrac{\alpha L}{4}$

12–129. $a = L - \left(\dfrac{72\Delta EI}{w_0}\right)^{1/4}$

12–130. $F_{CD} = 7.48$ kip

12–131. $M_{\max} = \dfrac{\pi^2 bt\gamma\omega^2 r^3}{108g}$

R12–1. $v = \dfrac{1}{EI}\left[-30x^3 + 46.25\langle x - 12\rangle^3 \right.$
$\left. - 11.7\langle x - 24\rangle^3 + 38{,}700x - 412{,}560\right]$ lb·in^3

R12–2. $v_1 = \dfrac{1}{EI}(4.44x_1^3 - 640x_1)$ lb·in^3,
$v_2 = \dfrac{1}{EI}(-4.44x_2^3 + 640x_2)$ lb·in^3

R12–3. $M_B = \dfrac{w_0 L^2}{30}, M_A = \dfrac{w_0 L^2}{20}$

R12–5. $(v_2)_{\max} = \dfrac{wL^4}{18\sqrt{3}EI}$

R12–6. $\theta_B = \dfrac{Pa^2}{4EI}, \Delta_C = \dfrac{Pa^3}{4EI}\uparrow$

R12–7. $B_y = 138$ N $\uparrow, A_y = 81.3$ N $\uparrow, C_y = 18.8$ N $\downarrow$

R12–9. $\Delta_C = 0.644$ in. $\downarrow$

Chapter 13

13–1. $P_{cr} = \dfrac{5kL}{4}$

13–2. $P_{cr} = kL$

13–3. Use $d = \dfrac{9}{16}$ in.

13–5. $P_{cr} = 1.84$ MN

13–6. $P_{cr} = 902$ kN

13–7. F.S. $= 1.87$

13–9. $P_{cr} = 1.30$ MN

13–10. $P_{cr} = 325$ kN

13–11. $P_{cr} = 20.4$ kip

13–13. $P = 42.8$ kN

13–14. $P_{cr} = 575$ kip

13–15. $P_{cr} = 70.4$ kip

13–17. $P_{cr} = 2.92$ kip

13–18. $P_{cr} = 5.97$ kip

13–19. $P = 17.6$ kip

13–21. Use $d_i = 1\dfrac{1}{8}$ in.

13–22. $W = 4.31$ kN

13–23. $W = 5.24$ kN, $d = 1.64$ m

13–25. $P = 62.3$ kip

13–26. $P = 2.42$ kip

13–27. Use $d_{AB} = 2\dfrac{1}{8}$ in., $d_{BC} = 2\dfrac{1}{4}$ in.

13–29. Use $d_{AB} = 1\dfrac{1}{2}$ in., $d_{BC} = 1\dfrac{3}{8}$ in.

13–30. $P = 129$ lb

13–31. $w = 1.17$ kN/m

13–33. $P = 14.8$ kN

13–34. Use $d = 62$ mm.

13–35. Use $d = 52$ mm.

13–37. $P = 37.5$ kip

13–38. $P = 5.79$ kip

13–39. Use $d = 1\dfrac{3}{4}$ in.

13–41. $M_{\max} = -\dfrac{wEI}{P}\left[\sec\left(\dfrac{L}{2}\sqrt{\dfrac{P}{EI}}\right) - 1\right]$

13–42. $M_{\max} = -\dfrac{F}{2}\sqrt{\dfrac{EI}{P}}\tan\left(\dfrac{L}{2}\sqrt{\dfrac{P}{EI}}\right)$

13–43. $P_{cr} = \dfrac{\pi^2 EI}{4L^2}$

13–46. $P = 31.4$ kN

13–47. $v_{\max} = 0.387$ in.

13–49. $P_{allow} = 7.89$ kN

13–50. $E_t = 14.6(10^3)$ ksi

13–51. $L = 8.34$ m

13–53. $P = 65.8$ kip

13–54. $P = 45.7$ kip

13–55. Yes

13–57. $P_{\max} = 61.2$ kip

13–58. $L = 21.2$ ft

13–59. $P_{\max} = 16.9$ kN

13–61. $L = 1.71$ m

13–62. $P = 174$ kN, $v_{\max} = 16.5$ mm

13–63. $P = 129$ kip

13–65. $P_{cr} = 83.5$ kN

13–66. $L = 2.53$ m

13–67. $d = 98.3$ mm

13–69. $P = 88.5$ kip

13–70. $P_{cr} = 1.32(10^3)$ kN

13–71. $P_{cr} = 5.29(10^3)$ kN

13–73. For $49.7 < KL/r < 99.3$,
$P/A = 200$ MPa

13–74. For $49.7 < L/r < 99.3$,
$P/A = 25$ ksi

13–75. $P_{cr} = 661$ kN

13–77. $P_{cr} = 1.35(10^3)$

13–78. $L = 8.99$ ft

13–79. Use W6 × 9.

13–81. $L = 20.3$ ft

13–82. Use W6 × 12.

13–83. $L = 18.0$ ft

13–85. $L = 33.7$ ft

13–86. Use W6 × 9.

13–87. $d = 1.42$ in.

13–89. Yes

13–90. Yes

13–91. $b = 0.704$ in.

13–93. $L = 1.92$ ft

13–94. $L = 3.84$ ft

13–95. $P_{\text{allow}} = 380$ kip

13–97. $P_{\text{allow}} = 129$ kip

13–98. $P_{\text{allow}} = 143$ kip

13–99. $P_{\text{allow}} = 109$ kip

13–101. $P_{\text{allow}} = 8.61$ kip

13–102. $L = 8.89$ ft

13–103. Use $a = 7\frac{1}{2}$ in.

13–105. $P_{\text{allow}} = 8.68$ kip

13–106. $P_{\text{allow}} = 27.7$ kip

13–107. $P = 8.83$ kip

13–109. $P = 18.4$ kip

13–110. $P = 5.93$ kip

13–111. $P = 32.7$ kip

13–113. $P = 0.967$ kip

13–114. $P = 0.554$ kip

13–115. The column is not adequate.

13–117. $P = 33.1$ kip

13–118. $P = 57.7$ kip

13–119. $P = 2.79$ kip

13–121. $P = 98.0$ kip

13–122. $P = 132$ kip

13–123. No

13–125. Yes

13–126. $P = 1.69$ kip

13–127. $P = 3.44$ kip

R13–1. $P = 5.76$ kip

R13–2. $P = 9.01$ kip

R13–3. $P_{\text{cr}} = 12.1$ kN

R13–5. $P = 12.5$ kip

R13–6. Use $d = 2\frac{1}{8}$ in.

R13–7. $t = 5.92$ mm

R13–9. $P_{\text{allow}} = 77.2$ kN

R13–10. It does not buckle or yield.

Chapter 14

14–1. $\dfrac{U_i}{V} = \dfrac{1}{2E}(\sigma_x^2 + \sigma_y^2 - 2\nu\sigma_x\sigma_y) + \dfrac{\tau_{xy}^2}{2G}$

14–3. $a = \sqrt{\dfrac{\pi}{2}}\, r$

14–5. (a) $U_a = \dfrac{N^2 L_1}{2AE}$, (b) $U_b = \dfrac{N^2 L_2}{2AE}$

Since $U_b > U_a$, i.e., $L_2 > L_1$, the design for case (b) is better able to absorb energy.

14–6. $U_i = 43.2$ J

14–7. $P = 375$ kN, $U_i = 1.69$ kJ

14–9. $U_i = 149$ J

14–10. $U_i = 0.0638$ J

14–11. $U_i = 64.4$ J

14–13. $P = 113$ kip, $U_i = 7.37$ in. · kip

14–15. $U_i = 0.125$ ft · kip

14–17. $U_i = \dfrac{17 w_0^2 L^5}{10\,080\, EI}$

14–18. $(U_i)_{sp} = 1.00$ J, $(U_i)_b = 0.400$ J

14–19. $U_i = 3.24$ in. · lb

14–21. $U_i = \dfrac{w_0^2 L^5}{504\, EI}$

14–22. $(U_i)_b = 0.477\,(10^{-3})$ J, $(U_i)_l = 0.0171$ J

14–23. $U_i = \dfrac{w^2 L^5}{40\, EI}$

14–25. $(\Delta_D)_v = \dfrac{3.50 PL}{AE}$

14–26. $(\Delta_C)_h = \dfrac{3 PL}{2AE}$

14–27. $(\Delta_A)_h = 0.0710$ in.

14–29. $(\Delta_C)_v = 13.3$ mm

14–30. $\Delta_B = 3.46$ mm

14–31. $\Delta_B = 11.7$ mm

14–33. $\Delta_B = 0.100$ mm

14–34. $\Delta_E = 5.46$ in.

14–35. $\theta_A = -\dfrac{M_0 L}{3EI}$

14–37. $\Delta_A = \dfrac{3\pi P r^3}{2EI}$

14–38. $\Delta_A = \dfrac{\pi P r^3}{2EI}$

14–39. $\Delta_B = 1.82$ in.

14–41. $\Delta_B = 15.2$ mm

14–42. (a) $U_i = 4.52$ kJ, (b) $U_i = 3.31$ kJ

14–43. $d = 5.35$ in.

14–45. (a) $\sigma_{\max} = 45.4$ ksi, (b) $\sigma_{\max} = 509$ psi, (c) $\sigma_{\max} = 254$ psi

14–46. $h = 1.75$ ft

14–47. $\sigma_{\max} = 20.3$ ksi

14–49. $L = 850$ mm

14–50. Yes

14–51. $h = 5.29$ mm

14–53. $\sigma_{\max} = 307$ MPa

14–54. $h = 95.6$ mm

14–55. Use $d = 2\frac{1}{8}$ in.

14–57. Yes, from any position

14–58. $(\Delta_A)_{\max} = 15.4$ in.

14–59. $\sigma_{\max} = 6.20$ ksi

14–61. $n = 19.7$

14–62. $h = 4.20$ ft

14–63. $\sigma_{\max} = 35.2$ ksi

14–65. $\Delta_B = 0.247$ in., $\sigma_{max} = 2.34$ ksi

14–66. $\sigma_{max} = 137$ MPa

14–67. $h = 8.66$ m

14–69. $h = 7.45$ in.

14–70. $(\Delta_B)_{max} = 0.661$ in.

14–71. $\Delta_{max} = 23.3$ mm, $\sigma_{max} = 4.89$ MPa

14–73. $(\Delta_B)_h = 0.223(10^{-3})$ in. $\leftarrow$

14–74. $(\Delta_B)_v = 0.00112$ in.$\downarrow$

14–75. $(\Delta_B)_v = 0.0132$ in.$\downarrow$

14–77. $(\Delta_B)_h = 0.699(10^{-3})$ in. $\rightarrow$

14–78. $(\Delta_B)_v = 0.0931(10^{-3})$ in. $\downarrow$

14–79. $(\Delta_B)_h = 0.367$ mm $\leftarrow$

14–81. $(\Delta_C)_h = 0.234$ mm $\leftarrow$

14–82. $(\Delta_D)_v = 1.16$ mm $\downarrow$

14–83. $(\Delta_A)_v = 3.18$ mm$\downarrow$

14–85. $(\Delta_D)_h = 4.12$ mm $\rightarrow$

14–86. $(\Delta_E)_h = 0.889$ mm $\rightarrow$

14–87. $\Delta_C = \dfrac{23Pa^3}{24EI}$

14–89. $\Delta_C = \dfrac{2Pa^3}{3EI}$

14–90. $\theta_C = -\dfrac{5Pa^2}{6EI}$

14–91. $\theta_A = \dfrac{Pa^2}{6EI}$

14–93. $\theta_B = -0.353°$

14–94. $\theta_C = 0.337°$

14–95. $\Delta_B = 47.8$ mm $\downarrow$

14–97. $\theta_A = 0.289°$

14–98. $\theta_B = 0.124°$

14–99. $\Delta_C = \dfrac{PL^3}{8EI}\downarrow$

14–101. $\theta_C = -\dfrac{13wL^3}{576EI}$

14–102. $\Delta_D = \dfrac{wL^4}{96EI}\downarrow$

14–103. $\theta_A = -1.28°$

14–105. $\Delta_C = \dfrac{PL^3}{48EI}\downarrow, \theta_B = \dfrac{PL^2}{16EI}$

14–106. $\Delta_C = 0.122$ in. $\downarrow$

14–107. $\theta_A = 0.232°$

14–109. $\theta_C = -1.14°$

14–110. $\Delta_D = 0.219$ in.$\uparrow$

14–111. $\Delta_{tot} = \left(\dfrac{w}{G}\right)\left(\dfrac{L}{a}\right)^2\left[\left(\dfrac{5}{96}\right)\left(\dfrac{L}{a}\right)^2 + \dfrac{3}{20}\right],$

$\Delta_b = \dfrac{5w}{96G}\left(\dfrac{L}{a}\right)^4$

14–113. $\theta_A = -\dfrac{5w_0 L^3}{192EI}$

14–114. $(\Delta_C)_h = \dfrac{640\,000\ \text{lb}\cdot\text{ft}^3}{EI}\leftarrow,$

$(\Delta_C)_v = \dfrac{1\,228\,800\ \text{lb}\cdot\text{ft}^3}{EI}\downarrow$

14–115. $\Delta_B = 43.5$ mm $\downarrow$

14–117. $\Delta_C = 17.9$ mm $\downarrow$

14–118. $\theta_A = -0.0568°$

14–119. $(\Delta_C)_h = \dfrac{5wL^4}{8EI}\rightarrow$

14–121. $(\Delta_B)_v = \dfrac{Pr^3}{4\pi EI}\left(\pi^2 - 8\right)\downarrow$

14–122. $(\Delta_A)_h = \dfrac{\pi Pr^3}{2EI}\leftarrow$

14–123. $(\Delta_B)_h = 0.223(10^{-3})$ in. $\leftarrow$

14–125. $(\Delta_B)_v = 0.0132$ in. $\downarrow$

14–126. $(\Delta_E)_v = 0.0149$ in. $\downarrow$

14–127. $(\Delta_B)_h = 0.699(10^{-3})$ in. $\rightarrow$

14–129. $(\Delta_C)_h = 0.234$ mm $\leftarrow$

14–130. $(\Delta_C)_v = 0.0375$ mm $\downarrow$

14–131. $(\Delta_D)_h = 4.12$ mm $\rightarrow$

14–133. $\theta_C = -\dfrac{5Pa^2}{6EI}$

14–134. $\theta_A = \dfrac{Pa^2}{6EI}$

14–135. $\Delta_C = 0.369$ in.

14–137. $\Delta_B = 47.8$ mm

14–138. $\Delta_D = 3.24$ mm

14–139. $\theta_A = 0.289°$

14–141. $\theta_A = \dfrac{wL^3}{24EI}$

14–142. $\Delta_C = \dfrac{5wL^4}{8EI}$

14–143. $\Delta_B = \dfrac{wL^4}{4EI}$

14–145. $\theta_B = \dfrac{wL^3}{8EI}$

R14–1. $U_i = 496\,\text{J}$

R14–2. $\sigma_{max} = 116$ MPa

R14–3. $h = 10.3$ m

R14–5. $\sigma_{max} = 43.6$ ksi

R14–6. $U_i = 45.5\,\text{ft}\cdot\text{lb}$

R14–7. $(\Delta_C)_v = 0.114$ in. $\downarrow$

R14–9. $\theta_B = \dfrac{M_0 L}{EI}$

R14–10. $\theta_B = \dfrac{M_0 L}{EI}$

R14–11. $\theta_C = -\dfrac{2\,wa^3}{3\,EI}, \Delta_C = \dfrac{5\,wa^4}{8\,EI}\downarrow$

INDEX